A Guide to Oracle8

Joline Morrison,
Mike Morrison
University of Wisconsin, Eau Claire

Thomson Learning™

ONE MAIN STREET, CAMBRIDGE, MA 02142

Australia • Canada • Denmark • Japan • Mexico • New Zealand • Philippines
Puerto Rico • Singapore • South Africa • Spain • United Kingdom • United States

A Guide to Oracle8 is published by Course Technology.

Senior Editor:	*Jennifer Normandin*	*Quality Assurance Manager:*	*John Bosco*
Associate Publisher:	*Kristen Duerr*	*Marketing Manager:*	*Susan Ogar*
Development Editor:	*Amanda Brodkin*	*Composition House:*	*GEX, Inc.*
Production Editors:	*Melissa Panagos and Debbie Masi*	*Text Designer:*	*GEX, Inc.*
Editorial Assistant:	*Amanda Young*	*Cover Designer:*	*Efrat Reis*

For more information contact:

Course Technology
1 Main Street
Cambridge, MA 02142

Or find us on the World Wide Web at: http://www.course.com.

Asia (excluding Japan)
Thomson Learning
60 Albert Street, #15-01
Albert Complex
Singapore 189969

Latin America
Thomson Learning
Seneca, 53
Colonia Polanco
11560 Mexico D.F. Mexico

Japan
Thomson Learning
Palaceside Building 5F
1-1-1 Hitotsubashi, Chiyoda-ku
Tokyo 100 0003 Japan

South Africa
Thomson Learning
Zonnebloem Building,
Constantia Square
526 Sixteenth Road
P.O. Box 2459
Halfway House, 1685
South Africa

Australia/New Zealand
Nelson/Thomson Learning
102 Dodds Street
South Melbourne, Victoria 3205

Canada
Nelson/Thomson Learning
1120 Birchmount Road
Scarborough, Ontario
Canada M1K 5G4

UK/Europe/Middle East
Thomson Learning
Berkshire House
168-173 High Holborn
London
WCIV 7AA United Kingdom

Business Press/Thomson Learning
Berkshire House
168-173 High Holborn
London WCIV 7AA United
Kingdom

Thomson Nelson & Sons LTD
Nelson House
Mayfield Road
Walton-on-Thames
KT12 5PL United Kingdom

Spain
Paraninfo/Thomson Learning
Calle Magallanes, 25
28015-MADRID
ESPANA

Distrubution Services
Thomson Learning
Ceriton House
North Way
Andover, Hampshire SP10 5BE

International Headquarters
Thomson Learning
International Division
290 Harbor Drive, 2nd Floor
Stamford, CT 06902-7477

ISBN 0-619-00027-9

Printed in Canada

2 3 4 5 6 7 8 9 WC 03 02 01 00 99

Contents

chapter 4

INTRODUCTION TO PL/SQL, TRIGGERS, AND PROCEDURE BUILDER *151*

chapter 8

USING GRAPHICS BUILDER *603*

chapter 9

CREATING AN INTEGRATED DATABASE APPLICATION *653*

A Guide to Oracle8 Preface

A Guide to Oracle8 is designed to provide a comprehensive guide for developing a relational database application using the Oracle8 database and application development utilities. The goal of this textbook is to support a database development course. It assumes that students already have experience with some other programming language and a basic understanding of relational database concepts.

Organization and Coverage

A Guide to Oracle8 contains an introductory chapter introducing client/server and relational database concepts. The next two chapters provide hands-on instruction for performing SQL command-line operations for creating tables, inserting data, and displaying data. Chapter 4 provides an in-depth introduction to PL/SQL, Oracle's procedural programming language, as well as instructions for creating and saving PL/SQL programs using a variety of approaches. Chapters 5 through 9 show students how to develop applications for entering and displaying database data and creating an integrated database project. Chapter 10 addresses the development of user database interfaces using dynamic Web pages.

When students complete this book, they will know how to create and modify database tables and add and modify database data using SQL*Plus, the Oracle command-line SQL environment. Students also will be able create queries using Query Builder, Oracle's graphical utility for creating complex queries that join multiple database tables. They will learn how to use Procedure Builder, a graphical utility for creating PL/SQL programs that interact with database tables. Specific instructions are included for creating stored program units, libraries, packages, and database triggers. Students will create a variety of different user interfaces, reports, and graphical charts for database applications, and develop an integrated database application using the Oracle Developer application set, which includes Form Builder, Report Builder, Graphics Builder, and Project Builder. They will create applications using the Wizards supplied in the Developer applications, and they will also create custom form applications using PL/SQL programs. Students also will learn how to program dynamic Web pages to display and manipulate database data using different approaches. This book emphasizes sound database design and development techniques and GUI design skills.

Approach

A Guide to Oracle8 distinguishes itself from other Oracle books because it is written and designed specifically for students and instructors in educational environments who are participating in the Oracle Academic Initiative. The Oracle client/server database software enables educators to illustrate multiuser and client/server database concepts, such as managing concurrent users and sharing database resources, and allows students to develop database applications in a production environment using the Developer utilities. The exercises in this book emphasize the issues that must be considered when developing applications using a client/server database. The book and the Instructor's Manual also address the unique issues that must be considered when installing and using the Oracle software utilities in student laboratory environments and creating and administering student accounts.

Features

A Guide to Oracle8 is a superior textbook because it also includes the following features:

- **"Read This Before You Begin" Page** This page supports Course Technology's unequaled commitment to helping instructors introduce technology into the classroom. Technical considerations and assumptions about hardware, software, and default settings are listed in one place to help instructors save time and eliminate unnecessary aggravation.
- **Case Approach** Two running cases address database-related problems that students could reasonably expect to encounter in business, followed by a demonstration of an application that could be used to solve the problem. Showing students the completed application before they learn how to create it is motivational and instructionally sound. By allowing the students to see the type of application they will create after completing the chapter, students will be more motivated to learn, because they can see how the programming concepts that they are about to learn can be used and, therefore, why the concepts are important. The databases referenced in the two ongoing cases represent realistic client/server applications with several database tables, and require supporting multiple users simultaneously at different physical locations. The Clearwater Traders database represents a standard sales order and inventory system, and the Northwoods University database illustrates a student registration system.
- **Step-by-Step Methodology** The unique Course Technology methodology keeps students on track. They click buttons or press keys always within the context of solving the problem posed in the chapter. The text constantly guides students and lets them know where they are in the process of solving the problem. The numerous illustrations include labels that direct students' attention to what they should look at on the screen.
- **HELP?** These paragraphs anticipate the problems students are likely to encounter and help them resolve these problems on their own. This feature facilitates independent learning and frees the instructor to focus on substantive issues rather than on common procedural errors.
- **TIPS** These notes provide additional information about a procedure—for example, an alternative method of performing the procedure.
- **Summaries** Following each chapter is a Summary that recaps the programming concepts and commands covered in the lesson.

- **Review Questions and Problem-Solving Cases** Each chapter concludes with meaningful, conceptual Review Questions that test students' understanding of what they learned in the chapter. Problem-Solving Cases provide students with additional practice of the skills and concepts they learned in the lesson. These exercises increase in difficulty and are designed to allow the student to explore the language and programming environment independently.

The Oracle Server and Client Software

This book was written using Oracle8i Enterprise Edition Server, Version 8.15, and Oracle Application Server, Version 4.0, installed on a Windows NT database server. The client software was Oracle Developer, Version 6.05, installed on Windows 95, Windows 98, and Windows NT client workstations. Later software versions might have slightly different features, but the core functionality is usually similar. Specific instructions for installation and configuring these applications on a Windows NT server and Windows 95/98/Windows NT clients are provided in the Instructor's Manual.

The Supplements

All of the supplements for this text are found in the Instructor's Resource Kit, which includes a printed Instructor's Manual and a CD-ROM.

- **Instructor's Manual** The authors wrote the Instructor's Manual and it was quality assurance tested. It is available in printed form and through the Course Technology Faculty Online Companion on the World Wide Web at www.course.com. (Call your customer service representative for the specific URL and your password.) The Instructor's Manual contains the following items:
  - Complete instructions for downloading the Java Development Kit from the Sun Web site or installing it from the CD-ROM included with the book.
  - Complete instructions for installing and configuring the server and client software.
  - Answers to all of the review questions and solutions to all of the problem-solving cases.
  - Teaching notes to help introduce and clarify the material presented in the chapters.
  - Technical notes that include troubleshooting tips.
- **Course Test Manager Version 1.1 Engine and Test Bank** Course Test Manager (CTM) is a cutting-edge Windows-based testing software program, developed exclusively for Course Technology, that helps instructors design and administer examinations and practice tests. This full-featured program allows students to generate practice tests randomly that provide immediate on-screen feedback and detailed study guides for incorrectly answered questions. Instructors can also use Course Test Manager to create printed and online tests. You can create, preview, and administer a test on any or all chapters of this textbook entirely over a local area network. Course

Test Manager can grade tests that students take automatically at the computer and can generate statistical information on individual as well as group performance. A CTM test bank has been written to accompany the textbook and is included on the CD-ROM. The test bank includes multiple-choice, true/false, short answer, and essay questions.

- **Solutions Files** Solutions Files contain every file students are asked to create or modify in the chapters and cases.
- **Student Disk Files** Student Disk Files, containing all of the data that students will use for the chapters and cases in this textbook, are provided through Course Technology's Online Companion, as well as on the Instructor's Resource Kit CD-ROM. A Readme file includes technical tips for lab management. See the inside covers of this textbook and the "Read This Before You Begin" page before Chapter 1 for more information on Student Files.

Acknowledgments

We would like to thank all of the people who helped to make this book a reality, especially Amanda Brodkin, our Development Editor, who displayed incredible persistence, patience, and good humor even after her first attempt to install the software crashed her computer. Thanks also to Kristen Duerr, Associate Publisher; Jennifer Normandin, Senior Editor; Melissa Panagos and Debbie Masi, Production Editors; and John Bosco and his Quality Assurance testing team.

We are grateful to the many reviewers who provided helpful and insightful comments during the development of this book, including Rocky Conrad, San Antonio College; Jeff Guan, University of Louisville; and Srini Srinivasan, University of Louisville. Thanks to our colleagues in the MIS department and Business College at the University of Wisconsin, Eau Claire, who gave us the opportunity to incorporate Oracle into the curriculum, and recognized the effort and contribution involved in creating a textbook like this. We would also like to express our appreciation to Carol Accola-Koich and her colleagues in our Computing and Network Services department, who have given us the freedom and support we've needed to install, administer, and use the Oracle Software in the public access laboratories. And thanks to all of the students who have taught us about teaching Oracle.

Joline Morrison
Mike Morrison

Read This Before You Begin

To the Student

Student Disks

You will need Student Disks that provide files that are needed to complete the chapters and exercises. You will be instructed to save the files you create on these Student Disks. Some of the completed files are very large and take a long time to load and execute if they are stored on a floppy diskette, so we recommend that you store the Student Disk files on a hard drive and save your work there. If you cannot do this, then you can save your work on floppy disks. The following paragraphs describe how to set up your Student Disks both for a hard disk and a floppy disk installation. If you are installing the Student Disk files on a hard disk, your instructor will tell you how to access the Student Disk files. If you are using floppy disks, your instructor will provide you with Student Disks or ask you to make your own.

Hard Disk Installation

If you are asked to make your Student Disks on a hard drive from a file server, your instructor will tell you which drive letter and folders contain the files you need. Start Windows Explorer, navigate to the folder, select all of the subfolders, and copy them to the folder on your hard drive where you want to store your Student Disk files. There is a folder for each of the chapters in the textbook. All of the folders contain files you will need to complete the chapters and exercises in the book, and some of the folders contain subfolders.

Floppy Disk Installation

If you are asked to make your own Student Disks, you will need ten blank, formatted high-density disks. You will need to copy a set of folders from a file server or standalone computer onto your disks. Your instructor will tell you which computer, drive letter, and folders contain the files you need. The following table shows you which folders go on each of your disks, so that you will have enough disk space to complete all the chapters and exercises:

When you begin each chapter, make sure you are using the correct Student Disk. See the inside front or inside back cover of this book for more information on Student Disk files, or ask your instructor or technical support person for assistance.

Student Disk	Write this on the disk label	Put these folders on the disk
1	Oracle8 Chapters 2, 3, 4, 5, 10	Chapter2 Chapter3 Chapter4 Chapter5 Chapter10
2	Oracle8 Chapter6 Disk 1	Chapter6
3	Oracle8 Chapter6 Disk 2	Chapter6
4	Oracle8 Chapter7 Disk 1	Chapter7
5	Oracle8 Chapter7 Disk 2	Chapter7
6	Oracle8 Chapter8 Disk 1	Chapter8
7	Oracle8 Chapter8 Disk 2	Chapter8
8	Oracle8 Chapter9 Disk1	Chapter9
9	Oracle8 Chapter9 Disk2	Chapter9
10	Oracle8 Chapter9 Disk3	Chapter9

To the Instructor

To complete the chapters in this book, your students must use a set of student files. These files are included in the Instructor's Resource Kit. They may also be obtained electronically through the Internet. See the inside front or back cover of this book for more details. Follow the instructions in the Readme file to copy the student files to your server or standalone computer. You can view the Readme file using a text editor such as WordPad or Notepad.

Once the files are copied, you can make Student Disks for the students yourself, or tell students where to find the files so they can make their own Student Disks. Make sure the files get copied correctly onto the Student Disks by following the instructions in the Student Disks section, which will ensure that students have enough disk space to complete all the chapters and exercises in this book.

Because Oracle can be installed with many variations, you or your database administrator should use the instructions provided in the *Installing and Configuring Oracle on the Client and Server Workstations* instructor's supplement for setting up student user names and accounts and for installing and initializing the Oracle DBMS and client utilities. You also should advise students of any differences that they will encounter in the lab, such as how to start the programs.

Course Technology Student Files

You are granted a license to copy the student files to any computer or computer network used by students who have purchased this book.

Visit Our World Wide Web Site

Additional materials designed especially for you might be available for your course on the World Wide Web. Go to **www.course.com**. Search for this book title periodically on the Course Technology Web site for more details.

CHAPTER 1

An Introduction to Client/Server Databases

objectives

- Identify the differences between client/server databases (such as Oracle) and other popular database environments
- Identify the components of the Oracle database development environment and understand their functions
- Review relational database concepts and terms
- Understand the Clearwater Traders sales order database
- Understand the Northwoods University student registration database

Introduction▶ In this chapter you will learn how a client/server database, such as Oracle, is used to create and maintain data in an organized way. The Oracle development environment includes a client/server database management system (DBMS) and several utilities for developing database applications. In this book you will learn about these utilities and practice using them to build databases for two fictitious organizations: Clearwater Traders and Northwoods University. When you have finished with this chapter, you will have the background you need to begin your exploration of Oracle.

Database Systems

When organizations first began converting from manual to computerized processing systems, each individual application had its own set of data files that were used only for that application. For example, in a bank's computer system, the checking account processing system would have its own data files, the auto loan system would have its own files, and the savings account system would have its own files. This file-based approach to data processing is shown in Figure 1-1.

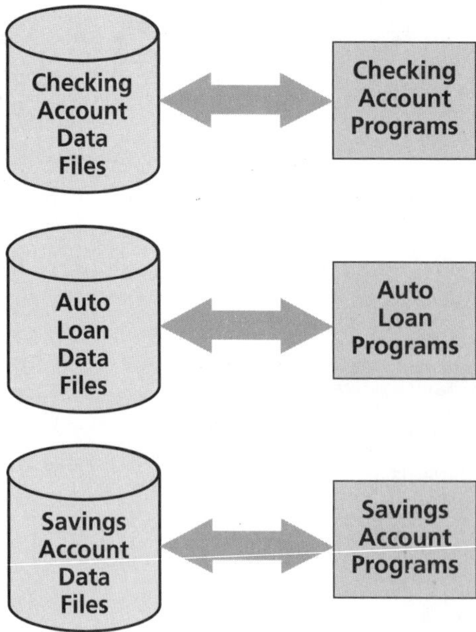

Figure 1-1: File-based approach to data processing

This approach had several problems. One problem was that many redundant programs had to be written and maintained for performing routine file operations such as adding new data, updating data, and deleting data. Another problem was that the same data might exist in multiple files. For example, the same information about a customer (such as the customer's name, address, and phone number) might be in several different files. This redundancy takes up extra storage space and causes a problem when information needs to be updated. For example, when a customer changes her address, the information might not get updated in every file, resulting in data files with inconsistent data.

To address this problem, database systems were developed that viewed data as an organizational resource. All data were stored in a central repository, and applications interfaced with the database, using the DBMS, which provided a central set of common functions for inserting, updating, retrieving, and deleting data. The database approach to data processing is illustrated in Figure 1-2.

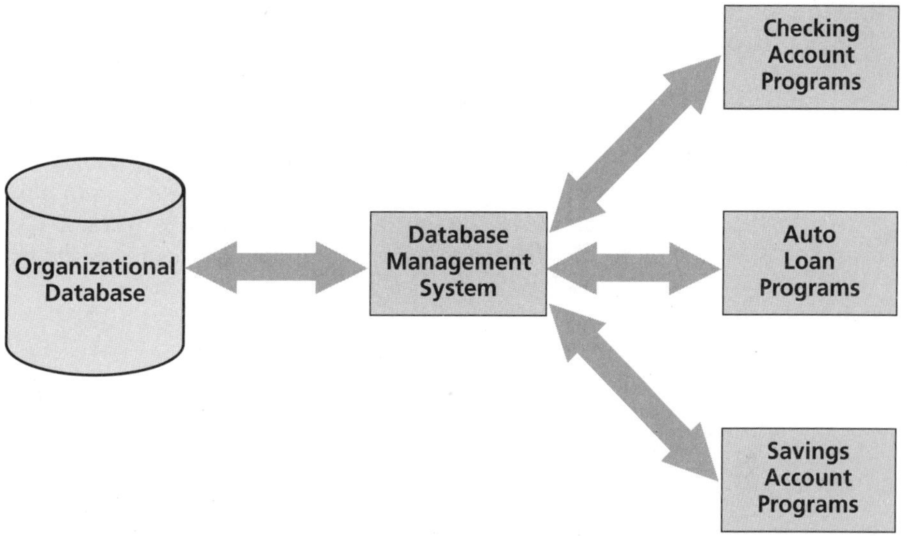

Figure 1-2: Database approach to data processing

The first databases were stored on large centralized mainframe computers that users accessed from **terminals**, which are devices that do not perform any processing but only send keyboard input to and display visual output from a central computer. As distributed computing and microcomputers became popular during the 1980s, two new kinds of databases emerged: **personal databases** and **client/server databases.**

Personal and Client/Server Databases

Personal database systems, such as Microsoft Access and FoxPro, are aimed toward single-user database applications that usually are stored on a single user's desktop computer, or a **client workstation**. When a personal DBMS is used for a multiuser application, the database application files are stored on a file server and transmitted to the individual users across a network, as shown in Figure 1-3. **Server** is a broad term for any computer willing to accept requests from other computers and to share some or all of its resources. **Resources** could include printers connected to a server, files stored on hard disks attached to a server, programs running in a server's main memory, and more. A **network** is an infrastructure of telecommunications hardware and software that enables computers to transmit messages to each other.

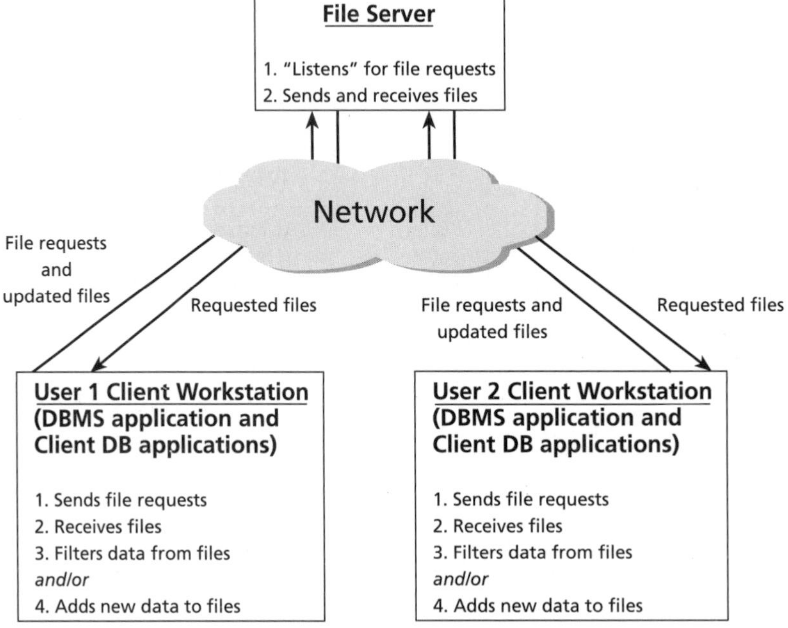

Figure 1-3: Personal database used for multiuser application

With a personal DBMS, each client workstation must load the entire database application into main memory along with the client database application in order to view, insert, update, or print data. A client request for a small amount of data from a large database might require the server to transmit the entire database (which might be hundreds of megabytes in size) to the client's workstation. Newer personal databases use indexed files that enable the server to send only part of the database, but in either case, these DBMSs put a heavy demand on client workstations and on the network. The network must be fast enough to handle the traffic generated when transferring database files to the client workstation and sending them back to the server for database additions and updates. Although system response time depends on the speed of the network, the size of the database, and the way the database is used, a personal database running on a server might handle 10 users making database transactions at the same time before becoming overloaded.

In contrast, **client/server databases,** such as Oracle, split the DBMS and the applications accessing the DBMS into a "process" running on the server and the applications running on the client, as shown in Figure 1-4. The **client application** sends data requests across the network. When the server receives a request, the server **DBMS process** retrieves the data from the database, performs the requested functions on the data (sorting, filtering, etc.), and sends *only* the final query result (not the entire database) back via the network to the client. As a result, multiuser client/server databases generate less network traffic than personal databases and are less likely to bog down due to an overloaded network.

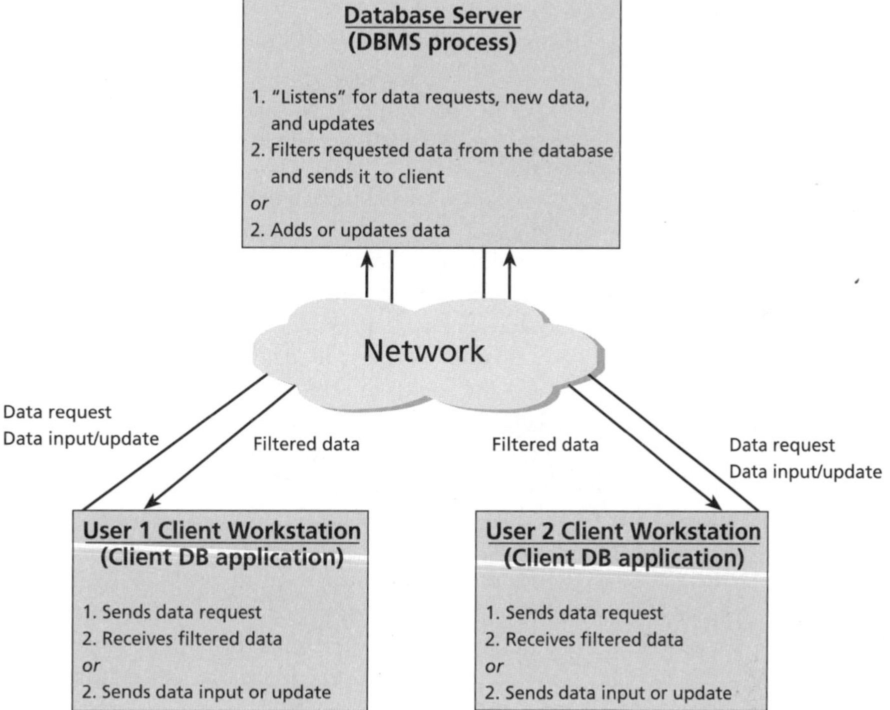

Figure 1-4: Client/server database

Another important difference between client/server and personal databases is how they handle client failures. In a personal database system, when a client workstation fails as a result of a software malfunction or power failure, the database is likely to become damaged due to interrupted updates, insertions, or deletions. Records in use at the time of the failure are **locked** by the failed client, which means they are unavailable to other users. The database might be reparable, but all users must log off during the repair process, which could take several hours. Updates, deletions, and insertions taking place at the time of the failure often cannot be reconstructed. If repair is not possible, the person responsible for installing, administering, and maintaining the database, who is called the **database administrator** or **DBA**, can restore the database to the last regular backup, but transactions that occurred since the backup are lost.

On the other hand, a client/server database is not affected when a client workstation fails. The failed client's in-progress transactions are lost, but the failure of a single client does not affect other users. In the case of a server failure, a central synchronized **transaction log,** which contains a record of all current database changes, enables in-progress transactions from all clients to be either fully completed or rolled back. **Rolling back** a database transaction means that the database is made to look as if the transaction never took place. Using the transaction log, a DBA can notify users with rolled-back transactions to resubmit them. Most client/server DBMS servers have extra features to minimize the chance of failure, and when they do fail, they have fast, powerful recovery mechanisms.

Client/server systems also differ from personal database systems in the way that they handle competing user transactions. Consider an airline seat ticketing system as an example. In a client/server system, such as Oracle, the transaction to sell a ticket causes the database to read the table and simultaneously lock all or part of the table that lists available seats prior to updating the table. Then the sold seat is marked as unavailable (or deleted from the table), and the table is unlocked. Without the lock, a second sales agent could read the table after the first agent reads it but before the first agent updates a certain seat as sold. The second agent might see that the given seat still is available and inadvertently sell the same seat to a second customer. By default, a personal database, such as Microsoft Access, uses **optimistic locking**: it hopes that two competing transactions will not take place at the same time. Optimistic locking, therefore, does not really lock the table. Access will notify the second agent that the table has been changed since he or she last read it, but then Access offers to proceed and save the update anyway, selling the already sold seat. Access *will* allow developers to write explicit code to lock tables, but it takes a lot more effort than a comparable Oracle application that handles locking automatically.

Another difference is in the way client/server systems and personal databases handle transaction processing. **Transaction processing** refers to grouping related database changes into batches that must either all succeed or all fail. For example, assume a customer writes a check from a checking account and deposits it into a money market account. The bank must ensure that the checking account is debited for the amount of the check and that the money market account is credited for the same amount. If any part of the transaction fails, then neither account balance should change. Microsoft Access provides procedures to group related changes, keep a record of these changes in main memory on the client workstation, and roll back the changes if the grouped transactions fail. However, if the client making the changes fails in the middle of a group of transactions, then the transaction log stored in the client's main memory is lost. Unlike Oracle, Microsoft Access doesn't have a file-based transaction log on which to base a rollback, and the partial changes cannot be reversed. Depending on the order of the transactions, a failed client could result in a depleted checking account and an unchanged money market account, or an enlarged money market account and an unchanged checking account.

Client/server databases are preferred for database applications that retrieve and manipulate small amounts of data from databases containing large numbers of records, because they minimize network traffic and improve response times. Client/server systems are essential for mission-critical applications because of their failure handling, recovery mechanisms, and transaction management control. They also have a rich set of database management and administration tools for handling large numbers of users. Some general guidelines are to use a client/server database if your database will have more than 10 simultaneous users, if your database is mission critical, or if you need a rich set of administration tools.

The Oracle8 Environment

The Oracle8 database development environment includes the Oracle8 client/server DBMS as well as several utilities for developing database systems. In this book, you will learn about the following tools:

- *SQL*Plus:* for creating and testing command-line SQL queries used to create, update, and delete database tables, views, and sequences
- *Query Builder:* for creating SQL queries using a visual environment
- *PL/SQL* and *Procedure Builder:* for creating procedural programs that process database data
- *Developer:* for developing database applications. Developer consists of the following tools:
  - *Form Builder:* for creating graphical forms and menus for user applications
  - *Report Builder:* for creating reports for displaying and printing data
  - *Graphic Builder:* for creating graphics charts based on database data
- *Enterprise Manager:* for managing and tuning the database. Enterprise Manager uses the following tools:
  - *Security Manager:* for creating and managing user accounts
  - *Storage Manager:* for creating and managing tablespaces
  - *Instance Manager:* for starting, shutting down, and tuning the database
- *Oracle Web Application Server:* for creating a World Wide Web site that allows users to access Oracle databases and create dynamic Web pages that serve as a database interface

Overview of Relational Databases

Early databases used a **hierarchical** structure, whereby all related data had a parent-to-child relationship: a "parent" data item (such as a customer) could have multiple "child" data items (such as a checking account, an auto loan, and a savings account). Relationships between related data were created using **pointers**, which are links to the physical locations where data are written on a disk. Figure 1-5 illustrates a hierarchical database structure, where the Customer data represent the parent records, and the Checking Account and Auto Loan data represent the child records. Each customer's checking account and auto loan data are linked using a pointer. This approach solved the problems of redundant applications and inconsistent data in file-based systems, but introduced a new problem: the data were physically dependent on their location on the storage media, and it was difficult to migrate a database to a new medium when the disk became full or new hardware was purchased. This, as well as other problems, led to the development of relational databases such as Oracle.

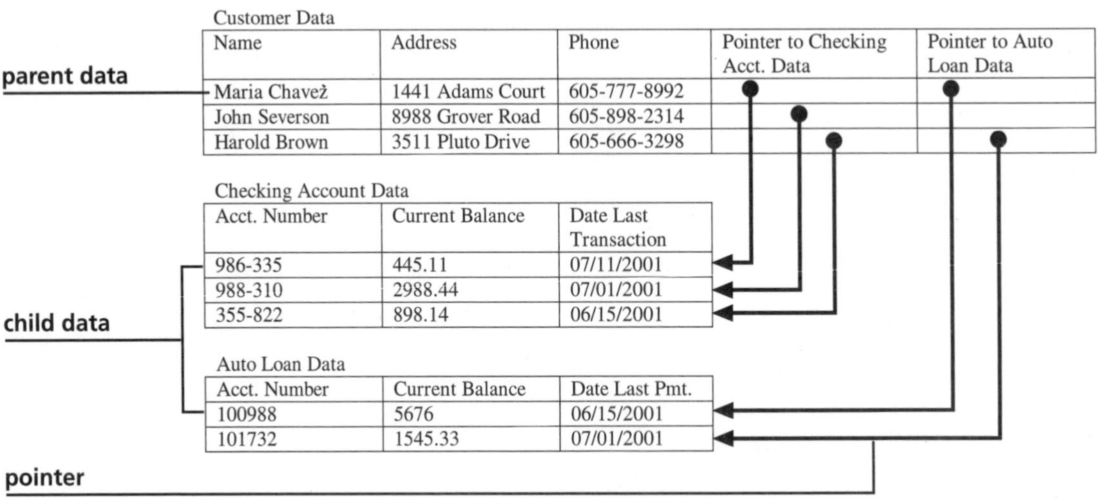

Name	Address	Phone	Pointer to Checking Acct. Data	Pointer to Auto Loan Data
Maria Chavež	1441 Adams Court	605-777-8992		
John Severson	8988 Grover Road	605-898-2314		
Harold Brown	3511 Pluto Drive	605-666-3298		

Acct. Number	Current Balance	Date Last Transaction
986-335	445.11	07/11/2001
988-310	2988.44	07/01/2001
355-822	898.14	06/15/2001

Acct. Number	Current Balance	Date Last Pmt.
100988	5676	06/15/2001
101732	1545.33	07/01/2001

Figure 1-5: Hierarchical database example

A relational database views data in **tables**, or matrixes with columns and rows. **Columns** represent different data categories, and **rows** contain the actual data values. Columns also are called **fields**, and rows also are called **records.** Figure 1-6 shows an example of a relational database that contains two tables.

PRODUCT table

PRODUCT_ID	DESCRIPTION	QUANTITY_ON_HAND
1	Plain Cheesecake	8
2	Cherry Cheesecake	10

PURCHASE table

ORDER_ID	PRODUCT_ID	ORDER_QUANTITY
100	1	2
100	2	2
101	2	1
102	1	3

Figure 1-6: A relational database with two tables

Individual records in relational database tables are identified using primary keys. A **primary key** is a field whose value must be unique for each record. Every record must have a primary key that is defined as such by the database programmer, and the primary key cannot be **NULL**, which means that its value is indeterminate or undefined. In the PRODUCT table shown in Figure 1-6, the PRODUCT_ID field is a good choice for the primary key because a unique value can be assigned for each product. The DESCRIPTION field might be another choice for the primary key, but there are two drawbacks to this option. First, products must always have unique descriptions, and second, the field is a text field and is therefore prone to typographical, spelling, and punctuation data entry errors. This is a problem because primary key values are used to create relationships with other database tables. When you are looking at a relational database table, you cannot tell which key is the primary key, but you can identify fields that could be used as the primary key. These fields are called **candidate keys**. Once you have selected the field that will be used as the primary key, you must specify this field as the primary key when you create the table. Other users can use special commands to identify the primary keys in existing database tables.

In this textbook, database table and field names are identified by typing them in all capital letters. The text reflects the style that Oracle uses to reference objects related to database tables or fields, which can vary depending on the Oracle application you are using.

Relationships among database tables are created by matching key values. For example, suppose Order 100 is for two plain cheesecakes and two cherry cheesecakes. Figure 1-7 shows how Products 1 and 2 in the PRODUCT table relate to Order 100 in the PURCHASE table. A field in one table that is a primary key in another table, and thus creates a relationship between the two tables, is called a **foreign key**. PRODUCT_ID is a foreign key in the PURCHASE table.

PRODUCT table

PRODUCT_ID	DESCRIPTION	QUANTITY_ON_HAND
1	Plain Cheesecake	8
2	Cherry Cheesecake	10

PURCHASE table

ORDER_ID	PRODUCT_ID	ORDER_QUANTITY
100	1	2
100	2	2
101	2	1
102	1	3

Figure 1-7: Creating a relationship using a foreign key

A foreign key value must exist as a primary key in the referenced table. For example, suppose you have a new record for ORDER_ID 103 that specifies that the customer ordered one unit of Product 3. There is no record for PRODUCT_ID 3 in the PRODUCT table, so the purchase record does not make sense. Foreign key values must match the value in the primary key table *exactly*. That is why it is not a good idea to use text fields, and risk typographical, punctuation, and spelling errors in the primary key.

Sometimes you have to combine multiple fields to create a unique primary key. Figure 1-8 shows an example of this situation. For ORDER_ID 100, the customer purchased two units of Product 1 and two units of Product 2, so the ORDER_ID value is not unique for each record. However, the combination of the ORDER_ID and PRODUCT_ID values is unique. When multiple fields are used to create a unique primary key, the key is called a **composite primary key**, or a **composite key**. Note that PRODUCT_ID, which is a foreign key, also can be part of a composite key. Sometimes when you are looking for a candidate key for a table, you find that you must use multiple fields to uniquely identify a record.

PURCHASE table

composite primary key

ORDER_ID	PRODUCT_ID	ORDER_QUANTITY
100	1	2
100	2	2
101	2	1
102	1	3

Figure 1-8: Example of a composite primary key

Sometimes a database table does not have a field that would make a good primary key. Figure 1-9 shows a CUSTOMER table with this problem. The LAST_NAME or FIRST_NAME field, or even the combination of the two fields, is not a good candidate key because many people have the same name. Multiple people can share the same address and phone number as well. Phone numbers are not good candidate keys because people often get new phone numbers, and if the phone number is updated in the table where it is the primary key, it must also be updated in every table where it is a foreign key. If the phone number is not updated, relationships are lost. A good database development practice is to create a **surrogate key**, which is a numerical value that usually is generated by the database, with the sole purpose of being the primary key identifier for a record. In Figure 1-6, PRODUCT_ID and ORDER_ID are examples of surrogate keys. Surrogate keys are unique identifiers that provide no unique information about the record. For the table in Figure 1-9, you probably would create a surrogate key called CUSTOMER_ID, and you might start the CUSTOMER_ID numbers at 1 or 100. The CUSTOMER_ID numbers will not change, and every customer gets a unique number.

CUSTOMER table

LAST_NAME	FIRST_NAME	ADDRESS	PHONE
Brown	John	101 Main Street	7155554321
Brown	John	3567 State Street	7155558901
Carlson	Mike	233 Water Street	7155557890
Carlson	Martha	233 Water Street	7155557890
Davis	Carol	1414 South Street	7155555566

Figure 1-9: Table lacking suitable primary key candidates

The Case Study Databases

This textbook uses chapter examples and end-of-chapter exercises that illustrate the Oracle utilities, using databases developed for Clearwater Traders and Northwoods University, which are two fictitious organizations. The focus of this textbook is on database development rather than on database design, so the text does not provide an explanation of the rationale behind the design of these database tables.

The described database systems require a client/server database application because they will have many simultaneous users accessing the system from different locations. Each database has data examples to illustrate the tasks that are addressed in this textbook, including the development of data entry and maintenance forms, output reports, chart displays, and a Web site application.

The Clearwater Traders Sales Order Database

Clearwater Traders markets a line of clothing and sporting goods via mail-order catalogs and a site on the World Wide Web. Clearwater Traders accepts customer orders via phone, mail, and fax. The company recently experienced substantial growth, and as a result, management decided to offer 24-hour customer order service. The existing microcomputer-based database system is not able to handle the current transaction volume that is generated by sales representatives processing incoming orders. Management is concerned that this system does not have the failure-handling and recovery capabilities needed to withstand any failures or downtime.

When a customer orders an item, the sales representative must check to see if the ordered item is in stock. If the item is in stock, the representative must update the available quantity on hand to reflect that the item has been sold. If the item is not in stock, the representative needs to advise the customer when the item will be available. When new inventory shipments are received, a receiving clerk must update the inventory to show the new quantities on hand. The system must produce invoices that can be included with customer shipments and reports showing inventory levels, possibly using charts or graphs. Marketing managers would like to be able to track each order's source (particular catalog number or Web site) to help plan future promotions.

The following data items have been identified:

- Customer name, address, and daytime and evening phone numbers
- Order date, payment method (check or credit card), order source (catalog number or Web site), and associated item numbers, sizes, colors, and quantities ordered
- Item descriptions, categories (women's clothing, outdoor gear, etc.), prices, quantities on hand, and next shipment information for out-of-stock items. Many clothing items are available in multiple sizes and colors. Sometimes the same item has different prices depending on the item size.
- Information about incoming product shipments and back-ordered shipments

Figure 1-10 shows sample data for Clearwater Traders. Five customer records are shown. Customer 107 is Paula Harris, who lives at 1156 Water Street, Apt. #3, Osseo, WI, and her ZIP code is 54705. Her daytime phone number is 715-555-8943, and her evening phone number is 715-555-9035. CUSTID has been designated as the table's primary key.

CUSTOMER

CUSTID	LAST	FIRST	MI	CADD	CITY	STATE	ZIP	DPHONE	EPHONE
107	Harris	Paula	E	1156 Water Street, Apt. #3	Osseo	WI	54705	7155558943	7155559035
232	Edwards	Mitch	M	4204 Garner Street	Washburn	WI	54891	7155558243	7155556975
133	Garcia	Maria	H	2211 Pine Drive	Radisson	WI	54867	7155558332	7155558332
154	Miller	Lee		699 Pluto St. NW	Silver Lake	WI	53821	7155554978	7155559002
179	Chang	Alissa	R	987 Durham Rd.	Sister Bay	WI	54234	7155557651	7155550087

CUST_ORDER

ORDERID	ORDERDATE	METHPMT	CUSTID	ORDERSOURCE
1057	5/29/2001	CC	107	152
1058	5/29/2001	CC	232	WEBSITE
1059	5/31/2001	CHECK	133	152
1060	5/31/2001	CC	154	153
1061	6/01/2001	CC	179	WEBSITE
1062	6/01/2001	CC	179	WEBSITE

Figure 1-10: Sample data for the Clearwater Traders database

ITEM

ITEMID	ITEMDESC	CATEGORY
894	Women's Hiking Shorts	Women's Clothing
897	Women's Fleece Pullover	Women's Clothing
995	Children's Beachcomber Sandals	Children's Clothing
559	Men's Expedition Parka	Men's Clothing
786	3-Season Tent	Outdoor Gear

INVENTORY

INVID	ITEMID	ITEMSIZE	COLOR	CURR_PRICE	QOH
11668	786		Sienna	259.99	16
11669	786		Forest	259.99	12
11775	894	S	Khaki	29.95	150
11776	894	M	Khaki	29.95	147
11777	894	L	Khaki	29.95	0
11778	894	S	Olive	29.95	139
11779	894	M	Olive	29.95	137
11780	894	L	Olive	29.95	115
11795	897	S	Teal	59.95	135
11796	897	M	Teal	59.95	168
11797	897	L	Teal	59.95	187
11798	897	S	Coral	59.95	0
11799	897	M	Coral	59.95	124
11800	897	L	Coral	59.95	112
11820	995	10	Blue	15.99	121
11821	995	11	Blue	15.99	111
11822	995	12	Blue	15.99	113
11823	995	1	Blue	15.99	121
11824	995	10	Red	15.99	148
11825	995	11	Red	15.99	137
11826	995	12	Red	15.99	134
11827	995	1	Red	15.99	123
11845	559	S	Navy	199.95	114
11846	559	M	Navy	199.95	17
11847	559	L	Navy	209.95	0
11848	559	XL	Navy	209.95	12

Figure 1-10 (continued): Sample data for the Clearwater Traders database

SHIPPING

SHIPID	INVID	DATE_EXPECTED	QUANTITY_EXPECTED	DATE_RECEIVED	QUANTITY_RECEIVED
211	11668	09/15/2001	25		
211	11669	09/15/2001	25		
212	11669	11/15/2001	25		
213	11777	06/25/2001	200		
214	11778	09/25/2001	200		
214	11779	09/25/2001	200		
215	11798	08/15/2001	100		
215	11799	08/15/2001	100		
216	11799	08/25/2001	100	08/25/2001	0
217	11800	08/15/2001	100		
218	11845	8/12/2001	50	8/15/2001	0
218	11846	8/12/2001	100	8/15/2001	100
218	11847	8/12/2001	100	8/15/2001	50
218	11848	8/12/2001	50	8/15/2001	50

BACKORDER

BACKORDERID	SHIPID	INVID	DATE_EXPECTED	QUANTITY_EXPECTED	DATE_RECEIVED	QUANTITY_RECEIVED
1	216	11799	09/01/2001	100	08/31/2001	100
2	218	11845	09/15/2001	50		
3	218	11847	09/15/2001	50		

ORDERLINE

ORDERID	INVID	ORDER_PRICE	QUANTITY
1057	11668	259.99	1
1057	11800	69.96	2
1058	11824	15.99	1
1059	11846	129.95	1
1059	11848	139.95	1
1060	11798	59.95	2
1061	11779	29.95	1
1061	11780	29.95	1
1062	11799	59.95	1
1062	11669	229.99	3

COLOR

COLOR
Sienna
Forest
Khaki
Olive
Teal
Coral
Blue
Red
Navy
Brown

ORDERSOURCE

ORDERSOURCE
99
122
123
145
146
151
152
153
211
WEBSITE

Figure 1-10 (continued): Sample data for the Clearwater Traders database

The CUST_ORDER table shows six customer orders. The first is 1057, dated 5/29/2001, method of payment CC (credit card), and ordered by customer 107, Paula Harris. Catalog 152 was the order source. ORDERID is the table's primary key, and CUSTID is a foreign key that creates a relationship to the CUSTOMER table.

The ITEM table contains five items: Women's Hiking Shorts, Women's Fleece Pullover, Children's Beachcomber Sandals, Men's Expedition Parka, and 3-Season Tent. Item 894, Women's Hiking Shorts, is in the Women's Clothing category. ITEMID is the primary key in this table, and it also is a surrogate key that was created because neither of the other table fields is a good primary key choice—ITEMDESC is a text field, and CATEGORY is also a text field and not unique for every record.

The INVENTORY table contains specific inventory numbers for specific item sizes and colors. It also shows the current price and quantity on hand (QOH) for each item. Items that are not available in different sizes contain NULL, or undefined values, in the ITEMSIZE column. Notice that some items have different prices for different sizes. INVID is the primary key of this table, and ITEMID is a foreign key that creates a relationship with the ITEM table.

The SHIPPING table contains a schedule of expected inventory item shipments that includes the inventory item number, the date and quantity of expected items, and the date and quantity of items delivered. As you can see in the table, a shipment can consist of one or more inventory items. The primary key of this table is a composite key made up of the combination of SHIPID and INVID. Remember that composite keys are required when no single field in the table uniquely identifies each record in the table.

Why is SHIPID needed in this table? First, without SHIPID, the primary key could be composed of both INVID and DATE because it would take both of these entries to identify a record in this table uniquely. But what if there are two separate shipments of the same item arriving on the same day? And what if both shipments have the same quantity? SHIPID provides a useful way to distinguish that there are *two* separate shipments for the same item on the same day.

The BACKORDER table shows back orders corresponding to shipments. Look at the records in the SHIPPING table for shipment 218. When the shipment arrived on 8/15/2001, none of the units for item 11845 arrived, and only 50 of the 100 units ordered for item 11847 arrived. Clearwater Traders was notified that the missing units were back-ordered and would be shipped on 9/15/2001. The BACKORDER records show the associated shipment ID, inventory ID, expected date, and expected quantity for these back-ordered items.

The ORDERLINE table represents the individual inventory items in a customer order. The first line of order 1057 is one Sienna-colored 3-Season Tent, with a price of 259.99. The second line of this order specifies two large Coral Women's Fleece Pullovers, with a price of 69.96 each. This information is used to create the printed customer order invoice and to calculate sales revenues. The order prices must be retained to show the actual price paid for an item on a particular order, because the price could have changed in the INVENTORY table (because of a markdown or discount) since the order was placed. Note that the primary key of this table is not ORDERID, because more than one record might have the same ORDERID. The primary key is a composite key made up of the combination of ORDERID and INVID. An order might have several different inventory items, but it will never have the same inventory item listed more than once. Along with being part of the primary key, INVID is also a foreign key in this table because it creates a relationship to the INVENTORY table.

The last two tables, COLOR and ORDERSOURCE, are lookup tables, or pick lists. A **lookup table** is a list of legal values for a field in another table. Notice the variety of colors shown in the INVENTORY table (Sienna, Forest, Khaki, Olive, Teal, etc.). If users are required to type these colors each time an inventory item is added to the table, typing errors might occur. For example, a query looking for sales of items with the Sienna color will not find instances where Sienna is misspelled as Siena, SIENNA, etc. Typically, a user entering a new inventory item will select a color from a pick list that displays values from the COLOR table, so the color is not typed directly, thus reducing errors. Small lists that are unlikely to change over time might be coded directly into an application, but lists with many items that might be added to over time usually are stored in a separate lookup table.

The Northwoods University Student Registration Database

Northwoods University has decided to replace its aging mainframe-based student registration system with a more modern client/server database system. School officials want to provide students with the capability to retrieve course availability information, register for courses, and print unofficial transcripts, using personal computers located in the student computer labs. Additionally, faculty members must be able to retrieve student course lists, drop and add students to courses, and record course grades. Faculty members must also be able to view records for the students they advise. Security is a prime concern, so student and course records must be protected by password access.

Students will log on to the Northwoods system using their name and PIN (personal identification number). They will be given the option of viewing current course listings or viewing information on courses they have completed. They can check which courses are available during the current term by viewing course information such as course names, call IDs (such as MIS 101), section numbers, days, times, locations, and the availability of open seats in the course. They also can view information about the courses they have taken in the past, and print a transcript report showing past course grades and grade point averages. Faculty members will log on to the system by entering their PINs. Then they can select from a list of the courses they are teaching in the current term, and retrieve a list of students enrolled in the selected course. A faculty member also can retrieve a list of his or her student advisees, select one, and then retrieve that student's past and current course enrollment information. The data items for the Northwoods database are:

1. Student name, address, phone number, class (freshman, sophomore, etc.), date of birth, and advisor ID
2. Course call number, name, credit, maximum enrollment, instructor, and term offered
3. Instructor name, office location, phone number, and rank
4. Student/course grades

Figure 1-11 shows sample data for the Northwoods database. Six student records are shown. Student 100 is Sarah Miller, who lives at 144 Windridge Blvd., Eau Claire, WI, and her ZIP code is 54703. Her phone number is 715-555-9876, she is a senior, her date of birth is 7/14/79, and her faculty advisor is Kim Cox. Note that the SPIN (student PIN) field has been added to store student personal identification numbers to control data access. Only the student or his faculty advisor is able to update student record data in this table. SID (student ID) is the table's primary key. FID (faculty ID) is a foreign key that refers to the FID field in the FACULTY table.

STUDENT

SID	SLNAME	SFNAME	SMI	SADD	SCITY	SSTATE	SZIP	SPHONE	SCLASS	SDOB	SPIN	FID
100	Miller	Sarah	M	144 Windridge Blvd.	Eau Claire	WI	54703	7155559876	SR	07/14/79	8891	1
101	Umato	Brian	D	454 St. John's Street	Eau Claire	WI	54702	7155552345	SR	08/19/79	1230	1
102	Black	Daniel		8921 Circle Drive	Bloomer	WI	54715	7155553907	JR	10/10/77	1613	1
103	Mobley	Amanda	J	1716 Summit St.	Eau Claire	WI	54703	7155556902	SO	9/24/78	1841	2
104	Sanchez	Ruben	R	1780 Samantha Court	Eau Claire	WI	54701	7155558899	SO	11/20/77	4420	4
105	Connoly	Michael	S	1818 Silver Street	Elk Mound	WI	54712	7155554944	FR	12/4/77	9188	3

FACULTY

FID	FLNAME	FFNAME	FMI	LOCID	FPHONE	FRANK	FPIN
1	Cox	Kim	J	53	7155551234	ASSO	1181
2	Blanchard	John	R	54	7155559087	FULL	1075
3	Williams	Jerry	F	56	7155555412	ASST	8531
4	Sheng	Laura	M	55	7155556409	INST	1690
5	Brown	Phillip	E	57	7155556082	ASSO	9899

Figure 1-11: Sample data for the Northwoods University database

LOCATION

LOCID	BLDG_CODE	ROOM	CAPACITY
45	CR	101	150
46	CR	202	40
47	CR	103	35
48	CR	105	35
49	BUS	105	42
50	BUS	404	35
51	BUS	421	35
52	BUS	211	55
53	BUS	424	1
54	BUS	402	1
55	BUS	433	1
56	LIB	217	2
57	LIB	222	1

TERM

TERMID	TDESC	STATUS
1	Fall 2000	CLOSED
2	Spring 2001	CLOSED
3	Summer 2001	CLOSED
4	Fall 2001	CLOSED
5	Spring 2002	CLOSED
6	Summer 2002	OPEN

COURSE

CID	CALLID	CNAME	CCREDIT
1	MIS 101	Intro. to Info. Systems	3
2	MIS 301	Systems Analysis	3
3	MIS 441	Database Management	3
4	CS 155	Programming in C++	3
5	MIS 451	Client/Server Systems	3

Figure 1-11 (continued): Sample data for the Northwoods University database

COURSE_SECTION

CSECID	CID	TERMID	SECNUM	FID	DAY	TIME	LOCID	MAXENRL	CURRENRL
1000	1	4	1	2	MWF	10:00	45	140	135
1001	1	4	2	3	TTh	9:30	51	35	35
1002	1	4	3	3	MWF	8:00	46	35	32
1003	2	4	1	4	TTh	11:00	50	35	35
1004	2	5	2	4	TTh	2:00	50	35	35
1005	3	5	1	1	MWF	9:00	49	30	25
1006	3	5	2	1	MWF	10:00	49	30	28
1007	4	5	1	5	TTh	8:00	47	35	20
1008	5	5	1	2	MWF	2:00	49	35	32
1009	5	5	2	2	MWF	3:00	49	35	35
1010	1	6	1	1	M-F	8:00	45	50	35
1011	2	6	1	2	M-F	8:00	50	35	35
1012	3	6	1	3	M-F	9:00	49	35	29

ENROLLMENT

SID	CSECID	GRADE
100	1000	A
100	1003	A
100	1005	B
100	1008	B
101	1000	C
101	1004	B
101	1005	A
101	1008	B
102	1000	C
102	1011	
102	1012	
103	1010	
103	1011	
104	1000	B
104	1004	C
104	1008	C
104	1012	
105	1010	
105	1011	

Figure 1-11 (continued): Sample data for the Northwoods University database

The FACULTY table describes five faculty members. The first record shows faculty member Kim Cox, whose office is located at BUS 424, and her phone number is 715-555-1234. She has the rank of associate professor, and her PIN is 1181. This FPIN (faculty PIN) will be used as a password to control a faculty member's access to records and ability to update specific student or course records. FID is the primary key, and LOCID is a foreign key that references LOCID in the LOCATION table.

LOCATION is another lookup table. In this case, users will see a sorted list that identifies building codes, room numbers, and room capacities. LOCID is the primary key. After a user makes a selection from the LOCATION table, the LOCID is inserted into the COURSE_SECTION table and FACULTY table. The user never needs to see the LOCID field in the LOCATION table.

The TERM table provides a textual description of each term along with an ID number that is used to link the semester to different course offerings, and a

STATUS field that shows whether enrollment is open or closed. The first record shows a TERMID of 1 for the Fall 2000 term, with enrollment status as CLOSED. TERMID is the primary key.

The COURSE table shows five courses. The first, CID 1, has the call ID MIS 101 and is named "Intro. to Info. Systems." It provides three credits. CID is the primary key, and there are no foreign keys.

The COURSE_SECTION table shows the course offerings for specific terms and includes fields for course ID, term, section number, ID of the instructor teaching the section, and course day, time, location, maximum allowable enrollment, and current enrollment. CSECID is the primary key, and CID, TERMID, FID, and LOCID are all foreign key fields. The FID field is an example of a foreign key field that can have a NULL value. The first record shows that CSECID 1000 is Section 1 of MIS 101. It is offered in the Fall 2001 term and is taught by John Blanchard. The section meets on Mondays, Wednesdays, and Fridays at 10:00 in room CR 101. It has a maximum enrollment of 140 students, and 135 students are currently enrolled.

The ENROLLMENT table shows students who currently are enrolled in each course section, and their associated grade if it has been assigned. The primary key for this table is a composite key of SID and CSECID.

SUMMARY

- Personal database management systems (DBMSs) are aimed toward single-user database applications that are usually stored on a single user's desktop computer, whereas a client/server DBMS runs on a network server and the user applications run on the individual client workstations.
- Personal databases download all of the data a user needs to the user's client workstation, manipulate the data, and then upload it back to the server. This process can be slow and cause network congestion.
- Client/server databases send data requests to the server and the results of data requests back to the client workstation. This process minimizes network traffic and congestion.
- Client/server databases have better failure recovery mechanisms than personal databases.
- Client/server databases automatically provide table and record locking.
- Client/server databases maintain a file-based transaction log that is not lost in the event of a client workstation failure.
- Client/server databases have utilities for managing systems with many users.
- The Oracle database development environment has utilities for creating database tables and queries; developing data forms, reports, and graphics based on database table data; performing database administration tasks; and creating database applications that run on the World Wide Web.
- A relational database views data in tables, or matrixes with columns and rows. Columns represent different data categories, and rows contain the actual data values.

- A primary key is a field that uniquely identifies every record in a database table. Primary key values must be unique within a table and cannot be NULL.
- A surrogate key can be created if no single existing data field uniquely identifies each record in a table or if existing primary key candidates are unsuitable because they are text fields or their values might change.
- A composite key is a primary key composed of the combination of two or more fields.
- Foreign keys link related records. A foreign key field in a table must exist as a primary key in another table.
- The Clearwater Traders sales order database includes the following data tables:
    - CUSTOMER
    - CUST_ORDER
    - ITEM
    - INVENTORY
    - SHIPPING
    - BACKORDER
    - ORDERLINE
    - COLOR
    - ORDERSOURCE
- The Northwoods University student registration database includes the following data tables:
    - STUDENT
    - FACULTY
    - LOCATION
    - TERM
    - COURSE
    - COURSE_SECTION
    - ENROLLMENT

REVIEW QUESTIONS

1. What is the main difference between the way that client/server and personal databases handle and process data?
2. Why can client/server databases handle client failures better than personal systems?

3. Give two examples of database applications that might be appropriate for a client/server system. Give two examples of database applications that might be more appropriate for a personal database system. (*Hint*: Personal database systems are appropriate for situations in which a client/server system is not required.)
4. In a relational database table, what is a field, and what is a record?
5. What is a primary key?
6. What is a foreign key?
7. What is a surrogate key?
8. What is a composite key?
9. What does it mean when a data value is NULL?

Answer the following questions, using the database tables shown in Figure 1-10. No computer work is necessary; you will learn how to do these queries using SQL queries with the Oracle database in another chapter.

10. Identify every primary and foreign key field in the Clearwater Traders database.
11. Find the names of every customer who placed orders using the Web site as a source.
12. Find the names of every item that currently is out of stock (QOH = 0).
13. Find the customer name, item description, size, color, quantity, extended total (quantity times order price), and total order amount for every item in order 1057.
14. Find the name and address of every customer who ordered items in the Women's Clothing category.
15. Find the total amount of every order generated from the Web site source.
16. Find the total amount of every order generated from the Catalog 152 source.

Answer the following questions, using the database tables shown in Figure 1-11. No computer work is necessary; you will learn how to do these queries using SQL queries with the Oracle database in another chapter.

17. Identify every primary and foreign key field in the Northwoods University database.
18. Find the name of every course offered during the Summer 2002 term.
19. List the name of every student that faculty member Kim Cox advises.
20. List the course name and section of every course that is filled to capacity (CURRENRL = MAXENRL).
21. List the course call ID, section number, term description, and grade for every course that student Sarah Miller is taking.
22. Calculate the total credits earned to date by student Brian Umato.
23. Calculate the total number of students taught by John Blanchard during the Spring 2002 term (use the CURRENRL figures).
24. Calculate the total number of student credits generated during the Spring 2002 term (student credits = CURRENRL * course credits).

CHAPTER

2

Creating and Modifying Database Tables

objectives

- Identify the properties of Oracle database tables
- Learn about the different data types used in Oracle database tables
- Learn how integrity and value constraints are defined in Oracle database tables
- Create and modify database tables using SQL*Plus
- Learn how to modify existing database tables

Introduction▶ Data in relational database tables are inserted, retrieved, and modified using commands called **queries**. Queries are performed using high-level **query languages** that use standard English commands such as INSERT, UPDATE, and DELETE. The standard query language for relational databases is **SQL (Structured Query Language)**. The structure and syntax of SQL are the same across most database management systems (DBMSs), although different platforms often support slightly different commands.

The first task in developing a database is to create the database tables. From a physical viewpoint, a database is a set of files stored on a disk. From a logical viewpoint, a database is a set of user accounts identified by unique user names and passwords. Each user account owns tables and other data objects. Within an individual user account, every table name must be unique. In this book, you will create the tables for both the Northwoods University and Clearwater Traders databases in your user account. You can create tables using SQL*Plus, the Oracle command-line SQL environment. This chapter will introduce SQL*Plus and show you how to create and modify Oracle database tables.

Table Names and Properties

A **table** stores data in a relational database. When you create an Oracle database table, you must specify the table name, the name of each data field, and the data type and size of each data field. You must also define constraints that specify whether the field is a primary key, whether it is a foreign key, whether NULL values are allowed, and/or whether data are restricted to certain values, such as an M or F for a gender field. Table and field names can be from one to 30 characters long and can consist only of alphanumeric characters and the special characters $, _, and #. Table and field names must begin with a letter and cannot contain blank spaces or hyphens. Figure 2-1 shows some invalid SQL table and field names and their descriptions.

Invalid Table or Field Name	Description
STUDENT TABLE	Spaces not permitted
STUDENT-TABLE	Hyphens not permitted
#CUST	Must begin with a letter
US_SOCIAL_SECURITY_NUMBERS_COLUMN	Cannot exceed 30 characters

Figure 2-1: Invalid SQL table and field names

Data Types

When you create a table, you must assign to each column a **data type** to specify what kind of data will be stored in the field. Data types are used for two primary reasons. First, assigning a data type provides a means for error checking. For example, you cannot store the character data "Chicago" in a field assigned a

DATE data type. Data types also cause storage space to be used more efficiently by optimizing the way specific types of data are stored. The main Oracle character data types are described next.

Note: A convention used in this book is that in SQL commands, all command words, also known as reserved words, appear in all uppercase letters, and all user-supplied variable names appear in lowercase letters (although Oracle displays them in uppercase letters). You should use this convention when entering your own queries, as well. While SQL*Plus commands are not case sensitive, this convention makes queries, as well as the examples in this book, easier to interpret. When a table or field is referenced outside of a command, its name will appear in all uppercase letters.

VARCHAR2, CHAR, NCHAR, and LONG Character Data Types

Alphanumeric fields containing text and numbers not used in calculations (such as telephone numbers and postal codes) are stored in character data fields. The main Oracle character data types, VARCHAR2, CHAR, NCHAR, and LONG, each store data in a different way.

VARCHAR2 Character Data Type The **VARCHAR2** data type stores variable-length character data up to a maximum of 4,000 characters in Oracle8, expanded from 2,000 characters in previous Oracle versions. When you declare a field using VARCHAR2, you must specify a field size. If entered data are smaller than the specified size, only the entered data are stored. Trailing blank spaces are not added to the end of the entry to make it fill the specified column length. If an entered data value is wider than the specified size, an error is returned. Because no trailing blank spaces are added, this data type is recommended for most character data fields. For example, the SLNAME field in the STUDENT table is defined as `slname VARCHAR2(30)`. This means that the field in which a student's last name is entered is a variable-length character field with a maximum of 30 characters. Examples of data stored in this field include Miller and Umato.

CHAR Character Data Type The **CHAR** data type holds fixed-length character data up to a maximum size of 255 characters. If the value stored in a CHAR field is less than the field size you specify, then trailing spaces are added. If no field size is specified, the default size value is one character. An example of a CHAR field definition is the SCLASS field in the STUDENT table in the Northwoods University database. The definition of the SCLASS field is `sclass CHAR(2).` The data that will be stored in the SCLASS field are SR, JR, SO, FR (senior, junior, sophomore, freshman).

You should use the CHAR data type only for fixed-length character fields that have a restricted set of values, such as the SCLASS field. Using the CHAR data type on fields that might not fill the column width forces the DBMS to add trailing blank spaces, which causes inconsistent query results in other Oracle applications.

The CHAR data type uses data storage space more efficiently than VARCHAR2 and can be processed faster. However, if there is any chance that all column spaces will not be filled, use the VARCHAR2 data type.

The next two data types, NCHAR and LONG, are advanced Oracle data formats that are not covered in this book. Definitions are included here for background purposes only.

NCHAR Character Data Type Typically, character data values are stored using American Standard Code for Information Interchange (ASCII) coding, which represents each character as an 8-digit (1-byte) binary value, and can represent a total of 256 different characters. To represent data in alternate alphabets, such as the Japanese Kanji character set, which has thousands of different characters, each character must be encoded using a 16-digit binary value. The **NCHAR** data type is similar to the CHAR data type, except that it supports 16-digit (2-byte) binary character codes.

LONG Character Data Type The **LONG** data type is used to store large amounts (up to 2 gigabytes) of variable-length character data. The LONG data type is appropriate when you need to store an indeterminate amount of unformatted textual data, such as that found in an invoice, letter, or report. You can include only one LONG field in a table. Suppose you need to store a student's transcript in the STUDENT table. This field would be declared as `s_transcript LONG`. Then the S_TRANSCRIPT field could store each student's transcript in a text format.

NUMBER Data Types

The **NUMBER** data type stores negative, positive, fixed, and floating point numbers between 10^{-130} and 10^{126} with precision up to 38 decimal places. The NUMBER data type is used for all numerical data and must be used for fields that will be used in mathematical operations. When you declare a NUMBER field, you can specify the **precision** (the total number of digits both to the left and to the right of the decimal point), and the **scale** (the number of digits to the right of the decimal point). There are three NUMBER data types: integer, fixed-point, and floating-point.

Integers An **integer** is a whole number with no digits to the right of the decimal point. For example, the SID field in the STUDENT table stores only integers (student identification numbers), so it is defined as `sid NUMBER(5)`. Examples of values in this field are 100, 101, 102, etc.

NUMBER(5) specifies that the SID field has a maximum length of five digits and has no digits to the right of the decimal point. When you are defining any field, it is important to allow room for growth. Northwoods University begins its SID numbers at 100, but there will be more than 999 students enrolled, so the data type is defined to store five digits, or up to 99,999 students.

Fixed-Point Numbers A **fixed-point number** contains a specific number of decimal places. An example is the CURR_PRICE field in the INVENTORY table in the Clearwater Traders database, which is defined as `curr_price NUMBER(5,2)`. Examples of such values are 259.99 and 59.99.

NUMBER(5,2) specifies that all values will have exactly two digits to the right of the decimal point, and that there is a total of five digits. Therefore, no merchandise prices can exceed $999.99. Note that the decimal point itself is *not* included in the precision value.

Floating-Point Numbers A **floating-point value** is a number with a variable number of decimal places. The decimal point can appear anywhere, from before the first digit to after the last digit, or can be omitted entirely. Floating-point values are defined in Oracle by *not* specifying either the precision or scale in the field declaration. While no floating-point values exist in either of the case study databases, a potential floating-point field in the Northwoods database could be student grade point average, which might be declared as `s_gpa NUMBER`. A student's GPA might include one or more decimal places, such as 2.7045, 3.25, or 4.0. A floating-point type allows you to store precise numbers.

DATE Data Type

The **DATE** data type stores dates from January 1, 4712 BC to December 31, 4712 AD. The DATE data type stores the century, year, month, day, hour, minute, and second. The default date format is DD-MON-YY to indicate the day of the month, a hyphen, the month (abbreviated using three capital letters), another hyphen, and the last two digits of the year. The default time format is HH:MI:SS A.M., to indicate the hours, minutes, and seconds using a 12-hour clock. If no time is specified in a date data value, the system default time is 12:00:00 A.M. If no date is specified when a time is entered in a DATE data value, the default date is the first day of the current month.

The SDOB field (student date of birth) in the STUDENT table is declared as `sdob DATE`, which stores a value such as 07-OCT-67 12:00:00 A.M. DATE fields are stored in a standard internal format in the database, so no length specification is required.

"DATE" is a reserved word, so it cannot be used as a field name.

Large Object Data Types

Sometimes you might need to store binary data such as digitized sounds or images, or references to binary files from a word processor or spreadsheet. In these cases, you can use one of the new large object (LOB) data types available in Oracle8. The four LOB data types are summarized in Figure 2-2. Since LOB data types are not yet readily supported in many Oracle tools, they will not be included in this book and are described here for background purposes only.

LOB Data Type	Description
BLOB	Binary LOB, storing up to 4 GB of binary data in the database
CLOB	Character LOB, storing up to 4 GB of character data in the database
BFILE	Binary File, storing a reference to a binary file that is located outside the database in a file maintained by the operating system
NCLOB	Character LOB that supports 2-byte character codes

Figure 2-2: LOB data types

Suppose you need to store an image containing photographs of items in the INVENTORY database table. You could store the actual raw (binary) image data using a BLOB data type, or you could store a reference to the location of an external image file using the BFILE data type.

Format Masks

Recall that the default date format is DD-MON-YY HH:MI:SS A.M. Suppose you have a date data field that currently contains the value 07/22/00 9:29:00 P.M. To display the date in a different format (such as July 22, 2000 or 22-JUL-00 9:29 P.M.), you would specify an alphanumeric character string called a **format mask** to specify the desired output format. Format masks only affect the way the data are displayed or printed; the full data value is stored in the database. Format masks use "9" as a default value to represent a number. Figure 2-3 lists how the numeric data value stored as 059783 is displayed using some common numerical data format masks.

Format Mask	Explanation	Sample Data
99999	Number of 9s determines display width	59783
099999	Displays leading zeros	059783
$99999	Prefaces the value with dollar sign	$59783
99999MI	Displays "-" before negative values	-59783
99999PR	Displays negative values in angle brackets	<59783>
99,999	Displays a comma in the indicated position	59,783
99999.99	Displays a decimal point in the indicated position	59783.00

Figure 2-3: Common numerical format masks

Figure 2-4 shows some common date format masks using the example date of 5:45:35 P.M., Friday, January 15, 2001.

Format Mask	Explanation	Sample Data
YYYY	Displays 4-digit year	2001
YYY or YY or Y	Displays last 3, 2, or 1 digits of year	001, 01, 1
RR	Last two digits of the year, but modified to store dates from different centuries using two digits. (If the current year's last two digits are 0 to 49, then years numbered 0 to 49 are assumed to belong to the current century, and years numbered 50 to 99 are assumed to belong to the previous century. If the current year's last two digits are from 50 to 99, then years numbered 0 to 49 are assumed to belong to the next century, and years numbered 50 to 99 are assumed to belong to the current century.)	01
MM	Displays month as digit (01–12)	01
MONTH	Displays name of month, spelled out (for months with fewer than 9 characters in their name, trailing blank spaces are added to pad the name to 9 characters)	JANUARY
DD	Displays day of month (01–31)	15
DDD	Displays day of year (01–366)	15
DAY	Displays day of week, spelled out	FRIDAY
DY	Displays name of day as a 3-letter abbreviation	FRI
AM, PM, A.M., P.M.	Meridian indicator (without or with periods)	PM
HH	Displays hour of day using 12-hour clock	05
HH24	Displays hour of day using 24-hour clock	17
MI	Displays minutes (0–59)	45
SS	Displays seconds (0–59)	35

Figure 2-4: Common date format masks

Front slashes, hyphens, and colons can be used as separators between different date elements. For example, the format mask MM/DD/YY would appear as 01/15/01, and the mask HH:MI:SS would appear as 05:45:35. Additional characters such as commas, periods, and blank spaces can also be included. For example, the format mask DAY, MONTH DD, YYYY would appear as FRIDAY, JANUARY 15, 2001.

You can embed characters within character data by preceding the format mask with the characters "FM" to specify that the user will not have to enter the embedded format characters and the double quotation marks. Figure 2-5 lists some commonly used character data format masks.

Data Item	Format Mask	Stored Value	Displayed Value
Phone number	FM"("999") "999"-"9999	7155551234	(715) 555-1234
Social Security	FM999"-"99"-"9999;	123456789	123-45-6789

Figure 2-5: Common character format masks

Format masks are not used when you create database tables, because date fields are specified simply using the DATE data type. You will use format masks in later chapters when you start inserting and retrieving data values.

Integrity Constraints

An **integrity constraint** allows you to define primary key fields and specify foreign keys and their corresponding table and column references. The general format for defining a primary key constraint is CONSTRAINT <constraint name> PRIMARY KEY. The **constraint name** is the internal name that Oracle uses to identify the constraint. Every constraint name must be unique for each database user—in other words, you cannot have two constraints with the same name in the database tables that you create. A convention for assigning unique constraint names is to use the general format <tablename where the constraint is being created>_<fieldname>_<constraint id>. The constraint ID is a two-character abbreviation that specifies the type of constraint. Commonly used constraint ID abbreviations are shown in Figure 2-6.

Constraint Type	Constraint ID Abbreviation
Primary Key	pk
Foreign Key	fk
Check Condition	cc
Not Null	nn

Figure 2-6: Common constraint ID abbreviations

An example of a primary key definition is the SID field in the STUDENT table. In the CREATE TABLE command for the STUDENT table, the SID field is defined as sid NUMBER(5) CONSTRAINT student_sid_pk PRIMARY KEY. This declaration defines a column named SID with a data type of NUMBER and precision 5 as the primary key of the STUDENT table.

If you omit the constraint variable name in a constraint declaration, Oracle will assign a system-generated name to the constraint. It is wise to name the constraint yourself, using the naming convention shown, so you can easily interpret the meaning of each constraint. Each constraint variable name must be unique for the entire database; that is, you cannot have two constraints named student_sid_pk. By using the naming convention, you ensure that each constraint variable name is unique.

The general format for specifying a foreign key constraint is CONSTRAINT <constraint variable name> REFERENCES <table where field is a primary key>(<name of field in table where it is the primary key>); An example of a foreign key is the LOCID field in the FACULTY table. The CREATE TABLE command for the FACULTY table includes the following definition:

```
locid NUMBER(5) CONSTRAINT faculty_locid_fk
REFERENCES location(locid)
```

Before you can create a foreign key reference, the table where the field is a primary key must already exist, and the field must be defined as a primary key. In this example, before creating the FACULTY table, you must create the LOCATION table and define LOCID as the primary key.

Figure 2-7 illustrates the components of this command. This declaration defines a field named LOCID as an integer field with precision 5. The declaration also specifies that the variable faculty_locid_fk identifies LOCID as a foreign key in the FACULTY table and references the LOCID field in the LOCATION table as well. When a new record is inserted into the FACULTY table, the DBMS checks to make sure that the value to be inserted for LOCID in the FACULTY record already exists in the LOCATION table.

locid NUMBER(5), CONSTRAINT faculty_locid_fk REFERENCES location(locid);

- **field name** — locid
- **data type** — NUMBER(5)
- **constraint name** — faculty_locid_fk
- **table where field is a primary key** — location
- **field name in table where field is a primary key** — (locid)

Figure 2-7: Declaring a foreign key reference

The constraints discussed so far have been **field constraints**, which specify the limitations for data values that can be inserted into an individual field. Sometimes you need to create a constraint that limits the values that can be inserted into multiple fields in a table record. To do this, you must create a **table constraint**. A common table constraint is one used for specifying a composite primary key, which is a primary key composed of two or more data fields. Usually this table constraint is declared after all of the table fields are declared, using the following general command: `CONSTRAINT <constraint name> PRIMARY KEY (<first field name>, <second field name>, ...)`. The constraint name should consist of the table name, the names of all fields involved in the primary key, and the identifier "pk."

In the ENROLLMENT table, the primary key consists of both the SID and the CSECID. SID and CSECID are also foreign keys in this table, and their associated values must also exist in the STUDENT and COURSE_SECTION tables. The command to create the ENROLLMENT table would include the following field and table constraint definitions:

```
CREATE TABLE enrollment
(sid NUMBER(5) CONSTRAINT enrollment_sid_fk REFERENCES student(sid),
csecid NUMBER(8) CONSTRAINT enrollment_csecid_fk REFERENCES
course_section(csecid),
CONSTRAINT enrollment_sid_csecid_pk PRIMARY KEY (sid, csecid))
```

Figure 2-8 illustrates the individual components of the composite key declaration.

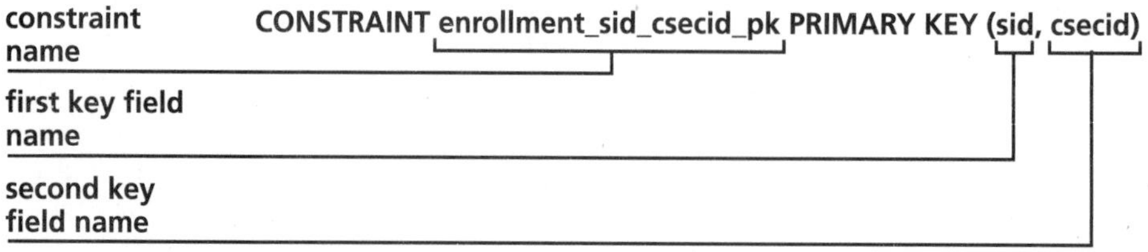

CONSTRAINT enrollment_sid_csecid_pk PRIMARY KEY (sid, csecid)

constraint name

first key field name

second key field name

Figure 2-8: Declaring a composite key

Value Constraints

Value constraints restrict what data can be entered into a given field, enable you to specify whether a field can or cannot be NULL, and allow you to specify a default value for a field. To restrict the data values that can be inserted into a given field, you use a value constraint called a **check condition**. For example, you can create a check condition to specify that numeric data must fall within a specific range (entries must be greater than zero but less than 1,000) or that character data must be from a set of specific values (entries must be SR, JR, SO, or FR). Specifying check conditions in table definitions must be done prudently, because once the table is populated with data, it is difficult or impossible to modify the constraint—all records must satisfy the constraint. Check conditions should be used only when the number of allowable values is limited and not likely to change. An example of an appropriate use of a check condition is for a gender field where the values are restricted to "M" or "F." An inappropriate use would be for the COLOR field in the INVENTORY table in the Clearwater Traders database—there are many possible values, and the values change constantly. A better approach is to create a lookup table, such as the one used for the COLOR table.

Each expression in a check condition must be able to be evaluated as true or false, and expressions can be combined using the logical operators AND and OR. When two expressions are joined by the AND operator, both expressions must be true for the expression to be true. When two expressions are joined by an OR operator, only one expression needs to be true for the expression to be true. A check condition can validate specific values, such as whether a gender field is equal to M or F, or validate ranges of allowable values. An example of a range check condition is used in the CCREDIT field when creating the COURSE table in the Northwoods University database to specify that course credits must be greater than 0 and less than 12. The definition is:

```
ccredit NUMBER(2) CONSTRAINT course_ccredit_cc
CHECK ((ccredit > 0) AND (ccredit < 12))
```

Note that both the expressions (ccredit > 0) and (ccredit < 12) can be evaluated as either true or false. The AND condition specifies that both conditions must be true for the check condition to be satisfied.

Another check condition in the database checks to ensure that the value entered in the SCLASS field of the STUDENT table is FR, SO, JR, or SR:

```
sclass CHAR(2) CONSTRAINT student_sclass_cc
CHECK ((sclass = 'FR') OR (sclass = 'SO')
OR (sclass = 'JR') OR (sclass = 'SR'))
```

Again, each of the expressions in parentheses can be evaluated as true or false. The OR condition specifies that if any one condition is true, then the check condition is satisfied.

The NOT NULL value constraint specifies whether a field must have a value entered for every record or whether it can be NULL (indeterminate or unknown). When a field is specified as a primary key, it automatically has a NOT NULL constraint. From a business standpoint, some fields, such as a customer's name, should not be NULL. Although foreign keys can be NULL (meaning there is no link between a given record and another table), sometimes foreign keys should not be NULL. For example, in the Northwoods University database, it doesn't make sense for a COURSE_SECTION record not to have an associated value for a term ID. Foreign key fields that are not allowed to be NULL must have an explicit NOT NULL constraint. The following example declares TERMID as a foreign key in the COURSE_SECTION table, and adds a NOT NULL constraint:

```
termid NUMBER(3) CONSTRAINT course_section_termid_fk REFERENCES
term(termid)
CONSTRAINT course_section_termid_nn NOT NULL
```

Finally, a DEFAULT value constraint specifies that a particular field will have a default value that is inserted automatically for every table record. For example, if most Northwoods University students live in Wisconsin, the default SSTATE field could be declared with a default value of WI, using the declaration sstate CHAR(2) DEFAULT 'WI'. The default will be overridden if the user inserts any value; the default will only be used if a null value is inserted into SSTATE.

Starting SQL*Plus

Starting an Oracle application is a two-step process. First you start the application on your client workstation, and then you log on to the Oracle database. Your instructor will advise you of the correct logon procedures for your lab. Now you will start SQL*Plus, the Oracle command-line SQL utility, and create some database tables.

To start SQL*Plus:

1. Click the **Start** button on the Windows taskbar, point to **Programs**, point to **Oracle for Windows 95**, and then click **SQL Plus 8.0**. After a few moments, Oracle SQL*Plus starts, opens the Log On dialog box, and requests your user name and password. You must also enter your host string (or connect string), as shown in Figure 2-9.

help

In most cases, Windows NT users will see "Windows NT" instead of "Windows 95" in the Start menu program names.

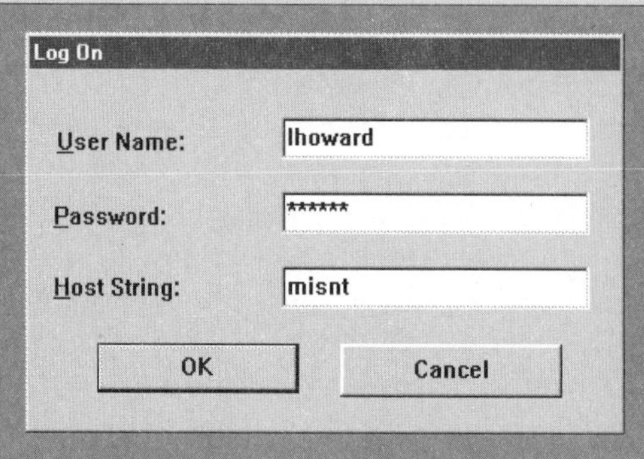

Figure 2-9: Log On dialog box

Another way to start SQL*Plus is to start Windows Explorer, change to the ORAWIN95\BIN (or ORANT\BIN) directory, and then double-click the plus80w.exe file.

2 Enter your user name, press the **Tab** key, enter your password, press the **Tab** key, type your host string, and then click **OK**. Now Oracle SQL*Plus is active, as shown in Figure 2-10.

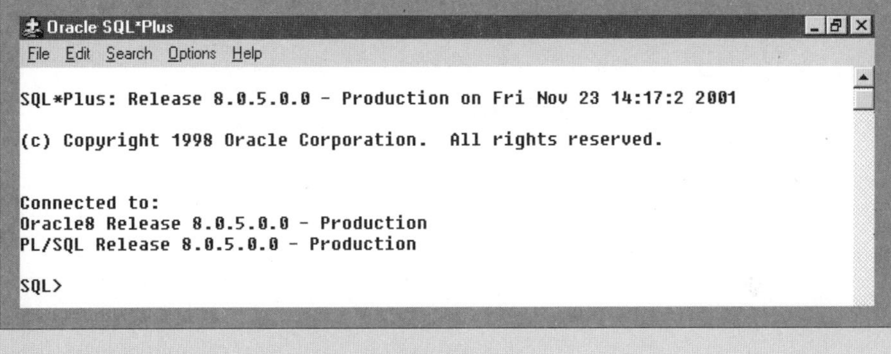

Figure 2-10: Oracle SQL* Plus program window

help If necessary, click the Maximize button on the Oracle SQL*Plus title bar to maximize the program window.

help Your version of SQL*Plus might have a different version number, but the steps in this book will work the same.

Creating a Database Table Using SQL*Plus

To create the STUDENT table, you will need to create the FACULTY table first because the FID field in the STUDENT table is a foreign key reference to the FID field in the FACULTY table. However, note that the FACULTY table also has a foreign key reference to LOCID in the LOCATION table. The LOCATION table does not have any foreign key references, so you will create it before creating the FACULTY and STUDENT tables.

The SQL command used to create a new table is the CREATE TABLE command. The general format of the CREATE TABLE command appears below. The constraint declarations are optional.

```
CREATE TABLE <tablename>
(<fieldname1> <data declaration>
      CONSTRAINT <integrity constraint declaration>
      CONSTRAINT <fieldname1 value constraint declaration>),
(<fieldname1> <data declaration>, ...;
```

The CREATE command is an example of a **data definition language (DDL)** statement in SQL. (The other commonly used DDL statements in SQL are DROP and ALTER, which you will learn later.) Data definition statements change the database structure and do not require an explicit COMMIT command. With a CREATE command, the database table is created as soon as the query executes. You will create the LOCATION table next.

To create the LOCATION table:

1 Type the query shown in Figure 2-11 to create the LOCATION table. Press the **Enter** key after typing each line to go to the next line. Do not type the line numbers because SQL*Plus adds line numbers to your query after you press the Enter key. Each line of the command defines a different column of the database. The line numbers can be used for referencing specific command lines using the SQL*Plus online editing facility, which you will learn about later.

ch2queries.txt - Notepad
File Edit Search Help
CREATE TABLE LOCATION
(locid NUMBER(5) CONSTRAINT location_locid_pk PRIMARY KEY,
bldg_code VARCHAR2(10) CONSTRAINT location_bldg_code_nn NOT NULL,
room VARCHAR2(6) CONSTRAINT location_room_nn NOT NULL,
capacity NUMBER(5) CONSTRAINT location_capacity_nn NOT NULL);

Figure 2-11: SQL query to create the LOCATION table

2 Type ; after the last line to end the command. The semicolon is the signal to the SQL*Plus compiler that it has reached the end of the SQL command.

3 Press the **Enter** key to execute the query. You should see the confirmation message "Table Created," as shown in Figure 2-11.

SQL is not case sensitive, and spaces between characters for formatting are ignored, as are line breaks. If you did not receive the "Table Created" confirmation message (or even if you did), proceed to the next section to learn how to edit and debug SQL commands.

Editing and Debugging SQL Commands

Many SQL commands are long and complex, and it is easy to make typing errors. There are two editing approaches in SQL*Plus: using the online editing facility or using an alternate text editor. The SQL*Plus environment has some limited online editing capabilities that can be used to make minor changes. For major changes, it is best to enter and edit your commands in an alternate editing environment, such as Notepad or any other Windows text editor, and then copy and paste your command into SQL*Plus.

Online Editing in SQL*Plus

When you press the Enter key to run a query, the query text is stored in a SQL*Plus memory area called the **edit buffer**. To list the current contents of the edit buffer, type the letter **L** (for list) at the SQL prompt. Figure 2-12 shows an example where a user has an error in her query. When she types the letter **L** at the SQL prompt, an asterisk marks the current line which is the line that is available for editing.

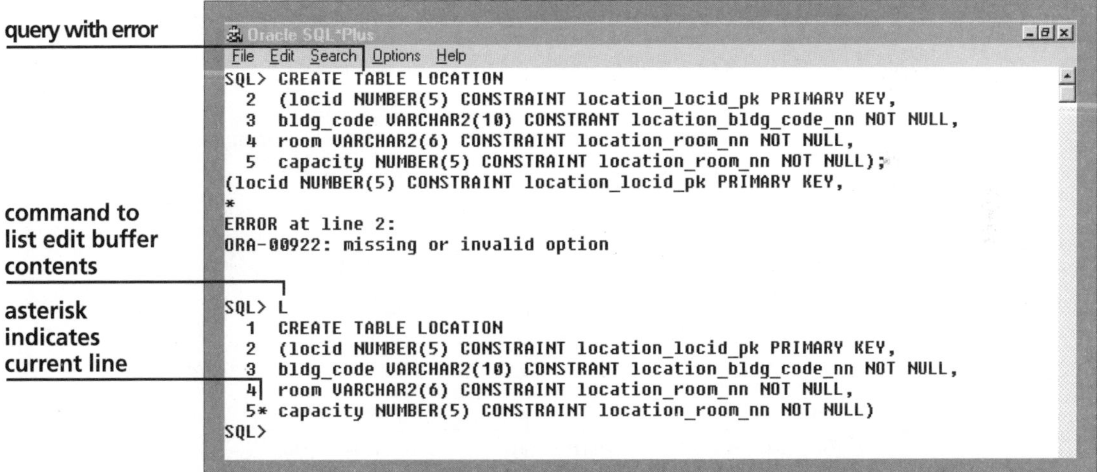

Figure 2-12: Listing the edit buffer contents

You can use the editing commands shown in Figure 2-13 at the SQL prompt to perform the indicated operations. The command is shown in boldface type. The descriptive word that corresponds to the command are shown in parentheses and the values you enter for line numbers or text are shown in angle brackets. (You don't type the parentheses or descriptive words—they are listed only to help you understand the meaning of the commands.)

Command	Operation
L (ist)	Shows the complete contents of the edit buffer
<line number>	Changes the current line number to the entered line number and shows the listed line number
A (ppend) **<text>**	Adds the entered text to the end of the current line
C (hange) **/old text/new text/**	Replaces an existing text string in the line (shown here as *old text* and delimited by front slashes) with a new text string (shown here as *new text* and ended with a front slash)
DEL (ete)	Deletes the current line
I (nsert) **<text>**	Creates a new line after the current line that contains the given input text
L (ist) **<line number>**	Shows the indicated line number text
/	Executes the contents of the edit buffer

Figure 2-13: SQL*Plus online editing commands

Figure 2-14 shows examples of how to use these commands. You will learn more about the specific error messages later in the chapter.

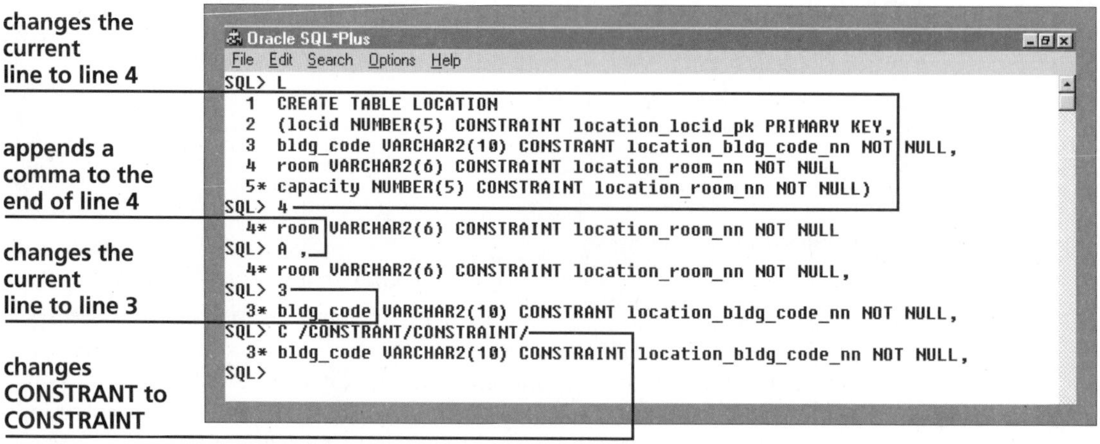

changes the current line to line 4

appends a comma to the end of line 4

changes the current line to line 3

changes CONSTRANT to CONSTRAINT

```
Oracle SQL*Plus
File Edit Search Options Help
SQL> L
  1  CREATE TABLE LOCATION
  2  (locid NUMBER(5) CONSTRAINT location_locid_pk PRIMARY KEY,
  3  bldg_code VARCHAR2(10) CONSTRANT location_bldg_code_nn NOT NULL,
  4  room VARCHAR2(6) CONSTRAINT location_room_nn NOT NULL
  5* capacity NUMBER(5) CONSTRAINT location_room_nn NOT NULL)
SQL> 4
  4* room VARCHAR2(6) CONSTRAINT location_room_nn NOT NULL
SQL> A ,
  4* room VARCHAR2(6) CONSTRAINT location_room_nn NOT NULL,
SQL> 3
  3* bldg_code VARCHAR2(10) CONSTRANT location_bldg_code_nn NOT NULL,
SQL> C /CONSTRANT/CONSTRAINT/
  3* bldg_code VARCHAR2(10) CONSTRAINT location_bldg_code_nn NOT NULL,
SQL>
```

Figure 2-14: Editing in SQL*Plus

Using an Alternate Text Editor with SQL*Plus

You can use an alternate editing environment in SQL*Plus, such as Notepad or any other Windows text editor, to type your commands. While SQL*Plus is running, start Notepad or any other text editor as a separate application, type the command, select and copy the command text, then switch to SQL*Plus and paste the selected command text, and execute the command. If the command fails to execute correctly, simply switch back to the text editor, edit the command, copy and paste the edited text back into SQL*Plus, and re-execute the command. You will do this next.

To use an alternate text editor with SQL*Plus:

1. Click the mouse pointer on the **C** in CREATE TABLE, and then drag the mouse pointer to select the **CREATE TABLE** command for the LOCATION table, as shown in Figure 2-15. Do not highlight the line numbers or error message, if one exists.

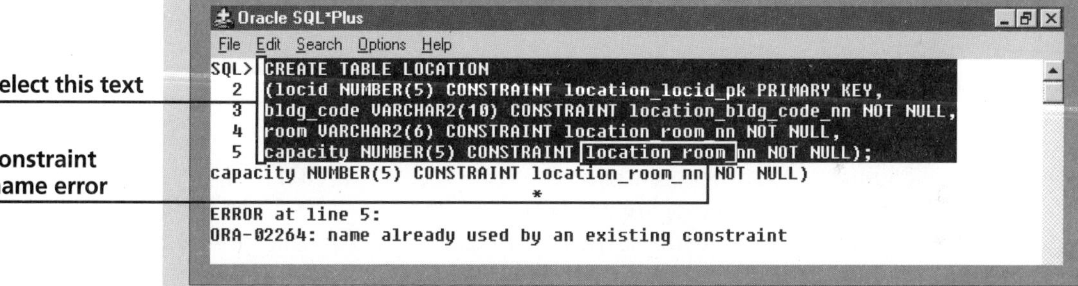

Figure 2-15: Selecting query command text in SQL*Plus

2. Click **Edit**, then click **Copy** to copy the highlighted text.
3. Click **Start** on the taskbar, then point to **Programs**, **Accessories**, then click **Notepad** to start Notepad.
4. Click **Edit**, then click **Paste** to copy the highlighted text into Notepad.

   Now you can edit your query to correct typographical errors if necessary. For example, note that in Figure 2-15, the constraint variable for capacity was entered incorrectly as location_room_nn. You can change the constraint variable for capacity to location_capacity_nn in Notepad and then copy your corrected command back and paste it into SQL*Plus. Your query text will be pasted into SQL*Plus, and then you can execute it by pressing the Enter key.

When you are creating database tables, it is a good idea to save the text of all of your CREATE TABLE queries in a single Notepad text file so you have a record of the original code, and you can easily re-create the tables if changes are required later. To save a text file in Notepad, click File on the menu bar, and then click Save and specify the drive, folder, and filename for the file. You can save more than one CREATE TABLE command in a text file, if you want. Just make sure that they are in the proper order so that foreign key references are made after their parent tables are created. Next, you will save your Notepad file so you have a record of your query text for creating the LOCATION table. You will continue to use this file to record and edit your SQL commands for the rest of this tutorial.

To save the Notepad file:

1. If necessary, click the **Untitled – Notepad** button on the task bar to make Notepad the active application.
2. Click **File**, then click **Save**. Save the file as **ch2queries.txt** in the Chapter2 folder on your Student Disk.
3. Minimize the Notepad window.

Using Online Help

Suppose that when you tried to create the LOCATION table, you got the error message shown in Figure 2-16. The error message shows the line and the position on the line where the error occurred, indicated by the position of the asterisk above the erroneous line, as well as the error code number and a brief description of the error.

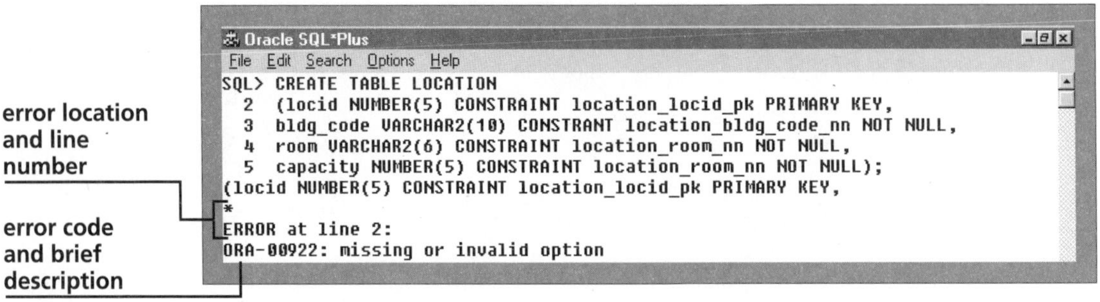

Figure 2-16: Query error message

The error message indicates that the error occurred on line 2, and the asterisk indicates that it was at the opening parenthesis of the column definitions. Line 2 looks correct, but the position of the asterisk indicates that the error could have occurred in any one of the table column declarations. The next step is to use online Help to get a more detailed error description. You will do this next.

To identify the error message:

1 Click the **Start** button, point to **Programs**, point to **Oracle for Windows 95**, and then click **Oracle8 Error Messages**. The Help Topics: Oracle8 Codes and Messages dialog box opens. Oracle does not have a central online Help application for all of its products, but has specific online Help applications for each individual product. This online Help application provides explanations for DBMS-generated (ORA-) error messages, while other Help applications provide other types of information.

Another way to start Oracle8 Error Messages is to start Windows Explorer, change to the ORAWIN95\MSHELP (or ORANT\MSHELP) directory, and then double-click the ora.hlp file.

2 Click the **Index** tab, and then type **ORA-00922** in the first text box, as shown in Figure 2-17.

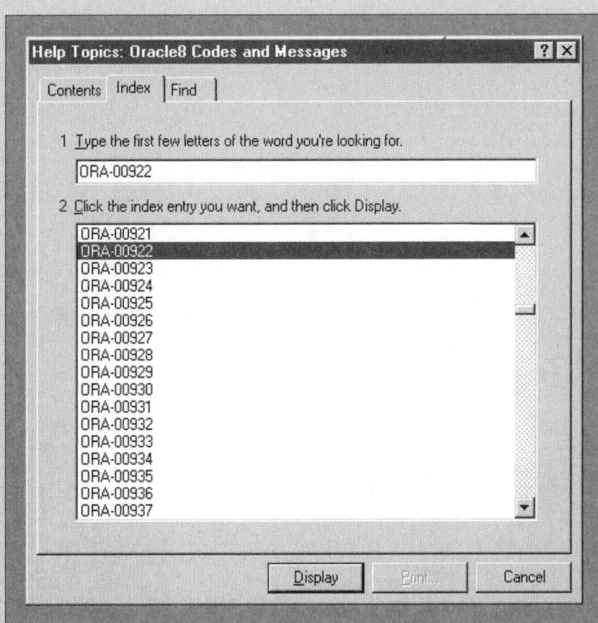

Figure 2-17: Searching for an Oracle error code in online Help

3 Click the **Display** button. Read the explanation of the error message that appears in the Oracle8 Messages and Codes window, as shown in Figure 2-18.

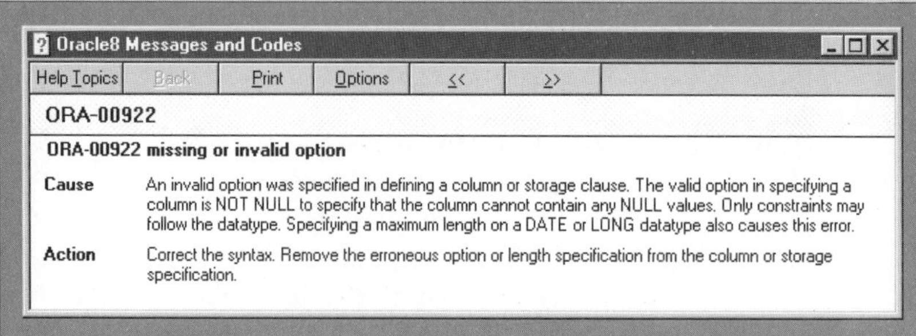

Figure 2-18: Oracle error code explanation

4 Close the Oracle8 Messages and Codes window by clicking its **Close** button.

5 If necessary, click **Help Topics** to go back to the index, then click the dialog box **Close** button to close Help.

This error explanation indicates that the error occurred in one of the column definitions, and suggests some possible causes. The key cause is "Only constraints may follow the datatype." Look carefully at the query and review each column definition. The problem is that "CONSTRAINT" was spelled as "CONSTRANT" on line 3.

Suppose you fix this typo using the C(hange) online editing command, as shown in Figure 2-19. When you run the query again, you get the new error message shown in the figure.

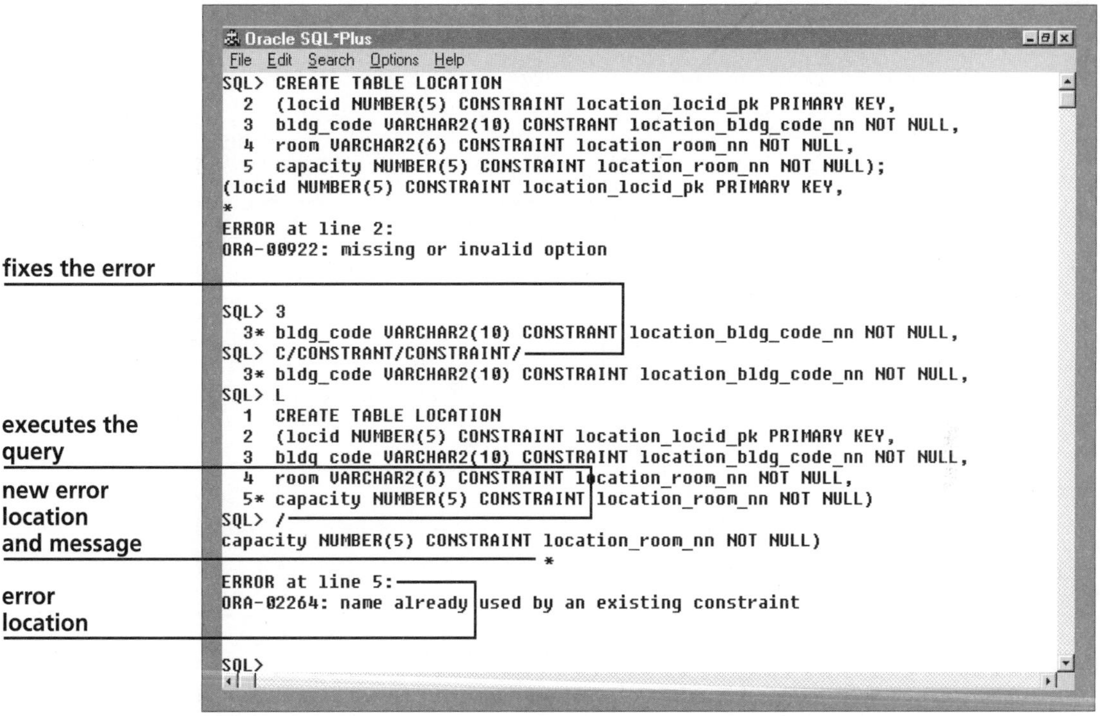

```
Oracle SQL*Plus
File Edit Search Options Help
SQL> CREATE TABLE LOCATION
  2  (locid NUMBER(5) CONSTRAINT location_locid_pk PRIMARY KEY,
  3  bldg_code VARCHAR2(10) CONSTRANT location_bldg_code_nn NOT NULL,
  4  room VARCHAR2(6) CONSTRAINT location_room_nn NOT NULL,
  5  capacity NUMBER(5) CONSTRAINT location_room_nn NOT NULL);
(locid NUMBER(5) CONSTRAINT location_locid_pk PRIMARY KEY,
*
ERROR at line 2:
ORA-00922: missing or invalid option

SQL> 3
  3* bldg_code VARCHAR2(10) CONSTRANT location_bldg_code_nn NOT NULL,
SQL> C/CONSTRANT/CONSTRAINT/
  3* bldg_code VARCHAR2(10) CONSTRAINT location_bldg_code_nn NOT NULL,
SQL> L
  1  CREATE TABLE LOCATION
  2  (locid NUMBER(5) CONSTRAINT location_locid_pk PRIMARY KEY,
  3  bldg_code VARCHAR2(10) CONSTRAINT location_bldg_code_nn NOT NULL,
  4  room VARCHAR2(6) CONSTRAINT location_room_nn NOT NULL,
  5* capacity NUMBER(5) CONSTRAINT location_room_nn NOT NULL)
SQL> /
capacity NUMBER(5) CONSTRAINT location_room_nn NOT NULL)
                              *
ERROR at line 5:
ORA-02264: name already used by an existing constraint

SQL>
```

Figure 2-19: Second SQL*Plus error message

This error message and the position of the asterisk suggest that the problem probably has to do with the location_room_nn NOT NULL constraint definition. Look at this definition carefully. Note that location_room_nn was entered as the constraint variable name for both ROOM and CAPACITY in lines 4 and 5. Remember that every constraint variable in a database must have a unique name. Figure 2-20 shows the process for changing the constraint variable name from location_room_nn to location_capacity_nn, executing the query again, and then receiving the message that the table was successfully created.

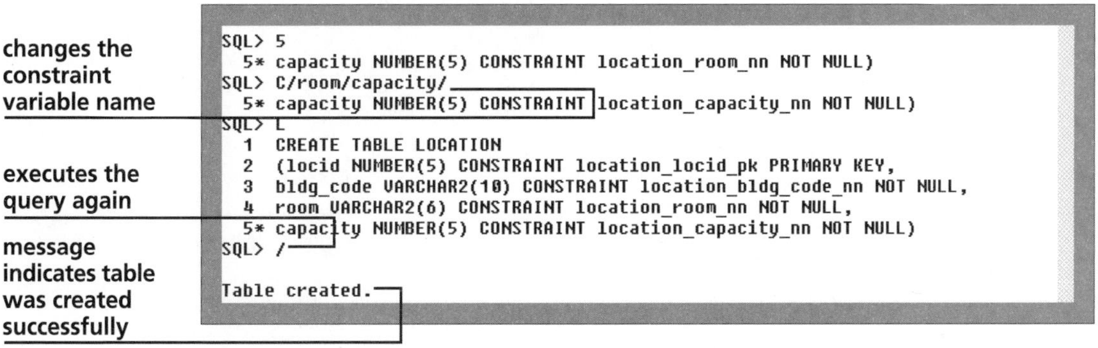

```
SQL> 5
  5* capacity NUMBER(5) CONSTRAINT location_room_nn NOT NULL)
SQL> C/room/capacity/
  5* capacity NUMBER(5) CONSTRAINT location_capacity_nn NOT NULL)
SQL> L
  1  CREATE TABLE LOCATION
  2  (locid NUMBER(5) CONSTRAINT location_locid_pk PRIMARY KEY,
  3  bldg_code VARCHAR2(10) CONSTRAINT location_bldg_code_nn NOT NULL,
  4  room VARCHAR2(6) CONSTRAINT location_room_nn NOT NULL,
  5* capacity NUMBER(5) CONSTRAINT location_capacity_nn NOT NULL)
SQL> /

Table created.
```

Figure 2-20: Correcting the constraint variable error

Some errors are harder to find than others. For example, the error caused by the misspelling of CONSTRAINT was identified at line 2. Line 2 was fine; the error was actually on line 3. However, the column declarations began on line 2, and the problem was in a subsequent column declaration. To debug SQL queries, always *start* looking for an error at the line referred to in the error message, and keep in mind that the error won't necessarily be on that line. If the referenced line is correct, examine the error message, and, if necessary, look it up in Online Help to identify what type of problem might cause the error. Then examine all the query lines to look for typographical errors, omitted or misplaced commas or parentheses, misspelled words, or repeated constraint variable names.

When you have an error that you cannot locate, a "last resort" debugging technique is to create the table multiple times and add one additional column declaration each time, until you find the declaration causing the error, as shown in Figure 2-21.

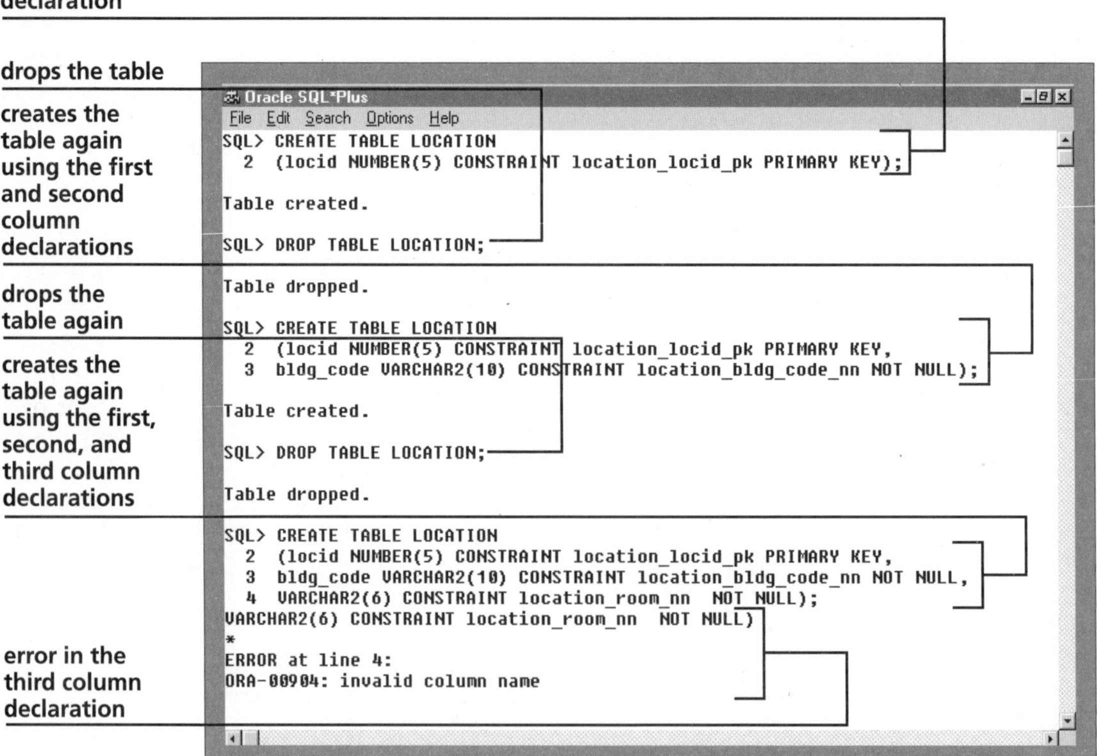

Figure 2-21: Creating, dropping, and re-creating a table to find the location of an error

First paste your nonworking query in a Notepad file and modify it so that it creates the table with only the first column declared. Copy the modified query, and paste it into SQL*Plus. If the table is created successfully, you now know that the error was not in the first column declaration. Delete the table using the DROP TABLE command, which has the following general format: `DROP TABLE <tablename>;`. Then modify the query in Notepad to create the table using only the first and second column declarations. If this works, you now know that the problem was not in either the first or second column declaration. Drop the table again, and modify the query so that the table is created using only the first, second, and third column declarations. Continue this process of adding one more column declaration to the CREATE statement until you locate the column declaration that is causing the error. Figure 2-21 shows how the LOCATION table is created and dropped repeatedly until the error message—not specifying a column name—is located on line 4.

Exiting SQL*Plus

There are three ways to exit SQL*Plus. You can type exit at the SQL prompt, or click File on the menu bar, and then click Exit, or you can click the Close button on the program window title bar. When you connect to a database on a server, your database connection disconnects automatically when you exit SQL*Plus. Use any one of these methods to exit SQL*Plus now.

Spooling SQL*Plus Commands and Output to a File

Sometimes it is useful to keep a record of all of your SQL commands and output in a text file. The file spooler is a feature of SQL*Plus that will automatically record all of the commands you type and all of the database output to a text file. When you invoke the file spooler, you are prompted to enter a filename. If the file does not exist yet, it will be created. If you select an existing file, it will be overwritten. The input commands and database output are not actually written to the file until you turn the spooler off. If you exit SQL*Plus without turning the spooler off, the spooled commands are automatically written to the spool file. Now you will start SQL*Plus again, invoke the file spooler, create the FACULTY table, and then turn the spooler off. Then you will view the spooled file in Notepad.

To start the file spooler, create the FACULTY table, and view the spooled file:

1. Start SQL*Plus, click **File** on the menu bar, point to **Spool**, then click **Spool File**. The Select File window opens. Navigate to the Chapter2 folder on your Student Disk, type **ch2spool** in the File name text box, then click **Save**.
2. Click on **ch2queries.txt – Notepad** on the taskbar to make Notepad the active application.

3 Place the mouse pointer at the end of the query creating the LOCATION table, and press the **Enter** key twice so there is a blank space between the end of the last query and the start of the next query.

4 Type the query shown in Figure 2-22 to create the FACULTY table.

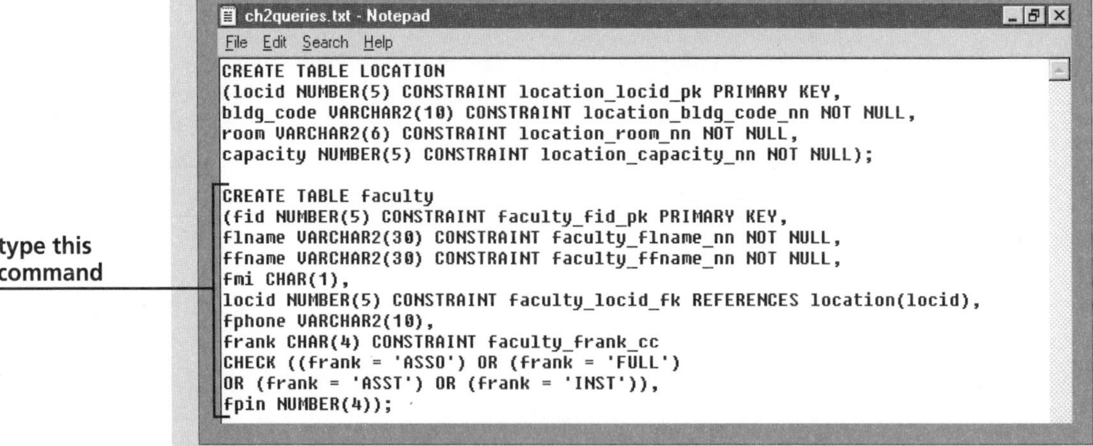

ch2queries.txt - Notepad
File Edit Search Help
CREATE TABLE LOCATION
(locid NUMBER(5) CONSTRAINT location_locid_pk PRIMARY KEY,
bldg_code VARCHAR2(10) CONSTRAINT location_bldg_code_nn NOT NULL,
room VARCHAR2(6) CONSTRAINT location_room_nn NOT NULL,
capacity NUMBER(5) CONSTRAINT location_capacity_nn NOT NULL);
CREATE TABLE faculty
(fid NUMBER(5) CONSTRAINT faculty_fid_pk PRIMARY KEY,
flname VARCHAR2(30) CONSTRAINT faculty_flname_nn NOT NULL,
ffname VARCHAR2(30) CONSTRAINT faculty_ffname_nn NOT NULL,
fmi CHAR(1),
locid NUMBER(5) CONSTRAINT faculty_locid_fk REFERENCES location(locid),
fphone VARCHAR2(10),
frank CHAR(4) CONSTRAINT faculty_frank_cc
CHECK ((frank = 'ASSO') OR (frank = 'FULL')
OR (frank = 'ASST') OR (frank = 'INST')),
fpin NUMBER(4));
type this command

Figure 2-22: Query to create the FACULTY table

5 Select the query text for creating the FACULTY table, copy it, paste it into SQL*Plus, and then press the **Enter** key. The "Table created" message indicates that your command contained no errors and that Oracle created the table. If you do not get this message, refer back to the section on editing and debugging SQL commands and debug your query until you create the FACULTY table successfully.

6 Click **File** on the menu bar, point to **Spool**, then click **Spool Off.** The spooled input and output are not written to the file until the spooler is turned off.

7 Click the **Start** button on the taskbar, point to **Programs**, point to **Accessories,** then click **Notepad** to start a second Notepad session. Open the **ch2spool.lst** file from the Chapter2 folder on your Student Disk. The CREATE TABLE command for the FACULTY table and the database output ("Table created") are displayed in the file.

Next you will start the spooler again using the same ch2spool.lst file. Then you will create the STUDENT table, and view the spooled file to confirm that the original file was overwritten.

To start the spooler again, create the STUDENT table, and overwrite the original spooled file:

1 Start the file spooler, and select the **ch2spool.lst** file from the Chapter2 folder on your Student Disk.

2 Click the **ch2queries.txt** file on the taskbar, and type the command shown in Figure 2-23 to create the STUDENT table below your other query commands.

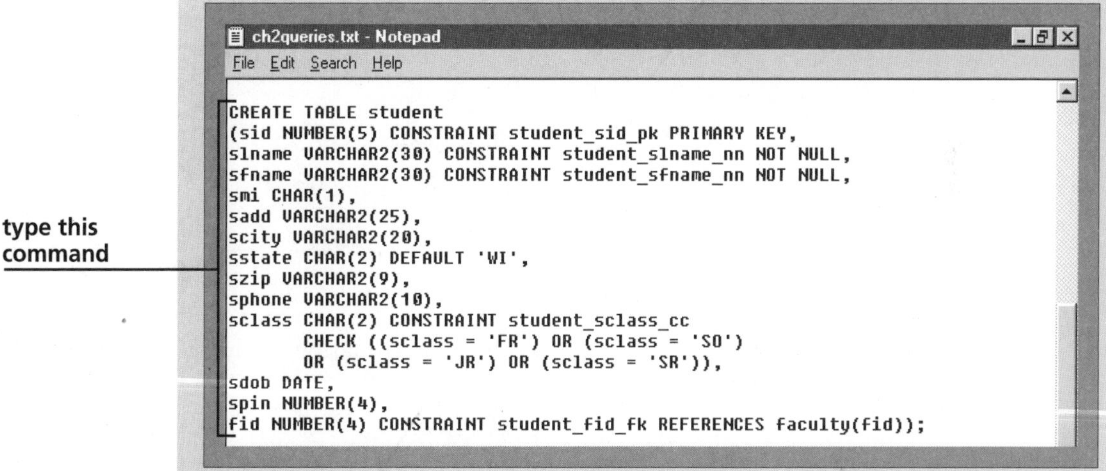

type this command

```
CREATE TABLE student
(sid NUMBER(5) CONSTRAINT student_sid_pk PRIMARY KEY,
slname VARCHAR2(30) CONSTRAINT student_slname_nn NOT NULL,
sfname VARCHAR2(30) CONSTRAINT student_sfname_nn NOT NULL,
smi CHAR(1),
sadd VARCHAR2(25),
scity VARCHAR2(20),
sstate CHAR(2) DEFAULT 'WI',
szip VARCHAR2(9),
sphone VARCHAR2(10),
sclass CHAR(2) CONSTRAINT student_sclass_cc
       CHECK ((sclass = 'FR') OR (sclass = 'SO')
       OR (sclass = 'JR') OR (sclass = 'SR')),
sdob DATE,
spin NUMBER(4),
fid NUMBER(4) CONSTRAINT student_fid_fk REFERENCES faculty(fid));
```

Figure 2-23: Query to create the STUDENT table

3 Select the query text, copy it, and then paste it into SQL*Plus. Press the **Enter** key to run the command. The "Table created" message indicates that your table was created successfully. Edit and debug your query if necessary.

4 Stop the file spooler.

5 Switch to the ch2spool.lst file on the taskbar, click **File** on the menu bar, click **Open**, then reopen the file to refresh your Notepad window. The spooled file displays only the CREATE TABLE query for the STUDENT table, showing that the original file was overwritten.

help

If you need to collect a long series of commands and outputs in a single spool file but find you don't have time to enter all of the commands in one SQL*Plus session, save the commands to multiple spool files, then consolidate the commands into a single file by copying and pasting.

Viewing Table Structure Information Using SQL*Plus

After you create your database tables and before you begin writing more advanced queries, you often need to review the table names, field and data type values, and constraint types. You can list the names of all of the tables in your user account to see which tables you have created already and which ones still need to be made. To do this, you will execute a query to view data in database tables. (You will learn more about queries in the next chapter.)

To list the tables you own:

1 In SQL*Plus, type the following command:

```
SELECT table_name FROM user_tables;
```

2 Press the **Enter** key to execute the command and list all of the tables in your account.

USER_TABLES is a system database table. **System database tables** are created and maintained by the DBMS for storing information about the database, such as tables and users. This table contains information about all users' data tables, and TABLE_NAME is one of its fields. The query that you just entered shows table names for the current user only. Your output should show the FACULTY, LOCATION, and STUDENT tables. Your output might display additional tables if you are using Personal Oracle.

Viewing Field Definitions

You also might need to view table field names and data types. To do this, you use the DESCRIBE command.

To view the field definitions for the STUDENT table:

1 Type **DESCRIBE student;** and then press the **Enter** key to list the STUDENT table's fields and field data types and sizes. See Figure 2-24.

```
Oracle SQL*Plus
File Edit Search Options Help
SQL> DESCRIBE student;
 Name                            Null?    Type
 ------------------------------- -------- ----
 SID                             NOT NULL NUMBER(5)
 SLNAME                          NOT NULL VARCHAR2(30)
 SFNAME                          NOT NULL VARCHAR2(30)
 SMI                                      CHAR(1)
 SADD                                     VARCHAR2(25)
 SCITY                                    VARCHAR2(20)
 SSTATE                                   CHAR(2)
 SZIP                                     VARCHAR2(9)
 SPHONE                                   VARCHAR2(10)
 SCLASS                                   CHAR(2)
 SDOB                                     DATE
 SPIN                                     NUMBER(4)
 FID                                      NUMBER(4)
```

Figure 2-24: Output of the DESCRIBE command

Viewing Constraint Information

You also might need to view information about the constraints that you entered for table fields. To do this, you will query the USER_CONSTRAINTS system database table.

To view the constraint information:

1 Enter and execute the query shown in Figure 2-25. The output shows a listing of all of the constraint variable names assigned to the STUDENT table. Based on the constraint variable naming convention, you can interpret exactly what constraints exist on specific table fields.

tip

Oracle stores table names in uppercase letters.

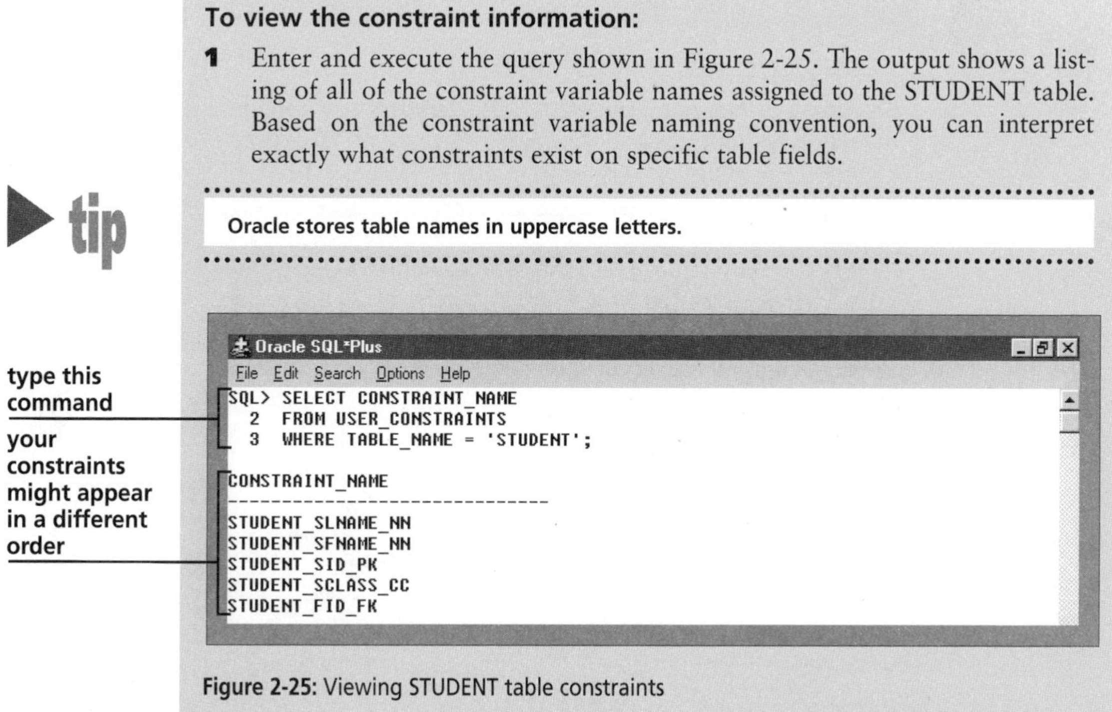

Figure 2-25: Viewing STUDENT table constraints

Modifying Tables Using SQL*Plus

After using the DESCRIBE command or selecting the USER_CONSTRAINTS information, you might find that you need to make changes to the column lengths or data types. You should plan your tables carefully to avoid having to make changes to the overall table structure, but inevitably, changes will need to be made. There are some parts of an Oracle database table that you can modify, and others that you cannot. Figure 2-26 shows the restricted and unrestricted actions for a table.

Restricted
Changing a column's data type, size, and default value is allowed only if there are no data in the column being modified
Adding a primary key is allowed only if current field values are unique (no duplicate entries)
Adding UNIQUE and CHECK CONDITION constraints to a column is allowed only if current field values match the added condition
Adding a foreign key is allowed only if current field values are NULL or exist in the referenced table
Changing a column name is not allowed
Deleting a column is not allowed
Unrestricted
Adding a new column to a table
Deleting a primary key constraint (also removes any foreign key references to the field in other tables)
Deleting a foreign key constraint

Figure 2-26: Restricted and unrestricted table changes

Sometimes you might find that you need to make changes that are not allowed in SQL, such as changing the name of a column. One approach is to drop a table so you can re-create it correctly.

Dropping and Renaming Existing Tables

When you drop a table from the database, you delete the table structure and all the data it contains. To delete a table, use the `DROP TABLE <tablename>`; command, where *tablename* is the name of the table that you want to drop. To rename a table, use the `RENAME <oldtablename> TO <newtablename>;` command, and replace the markers *oldtablename* and *newtablename* with the appropriate names.

Adding Fields to Existing Tables

The basic format of the statement to add a new field to a table is:

```
ALTER TABLE <tablename>
ADD <fieldname><data declaration>
CONSTRAINT <integrity constraints>
CONSTRAINT <value constraints>;
```

Suppose that you need to add a field to the FACULTY table that specifies each faculty member's employment start date. The ALTER TABLE command adds a column to an existing table. You will add the STARTDATE field to the FACULTY table next.

To add the STARTDATE field to the FACULTY table:

1 Type **ALTER TABLE faculty ADD (startdate DATE);**, and then press the **Enter** key. The "Table altered" confirmation message appears, as shown in Figure 2-27. 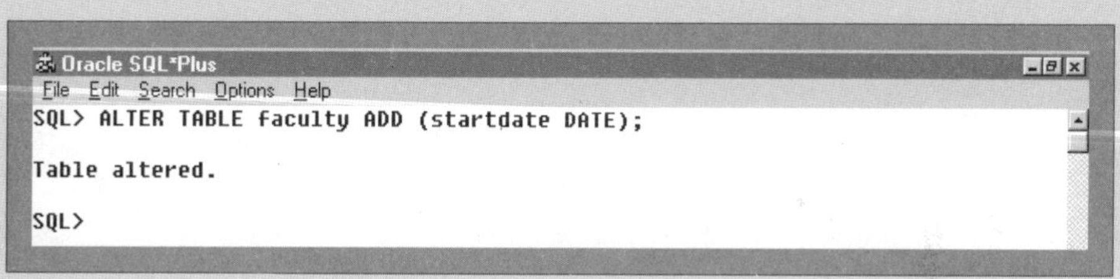

```
Oracle SQL*Plus
File Edit Search Options Help
SQL> ALTER TABLE faculty ADD (startdate DATE);

Table altered.

SQL>
```

Figure 2-27: ALTER TABLE command

Modifying Existing Fields

The general format of the command to modify an existing field's data declaration is:

```
ALTER <tablename>
MODIFY <fieldname><new data declaration>;
```

Suppose that you decide to modify the FACULTY table's FRANK field to a data type of VARCHAR2 with a length of six.

To change the data type and size of the FRANK field:

1 Type **ALTER TABLE faculty MODIFY (frank VARCHAR2 (6));**, and then press the **Enter** key to change the data type and size of the FRANK field in the FACULTY table.

Although you modified the field's data type and size, the check condition still exists for the field and does not require any modification.

To modify a NOT NULL value constraint, you use the MODIFY command. Next, you will modify the SPIN field in the STUDENT table so it is NOT NULL.

To modify the SPIN field in the STUDENT table to NOT NULL:

1. Type **ALTER TABLE student MODIFY (spin NUMBER(4) CONSTRAINT student_spin_nn NOT NULL);**, and then press the **Enter** key to add a NOT NULL constraint to an existing field.

To change an existing check condition value constraint on a field, you must first drop the existing check condition, using the DROP clause, and then add the new check condition. Suppose you decide to change the check condition in the SCLASS field in the STUDENT table to allow a GR value for graduate students.

To add the check conditions to the STUDENT table:

1. Type **ALTER TABLE student DROP CONSTRAINT student_sclass_cc;**, and then press the **Enter** key to modify the existing SCLASS field so it no longer has the check constraint.
2. Type the following code to add the new constraint:

   ```
   ALTER TABLE student
   ADD CONSTRAINT student_sclass_cc
   CHECK ((sclass = 'FR') OR (sclass = 'SO')
   OR (sclass = 'JR') OR (sclass = 'SR')
   OR (sclass = 'GR'));
   ```

3. Press the **Enter** key to make the change. Next, you would like to make FR (freshman) the default value for the SCLASS field.
4. Type **ALTER TABLE student MODIFY (sclass CHAR(2) DEFAULT 'FR');**, and then press the **Enter** key.

   A quirk of the ALTER TABLE command is that the MODIFY clause can be used only to change the data type, size, default values, and the NOT NULL check condition. For all other check condition modifications, you must drop the existing column constraint and then use the ADD clause to define the new constraint. Figure 2-28 shows when to use the ALTER TABLE/ADD and the ALTER TABLE/MODIFY clauses. Remember that you must drop any existing check condition constraints on the column first before using the ALTER TABLE/ADD clause.
5. Exit SQL*Plus.
6. Switch to Notepad, save your file and exit Notepad.

Use ALTER TABLE/ADD when:	Use ALTER TABLE/MODIFY when:
Adding a new column	Modifying a column's data type or size
Adding a new foreign key or check condition constraint	Modifying a column to add a NOT NULL constraint
	Modifying a column to add a default value

Figure 2-28: Summary of ALTER TABLE command usage

SUMMARY

- Data in relational database tables are inserted, retrieved, and modified using commands, called queries, such as INSERT, UPDATE, and DELETE.
- SQL is a high-level language that is used to query relational databases.
- Reserved words are SQL commands that cannot be used as table or field names.
- Oracle table and field names can be 1 to 30 characters in length and consist of only alphanumeric characters and the special characters $, _, and #. Table and field names must begin with a letter and cannot contain spaces.
- When you create a table, you must assign to each column a data type to specify what kind of data will be stored in the field. Data types provide error checking by making sure the correct data values are entered, and data types also cause storage space to be used more efficiently by optimizing the way in which specific types of data are stored.
- Character data can be stored using the CHAR, VARCHAR2, NCHAR, and LONG data types.
- Numerical data can be stored using the NUMBER data type. The precision of a NUMBER data type declaration specifies the total number of digits both to the left and to the right of the decimal point, and the scale specifies the number of digits to the right of the decimal point.
- The DATE data type stores date and time data.
- The Large Object (LOB) data type stores large amounts of binary or character data, or references to external binary files.
- A format mask is an alphanumeric character string that specifies how data output is formatted.
- In an Oracle database, integrity constraints are enforced through primary key specifications and foreign key references.

- In an Oracle database, value or check constraints on fields specify which values can be entered into those fields, and whether the values can be NULL.
- A convention for naming constraints to ensure that every constraint in the database has a unique name is to use the general format <table name where the constraint is being created>_<fieldname>_<constraint identifier>.
- In an Oracle database, you can specify whether a field has a default value that will be inserted automatically for every record.
- To create a new database table in SQL*Plus, use the CREATE TABLE command followed by the new table name, an opening parenthesis, and then a list of each of the field names, their data type declarations, and constraint declarations, followed by a closing parenthesis.
- The CREATE, ALTER, and DROP commands are examples of data definition SQL statements that change the database structure and do not require an explicit COMMIT command.
- SQL is not case sensitive, and spaces and line breaks for formatting are ignored.
- You can edit SQL*Plus commands using command-line editing or an alternate text editor. The edit buffer is a memory area in SQL*Plus that stores the text of the last executed query.
- Oracle does not have a central online Help application for all of its products, but has specific online Help applications for each different product. Detailed descriptions of Oracle error message codes can be obtained by searching for the code prefix (such as ORA) followed by a hyphen and the error code number (such as -00922) in the Oracle8 error messages help system.
- You can change a column's data type, size, or default value only if there are no data in the column. You can add a primary key constraint to a column only if all of the existing column values are unique. You can add a foreign key constraint to a column only if the existing column values are NULL or if they exist in the referenced table.
- A constraint variable is the internal name used to identify integrity and value constraints. Every constraint variable must have a unique name.
- A system table is a database created and maintained by the DBMS for storing data about database objects such as tables and users.
- To list the names of the tables you have made, use the following query: SELECT table_name FROM user_tables;.
- To list the fields and data types in a particular table, use the following query: DESCRIBE <table name>;.
- You cannot change a column name or delete a column.

REVIEW QUESTIONS

1. What is SQL?
2. Which of the following declarations are legal Oracle table and field names?
   a. SALES-REP-NAME
   b. INVENTORY_TABLE
   c. CLEARWATER_TRADERS_INVENTORY_TABLE
   d. SalesRepID
   e. _ITEM_PRICE
   f. SALESREP TABLE
3. What is the purpose of a data type?
4. List which data type you would recommend for the following data items:
   a. Social Security number
   b. gender (M or F)
   c. telephone number
   d. an image of a city map
   e. sales representative last name
   f. sales representative date of birth
   g. a pointer to a file containing a spreadsheet
5. Write the data field declarations for the following variables. Include the data type and length. Do not include value constraints.
   a. cust_age, containing integer values ranging from 1 to 99
   b. kilos_purchased, containing numeric values ranging from 0 to 100, and having a variable number of decimal places
   c. product_price, containing numeric values ranging from 0 to 999.99, and rounded to two decimal places
   d. appt_start_time, containing times ranging from 8 a.m. to 5 p.m.
6. What is a format mask? List the correct format masks for the following values:

Stored Data Value	Displayed Value	Format Mask
0123456789	123,456,789	____________
0123456789	01234567.89	____________
888888888	888-88-8888	____________
8888888888	(888)-888-8888	____________
15-JUL-2000	15/07/00	____________
6:33 PM 15-JUL-2000	18:33:00 PM	____________
6:33 PM 15-JUL-2000	6 PM	____________

7. What are the two types of integrity constraints? What happens when you omit the constraint name in a data declaration?
8. List the order in which the tables in the Clearwater Traders database must be created so that all fields that are referenced as foreign keys are first created as primary keys.

9. What is a value constraint? Give an example using one of the fields from the Clearwater Traders database.
10. Using the general constraint variable naming convention, write valid names for the following constraints:
    a. primary key in the COURSE_SECTION table
    b. SID foreign key reference in the ENROLLMENT table
    c. check condition requiring the enrollment column in the TERM table to have a value of either OPEN or CLOSED
    d. constraint requiring the TDESC column in the TERM table to not be NULL
11. Write the data field declarations for the following variables. Include the data type, length, and appropriate value constraints.
    a. cust_gender, where the values are restricted to M or F
    b. cust_name, where the value can be a maximum of 30 characters long and cannot be NULL
    c. sale_qty, which is an integer value ranging from 1 to 99. The value can be NULL, but cannot be 0.
12. Which properties of an Oracle database table can you always change?
13. Which properties of an Oracle database table can you never change?

PROBLEM-SOLVING CASES

1. Use Figure 1-11 in Chapter 1 as a reference to write the SQL commands to create the TERM, COURSE, COURSE_SECTION, and ENROLLMENT tables in the Northwoods University database. Save your commands in a Notepad file as Ch2Ex1.sql in the Chapter2 folder on your Student Disk. Test your queries in SQL*Plus to make sure they are correct, and spool the output to a file named Ch2spoolEx1.lst, saved in the Chapter2 folder on your Student Disk. Do not enter any data into the tables.
2. In a Notepad file, write the SQL*Plus commands in a Notepad file that make the following modifications to your database tables. Save the file as Ch2Ex2.sql in the Chapter2 folder on your Student Disk. Test each command by copying and pasting it to the SQL*Plus command line, and as you run the commands, spool the commands to a file named Ch2spoolEx2.lst. Make sure each command is modifying the database table as intended. (Depending on the kind of change you are making, you can check either by using the DESCRIBE command or by querying the USER_CONSTRAINTS table.) When all of the commands are working properly, save the Ch2Ex2.sql file.
    a. Add a check condition on the GRADE field in the ENROLLMENT table so that it allows only values of A, B, C, D, or F.

b. Modify the STUDENT table so that it includes fields for each student's permanent address, city, state, ZIP code, and telephone number, using the following field names, data types, and constraints:

Column Name	Data Type and Size	Constraints
PERMADD	VARCHAR2(30)	NOT NULL
PERMCITY	VARCHAR2(30)	NOT NULL
PERMSTATE	CHAR(2)	NOT NULL
PERMZIP	VARCHAR2(9)	NOT NULL
PERMPHONE	VARCHAR2(10)	NULL OK

c. Modify the COURSE table so that the default value for CCREDIT is 3.

d. Modify the COURSE_SECTION table so that the default value for CURRENRL is 0 (zero).

3. On a piece of paper, write the column names, data types, sizes, and constraints to use for each of the tables in the Clearwater Traders database shown in Figure 1-10 in Chapter 1. Use the following format:

TABLE NAME

Column Name	Data Type and Size	NOT NULL?	Primary Key?	Foreign Key Reference

4. Use Figure 1-10 in Chapter 1 to write the SQL commands to create the CUSTOMER, CUST_ORDER, ITEM, INVENTORY, ORDERSOURCE, and COLOR tables in the Clearwater Traders database. Save your commands in a Notepad file as Ch2Ex4.sql in the Chapter2 folder on your Student Disk. Test your queries in SQL*Plus to make sure they are correct, and spool the output to a file named Ch2spoolEx4.lst, saved in the Chapter2 folder on your Student Disk. Do not enter any data into the tables.

5. Use Figure 1-10 in Chapter 1 as a reference to create the SHIPPING, BACKORDER, and ORDERLINE tables in the Clearwater Traders database. Save your commands in a Notepad file as Ch2Ex5.sql in the Chapter2 folder on your Student Disk. Test your queries in SQL*Plus to make sure they are correct, and spool the output to a file named Ch2spoolEx5.lst, saved in the Chapter2 folder on your Student Disk. Do not enter any data into the tables.

CHAPTER

3

Using Oracle to Add, View, and Update Data

Introduction▶ The next step in developing a database is to insert, select, modify, and delete data records. Ultimately, the database users will perform these operations using forms and reports that automate the data entry, modification, and summarization processes. To make these forms and reports operational, the database developers often must translate the user input into SQL queries that will be submitted to the database. As a result, the developers must be very proficient with SQL command-line operations. Therefore, it is important that you know how to add, view, and update data records using SQL*Plus. This chapter will provide the foundation for the SQL commands that you will use in later chapters in this book—it is not intended as a complete SQL reference.

You also will learn how to use Query Builder, which is a visual tool that creates queries and generates the corresponding SQL commands. Finally, you will learn how to grant the privileges needed to allow other users to manipulate data in your database tables using SQL*Plus.

LESSON A

objectives

- **Automate SQL commands using scripts**
- **Insert data into database tables using SQL*Plus**
- **Commit data to the database and create database transactions**
- **Update and delete database records**

Running a SQL Script

A **script** is a text file that contains a sequence of SQL commands that can be executed in SQL*Plus. You created a script in Chapter 2 when you typed a series of CREATE TABLE commands, one after another, in your Notepad text file. Usually, SQL scripts have an .sql file extension.

To run a script, you type the START command at the SQL prompt, followed by a space, and then the full path and filename of the script text file. For example, to start a script saved in the text file A:\Chapter3\myscript.sql, you would enter the following command: `START a:\Chapter3\myscript.sql`. The pathname and text filename and extension can be any legal Windows 95 filename, but they *cannot* contain any blank spaces.

Before you can complete the exercises in this chapter, you need to run a script named emptynu.sql that will delete all of the Northwoods University database tables that you already have created and then re-create the tables.

To run the script file:

1 Start SQL*Plus and log on to the database.

2 Run the emptynu.sql script by typing the following command at the SQL prompt: **START A:\Chapter3\emptynu.sql**.

help Don't worry if you receive an error message in the DROP TABLE command, as shown in Figure 3-1. This error message indicates that the script is trying to drop a table that does not exist. The script is written so it drops all existing Northwoods University database table definitions before creating the new tables. If a script tries to create a table that already exists, SQL*Plus will generate an error and will not re-create the table.

```
Oracle SQL*Plus
File Edit Search Options Help
SQL> START a:\chapter3\emptynu.sql;
DROP TABLE enrollment
           *
ERROR at line 1:
ORA-00942: table or view does not exist


DROP TABLE course_section
           *
ERROR at line 1:
ORA-00942: table or view does not exist
```

Figure 3-1: Error in the script DROP TABLE command

help

If you receive an error message in a DROP TABLE command as shown in Figure 3-2, it probably means that you created a table in the Northwoods University database using a name other than the ones shown in Figure 1-11 in Chapter 1, when you did the Chapter 2 tutorial or exercises. This improperly named table has a foreign key reference to another table—in this case, to the LOCATION table. The table with the foreign key reference cannot be dropped until the table that is referencing it is dropped. You will need to view all of your database tables using the command SELECT table_name FROM USER_TABLES. Delete all of the tables manually, using the DROP TABLE command, and then rerun the script.

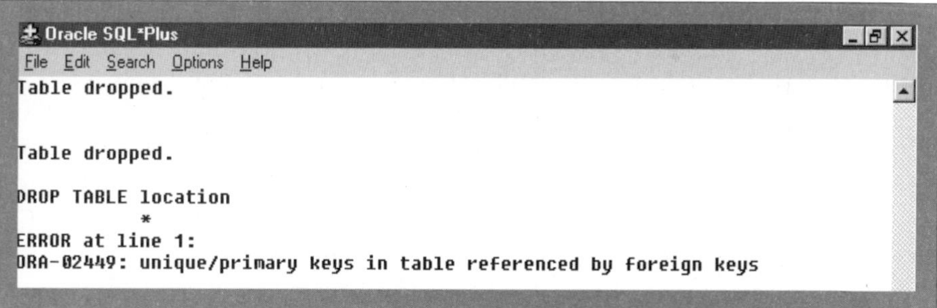

Oracle SQL*Plus
File Edit Search Options Help
Table dropped.
Table dropped.
DROP TABLE location
*
ERROR at line 1:
ORA-02449: unique/primary keys in table referenced by foreign keys

Figure 3-2: Error in script indicating a table has been created incorrectly and has a foreign key reference to the LOCATION table

Using SQL*Plus to Insert Data

After successfully running the script to create the tables, you are ready to begin adding data to them. In a business setting, programs called **forms** are used to automate the data entry process. Program developers who create forms often use the SQL INSERT statement in the form program code to insert data into tables.

The INSERT statement can be used two ways: to insert all fields into a record at once, or to insert a few selected fields. The basic format of the INSERT statement for inserting all table fields is:

```
INSERT INTO <tablename>
VALUES (column1 value, column2 value, …);
```

When you are inserting a value for all record fields, the VALUES clause of the INSERT statement must contain a value for each column in the table, or the word NULL instead of a data value if the data value is currently unknown or undefined. Column values must be listed in the same order in which the columns were defined in the CREATE TABLE command.

The basic format of the INSERT statement for inserting selected table fields is:

```
INSERT INTO <tablename> (column1 name, column2 name, …)
VALUES (column1 value, column2 value, …);
```

The column names can be listed in any order. The data values must be listed in the same order as their associated columns, and they must be the correct data type. A NULL value will automatically be inserted for columns omitted from the INSERT INTO clause.

Before you can insert a new data record, you must ensure that all of the foreign key records referenced by the new record have been added. Looking at the Northwoods database (Figure 1-11), you see that in the first STUDENT record, Sarah Miller's FID is 1. This refers to FID 1 (Kim Cox) in the FACULTY table. The FACULTY record must be added before you can add the first STUDENT record, or you will get a foreign key reference error. Look at Kim Cox's FACULTY record, and note that it has a foreign key value of LOCID 53. Similarly, this LOCATION record must be added before you can add the FACULTY record. Thankfully, the LOCID 53 record in the LOCATION table has no foreign key values to reference. Therefore, you can insert LOCID 53 in the LOCATION table, and then insert the associated FACULTY record, and finally add the STUDENT record.

To insert a record:

1 Start Notepad, and type the SQL command shown in Figure 3-3. Save the Notepad file as **chapter3queries.txt** in the Chapter 3 folder of your Student Disk.

2 Copy the query text and paste it into SQL*Plus, and then press the **Enter** key. The message "1 row created" should appear, indicating that the record has been added to the LOCATION table. 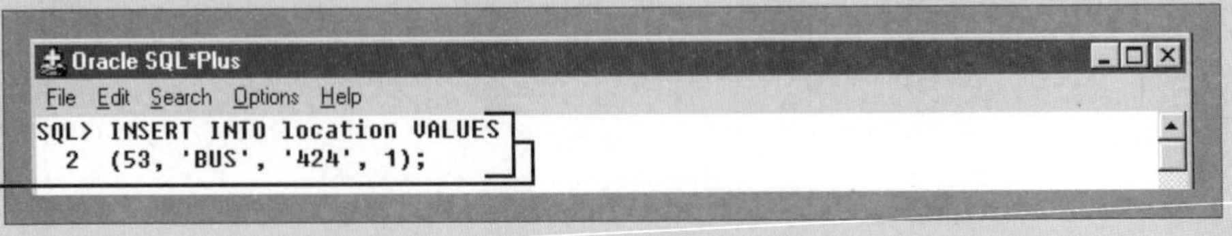

type this query

```
Oracle SQL*Plus
File Edit Search Options Help
SQL> INSERT INTO location VALUES
  2  (53, 'BUS', '424', 1);
```

Figure 3-3: Query text to insert a record into the LOCATION table

Note: For the remainder of the tutorial, assume that you will enter the query text into your Notepad text file, and then copy and paste it into SQL*Plus.

Note that the LOCATION table has four columns—LOCID, BLDG_CODE, ROOM, and CAPACITY—and that their associated data types are NUMBER, VARCHAR2, VARCHAR2, and NUMBER, respectively. The NUMBER fields are entered as digits, while the VARCHAR2 fields are enclosed within single quotation marks ('). Data stored in CHAR and VARCHAR2 fields must be enclosed in single quotation marks, and the text within the single quotation marks is case sensitive.

The field values in the INSERT statement are entered in the same order as the fields in the table, and a value must be included for every column in the table. When you insert a record into the LOCATION table, the DBMS expects a NUMBER data type, then a VARCHAR2, another VARCHAR2, and then another NUMBER. An error is displayed if you try to insert the values in the wrong order or if you omit a column value. If you cannot remember the order of the columns or their data types, use the DESCRIBE command to verify the table's structure.

The next step is to insert Kim Cox's record into the FACULTY table. To practice the command for inserting selected fields instead of inserting all of the table fields, you will enter values only for the FID, FLNAME, FFNAME, FMI, and LOCID columns, and omit the values for FPHONE, FRANK, and FPIN. This method of inserting data is useful when you are inserting data into only a few selected fields in a record.

To enter the record for faculty member Kim Cox:

1 Insert the first record into the FACULTY table by typing the SQL query shown in Figure 3-4. The message "1 row created" indicates that the record has been added to the table.

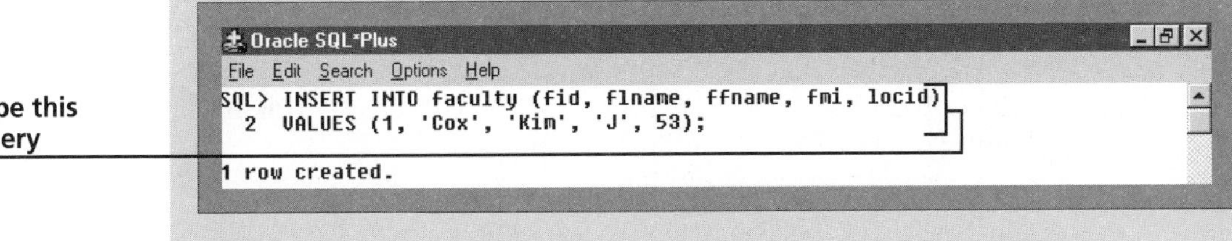

Figure 3-4: Inserting a record with specified columns into the FACULTY table

Next you will add the record for John Blanchard. First, you must insert the associated LOCATION record.

To insert the record for John Blanchard:

1 Type the queries shown in Figure 3-5 to insert the records into the LOCATION and FACULTY tables.

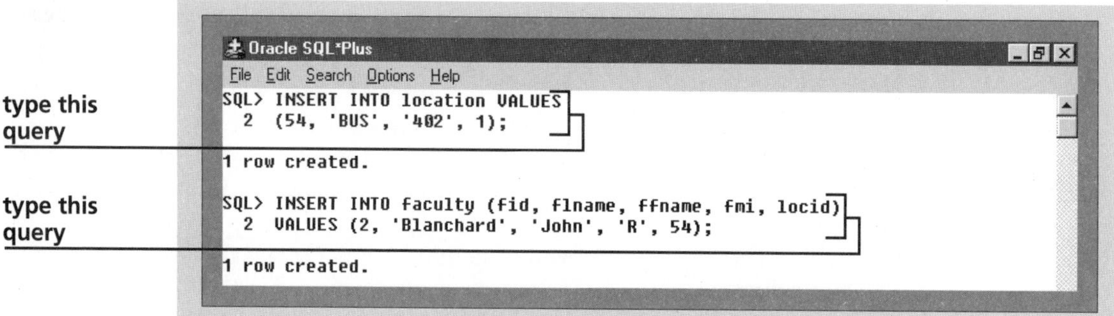

type this query

type this query

```
SQL> INSERT INTO location VALUES
  2  (54, 'BUS', '402', 1);

1 row created.

SQL> INSERT INTO faculty (fid, flname, ffname, fmi, locid)
  2  VALUES (2, 'Blanchard', 'John', 'R', 54);

1 row created.
```

Figure 3-5: Adding another LOCATION and FACULTY record

Now you can start adding the STUDENT records. The STUDENT table contains the SDOB field, which has a DATE data type. To insert a date into an Oracle database using SQL*Plus, you will enter the date as a character string, and instruct the DBMS to convert this character string to a DATE, using the TO_DATE function. The general format of the TO_DATE function is:

```
TO_DATE('<date that you are inserting>',
'<Oracle date format mask used for the date you are inserting>')
```

The different Oracle date format masks were summarized in Chapter 2. Recall that these date formats can be combined using embedded characters such as -, /, and :. For example, the character string '08/24/92' is converted to a DATE data type using the following command:

```
TO_DATE('08/24/92', 'MM/DD/YY')
```

To enter the same date as 24-AUG-1992, use the following command:

```
TO_DATE('24-AUG-1992', 'DD-MON-YYYY')
```

Remember that the DATE data type stores times as well. To convert the 10:00 A.M. start time to a DATE for CSECID 1000 in the COURSE_SECTION table, you would use the following command:

```
TO_DATE('10:00 A.M.', 'HH:MI AM')
```

To add the STUDENT records:

1 Enter the first record into the STUDENT table by typing the query shown in Figure 3-6. The message "1 row created" indicates that one record has been added to the table.

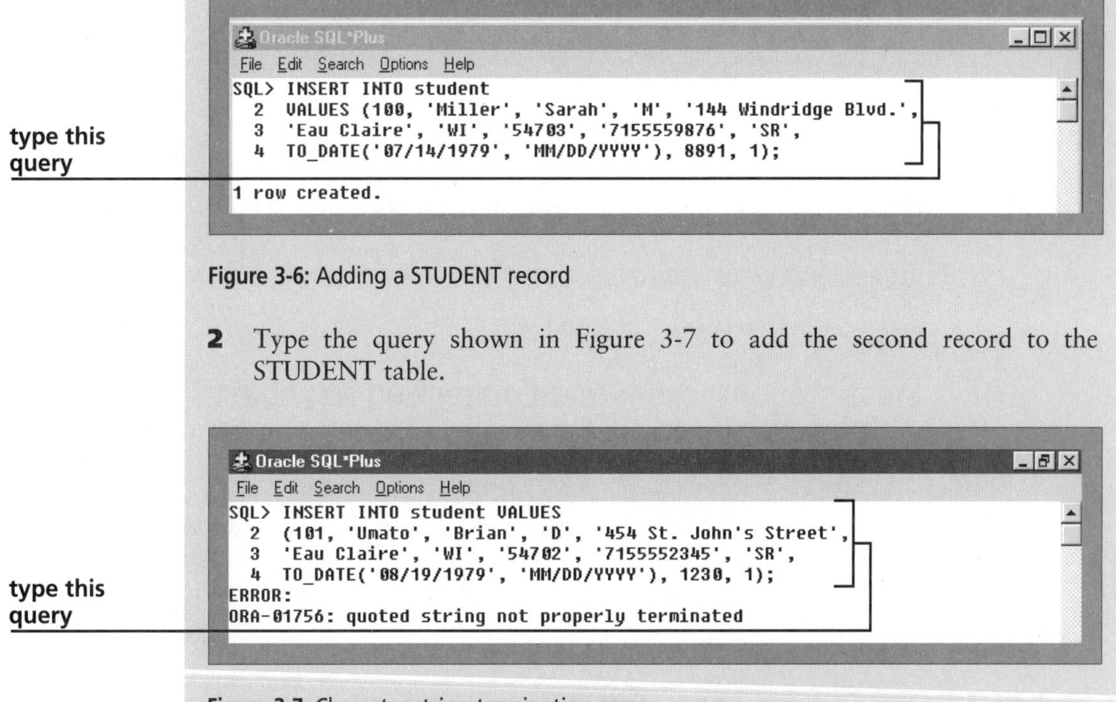

type this query
Oracle SQL*Plus
File Edit Search Options Help
SQL> INSERT INTO student
2 VALUES (100, 'Miller', 'Sarah', 'M', '144 Windridge Blvd.',
3 'Eau Claire', 'WI', '54703', '7155559876', 'SR',
4 TO_DATE('07/14/1979', 'MM/DD/YYYY'), 8891, 1);
1 row created.
type this query
Oracle SQL*Plus
File Edit Search Options Help
SQL> INSERT INTO student VALUES
2 (101, 'Umato', 'Brian', 'D', '454 St. John's Street',
3 'Eau Claire', 'WI', '54702', '7155552345', 'SR',
4 TO_DATE('08/19/1979', 'MM/DD/YYYY'), 1230, 1);
ERROR:
ORA-01756: quoted string not properly terminated

Figure 3-6: Adding a STUDENT record

2 Type the query shown in Figure 3-7 to add the second record to the STUDENT table.

Figure 3-7: Character string termination error

The error message shown in Figure 3-7 appears. What happened? Notice that the SADD value '454 St. John's Street' has a single quotation mark within the text. When the DBMS reached this single quotation mark, it assumed that this was the end of the SADD field, and then it expected a comma. When it found the letter "s" instead, it generated an error. To add text strings with embedded single quotation marks, you need to enter the single quotation mark twice. You will add the correct record next.

To enter text that contains a single quotation mark:

1 Change the SADD text to **'454 St. John''s Street'**, and then execute the command again. The message "1 row created" indicates that the record has been added to the table. Note that the '' is two single quotation marks, not a double quotation mark (").

2 Enter the record for SID 102, as shown in Figure 3-8. Daniel Black does not have a middle initial, so the word "NULL" is inserted for SMI because a value must be included in the INSERT statement for every column in the table.

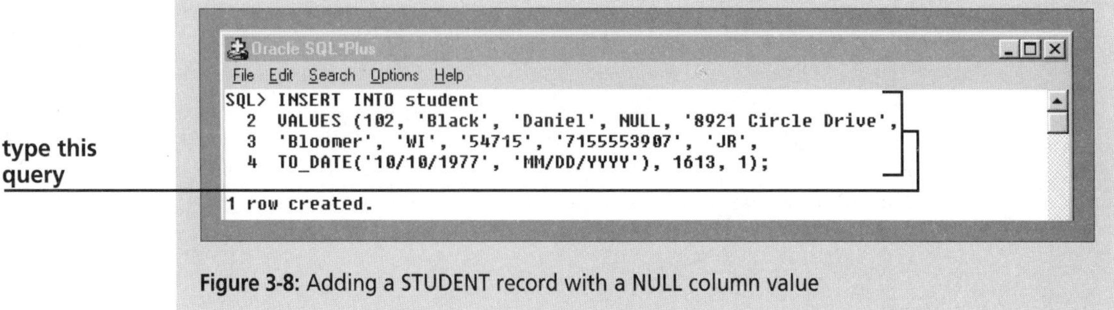

Figure 3-8: Adding a STUDENT record with a NULL column value

Creating Transactions and Committing New Data

When you create a new table or update the structure of an existing table, the change is effective immediately at the central database on the server. However, when you insert, update, or delete database records, the database on the server is not changed until you explicitly **commit** your changes to make them permanent. To commit changes during your current SQL*Plus session, you use the COMMIT command, as you will see next.

To commit your changes:

1 Type **COMMIT;**. The message "Commit complete" indicates that your changes are made in the central database.

Prior to issuing the COMMIT command, your changes were made to the **database buffer**, which is a memory location that stores the results of changes that add, modify, or delete database records. Other users cannot see the records you added or access the rows you modified, because these rows are locked until you commit your changes. When a record is locked, other users cannot change the record until the user who locked the record commits the change and releases the lock. SQL*Plus automatically commits your changes when you exit the program. However, it is a good idea to commit your changes often so that the records will be available to other users, and your changes will be saved if you do not exit normally because of a power failure or workstation malfunction.

A **transaction** is one or more related SQL commands that constitute a logical unit of work. An example of a transaction is the entry of a new customer order at Clearwater Traders. The associated CUST_ORDER record must be inserted to record the order ID, date, payment method, customer ID, and order source. Then one or more ORDERLINE records must be inserted to record the inventory ID, order price, and quantity ordered for each item ordered. If all parts of the transaction are not completed, the database will contain inconsistent data. For example, suppose you successfully insert the ORDER record, but before you can insert the ORDERLINE record, your workstation malfunctions. The ORDER record exists but has no associated order line information.

SQL has a process called a **rollback** that enables you to return the database to its original state by undoing the effects of all of the commands since the last commit occurred. An example of a rolled back transaction is shown in Figure 3-9. A new record is inserted into the CUST_ORDER table, and then the ROLLBACK command is issued. The final SELECT statement shows that the record was not inserted into the CUST_ORDER table, and that the INSERT command was effectively rolled back.

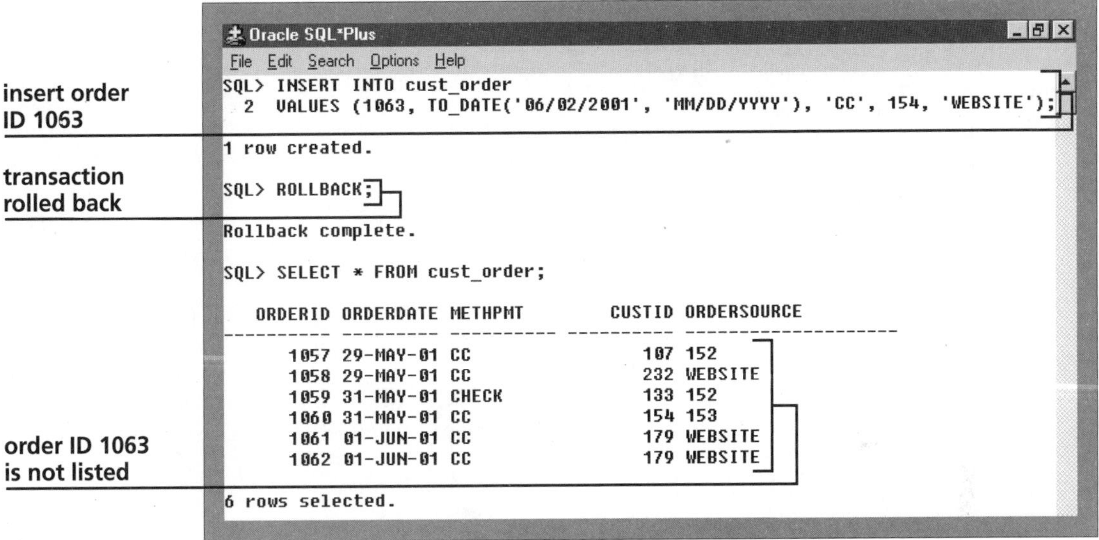

Figure 3-9: Using the ROLLBACK command

Rollbacks can be used with **savepoints** that mark the beginning of individual sections of a transaction. By using savepoints, you can roll back part of a transaction without having to process all of the transaction commands again. Figure 3-10 shows how to create a savepoint and then roll back a SQL*Plus session to that savepoint. A savepoint named ORDER_SAVE is created. A record is inserted for CUST_ORDER 1063, and an associated ORDERLINE record is inserted. Finally a rollback command to the savepoint is issued, and neither record is saved in the database.

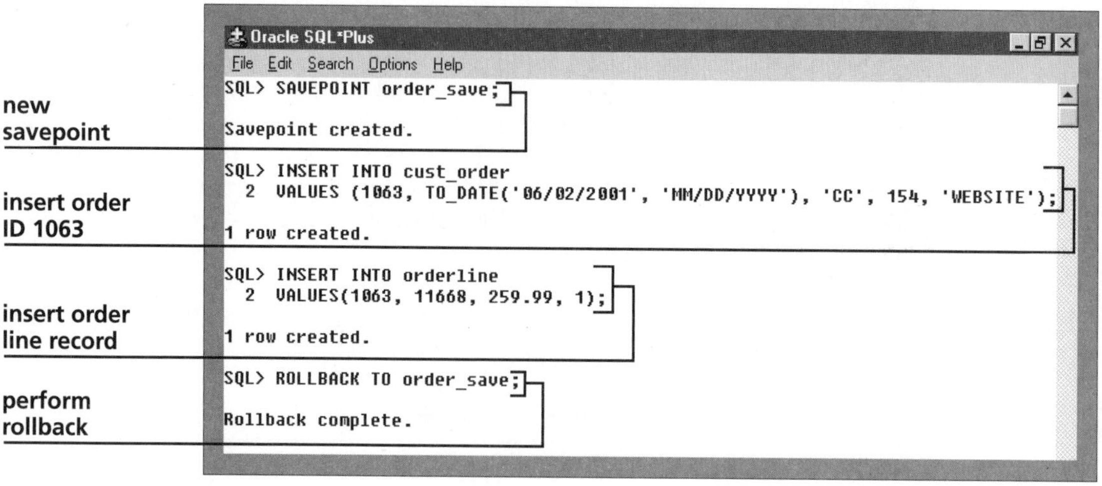

Oracle SQL*Plus

File Edit Search Options Help

```
SQL> SAVEPOINT order_save;

Savepoint created.

SQL> INSERT INTO cust_order
  2  VALUES (1063, TO_DATE('06/02/2001', 'MM/DD/YYYY'), 'CC', 154, 'WEBSITE');

1 row created.

SQL> INSERT INTO orderline
  2  VALUES(1063, 11668, 259.99, 1);

1 row created.

SQL> ROLLBACK TO order_save;

Rollback complete.
```

new savepoint

insert order ID 1063

insert order line record

perform rollback

Figure 3-10: Using the ROLLBACK command with a savepoint

Updating Existing Records in a Table

An important data maintenance operation is the updating of existing data records. Student addresses and phone numbers change often, and every year students (hopefully) move up to the next class. The general format of the UPDATE statement is:

```
UPDATE <tablename>
SET <column> = <new data value>
WHERE <search condition>;
```

You can update records in only one table at a time using a single UPDATE command. Search conditions are listed in the WHERE clause to make the command update specific records. The general format of a search condition is:

```
WHERE <expression> <comparison operator> <expression>
```

Every search condition must be able to be evaluated as either TRUE or FALSE. Expressions usually are field names or constants. Figure 3-11 lists some common comparison operators that are used in SQL expressions.

Operator	Description
=	Equal to
>	Greater than
<	Less than
>=	Greater than or equal to
<=	Less than or equal to
< >	Not equal to

Figure 3-11: Common search condition comparison operators

For example, the condition that finds all locations with a capacity greater than or equal to 50 is capacity >= 50. Comparison operators also can compare CHAR and VARCHAR2 values. For example, the expression that finds student records where the SCLASS value is equal to SR is sclass = 'SR'.

Oracle pads CHAR values with blank spaces if an entered data value does not fill all of the declared variable's size. For example, suppose you want to declare SCLASS as a CHAR field with size 2 and then enter a data value of 'U' (for unclassified). To search for this value in the WHERE clause, you would have to type SCLASS = 'U ' (with a blank space after the U).

Don't forget that values enclosed within single quotation marks are case sensitive. If you type SCLASS = 'sr', you will not retrieve rows in which the SCLASS value is 'SR'.

To use the UPDATE command:

1 Type the query shown in Figure 3-12 to update student Daniel Black's SCLASS value to SR. The "1 row updated" message indicates that the row was updated. You also can update multiple fields in a record using a single UPDATE command.

type this query

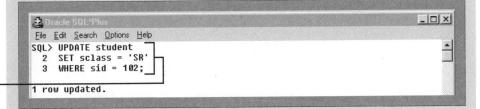

Figure 3-12: Updating a single field in a record

2 Type the query shown in Figure 3-13 to change Daniel Black's SCLASS back to JR and update his FID to 2.

type this query

```
Oracle SQL*Plus
File Edit Search Options Help
SQL> UPDATE student
  2  SET sclass = 'JR', fid = 2
  3  WHERE sid = 102;

1 row updated.
```

Figure 3-13: Updating multiple fields in a record

You can update multiple records in a table using a single UPDATE command by specifying a search condition that matches multiple records and uses the greater than (>) or less than (<) mathematical operators, or the AND and OR operators. You also can combine multiple search conditions using the AND and OR operators. When you use the **AND** operator to connect two search conditions, both conditions must be true for the row to satisfy the search conditions. If no data rows exist that match both conditions, then no data are retrieved. For example, the following search condition would find all records where the BLDG_CODE is 'CR' and the capacity is greater than 50: `WHERE bldg_code = 'CR' AND capacity > 50`.

When you use the **OR** operator to connect two search conditions, only one of the conditions must be true for the row to satisfy the search conditions. For example, the following search condition would find all course section records that meet either on Tuesday and Thursday or on Monday, Wednesday, and Friday: `WHERE day = 'TTh' OR day = 'MWF'`.

To update multiple records using a single UPDATE command:

1 Type the query shown in Figure 3-14 to change the capacity of LOCIDs 52 through 57 to 2. The WHERE clause specifies that the LOCID records to change must be greater than or equal to 52 and less than or equal to 57. Note that your confirmation message is "2 rows updated" because only LOCIDs 52 and 54 have been inserted into the database at this point.

type this query

```
Oracle SQL*Plus
File Edit Search Options Help
SQL> UPDATE location
  2  set capacity = 2
  3  WHERE locid >= 52 AND locid <= 57;

2 rows updated.
```

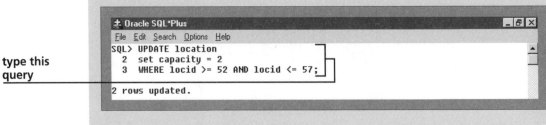

Figure 3-14: Updating multiple records

Deleting Records

Another table maintenance operation is deleting records. The general format of the DELETE command is:

```
DELETE FROM <tablename>
WHERE <search condition>;
```

Always include a WHERE clause when deleting a record from a table, to ensure that the correct record is deleted. Otherwise, all table records will be deleted.

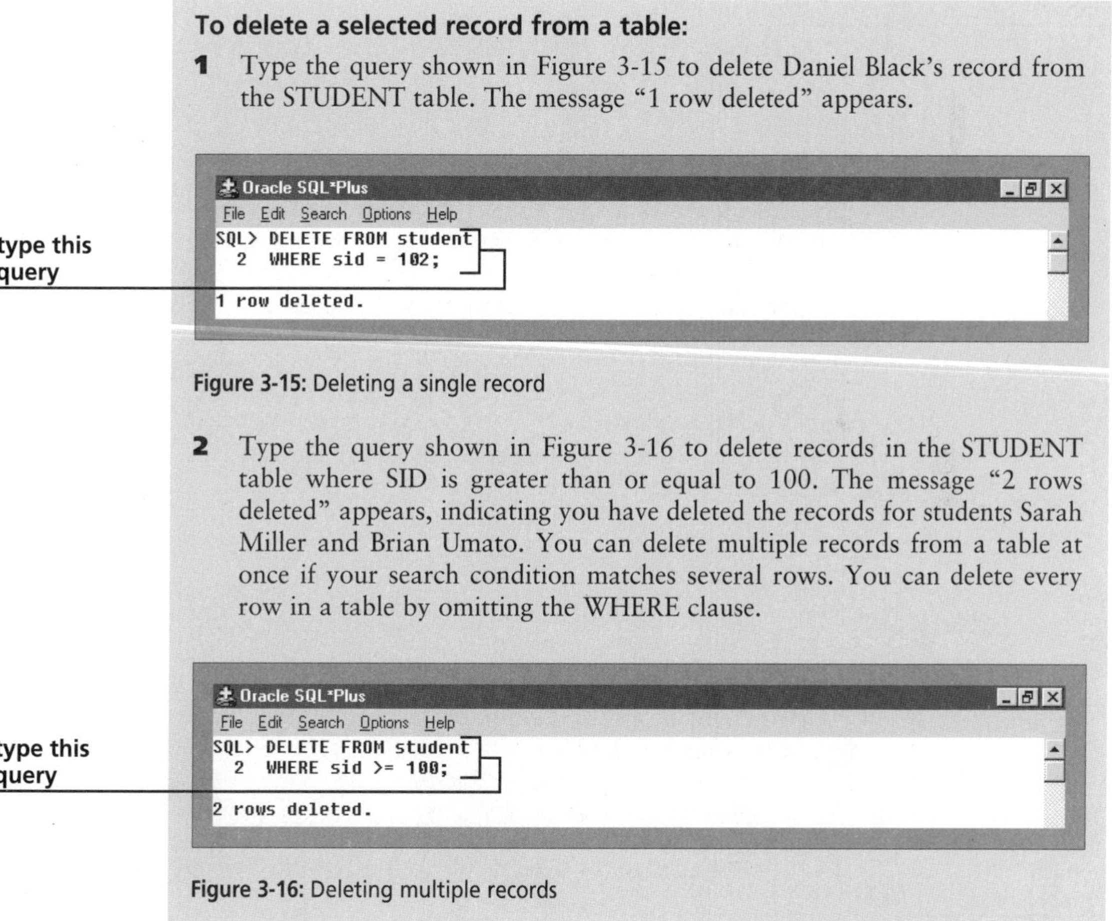

To delete a selected record from a table:

1 Type the query shown in Figure 3-15 to delete Daniel Black's record from the STUDENT table. The message "1 row deleted" appears.

Figure 3-15: Deleting a single record

2 Type the query shown in Figure 3-16 to delete records in the STUDENT table where SID is greater than or equal to 100. The message "2 rows deleted" appears, indicating you have deleted the records for students Sarah Miller and Brian Umato. You can delete multiple records from a table at once if your search condition matches several rows. You can delete every row in a table by omitting the WHERE clause.

Figure 3-16: Deleting multiple records

You cannot delete a record if it is a foreign key reference to another record, as you will see next.

To try to delete a record with a foreign key reference:

1 Type the command shown in Figure 3-17 to delete LOCID 53 (Kim Cox's office) from the LOCATION table. Note the error message that appears.

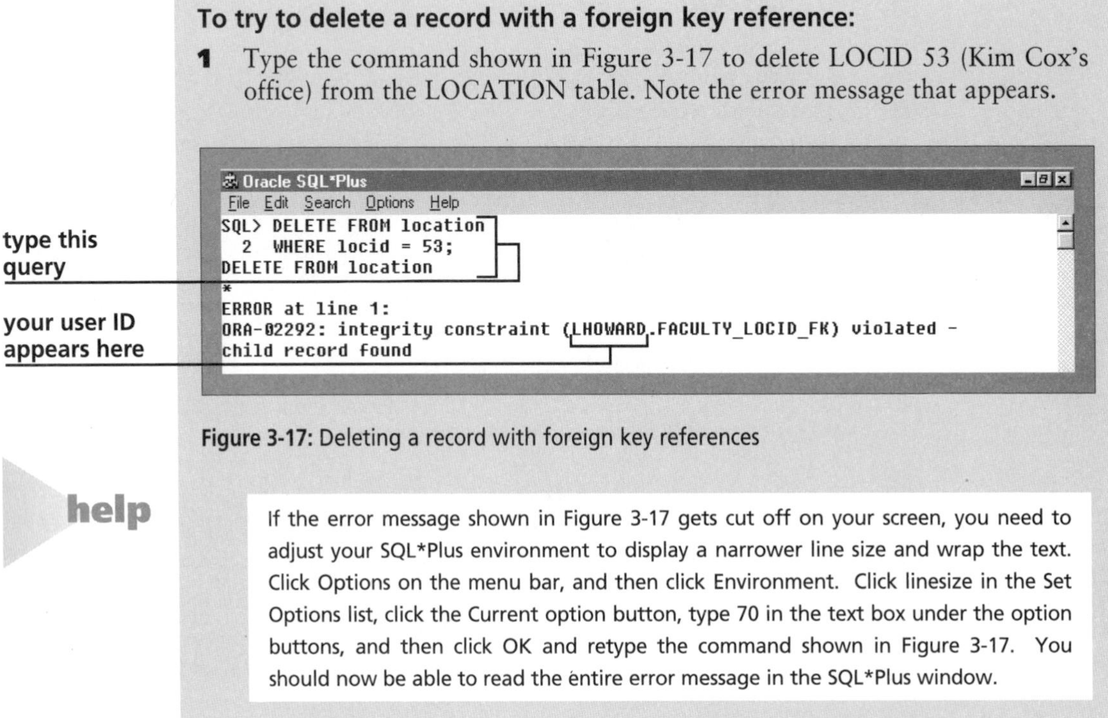

Figure 3-17: Deleting a record with foreign key references

help If the error message shown in Figure 3-17 gets cut off on your screen, you need to adjust your SQL*Plus environment to display a narrower line size and wrap the text. Click Options on the menu bar, and then click Environment. Click linesize in the Set Options list, click the Current option button, type 70 in the text box under the option buttons, and then click OK and retype the command shown in Figure 3-17. You should now be able to read the entire error message in the SQL*Plus window.

The value for LOCID 53 is a foreign key in the FACULTY table for Kim Cox's data record. If you want to delete LOCID 53, you have to delete Kim Cox's FACULTY record first. Now, you will delete all of the records you inserted into the FACULTY and LOCATION tables.

To delete the records you inserted:

1 Type the queries shown in Figure 3-18 to delete the records in the FACULTY and LOCATION tables. The message "2 rows deleted" should appear as shown.

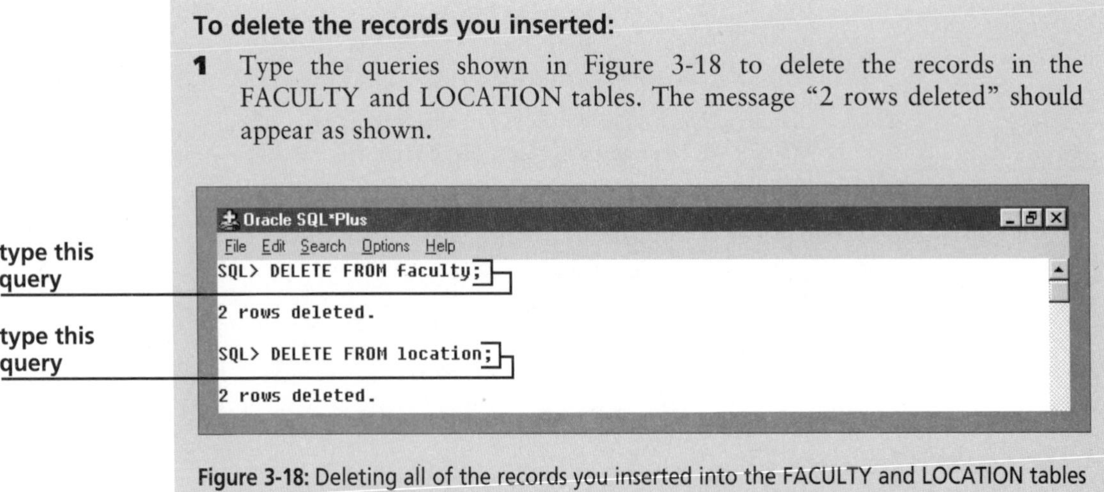

Figure 3-18: Deleting all of the records you inserted into the FACULTY and LOCATION tables

SUMMARY

- Database developers must have a complete understanding of SQL commands for selecting, inserting, updating, and deleting data, because these commands are used extensively in Oracle database applications.
- The INSERT command is used to insert new data records into a table. The INSERT command can be used to insert all of the data fields into a record, or to insert only specific fields.
- Before you can add a new data record, you must ensure that all of the foreign key records that it references have been previously added.
- Character fields in Oracle are case sensitive, and values must be enclosed within single quotation marks when they are inserted.
- To add text strings with embedded single quotation marks, type the single quotation mark twice.
- DATE fields are inserted as characters and then converted to dates using the TO_DATE function. DATE fields also store time values.
- A transaction is one or more related SQL commands that constitute a logical unit of work. All parts of the transaction must be completed, or the database will contain inconsistent data.
- When you create a new table or update the structure of an existing table, the change is effective immediately at the central database on the server. When you insert database records, the database on the server is not changed until you explicitly commit your changes using the COMMIT command or until you exit the program. Other users cannot see the records you have added or access the rows you have modified, because these rows are locked until you commit your changes.
- It is a good idea to commit your changes often (just like saving work in other applications) so that the records will be available to other users and so that your changes will be saved if you do not exit normally, because of a power failure or workstation malfunction.
- A rollback enables you to return the database to its original state by undoing the effects of all the commands since the last commit.
- A savepoint marks the beginnings of individual sections of a transaction. You can roll back transactions to particular savepoints.
- You can update records in only one table at a time using a single UPDATE command.
- You can update multiple records in a table at a time using a single UPDATE command by specifying a search condition that matches multiple records.
- Always include a WHERE clause when deleting a record from a table, because otherwise all table records will be deleted.
- You cannot delete a record if it is a foreign key reference to another record.

REVIEW QUESTIONS

1. What is a script? What is the purpose of creating a script?
2. In an INSERT statement, in what order do the column names in the VALUES clause have to be?
3. Write the INSERT statement to insert CSECID 1000 into the COURSE_SECTION table of the Northwoods University database. What other records must be inserted prior to running this command?
4. Write the TO_DATE function that converts the value 09/16/99 to a date using the format DD-MON-YYYY.
5. Write a TO_DATE function to convert the text value 10:45:07 P.M. to a date.
6. What is the purpose of the COMMIT command?
7. What is the purpose of a savepoint?
8. Write the SQL query to delete all records from the FACULTY table where FRANK = 'ASSO'.
9. Write the SQL query to update the SCLASS field to 'SO' for all records in the STUDENT table where SCLASS = 'FR'.

PROBLEM-SOLVING CASES

1. Run the emptynu.sql script in the Chapter3 folder on your Student Disk. Using Notepad or another text editor, write a script named Ch3aEx1.sql that contains all of the commands required to insert CSECID 1000 into the COURSE_SECTION table of the Northwoods University database. Be sure that the commands are in the correct order so that all foreign key references are inserted first. Test your script by running it in SQL*Plus to make sure that it is working correctly.
2. Modify the script you created in Exercise 1 so that after the records are inserted, they are then deleted in the correct order. Name the modified file Ch3aEx2.sql. Test your script by running it in SQL*Plus to make sure that it is working correctly.
3. Run the emptynu.sql script in the Chapter3 folder on your Student Disk. Write a script named Ch3aEx3.sql that inserts the first three records into the ENROLLMENT table of the Northwoods University database, along with all of the foreign key referenced fields, and then deletes all of the inserted records. Test your script by running it in SQL*Plus to make sure that it is working correctly.

4. Run the emptynu.sql script in the Chapter3 folder on your Student Disk. Write a script named Ch3aEx4.sql that inserts all of the records into the STUDENT table of the Northwoods University database, and then updates the SCLASS field in each record as follows:

Current SCLASS value	Updated SCLASS value
FR	SO
SO	JR
JR	SR

Be sure to insert records for all necessary foreign key references. Do not update any record more than once.

5. Run the emptycw.sql script in the Chapter3 folder on your Student Disk. Using Notepad or another text editor, write a script named Ch3aEx5.sql that contains the commands required to insert the first two records into the CUST_ORDER table of the Clearwater Traders database, and then update the ORDERSOURCE field of these records to 155 for all records where the order source is currently 152. Include the commands to insert all required foreign key references. Test your script by running it in SQL*Plus to make sure that it is working correctly. (*Hint*: You will need to insert the record for ORDERSOURCE 155.)

6. Run the emptycw.sql script in the Chapter3 folder on your Student Disk. Modify the script you created in Exercise 5 so that it deletes all of the records that were inserted. Save the modified script file as Ch3aEx6.sql. Test your script by running it in SQL*Plus to make sure that it is working correctly.

7. Run the emptycw.sql script in the Chapter3 folder on your Student Disk. Write a script named Ch3aEx7.sql that inserts the first two records into the ORDERLINE table of the Clearwater Traders database, along with all of the foreign key referenced fields, and then deletes all of the inserted records. Test your script by running it in SQL*Plus to make sure that it is working correctly.

8. Run the emptycw.sql script in the Chapter3 folder on your Student Disk. Write a script named Ch3aEx8.sql that inserts the three records into the BACKORDER table of the Clearwater Traders database. Also, insert the first three records into the BACKORDER table of the Clearwater Traders database. Test your script by running it in SQL*Plus to make sure that it is working correctly.

9. Run the emptycw.sql script in the Chapter3 folder on your Student Disk. Write a script named Ch3aEx9.sql that inserts all of the records into the ITEM table of the Clearwater Traders database as well as all referenced foreign key records, and then updates all records where CATEGORY is 'Children's Clothing' to 'Kid's Clothing'.

LESSON B

objectives

- **Write SQL queries to retrieve data records from a single database table**
- **Sort query output**
- **Use mathematical calculations, number functions, and date arithmetic in queries**
- **Create a column alias**
- **Group query output**

Now that you have learned how to insert, update, and delete data records using SQL*Plus, the next step is to learn how to retrieve data. SQL enables you to view data in relational database tables, sort the output, and perform calculations on data, such as calculating a person's age from data stored in a date of birth column or calculating a total for an invoice. You can also group retrieved records and perform group functions, such as finding the sum or average of a set of retrieved records.

Retrieving Data from a Single Table

The basic format of a SQL query for retrieving data is:

```
SELECT <column1, column2, etc.>
FROM <ownername.table1 name, ownername.table2 name, …>
WHERE <search condition>;
```

The SELECT clause lists the columns that you want to display in your query. The FROM clause lists the name of each table involved in the query. If you are querying a table that you did not create, then you must preface it with the creator's user name, and the creator must have given you the privilege to SELECT data from that table (you will learn how to grant user privileges in Lesson C of this chapter). To query another user's database tables, you must use the format "ownername.tablename." For example, if you want to query user LHOWARD's LOCATION table, you would write the table name as LHOWARD.LOCATION. The WHERE clause is an optional clause that is used to identify which rows to display by applying a search condition. For example, if you want to display only the records associated with the CR building in the LOCATION table, you would assign the search condition WHERE bldg_code = 'CR'.

The SELECT queries in this section will be performed using fully populated Northwoods University and Clearwater Traders database tables. You will create these tables by running the northwoo.sql and clearwat.sql scripts on your Student Disk. You will do this next.

To run the script file:

1. If necessary, start "SQL*Plus and log on to the database. Run the northwoo.sql script by typing the following command at the SQL prompt: **START A:\Chapter3\northwoo.sql**.
2. Run the clearwat.sql script by typing the following command at the SQL prompt: **START A:\Chapter3\clearwat.sql.**

To retrieve every row in a table, the data do not need to satisfy a search condition, so you omit the WHERE clause. You will retrieve the first name, middle initial, and last name from every row in the STUDENT table in the Northwoods University database next.

To retrieve every row from the STUDENT table:

1. Type the query shown in Figure 3-19 to retrieve every row from the STUDENT table.

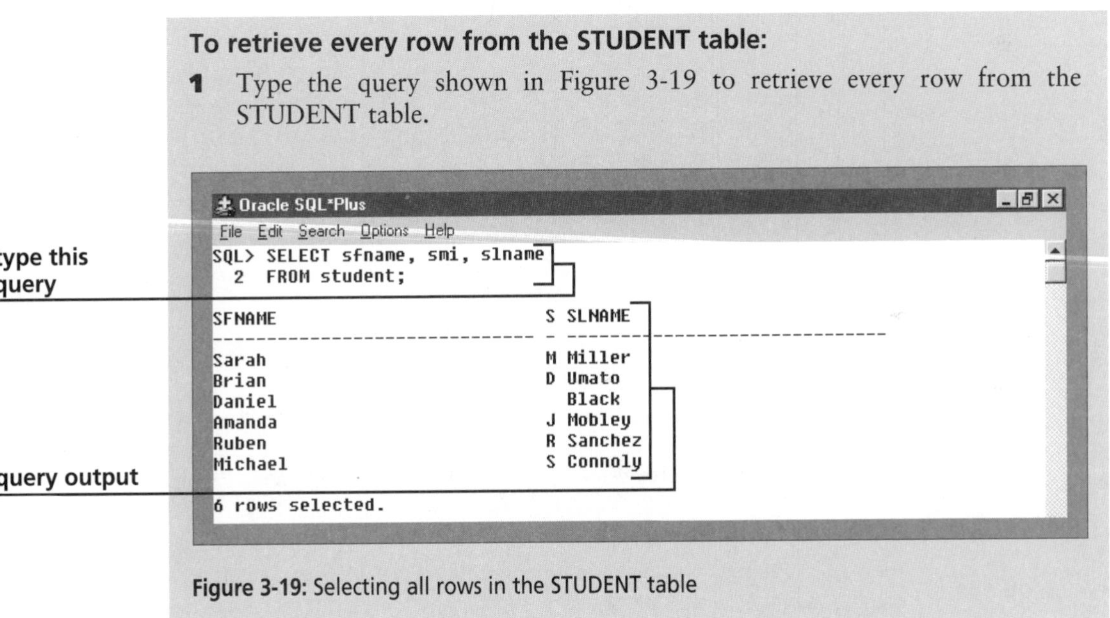

Figure 3-19: Selecting all rows in the STUDENT table

Your output should look like Figure 3-19. Note that the column names appear as column headings. Only the columns included in the SELECT statement are displayed, and the columns appear in the same order in which they are listed in the SQL command.

If you want to retrieve all of the columns in a table, you can use an asterisk (*) as a wildcard character in the SELECT statement instead of typing every column name.

To retrieve every row and column from a table:

1 Type the query shown in Figure 3-20 to select all rows and columns from the LOCATION table.

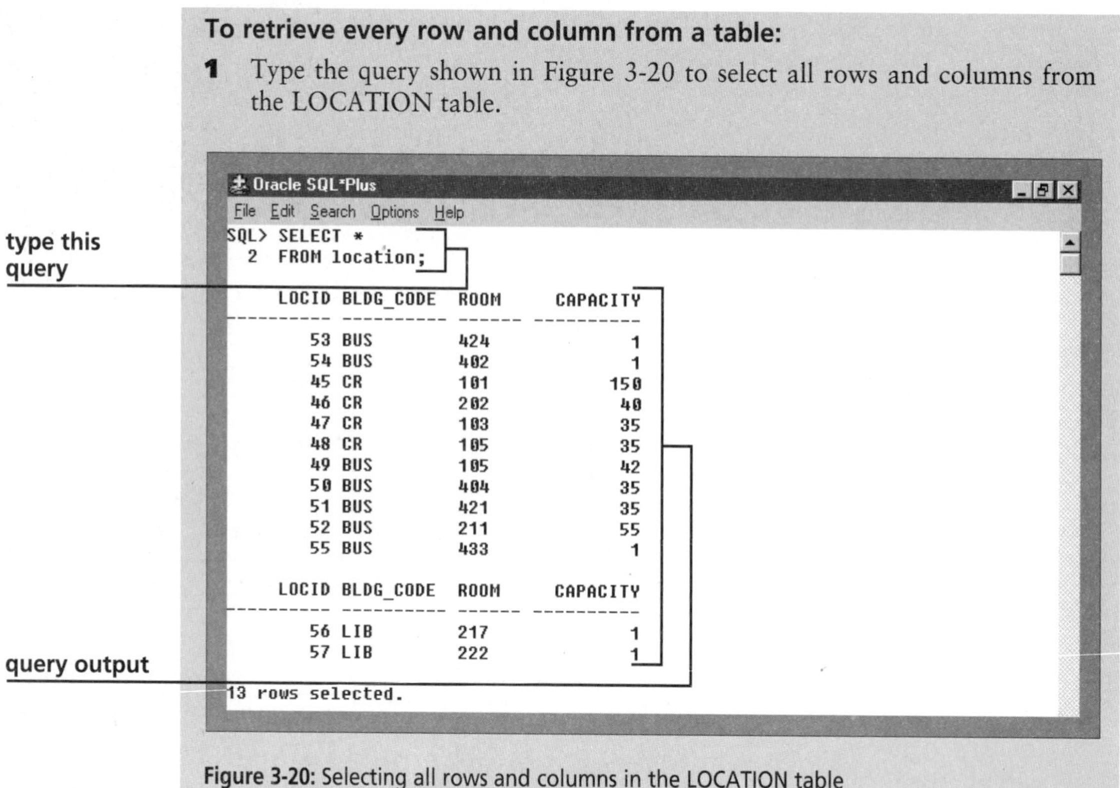

Figure 3-20: Selecting all rows and columns in the LOCATION table

help The SQL*Plus environment is set up to show data output one screen at a time. If the query output is wider than the current SQL*Plus line width, then the output will wrap to the next line. If the query output is longer than the current SQL*Plus page size, then the column headings will be redisplayed, and the remaining output will be displayed under the repeated column headings. To customize the SQL*Plus environment to match your screen resolution, click Options on the menu bar, then click Environment. To change the default line width, select linesize in the Set Options list, click the Current option button, then change the line size to the desired value. To change the page length, select pagesize in the Set Options list, click the Current option button, and change the page length to the desired value.

Your output should look like Figure 3-20. Sometimes a query will retrieve duplicate rows. For example, suppose you want to see the different ranks for faculty members.

To retrieve duplicate rows:

1 Type the command shown in Figure 3-21.

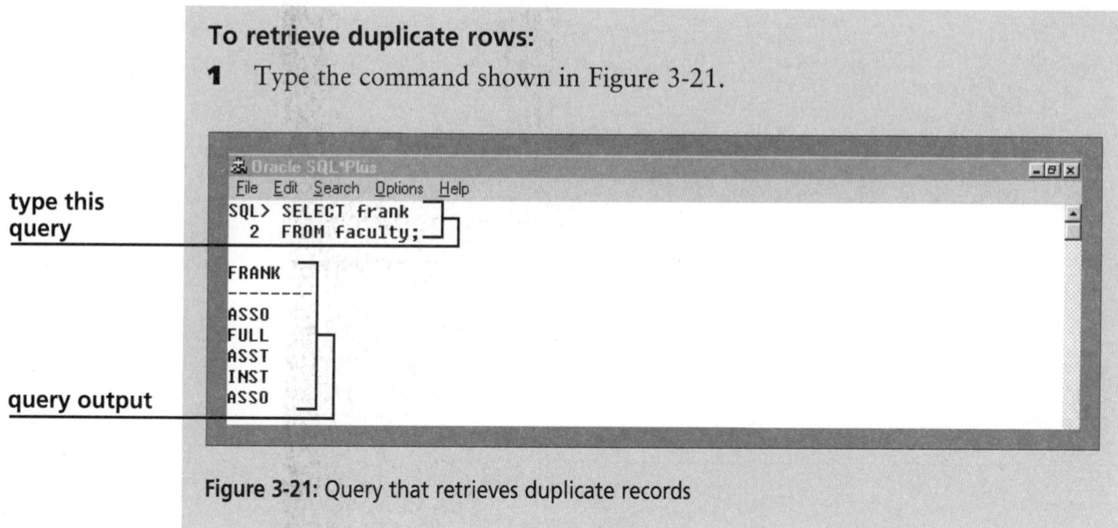

Figure 3-21: Query that retrieves duplicate records

Your output should look like Figure 3-21. The ASSO value is listed twice because there are two records in the table with the ASSO value in the FRANK field. To suppress duplicate values, use the DISTINCT qualifier immediately after the SELECT command. This tells SQL to display each value only once.

To use the DISTINCT qualifier:

1 Type the query shown in Figure 3-22.

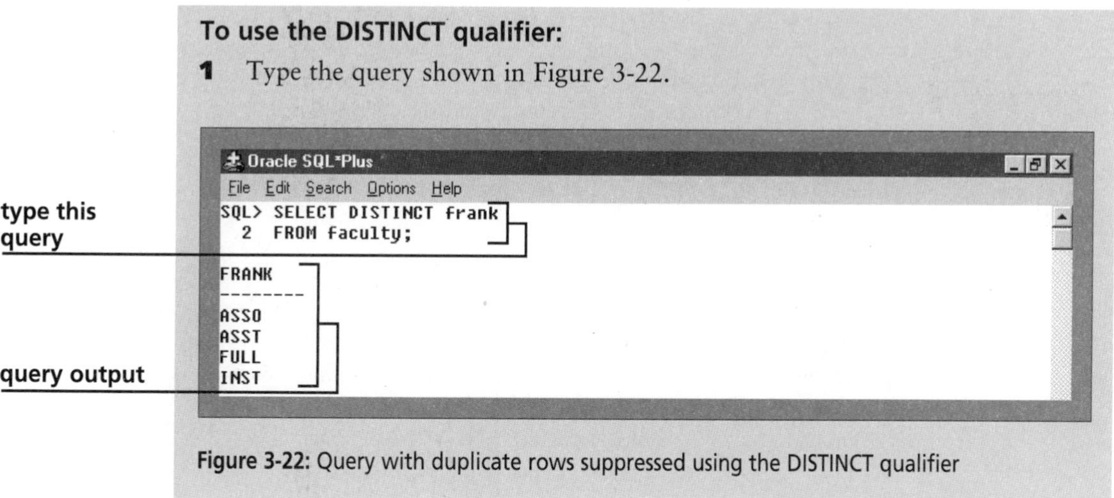

Figure 3-22: Query with duplicate rows suppressed using the DISTINCT qualifier

Your output should look like Figure 3-22, with the duplicate rows suppressed.

Writing Queries that Retrieve Specific Records

Next, you will use a search condition to retrieve specific records in the FACULTY table. Search conditions in the SELECT command work just the same as they do in the UPDATE and DELETE commands. Recall that search conditions use comparison operators, and each search condition must be able to be evaluated as TRUE or FALSE. Search conditions can be combined using the AND and OR operators.

To use a search condition in a SQL command:

1 Type the query shown in Figure 3-23. The query output lists the first name, middle initial, last name, and rank of all faculty members with a rank of ASSO.

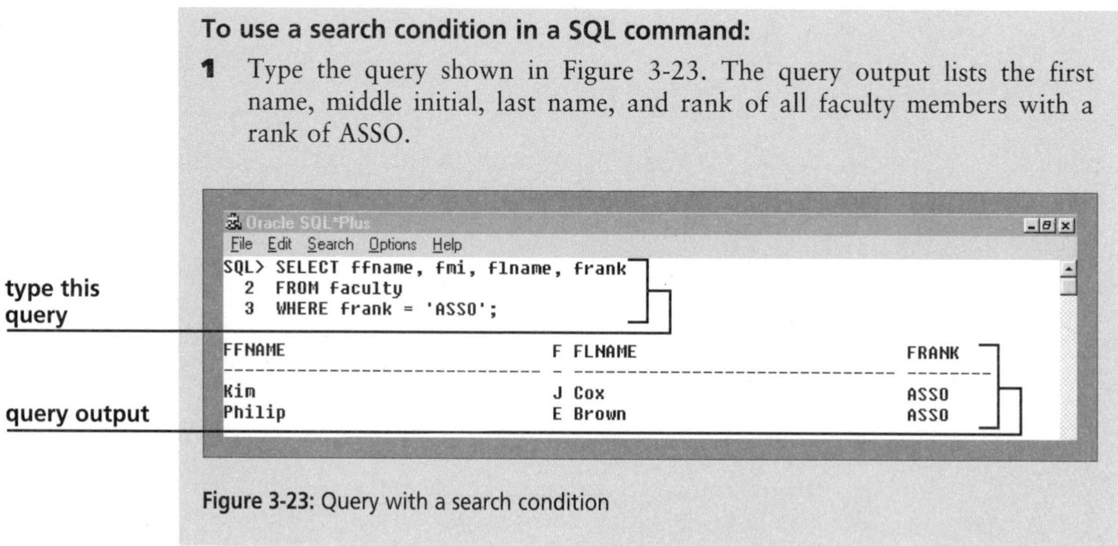

Figure 3-23: Query with a search condition

You also can combine multiple search conditions using the AND and OR operators. Recall that when you use the AND operator to connect two search conditions, both conditions must be true for the row to appear in the query outputs. If no data rows exist that match both conditions, then no data are retrieved. When you use the OR operator to connect two search conditions, only one of the conditions must be true for the row to appear in the query outputs. You will use the AND operator in the next query.

To use the AND operator in a search condition:

1 Type the query shown in Figure 3-24 to find the room numbers of all rooms in the BUS building that have a capacity greater than or equal to 40. The output displays values that match both conditions in the query.

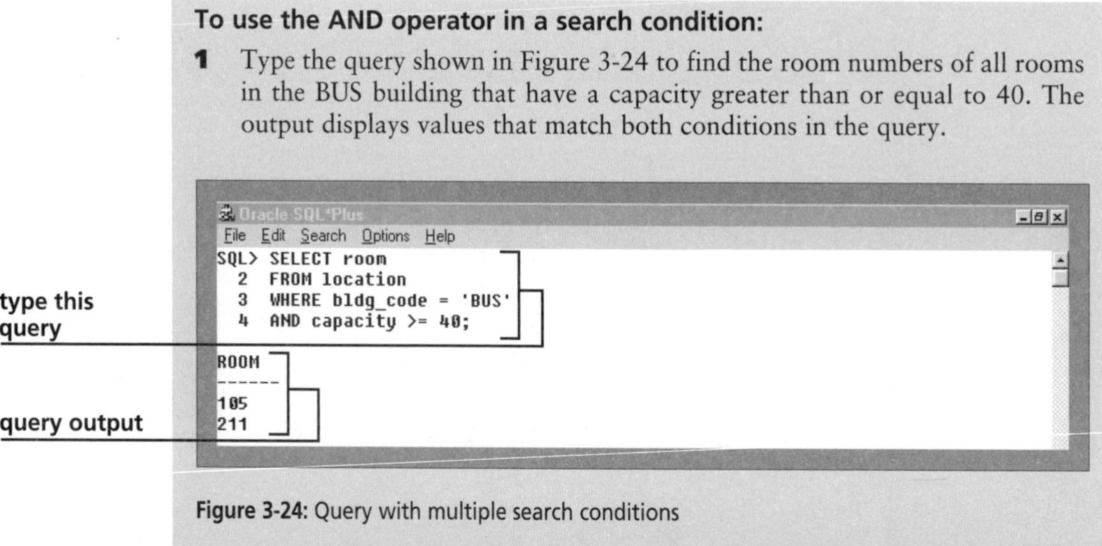

Figure 3-24: Query with multiple search conditions

2 Type the query shown in Figure 3-25 to list the first and last names of all students who are freshmen and who were born before January 1, 1977. Note that whenever you use a data field with the DATE data type in a search condition, you must convert the date from a character string to a DATE data type, using the TO_DATE function. No STUDENT records match both of the conditions, so no rows are listed in the results, as indicated by the "no rows selected" message.

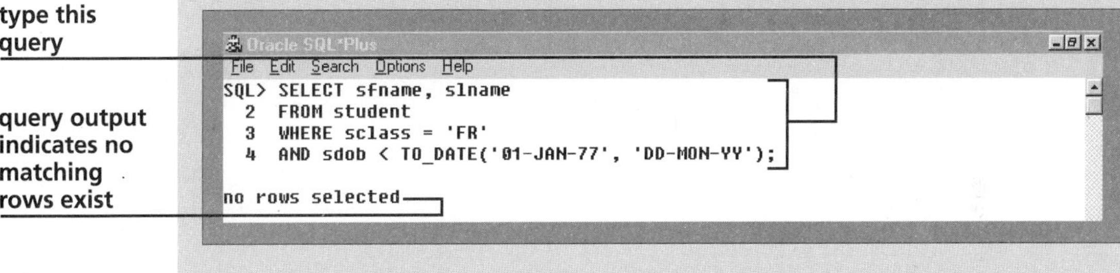

Figure 3-25: Multiple search condition query with no rows selected

When the OR operator is used to connect two conditions, a row is returned if *either* the first *or* second condition is true. If neither condition is true, no data are returned.

To use the OR operator in a condition:

1 Type the query shown in Figure 3-26 to find the first and last name of every student who was born after January 1, 1978, or who is a freshman. The query output lists one student who is a freshman (Michael Connoly), along with those who were born after January 1, 1978, but are not freshmen (Sarah Miller, Brian Umato, and Amanda Mobley).

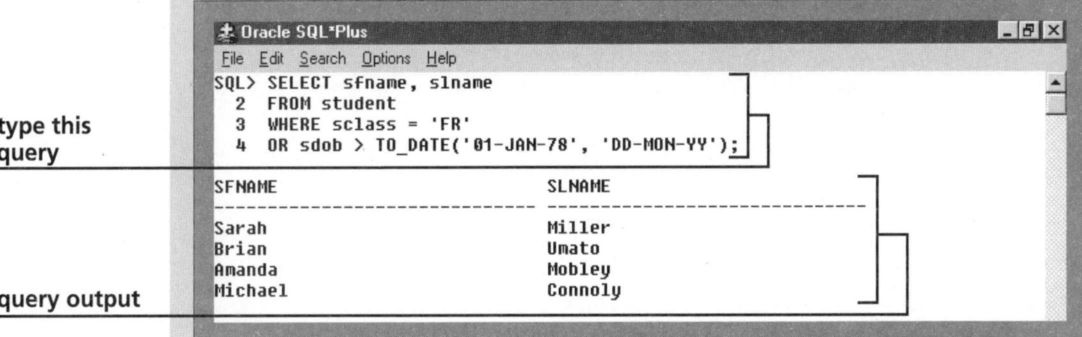

Figure 3-26: Multiple search condition query using the OR operator

You can combine the AND and OR operators in a single query. This is a very powerful operation, but it can be tricky to use. SQL evaluates AND conditions first, and then SQL evaluates the result against the OR condition.

To use the AND and OR operators in a single condition:

1 Type the query shown in Figure 3-27 to find the location ID, building code, room number, and capacity of every room in the BUS or CR building whose capacity is greater than 35.

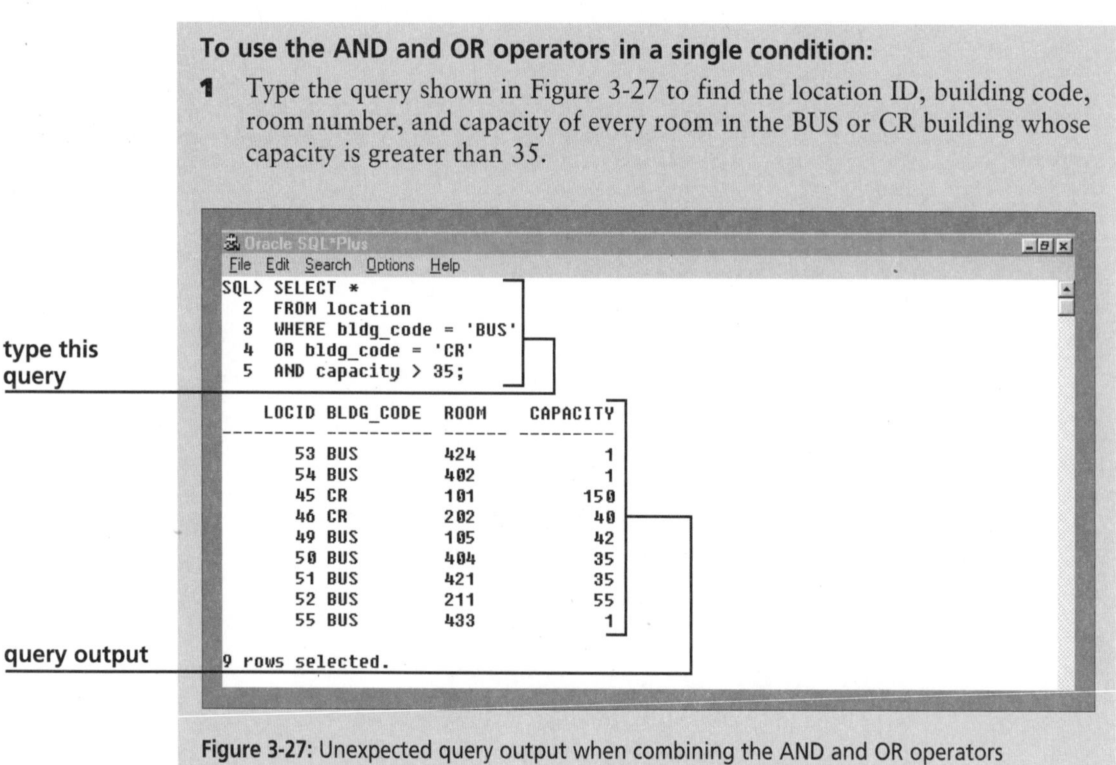

Figure 3-27: Unexpected query output when combining the AND and OR operators

The query output shown in Figure 3-27 is puzzling. Why were rooms with capacity of less than 35 returned? SQL first evaluates the AND condition (bldg_code = 'CR' AND capacity > 35), and then returns rows that contain LOCIDs of 45 and 46. Then SQL evaluates the first half of the OR condition (bldg_code = 'BUS'), and returns all rows with bldg_code = 'Bus' (LOCIDs of 49, 50, 51, 52, 53, 54, and 55). Finally, SQL combines these results with the result of the AND condition (LOCIDs 45 and 46). Therefore, it returns rows of 45, 46, and 49 through 55, which is not the result you wanted.

The best way to overcome AND/OR ordering problems is to put the operation that should be performed first in parentheses. SQL always evaluates operations in parentheses first—regardless of whether or not they contain an AND or OR operator.

To use the AND and OR operators correctly in a single condition:

1 Type the query shown in Figure 3-28. Now the query output is correct. SQL performs the OR operation first (by returning all LOCIDs either in CR or BUS), and then evaluates the AND condition by returning only those locations that have a capacity of greater than 35.

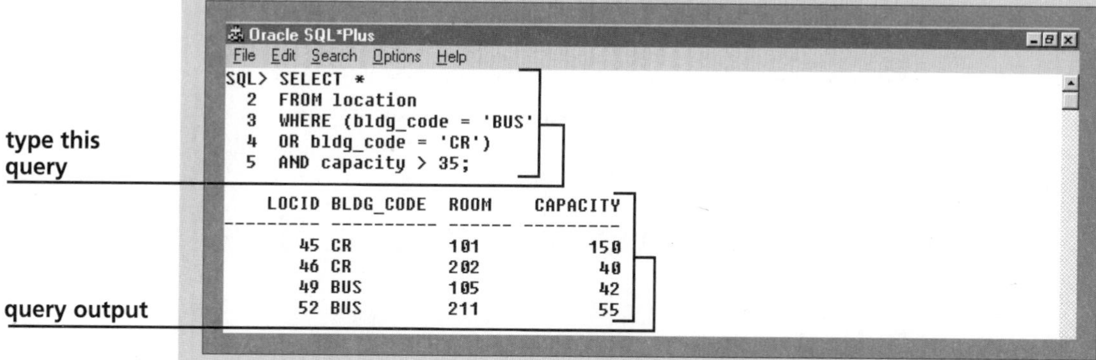

Figure 3-28: Using parentheses when combining the AND and OR operators

You also can use the **NOT** operator in a condition to express negative conditions, as you will see next.

To use the NOT operator in a condition:

1 Type the query shown in Figure 3-29 to list the first and last name of every student who is not a senior. The query output lists students who are not seniors.

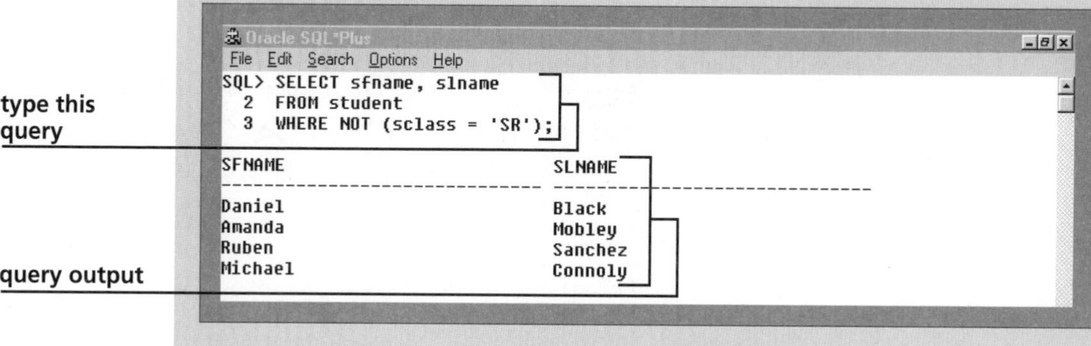

Figure 3-29: Using the NOT operator in a search condition

Sometimes you need to create a query to return records where the value of a particular field is NULL. To do this, you use the following general format for the search condition: `WHERE <field name> IS NULL`. Similarly, to return records where the value of a particular field is not NULL, you use the format `WHERE <field name> IS NOT NULL`. Next, you will create a query to find all records in the ENROLLMENT table where the grade field has not been assigned and is currently NULL.

To create a query using an IS NULL search condition:

1 Type the query shown in Figure 3-30. Note that the returned records show all enrollment records where the grade value has not yet been entered.

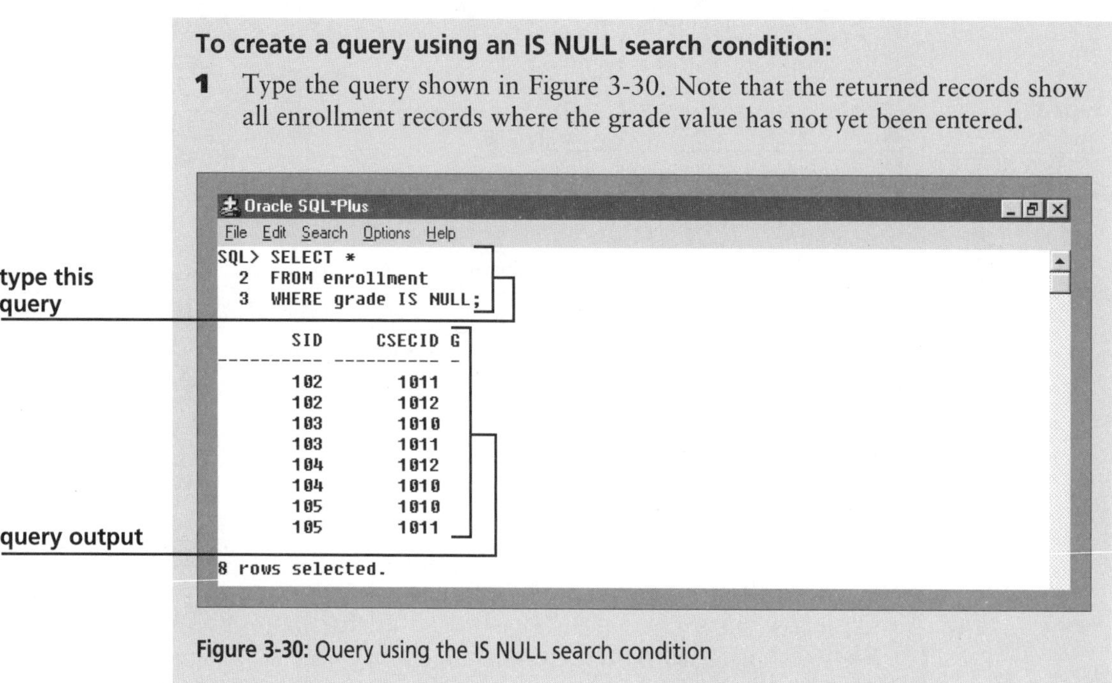

Figure 3-30: Query using the IS NULL search condition

Now, you will create a query to find all records in the ENROLLMENT table where the grade field has been assigned, using the IS NOT NULL search condition.

To create a query using an IS NOT NULL search condition:

1 Type the query shown in Figure 3-31. Note that the returned records show all enrollment records where the grade value has been entered.

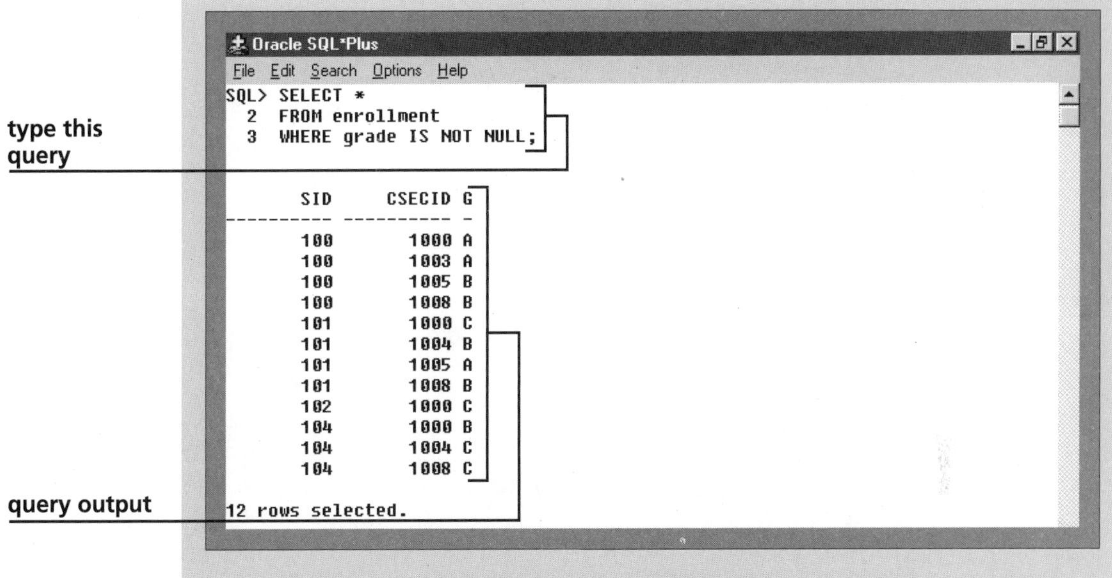

Figure 3-31: Query using the IS NOT NULL search condition

Sorting the Query Output

The query output you have seen so far has not displayed data in any particular order. You can sort query output by using the ORDER BY clause and specifying the **sort key**, which is the column SQL will use as a basis for ordering the data. The format for the ORDER BY clause is `ORDER BY <column name>`. By default, records are sorted in numerical ascending order if the sort key is a NUMBER column, and in alphabetical ascending order if the sort key is a CHAR or VARCHAR2 column. To sort the records in descending order, insert the DESC command at the end of the ORDER BY clause.

To use the ORDER BY clause to sort data:

1 Type the query shown in Figure 3-32 to list the building code, room number, and capacity for every room with a capacity that is greater than or equal to 40, sorted in ascending order by capacity.

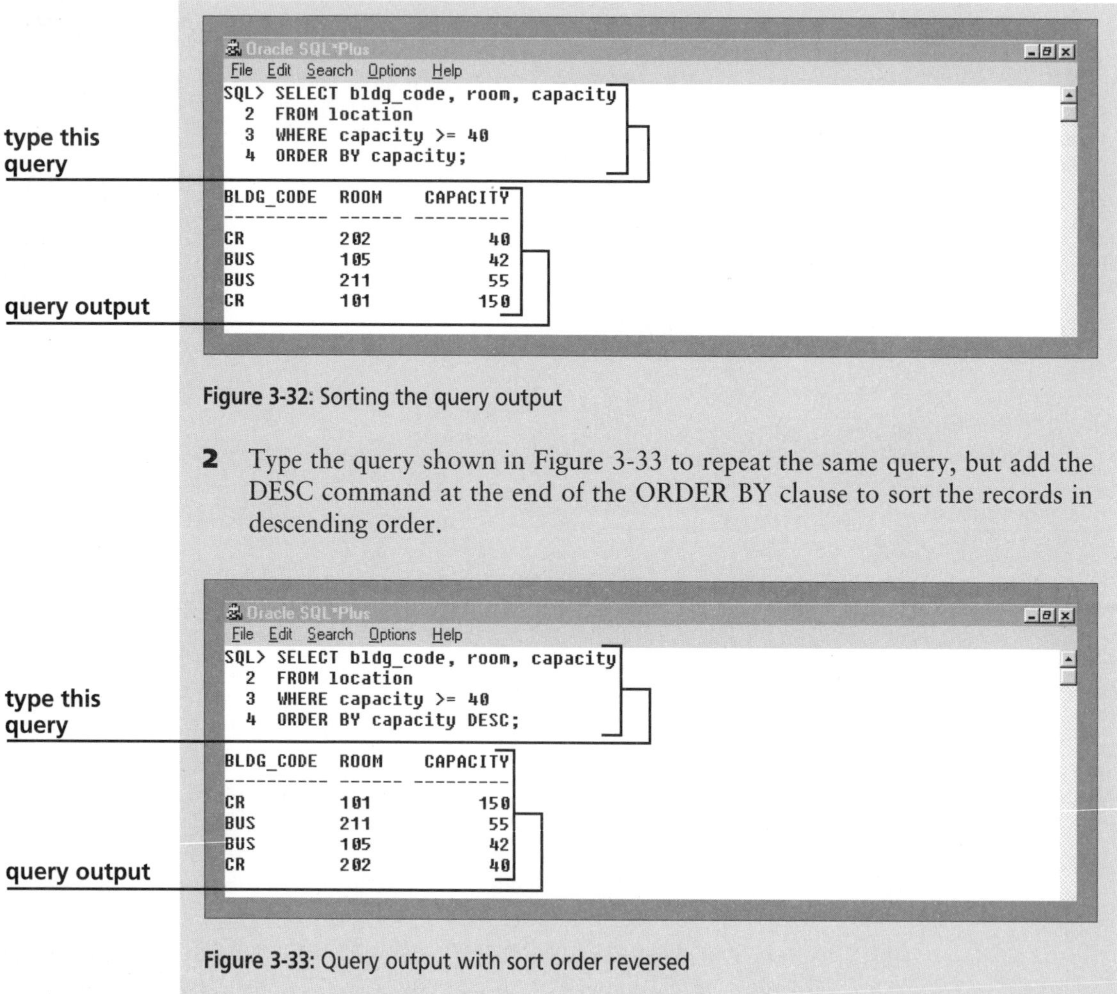

Figure 3-32: Sorting the query output

2 Type the query shown in Figure 3-33 to repeat the same query, but add the DESC command at the end of the ORDER BY clause to sort the records in descending order.

Figure 3-33: Query output with sort order reversed

You also can specify multiple sort keys to sort query outputs on the basis of multiple columns. You must specify which column gets sorted first, second, etc. The next query lists all building codes, rooms, and capacities, sorted first by building code, and then by room number.

To sort data on the basis of multiple columns:

1 Type the query shown in Figure 3-34. The query output lists every row in the LOCATION table, first sorted alphabetically by building code and then sorted within building codes by ascending room numbers.

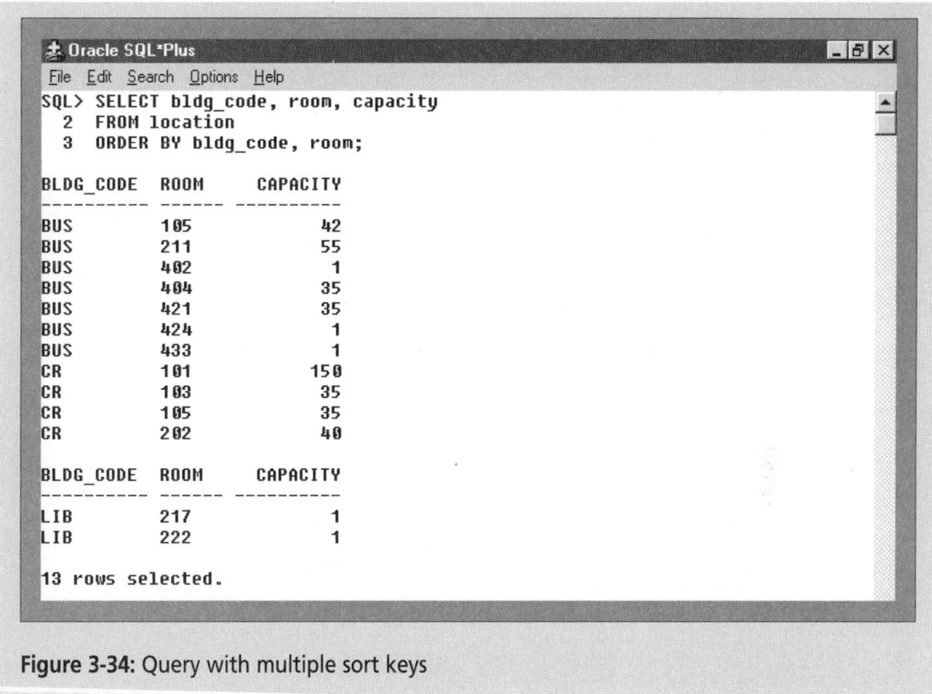

Figure 3-34: Query with multiple sort keys

Using Calculations in a Query

You can perform basic mathematical calculations on retrieved data. Figure 3-35 lists the operations and their associated SQL operators.

SQL Operator	Description
+	Addition
-	Subtraction
*	Multiplication
/	Division

Figure 3-35: SQL operators used in calculations

For example, suppose that you want to display the course section ID, maximum enrollment, current enrollment, and the difference between the maximum enrollment and the current enrollment for each course section.

To calculate data using a SQL query:

1 Type the query shown in Figure 3-36. The query output lists the calculated value by subtracting the CURRENRL value from the MAXENRL value.

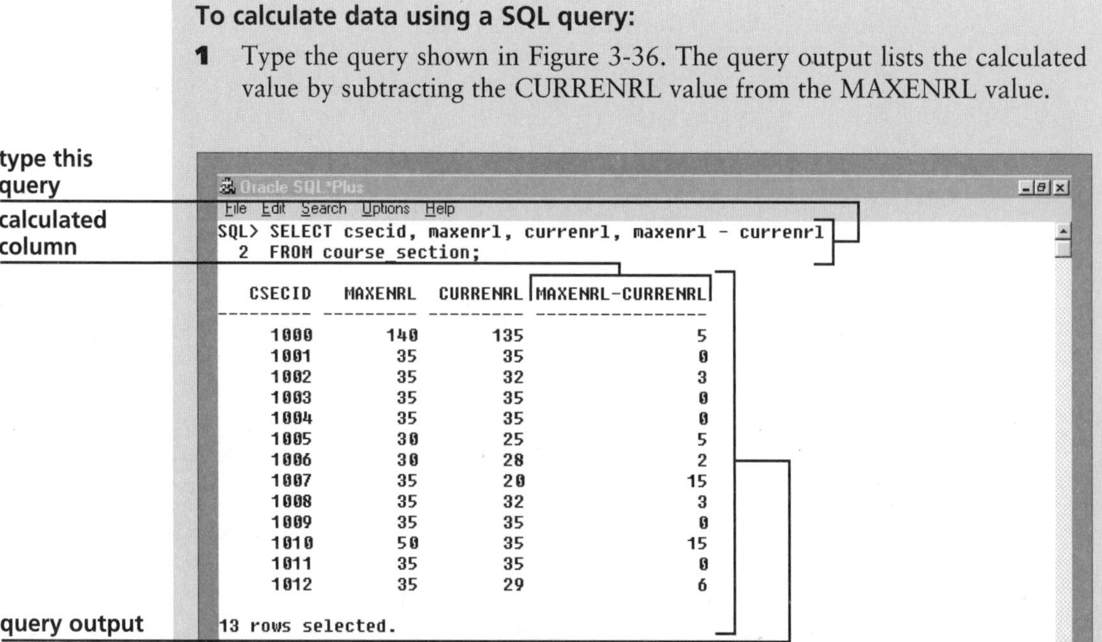


```
SQL> SELECT csecid, maxenrl, currenrl, maxenrl - currenrl
  2  FROM course_section;

   CSECID    MAXENRL   CURRENRL MAXENRL-CURRENRL
--------- ---------- ---------- ----------------
     1000        140        135                5
     1001         35         35                0
     1002         35         32                3
     1003         35         35                0
     1004         35         35                0
     1005         30         25                5
     1006         30         28                2
     1007         35         20               15
     1008         35         32                3
     1009         35         35                0
     1010         50         35               15
     1011         35         35                0
     1012         35         29                6

13 rows selected.
```

Figure 3-36: Query with calculated column

The next query uses the SYSDATE function, which returns the current system date and time from the client workstation. To form the query, you use the following format: `SELECT SYSDATE FROM DUAL`. DUAL is a system table that is created when the database is installed. It belongs to the system, but all users have SELECT privileges on it. DUAL is used by the SQL compiler for processing values that involve constant values that are not retrieved from the database, such as numeric constants and the system date. To retrieve the current system date, use SYSDATE in the SELECT clause, and then specify DUAL in the FROM clause.

To use the SYSDATE function:

1 Type the query shown in Figure 3-37 to list the student ID, last name, and age for each student. The student age is calculated by subtracting the student's date of birth from the current date.

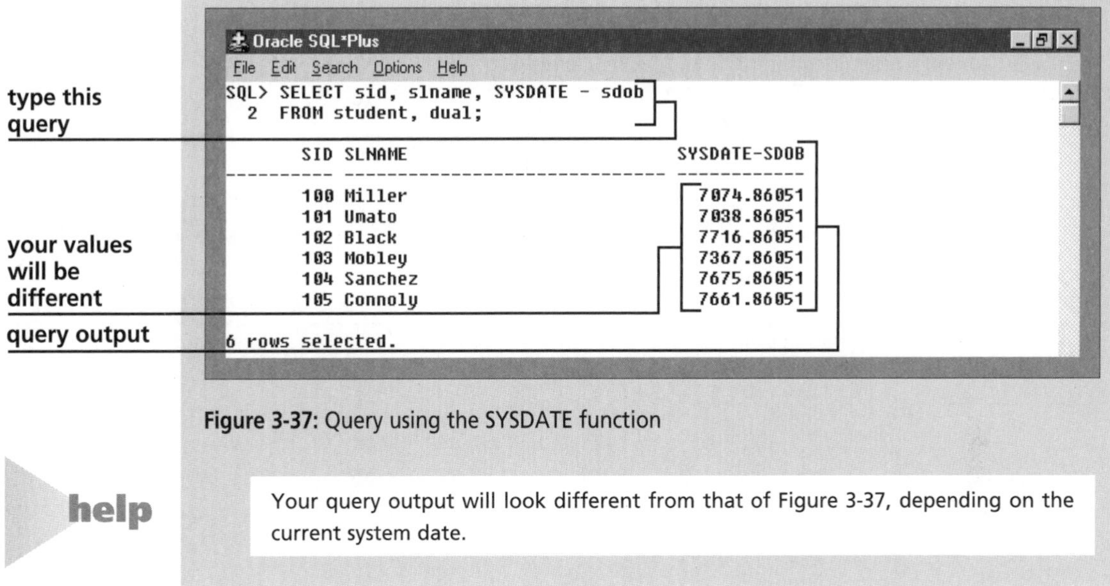

Figure 3-37: Query using the SYSDATE function

help

Your query output will look different from that of Figure 3-37, depending on the current system date.

The query output lists the calculated ages in days rather than in years. One solution to this problem is to divide this value by the number of days in a year, which is approximately 365.25 (including leap years). To display the query output in years, you will have to use the following expression: SYSDATE – sdob/365.25. You can do this in SQL by combining multiple arithmetic operations in a single query.

In mathematics and in programming languages, expressions that contain more than one operator must be evaluated in a specific order. SQL evaluates division and multiplication operations first, and addition and subtraction operations last. Recall that SQL always evaluates expressions enclosed within parentheses first. Therefore, the previous expression is evaluated as SYSDATE – (SDOB/365.25), which is not what you want, because the division operation is evaluated before the subtraction operation. However, your intent is to calculate the difference between the current date and the student's date of birth first, and then divide that result by 365.25. You can use parentheses to indicate that the subtraction operation should be evaluated first.

To combine multiple arithmetic operations in a single query:

1. Type the query shown in Figure 3-38. The query output lists the students' ages in years.

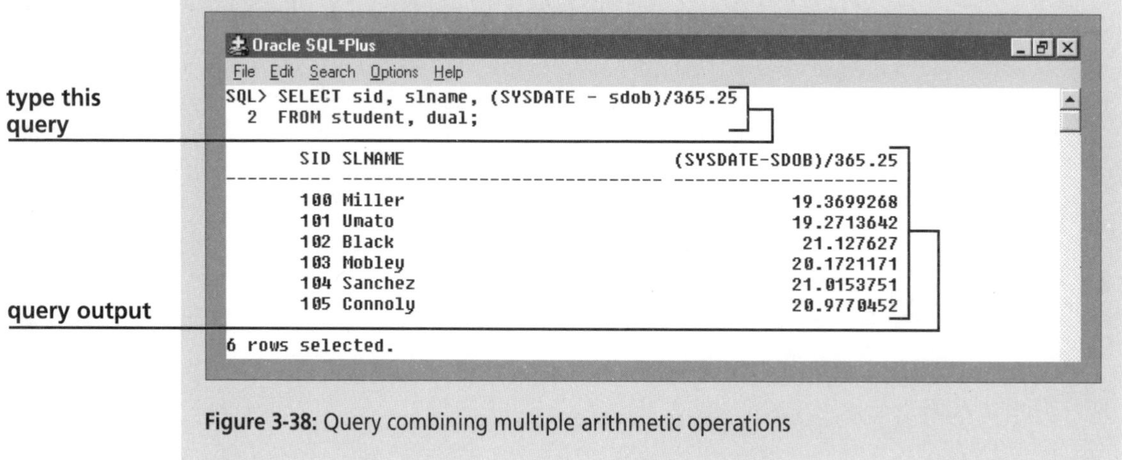

Figure 3-38: Query combining multiple arithmetic operations

Using Number Functions in a Query

In Figure 3-38, each student's age is shown in years with a fraction that represents the time since the student's last birthday. To remove this fraction, you will need to use a SQL number function. SQL has several **number functions** that can be used to manipulate retrieved data. Figure 3-39 summarizes some of the commonly used SQL number functions.

SQL Function	Description	Example	Result
ABS(*n*)	Returns the absolute value of *n*	SELECT ABS(-1) FROM DUAL;	1
POWER(*n*,power)	Returns *n* raised to the specified power	SELECT POWER(3,2) FROM DUAL;	9
ROUND(*n*, precision)	Returns *n* rounded to the specified precision	SELECT ROUND (123.476, 1) FROM DUAL;	123.5
TRUNC(*n*, precision)	Returns *n* truncated to the specified precision	SELECT TRUNC (123.476, 1) FROM DUAL;	123.4

Figure 3-39: SQL number functions

To use a SQL number function, list the function name followed by the required parameter (or parameters) in parentheses. If parameters are actual numeric values (such as rounding the number 123.476), then the general query format is SELECT<function name (parameter)> FROM DUAL;. If the parameter is a retrieved or calculated database value (such as truncating the fraction from the retrieved student ages), the general query format is SELECT<function name>(<field name>) FROM <tablename> WHERE <search condition>;. The next query demonstrates how to use a number function with a calculated database value. You will use the TRUNC function to truncate the fraction portion of the calculated student ages.

To list the students' ages in years without fractions:

1 Type the query shown in Figure 3-40. The query output shows the students' ages in years without fractional values. Your values might be different due to different system dates.

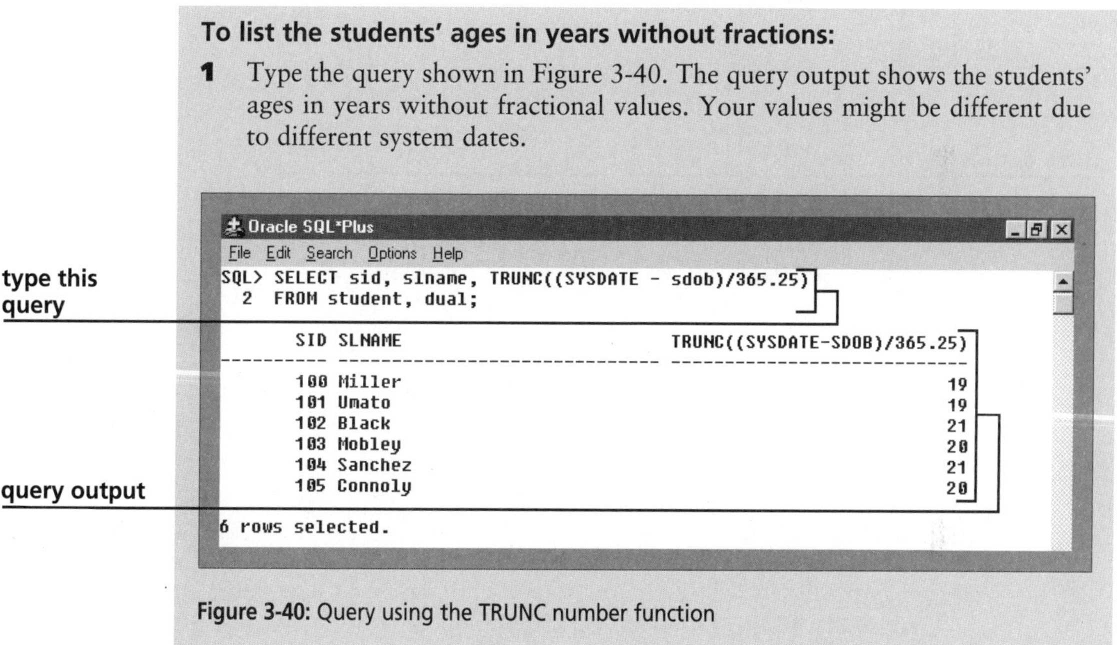

Figure 3-40: Query using the TRUNC number function

Creating a Column Alias

In Figure 3-40, the heading for the output column showing the students' ages includes the formula used to calculate the age, along with the TRUNC number function. When you are creating a query with a complex arithmetic operation or number function like this, it is desirable to create an **alias**, which is an alternate column name that describes the column data and hides the underlying complexity of the query. The general format for creating an alias is: SELECT <field name> AS <alias name>. The alias name must obey the same rules as for naming database columns—it must be between 1 and 30 characters, begin with a letter, not be a reserved word, and not contain spaces or hyphens. You will now repeat the query to calculate the students' ages, but modify it to create an alias for the age column.

To create an alias:

1 Type the query shown in Figure 3-41. The alias appears as the new column heading.

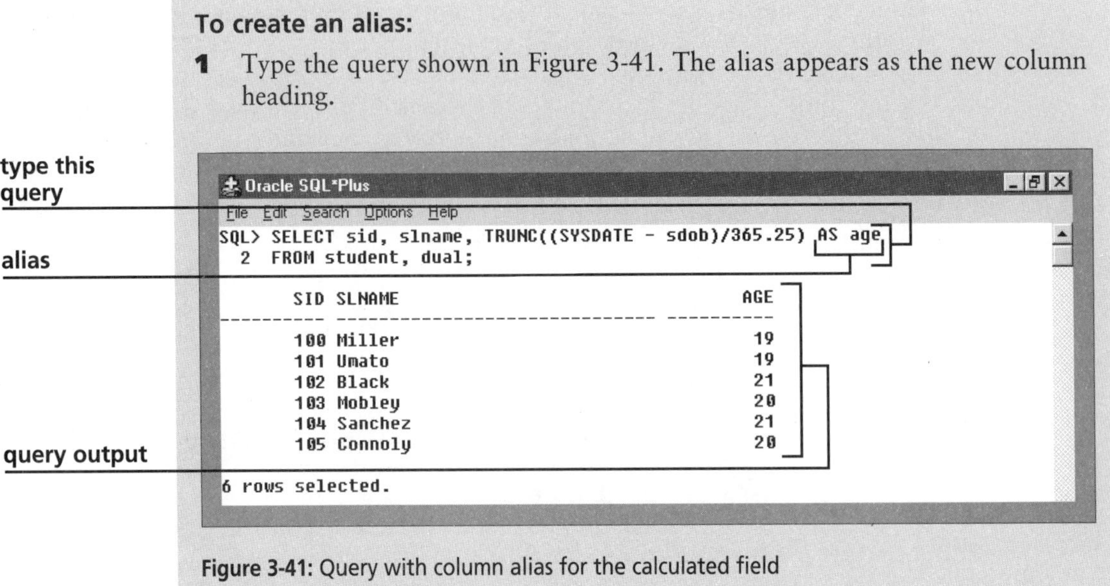

Figure 3-41: Query with column alias for the calculated field

Using Date Arithmetic

Often when creating database reports, it is useful to be able to perform arithmetic calculations on dates for retrieving records corresponding to a date that is within the next week, next month, or next year. For example, you might need to know quantities of shipments that are scheduled to arrive within a certain period of time.

First, you will learn how to specify a date that is within a specific number of days before or after a known date. Suppose today is August 10, 2001, and you would like to determine all shipments that are expected to arrive at Clearwater Traders within the next seven days. To do this, you add the number of days to the known date.

To write a query that specifies a date after a known date, using date arithmetic:

1 Type the query shown in Figure 3-42. Note that the query output specifies shipments that are expected to arrive in the period from 8/11/2001 through 8/17/2001. The query will retrieve all dates after (greater than) 8/10/2001, and all dates before (less than) or equal to 8/10/2001 plus seven days, or 8/17/2001.

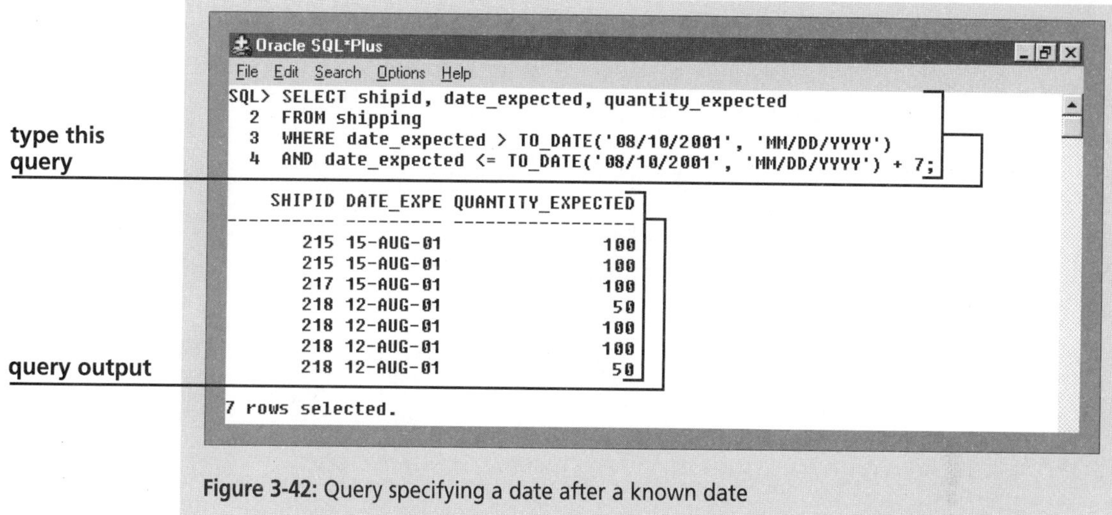

Figure 3-42: Query specifying a date after a known date

To write a query that specifies a date that occurs a specific number of days before a known date, you subtract the number of days from the date. Suppose today is August 31, 2001, and you would like to view all shipments that were received within the past seven days. You will write the query to do this next.

To write a query that specifies a date before a known date, using date arithmetic:

1 Type the query shown in Figure 3-43. Note that the query output specifies shipments that were received in the period from 8/25/2001 through 8/31/2001.

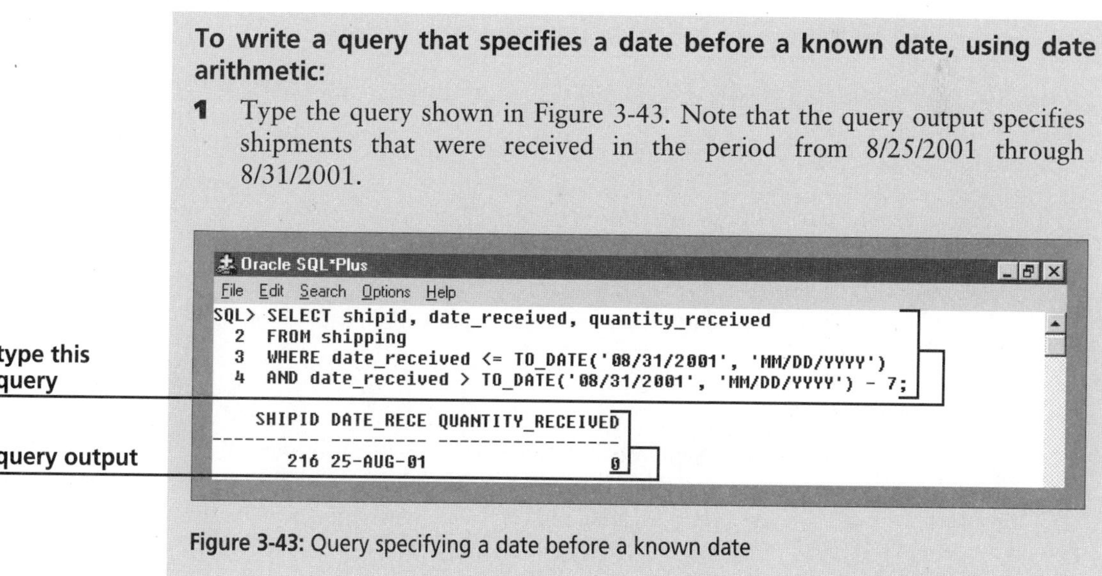

Figure 3-43: Query specifying a date before a known date

To specify a date that is within a specific number of months of a given date, you use the ADD_MONTHS function. The general format of this function is `ADD_MONTHS(<known date>, <count>)`, where count is the number of months before or after the known date. To specify a date that is an exact number of months after a known date, use a positive value for count; to specify an exact number of months before a known date, use a negative value for count. Suppose today is September 1, 2001, and you would like a listing of all shipments that are expected to be received during the next two months. You will write the query to do this next, using the ADD_MONTHS function.

To write a query using the ADD_MONTHS function:

1 Type the query shown in Figure 3-44. Note that you can combine the TO_DATE function with the ADD_MONTHS function. The query output specifies shipments expected to be received during the two-month period following September 1, 2001.

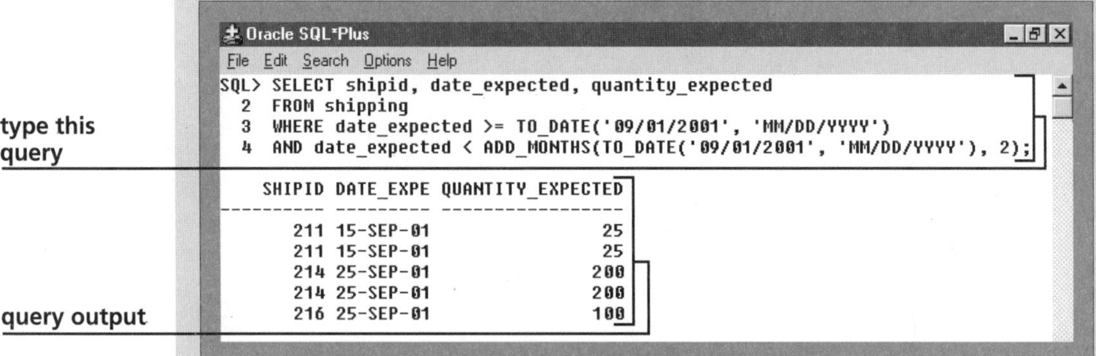

Figure 3-44: Query specifying a period after a known date using the ADD_MONTHS function

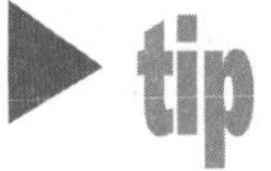

Using the < sign in line 4 of the query shown in Figure 3-44 returns all dates including today (September 1) and less than two months from today (November 1), thus returning September 1 through October 31. If you replace the < with <=, the query will also return November 1.

Similarly, you use a negative value for count to specify a period in the past. Suppose today is September 1, 2001, and you would like a listing of all shipments that were received during the past two months. You will write the query to do this next, using the ADD_MONTHS function, with a negative value for count.

To write a query using the ADD_MONTHS function to specify a past date:

1 Type the query shown in Figure 3-45. The query output specifies shipments that were received during the two-month period before September 1, 2001.

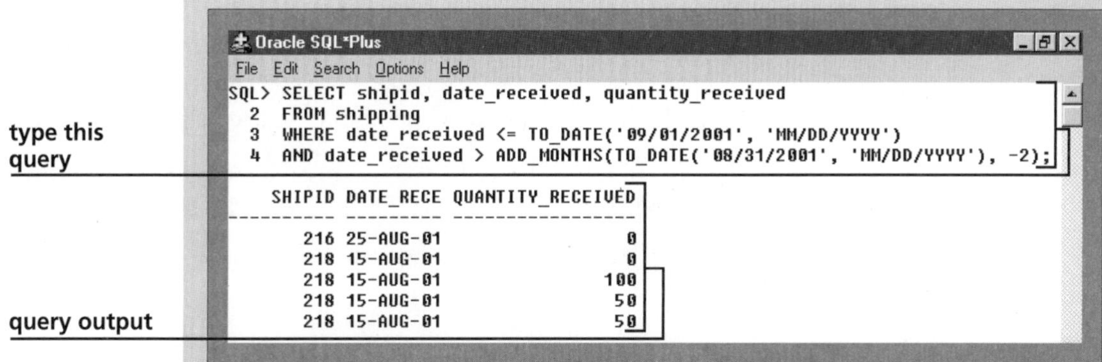

Figure 3-45: Query specifying a period before a known date using the ADD_MONTHS function

Another useful date arithmetic function is MONTHS_BETWEEN, which returns the number of months between two given dates. The general format of the function is: `MONTHS_BETWEEN(<date1>, <date2>)`. This function returns the number of months between the two dates as a whole number, as well as a fractional portion representing additional days. In the next query, you will calculate the number of days between the date received and date expected for inventory item 11845 in shipment 218, using the MONTHS_BETWEEN function, with an alias of "Months Late" for the output.

To use the MONTHS_BETWEEN function:

1 Type the query shown in Figure 3-46. The query output specifies the fractional value of the months between the two dates.

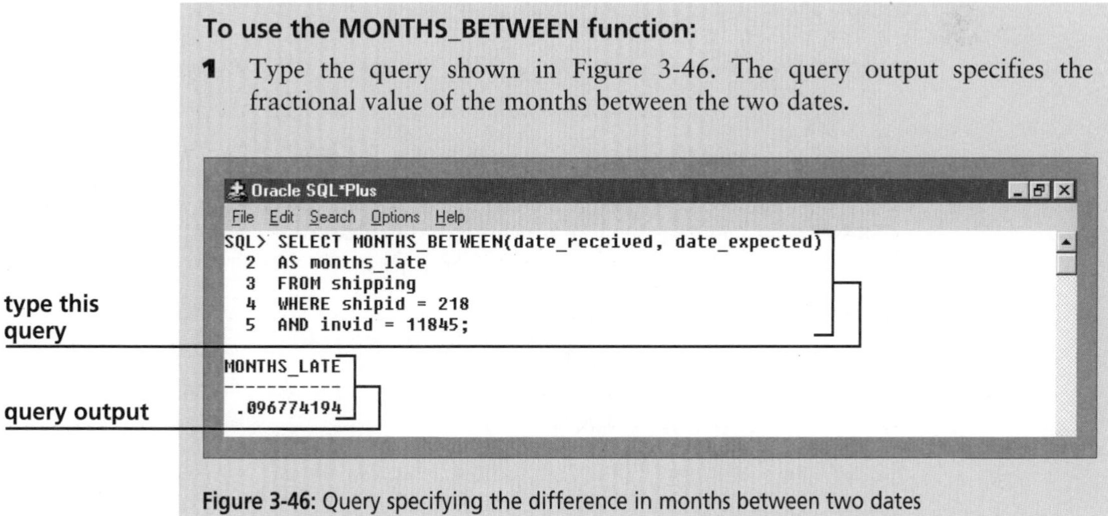

Figure 3-46: Query specifying the difference in months between two dates

2 To convert the result of the previous query to days, you will need to multiply the result by 30.4 (the average number of days in a month), and then use the TRUNC function to remove the fractional remainder. Type the query shown in Figure 3-47 to display the number of days between two dates.

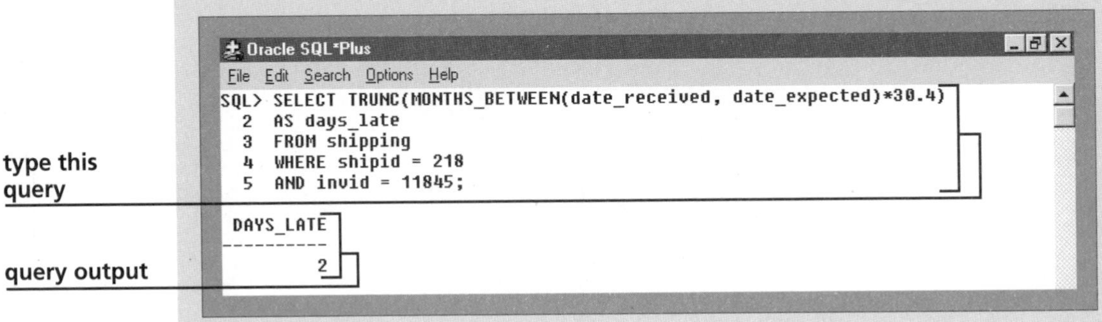

type this query

```
SQL> SELECT TRUNC(MONTHS_BETWEEN(date_received, date_expected)*30.4)
  2  AS days_late
  3  FROM shipping
  4  WHERE shipid = 218
  5  AND invid = 11845;

 DAYS_LATE
----------
         2
```

query output

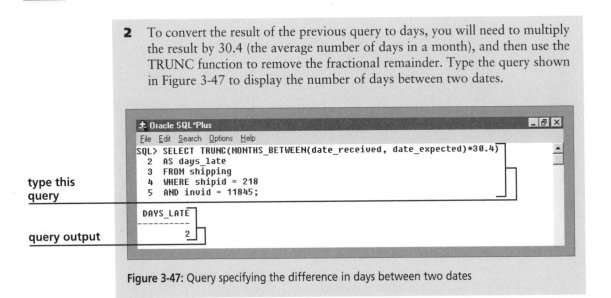

Figure 3-47: Query specifying the difference in days between two dates

Using Group Functions in Queries

A SQL **group function** performs an operation on a group of queried rows and returns a single result, such as a column sum. You might use this function to sum the total capacity of all rooms in the CR building. Figure 3-48 describes five commonly used SQL group functions.

Function	Description	Example
AVG	Returns the average value of a numeric column's returned values	AVG(capacity)
COUNT	Returns an integer representing a count of the number of returned rows	COUNT(grade)
MAX	Returns the maximum value of a numeric column's returned values	MAX(currenrl)
MIN	Returns the minimum value of a numeric column's returned values	MIN(currenrl)
SUM	Sums a numeric column's returned values	SUM(ccredit)

Figure 3-48: SQL group functions

To use a group function in a SQL query, list the function name followed by the column name on which to perform the calculation in parentheses, using the general format `<function name>(<column name>)`. Next, you will enter a query using a group function.

To use a group function in a query:

1 Type the query shown in Figure 3-49 to sum the total current enrollment and calculate the average, maximum, and minimum current enrollment for each course section for the Summer 2002 term (TERMID = 6).

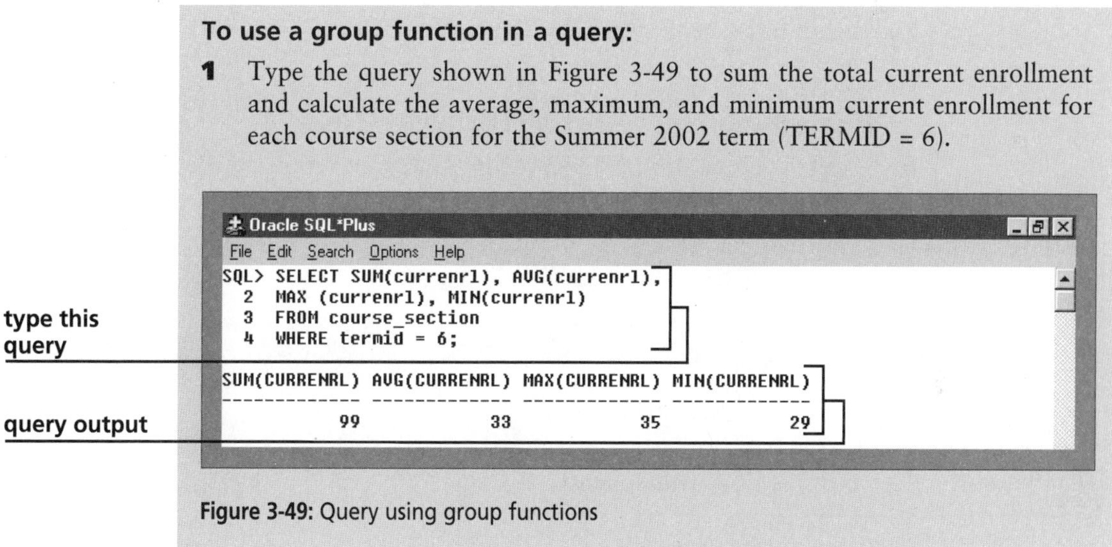

Figure 3-49: Query using group functions

Sometimes you need to divide the query output rows into groups that have matching values and then apply a group function to the grouped data. To do this, use the GROUP BY clause.

To use the GROUP BY clause to group rows:

1 Type the query shown in Figure 3-50 to list the building code name with the total, average, and maximum room capacities (excluding private offices with a capacity of less than five) of all buildings in the LOCATION table.

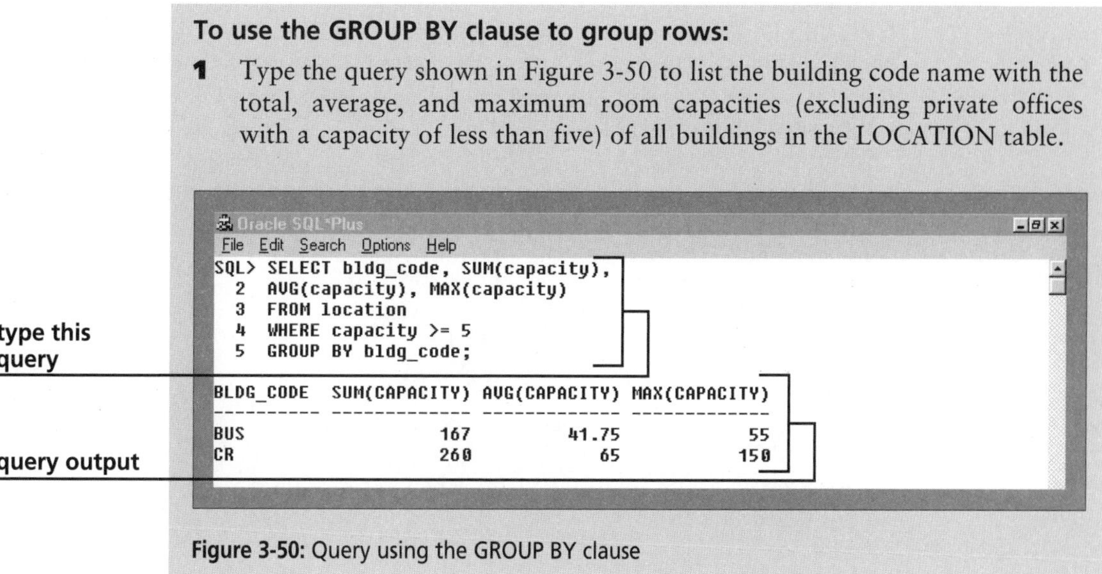

Figure 3-50: Query using the GROUP BY clause

If you create a query where one or more of the columns in the SELECT clause involve group functions and one or more columns do not, then the columns that are not in the group function must be included in the GROUP BY clause. Next you will discover what happens when you repeat this query to return building codes and the sums of their capacities but omit the GROUP BY clause.

To repeat the query without the GROUP BY clause:

1 Type the query shown in Figure 3-51 to list the building code name with the total, average, and maximum room capacities (excluding private offices with a capacity of less than five) of all buildings in the LOCATION table, omitting the GROUP BY clause.

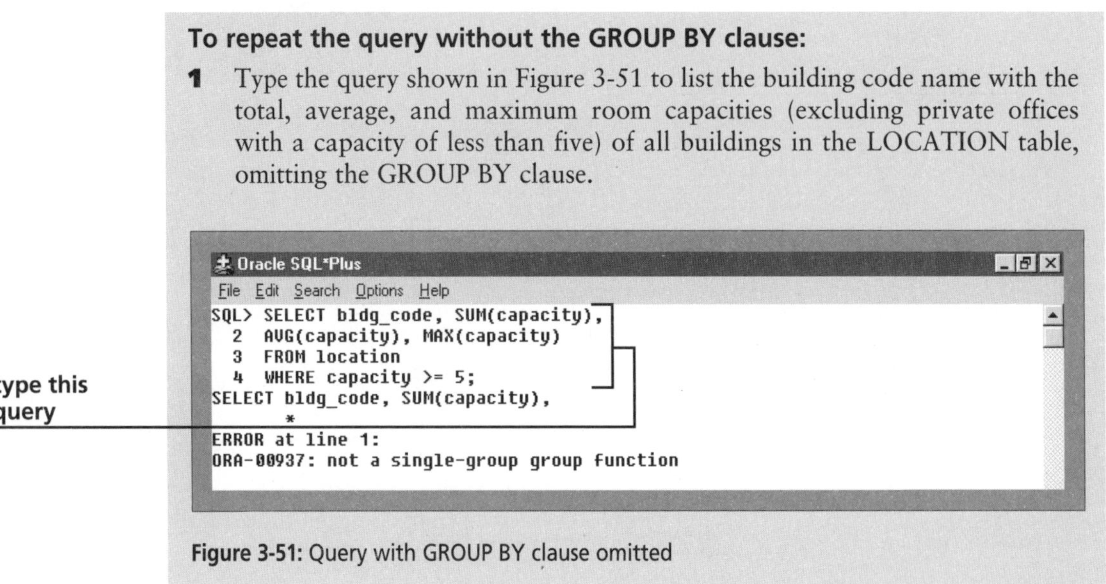

Figure 3-51: Query with GROUP BY clause omitted

The error message "ORA-00937: not a single-group group function" indicates that a SELECT clause cannot contain both a group function and an individual column expression unless the individual column expression is in a GROUP BY clause. To solve this problem, you would include BLDG_CODE in the GROUP BY clause, as in the query shown in Figure 3-50.

When you create a query with one or more group functions, you can group the output by only one column, and you cannot list any columns in the SELECT clause except the grouping column. Suppose you want to display room numbers along with building codes in the previous query. Now you will type the query again, but include ROOM in the SELECT clause.

To display room numbers in the group function query:

1 Type the query shown in Figure 3-52, adding ROOM to the SELECT clause. The ORA-00937 error is displayed again, indicating that all SELECT clause columns must either be in group functions or be included in the GROUP BY clause.

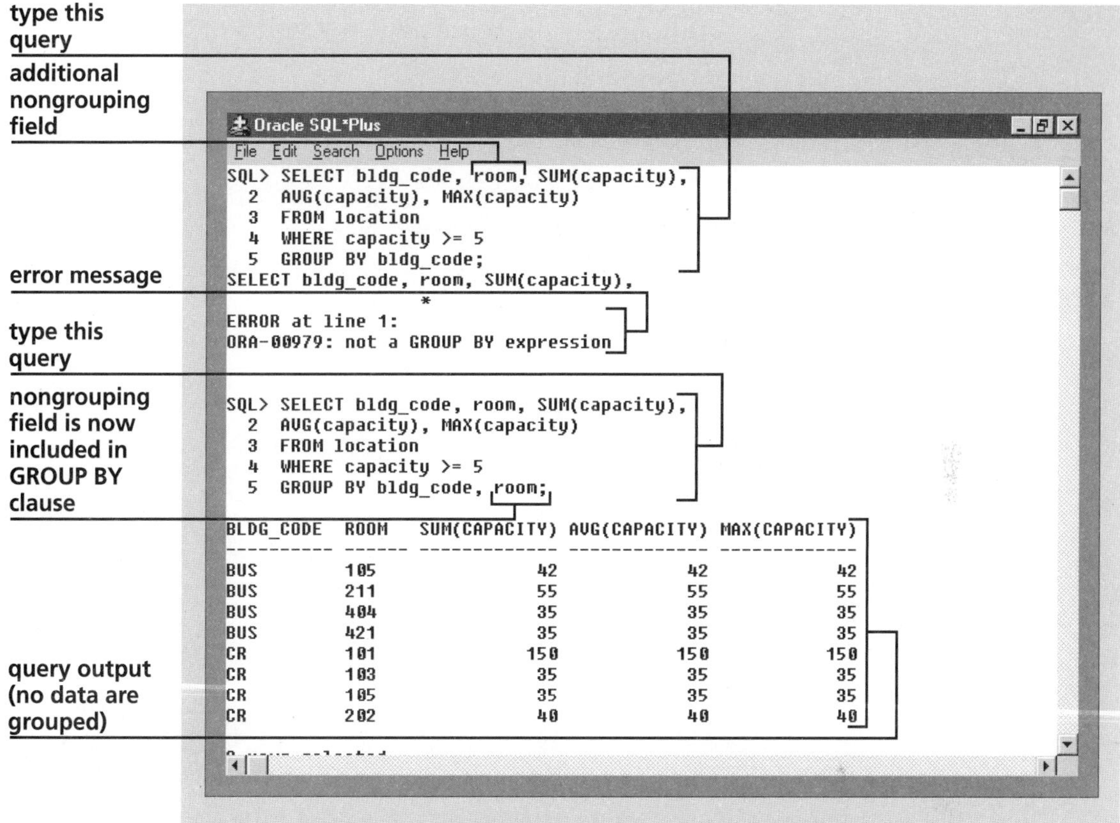

Figure 3-52: Query with additional nongrouping field in SELECT clause

2 Type the query again as shown in Figure 3-52, this time including ROOM in the GROUP BY clause.

Adding ROOM to the GROUP BY clause enables the query to run, but the output is not correct. The DBMS is instructed to group the data by both the building code and room, but there are no records that have the same building code and room. Therefore the group functions operate on each record individually, and the sum, average, and maximum capacities for each room are the same as each room's actual capacity.

The COUNT group function returns an integer that represents the number of records that are returned by a given query. The COUNT(*) version of this function calculates the total number of rows in a table that satisfy a given search condition. The COUNT(<column name>) version calculates the number of rows in a table that satisfy a given search condition and also contain a non-null value for the given column. You will use both versions of the COUNT function next.

To use the COUNT group function:

1 Type the query shown in Figure 3-53 to count the total number of students enrolled in CSECID 1010.

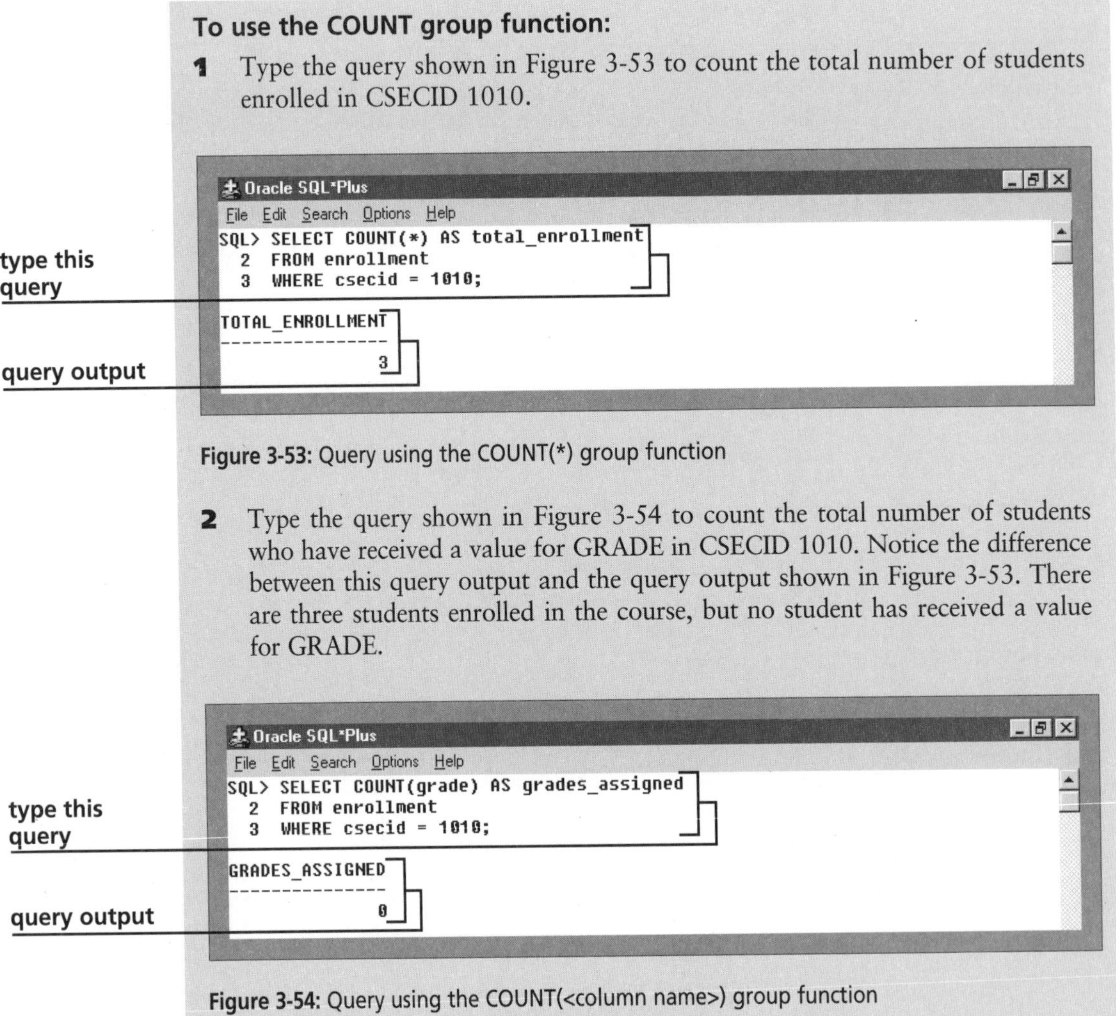

Figure 3-53: Query using the COUNT(*) group function

2 Type the query shown in Figure 3-54 to count the total number of students who have received a value for GRADE in CSECID 1010. Notice the difference between this query output and the query output shown in Figure 3-53. There are three students enrolled in the course, but no student has received a value for GRADE.

Figure 3-54: Query using the COUNT(<column name>) group function

SUMMARY

- In a SELECT command, the SELECT clause lists the columns that you want to display in your query, the FROM clause lists the name of each table involved in the query, and the WHERE clause specifies a search condition.
- If you are querying a table that you did not create, then you must preface it with the creator's user name, and the creator must have given you the privilege to SELECT data from that table.

- To retrieve every row in a table, the data do not need to satisfy a search condition, so you omit the WHERE clause.
- If you want to retrieve all of the columns in a table, you can use an asterisk (*) as a wildcard character in the SELECT statement instead of typing every column name.
- The SELECT DISTINCT command is used to suppress duplicate records in query output.
- To return records where the value of a particular field is NULL, you use the search condition WHERE <field name> IS NULL. To return records where the value of a particular field is not NULL, you use the search condition WHERE <field name> IS NOT NULL.
- Query outputs can be sorted using the ORDER BY clause.
- SQL has operations for adding, subtracting, multiplying, and dividing query values.
- The SYSDATE function returns the current system (server) date and time.
- SQL has number functions that allow you to take the absolute value of a returned value, raise a returned value to an exponential power, round a value, or truncate a value.
- SQL has date arithmetic functions for retrieving records corresponding to a date that is within a given number of days or months of a known date, and for calculating the number of months between two dates.
- SQL has group functions for summing, averaging, finding the maximum or minimum value of, or counting the number of records returned by a query.
- An alias is an alternate output column name that describes the column data and hides the underlying complexity of the query.

REVIEW QUESTIONS

1. Write a SELECT statement to retrieve all of the fields and rows from the STUDENT table.
2. Write a SELECT statement to retrieve the SFNAME and SLNAME from the STUDENT table for SID 105.
3. Write a SELECT statement to retrieve the STATUS field from all rows in the TERM table. Suppress duplicate output rows.
4. Write a query to return every field in the BACKORDER table for every row where the DATE_RECEIVED field is NULL.
5. In what order are AND and OR operations evaluated? How can you force a different order?

6. List the output of the following sorted queries:
   a. SELECT DISTINCT category FROM item ORDER BY category;
   b. SELECT sfname, slname FROM student ORDER BY fid DESC;
7. In what order are arithmetic operations evaluated in SQL?
8. Write a SELECT command that returns the current system date.
9. List the output of the following queries that use number functions:
   a. SELECT ABS(qoh) FROM inventory WHERE invid = 11777;
   b. SELECT POWER(ccredit, 2) FROM course WHERE cid = 3;
   c. SELECT ROUND(curr_price, 1) FROM inventory WHERE invid = 11668;
   d. SELECT TRUNC(curr_price) FROM inventory WHERE invid = 11668;
10. List the output of the following queries that use data arithmetic functions:
    a. SELECT slname, sfname
       FROM student
       WHERE sdob > TO_DATE('01/01/77', 'MM/DD/YY')
       AND sdob <= ADD_MONTHS(TO_DATE('01/01/77', 'MM/DD/YY'), 11);
    b. SELECT orderid, orderdate
       FROM cust_order
       WHERE orderdate < TO_DATE('06/01/2001', 'MM/DD/YYYY')
       AND orderdate >= TO_DATE('06/01/2001', 'MM/DD/YYYY') – 2;
11. List the output of the following queries that use group functions:
    a. SELECT AVG(capacity) FROM location WHERE bldg_code = 'LIB';
    b. SELECT COUNT(*) FROM location WHERE bldg_code = 'CR';
    c. SELECT COUNT(frank) FROM faculty;
    d. SELECT MAX(currenrl) FROM course_section WHERE termid = 4;
    e. SELECT MIN(capacity) FROM location WHERE bldg_code = 'LIB';
    f. SELECT SUM(capacity) FROM location WHERE bldg_code = 'CR' OR bldg_code = 'LIB';

PROBLEM-SOLVING CASES

Save all the following SQL*Plus queries in a text file named Ch3bNU.sql in the Chapter3 folder on your Student Disk.

1. Write a query that returns all fields in the TERM table where status = 'CLOSED'. Use 'CLOSED' as the search condition in the query's WHERE clause.
2. Write a query that returns the CALLID and CNAME for all rows in the COURSE table.
3. Write a query that returns the SCLASS field for all rows in the STUDENT table, and suppresses duplicate outputs.

4. WRITE a query that returns the SFNAME, SLNAME, and SMI fields for every student who does not have a value for SMI. Write a second SELECT command that returns the SFNAME, SLNAME, and SMI fields for students who have a value for SMI.
5. Write a query that returns the SFNAME, SLNAME, and SDOB for all students whose SCLASS is either 'JR' or 'SR'.
6. Write a query that returns the SFNAME, SLNAME, and SDOB of all students who were born during 1977, using a search condition that contains the ADD_MONTHS function.
7. Write a query that returns the SID, CSECID, and GRADE for all records in the ENROLLMENT table where CSECID = 1000. Order the records by GRADE, showing the A's first, then the B's, etc.
8. Write a query that calculates the difference between the current SYSDATE and student Michael Connoly's SDOB in days. Display the output using the column alias AGE_IN_DAYS. Use 'Michael' and 'Connoly' as the search condition in the query's WHERE clause.
9. Modify the query in Exercise 8 to display the SDOB in months. Display the output using the column alias AGE_IN_MONTHS, and truncate the output to the previous month (don't show fractions of months).
10. Modify the query in Exercise 8 to display the SDOB in years. Display the output using the column alias AGE_IN_YEARS, and truncate the output to the previous year (don't show fractions of years).
11. Write a query that calculates the average current enrollment for all sections of CID 1 during TERMID 4.
12. Write a query that calculates the total number of students advised by FID 1.
13. Write a query that calculates the total capacity of all rooms in the 'BUS' building. Use 'BUS' as the search condition in the query's WHERE clause.

Save the following SQL*Plus queries in a text file named Ch3bCT.sql in the Chapter3 folder on your Student Disk.

14. Write a query that returns all fields from the CUST_ORDER table where the order source is the Web site. Use 'WEBSITE' as the search condition in the query's WHERE clause.
15. Write a query that returns the INVID, COLOR, and CURR_PRICE for all rows in the INVENTORY table where the color is Navy. Use 'Navy' as the search condition in the query's WHERE clause.
16. Write a query that returns the COLOR field from every row in the INVENTORY table, and suppresses duplicate outputs.
17. Write a query that returns the item ID, size, color, and the value of the inventory (QOH * CURR_PRICE) for ITEMID 897, size S, in the INVENTORY table.
18. Write a query that returns every category in the ITEM table, sorted in alphabetical order, with duplicate outputs suppressed.
19. Write a query that returns all of the fields in all of the records from the CUST_ORDER table, ordered by ORDERDATE with the most recent orders listed first.

20. Write a query that returns the SHIPID, INVID, and DATE_EXPECTED for all shipments expected to arrive during the month of August 2001, using the ADD_MONTHS function. Do not include shipments that have already been received. (*Hint*: Look for records where DATE_RECEIVED is NULL in your search condition.)
21. Write a query that calculates the number of days between the current SYSDATE and the DATE_EXPECTED for back order ID 1. Truncate the output to the nearest day, and display the output using a column alias named DIFFERENCE_IN_DAYS.
22. Modify the query in Exercise 21 to calculate the number of months between the current SYSDATE and the DATE_EXPECTED for back order ID 1. Truncate the output to the nearest month, and display the output using a column alias named DIFFERENCE_IN_MONTHS.
23. Modify the query in Exercise 21 to calculate the number of years between the current SYSDATE and the DATE_EXPECTED for back order ID 1. Truncate the output to two decimal places, and display the output using a column alias named DIFFERENCE_IN_YEARS.
24. Write a query that calculates the total quantity on hand of ITEMID 995 in the INVENTORY table.
25. Write a query that returns the total number of customer orders that have WEBSITE as the order source. Use 'WEBSITE' as the search condition in the query's WHERE clause.
26. Write a query that returns the ORDERID and total quantity of each order in the ORDERLINE table. (*Hint*: You will need to use the GROUP BY function.)

LESSON C

objectives

- Write SQL queries to join multiple database tables
- Combine query results using set operators
- Select records for update
- Create database views
- Grant table privileges to other users using SQL*Plus
- Create sequences to automatically generate surrogate primary keys
- Retrieve data and create SQL queries using Query Builder

Using Queries to Join Multiple Tables

All of the queries you have seen so far have retrieved data from a single table. However, one of the strengths of SQL is its ability to **join**, or combine, data from multiple database tables using foreign key references. The general format of a SELECT statement with a join operation is:

```
SELECT <column1, column2, …>
FROM <table1 name, table2 name>
WHERE <table1 name.joincolumn name> = <table1 name.joincolumn name>;
```

The SELECT clause contains the names of the columns to display in the query output. If you display a column that exists in more than one of the tables in the FROM clause, you must write the table name, followed by a period, and then write the column name with the name of one of the tables. Otherwise, the DBMS will issue an error. Listing the table name before a column name is known as **qualifying** the column name.

The FROM clause contains the name of each table involved in the join operation. The WHERE clause contains the **join condition**, which specifies the table and column names on which to join the tables. The join condition contains the foreign key reference in one table, and the primary key in the other table. Additional search conditions are listed in the WHERE clause, using the AND and OR operators. Next, you will retrieve student last and first names along with each student's advisor's name. This requires you to retrieve data from both the STUDENT and FACULTY tables, and to join the tables on the FID column, which is also the FID in the FACULTY table.

To retrieve records from two tables by joining them through a foreign key reference:

1 If necessary, start SQL*Plus and log on to the database. Type the query shown in Figure 3-55 to list the student ID, last name, and advisor last name.

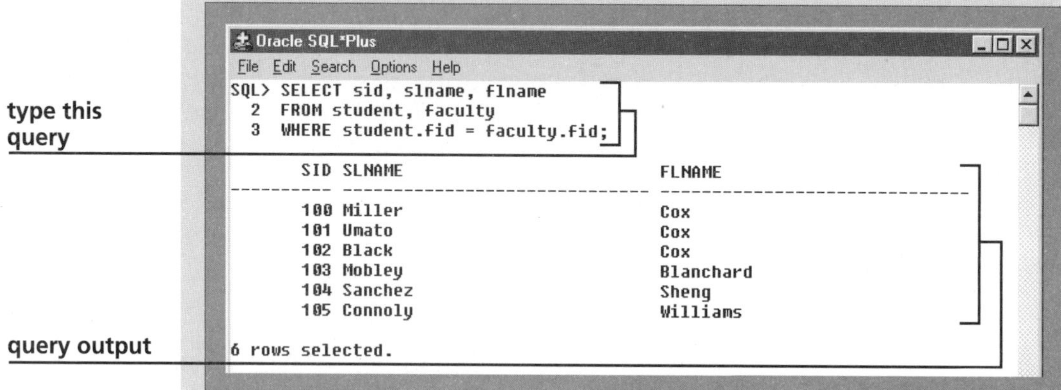

Figure 3-55: Query joining two tables

2 You can add additional search conditions in the WHERE clause using the AND operator, as demonstrated next. Type the query shown in Figure 3-56 to list the location ID, building code, and room number for faculty member Laura Sheng. Note that in the SELECT clause LOCID had to be prefaced with a table name, because the LOCID field exists in both the FACULTY and LOCATION tables. You could have prefaced LOCID with either LOCATION or FACULTY, since the value is the same in both tables for the joined records.

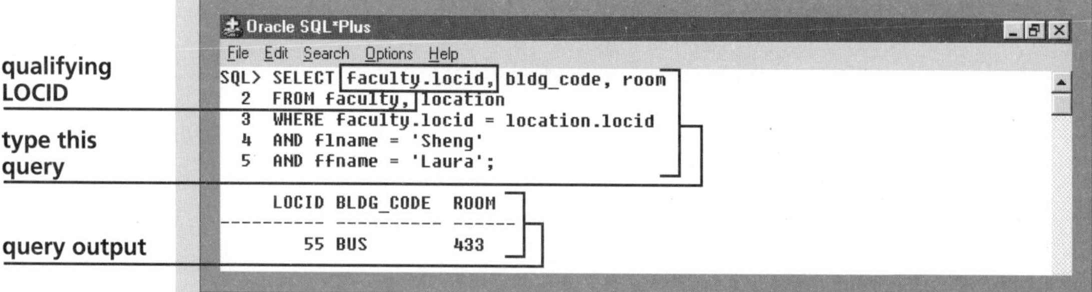

Figure 3-56: Join query with search conditions

In the previous examples, you joined two tables, but you can join any number of tables in a SELECT command. When you join tables, each table involved in the query must be listed in the FROM clause, and each table in the FROM clause must be listed in the WHERE clause in a join condition. Suppose that you want to create a query to display the call ID and grade for each of Sarah Miller's courses. This query requires you to join four tables: STUDENT (to search for SFNAME and SLNAME), ENROLLMENT (to display GRADE), COURSE (to display CALLID), and COURSE_SECTION (to join CALLID in COURSE to CSECID in ENROLLMENT). For complex queries like this, it is often helpful to draw a diagram like the one shown in Figure 3-57, to show the columns and associated tables that you need to display and search, as well as the required joining columns and their links.

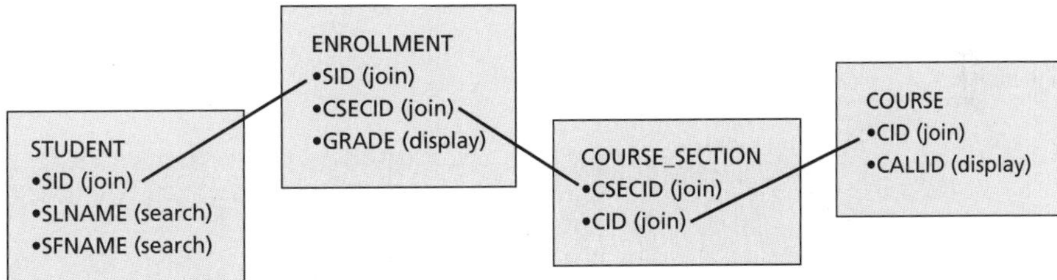

Figure 3-57: Join query design diagram

You can derive your query from the diagram by following these steps:

- Place the display fields in the SELECT clause
- List all of the tables in the FROM clause
- Include the links in join conditions in the WHERE clause
- Include all of the search fields in the WHERE clause

Note that you must always have one fewer join condition than the total number of tables joined in the query. In this query, you are joining four tables, and you have three join conditions.

You will create a SQL query to join four tables next.

To join four tables in a single query:

1 Type the query shown in Figure 3-58.

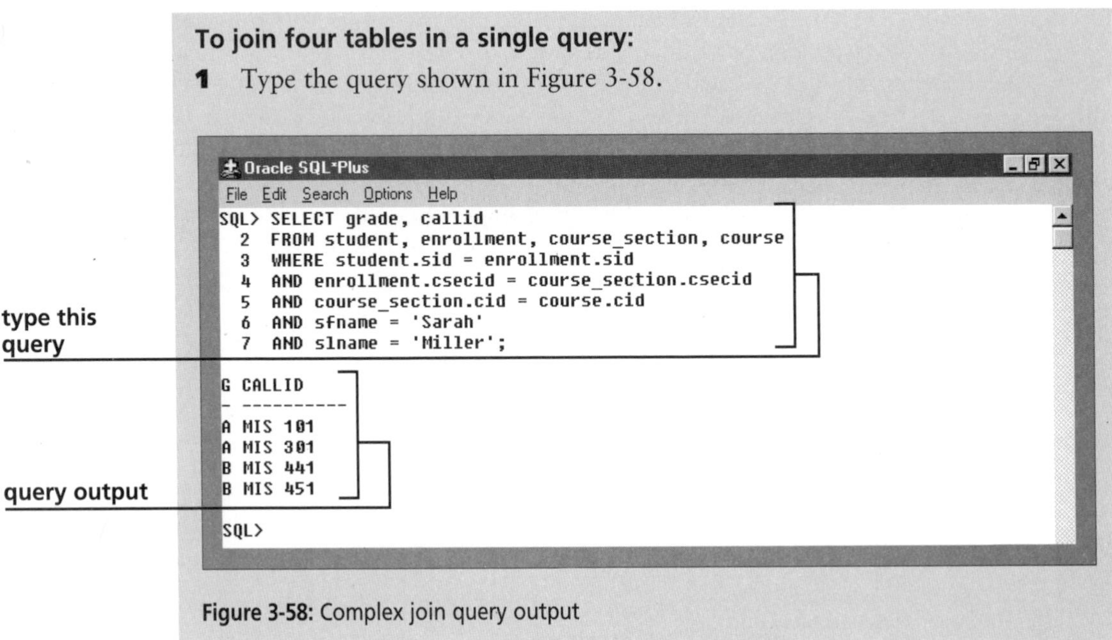

Figure 3-58: Complex join query output

If you accidentally omit a join condition in a multiple-table query, the result is a **Cartesian product**, whereby every row in one table is joined with every row in the other table. For example, suppose you repeat the query to show each student record along with each student's advisor, but you omit the join condition. Every row in the STUDENT table (six rows) is joined with every row in the FACULTY table (five rows). The result is 6 * 5 rows, or 30 rows. You will do this next.

To create a Cartesian product by omitting a join condition:

1 Type the query shown in Figure 3-59. Each row in the STUDENT table is first joined with the first row in the FACULTY table (FLNAME 'Cox'). Next, each row in the STUDENT table is joined with the second row in the FACULTY table (FLNAME 'Blanchard'). This continues until each STUDENT row is joined with each FACULTY row, for a total of 30 rows returned.

When a multiple-table query returns more records than you expect, look for missing join condition statements.

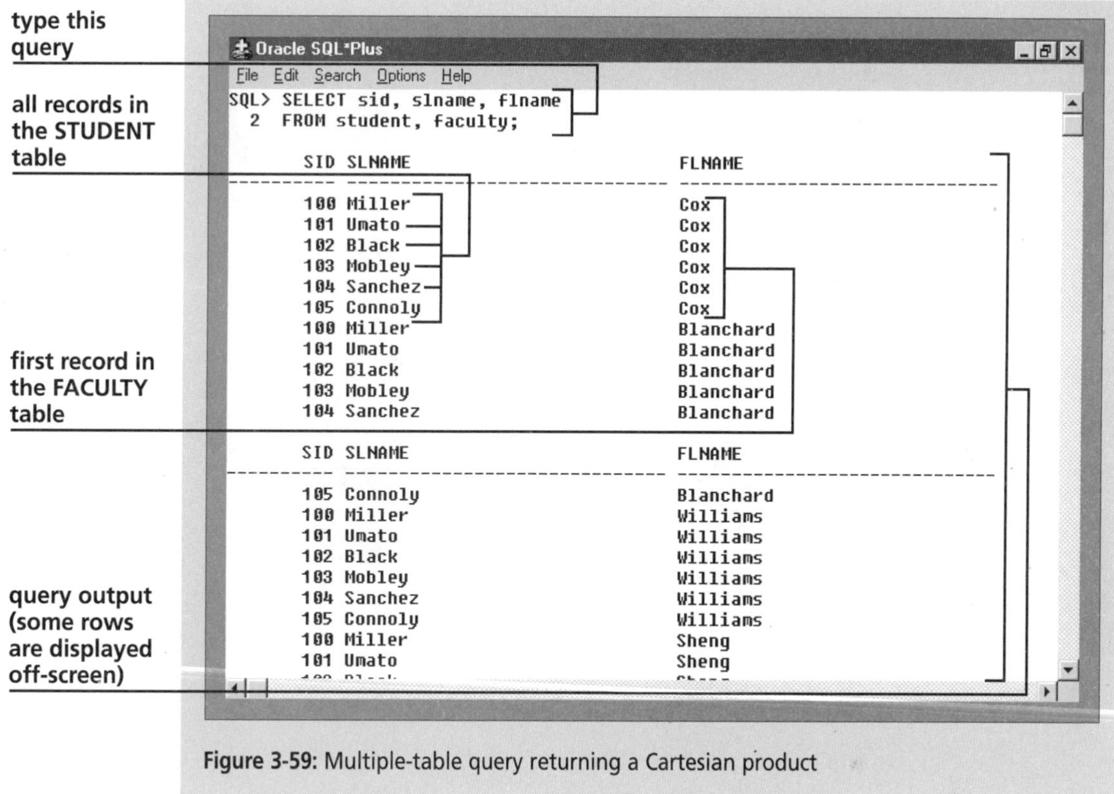

Figure 3-59: Multiple-table query returning a Cartesian product

Using Set Operators to Combine Query Results

You can use set operators to combine the results of two separate queries into a single result. Figure 3-60 lists the set operators and their purpose.

Operator	Purpose
UNION	Returns all rows from both queries, but only displays duplicate rows once
UNION ALL	Returns all rows from both queries, and displays all duplicate rows
INTERSECT	Returns all rows that are returned by both queries
MINUS	Returns the rows returned by the first query minus the matching rows returned by the second query

Figure 3-60: SQL set operators

UNION and UNION ALL

A **UNION** query combines the output of two unrelated queries into a single output result. Suppose you need to create a query to list the last name, first name, and phone number of every student and faculty member in the Northwoods University database, to create a phone directory. If you enter the SELECT statement shown in Figure 3-61, the output is a product of the FACULTY and STUDENT tables that was caused by omitting a join clause. However, no join clause was included, because the relationship between the STUDENT and FACULTY records is not relevant for this query. A list of all student and faculty names and phone numbers has nothing to do with student/advisor relationships. A single SELECT command cannot return data from two unrelated queries as a single output. Instead, you must use a UNION.

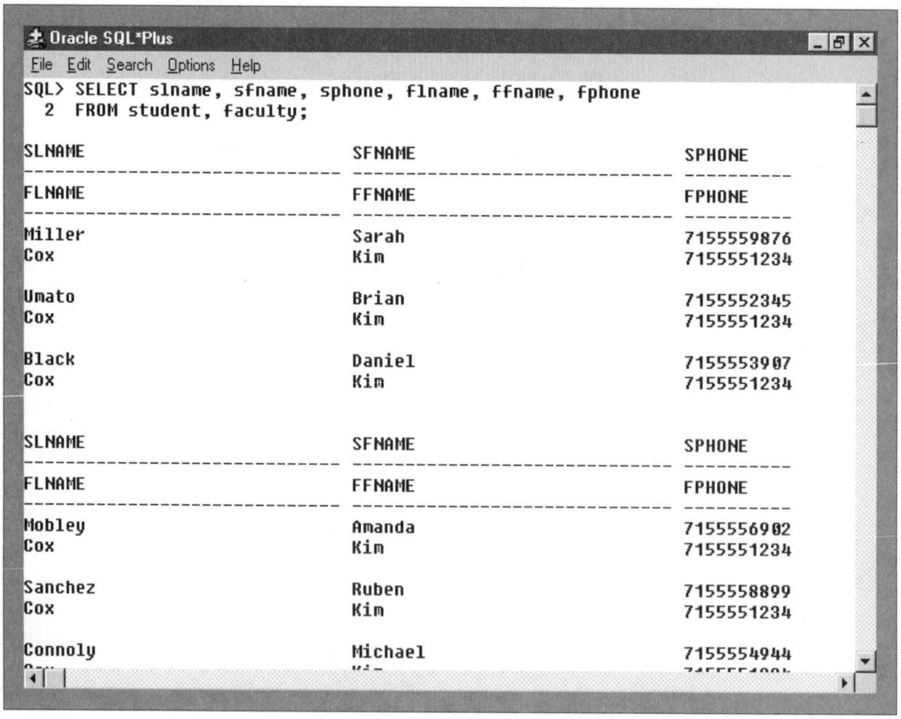

```
Oracle SQL*Plus
File Edit Search Options Help
SQL> SELECT slname, sfname, sphone, flname, ffname, fphone
  2  FROM student, faculty;

SLNAME                         SFNAME                         SPHONE
------------------------------ ------------------------------ ----------
FLNAME                         FFNAME                         FPHONE
------------------------------ ------------------------------ ----------
Miller                         Sarah                          7155559876
Cox                            Kim                            7155551234

Umato                          Brian                          7155552345
Cox                            Kim                            7155551234

Black                          Daniel                         7155553907
Cox                            Kim                            7155551234


SLNAME                         SFNAME                         SPHONE
------------------------------ ------------------------------ ----------
FLNAME                         FFNAME                         FPHONE
------------------------------ ------------------------------ ----------
Mobley                         Amanda                         7155556902
Cox                            Kim                            7155551234

Sanchez                        Ruben                          7155558899
Cox                            Kim                            7155551234

Connoly                        Michael                        7155554944
```

Figure 3-61: Example of output that is a Cartesian product of the STUDENT and FACULTY tables

A union requires that both queries have the same number of display columns in the SELECT statement, and that each column in the first query has the same data type as the corresponding column in the second query. The general format of the UNION query is <query 1> UNION <query 2>. You will create a UNION query next.

To create a UNION query:

1 Type the query shown in Figure 3-62. The query output shows the student and faculty names and phone numbers in a single list. Note that by default the column titles are taken from the column names in the first query's SELECT statement, and that output results appear in order, based on the first column of the first SELECT statement.

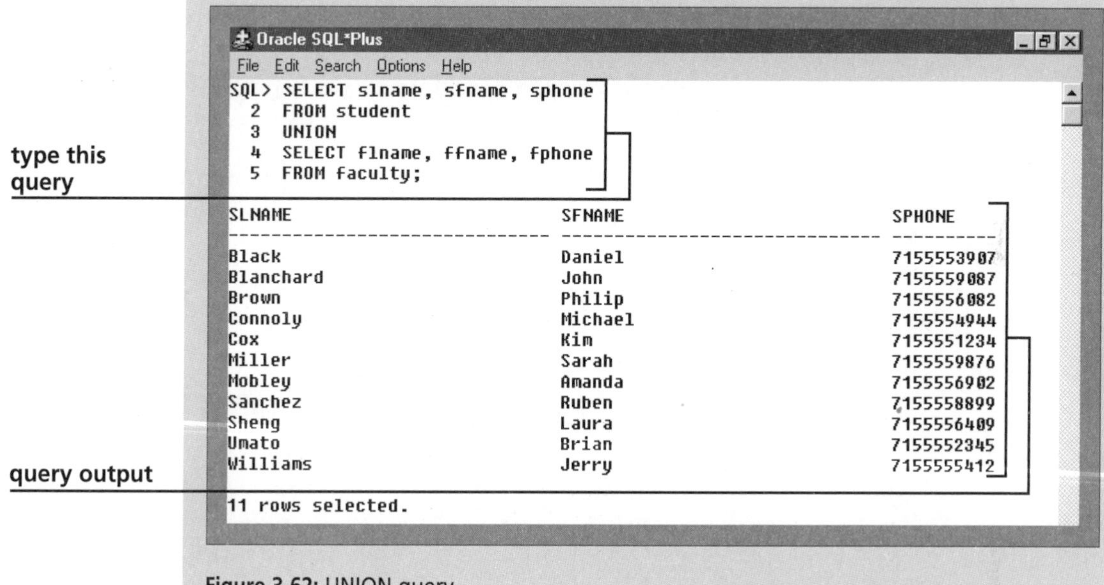

Figure 3-62: UNION query

If one record in the first query exactly matches a record in the second query, the UNION output only shows the duplicate record once. To display duplicate outputs, you must use the UNION ALL operator. Suppose you need to create a query that displays the shipment ID and quantity expected for all shipments or back orders expected to arrive on 09/15/2001. When you examine the SHIPPING and BACKORDER tables in the Clearwater Traders database, you see that the first query (from the SHIPPING table) should return shipment 211 (inventory ID 11668, quantity 25) and shipment 211 (inventory ID 11669, quantity 25). The second query (from the BACKORDER table) should return shipment 218 (inventory ID 11845, quantity 50) and shipment 218 (inventory ID 11847, quantity 50). If only the SHIPID and QUANTITY fields are included in the SELECT statement, the queries will return duplicate records. If you create a UNION of these two queries, the duplicate records will each be displayed only once. Now you will enter the query first with the UNION operator and then with the UNION ALL operator, and examine the results.

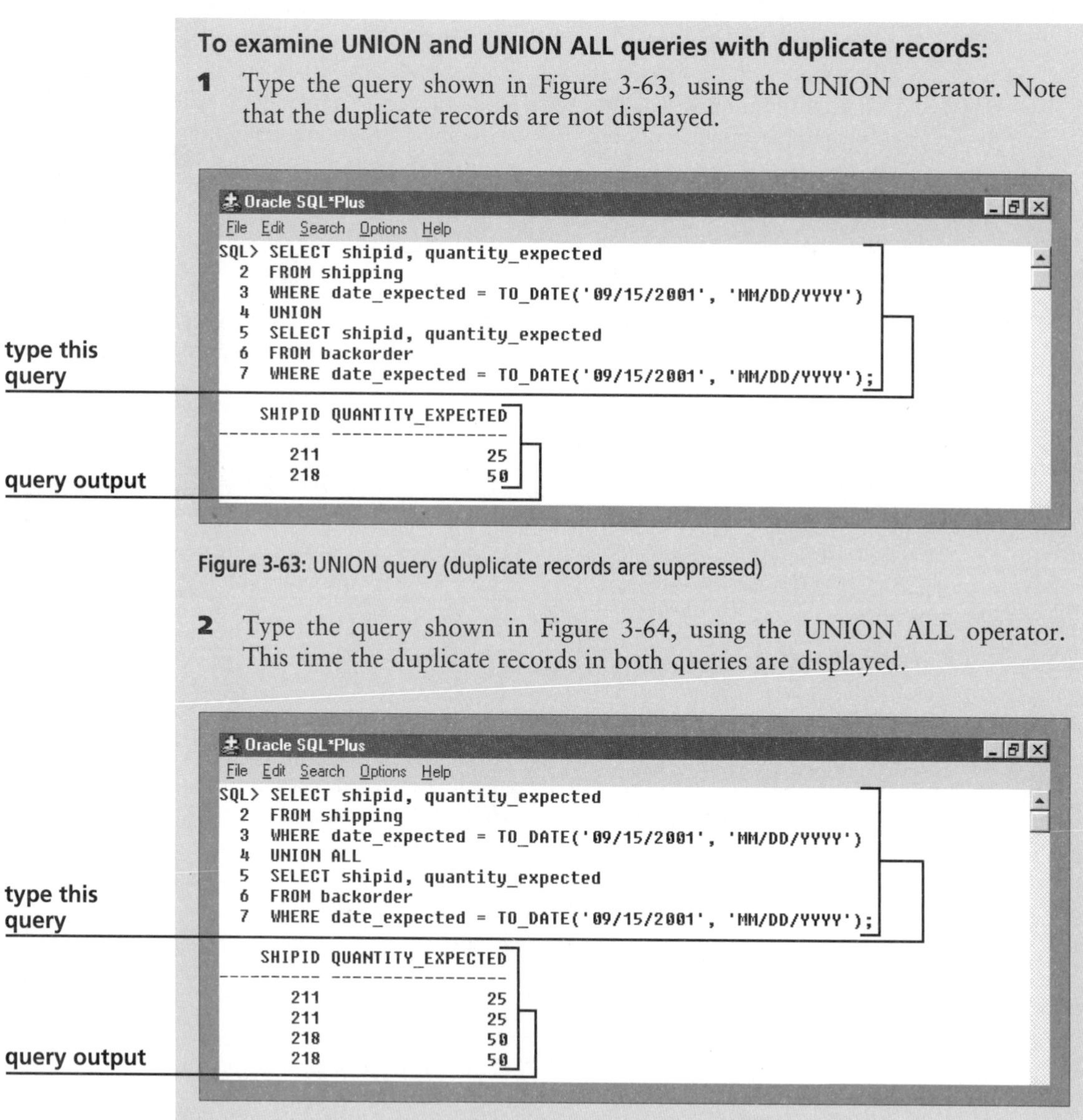

To examine UNION and UNION ALL queries with duplicate records:

1 Type the query shown in Figure 3-63, using the UNION operator. Note that the duplicate records are not displayed.

```
Oracle SQL*Plus
File Edit Search Options Help
SQL> SELECT shipid, quantity_expected
  2  FROM shipping
  3  WHERE date_expected = TO_DATE('09/15/2001', 'MM/DD/YYYY')
  4  UNION
  5  SELECT shipid, quantity_expected
  6  FROM backorder
  7  WHERE date_expected = TO_DATE('09/15/2001', 'MM/DD/YYYY');

   SHIPID QUANTITY_EXPECTED
--------- -----------------
      211                25
      218                50
```

Figure 3-63: UNION query (duplicate records are suppressed)

2 Type the query shown in Figure 3-64, using the UNION ALL operator. This time the duplicate records in both queries are displayed.

```
Oracle SQL*Plus
File Edit Search Options Help
SQL> SELECT shipid, quantity_expected
  2  FROM shipping
  3  WHERE date_expected = TO_DATE('09/15/2001', 'MM/DD/YYYY')
  4  UNION ALL
  5  SELECT shipid, quantity_expected
  6  FROM backorder
  7  WHERE date_expected = TO_DATE('09/15/2001', 'MM/DD/YYYY');

   SHIPID QUANTITY_EXPECTED
--------- -----------------
      211                25
      211                25
      218                50
      218                50
```

Figure 3-64: UNION ALL query (duplicate records are displayed)

INTERSECT

Some queries require an output that finds the intersection, or matching rows, in two unrelated queries. For example, suppose you need to find a list of faculty members whose offices are in the BUS building and who have also taught a course in the BUS building. Like a UNION, an INTERSECT query requires that both queries have the same number of display columns in the SELECT statement, and that each column in the first query has the same data type as the corresponding column in the second query. First you will type the individual queries and examine the results, then you will combine the queries using the INTERSECT set operator to find the names that are returned by both queries.

To create a query using the INTERSECT operator:

1 Type the queries shown in Figure 3-65. Faculty members Cox, Blanchard, and Sheng each have an office in BUS, while faculty members Sheng, Cox, Blanchard, and Williams have taught courses in BUS.

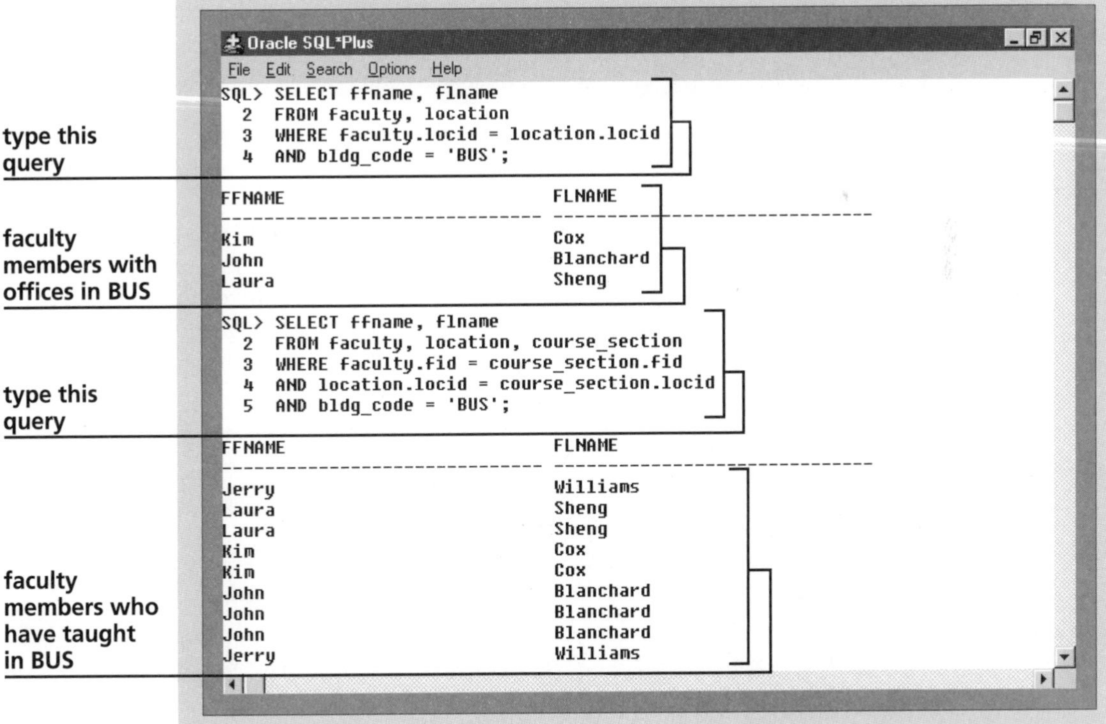

```
Oracle SQL*Plus
File Edit Search Options Help
SQL> SELECT ffname, flname
  2  FROM faculty, location
  3  WHERE faculty.locid = location.locid
  4  AND bldg_code = 'BUS';

FFNAME                         FLNAME
------------------------------ ------------------------------
Kim                            Cox
John                           Blanchard
Laura                          Sheng

SQL> SELECT ffname, flname
  2  FROM faculty, location, course_section
  3  WHERE faculty.fid = course_section.fid
  4  AND location.locid = course_section.locid
  5  AND bldg_code = 'BUS';
FFNAME                         FLNAME
------------------------------ ------------------------------
Jerry                          Williams
Laura                          Sheng
Laura                          Sheng
Kim                            Cox
Kim                            Cox
John                           Blanchard
John                           Blanchard
John                           Blanchard
Jerry                          Williams
```

type this query

faculty members with offices in BUS

type this query

faculty members who have taught in BUS

Figure 3-65: Individual query results

2 Type the query shown in Figure 3-66, which uses the INTERSECT operator to find the names on both lists. Note that when you use the INTERSECT operator, duplicate records are suppressed.

type this query

query result

```
Oracle SQL*Plus
File Edit Search Options Help
SQL> SELECT ffname, flname
FROM faculty, location
WHERE faculty.locid = location.locid
AND bldg_code = 'BUS'
INTERSECT
SELECT ffname, flname
FROM faculty, location, course_section
WHERE faculty.fid = course_section.fid
AND location.locid = course_section.locid
AND bldg_code = 'BUS';

FFNAME                         FLNAME
------------------------------ ------------------------------
John                           Blanchard
Kim                            Cox
Laura                          Sheng
```

Figure 3-66: Combined query result using INTERSECT operator

MINUS

The MINUS operator allows you to find the difference between two unrelated query result lists. For example, suppose you need to find the names of all faculty members who have taught courses in BUS, but whose offices are not located in BUS. You will create a query that returns the names of all faculty members who have taught in BUS, then subtracts the names of the faculty members whose offices are in BUS.

To create a query using the MINUS operator:

1 Type the query shown in Figure 3-67. The query output shows that the only faculty member whose name was not on both lists in Figure 3-65 was Jerry Williams. Note that duplicate outputs are suppressed—Jerry Williams's name appears twice in the second query in Figure 3-65, but only once in the MINUS query.

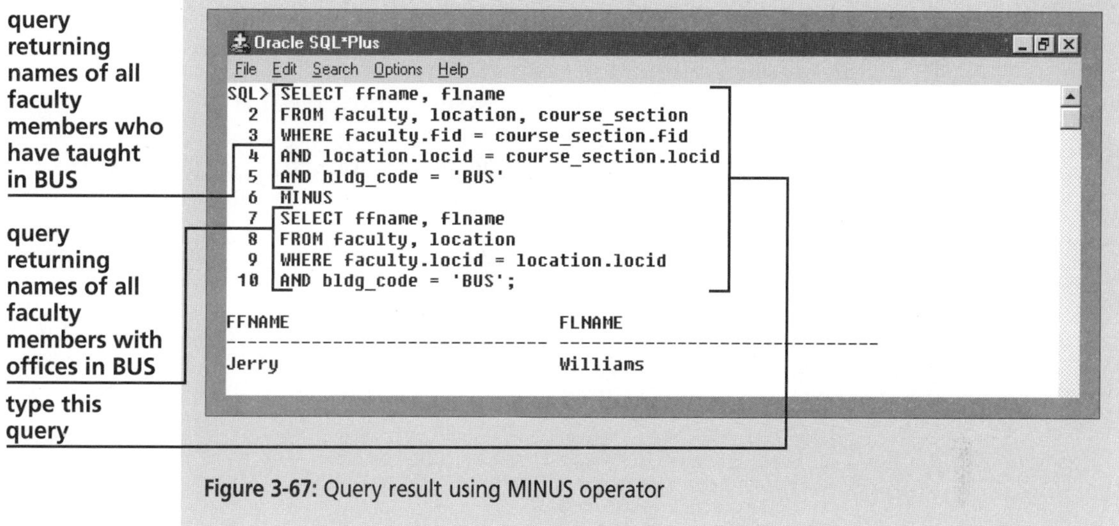

Figure 3-67: Query result using MINUS operator

Selecting Records for Update

When you enter a SELECT command to query database records, no locks are placed on the selected records, and other database users can view and update these records at the same time you are viewing them. This is necessary for databases with many concurrent users, because otherwise many or all of the database records would be locked most of the time, and work would grind to a halt. However, records that have been changed with an INSERT or UPDATE statement are locked until the user holding the lock releases it by issuing a COMMIT command.

Sometimes you might want to view a record and then update it in the same transaction. For example, when a Clearwater Traders customer wants to order an item, the sales representative must first determine if the item is in stock before placing the order. Suppose that a particular item is in stock and the sales representative verbally confirms this fact to the customer. But, before the sales representative can place the order and commit the update, another sales representative sells the entire inventory on hand. To avoid this situation, you must be able to view the quantity on hand and then update it in a single transaction. You can do this in Oracle using the SELECT … FOR UPDATE command.

The general format of the SELECT … FOR UPDATE command is:

```
SELECT <column names>
FROM <table names>
WHERE <search conditions>
FOR UPDATE OF <column names to be updated>
NOWAIT;
```

The column names listed in the FOR UPDATE OF command do not restrict which columns can be updated in the record, because the entire record is locked.

However, listing these column names helps to document which fields are to be updated. The NOWAIT command causes the system to generate an error message immediately if another user has previously locked the selected records. If the NOWAIT command is omitted, the system forces the user to "wait" until the requested records are unlocked, and the user can do no further processing.

Next you will select item 11668 using the SELECT ... FOR UPDATE command to select and lock an inventory item.

To select and lock an inventory item in a single transaction:

1 Type the command shown in Figure 3-68 to select and lock the record for INVID 11668.

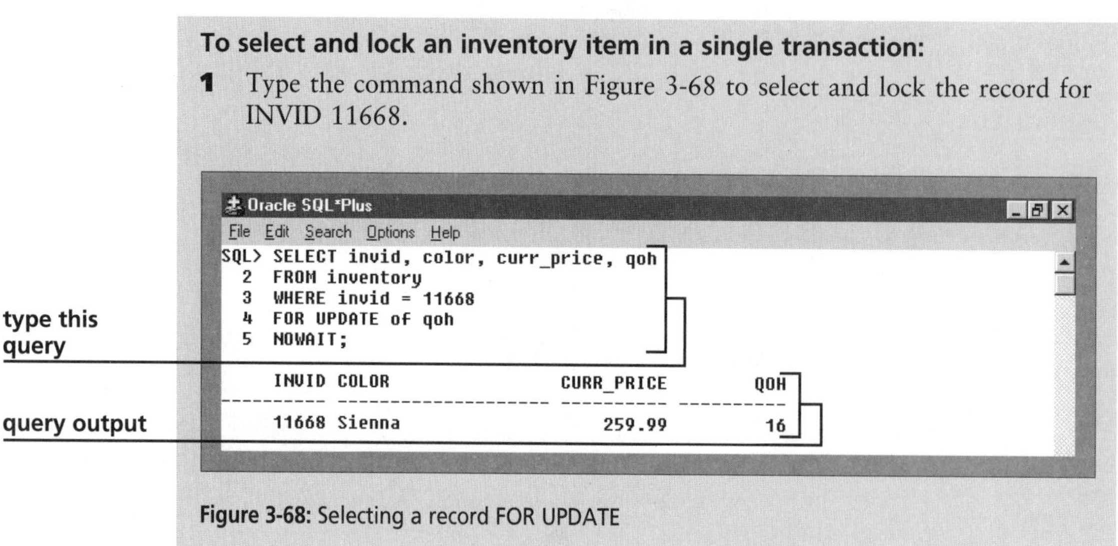

type this query
query output
Oracle SQL*Plus
File Edit Search Options Help
SQL> SELECT invid, color, curr_price, qoh
2 FROM inventory
3 WHERE invid = 11668
4 FOR UPDATE of qoh
5 NOWAIT;
INVID COLOR CURR_PRICE QOH
11668 Sienna 259.99 16

Figure 3-68: Selecting a record FOR UPDATE

Currently, your SQL*Plus session has the record for INVID 11668 locked. No other users can update or delete this record until you release the record with a COMMIT command. To confirm this, you will start a second SQL*Plus session and attempt to update the record.

To attempt to update the record in a second SQL*Plus session:

1 Click **Start** on the Windows taskbar, point to **Programs, Oracle for Windows 95** or **Windows NT**, and click **SQL*Plus 8.0** to start a new SQL*Plus session. Log on to SQL*Plus in the usual way.

2 Type the UPDATE query shown in Figure 3-69, and then press the **Enter** key. Your SQL*Plus query "waits" because it cannot be processed, since the current record is locked by your first SQL*Plus session.

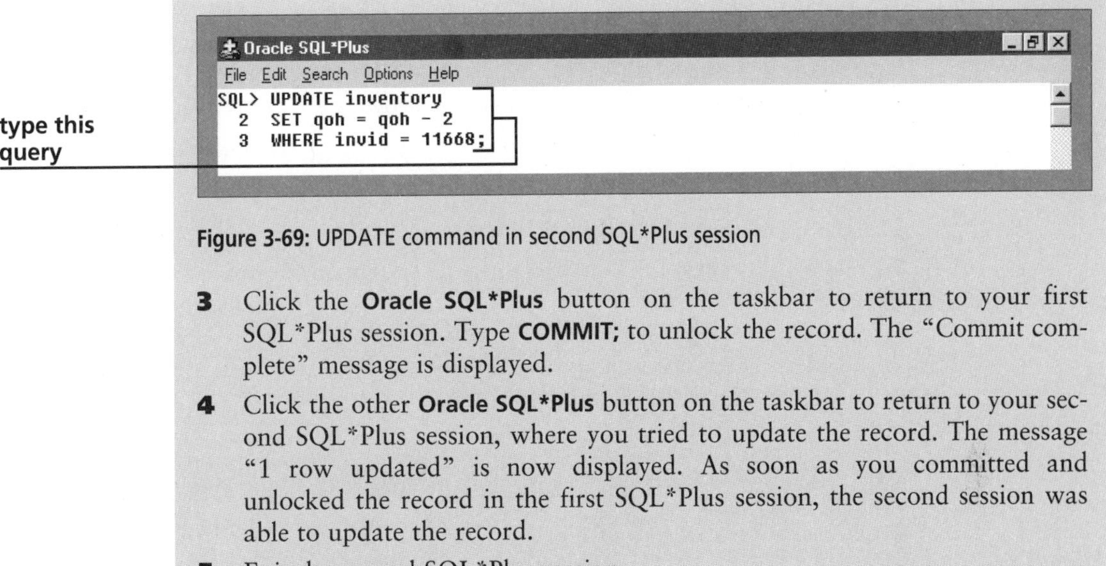
type this query
Oracle SQL*Plus
File Edit Search Options Help
SQL> UPDATE inventory
2 SET qoh = qoh - 2
3 WHERE invid = 11668;

Figure 3-69: UPDATE command in second SQL*Plus session

3 Click the **Oracle SQL*Plus** button on the taskbar to return to your first SQL*Plus session. Type **COMMIT;** to unlock the record. The "Commit complete" message is displayed.

4 Click the other **Oracle SQL*Plus** button on the taskbar to return to your second SQL*Plus session, where you tried to update the record. The message "1 row updated" is now displayed. As soon as you committed and unlocked the record in the first SQL*Plus session, the second session was able to update the record.

5 Exit the second SQL*Plus session.

Sometimes you might inadvertently lock your records from yourself if you are multitasking between different Oracle applications. If a query "waits" and will not execute while you are working, determine if you have an uncommitted INSERT, UPDATE, or SELECT … FOR UPDATE command in any other Oracle application, and then issue a commit.

Database Views

A database **view** is a logical table based on a query. It does not store data, but presents it in a format different from the one in which it is stored in the underlying tables. To an application, a view looks and acts like a table, and can be queried like a table. Sometimes applications are not able to display data from multiple tables, but you can create a view to join these tables into what looks like a single table. For example, suppose you are creating a report to list the item description, size, color, price, and quantity ordered for all items associated with ORDERID 1057. This involves joining four tables (CUST_ORDER, ORDERLINE, INVID, and ITEMID). However, the application you are using to generate the report will only display data from a single table. You can create a view, and then attach the view to the application—and it will work just like a table.

A view does not physically exist in the database—it is derived from other database tables. When the data in its source tables are updated, the view reflects the updates as well. Views are useful because you do not have to reenter complex query commands that are used frequently. They also can be used by the DBA to enforce database security and enable certain users to view only selected table fields or records.

The general format for creating a view is `CREATE VIEW <view name> AS <view query specification>`. The following example demonstrates how to create a view named FACULTY_VIEW based on the FACULTY table. This view contains all of the FACULTY columns except the FPIN. The columns will be sorted by FLNAME.

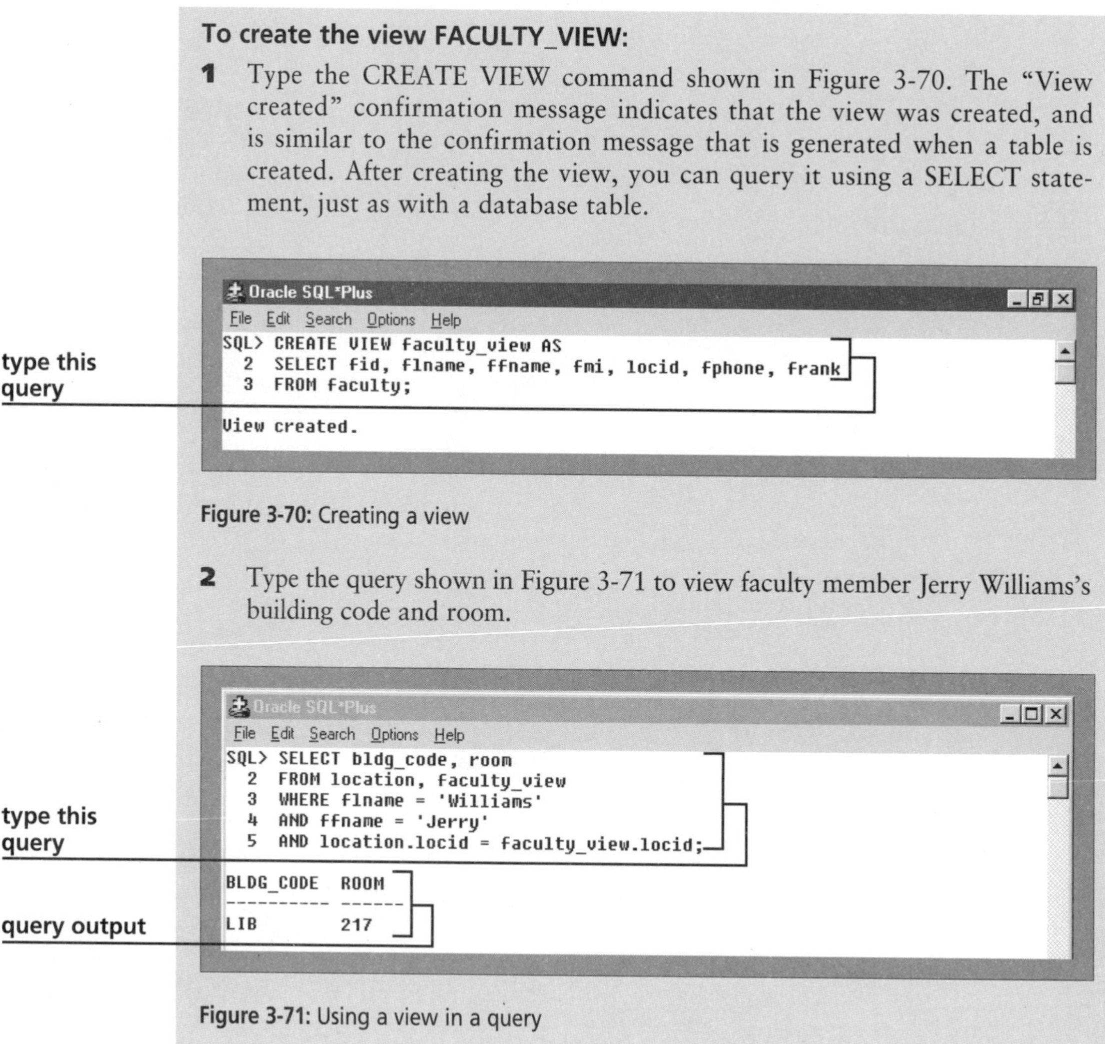

To create the view FACULTY_VIEW:

1 Type the CREATE VIEW command shown in Figure 3-70. The "View created" confirmation message indicates that the view was created, and is similar to the confirmation message that is generated when a table is created. After creating the view, you can query it using a SELECT statement, just as with a database table.

Figure 3-70: Creating a view

2 Type the query shown in Figure 3-71 to view faculty member Jerry Williams's building code and room.

Figure 3-71: Using a view in a query

There are two ways you can list the views in the database. Recall that system database tables maintain information about what is stored in the database. The ALL_VIEWS system table contains information about views created by all database users, and the USER_VIEWS system table contains information about only a specific user's views. You can query the ALL_VIEWS table using a specific user name as a search condition, and you can query the USER_VIEWS table to view your views only. Now you will write a query to display the names of another user's views and your own views.

To display the names of views:

1 Type the query shown in Figure 3-72 to query the ALL_VIEWS table and show the views that are created by SYSTEM. You could substitute any other user name for SYSTEM to see the names of the views created by another database user.

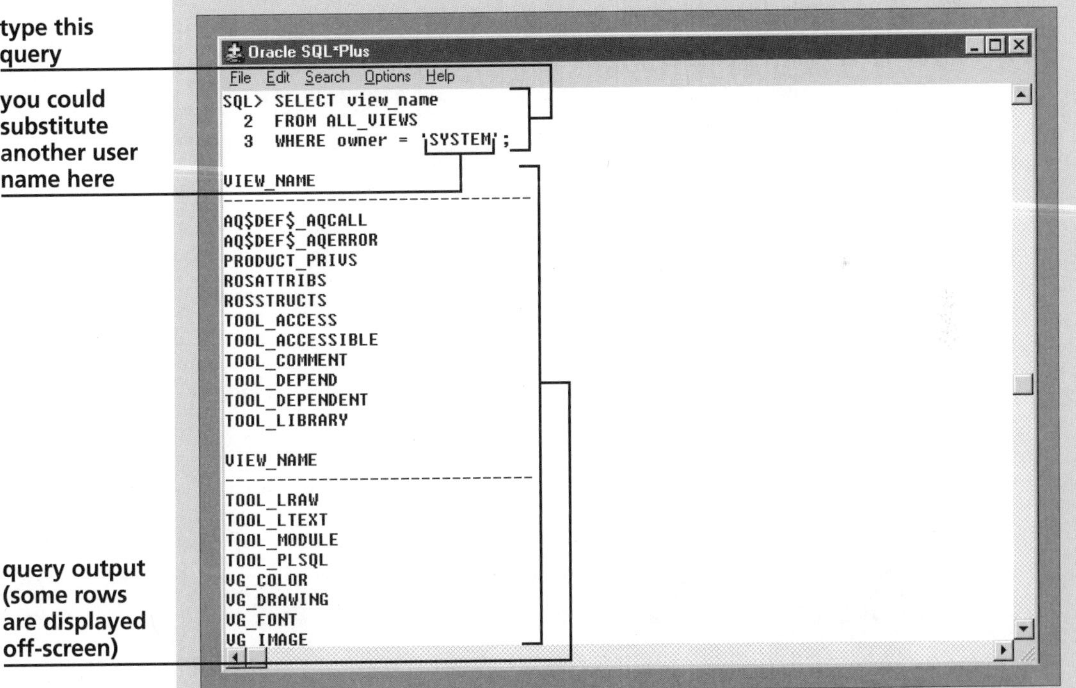

Figure 3-72: Displaying the names of the views created by SYSTEM

2 Type the query shown in Figure 3-73 using the USER_VIEWS system table to display the names of the views you have created.

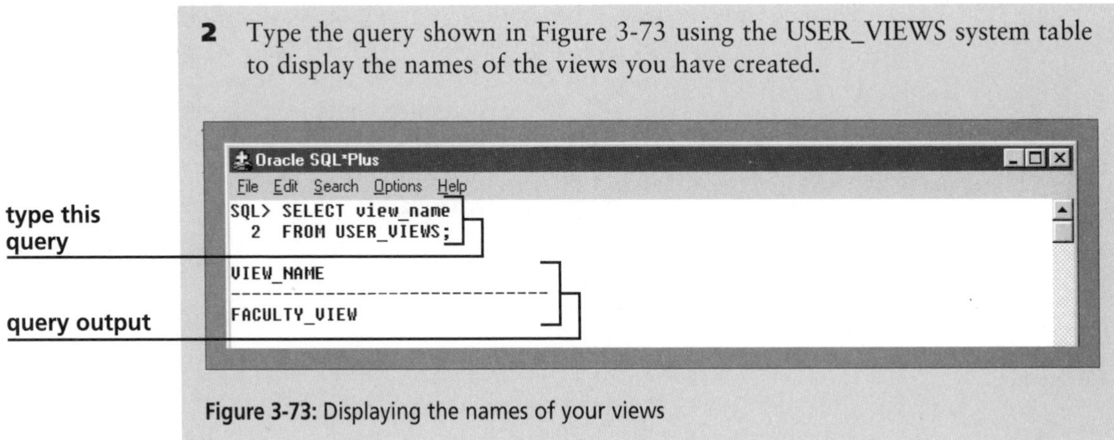

Figure 3-73: Displaying the names of your views

You use the DROP VIEW command to delete a view from the database. Remember that a view is based on a query that executes to display the requested data from the underlying base table. When you drop a view, you do not drop the data that appear in the view—only the view definition is dropped.

To drop the FACULTY_VIEW view:

1 Type **DROP VIEW faculty_view;** to drop FACULTY_VIEW. The "View dropped" confirmation message indicates that the view was dropped.

Granting Table Privileges to Other Users

When you create a database table in Oracle, other users cannot modify your tables or view or change your data records unless you give them explicit privileges to do so. Figure 3-74 lists some commonly used table privileges and their descriptions.

Command	Description
ALTER	Allows the user to change a table's structure using the ALTER TABLE command
DELETE	Allows the user to delete records from a table using the DELETE command

Figure 3-74: Commonly used table privileges

Command	Description
INSERT	Allows the user to insert records into a table using the INSERT command
REFERENCES	Allows the user to reference table fields as foreign keys using the REFERENCED BY command
SELECT	Allows the user to view data using the SELECT command
UPDATE	Allows the user to modify data using the UPDATE command
ALL	Grants all privileges to the user

Figure 3-74 (continued): Commonly used table privileges

Table privileges are granted to other users using the SQL GRANT command. The general format of the SQL GRANT command is:

```
GRANT <privilege1, privilege2, …>
ON <table name>
TO <user1, user2, …>;
```

To grant a privilege:

1 Type the query shown in Figure 3-75 to grant SELECT and ALTER privileges on your STUDENT table to two other students in your class by substituting the students' user names for MORRISJP and MORRISCM in the query. When your query executes successfully, the confirmation message "Grant succeeded" is displayed. Note that you can grant privileges for only one table at a time, but that you can grant privileges to many users at once. If you want to grant privileges to every database user, you can use the word PUBLIC in the TO clause.

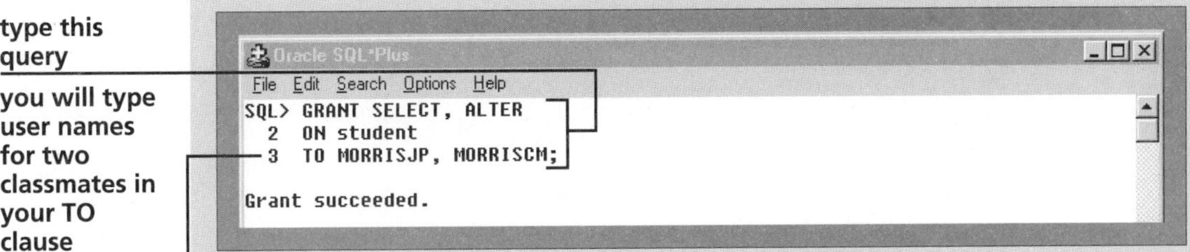
type this query
you will type user names for two classmates in your TO clause
Oracle SQL*Plus
File Edit Search Options Help
SQL> GRANT SELECT, ALTER
2 ON student
3 TO MORRISJP, MORRISCM;
Grant succeeded.

Figure 3-75: Granting SELECT and ALTER privileges to two selected users

2 Type the command shown in Figure 3-76 to grant all privileges on your FACULTY table to all database users.

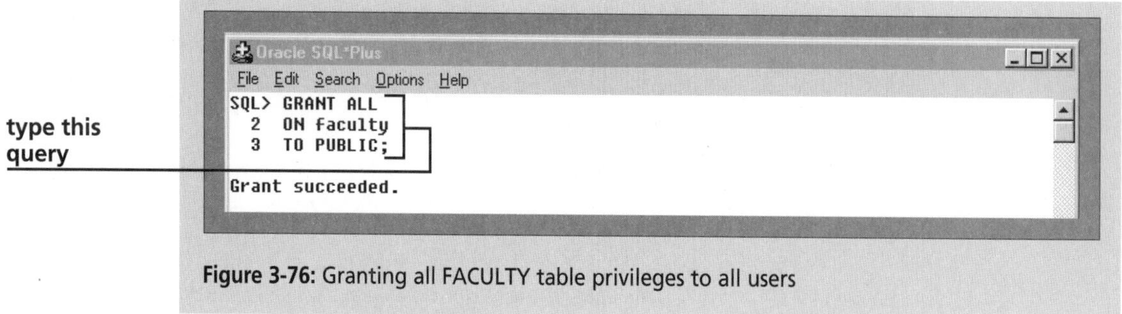
type this query
Oracle SQL*Plus
File Edit Search Options Help
SQL> GRANT ALL
2 ON faculty
3 TO PUBLIC;
Grant succeeded.

Figure 3-76: Granting all FACULTY table privileges to all users

Revoking Table Privileges

To cancel a privilege to a user, you use the SQL REVOKE command. The general format of the REVOKE command is:

```
REVOKE <privilege1, privilege2, etc.>
ON <table name>
FROM <user1, user2, etc.>;
```

To revoke a privilege:

1 Type the command shown in Figure 3-77 to revoke the SELECT and ALTER privileges that you granted to your two classmates on the STUDENT table. Remember that you will have to change the user names from MORRISJP and MORRISCM to the names of the users to whom you granted these privileges in Figure 3-75.

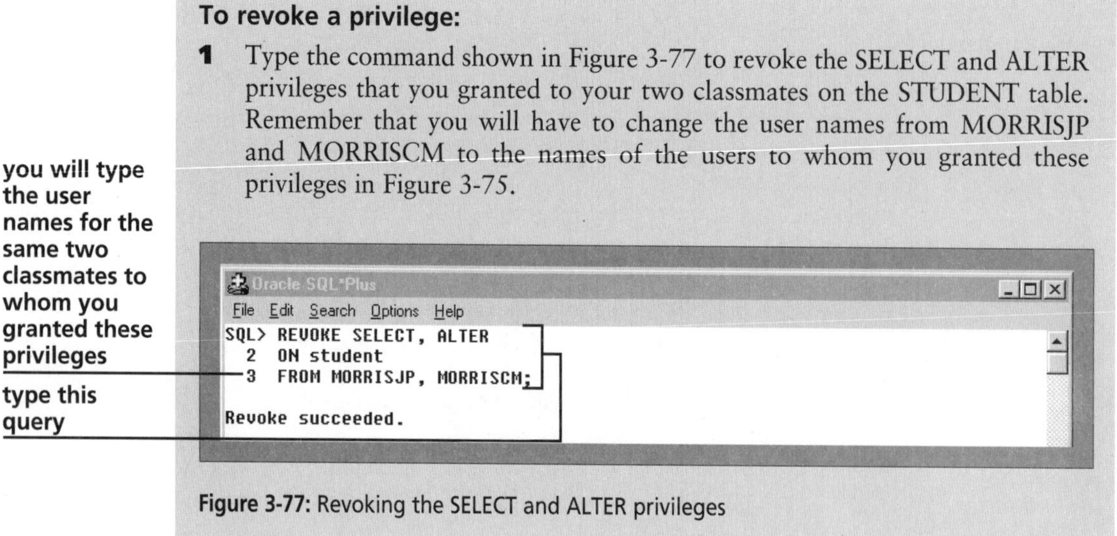
you will type the user names for the same two classmates to whom you granted these privileges
type this query
Oracle SQL*Plus
File Edit Search Options Help
SQL> REVOKE SELECT, ALTER
2 ON student
3 FROM MORRISJP, MORRISCM;
Revoke succeeded.

Figure 3-77: Revoking the SELECT and ALTER privileges

Sequences

Oracle **sequences** are sequential lists of numbers that are generated automatically by the database. Sequences are useful for creating unique surrogate key values for primary key fields when no field exists to use as the primary key. For example, you might assign a sequence for a customer ID field so that each record is unique.

Fields using sequences must have a NUMBER data type. In the following exercises, you will become familiar with creating and manipulating sequences using SQL*Plus.

Creating New Sequences

The general format for creating a new sequence is shown below, with optional commands enclosed in square brackets:

```
CREATE SEQUENCE <sequence name>
[INCREMENT BY <number>]
[START WITH <start value number>]
[MAXVALUE <maximum value number>]
[NOMAXVALUE]
[MINVALUE <minimum value number>]
[CYCLE]
[NOCYCLE]
[CACHE <number of sequence values to cache>]
[NOCACHE]
[ORDER]
[NOORDER];
```

Every sequence must have a unique name, and the naming rules are the same as for Oracle database tables and fields. The CREATE SEQUENCE command is the only command required to create a sequence; the rest of the commands are optional. Figure 3-78 describes these optional commands and shows their default value if the parameter specification is omitted.

Command	Description	Default Value
INCREMENT BY	Specifies the value by which the sequence is incremented	1
START WITH	Specifies the sequence start value. It is necessary to specify a start value if you already have data in the fields where the sequence will be used to generate primary key values. For example, in the ITEM table, the highest existing ITEMID is 995, so you would start the sequence with 996.	1
MAXVALUE	Specifies the maximum value to which the sequence can be incremented. If you don't specify a maximum value, the maximum allowable value for a sequence variable is $1 * 10^{27}$.	NOMAXVALUE, which specifies that the sequence will keep incrementing until the maximum allowable value is reached

Figure 3-78: Sequence optional parameters

Command	Description	Default Value
MINVALUE	Specifies the minimum value of the sequence for a decrementing sequence	NOMINVALUE
CYCLE	Specifies that, when the sequence reaches its MAXVALUE, it cycles back and starts again at the MINVALUE. For example, if you specify a sequence with a maximum value of 10 and a minimum value of 5, the sequence will increment up to 10 and then start again at 5.	NOCYCLE, which specifies that the sequence will continue to generate values until it reaches MAXVALUE and will not cycle
CACHE	Specifies that, whenever you request a sequence value, the Oracle database server automatically generates several sequence numbers and stores them in a server memory area called a cache to improve system performance. The default number of sequence values stored in the cache is 20. To specify a different number of sequence values to be cached, use the CACHE command along with the number of sequence values you want to generate and store.	20 sequence numbers are cached
NOCACHE	Directs the server not to cache any sequence values	n/a
ORDER	Ensures that the sequence numbers are granted in the exact chronological order in which they are requested. For example, the first user who requests a number from the sequence would be granted sequence number 1, the second user who requests a number from the sequence would be granted sequence number 2, and so forth. This is useful for tracking the order in which specific transactions occur.	NOORDER, which specifies that the sequence values are not necessarily granted in chronological order; therefore, although users are guaranteed to get unique sequence numbers, the order of the values might not correspond with the order in which the sequence numbers were requested

Figure 3-78 (continued): Sequence optional parameters

Next, you will create a sequence for the ITEMID field in the ITEM table in the Clearwater Traders database, named ITEMID_SEQUENCE, which starts with 996 and has no maximum value. The sequence will increment by 1, so you will accept the default value and not use the INCREMENT BY command. You do not want the sequence to CYCLE, so you will omit the CYCLE command and accept the NOCYCLE default. You will not have the server cache sequence values, which means that you will include the NOCACHE command.

To create the ITEMID_SEQUENCE:

1 Type the SQL query shown in Figure 3-79. The sequence will start at 996, and it has no maximum value. The confirmation message "Sequence created" indicates that the ITEMID_SEQUENCE sequence was created.

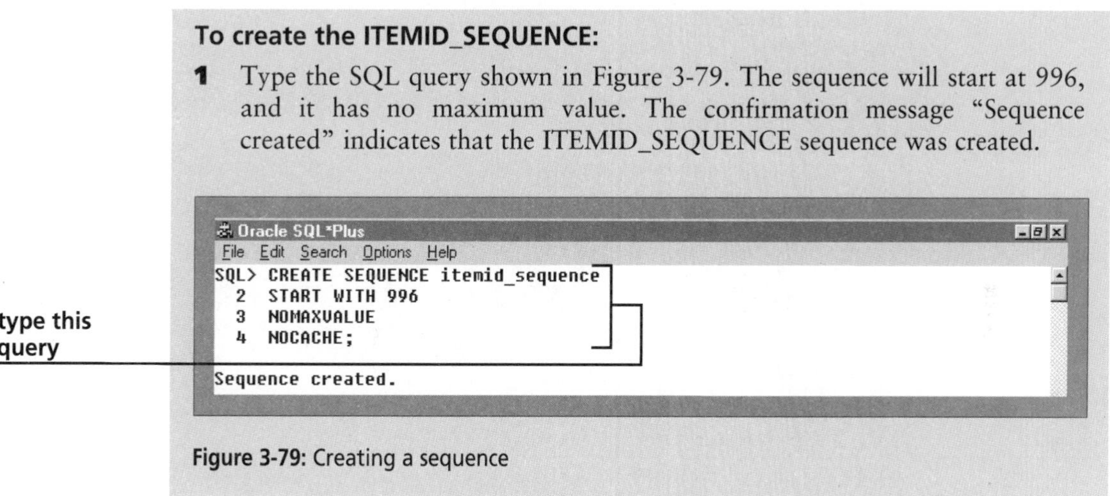
type this query
Oracle SQL*Plus
File Edit Search Options Help
SQL> CREATE SEQUENCE itemid_sequence
2 START WITH 996
3 NOMAXVALUE
4 NOCACHE;
Sequence created.

Figure 3-79: Creating a sequence

Accessing the Next Sequence Value

To access the next value in a sequence and use that value when you insert a new data record, use the following general command:

```
INSERT INTO <table name>
VALUES (<owner user name>.<sequence name>.NEXTVAL, <field2
data value>, <field3 data value>, …);
```

tip

This command assumes that the primary key associated with the sequence is the first table field.

Next, you will insert a new item record in the Clearwater Traders database for an item with the description "Heavy Duty Day Pack" in the Outdoor Gear category.

To insert a new ITEM record using a sequence:

1 Type the SQL query shown in Figure 3-80. The confirmation message "1 row created" indicates that you inserted one record. Next you will see what was inserted for ITEMID.

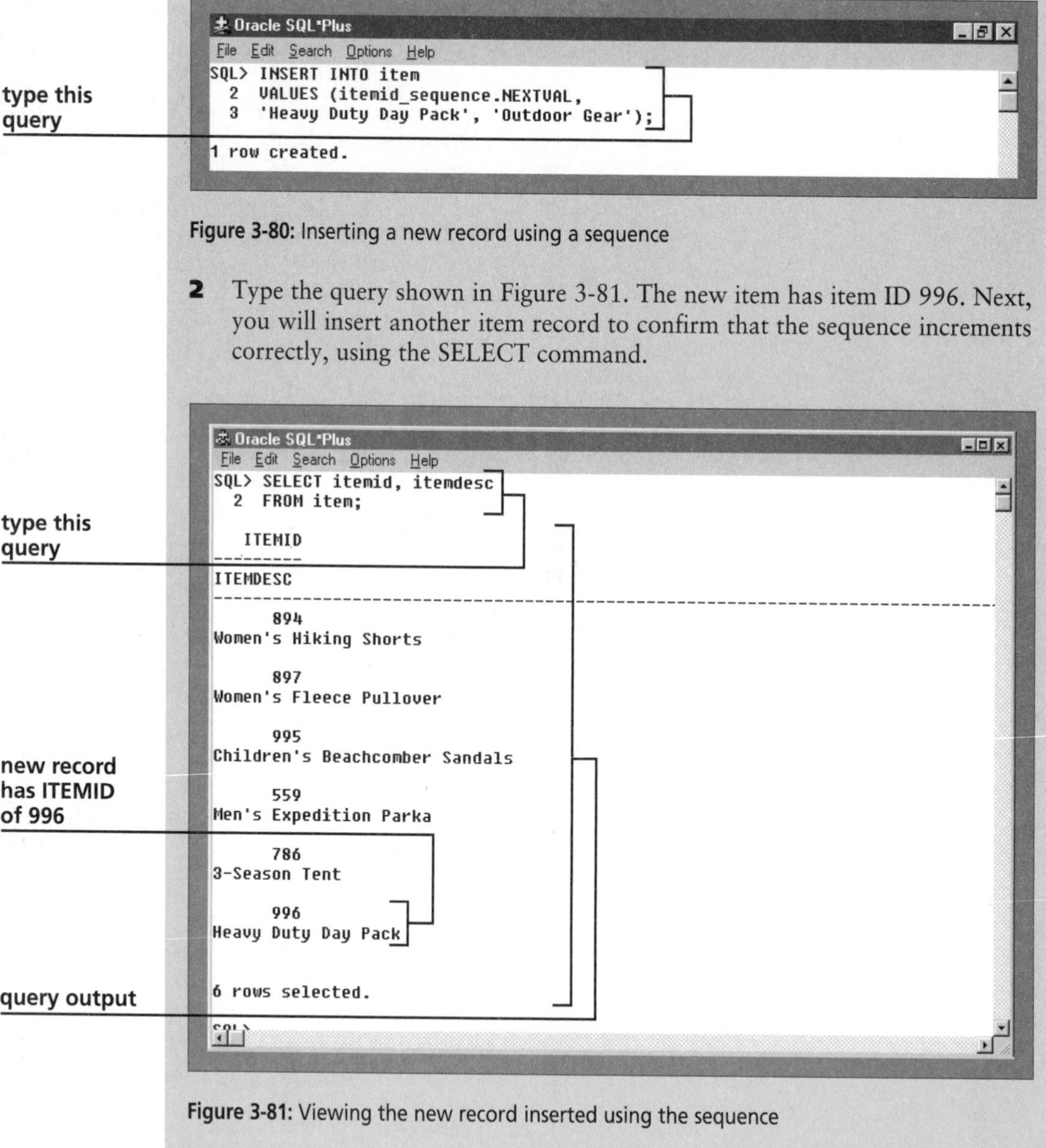
Oracle SQL*Plus
File Edit Search Options Help
SQL> INSERT INTO item
2 VALUES (itemid_sequence.NEXTVAL,
3 'Heavy Duty Day Pack', 'Outdoor Gear');
1 row created.
type this query
Oracle SQL*Plus
File Edit Search Options Help
SQL> SELECT itemid, itemdesc
2 FROM item;
ITEMID
ITEMDESC
894
Women's Hiking Shorts
897
Women's Fleece Pullover
995
Children's Beachcomber Sandals
559
Men's Expedition Parka
786
3-Season Tent
996
Heavy Duty Day Pack
6 rows selected.
type this query
new record has ITEMID of 996
query output

Figure 3-80: Inserting a new record using a sequence

2 Type the query shown in Figure 3-81. The new item has item ID 996. Next, you will insert another item record to confirm that the sequence increments correctly, using the SELECT command.

Figure 3-81: Viewing the new record inserted using the sequence

3 Type the SQL query shown in Figure 3-82. You should receive the "1 row created" message again.

type this query
Oracle SQL*Plus
File Edit Search Options Help
SQL> INSERT INTO item
2 VALUES (itemid_sequence.NEXTVAL,
3 'Mountain Parka', 'Men''s Clothing');
1 row created.

Figure 3-82: Inserting another record using the sequence

4 Type the SQL query shown in Figure 3-83 to view the new record. The Mountain Parka should have item ID 997.

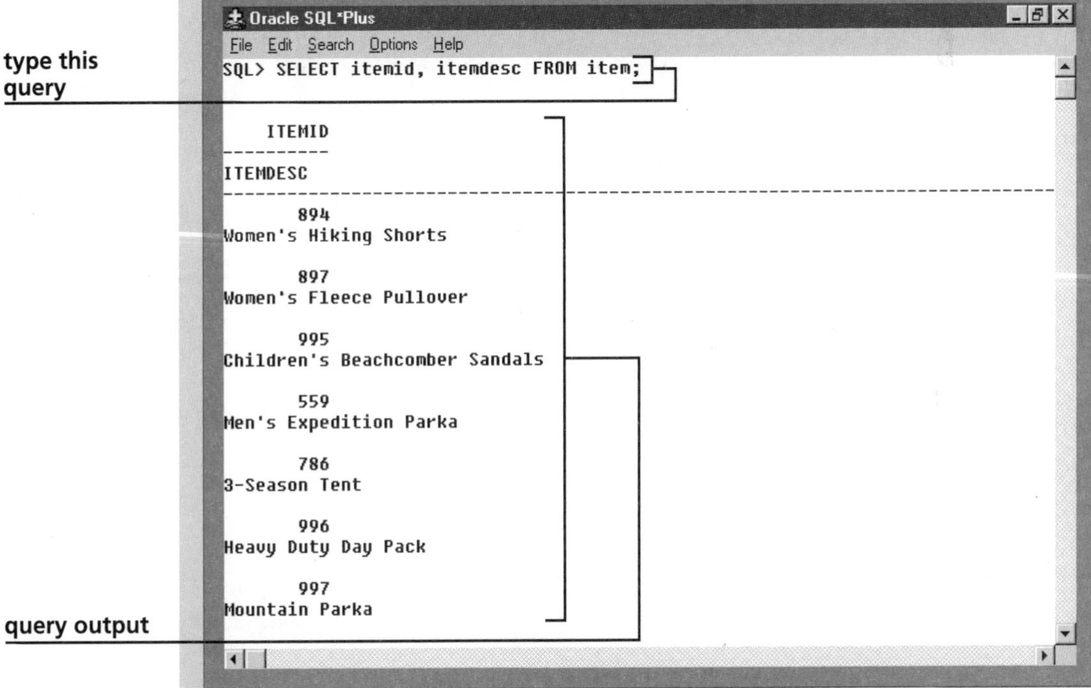
type this query
query output
Oracle SQL*Plus
File Edit Search Options Help
SQL> SELECT itemid, itemdesc FROM item;
ITEMID
ITEMDESC
894
Women's Hiking Shorts
897
Women's Fleece Pullover
995
Children's Beachcomber Sandals
559
Men's Expedition Parka
786
3-Season Tent
996
Heavy Duty Day Pack
997
Mountain Parka

Figure 3-83: Viewing the next record inserted using the sequence

help

If your query output prints beyond the length of the SQL*Plus window page size, the column headings will print again in the output. You can increase the size of the SQL*Plus window page by clicking Environment on the Options menu, selecting pagesize from the Set Options list, clicking the Current option button, entering a higher value in the text box under the option buttons, and then clicking OK.

Sometimes you need to access the next value of a sequence, but you don't want to insert a new record. For example, suppose you want to create a new customer order and display the ORDERID on the order form, but you need to have the user enter more information before you actually insert the new record. To do this, you use the SELECT command with the Oracle system database table DUAL. Recall that **DUAL** is a system database table that is used with a variety of SQL commands.

To use the SELECT command to access the next value in a sequence:

1 Type the SQL query shown in Figure 3-84. Your query result should show 998 as the NEXTVAL in the ITEMID_SEQUENCE. Now that it has been accessed, it is the current value of the sequence, and the next value will be 999.

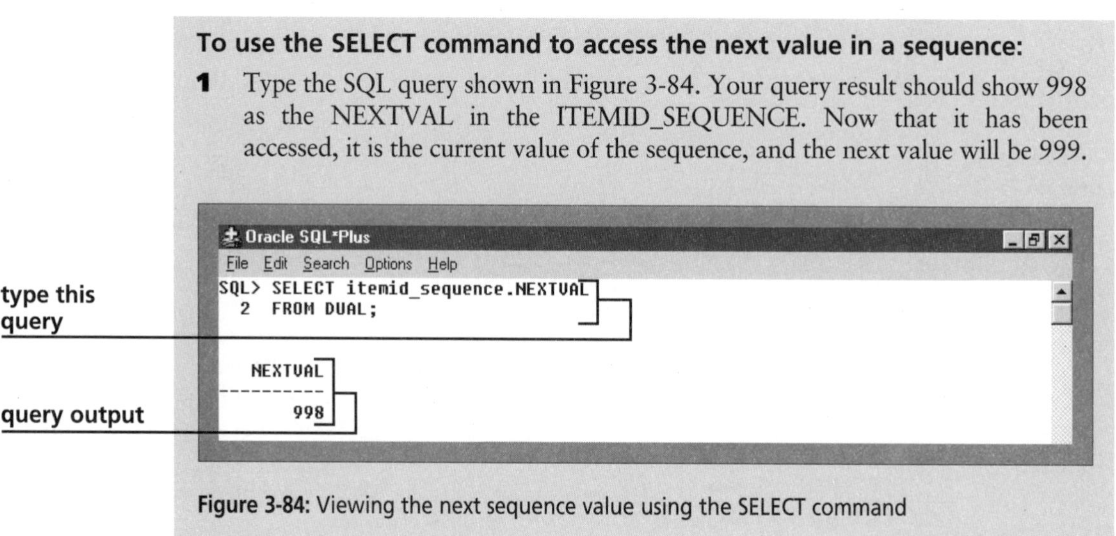
type this query
query output
Oracle SQL*Plus
File Edit Search Options Help
SQL> SELECT itemid_sequence.NEXTVAL
2 FROM DUAL;
NEXTVAL

998

Figure 3-84: Viewing the next sequence value using the SELECT command

You can confirm that 999 is the next value by accessing it, as you will see next.

To make sure that the sequence increments correctly:

1 Confirm that 999 is the next value by typing **L** and pressing the **Enter** key at the SQL prompt to list your previous SQL query. Then type **/** and press the **Enter** key to execute the command. Your query output should look like Figure 3-85.

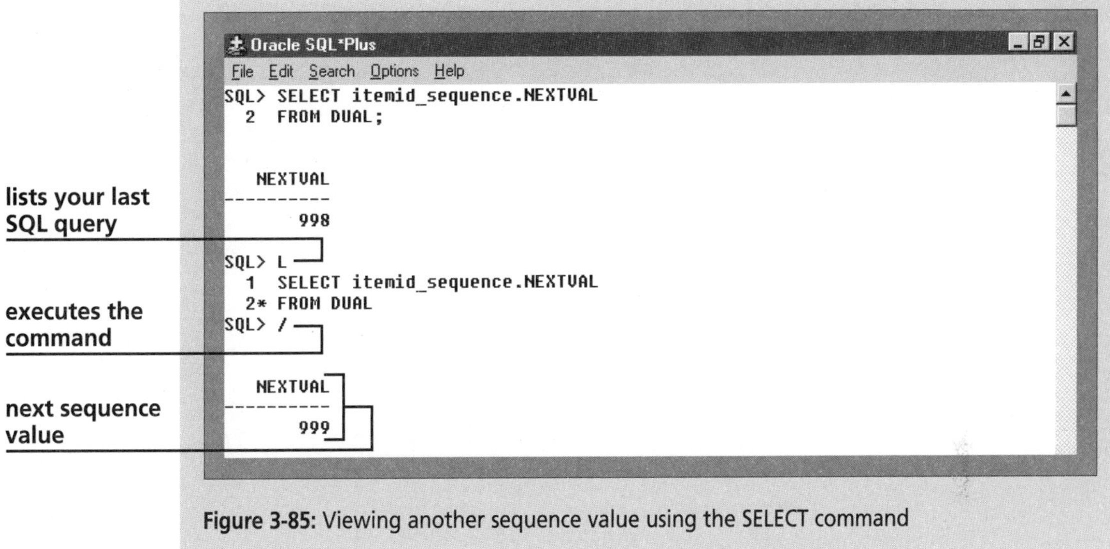

Figure 3-85: Viewing another sequence value using the SELECT command

Once you move on to the next value, the previous sequence value cannot be accessed using the sequence commands. This prevents you (or anyone else) from accidentally using the same sequence value as the primary key for two different records. In the previous example, 998 is "lost" as a sequence value, meaning that there will be no item in the table with 998 as an ID number. But that's okay—it is better to lose a sequence value than to use the same value as the primary key for two different records.

Accessing the Current Sequence Value

While you cannot access past sequence values, you can access the current value using the CURRVAL command.

To use the CURRVAL command:

1 Confirm that 999 is the current sequence value by typing the SQL query shown in Figure 3-86.

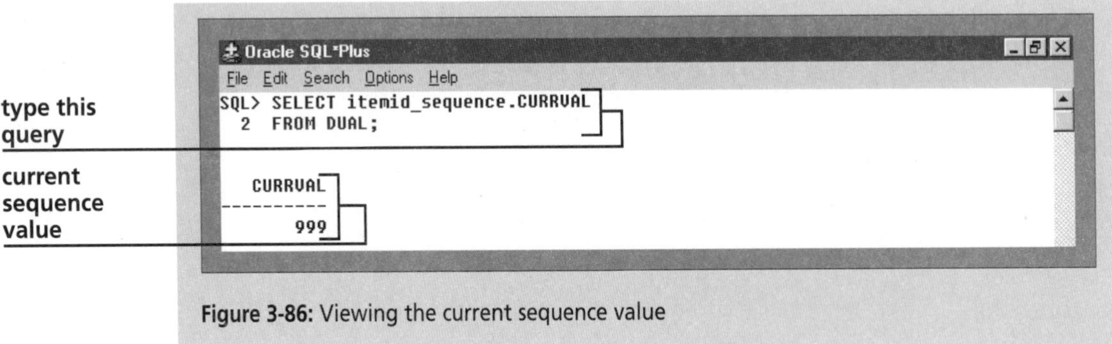

Figure 3-86: Viewing the current sequence value

The tricky thing about using the CURRVAL command is that it can be used only in the same database session and immediately after using the NEXTVAL command. This prevents two database users or processes from assigning the same sequence value to two different records. Next, you will confirm this by exiting SQL*Plus, starting it again, and then trying to use the CURRVAL command.

To make sure that the CURRVAL command works properly:

1. Exit SQL*Plus.
2. Start SQL*Plus again and log on to the database.
3. Type the SQL query shown in Figure 3-87. The error message indicates that the CURRVAL command will not select a value because you exited SQL*Plus and then started a new SQL*Plus session.

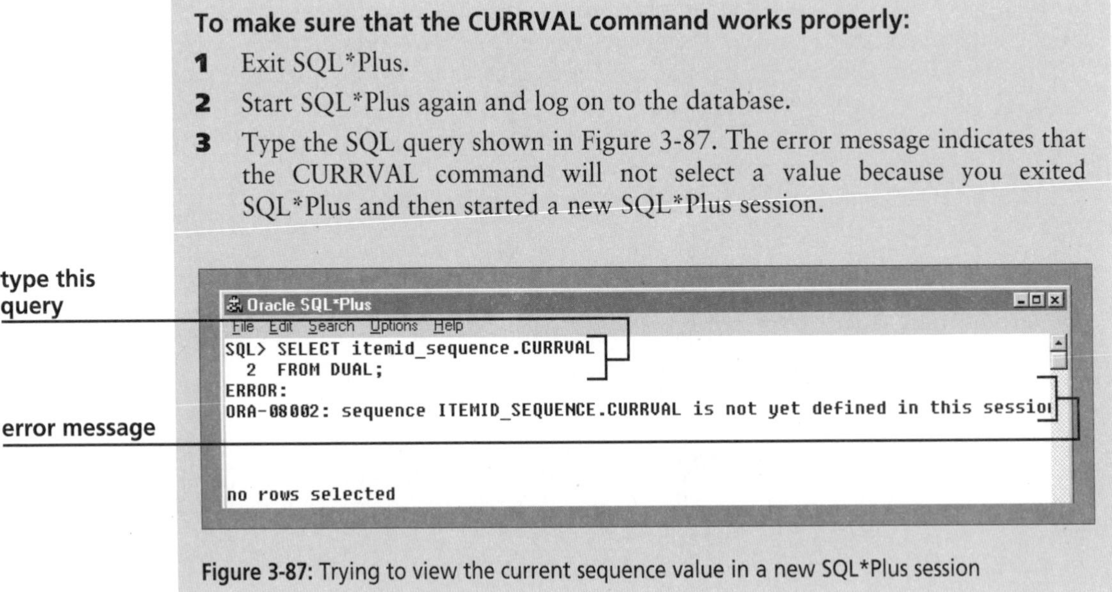

Figure 3-87: Trying to view the current sequence value in a new SQL*Plus session

Granting Sequence Privileges to Other Users

In order for other users to use your sequences, you must grant them explicit privileges to do so. You grant sequence privileges using the following general command format: GRANT SELECT ON <sequence name> TO <user name 1, user name 2, ...>. Next, you will grant privileges on your ITEMID_SEQUENCE to user PUBLIC.

To grant sequence privileges using the GRANT command:

1 Type the SQL query shown in Figure 3-88. The confirmation message "Grant succeeded" enables all database users to use the NEXTVAL, CURRVAL, and SELECT commands with your ITEMID_SEQUENCE. If you wanted to grant this privilege to only a small group of users, you would enter individual user names (separated by commas) instead of PUBLIC.

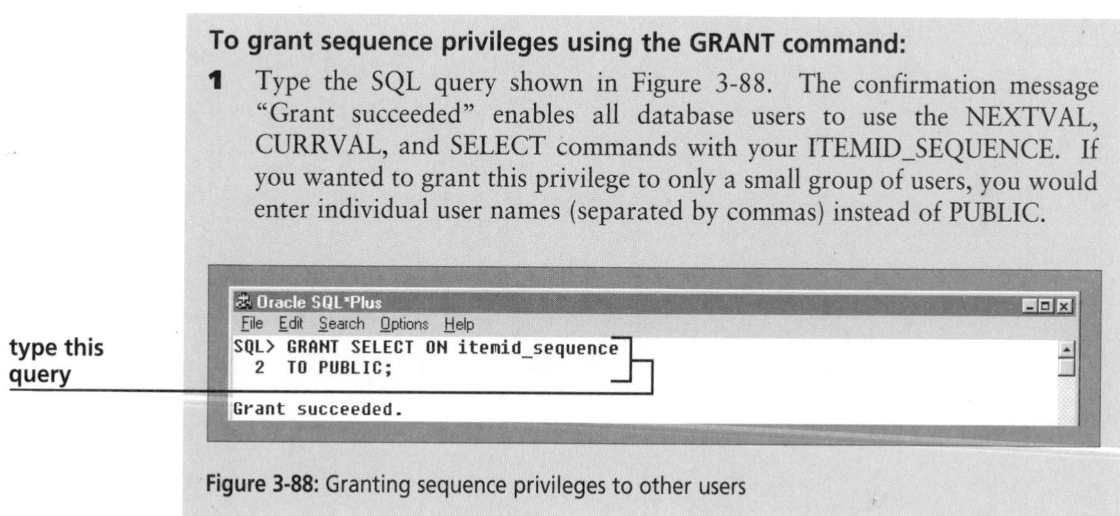

Figure 3-88: Granting sequence privileges to other users

Viewing and Deleting Sequences

Sometimes you might need to review the names and properties of your sequences after you create them. You can do this by querying the USER_SEQUENCES system table, using a SELECT command.

To review the properties of your sequences:

1 Type the SQL query shown in Figure 3-89. The query output lists your current sequence names and properties, including the minimum and maximum allowable values that were established when you created the sequence, and the value by which the sequence increments. The CYCLE and ORDER parameter values of "N" indicate that cycling and ordering are not used. The cache size is shown as 0, because you specified no cache sequence values using the NOCACHE command. The next value of the sequence is 1000.

maximum allowable value
minimum allowable value
sequence name
cache size
next value
increment value
CYCLE parameter
ORDER parameter

```
Oracle SQL*Plus
File Edit Search Options Help
SQL> SELECT *
  2  FROM user_sequences;

SEQUENCE_NAME                  MIN_VALUE MAX_VALUE INCREMENT_BY C O
------------------------------ --------- --------- ------------ - -
CACHE_SIZE LAST_NUMBER
---------- -----------
ITEMID_SEQUENCE                        1 1.000E+27            1 N N
         0        1000

SQL>
```

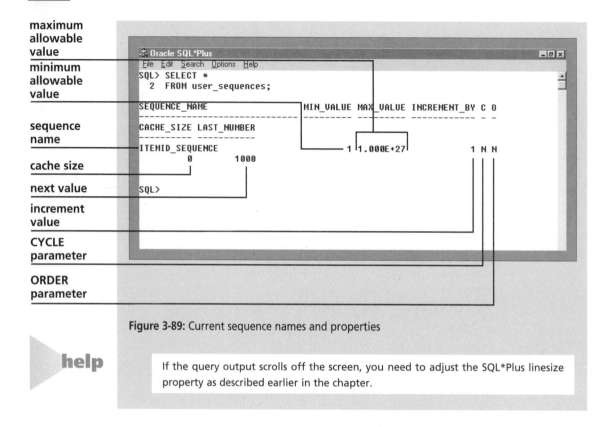

Figure 3-89: Current sequence names and properties

help If the query output scrolls off the screen, you need to adjust the SQL*Plus linesize property as described earlier in the chapter.

You might need to delete a sequence if you want to change its name or one of its other properties. When you need to delete a sequence, use the DROP SEQUENCE command, which has the following general format: `DROP SEQUENCE <sequence name>;`. Next, you will drop the ITEMID_SEQUENCE that you created.

To drop a sequence:

1 Type the query shown in Figure 3-90. The message "Sequence dropped" confirms that the sequence was dropped.

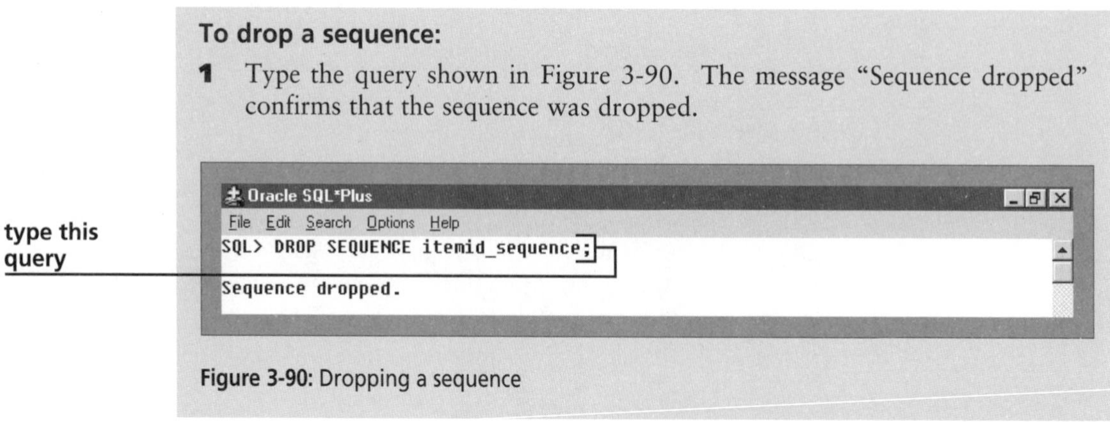

Figure 3-90: Dropping a sequence

Using Oracle Query Builder to Retrieve Data

Query Builder is a graphical environment that makes creating SQL SELECT statement queries easier. Rather than having to write the query manually, Query Builder allows you to select the tables you want to query, and it displays the tables and their associated columns on the screen. Foreign key links are shown as lines between the columns in the related tables. You can choose the query display columns by clicking on them. You can optionally specify search conditions by clicking on the field or fields you want to search on, and typing the search conditions. You can also easily specify group functions, such as summing, grouping, or ordering. Query Builder takes care of specifying the join conditions.

First, you need to refresh your Northwoods University database tables by running the northwoo.sql script on your Student Disk.

To run the script file and start Query Builder:

1. Run the northwoo.sql script by typing the following command: **START A:\Chapter3\northwoo.sql.**
2. Click the **Start** button on the Windows taskbar, point to **Programs**, point to **Oracle Developer 6.0**, and then click **Query Builder.**
3. Log on to Query Builder using your user name, password, and connect string, and then click **Connect.**

   The Start Query Builder dialog box opens. You can use this dialog box to create a new query, open an existing query from the database, or open an existing query from the file system. Query Builder enables you to create queries and then save them in the database or in the file system for later retrieval.

help

If a message box displaying Oracle error codes is displayed, it means that some of the scripts used with Query Builder were not run when Query Builder was installed on your workstation. Query Builder will still work correctly in the tutorial, so click OK.

4. Make sure that the **Create New Query** option button is selected, and then click **OK.** The Query Builder program window opens, and then the Select Data Tables dialog box opens. The Show check boxes specify what types of objects are displayed: Tables, Views, Queries, Snapshots, or Synonyms. The center list box allows you to choose whose objects are currently displayed—your user name will be displayed. The bottom list box lists your tables, views, and queries.

Creating a Query Using Query Builder

Remember the SQL query you created earlier to display the call IDs and grades for all of Sarah Miller's courses? Next, you will re-create that query using Query Builder.

To create a query using Query Builder:

1. Click the **COURSE** table in the Select Data Tables list box, press and hold down the **Ctrl** key, and then click **COURSE_SECTION**, **ENROLLMENT**, and **STUDENT**.
2. Click the **Include** button, wait a moment (the pointer will change from to), and then click the **Close** button. If necessary, click the Maximize button to maximize the Query Builder window.
3. Click the **Maximize** button on the Untitled1: Query window. The Query window has two panels: the Condition panel and the Datasource panel. The Datasource panel shows the tables you selected and the links between the tables created by foreign key relationships. The Condition panel is used to specify query search conditions.

   To make the display easier to interpret, you can drag and drop the boxes so the foreign key relationships are displayed more clearly. You can also resize the boxes that display the table fields by clicking the table heading to select it, then placing the pointer on the lower-right corner until the pointer changes to , and dragging the box to the size you want.
4. Resize and move the tables so they are positioned like the ones shown in Figure 3-91. All tables should be displayed on the screen, and all of the foreign key links should be visible.

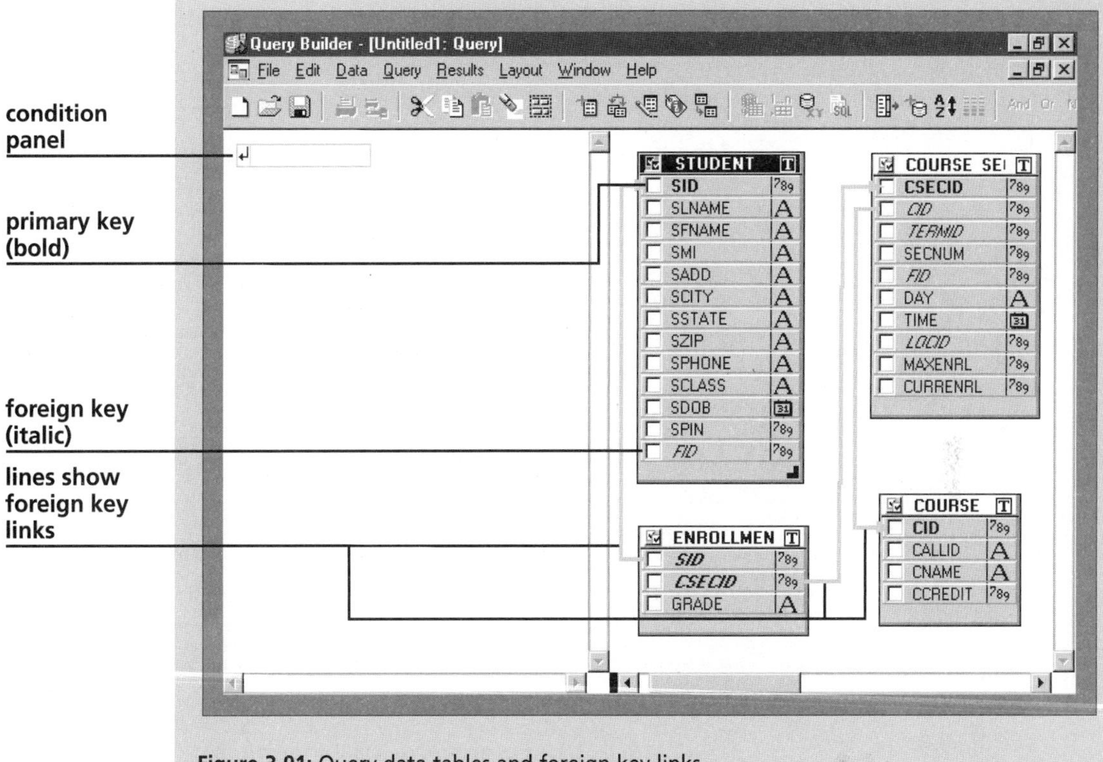

Figure 3-91: Query data tables and foreign key links

This display provides a useful tool for viewing the database structure and composing queries. Primary keys are displayed in bold, and foreign keys are displayed in italic. Links from primary keys to associated foreign keys are displayed as connecting lines. NUMBER data fields are designated with 789, character data fields with A, and DATE data fields with 31.

The next step in creating a query is to select the columns that will be displayed in the query output. To select a column for display, check the box next to the column name in the table, and a check mark appears in front of the column name to indicate that it is selected. The order in which you select the column names will specify the order in which they appear in the query output. Now, you will select the query display columns.

To select the query display columns:

1 Click the check box next to **CALLID** (in the COURSE table) and **GRADE** (in the ENROLLMENT table) to specify them as the display columns. A check mark appears in front of the column names to indicate that they are selected. To deselect a column, clear the check box.

Next, you will specify the search conditions. Currently, the search **Condition panel** is on the left side of the Query window (see Figure 3-91). It is easier to view the search conditions if you configure the Query window so that it splits horizontally and the Condition panel appears at the top of the window. Next, you will reconfigure the Condition panel and specify the search condition.

To reconfigure the Condition panel and specify the search condition:

1. Click **Edit** on the menu bar, and then click **Preferences.** Click **Query Window** in the Preferences for list box, click the **All Documents** option button, and then change the Panel Split Direction to Vertical and change the Condition Panel Size value to **20.**
2. Click **OK.** Figure 3-92 shows the reconfigured Condition panel. Adjust the table sizes and positions if necessary.
3. Click the **Condition panel** box to activate the Condition panel, as shown in Figure 3-92, and then click **SLNAME** in the STUDENT table to add it to the Condition panel.

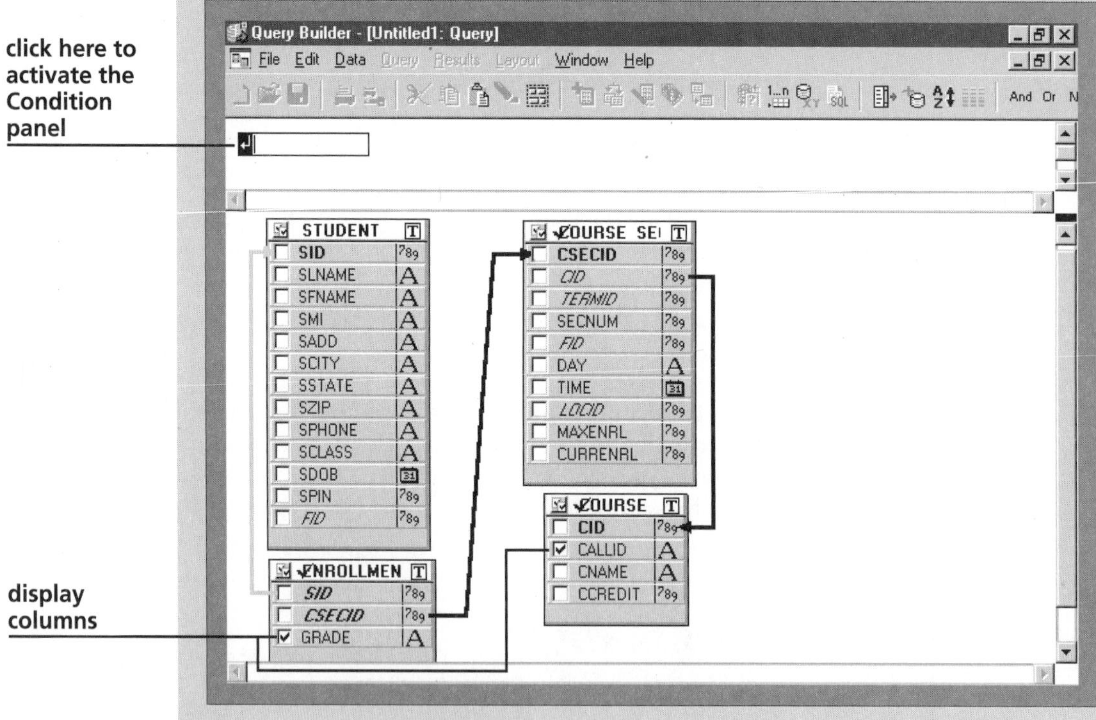

Figure 3-92: Activating the Condition panel

4 Immediately after STUDENT.SLNAME, type **='Miller'** as the first search condition. Make sure that you type the single quotation marks and use the same case as indicated.

5 Click the **And** button [And] on the toolbar to add the AND operator to your expression in the Condition panel.

6 Click **SFNAME** in the STUDENT table to add it to the Condition panel.

7 Click the Condition panel immediately after STUDENT.SFNAME, and then type =**'Sarah'** as the second search condition, as shown in Figure 3-93.

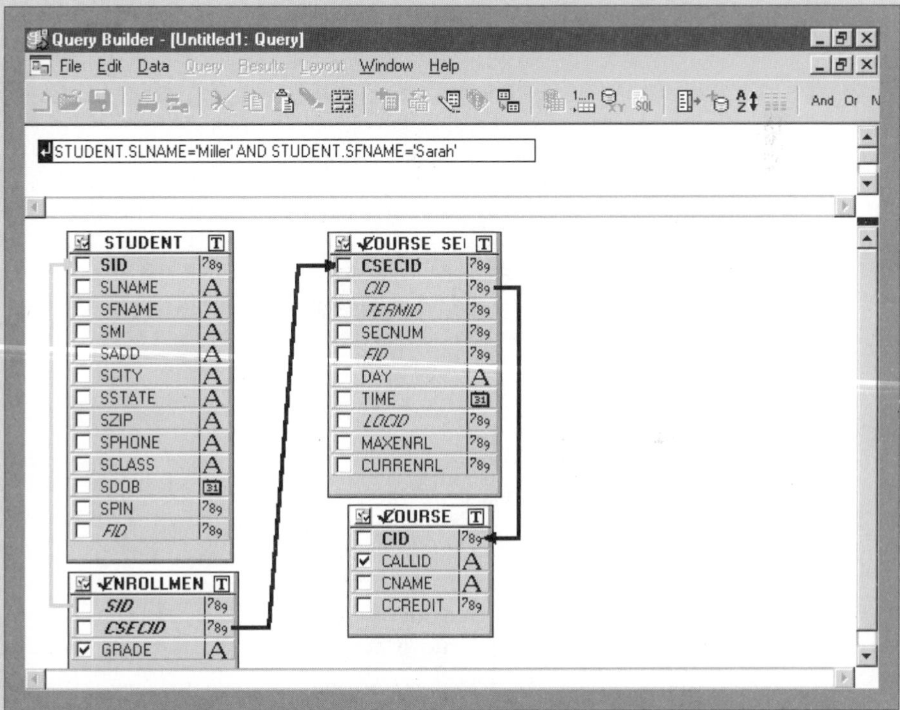

Figure 3-93: Specifying the query search condition

8 Click the **Accept** button on the toolbar to accept the search conditions, and then click the **Execute Query** button on the toolbar to execute the query. The query output is shown in Figure 3-94.

help

If the ACCEPT button is not visible on the toolbar, press the Enter key on the keyboard to accept the search condition.

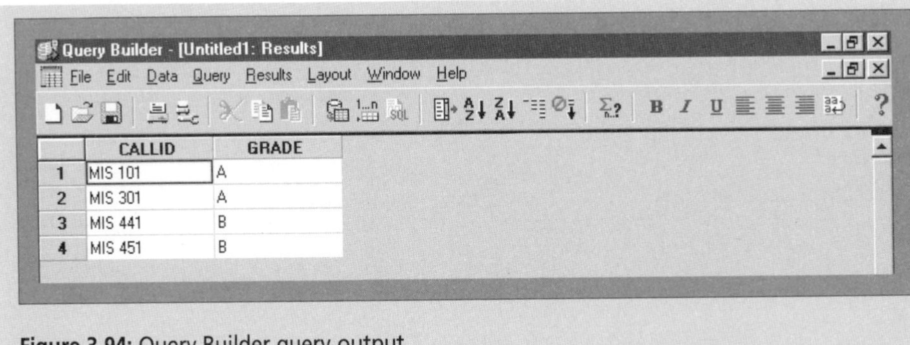

Figure 3-94: Query Builder query output

help

If the Execute Query button or menu commands are not available to you, you probably forgot to select the display fields. Double-click the fields you want the query to display, so that a check mark appears in front of their names.

help

If your query finds no records, check to make sure that you enclosed the character values within single quotation marks, and that the case of the character strings matches the way the values are entered in the database. If you still have problems, ask your instructor or technical support person for help.

Viewing the SQL Command Code

Query Builder simplifies the query process, but some knowledge of table relationships and search condition logic is still needed, so Query Builder probably will not be used by end users as a query tool. However, this utility is useful to create queries like this one that involve several tables, because it can generate SQL syntax that can be copied into other Oracle applications, such as SQL*Plus.

To view the SQL command code and copy it to SQL*Plus:

1 Click **Query** on the menu bar, and then click **Show SQL**. The SQL syntax for the query appears, as shown in Figure 3-95.

your user name will appear throughout the command instead of LHOWARD

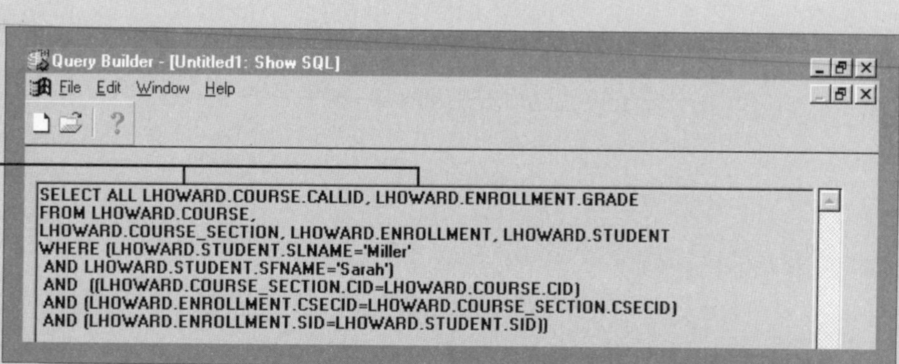

Figure 3-95: Viewing the SQL command created by Query Builder

2 Highlight the SQL command text, click **Edit** on the menu bar, and then click **Copy**.

3 Click the **Oracle SQL*Plus** command button on the Windows taskbar to return to SQL*Plus.

4 Click **Edit** on the menu bar, and then click **Paste**.

5 Type **;** at the end of the query, and then press the **Enter** key. The query output is shown in Figure 3-96.

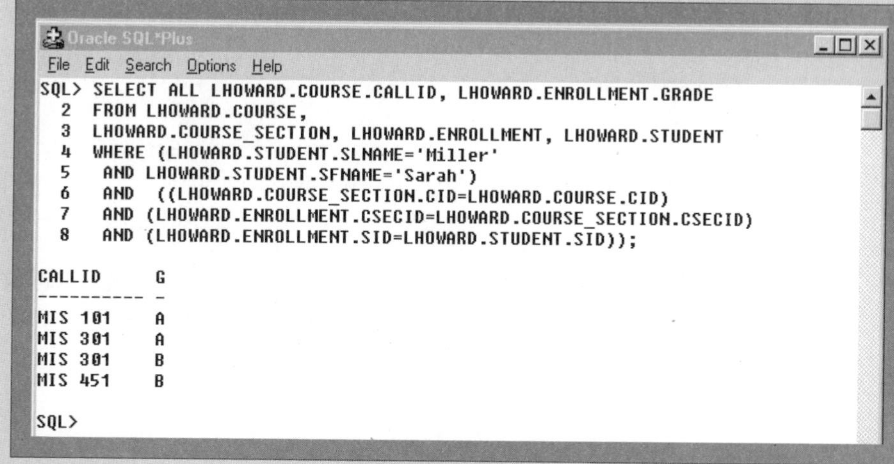

Figure 3-96: Running the Query Builder query in SQL*Plus

Saving Your Query Builder Query

It is useful to be able to save Query Builder queries so they can be modified later if needed, or so you can run them again and view the results as the data in the database change. You can save your queries in the database, to your file system, or as SQL text files. If queries are saved in the database or file system as Query Builder files, they can be retrieved only by using Query Builder. Queries saved as text files can be copied and then pasted into SQL*Plus or other Oracle applications or saved as SQL scripts.

To save your Query Builder query:

1. Click the **Query Builder** button on the taskbar to activate Query Builder.
2. Click **Window** on the menu bar. There are three windows currently open in your Query Builder environment: Untitled1: Query (in which you create and edit query specifications), Untitled1: Results (which displays the query output when you run the query), and Untitled1: Show SQL (which shows the SQL command syntax). Click **Untitled1: Query** to activate the Query window.
3. Click the **Save** button on the toolbar to open the Save As dialog box. You can save the query as a Query Builder query in the database, or in the file system as a file with a .brw extension. The final option is to save to the file system as a .qxf (Oracle Data Query) file. Oracle Data Query is an end-user query-building tool that is not covered in this book.
4. Click the **File System** option button, and then click **OK**. The Windows Save As dialog box opens.
5. Navigate to the Chapter3 folder on your Student Disk, and then save the file as **miller.brw**.
6. Click **File** on the menu bar, and then click **Save SQL** to open the Save As dialog box. Save the file as **miller.sql** in the Chapter3 folder on your Student Disk. This saves the query as a text file with a .sql extension that can be opened and modified using any text editor.
7. Click **File** on the menu bar, and then click **Close** to close the Query window.

Other Query Builder Functions

Query Builder has capabilities for creating queries that involve sorting, grouping, or summing data. In the next query, you will list every term description, the call IDs for all courses offered during that term, the number of students who were enrolled or currently are enrolled for each course, and the total enrollment for each course and for each term. Figure 3-97 shows the query design diagram that includes the required tables, display fields, search fields, and join fields for this query.

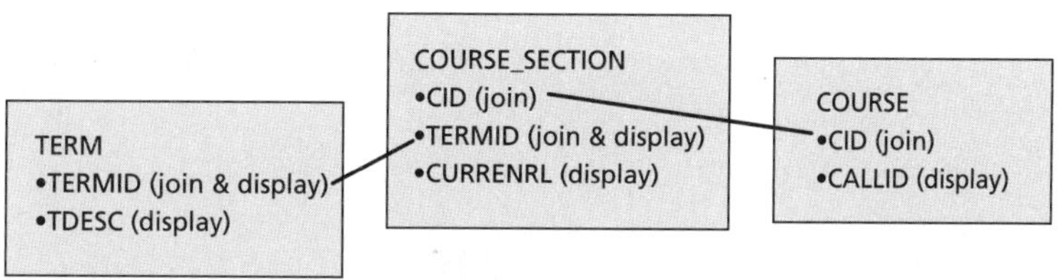

Figure 3-97: Query design diagram for Query Builder query

To create the query:

1 Click the **New** button on the toolbar to create a new Query Builder query. The Select Data Tables dialog box opens.

2 Click the **COURSE** table, press and hold down the **Ctrl** key, and then click the **COURSE_SECTION** and **TERM** tables. Click the **Include** button, wait a moment, and then click **Close**.

3 If necessary, click the **Maximize** button on the Query window, and then arrange the tables so you can clearly see the foreign key links.

4 Check the display fields in the following order: **TERMID** (in the Term table), **TDESC**, **CALLID**, and **CURRENRL**.

Remember that the fields will be displayed (and sorted) in the same order as they are selected.

Next, you want to specify the record order. The records need to be sorted by term and by call ID. You will specify how the records will be sorted next.

To create the record groupings in Query Builder:

1. Click **Results** on the menu bar, and then click **Group** to open the Group dialog box. Click **<your user name>.TERM.TERMID** as the first sort group, and then click the **Copy** button. Click **COURSE.CALLID** as the second sort group, and then click the **Copy** button again. TERMID is qualified by the user name and table name because it appears in multiple tables in the query. CALLID appears only in the COURSE table, so it does not need to be qualified by its user name and table name.
2. Click **OK** to close the Group dialog box.
3. Click the **Execute Query** button on the toolbar to run the query to check that the sorting is correct, as shown in Figure 3-98. The data are first sorted by TERMID, and then sorted by CALLID within TERMID.

records sorted by TERMID

records sorted by CALLID within TERMID

Query Builder - [Untitled3: Results]

File Edit Data Query Results Layout Window Help

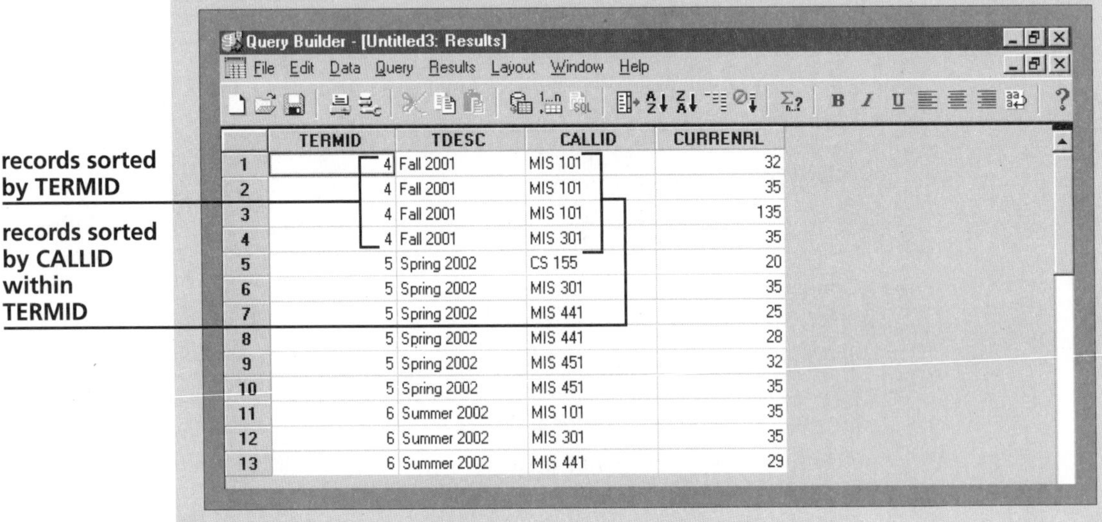

	TERMID	TDESC	CALLID	CURRENRL
1	4	Fall 2001	MIS 101	32
2	4	Fall 2001	MIS 101	35
3	4	Fall 2001	MIS 101	135
4	4	Fall 2001	MIS 301	35
5	5	Spring 2002	CS 155	20
6	5	Spring 2002	MIS 301	35
7	5	Spring 2002	MIS 441	25
8	5	Spring 2002	MIS 441	28
9	5	Spring 2002	MIS 451	32
10	5	Spring 2002	MIS 451	35
11	6	Summer 2002	MIS 101	35
12	6	Summer 2002	MIS 301	35
13	6	Summer 2002	MIS 441	29

Figure 3-98: Query output to check sorting

You need to sum the CURRENRL fields so that they show the total enrollment for each term and for each course. To do this, you will have to sum CURRENRL, and group the data on the TERMID and CALLID fields. To do this, you will create a field break. A **break** is similar to a SQL GROUP BY statement, and signifies that you want to perform a group function on a particular field.

To sum the CURRENRL field and create the required field GROUP BY breaks:

1 Click **Results** on the menu bar, and then click **Totals** to open the Totals dialog box. Click **CURRENRL** as the Select Column, and click the **Total** Summary Operations check box.

2 Click **OK** to close the Totals dialog box. The sums for the CURRENRL column should appear, as shown in Figure 3-99. Next, you can create the field breaks that show enrollments for class and term.

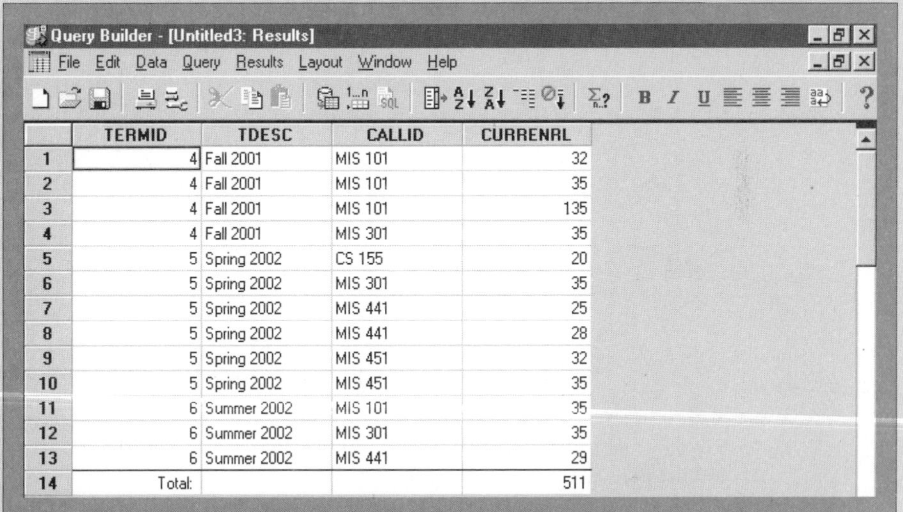

	TERMID	TDESC	CALLID	CURRENRL
1	4	Fall 2001	MIS 101	32
2	4	Fall 2001	MIS 101	35
3	4	Fall 2001	MIS 101	135
4	4	Fall 2001	MIS 301	35
5	5	Spring 2002	CS 155	20
6	5	Spring 2002	MIS 301	35
7	5	Spring 2002	MIS 441	25
8	5	Spring 2002	MIS 441	28
9	5	Spring 2002	MIS 451	32
10	5	Spring 2002	MIS 451	35
11	6	Summer 2002	MIS 101	35
12	6	Summer 2002	MIS 301	35
13	6	Summer 2002	MIS 441	29
14	Total:			511

Figure 3-99: Query output showing CURRENRL sum

3 Click **Results** on the menu bar, and then click **Break**. Click **<your user name>.TERM.TERMID** as the first break field, click the **Copy** button, click **CALLID** as the second break field, click the **Copy** button, and then click **OK**. The query output should look like Figure 3-100.

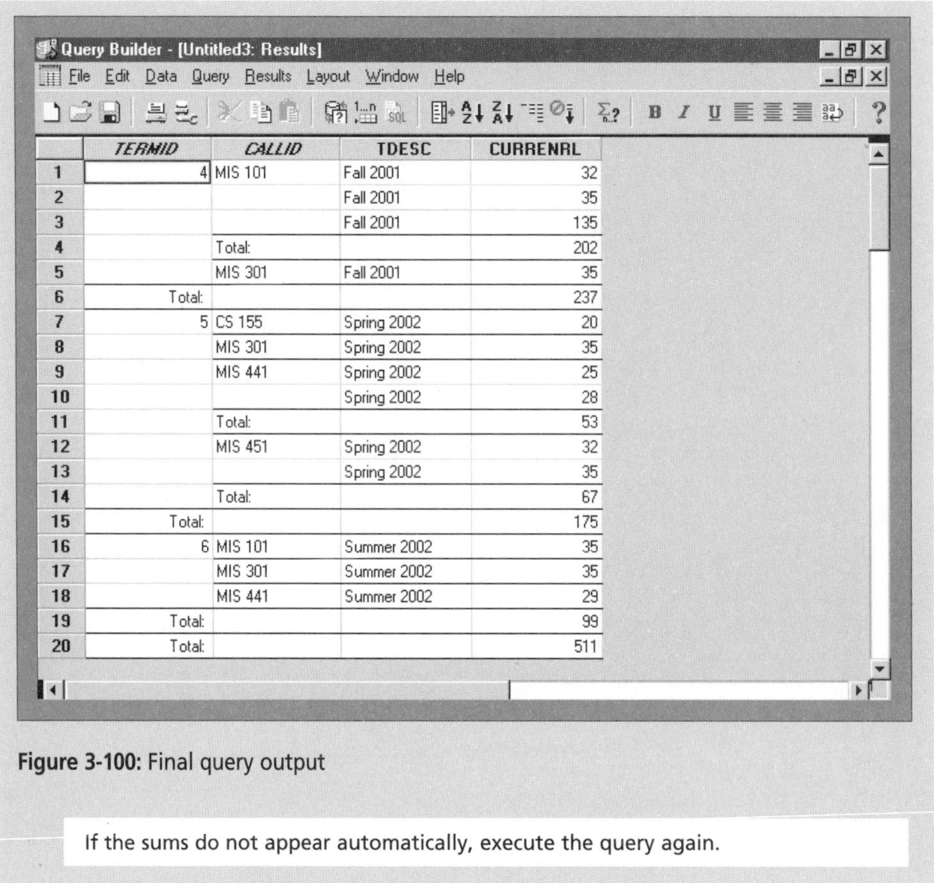

	TERMID	CALLID	TDESC	CURRENRL
1	4	MIS 101	Fall 2001	32
2			Fall 2001	35
3			Fall 2001	135
4		Total:		202
5		MIS 301	Fall 2001	35
6	Total:			237
7	5	CS 155	Spring 2002	20
8		MIS 301	Spring 2002	35
9		MIS 441	Spring 2002	25
10			Spring 2002	28
11		Total:		53
12		MIS 451	Spring 2002	32
13			Spring 2002	35
14		Total:		67
15	Total:			175
16	6	MIS 101	Summer 2002	35
17		MIS 301	Summer 2002	35
18		MIS 441	Summer 2002	29
19	Total:			99
20	Total:			511

Figure 3-100: Final query output

help If the sums do not appear automatically, execute the query again.

Although Query Builder allows you to create queries using functions such as sorting and grouping, it includes only some—not all—of these functions in the corresponding SQL code. Next, you will view the SQL code for your Query Builder query to see the group functions that are included in the SQL command.

To view the Query Builder SQL command:

1 Click **Query** on the menu bar, then click **Show SQL**. Note that the GROUP BY and ORDER BY commands are included in the SQL code, but the SUM command is not.

Now you will save your query and exit Query Builder and SQL*Plus.

To save your query and exit Query Builder:

1. Click the **Close** button to close the Show SQL window.
2. Click the **Save** button on the toolbar, and then save the query in the Chapter3 folder on your Student Disk as **term.brw**.
3. Click **File** on the menu bar, and then click **Exit** to exit Query Builder.
4. Close SQL*Plus.

SUMMARY

- To join any number of tables in a SELECT command, you must include every involved table in the FROM clause, and include a join condition for every link between tables in the WHERE clause.
- If you display a column in a multiple-table query that exists in more than one of the tables, you must qualify the column name by writing the table name, followed by a period, and then the column name.
- If you accidentally omit a join condition in a multiple-table query, it results in a Cartesian product, whereby every row in one table is joined with every row in the other table.
- A UNION query combines the results of two queries that are not related through foreign key relationships, and suppresses duplicate rows. A UNION ALL query displays duplicate rows. A union requires that both queries have the same number of display columns in the SELECT clause, and each column in the first query must have the same data type as the corresponding column in the second query.
- An INTERSECT query returns the intersection, or matching rows, in two unrelated queries.
- A MINUS query allows you to find the difference between two unrelated query result outputs.
- The SELECT ... FOR UPDATE command enables you to view and lock a record with a single command. You must COMMIT the transaction to unlock the record.
- A database view looks and acts like a table. It is created from a query and does not physically exist in the database, but it is derived from other database tables.
- Other users cannot view or modify your database tables unless you give them explicit privileges to do so. Table privileges can be given using the SQL GRANT command, and revoked using the REVOKE command.

- Sequences are sequential lists of numbers that are generated automatically by the database for creating unique surrogate key values for primary key fields.
- Once you move on to the next sequence value, the previous value cannot be accessed using the sequence commands. This prevents two users from accidentally using the same sequence value for a primary key.
- You have to grant explicit privileges to other users in order for them to use your sequences.
- Query Builder provides a graphical Oracle environment for creating SELECT queries.
- Query Builder queries can be saved in the file system or database, and they can be used to generate SQL commands.

REVIEW QUESTIONS

1. What is a join condition?
2. If the result of a query joining multiple tables is many more records than you expected, what probably happened?
3. If you are creating a query that joins four tables, how many join conditions will you need in the WHERE command?
4. Create a query design diagram like the one shown in Figure 3-57 for the following queries:
   a. Display CALLID, CNAME, SECNUM, DAY, and TIME for every course in the Spring 2002 term.
   b. Display FLNAME, FFNAME, FPHONE, DAY, and TIME for every faculty member who teaches in BUS Room 105 during the Spring 2002 term.
   c. Display the ITEMDESC, ITEMSIZE, and COLOR of every item ever ordered by customer Alissa Chang.
   d. Display the ITEMDESC, ITEMSIZE, COLOR, QUANTITY_EXPECTED, and DATE_EXPECTED for all unreceived shipments (DATE_RECEIVED is NULL) of every item where QOH = 0 in the INVENTORY table.
5. When do you need to use a union?
6. What is the difference between an INTERSECT and a MINUS query?
7. Give two reasons for creating database views.
8. When do you need to use the SELECT ... FOR UPDATE command?
9. What is the purpose of the NOWAIT statement in the SELECT ... FOR UPDATE command?
10. What is a sequence, and what is it used for?
11. Write the SQL command to create a sequence named PRACTICE_SEQUENCE that begins with 1 and has no maximum value.
12. Write the SQL command to access the next value in PRACTICE_SEQUENCE.

13. Which SQL group functions are translated into SQL code in Query Builder? Which function is not?
14. How can you specify that a privilege is to be granted to all database users?

PROBLEM-SOLVING CASES

Save all SQL*Plus queries in a text file named Ch3cNU.sql in the Chapter3 folder on your Student Disk. Save Query Builder queries in the Chapter3 folder on your Student Disk, using the specified filename with a .brw extension.

1. Use SQL*Plus to write a query to find the course names of all courses offered during the Summer 2002 term. (*Hint*: Use TDESC = 'Summer 2002' as a search condition.)
2. Use SQL*Plus to write a query to list the first and last name of every student whom faculty member Kim Cox advises. (*Hint*: Use FLNAME and FFNAME in your search conditions.)
3. Use SQL*Plus to write a query to list the course call ID, section number, term description, and grade for every course taken by student Ruben Sanchez for which a grade has been assigned. (*Hint*: Use SLNAME and SFNAME in your search conditions, and search for records where GRADE is NOT NULL.)
4. Use SQL*Plus to write a query to calculate the total number of students taught by John Blanchard during the Spring 2002 term. Sum the CURRENRL figures, and include FLNAME, FFNAME, and TDESC in your search condition.
5. Use SQL*Plus to create a query to calculate the total number of student credits generated during the Spring 2002 term (student credits = CURRENRL * course credits). Use DESC in your search condition.
6. Use SQL*Plus to create a query that lists the BLDG_CODE and ROOM of every room that is either currently in use as a faculty office or in use as a classroom during the Summer 2002 term. (*Hint*: You will need to use a UNION.)
7. Use SQL*Plus to create a view named TERM_VIEW that contains the CALLID, SECNUM, DAY, TIME, BLDG_CODE, and ROOM for all courses offered during the Summer 2002 term. Use 'Summer 2002' as a search condition. Then, write a query to view all of the fields in all of the view records.
8. Use SQL*Plus to write the commands to grant ALTER, DELETE, INSERT, UPDATE, and SELECT privileges to PUBLIC for your ENROLLMENT and COURSE_SECTION tables, and then write the commands to revoke all of these privileges.
9. Use Query Builder to create a query that lists the call ID, course name, and section number of all courses that were filled to capacity (CURRENRL = MAXENRL) during the Spring 2002 term. Use TDESC in your search condition. Save the query as Ch3cNU9.brw.
10. Use Query Builder to calculate the total credits earned so far by student Brian Umato. (*Hint:* Use SLNAME and SFNAME in your search conditions, and only return records where GRADE is not NULL.) Save the query as Ch3cNU10.brw.

Save all SQL*Plus queries in a text file named Ch3cCT.sql in the Chapter3 folder on your Student Disk. Save Query Builder queries in the Chapter3 folder on your Student Disk, using the specified filename with a .brw extension.

11. Use SQL*Plus to write a query to find the first and last name of every customer who has placed an order with Clearwater Traders using the Web site as a source. Use ORDERSOURCE in your search condition, and do not return any duplicate records.
12. Use SQL*Plus to write a query to find the ITEMID, ITEMDESC, ITEMSIZE, and COLOR of every item that is currently out of stock (QOH = 0).
13. Use SQL*Plus to write a query to find the ITEMDESC, ITEMSIZE, COLOR, DATE_EXPECTED, and QUANTITY_EXPECTED of every shipment that has not yet been received. (*Hint*: Use DATE_RECEIVED IS NULL as a search condition.)
14. Use SQL*Plus to write a query that finds the total revenue (ORDER_PRICE * QUANTITY) generated by all orders where the ORDERSOURCE is WEBSITE.
15. Use SQL*Plus to write a query that lists the INVID, ITEMDESC, ITEMSIZE, and COLOR of all items where QOH is currently 0, or items that have been ordered by customer Alissa Chang. Use CUSTOMER.FIRST and CUSTOMER.LAST in your search condition. (*Hint*: You will need to use a UNION.)
16. Create a view named INVENTORY_VIEW that contains CATEGORY, INVID, ITEMDESC, ITEMSIZE, COLOR, CURR_PRICE, and QOH for all Clearwater Traders inventory items. Then, write a query to display the item description, size, color, and quantity on hand for all records in the view.
17. Write the commands to grant ALTER, DELETE, INSERT, UPDATE, and SELECT privileges to PUBLIC for the ORDERLINE and COLOR tables, then write the commands to REVOKE these privileges.
18. Use Query Builder to create a query that lists the first and last names, addresses, cities, states, and ZIP codes of every customer who ordered items in the Women's Clothing category. Save your query as Ch3cCT8.brw.
19. Use Query Builder to write a query that displays the ORDERDATE, ITEMDESC, and QUANTITY of every item ordered by customer Alissa Chang. Use CUSTOMER.FIRST and CUSTOMER.LAST in your search condition. Save your query as Ch3cCT9.brw.
20. Use Query Builder to write a query that displays the INVID, ITEMDESC, ITEMSIZE, COLOR, and QUANTITY_EXPECTED for all shipments expected to arrive on 8/15/2001. (*Hint*: Use the TO_DATE function in the search condition.) Save your query as Ch3cCT10.brw.

CHAPTER 4

Introduction to PL/SQL, Triggers, and Procedure Builder

Introduction▶ A **procedural programming language** is a programming language that uses detailed, sequential instructions to process data. PL/SQL is the procedural language used to write procedures and functions for Oracle applications. A PL/SQL program combines SQL commands (such as SELECT and UPDATE) with procedural commands for tasks such as manipulating variable values, evaluating IF/THEN logic structures, and looping. This chapter presents an overview of the PL/SQL programming language. It assumes you have already used a procedural programming language such as Pascal, BASIC, COBOL, or FORTRAN.

LESSON A

objectives

- Learn about PL/SQL variables and data types
- View the structure of PL/SQL program blocks and their basic operations
- Concatenate and parse strings in PL/SQL
- Perform comparisons using IF/THEN selection structures

PL/SQL Variables and Data Types

Like SQL, PL/SQL variable names must begin with a letter and cannot contain more than 30 characters. Variables can contain letters, numbers, and the symbols $, _, and #. Hyphens and blank spaces are not permitted. Variables cannot be reserved words such as NUMBER, VALUES, BEGIN, or other words used in SQL or PL/SQL commands. Variable names cannot be the same as database table names. Variable names should be as descriptive as possible (such as StudentSID rather than X). Variable names are normally expressed in mixed uppercase and lowercase letters. PL/SQL capitalization styles are summarized in Figure 4-1.

Item Type	Capitalization	Example
Reserved words	Uppercase	BEGIN, DECLARE
Built-in functions	Uppercase	COUNT, TO_DATE
Predefined data types	Uppercase	VARCHAR2, NUMBER
SQL commands	Uppercase	SELECT, INSERT
Database objects	Lowercase	student, fid
Variable names	Mixed case	StudentFName, FacultyID

Figure 4-1: PL/SQL capitalization styles

PL/SQL data types and sample declarations are summarized in Figure 4-2.

Data Type	Usage	Sample Declaration
VARCHAR2	Variable-length character strings	StudentName VARCHAR2(30);
CHAR	Fixed-length character strings	StudentGender CHAR(1);
NUMBER	Floating, fixed-point, or integer numbers	CurrentPrice NUMBER(5, 2);
BINARY_INTEGER	Integers	CustID BINARY_INTEGER;
DATE	Dates	TodaysDate DATE;
BOOLEAN	True/False values	OrderFlag BOOLEAN;
%TYPE	Assumes the data type of a database field	CustAddress customer.cadd%TYPE;
%ROWTYPE	Assumes the data type of a database row	CustOrderRecord cust_order% ROWTYPE;

Figure 4-2: PL/SQL data types

The PL/SQL VARCHAR2 data type holds variable-length string data up to a maximum length of 32,767 characters. VARCHAR2 is different from its database counterpart, which can only hold a maximum of 4,000 characters when it is used as an Oracle8 database field. When you declare a VARCHAR2 variable in PL/SQL, you must also specify the maximum field width.

The CHAR data type is used for fixed-length character strings, to a maximum length of 32,767 characters. When you declare a CHAR variable, you must specify the maximum field width. When the variable is assigned a data value, if the entire field width is not filled by characters, the remainder is padded with blank spaces.

The NUMBER data type is identical to the Oracle database NUMBER data type, with the following general format: `NUMBER(<precision>, <scale>);`. The precision specifies the total length of the number, including decimal places; the scale specifies the number of digits to the right of the decimal place. When you declare a NUMBER variable, you specify the precision only for integer values, and both the precision and scale for fixed-point values. For floating-point values, you omit both the precision and scale specifications.

The BINARY_INTEGER data type can also be used to represent integer values. Data values are stored internally in binary format, which takes slightly less storage space than the NUMBER data type, and calculations can be performed on BINARY_INTEGER data values more quickly than on integer NUMBER values. The DATE data type is the same as its Oracle database counterpart, and stores both date and time values. The BOOLEAN data type is used to store True/False values, and is usually used as a flag. When a BOOLEAN variable is declared, it has a value of NULL until it is assigned a value of TRUE or FALSE.

Two other data types that are often used in PL/SQL are the reference data types %TYPE and %ROWTYPE, which assume the same data type as a referenced database column or row. The general format for a %TYPE data declaration is <variable name> <table name>.<field name>%TYPE;. Suppose you want to declare a variable named LName with the same data type as the FLNAME field in the FACULTY table in the Northwoods University database. The declaration would be written as LNAME FACULTY.FLNAME%TYPE;, and the field would assume a data type of VARCHAR2(30), as was used when the FLNAME field was declared in the CREATE TABLE command for the FACULTY table.

The ROWTYPE reference data type is useful when you want to declare a variable to hold an entire row of data. The general format for a %ROWTYPE data declaration is <row variable name> <table name>%ROWTYPE;. The following code will declare a variable that holds an entire row of data that was retrieved from the FACULTY table: FacRow FACULTY%ROWTYPE;. This variable would consist of all eight fields from the FACULTY table, and each would have the same data type as the associated database field.

PL/SQL Program Blocks

PL/SQL programs are structured in **blocks** with the following format:

```
DECLARE
  <variable declarations>
BEGIN
  <body containing procedure or function steps>
EXCEPTION
  <error-handling statements>
END;
```

PL/SQL is a **strongly typed language**, which means that all variables must be declared prior to use. Strong typing means assignments and comparisons can be performed only between variables with the same data type. PL/SQL program variables are declared in the program's DECLARE section. The general format for declaring a variable is <variable name> <data type>;. For example, suppose you want to create a variable named StudentID of data type NUMBER. The declaration would be written as StudentID NUMBER;. The **body** of a PL/SQL block consists of program statements (which can be assignments, conditional statements, looping statements, etc.) that lie between the BEGIN and EXCEPTION statements. The **exception section** is used for error handling. Each PL/SQL code statement must end with a semicolon. PL/SQL program lines can span many lines in a text editor, but when the code is compiled, the compiler considers everything up to an ending semicolon to be part of the same line of code. Forgetting to include a semicolon at the end of a code line will cause a compile error.

Comment statements are text comments within a computer program that do not contain program commands, but are used to explain or document a program step or series of steps. Comment statements must be marked, or **delimited**, so that the compiler does not try to interpret them as program commands. PL/SQL program comment statements can be delimited in two ways. A **block** of comments that spans several lines can begin with the symbols **/*** and end with the symbols ***/**. For example, the following code would be read by the compiler as a comment block:

```
/* Script:      orcl_cold_backup
   Purpose:     To perform a complete cold backup on the ORCL database
                instance
   Revisions: 9/8/2001 JM Script */
```

If a comment statement appears on a single line, you can delimit it by typing two hyphens at the beginning of the line, as shown in the following example:

```
DECLARE
--variable to hold current value of SID
StudentID    NUMBER;
```

PL/SQL Arithmetic Operators

The arithmetic operators used in PL/SQL are similar to those used in most programming languages. Figure 4-3 describes the PL/SQL arithmetic operators in the order in which they are evaluated. As with most programming languages, you can force operators to be evaluated in a specific order by placing the operation in parentheses.

Operator	Meaning	Example	Result
**	Exponentiation	2**3	8
*	Multiplication	2 * 3	6
/	Division	9/2	4.5
+	Addition	3 + 2	5
–	Subtraction	3 – 2	1
–	Negation	–5	negative 5

Figure 4-3: PL/SQL arithmetic operators

PL/SQL Assignment Statements

In programming, an assignment statement assigns a value to a variable. In PL/SQL, the assignment operator is **:=**. The variable that is being assigned the new value is placed on the left side of the assignment operator, and the new value is placed on the right side of the operator. The value can be a **literal**, which is an actual data value such as 'John', or another variable that has been assigned the desired value previously. Examples of both types of assignments are shown in Figure 4-4.

Description	Example
Variable assigned to literal value	SName := 'John';
Variable assigned to another variable	SName := CurrentStudentName;

Figure 4-4: PL/SQL assignment statements

You can use an assignment statement within a variable declaration to assign an initial value to a new variable. For example, to declare a variable named StudentID with data type NUMBER and assign it an initial value of 100, you would write the variable declaration as `StudentID NUMBER := 100;`.

PL/SQL Interactive Output

Normally you use PL/SQL within another Oracle development environment, such as Form Builder, which enables you to develop Windows applications that allow users to easily insert, update, view, and delete data. In the SQL*Plus environment, you can use the DBMS_OUTPUT.PUT_LINE function to display output within a PL/SQL program. Whenever you start a new SQL*Plus session and plan to use this function, you must first issue the command `SET SERVEROUTPUT ON`, which activates the **internal buffer**, or memory area that stores the values that are input and output. If you don't do this, the DBMS_OUTPUT commands will not work. The general format of this command is `DBMS_OUTPUT.PUT_LINE(<text to be displayed>);`. This command should just be used for illustration and debugging purposes, and is not meant as a way to create user applications or reports. Keep in mind that the PUT_LINE command can handle only one line (maximum 255 characters) of text data.

Writing a PL/SQL Program

Now that you have learned the basics of the PL/SQL programming language, you are ready to write a simple PL/SQL program that will be executed in SQL*Plus. First, you will start SQL*Plus, refresh your database tables, and enable the DBMS_OUTPUT function.

To start SQL*Plus, refresh your database tables, and enable the DBMS_OUTPUT function:

1. Start SQL*Plus.
2. Run the clearwat.sql and northwoo.sql scripts from the Chapter4 folder on your Student Disk to refresh your database tables.
3. Type **SET SERVEROUTPUT ON** at the SQL prompt to enable the DBMS_OUTPUT function.

Now you will write a PL/SQL program that will declare a DATE variable named TodaysDate. The program will then set TodaysDate equal to the current system date, and display the current date as output. It is a good programming practice to place the DECLARE, BEGIN, and END commands flush with the left edge of the window, and then indent the commands within each section. The blank spaces do not affect the program's functionality, but make it easier to read and understand. You can indent the lines by pressing the Tab key, or by pressing the spacebar to add blank spaces.

To write a PL/SQL program:

1. Start Notepad and type the program code shown in Figure 4-5. Save your Notepad file as Ch4aPLSQL1.txt in the Chapter4 folder on your Student Disk.

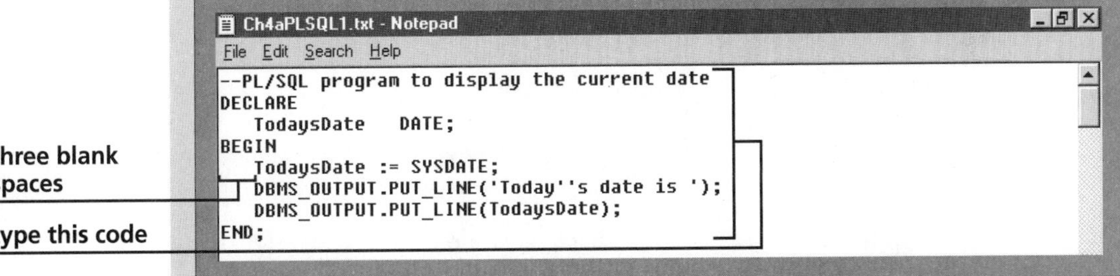

Figure 4-5: PL/SQL program code

2. Copy the text, and then paste it into SQL*Plus. Press the **Enter** key after the last line of the program, type **/**, and then press the **Enter** key again to execute the program. The output is shown in Figure 4-6. If your program doesn't run correctly, go to the section on Debugging PL/SQL Programs later in this lesson.

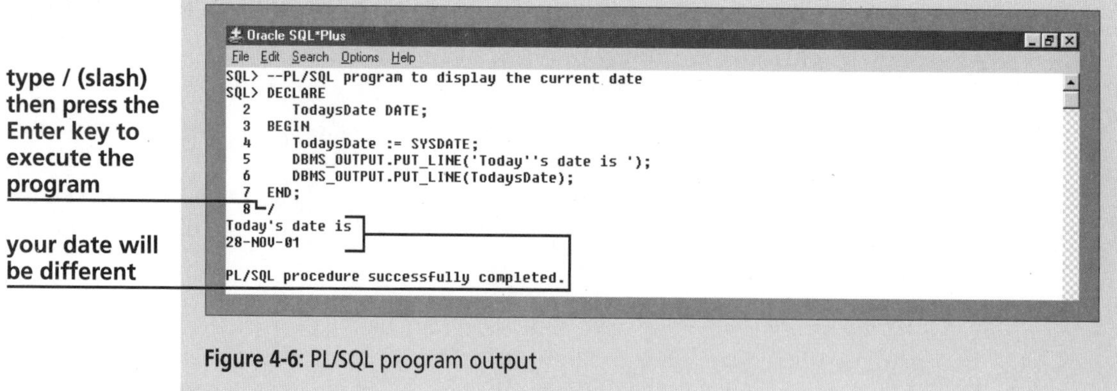

Figure 4-6: PL/SQL program output

Data Type Conversion Functions

Sometimes you need to convert a variable from one data type to another data type. For example, if the value '2' is input as a character variable, it must be converted to a number before you can use it to perform arithmetic calculations. PL/SQL performs **implicit data conversions**, and automatically converts one data type to another when it makes sense to do so. For example, in the statement `DBMS_OUTPUT.PUT_LINE(TodaysDate);` in your previous PL/SQL program, the DATE variable TodaysDate was automatically converted to a text string for output. However, it is very risky to rely on implicit conversions, because their output is unpredictable. Therefore, you should always perform **explicit data conversions**, which are data conversions performed using specific built-in conversion functions. The PL/SQL data conversion functions are listed in Figure 4-7.

Function	Description	Example
TO_DATE	Converts a character string to a date	TO_DATE('07/14/01', 'MM/DD/YY');
TO_NUMBER	Converts a character string to a number	TO_NUMBER('2');
TO_CHAR	Converts either a number or a date to a character string	TO_CHAR(2); TO_CHAR(SYSDATE);

Figure 4-7: PL/SQL data conversion functions

To convert a date to a character string, you use the TO_CHAR function, which has the following general format: TO_CHAR(<date value>, '<date format mask >'). For example, to convert the current SYSDATE to a character string with the format MM/DD/YYYY, you would use the following command:

```
TO CHAR (sysdate, 'MM/DD/YYYY');
```

Recall that to convert a character string to a date, you use the TO_DATE function, which has the following general format: TO_DATE(<character string representing date>, '<date format mask>'). Common date format masks are listed in Chapter 3.

Handling Character Strings

Manipulating **character strings**, which are character data values that consist of more than one character, is an important topic in any programming language. For example, you might want to create a single data field by **concatenating**, or joining, two separate character strings. Or, you might want to take a single character string consisting of two data items separated by commas and **parse**, or separate, it into two individual character strings. SQL*Plus has a variety of string-handling functions to perform these tasks.

Concatenating Character Strings

Suppose you are working on a PL/SQL program and have the following variable names and values:

Variable Name	Value
SFirstName	Sarah
SLastName	Miller

You would like to combine these two values into one variable named SFullName that has the value "Sarah Miller." To concatenate, or join, two character strings in PL/SQL, you use the double bar (||) operator. The command to concatenate these values is `SFullName := SFirstName || SLastName;`.

That was pretty simple, but there is a catch. This command puts the two character strings together with no spaces between them, so the value of SFullName is now "SarahMiller," which is not what you wanted. You need to insert a blank space between the first and last name. To do this, use the same command, but concatenate a character string consisting of a blank space between the two variable names: `SFullName := SFirstName || ' ' || SLastName;`.

Literals are unchanging data values coded directly into a program, such as 3.14159 for the value of pi, or 'Customer Name:' for a column heading. You must always enclose a literal value that is a character string, or a string literal, within single quotation marks. A string literal can enclose any combination of valid PL/SQL characters ('Student Name:' for example). If a single quotation mark is used within a string literal, you must type two single quotation marks (for example, 'Sarah''s Computer'). Suppose you saved the values for a building code, room, and capacity in three variables, as shown next.

Variable Name	Data Type	Value
BuildingCode	VARCHAR2	CR
RoomNum	VARCHAR2	101
RoomCapacity	NUMBER	150

You need to display a message that reads "CR Room 101 has 150 seats." However, the ROOM_CAPACITY value is a NUMBER data type and not a CHARACTER data type. You must use the TO_CHAR function to convert the number to a character before you can use it in a concatenation operation. The required code to create this character string (which is saved in a variable named RoomMessage) is:

```
RoomMessage := BuildingCode || ' Room '|| RoomNum || ' has ' ||
TO_CHAR(RoomCapacity) ||' seats.';
```

Another tricky part of this operation is remembering to add spaces before and after the variable values so that the strings don't run together. Now you will modify the PL/SQL program you wrote so that it displays the header message 'Today is' and the date as a single concatenated string. You will need to convert the TodaysDate variable to a character data type.

To modify the PL/SQL program to display the date as a single string:

1 Switch back to Notepad, and change your program code so it looks like the code shown in Figure 4-8. Do not type the line numbers, because they will be inserted when you copy the code into SQL*Plus. Save your modified file as Ch4aPLSQL2.txt.

2 Copy the text, and then paste it into SQL*Plus. Press the **Enter** key after the last line of the program, type **/**, and then press the **Enter** key again to execute the program. The output is shown in Figure 4-8. If your program does not execute correctly, go to the section on Debugging PL/SQL Programs later in the lesson.

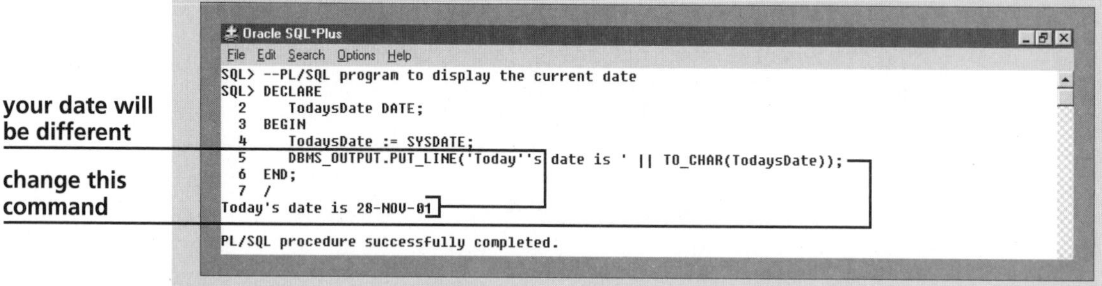

your date will be different

change this command

```
Oracle SQL*Plus
File Edit Search Options Help
SQL> --PL/SQL program to display the current date
SQL> DECLARE
       TodaysDate DATE;
     BEGIN
       TodaysDate := SYSDATE;
       DBMS_OUTPUT.PUT_LINE('Today''s date is ' || TO_CHAR(TodaysDate));
     END;
     /
Today's date is 28-NOV-01

PL/SQL procedure successfully completed.
```

Figure 4-8: PL/SQL program with output concatenated on a single line

Placing String Output on a New Line

Now suppose you are formatting several room records and you want each new room location to print on a new line. The ASCII characters for carriage return (ASCII 13) and line feed (ASCII 10) signal the beginning of a new line. The CHR function is used to convert ASCII numerical codes to characters that can be inserted into text strings using the || concatenation operator. If you concatenate the string `CHR(13) || CHAR(10)` to the end of a line, the DBMS_OUTPUT.PUT_LINE function will place the output of the next line on a new line on the screen. The previous room capacity example, modified to start each room on a new line, would be written as:

```
RoomMessage := BuildingCode || ' Room ' || RoomNum || ' has ' ||
TO_CHAR(RoomCapacity) ||' seats.' || CHR(13) || CHR(10);
```

You cannot use the CHR(13) || CHR(10) new line characters in SQL*Plus, because the DBMS_OUTPUT.PUT_LINE function only supports one line of output per command. However, you can use this function in other Oracle development environments.

Removing Blank Trailing Spaces from Strings

Sometimes when you store retrieved data values in variables or convert numeric or date values to characters, the data values contain blank spaces that pad out the value to its maximum column width. Suppose you have a variable named CustAddress, which is a CHAR data type of size 20. Its current value is '2103 First St   ' (13 characters followed by seven spaces). To remove all spaces from the right side of the variable named CustAddress, use the RTRIM function, as follows:

```
CustAddress := RTRIM(CustAddress);
```

Finding the Length of Character Strings

The LENGTH function returns an integer that is the length of a character string. Suppose you have a variable named BuildingCode, and its current value is 'CR'. To find the length of the string, you first declare an integer variable named CodeLength, and then enter the command `CodeLength := LENGTH(BuildingCode)`. Since the value of BuildingCode is 'CR', the LENGTH function would return the number 2 to the CodeLength variable. If the variable contains additional padded spaces to the right, these also will be counted in the length. For example, if BuildingCode is retrieved from a CHAR type database field of size six, the value of BuildingCode will be 'CR   ' (CR followed by four spaces), and the LENGTH function will return the number 6.

Character String Case Functions

PL/SQL has functions to convert character strings from lowercase to uppercase, and vice versa. The UPPER function converts lowercase or mixed-case characters to all uppercase characters. For example, if a variable named SFullName stores the value 'Sarah Miller', the following code will then convert the SFullName value to all uppercase characters: `SFullName := UPPER(SFullName);` and change the variable value to 'SARAH MILLER'.

Similarly, the LOWER function converts uppercase or mixed-case characters to all lowercase characters. Using the previous example's data values, `SFullName := LOWER(SFullName);` would change the variable value to 'sarah miller'.

Next, you will add the LOWER and LENGTH functions to your PL/SQL program so that it displays the current day of the week, and the number of characters in the name of the current day. You will declare a VARCHAR2 variable named Today that is assigned to the current day of the week, and another variable named DayLength that will be assigned to the length of the current day's character string. The program will then convert the day of the week to lowercase characters, trim the blank trailing spaces from the day using the RTRIM function, and display the day and the length of the name of the day.

To modify your PL/SQL program to use the LOWER and LENGTH functions:

1. Switch back to Notepad, and change your program code so it looks like the code shown in Figure 4-9. Save your modified file as Ch4aPLSQL3.txt.
2. Copy the text, and then paste it into SQL*Plus. Press the **Enter** key after the last line of the program, type /, and then press the **Enter** key again to execute the program. The output is shown in Figure 4-9. If your program does not execute correctly, go to the section on Debugging PL/SQL Programs later in the lesson.

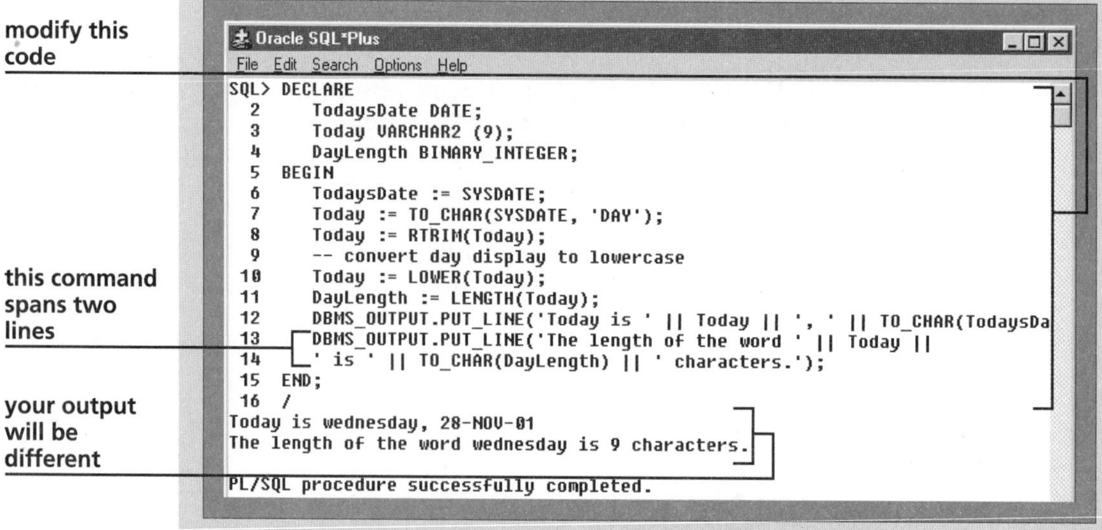

modify this code

this command spans two lines

your output will be different

Oracle SQL*Plus

File Edit Search Options Help

```
SQL> DECLARE
       TodaysDate DATE;
       Today VARCHAR2 (9);
       DayLength BINARY_INTEGER;
     BEGIN
       TodaysDate := SYSDATE;
       Today := TO_CHAR(SYSDATE, 'DAY');
       Today := RTRIM(Today);
       -- convert day display to lowercase
       Today := LOWER(Today);
       DayLength := LENGTH(Today);
       DBMS_OUTPUT.PUT_LINE('Today is ' || Today || ', ' || TO_CHAR(TodaysDa
       DBMS_OUTPUT.PUT_LINE('The length of the word ' || Today ||
       ' is ' || TO_CHAR(DayLength) || ' characters.');
     END;
     /
Today is wednesday, 28-NOV-01
The length of the word wednesday is 9 characters.

PL/SQL procedure successfully completed.
```

Figure 4-9: PL/SQL program using LOWER, LENGTH and RTRIM functions

The INSTR and SUBSTR String Functions

The INSTR and SUBSTR functions often are used together to process and modify strings. The INSTR function searches one string and looks for a matching substring. If it finds a matching substring, the function returns the starting position of the substring within the original string as an integer. If a matching substring is not found, the function returns the value 0.

The general format of the INSTR function is: <string starting position)>:= INSTR(<string being searched>, <string being looked for>);. The following code example uses the INSTR function to return the starting position of the single space in the SFullName variable that currently contains the value 'Sarah Miller', and returns the value to an integer number variable named BlankPosition: BlankPosition := INSTR(SFullName, ' ');. If the SFullName string variable contains the value 'Sarah Miller', BlankPosition will be set to 6.

The SUBSTR function extracts a specific number of characters from a character string, starting at a given point. The general format of the SUBSTR function is: <extracted string> := SUBSTR(<string being searched>, <starting point of string to be extracted>, <number of characters to extract>);. To extract the string 'Sarah' from the string of the SFullName variable and set it equal to a character variable named SFirstName, you would use the following code: SFirstName := SUBSTR(SFullname, 1, 5);. The 1 represents the starting place (the first character in the string), and the 5 represents the number of characters to extract (five, one for every letter in 'Sarah').

To use the INSTR and SUBSTR functions together to return the student's first name (up to the blank space) in a single command, you could use the INSTR function within the SUBSTR function. You must subtract 1 from the result of the INSTR function, since you only want to return the characters up to, but not including, the blank space. You would use the following code: SFirstName := SUBSTR(SFullname, 1, (INSTR(SFullname, ' ')- 1));.

Now, you will modify your PL/SQL program so that it will use the INSTR function to display the position in the name of the day of the week where the substring 'day' starts. It will also use the SUBSTR function to extract the characters in the day of the week that occur before the substring 'day'. The SUBSTR function will start on the first character of the string. The length of the string to be extracted will be the length of the string minus 3 (which is the number of characters in 'day'). You will declare a variable named PositionOfDay to hold the position where the substring 'day' starts, and a variable named StringBeforeDay to contain the substring of the name of the day of the week.

To modify your PL/SQL program to use the INSTR and SUBSTR functions:

1. Switch back to Notepad, and change your program code so it looks like the code shown in Figure 4-10. Save your modified file as Ch4aPLSQL4.txt.
2. Copy the text, and then paste it into SQL*Plus. Press the **Enter** key after the last line of the program, and then type /, and then press the **Enter** key again to execute the program. The output is shown in Figure 4-10. If your program does not execute correctly, go to the section on Debugging PL/SQL Programs later in the lesson.

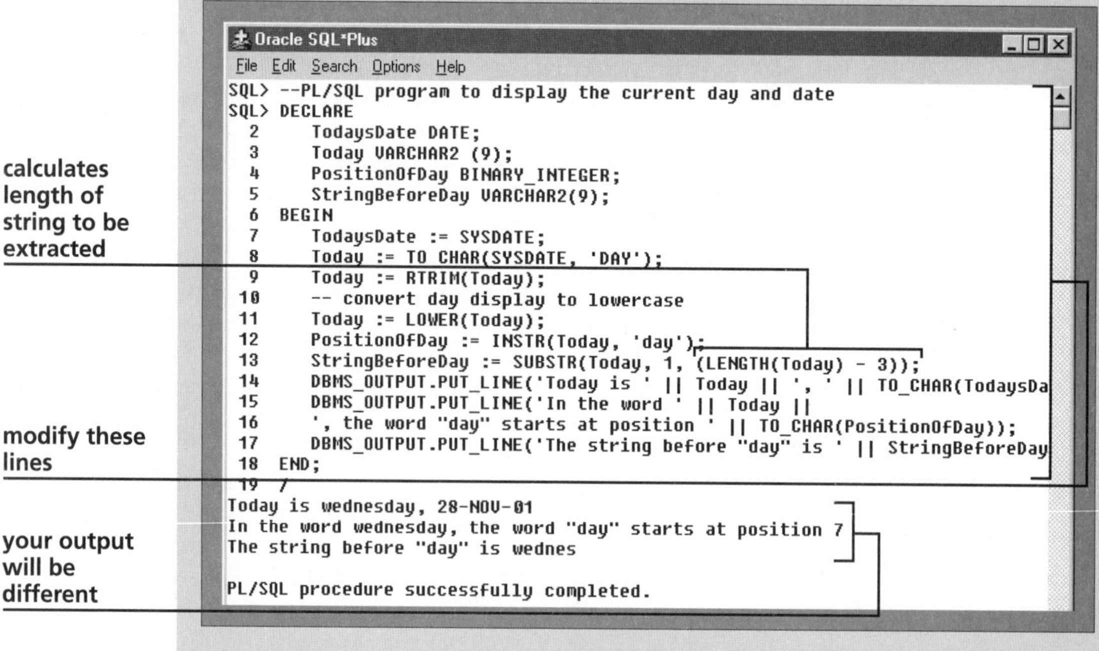

Oracle SQL*Plus

File Edit Search Options Help

```
SQL> --PL/SQL program to display the current day and date
SQL> DECLARE
       TodaysDate DATE;
       Today VARCHAR2 (9);
       PositionOfDay BINARY_INTEGER;
       StringBeforeDay VARCHAR2(9);
     BEGIN
       TodaysDate := SYSDATE;
       Today := TO_CHAR(SYSDATE, 'DAY');
       Today := RTRIM(Today);
       -- convert day display to lowercase
       Today := LOWER(Today);
       PositionOfDay := INSTR(Today, 'day');
       StringBeforeDay := SUBSTR(Today, 1, (LENGTH(Today) - 3));
       DBMS_OUTPUT.PUT_LINE('Today is ' || Today || ', ' || TO_CHAR(TodaysDa
       DBMS_OUTPUT.PUT_LINE('In the word ' || Today ||
       ', the word "day" starts at position ' || TO_CHAR(PositionOfDay));
       DBMS_OUTPUT.PUT_LINE('The string before "day" is ' || StringBeforeDay
     END;
     /
Today is wednesday, 28-NOV-01
In the word wednesday, the word "day" starts at position 7
The string before "day" is wednes

PL/SQL procedure successfully completed.
```

calculates length of string to be extracted

modify these lines

your output will be different

Figure 4-10: PL/SQL program using INSTR and SUBSTR functions

Debugging PL/SQL Programs

Programming errors usually fall into one of two categories: syntax and logic. A **syntax error** is an error that does not follow the guidelines of the programming language, and results in a compile error message. **Logic errors** are errors in the program logic that do not stop a program from compiling, but cause incorrect output. For example, in the previous exercise, you used the following program line to find the string before 'day' in the day of the current week:

```
StringBeforeDay := SUBSTR(Today, 1, (LENGTH(Today) - 3));
```

Suppose that for the last argument in the SUBSTR function, which is the number of characters to extract from the string, you used the expression `(LENGTH(Today – 2));`. Instead of returning the string representing the current day minus 'day', the SUBSTR function would return the current day minus 'ay', and the output would look like Figure 4-11. The program compiled and ran correctly, but gave an incorrect output.

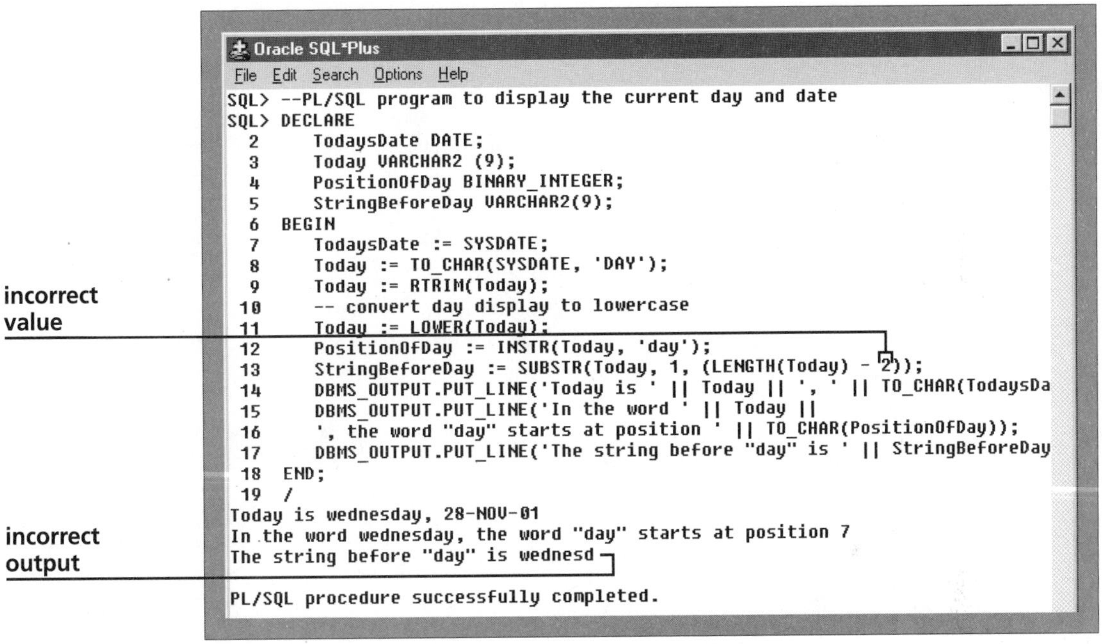

Oracle SQL*Plus

File Edit Search Options Help

```
SQL> --PL/SQL program to display the current day and date
SQL> DECLARE
       TodaysDate DATE;
       Today VARCHAR2 (9);
       PositionOfDay BINARY_INTEGER;
       StringBeforeDay VARCHAR2(9);
     BEGIN
       TodaysDate := SYSDATE;
       Today := TO_CHAR(SYSDATE, 'DAY');
       Today := RTRIM(Today);
       -- convert day display to lowercase
       Today := LOWER(Today);
       PositionOfDay := INSTR(Today, 'day');
       StringBeforeDay := SUBSTR(Today, 1, (LENGTH(Today) - 2));
       DBMS_OUTPUT.PUT_LINE('Today is ' || Today || ', ' || TO_CHAR(TodaysDa
       DBMS_OUTPUT.PUT_LINE('In the word ' || Today ||
       ', the word "day" starts at position ' || TO_CHAR(PositionOfDay));
       DBMS_OUTPUT.PUT_LINE('The string before "day" is ' || StringBeforeDay
     END;
     /
Today is wednesday, 28-NOV-01
In the word wednesday, the word "day" starts at position 7
The string before "day" is wednesd

PL/SQL procedure successfully completed.
```

incorrect value

incorrect output

Figure 4-11: Logic error example

The following paragraphs present examples of common errors, and tips for debugging.

Syntax Errors

Syntax errors, also called compile errors, might involve misspelling a reserved word; omitting a required character such as a parenthesis, single quote, or semicolon; or using a built-in function improperly. The line number and character location of these errors are flagged by the compiler, and an error code and message are displayed. Figure 4-12 shows an example of a compile error where the semicolon was omitted from the Today variable declaration. The ORA-06550 error message indicates that a compile error has occurred, and reports the line number and location. The PLS-00103 error message is a specific PL/SQL error message indicating that the compiler was expecting the semicolon end-of-line marker. You can find explanations and solution strategies for PLS error codes just as you do for ORA- error codes: click the Start button on the Windows taskbar, point to Programs, point to Oracle for Windows 95 or Windows NT, and then click Oracle8 Error Messages.

Recall that another way to start Oracle8 Error Messages is to start Windows Explorer, change to the ORAWIN95\MSHELP (or ORANT\MSHELP) directory, and then double-click the ora.hlp file.

Sometimes the location flagged for a syntax error is not the line where the error actually occurs. For example, Figure 4-12 shows an error where a semicolon is omitted after the Today variable declaration. The error is flagged on Line 4, but the error actually occurs on Line 3. When you receive an error message and the flagged line appears correct, look on lines preceding the flagged line.

actual error on Line 3

error flagged on Line 4

Oracle SQL*Plus

File Edit Search Options Help

```
SQL> --PL/SQL program to display the current day and date
SQL> DECLARE
  2     TodaysDate DATE;
  3     Today VARCHAR2 (9)
  4     DayLength BINARY_INTEGER;
  5  BEGIN
  6     TodaysDate := SYSDATE;
  7     Today := TO_CHAR(SYSDATE, 'DAY');
  8     -- convert day display to lowercase
  9     Today := LOWER(Today);
 10     DayLength := LENGTH(Today);
 11     DBMS_OUTPUT.PUT_LINE('Today is ' || Today || ', ' || TO_CHAR(TodaysDate));
 12     DBMS_OUTPUT.PUT_LINE('The length of the word ' || Today ||
 13     ' is ' || TO_CHAR(DayLength) || ' characters.');
 14  END;
 15  /
   DayLength BINARY_INTEGER;
   *
ERROR at line 4:
ORA-06550: line 4, column 4:
PLS-00103: Encountered the symbol "DAYLENGTH" when expecting one of the
following:
:= ; not null default character
The symbol ";" was substituted for "DAYLENGTH" to continue.
```

Figure 4-12: Compile error flagged on line after actual error

Another common error message is the one seen in Figure 4-13: 'Encountered the symbol ";" when expecting one of the following: ...'. This indicates that the compiler was expecting another character when it encountered the end-of-statement semicolon. In this case, the closing parenthesis around the DBMS_OUTPUT.PUT_LINE command was omitted. You must carefully examine the line where the error was flagged and determine which character was omitted.

should have two closing parentheses

error message indicating missing character

```
Oracle SQL*Plus
File Edit Search Options Help
SQL> --PL/SQL program to display the current day and date
SQL> DECLARE
  2     TodaysDate DATE;
  3     Today VARCHAR2 (9);
  4     DayLength BINARY_INTEGER;
  5  BEGIN
  6     TodaysDate := SYSDATE;
  7     Today := TO_CHAR(SYSDATE, 'DAY');
  8     -- convert day display to lowercase
  9     Today := LOWER(Today);
 10     DayLength := LENGTH(Today);
 11     DBMS_OUTPUT.PUT_LINE('Today is ' || Today || ', ' || TO_CHAR(TodaysDate);
 12     DBMS_OUTPUT.PUT_LINE('The length of the word ' || Today ||
 13     ' is ' || TO_CHAR(DayLength) || ' characters.');
 14  END;
 15  /
   DBMS_OUTPUT.PUT_LINE('Today is ' || Today || ', ' || TO_CHAR(TodaysDate);
                                                                           *
ERROR at line 11:
ORA-06550: line 11, column 76:
PLS-00103: Encountered the symbol ";" when expecting one of the following:
. ( ) , * % & | = - + < / > in mod not rem => ..
an exponent (**) <> or != or ~= >= <= <> and or like between
using is null is not || is dangling
The symbol ")" was substituted for ";" to continue.
```

Figure 4-13: Missing character compile error

Sometimes errors are very difficult to find visually, and the compiler does not specify the exact error location. For example, Figure 4-14 displays the error message "quoted string not properly terminated," but does not specify the line or column location.

```
Oracle SQL*Plus
File Edit Search Options Help
SQL> --PL/SQL program to display the current day and date
SQL> DECLARE
  2     TodaysDate DATE;
  3     Today VARCHAR2 (9);
  4     DayLength BINARY_INTEGER;
  5  BEGIN
  6     TodaysDate := SYSDATE;
  7     Today := TO_CHAR(SYSDATE, 'DAY');
  8     -- convert day display to lowercase
  9     Today := LOWER(Today);
 10     DayLength := LENGTH(Today);
 11     DBMS_OUTPUT.PUT_LINE('Today is ' || Today || ', ' || TO_CHAR(TodaysDate));
 12     DBMS_OUTPUT.PUT_LINE('The length of the word ' || Today ||
 13      is ' || TO_CHAR(DayLength) || ' characters.');
 14  END;
 15  /
ERROR:
ORA-01756: quoted string not properly terminated
```

Figure 4-14: Compile error with location not specified

If you cannot find a compile error visually, or if the location is not specified, you need to try to systematically determine which line is generating it. A useful debugging technique for isolating program errors is to **comment out** a program line, which means to change it so the compiler treats it as a comment statement and does not compile or execute it. In the program in Figure 4-14, only character strings are used on the DBMS_OUTPUT commands, so chances are the error is occurring on one of them. We will show an example of how to comment out the second DBMS_OUTPUT command to determine if the first or second DBMS_OUTPUT command is causing the error. Figure 4-15 shows the program with the second DBMS_OUTPUT command commented out. No error message is generated, so this isolates the error—it is occurring somewhere in the second DBMS_OUTPUT command, on Line 12 or Line 13.

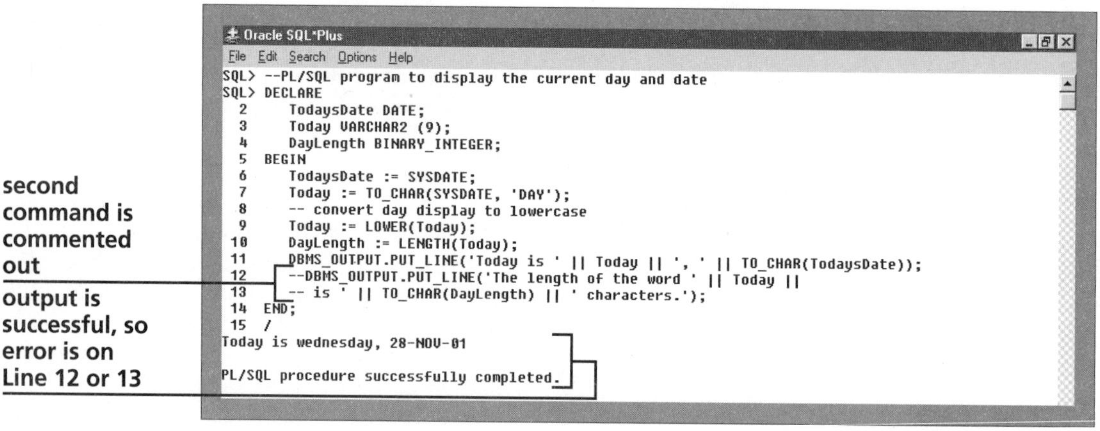

second command is commented out

output is successful, so error is on Line 12 or 13

```
Oracle SQL*Plus
File Edit Search Options Help
SQL> --PL/SQL program to display the current day and date
SQL> DECLARE
  2     TodaysDate DATE;
  3     Today VARCHAR2 (9);
  4     DayLength BINARY_INTEGER;
  5  BEGIN
  6     TodaysDate := SYSDATE;
  7     Today := TO_CHAR(SYSDATE, 'DAY');
  8     -- convert day display to lowercase
  9     Today := LOWER(Today);
 10     DayLength := LENGTH(Today);
 11     DBMS_OUTPUT.PUT_LINE('Today is ' || Today || ', ' || TO_CHAR(TodaysDate));
 12     --DBMS_OUTPUT.PUT_LINE('The length of the word ' || Today ||
 13     -- is ' || TO_CHAR(DayLength) || ' characters.');
 14  END;
 15  /
Today is wednesday, 28-NOV-01

PL/SQL procedure successfully completed.
```

Figure 4-15: Isolating the error by commenting out the second DBMS_OUTPUT command

Now we need to determine if the error is occurring on Line 12 or on Line 13. We will break the DBMS_OUTPUT command into two lines to determine if the error is occurring in the first or second line. Figure 4-16 shows how the command on Line 12 is ended properly by deleting the concatenation (||) operator and adding the end-of-function parenthesis and a semicolon. Line 13 remains commented out. The program again executes successfully, so the error is on Line 13.

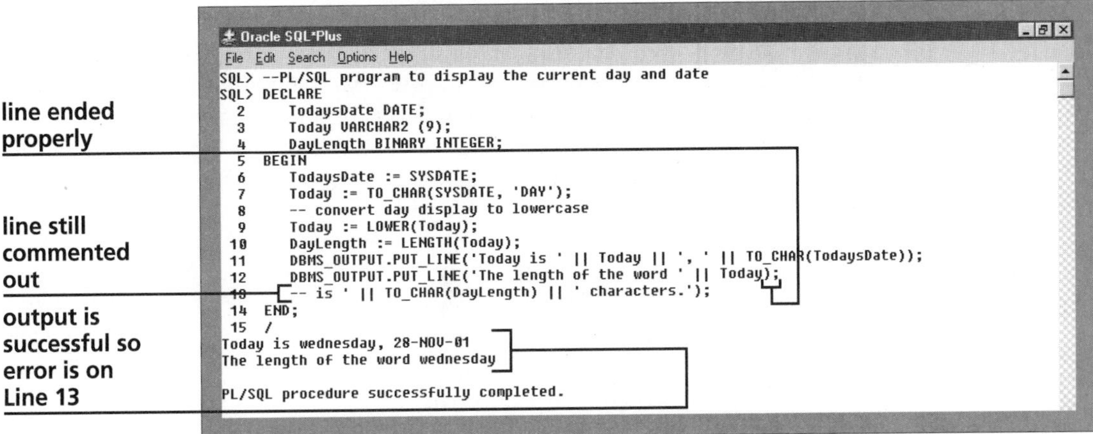

Figure 4-16: Isolating the error further by dividing the second DBMS_OUTPUT command

A careful examination of Line 13 locates the error: the initial single quote on the word 'is' was omitted at the beginning of the line.

Figure 4-17 shows an example of an error that is generated when a built-in function is used improperly. The TO_CHAR data conversion function is applied to the Today variable, but Today is already a character data field, so it cannot be converted, and the resulting error message is displayed.

line containing error

error location

```
Oracle SQL*Plus
File Edit Search Options Help
SQL> --PL/SQL program to display the current day and date
SQL> DECLARE
  2     TodaysDate DATE;
  3     Today VARCHAR2 (9);
  4     DayLength BINARY_INTEGER;
  5  BEGIN
  6     TodaysDate := SYSDATE;
  7     Today := TO_CHAR(SYSDATE, 'DAY');
  8     Today := TO_CHAR(Today);
  9     -- convert day display to lowercase
 10     Today := LOWER(Today);
 11     DayLength := LENGTH(Today);
 12     DBMS_OUTPUT.PUT_LINE('Today is ' || Today || ', ' || TO_CHAR(TodaysDate));
 13     DBMS_OUTPUT.PUT_LINE('The length of the word ' || Today ||
 14     ' is ' || TO_CHAR(DayLength) || ' characters.');
 15  END;
 16  /
DECLARE
*
ERROR at line 1:
ORA-06550: line 8, column 13:
PLS-00307: too many declarations of 'TO_CHAR' match this call
ORA-06550: line 8, column 4:
PL/SQL: Statement ignored
```

Figure 4-17: Error resulting from using the TO_CHAR function improperly

Sometimes one compile error can generate many more errors. For example, Figure 4-18 shows a program where DECLARE is misspelled. As a result, none of the variable declarations are processed, and several compile errors are generated. A good debugging strategy is to locate and fix the first compile error, and then rerun the program. Don't try to fix all of the compile errors before rerunning the program, because one error might generate several others, and fixing that one error may correct the others as well.

spelling error

additional errors generated as a result of first error

```
Oracle SQL*Plus
File Edit Search Options Help
SQL> --PL/SQL program to display the current day and date
SQL> DECALRE
unknown command "DECALRE" - rest of line ignored.
SQL>    TodaysDate DATE;
unknown command beginning "TodaysDate..." - rest of line ignored.
SQL>    Today VARCHAR2 (9);
unknown command beginning "Today VARC..." - rest of line ignored.
SQL>    DayLength BINARY_INTEGER;
unknown command beginning "DayLength ..." - rest of line ignored.
For a list of known commands enter HELP
and to leave enter EXIT.
SQL> BEGIN
  2     TodaysDate := SYSDATE;
  3     Today := TO_CHAR(SYSDATE, 'DAY');
  4     Today := TO_CHAR(Today);
  5     -- convert day display to lowercase
  6     Today := LOWER(Today);
  7     DayLength := LENGTH(Today);
  8     DBMS_OUTPUT.PUT_LINE('Today is ' || Today || ', ' || TO_CHAR(TodaysDate));
  9     DBMS_OUTPUT.PUT_LINE('The length of the word ' || Today ||
 10     ' is ' || TO_CHAR(DayLength) || ' characters.');
 11  END;
 12  /
BEGIN
*
ERROR at line 1:
ORA-06550: line 2, column 4:
PLS-00201: identifier 'TODAYSDATE' must be declared
ORA-06550: line 2, column 4:
PL/SQL: Statement ignored
ORA-06550: line 3, column 4:
PLS-00201: identifier 'TODAY' must be declared
ORA-06550: line 3, column 4:
PL/SQL: Statement ignored
ORA-06550: line 4, column 4:
PLS-00201: identifier 'TODAY' must be declared
```

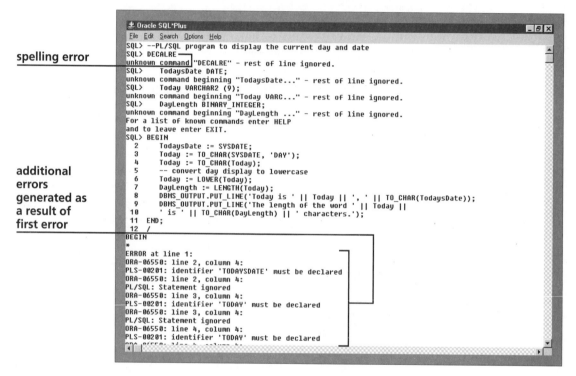

Figure 4-18: Compile error that generates several other errors

Keep in mind the following when attempting to resolve syntax errors:

- Error locations might not correspond to the line of the actual error.
- Try to isolate the error location systematically by commenting out and/or modifying suspect lines.
- One error might result in several other error messages.

Logic Errors

Logic errors causing incorrect program output can be caused by many things: not using the proper order of operations in arithmetic functions, passing incorrect parameter values to built-in functions, nonterminating loops, unexpected data values that are out of range or not of the right data type, as well as a long list of many other indiscretions. In complex commercial software applications, logic errors are located by legions of professional testers during long beta-testing processes. For the programs you will write in this book, you will usually be able to locate logic errors by determining what the output should be and comparing it to the actual program output. The best way to locate logic errors is to view variable values during program execution. PL/SQL does not have a built-in **debugger** in the SQL* Plus environment, which is a program that automatically tracks variable values during program execution. However, you can track variable values using DBMS_OUTPUT statements. Figure 4-19 shows the program that is supposed to display the string that is before 'day' in the name of the current day of the week. However, the program is currently displaying the string 'we' instead of 'wednes'.

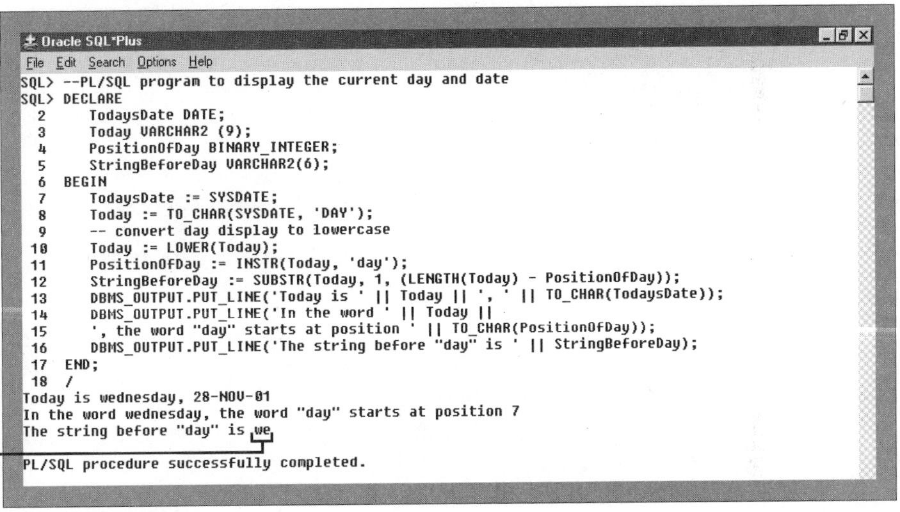

```
Oracle SQL*Plus
File Edit Search Options Help
SQL> --PL/SQL program to display the current day and date
SQL> DECLARE
  2     TodaysDate DATE;
  3     Today VARCHAR2 (9);
  4     PositionOfDay BINARY_INTEGER;
  5     StringBeforeDay VARCHAR2(6);
  6  BEGIN
  7     TodaysDate := SYSDATE;
  8     Today := TO_CHAR(SYSDATE, 'DAY');
  9     -- convert day display to lowercase
 10     Today := LOWER(Today);
 11     PositionOfDay := INSTR(Today, 'day');
 12     StringBeforeDay := SUBSTR(Today, 1, (LENGTH(Today) - PositionOfDay));
 13     DBMS_OUTPUT.PUT_LINE('Today is ' || Today || ', ' || TO_CHAR(TodaysDate));
 14     DBMS_OUTPUT.PUT_LINE('In the word ' || Today ||
 15     ', the word "day" starts at position ' || TO_CHAR(PositionOfDay));
 16     DBMS_OUTPUT.PUT_LINE('The string before "day" is ' || StringBeforeDay);
 17  END;
 18  /
Today is wednesday, 28-NOV-01
In the word wednesday, the word "day" starts at position 7
The string before "day" is we

PL/SQL procedure successfully completed.
```

output is incorrect

Figure 4-19: Program with a logic error

The variable value with the error is StringBeforeDay. The error is most likely occurring in the assignment statement on Line 12, where StringBeforeDay is assigned the value resulting from the SUBSTR function. To debug the program, we will place DBMS_OUTPUT statements before the line that uses the SUBSTR so we can view the variables being passed to the SUBSTR function. When you put debugging statements in your code, always include strings that label the variables, so you do not become confused about which variable values you are viewing. Figure 4-20 shows the DBMS_OUTPUT debugging statements, and their resulting output.

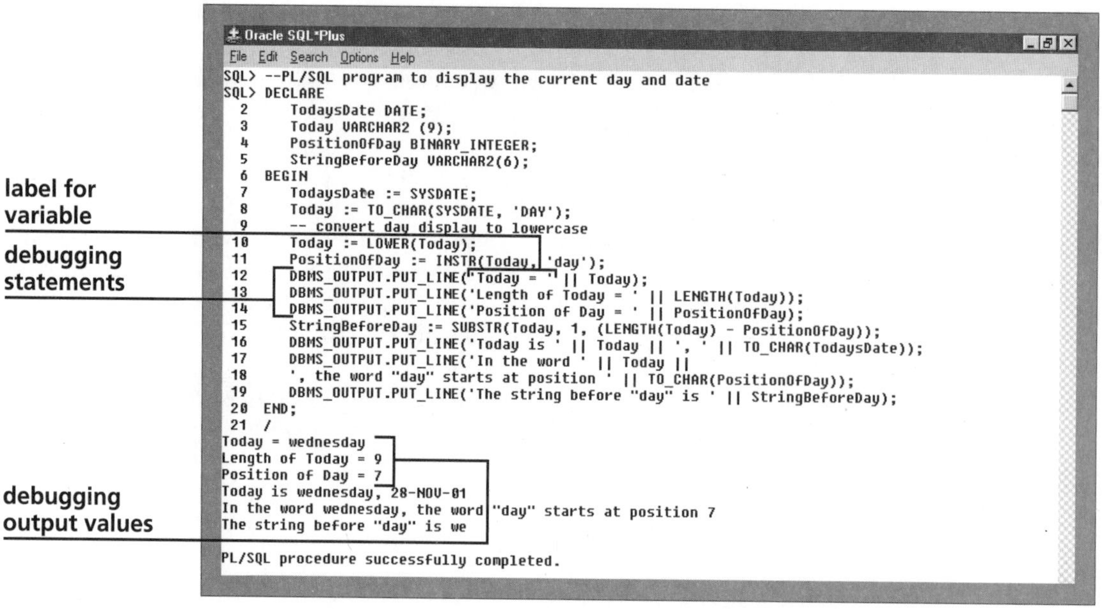

Oracle SQL*Plus

File Edit Search Options Help

```
SQL> --PL/SQL program to display the current day and date
SQL> DECLARE
       TodaysDate DATE;
       Today VARCHAR2 (9);
       PositionOfDay BINARY_INTEGER;
       StringBeforeDay VARCHAR2(6);
     BEGIN
       TodaysDate := SYSDATE;
       Today := TO_CHAR(SYSDATE, 'DAY');
       -- convert day display to lowercase
       Today := LOWER(Today);
       PositionOfDay := INSTR(Today, 'day');
       DBMS_OUTPUT.PUT_LINE('Today = ' || Today);
       DBMS_OUTPUT.PUT_LINE('Length of Today = ' || LENGTH(Today));
       DBMS_OUTPUT.PUT_LINE('Position of Day = ' || PositionOfDay);
       StringBeforeDay := SUBSTR(Today, 1, (LENGTH(Today) - PositionOfDay));
       DBMS_OUTPUT.PUT_LINE('Today is ' || Today || ', ' || TO_CHAR(TodaysDate));
       DBMS_OUTPUT.PUT_LINE('In the word ' || Today ||
       ', the word "day" starts at position ' || TO_CHAR(PositionOfDay));
       DBMS_OUTPUT.PUT_LINE('The string before "day" is ' || StringBeforeDay);
     END;
     /
Today = wednesday
Length of Today = 9
Position of Day = 7
Today is wednesday, 28-NOV-01
In the word wednesday, the word "day" starts at position 7
The string before "day" is we

PL/SQL procedure successfully completed.
```

label for variable

debugging statements

debugging output values

Figure 4-20: Program with DBMS_OUTPUT debugging statements

The output shows that the SUBSTR function receives the following parameters: ('wednesday', 1, (9 – 7)). As a result, the function returns 9 – 7, or two characters ('we'), starting with the first character of the string. However, the function should return six characters ('wednes'), starting with the first character of the string. This reveals the logic error: the third parameter in the SUBSTR function should be the length of Today less 3, not the length of Today less the PositionofDay variable value, which is 7.

Here is a checklist to use when trying to isolate a logic error:

1. Identify the output variable(s) that have the error.
2. Identify the inputs and calculations that contribute to the invalid output.
3. Find the values of the inputs that are contributing to the invalid output, using DBMS_OUTPUT statements.
4. If you still can't locate the problem, take a break and look at it again later.
5. If you still can't locate the problem, ask a fellow student for help.
6. If you still can't locate the problem, ask your instructor for help.

The PL/SQL Selection Structure

The programs you have written so far use sequential processing, in which statements are processed one after another. However, most programs require selection structures in the form of IF/THEN/ELSE statements that make the processing order change, depending on the values of certain variables.

The IF/THEN Structure

The PL/SQL IF/THEN selection structure has the following general format:

```
IF <condition> THEN
    <program statements that execute when condition is TRUE>;
END IF;
```

Every IF must have a corresponding END IF. The <condition> has to be able to be evaluated as either TRUE or FALSE. Figure 4-21 shows the PL/SQL relational operators, and examples of their usage.

Operator	Description	Example
=	Equal	Count = 5
<>	Not equal	Count <> 5
!=	Not equal	Count = 5
>	Greater than	Count > 5
<	Less than	Count < 5
>=	Greater than or equal to	Count >= 5
<=	Less than or equal to	Count <= 5

Figure 4-21: PL/SQL relational operators

Remember that the relational condition operator that compares two values is =, while the assignment operator that assigns a value to a variable is :=.

If the condition in the IF/THEN structure evaluates as TRUE, one or more program statements are executed. If the condition evaluates as FALSE, the program statements are skipped. It is a good programming practice to format IF/THEN structures by indenting the program statements that are executed if the condition is TRUE, so that the structure is easier to read and understand. You can indent these program statements by pressing the Tab key or by pressing the spacebar to add blank spaces.

Next, you will add an IF/THEN structure to your PL/SQL program so that it displays the current day of the week if today happens to be Friday. For the comparison to work properly, you will need to apply the RTRIM function to the Today variable to remove blank spaces that are added during the TO_CHAR conversion.

To add an IF/THEN structure to your PL/SQL program:

1 Switch to Notepad, and type the code as shown in Figure 4-22. Do not include the line numbers. Save your file as Ch4aPLSQL5.txt.

indent program line that is executed if condition is TRUE

type this code

```
Oracle SQL*Plus
File Edit Search Options Help
SQL> --PL/SQL program to display the current day
SQL> DECLARE
       Today VARCHAR2(9);
     BEGIN
       Today := TO_CHAR(SYSDATE, 'DAY');
       Today := RTRIM(Today);
      --add IF/THEN statement to determine if current day is Friday
      IF Today = 'FRIDAY' THEN
       DBMS_OUTPUT.PUT_LINE('Today is Friday');
      END IF;
     END;
     /

PL/SQL procedure successfully completed.
```

Figure 4-22: PL/SQL program with IF/THEN structure testing if current day is Friday

2 Copy the code into SQL*Plus, and execute the program. Unless it happens to be Friday (which it isn't in our example), the condition evaluates to FALSE, and no output is generated.

3 Modify your program code as shown in Figure 4-23 so the condition will evaluate to TRUE if it is not Friday, and save your file as Ch4aPLSQL6.txt. Unless you were one of the lucky ones whose condition evaluated to TRUE in the previous step, your output should now look like the output in Figure 4-23.

type this code

```
Oracle SQL*Plus
File Edit Search Options Help
SQL> --PL/SQL program to display the current day
SQL> DECLARE
       Today VARCHAR2(9);
     BEGIN
       Today := TO_CHAR(SYSDATE, 'DAY');
       Today := RTRIM(Today);
      --add IF/THEN statement to determine if current day is Friday
      IF Today != 'FRIDAY' THEN
        DBMS_OUTPUT.PUT_LINE('Today is not Friday');
      END IF;
     END;
     /
Today is not Friday

PL/SQL procedure successfully completed.
```

Figure 4-23: PL/SQL program with IF/THEN structure testing if current day is not Friday

The IF/THEN/ELSE Structure

The previous example suggests the need for a selection structure that executes alternate program statements when the condition evaluates to FALSE. This is usually called an IF/THEN/ELSE structure. In PL/SQL, the IF/THEN/ELSE structure has the following general format:

```
IF <condition> THEN
   <program statements that execute when condition is TRUE>;
ELSE
   <alternate program statements that execute when condition is FALSE>;
END IF;
```

Next, you will modify your program using an IF/THEN/ELSE structure so that it displays one output if the current day is Friday, and a different output if the current day is not Friday.

To modify your PL/SQL program using an IF/THEN/ELSE structure:

1. Switch to Notepad, and modify your PL/SQL code, as shown in Figure 4-24, to include the IF/THEN/ELSE structure. (Do not include the line numbers.) Save your file as Ch4aPLSQL7.txt.
2. Copy the code into SQL*Plus, and execute the program. Depending on the current day, the appropriate output is displayed.

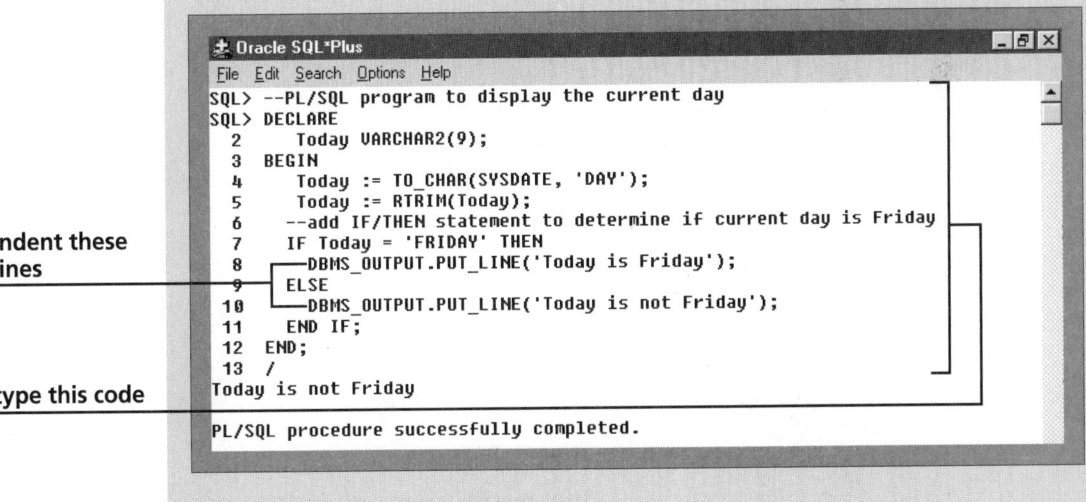

Figure 4-24: PL/SQL IF/THEN/ELSE structure

Nesting IF/THEN/ELSE Statements

IF/THEN/ELSE structures can be **nested**, which means that one or more additional IF/THEN/ELSE statements are included within the program statements, below either the IF or ELSE. It is especially important to properly indent the program lines following the THEN and ELSE commands in nested IF/THEN/ELSE structures—correct formatting enables you to better understand the program logic and spot syntax errors, such as missing END IF commands. Next, you will modify your PL/SQL program so that if the current day is not Friday, the program also tests to see if the current day is Saturday and displays the appropriate output.

To modify your PL/SQL program using a nested IF/THEN/ELSE structure:

1. Switch to Notepad, and modify your PL/SQL code, as shown in Figure 4-25, to include the nested IF/THEN/ELSE structure. Save your file as Ch4aPLSQL8.txt.
2. Copy the code into SQL*Plus, and execute the program. Depending on the current day, the appropriate output is displayed.

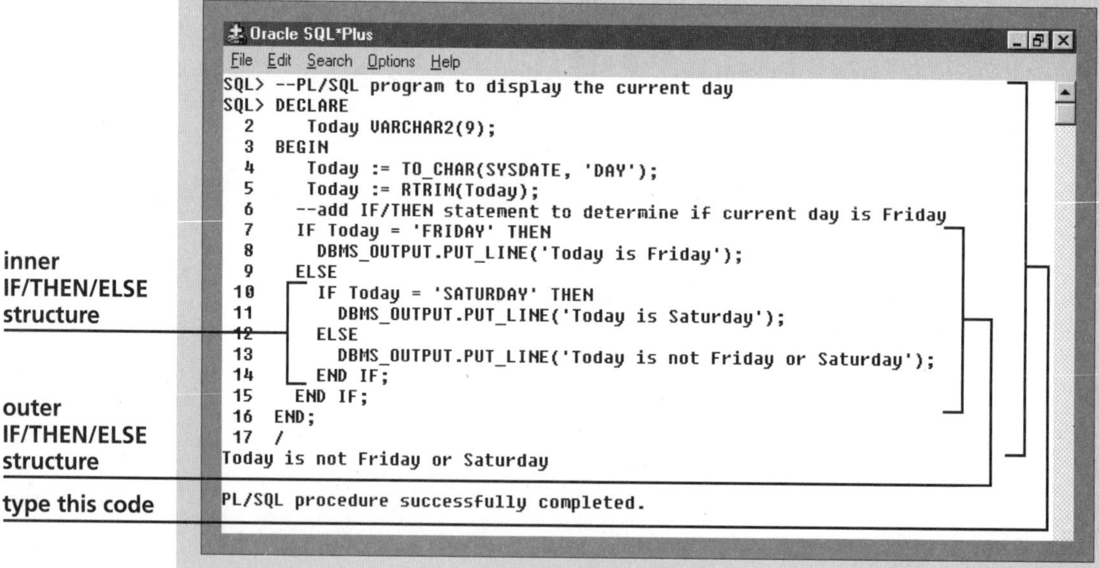

Oracle SQL*Plus

File Edit Search Options Help

```
SQL> --PL/SQL program to display the current day
SQL> DECLARE
       Today VARCHAR2(9);
     BEGIN
       Today := TO_CHAR(SYSDATE, 'DAY');
       Today := RTRIM(Today);
      --add IF/THEN statement to determine if current day is Friday
      IF Today = 'FRIDAY' THEN
        DBMS_OUTPUT.PUT_LINE('Today is Friday');
      ELSE
        IF Today = 'SATURDAY' THEN
          DBMS_OUTPUT.PUT_LINE('Today is Saturday');
        ELSE
          DBMS_OUTPUT.PUT_LINE('Today is not Friday or Saturday');
        END IF;
      END IF;
     END;
     /
Today is not Friday or Saturday

PL/SQL procedure successfully completed.
```

inner IF/THEN/ELSE structure

outer IF/THEN/ELSE structure

type this code

Figure 4-25: Nested PL/SQL IF/THEN/ELSE structure

The IF/ELSIF Structure

The IF/ELSIF structure allows you to test for many different conditions, and is similar to the CASE or SELECT CASE structure used in other programming languages. The general format for this structure is as follows:

```
IF <condition1> THEN
   <program statements that execute when condition1 is TRUE>;
ELSIF <condition2> THEN
   <program statements that execute when condition2 is TRUE>;
ELSIF <condition3> THEN
   <program statements that execute when condition3 is TRUE>;
...
ELSE
   <program statements that execute when none of the conditions are TRUE>;
END IF;
```

If the first condition is true, then its program statement(s) is executed, and the IF/THEN structure is exited. If an ELSIF condition is true, then its associated program statement(s) is executed, and the IF/THEN structure is exited. If all conditions are false, then the ELSE program statement(s) is executed. Next, you will modify your PL/SQL program so that it tests for each day of the week and displays the appropriate output. Note that the program statements following all of the ELSIF commands and the program statement following the ELSE command are all indented to make the structure easier to read and interpret.

To modify your PL/SQL program to use the IF/ELSIF structure:

1. Switch to Notepad, and modify your PL/SQL code, as shown in Figure 4-26, to include the IF/ELSIF structure. Save your file as Ch4aPLSQL9.txt.
2. Copy the code into SQL*Plus, and execute the program. Depending on the current day, the appropriate output is displayed.

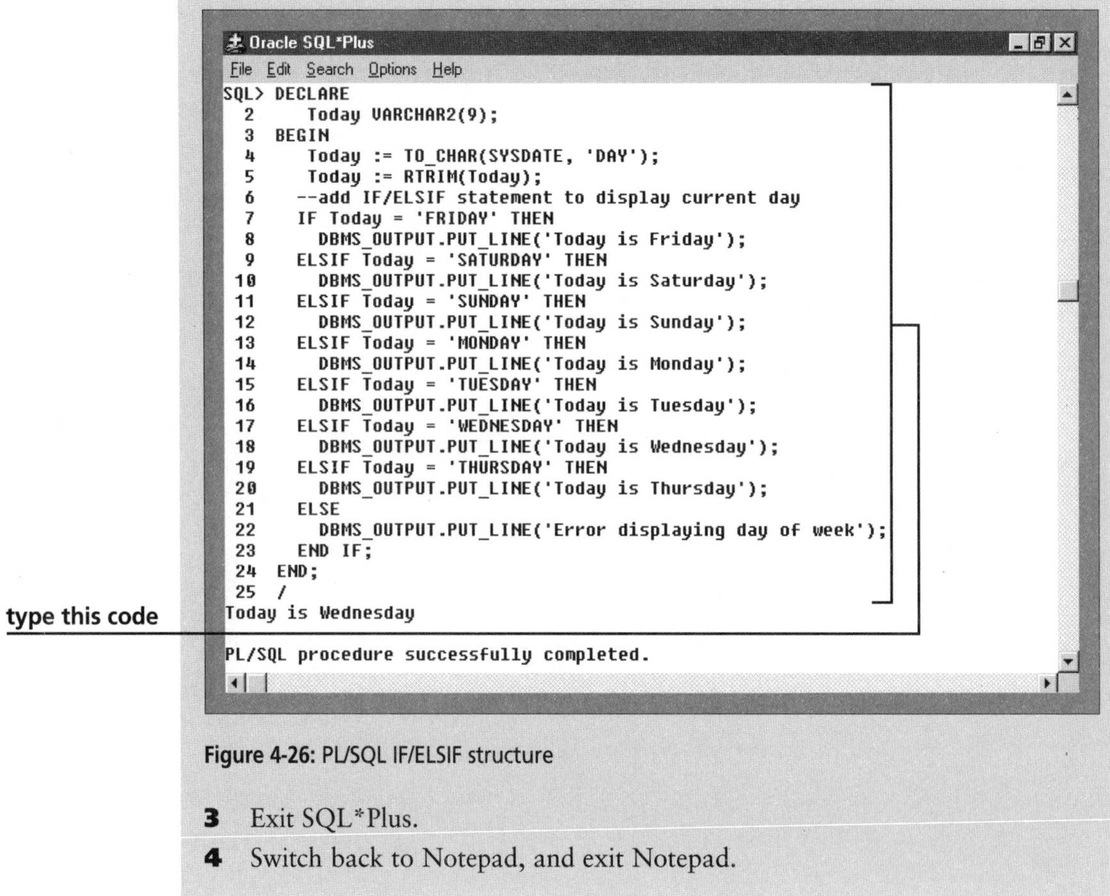

Figure 4-26: PL/SQL IF/ELSIF structure

3 Exit SQL*Plus.

4 Switch back to Notepad, and exit Notepad.

SUMMARY

- PL/SQL is the procedural language used to write procedures and functions for Oracle applications, and is used in most Oracle database application development tools.
- PL/SQL variables must begin with a character, and cannot contain more than 30 characters or be reserved words. Variables can contain letters, numbers, and the symbols $, _, and #.
- The main PL/SQL data types are VARCHAR2, CHAR, NUMBER, BINARY_INTEGER, DATE, and BOOLEAN.
- Reference data types assume the same data type as a referenced database column or row.

- A PL/SQL program consists of the following sections: declarations, where variables are defined or described; body, where program statements are placed; and exception, where error handling is done.
- In PL/SQL, all variables must be declared prior to use, and assignment and comparisons can be performed only between variables with the same type.
- In PL/SQL, a block of comment statements can be delimited by beginning the block with the characters /* and ending the block with the characters */. A single comment line is delimited by beginning it with two hyphens (--).
- In PL/SQL, values are assigned to variables using the := assignment operator. The variable name is placed on the left side of the operator, and the value is placed on the right side of the operator.
- The DBMS_OUTPUT.PUT_LINE function is used to display interactive output in PL/SQL programs in the SQL*Plus environment.
- Before using the DBMS_OUTPUT.PUT_LINE function, you must type SET SERVER OUTPUT ON at the SQL> prompt.
- PL/SQL performs implicit data conversions, but the output is unpredictable, so you should always use the explicit data conversion functions.
- To concatenate two character strings, use the double bar (||) operator in PL/SQL.
- A syntax error does not follow the guidelines of the programming language, and results in a compile error message. A logic error does not stop a program from compiling, but results in incorrect output.
- The IF/THEN structure evaluates a given condition. Every IF statement must have a corresponding END IF line in the program.
- The IF/THEN/ELSE structure is similar to the IF/THEN structure, except that when the condition evaluates to FALSE, one or more alternate program statements execute. The IF/THEN/ELSIF structure allows you to test for many different conditions.

REVIEW QUESTIONS

1. You have the following variable names and associated values:

Variable Name	Value	Data Type
City	Cheyenne	VARCHAR2
State	Wyoming	VARCHAR2
StateNickname	Equality State	VARCHAR2
CityNickname	Magic City of the Plains	VARCHAR2
CityPopulation	55,000	BINARY_INTEGER
StatehoodDate	07/10/1890	DATE

Write a PL/SQL statement that will concatenate the variables with the necessary literal values to give the following output:

a. Cheyenne is the largest city in Wyoming.
b. The population of Cheyenne is about 55,000 people.
c. Wyoming's nickname is the Equality State.
d. The nickname of Cheyenne, Wyoming, is "Magic City of the Plains."
e. Wyoming became a state on July 10, 1890.

2. Write the PL/SQL declaration for the following variables:

Variable Name	Data Description
MysteryText	Up to 2,000 characters of text
CourseCredit	An integer with maximum value of 99
OrderStatus	TRUE or FALSE
OrderPrice	A fixed-point number with two decimal places, with a maximum value of 999.99
DateReceived	Date
StudentClass	Text field that always contains two characters

3. What is the difference between the = and the := operators?
4. When should you issue the command SET SERVEROUTPUT ON?
5. What are the two approaches for delimiting comment blocks in a PL/SQL program?
6. What should you do before asking your instructor for help with a logic programming error?
7. You have a variable named TermDesc, and it currently contains the value 'Fall 2000'.
   a. Write the SUBSTR function that will return the value '2000' to a variable named Year.
   b. Write the INSTR function that will find the position of the blank space and assign the value to a variable named BlankPosition.
   c. Write the function that will convert the value stored in TermDesc to 'FALL 2000'.
   d. Write the commands that will find the characters before the blank space (in this case, 'Fall'), and assign them to a character variable named TermSeason. Write the command that will find the characters after the blank space (in this case, '2000'), and assign them to a DATE variable named TermYear. (*Hint*: You will need to use the INSTR and SUBSTR functions. Create new variables if needed.)

PROBLEM-SOLVING CASES

1. Using Notepad, write a PL/SQL program that declares a numeric variable named Counter and a string variable named MyString. Assign the value 100 to COUNTER and the text 'Hello world' to MyString, and display the values on two separate lines using the DBMS_OUTPUT function. Save your file as Ch4aEx1.txt.

2. Using Notepad, write a PL/SQL program that declares the following variables and assigns them the given values:

Variable Name	Data Type	Value
InventoryID	numeric	11668
InventoryColor	character	Sienna
InventoryPrice	numeric	259.99
InventoryQOH	numeric	16

Write the program statements to have the program display the following output, using the DBMS_OUTPUT function and the variable values. (Do not insert, or hard-code, the actual data values in the output function).

Inventory ID: 11668
Color: Sienna
Price: $259.99
Quantity on Hand: 16

Save your file as Ch4aEx2.txt.

3. Using Notepad, write a PL/SQL program that declares the following variables and assigns them the given values:

Variable Name	Data Type	Value
StudentLastName	character	Miller
StudentFirstName	character	Sarah
StudentMI	character	M
StudentAddress	character	144 Windridge Blvd.
StudentCity	character	Eau Claire
StudentState	character	WI
StudentZip	character	54703

Using the DBMS_OUTPUT function, write the program statements to display the following output (do not insert, or hard-code, the actual data values in the output function):

Sarah M. Miller
144 Windridge Blvd.
Eau Claire, WI 54703

Save your file as Ch4aEx3.txt.

4. Using Notepad, write a PL/SQL program that declares character variables named FacultyLastName, FacultyFirstName, and FacultyPhone. Assign the value 'Cox' to FacultyLastName, 'Kim' to FacultyFirstName, and '7155551234' to FacultyPhone. Write the program commands so that the program displays "Kim Cox's phone number is 715 555 1234" on one line, using the DBMS_OUTPUT function. (Do not insert, or hard-code, the actual data values in the output function.) Save your file as Ch4aEx4.txt.

5. Using Notepad, write a PL/SQL program that declares a date variable named Today, and assigns SYSDATE to it. Depending on the day number of the month, your program should display the following output:

Day	Output
1-10	It is the <day number> day of <month name>. It is early in the month.
11-20	It is the <day number> day of <month name>. It is the middle of the month.
21-31	It is the <day number> day of <month name>. It is nearly the end of the month.

For example, if it is currently November 30, your program should display "It is the 30 day of NOVEMBER. It is nearly the end of the month." (*Hint*: Do not worry about day suffixes such as the 30th day of November.) Save your file as Ch4aEx5.txt.

6. PL/SQL has a function named NEW_TIME that converts an input date and time from one time zone to another. The general format of the function is NEW_TIME(<input date and time>, <current_time_zone>, <desired_time_zone>). The time zone parameters require the following abbreviations:

CST	Central Standard Time
EST	Eastern Standard Time
GMT	Greenwich Mean Time
HST	Alaska/Hawaii Standard Time
MST	Mountain Standard Time
PST	Pacific Standard Time
YST	Yukon Standard Time

To convert the current time from Central Standard Time to Pacific Standard Time, you would use the following function: NEW_TIME(SYSDATE, 'CST', 'PST');.

Write a program that declares a character variable named TimeZone, and assigns it to the abbreviation corresponding to your time zone. (If your time zone is not listed, use 'EST'.) Then, write the code to display the following output:

The current time in New York City is <your current time converted to EST>.
The current time in Chicago is <your current time converted to CST>.
The current time in Honolulu is <your current time converted to HST>.
The current time in the Yukon is <your current time converted to YST>.
The current time in London is <your current time converted to GMT>.

For example, if it is currently 2:03 P.M. EST, the program will display "The current time in New York City is 2:03 P.M." for the first output line, and "The current time in Chicago is 1:03 P.M." for the second output line. Use the variable value (not the literal value) for your time zone in the NEW_TIME function. Save your file as Ch4aEx6.txt.

LESSON B

objectives

- Use SQL commands in PL/SQL programs
- Create loops in PL/SQL programs
- Use cursors to retrieve database data into PL/SQL programs
- Use the exception section to handle errors in PL/SQL programs

Using SQL Commands in PL/SQL Programs

SQL commands can be classified into different categories. **Data Definition Language (DDL)** commands, such as CREATE, ALTER, and DROP, change the database structure. **Data Manipulation Language (DML)** commands, such as SELECT and INSERT, query or manipulate the data in database tables. **Transaction control** commands, such as COMMIT, are used to organize commands into logical transactions and commit them to the database or roll them back. DDL commands cannot be used in PL/SQL programs, but both DML and transaction control commands can be used in PL/SQL programs. Figure 4-27 summarizes these SQL command categories, and shows which can and cannot be used in PL/SQL programs.

Category	Purpose	Examples of Commands	Can Be Used in PL/SQL
Data Definition Language (DDL)	Changes the database structure	CREATE, ALTER, DROP, GRANT, REVOKE	No
Data Manipulation Language (DML)	Queries or changes the data in the database tables	SELECT, INSERT, UPDATE, DELETE	Yes
Transaction control commands	Organize commands into logical transactions	COMMIT, ROLLBACK, SAVEPOINT	Yes

Figure 4-27: Categories of SQL commands

Now you will write a PL/SQL program that inserts the first three records into the TERM table of the Northwoods University database, using DML and transaction control commands. First, you will start SQL*Plus, and run a script to delete your current database tables and re-create the Northwoods University database tables without any data in them. Then, you will check to make sure the TERM table is empty.

To run the script to create the database tables and check the TERM table:

1 If necessary, start SQL*Plus and type **SET SERVEROUTPUT ON** to enable interactive output.

2 Run the emptynu.sql script from the Chapter4 folder on your Student Disk to drop your existing Northwoods University tables and create the empty Northwoods University database tables.

3 Type the following command at the SQL> prompt: **SELECT * FROM term;**. The message "no rows selected" is displayed, indicating that the table is empty.

Next, you will write a PL/SQL program to insert the first three records into the TERM table. You will declare variables corresponding to each of the table fields, assign the variables to the appropriate values, insert the records, and then commit the records.

To write the program to insert the TERM records:

1 Start Notepad, and type the PL/SQL program commands shown in Figure 4-28. Do not type the line number.

type this code

```
Oracle SQL*Plus
File Edit Search Options Help
SQL> DECLARE
       TermID BINARY_INTEGER;
       TermDesc VARCHAR2(20);
       TermStatus VARCHAR2(20);
     BEGIN
       TermID := 1;
       TermDesc := 'Fall 2000';
       TermStatus := 'CLOSED';
       -- insert the records
       INSERT INTO term VALUES (TermID, TermDesc, TermStatus);
       TermID := TermID + 1;
       TermDesc := 'Spring 2001';
       INSERT INTO term VALUES (TermID, TermDesc, TermStatus);
       TermID := TermID + 1;
       TermDesc := 'Summer 2001';
       INSERT INTO term VALUES (TermID, TermDesc, TermStatus);
       COMMIT;
     END;
     /

PL/SQL procedure successfully completed.
```

output when program successfully executes

Figure 4-28: PL/SQL program using INSERT command

2 Copy your text into SQL*Plus, and run the program. The message "PL/SQL procedure successfully completed" is displayed when your program runs correctly. If a compile error is displayed, debug and rerun your program until it successfully executes.

3 Type the query shown in Figure 4-29 at the SQL*Plus SQL> prompt. The output shows that the records were inserted correctly.

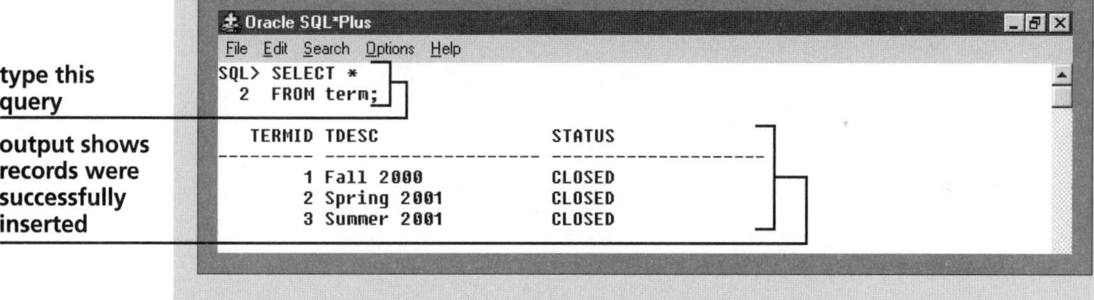

Figure 4-29: Contents of TERM table after running PL/SQL program

Loops

Programs use a repetition structure called a **loop** to repeat an action multiple times until an ending condition is reached. PL/SQL has five looping structures: LOOP ... EXIT, LOOP ... EXIT WHEN, WHILE ... LOOP, numeric FOR loops, and cursor FOR loops. The first four loop structures will be discussed in the following paragraphs. Cursor FOR loops will be discussed in the section on cursors later in this lesson. To illustrate the different types of loops, you will create a database table named COUNT_TABLE that has one data field named COUNTER. You will create PL/SQL programs that use the different loop structures to automatically insert the numbers 1, 2, 3, 4, and 5 into the COUNTER field in COUNT_TABLE, as shown in Figure 4-30.

COUNTER
1
2
3
4
5

Figure 4-30: COUNT_TABLE

You will now create COUNT_TABLE in SQL*Plus.

To create COUNT_TABLE:

1 Type the following command at the SQL*Plus SQL> prompt:

```
CREATE TABLE count_table
(counter    NUMBER(2));
```

The message "Table created." is displayed if the table is successfully created.

The LOOP ... EXIT Loop

The basic format of a LOOP ... EXIT loop is:

```
LOOP
     <program statements>
     IF <condition> THEN
       EXIT;
     END IF;
     <more program statements>
END LOOP
```

When the IF/THEN condition is true, the program statement EXIT directs the program to exit the loop and resume execution after the END LOOP statement. Repeating program statements can be placed before or after the IF/THEN condition. As a result, the loop can be either a **pretest loop** (the condition is tested before the program statements are executed) or a **posttest loop** (the condition is tested after the program statements are executed, and the program statements are always executed at least once). If the program statements might never be executed, use a pretest loop. If the program statements will always be executed at least once, use a posttest loop. Now, you will write a program to insert records 1 through 5 into COUNT_TABLE using a LOOP...EXIT loop. The loop in this program uses a pretest condition, and the value of LoopCount is tested before the record is inserted.

To insert the COUNT_TABLE records using a LOOP...EXIT loop:

1 Switch to Notepad, and type the code shown in Figure 4-31. It is a good programming practice to format the loop as shown by indenting the program lines between the LOOP and END LOOP commands.

2 Paste your code into SQL*Plus, run it, and debug it if necessary.

3 Type the query shown in Figure 4-31 to view the records in COUNT_TABLE and confirm that the records were inserted correctly.

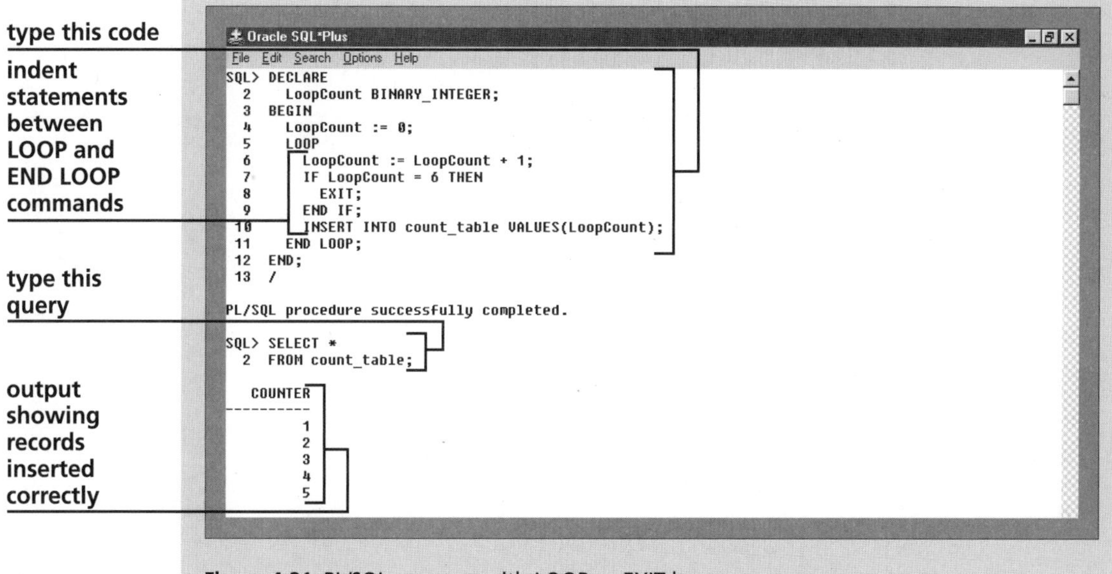

type this code
indent statements between LOOP and END LOOP commands
type this query
output showing records inserted correctly
Oracle SQL*Plus
File Edit Search Options Help
SQL> DECLARE
2 LoopCount BINARY_INTEGER;
3 BEGIN
4 LoopCount := 0;
5 LOOP
6 LoopCount := LoopCount + 1;
7 IF LoopCount = 6 THEN
8 EXIT;
9 END IF;
10 INSERT INTO count_table VALUES(LoopCount);
11 END LOOP;
12 END;
13 /
PL/SQL procedure successfully completed.
SQL> SELECT *
2 FROM count_table;
COUNTER

1
2
3
4
5

Figure 4-31: PL/SQL program with LOOP ... EXIT loop

The LOOP ... EXIT WHEN Loop

The general format of the LOOP ... EXIT WHEN loop is:

```
LOOP
     <program statements>
     EXIT WHEN <condition>;
END LOOP;
```

This loop executes the program statements, and then tests for the exit condition, using the EXIT WHEN command. This is a posttest loop, and the program statements are always executed at least once. Now, you will delete the records you inserted into COUNT_TABLE, and then write a PL/SQL program to insert the records using a LOOP ... EXIT WHEN loop.

To delete the records and write the program using a LOOP ... EXIT WHEN loop:

1. Delete the existing COUNT_TABLE records by typing the following command at the SQL> prompt in SQL*Plus: **DELETE FROM count_table;**. The message "5 rows deleted" is displayed.
2. Type the code shown in Figure 4-32. Save your file as Ch4bPLSQL3.txt in the Chapter4 folder on your Student Disk. Note that the program lines between the LOOP and END LOOP commands are indented.
3. Paste your code into SQL*Plus, run it, and debug it if necessary.

4 Type the SELECT query shown in Figure 4-32 to view the records in COUNT_TABLE and confirm that the records were inserted correctly.

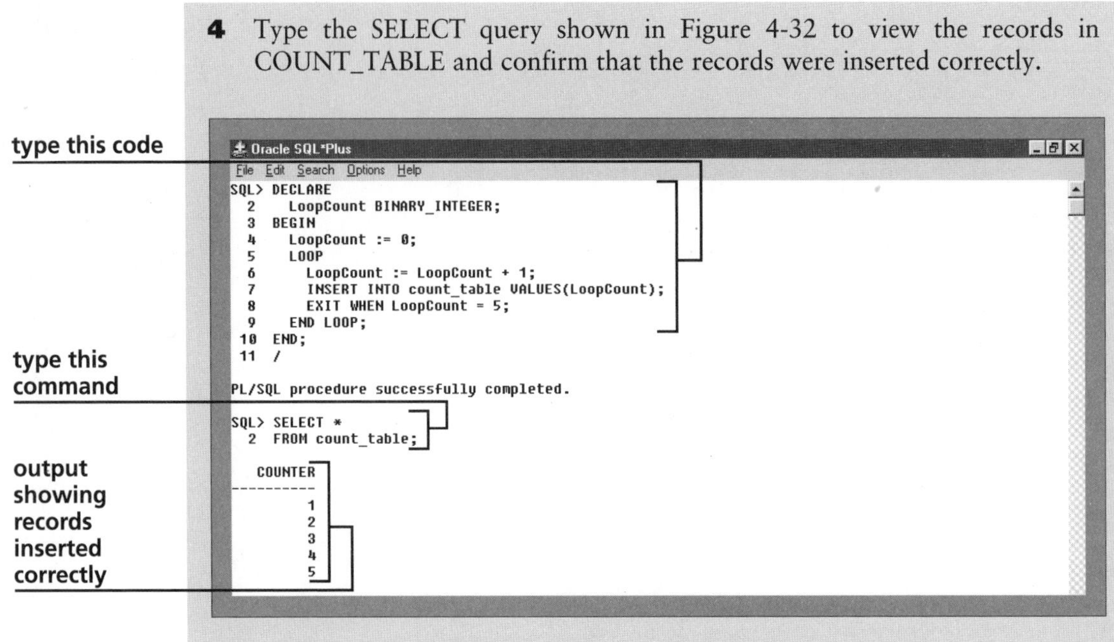
type this code
type this command
output showing records inserted correctly
Oracle SQL*Plus
File Edit Search Options Help
SQL> DECLARE
2 LoopCount BINARY_INTEGER;
3 BEGIN
4 LoopCount := 0;
5 LOOP
6 LoopCount := LoopCount + 1;
7 INSERT INTO count_table VALUES(LoopCount);
8 EXIT WHEN LoopCount = 5;
9 END LOOP;
10 END;
11 /
PL/SQL procedure successfully completed.
SQL> SELECT *
2 FROM count_table;
COUNTER
1
2
3
4
5

Figure 4-32: PL/SQL program with LOOP ... EXIT WHEN loop

The WHILE ... LOOP Loop

The WHILE ... LOOP is always a pretest loop, where the condition is evaluated before any program statements are executed. The general format of the WHILE ... LOOP loop is:

```
WHILE <condition>
LOOP
     <program statements>
END LOOP;
```

You will again delete the records you inserted into COUNT_TABLE, and then write a PL/SQL program to insert the records using the WHILE ... LOOP structure.

To delete the records and write the program using the WHILE ...LOOP:

1 Delete the existing COUNT_TABLE records by typing the following command at the SQL> prompt in SQL*Plus: **DELETE FROM count_table;**. The message "5 rows deleted" is displayed.

2 Type the code shown in Figure 4-33. Again, note that the program lines between the LOOP and END LOOP commands are indented.

3 Paste your code into SQL*Plus, run it, and debug it if necessary.

4 Type the SELECT query shown in Figure 4-33 to confirm that the records were inserted correctly.

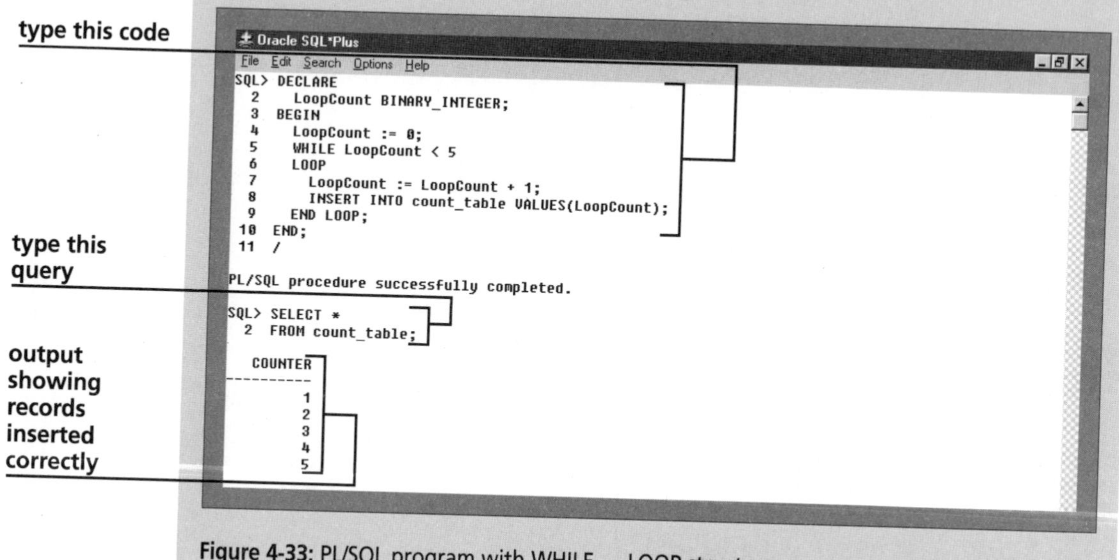
type this code
type this query
output showing records inserted correctly
Oracle SQL*Plus
File Edit Search Options Help
SQL> DECLARE
2 LoopCount BINARY_INTEGER;
3 BEGIN
4 LoopCount := 0;
5 WHILE LoopCount < 5
6 LOOP
7 LoopCount := LoopCount + 1;
8 INSERT INTO count_table VALUES(LoopCount);
9 END LOOP;
10 END;
11 /
PL/SQL procedure successfully completed.
SQL> SELECT *
2 FROM count_table;
COUNTER

1
2
3
4
5

Figure 4-33: PL/SQL program with WHILE ... LOOP structure

The Numeric FOR Loop

In all of the loops you have done so far, you have had to declare a counter variable and manually increment it within your program statements. The numeric FOR loop is different, because you do not need to declare the counter variable. The counter is defined in a series of numbers in the FOR statement, and is automatically incremented each time the loop repeats. The general format of the numeric FOR loop is:

```
FOR<counter variable> IN <start value> .. <end value>
LOOP
    <program statements>
END LOOP;
```

The start and end values must be integers, and are always incremented by one. Now, you will delete the records and reinsert the records using a numeric FOR loop.

To insert the records using a numeric FOR loop:

1. Delete the existing COUNT_TABLE records by typing the following command at the SQL> prompt in SQL*Plus: **DELETE FROM count_table;**. The message "5 rows deleted" is displayed.
2. Type the code shown in Figure 4-34. Again, the command between the LOOP and END LOOP commands is indented.
3. Paste your code into SQL*Plus, run it, and debug it if necessary.
4. Type the SELECT query shown in Figure 4-34 to confirm that the records were inserted correctly.

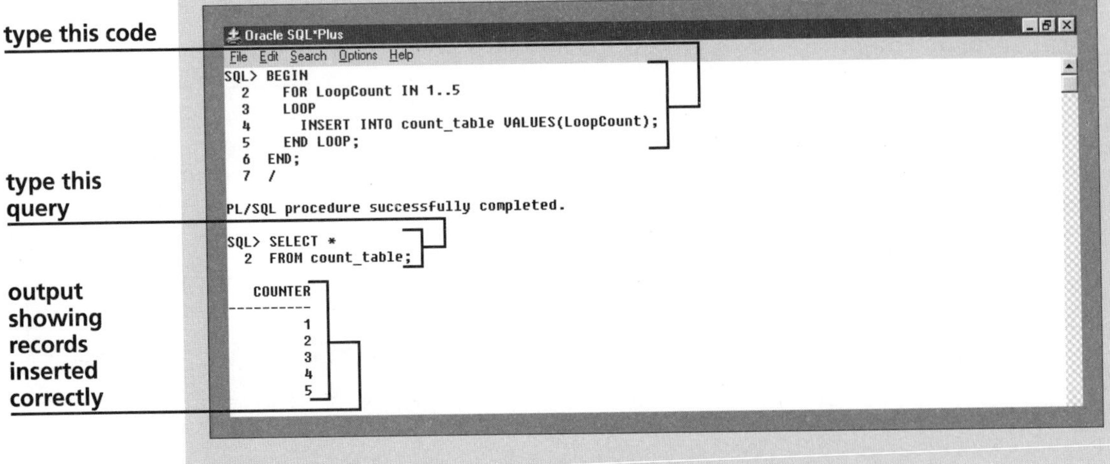
type this code
type this query
output showing records inserted correctly
Oracle SQL*Plus
File Edit Search Options Help
SQL> BEGIN
2 FOR LoopCount IN 1..5
3 LOOP
4 INSERT INTO count_table VALUES(LoopCount);
5 END LOOP;
6 END;
7 /
PL/SQL procedure successfully completed.
SQL> SELECT *
2 FROM count_table;
COUNTER

1
2
3
4
5

Figure 4-34: PL/SQL program with numeric FOR loop

Cursors

When Oracle processes a SQL command, it allocates a memory location in the database server's memory known as the **context area**. This memory location contains information about the command, such as the number of rows processed by the statement, a parsed (machine-language) representation of the statement, and in the case of a query that returns data, the **active set**, which is the set of data rows returned by the query. A **cursor** is a **pointer**, which is a variable that contains the address of the memory location that contains the SQL command's context area. Figure 4-35 illustrates a cursor that points to a context area for a query that returns all of the rows from the COURSE table in the Northwoods University database.

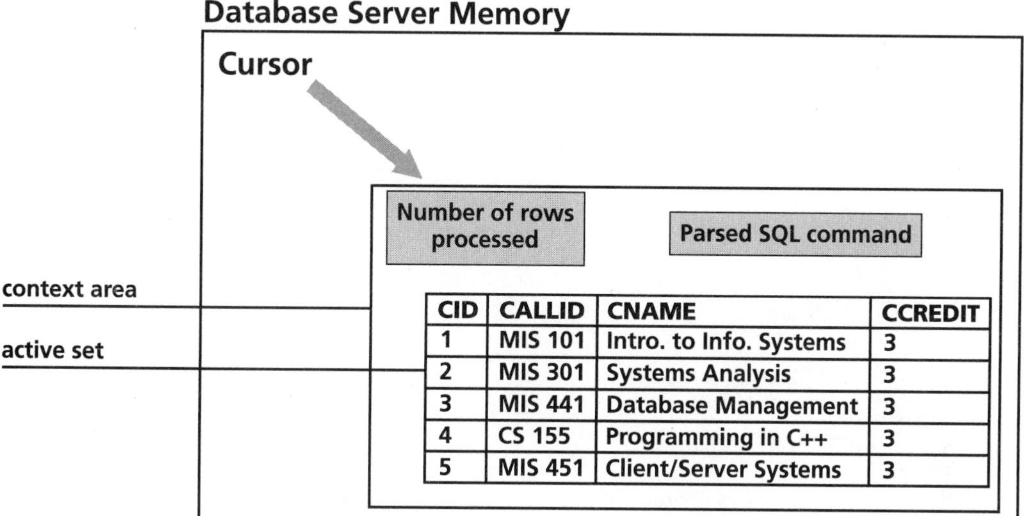

CID	CALLID	CNAME	CCREDIT
1	MIS 101	Intro. to Info. Systems	3
2	MIS 301	Systems Analysis	3
3	MIS 441	Database Management	3
4	CS 155	Programming in C++	3
5	MIS 451	Client/Server Systems	3

Figure 4-35: Cursor to return all rows in COURSE table

There are two kinds of cursors: implicit and explicit. An **implicit cursor** is created automatically within the Oracle environment, and does not need to be declared in the DECLARE section of a program. Oracle creates an implicit cursor every time you issue an INSERT, UPDATE, or DELETE command. You can use an implicit cursor when you want to assign the output of a SELECT query to a PL/SQL variable and you are sure that the query will return one—and only one—record. If the query returns more than one record, or does not return any records, an error message is generated. Implicit cursors have the following general format:

```
SELECT<data field(s)>
INTO<declared variable name(s)>
FROM <table name(s)>
WHERE<search condition that will return
a single record>;
```

The variables receiving the data from the query in the INTO clause must be the same data types as the returned field(s). Now, you will create an implicit cursor that retrieves a specific faculty member's last and first names into the declared variables LastName and FirstName from the FACULTY table in the Northwoods University database, and then displays the retrieved values using the DBMS_OUTPUT function. First, you will run a script to refresh your Northwoods University database tables.

To refresh your Northwoods University database tables and create a program with an implicit cursor:

1 Run the northwoo.sql script form the Chapter4 folder of your Student Disk.

2 Type the code shown in Figure 4-36 in Notepad. Note that a blank space is inserted before the faculty member's first name, and between the faculty member's first and last names, using single quotes. Save your file as Ch4bPLSQL6.txt.

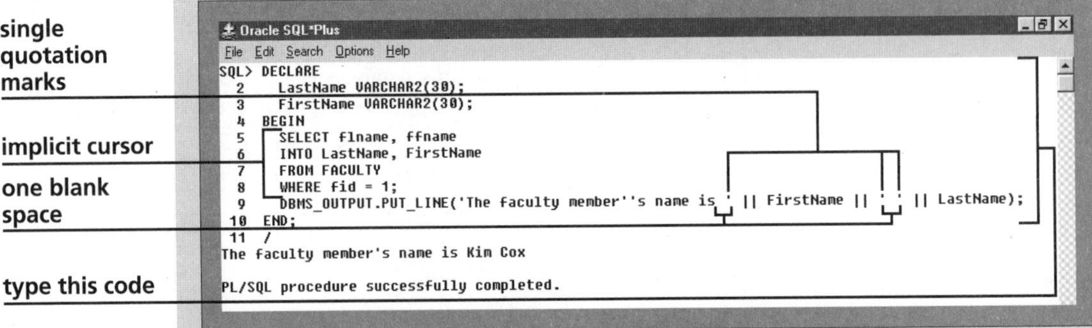
single quotation marks
implicit cursor
one blank space
type this code
Oracle SQL*Plus
File Edit Search Options Help
SQL> DECLARE
LastName VARCHAR2(30);
FirstName VARCHAR2(30);
BEGIN
SELECT flname, ffname
INTO LastName, FirstName
FROM FACULTY
WHERE fid = 1;
DBMS_OUTPUT.PUT_LINE('The faculty member''s name is ' || FirstName || ' ' || LastName);
END;
/
The faculty member's name is Kim Cox
PL/SQL procedure successfully completed.

Figure 4-36: PL/SQL program with implicit cursor

3 Paste your code into SQL*Plus, run it, and debug it if necessary. The output showing the retrieved data values is displayed.

Figure 4-37 shows the error message that is generated when an implicit cursor retrieves multiple records. The search condition was deleted in the previous implicit cursor query, so the query returned all five records in the FACULTY table, and the resulting error message is displayed.

the cursor search condition was modified so multiple records were retrieved

error message showing multiple records were retrieved

```
Oracle SQL*Plus
File Edit Search Options Help
SQL> DECLARE
  LastName VARCHAR2(30);
  FirstName VARCHAR2(30);
BEGIN
  SELECT flname, ffname
  INTO LastName, FirstName
  FROM FACULTY;
  DBMS_OUTPUT.PUT_LINE('The faculty member''s name is ' || FirstName || ' ' || LastName);
END;
/
DECLARE
*
ERROR at line 1:
ORA-01422: exact fetch returns more than requested number of rows
ORA-06512: at line 5
```

Figure 4-37: Implicit cursor that returns multiple records

Figure 4-38 shows the result of an implicit cursor that returns no records. The search condition was modified so the query searched for FID 6, but the database does not contain a record where FID = 6. An error message stating "no data found" is generated.

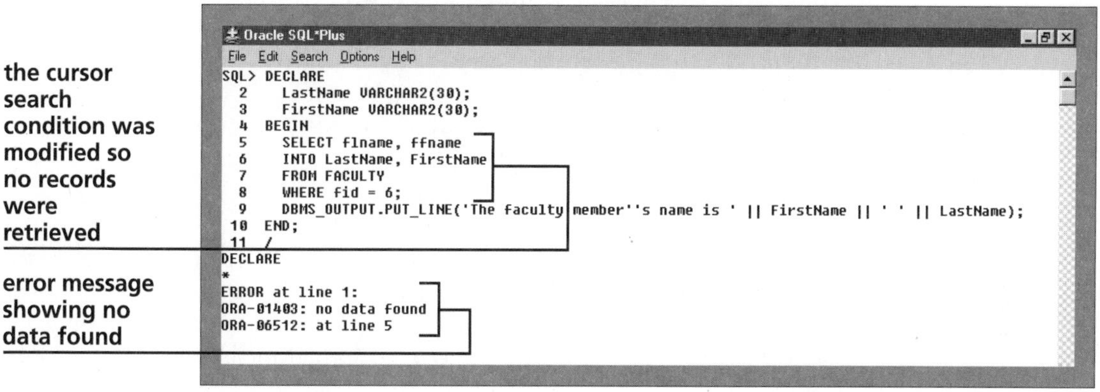

the cursor search condition was modified so no records were retrieved

error message showing no data found

```
Oracle SQL*Plus
File Edit Search Options Help
SQL> DECLARE
  2    LastName VARCHAR2(30);
  3    FirstName VARCHAR2(30);
  4  BEGIN
  5    SELECT flname, ffname
  6    INTO LastName, FirstName
  7    FROM FACULTY
  8    WHERE fid = 6;
  9    DBMS_OUTPUT.PUT_LINE('The faculty member''s name is ' || FirstName || ' ' || LastName);
 10  END;
 11  /
DECLARE
*
ERROR at line 1:
ORA-01403: no data found
ORA-06512: at line 5
```

Figure 4-38: Implicit cursor that returns no records

Explicit cursors must be used with SELECT statements that might retrieve a variable number of records or that might return no records. Explicit cursors are declared in the DECLARE section and processed in the program body. The steps for using an explicit cursor are:

1. Declare the cursor.
2. Open the cursor.
3. Fetch the cursor results into PL/SQL program variables.
4. Close the cursor.

Declaring an Explicit Cursor

Declaring an explicit cursor names the cursor and defines the query associated with the cursor. The general format for declaring an explicit cursor is:

```
CURSOR <cursorname> IS <SELECT statement>;
```

The cursor name can be any valid PL/SQL variable name. The SELECT statement is the query that will return the desired data values. You can use any legal SQL SELECT statement except one containing the UNION or MINUS operators. The query search condition can contain PL/SQL variables that have been assigned values, as long as the variables are declared before the cursor is declared. For example, suppose you want to create an explicit cursor named LocationCursor that retrieves all of the LOCATION records for a particular BLDG_CODE in the Northwoods University database. The BLDG_CODE that you want to retrieve will be stored in a character variable named CurrentBldgCode, and its value will be assigned before the cursor is processed. You would declare the variable and then the cursor as follows:

```
DECLARE
  CurrentBldgCode       VARCHAR2(5);
  CURSOR LocationCursor IS
       SELECT locid, room, capacity
       FROM location
       WHERE bldg_code = CurrentBldgCode;
```

Opening an Explicit Cursor

Opening the cursor causes the SQL compiler to **parse** the SQL query, which means to examine its individual components, and check them for syntax errors. The parsed query is then stored in the context area in an internal machine-language format, and the active set is defined. The general format of the command to open an explicit cursor is:

```
OPEN <cursor name>;
```

The OPEN command causes the cursor to identify the data rows that satisfy the SELECT query. However, the data values are not actually retrieved yet. The code for opening the LocationCursor you declared before is:

```
OPEN LocationCursor;
```

Fetching Data into an Explicit Cursor

The FETCH command retrieves the query data from the database into the active set, one row at a time. Since a query might return several rows, the FETCH command is usually executed within a loop. The general format of the FETCH command is:

```
FETCH <cursor name> INTO <record variable(s)>;
```

The record variable is either a single variable or a list of variables that will receive data from the field or fields currently being processed. Usually, this variable is declared using one of the reference variable data types that are declared using %TYPE or %ROWTYPE.

Recall that the %TYPE data type assumes the same data type as a specific database table field, and it is declared using the format `<tablename>.<field name>%TYPE`. Now we will look at the code for declaring and then processing a cursor that returns the CAPACITY field from the LOCATION table. The cursor output is fetched and stored into a variable named RoomCapacity that is declared to be the same data type as the CAPACITY field in the LOCATION table, using the reference data type %TYPE.

```
DECLARE
     CURSOR LocationCursor IS
          SELECT capacity
          FROM location;
     RoomCapacity     location.capacity%TYPE;
BEGIN
     OPEN LocationCursor;
     FETCH LocationCursor INTO RoomCapacity;
     <additional processing statements>
END;
```

If a cursor returns multiple data fields, the output is fetched into a variable that is declared using the %ROWTYPE data type. Recall that the %ROWTYPE data type assumes the same data type as a specific database table record when it is declared using the format `<table name>%ROWTYPE`. The %ROWTYPE data type can also assume the same data types as the data fields that a cursor returns when it is declared, using the format `<row variable name> <cursor name>%ROWTYPE;`.

Here is the code for declaring and then processing a cursor that returns three fields from the LOCATION table. The cursor output is fetched and stored into a variable named LocationRow that is declared using the reference data type `<cursor name>%ROWTYPE`. This variable automatically has the same number of fields and data types as a row returned by the cursor.

```
DECLARE
     CURSOR LocationCursor IS
          SELECT bldg_code, room, capacity
          FROM location;
     LocationRow      LocationCursor%ROWTYPE;
BEGIN
     OPEN LocationCursor;
     FETCH LocationCursor INTO LocationRow;
     <additional processing statements>
END;
```

The individual fields in the row variable can be referenced using the format `<row variable name>.<database field name>`. For instance, in the example, the current value of the BLDG_CODE variable in the cursor would be accessed using the variable name `LocationRow.bldg_code`.

Closing the Cursor

A cursor should be closed after processing is completed, so that its memory area and resources can be made available to the system for other tasks. The general format for the cursor CLOSE command is:

```
CLOSE <cursor name>;
```

If you forget to close a cursor, it will automatically be closed when the program where the cursor is declared ends.

Processing an Explicit Cursor

Explicit cursors can be processed using a loop that terminates when all rows are processed. Two different looping structures may be used: the LOOP ... EXIT WHEN loop, or the cursor FOR loop. When processing a cursor, the general format of the LOOP ... EXIT WHEN loop is:

```
BEGIN
     OPEN<cursor name>;
     LOOP
          FETCH <cursor name> INTO <cursor variable(s)>;
          EXIT WHEN <cursor name>%NOTFOUND;
          <additional program statements to process cursor fields>
     END LOOP;
     CLOSE <cursor name>;
END;
```

The EXIT WHEN command automatically exits the loop when the empty record that is after the last cursor record is fetched. Notice that the cursor processing steps are placed after the EXIT WHEN command. Otherwise, the last record will be processed twice.

Now you will write a program to create an explicit cursor and display its output using a LOOP ... EXIT WHEN processing structure. This cursor will retrieve and display the records in the LOCATION table where BLDG_CODE is 'LIB'. The search condition string 'LIB' will be assigned and referenced as a variable value.

To create and process an explicit cursor using a LOOP ... EXIT WHEN structure:

1. Type the code shown in Figure 4-39.
2. Run the program, and debug it if necessary. The formatted output should be displayed as shown.

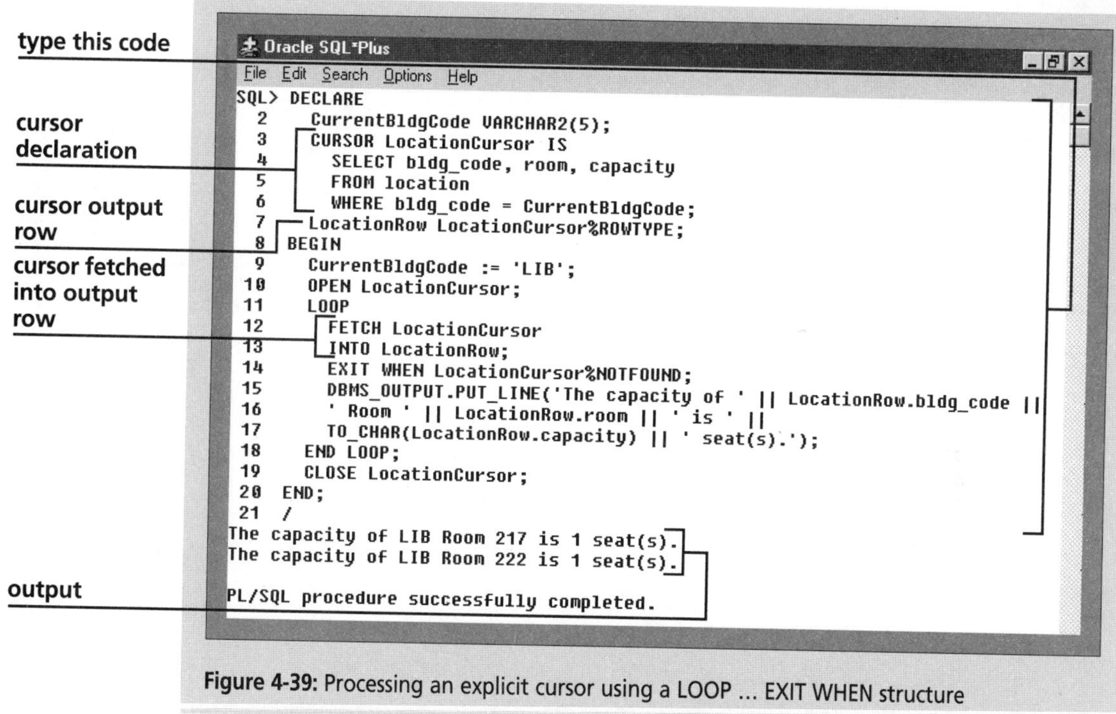
type this code
cursor declaration
cursor output row
cursor fetched into output row
output
Oracle SQL*Plus
File Edit Search Options Help
SQL> DECLARE
CurrentBldgCode VARCHAR2(5);
CURSOR LocationCursor IS
SELECT bldg_code, room, capacity
FROM location
WHERE bldg_code = CurrentBldgCode;
LocationRow LocationCursor%ROWTYPE;
BEGIN
CurrentBldgCode := 'LIB';
OPEN LocationCursor;
LOOP
FETCH LocationCursor
INTO LocationRow;
EXIT WHEN LocationCursor%NOTFOUND;
DBMS_OUTPUT.PUT_LINE('The capacity of ' || LocationRow.bldg_code ||
' Room ' || LocationRow.room || ' is ' ||
TO_CHAR(LocationRow.capacity) || ' seat(s).');
END LOOP;
CLOSE LocationCursor;
END;
/
The capacity of LIB Room 217 is 1 seat(s).
The capacity of LIB Room 222 is 1 seat(s).
PL/SQL procedure successfully completed.

Figure 4-39: Processing an explicit cursor using a LOOP ... EXIT WHEN structure

The cursor FOR loop provides an easier way to process explicit cursors, because you do not need to explicitly open, fetch the rows, or close the cursor. By using the cursor FOR loop, Oracle implicitly performs these operations. The general format of a cursor FOR loop is:

```
BEGIN
    FOR <cursor variable(s)> IN <cursor name> LOOP
        <additional program statements>
    END LOOP;
END;
```

Now you will write a program to use a cursor FOR loop for processing the LocationCursor. You do not need to explicitly open, fetch into, or close the cursor, so the processing requires less code.

To process a cursor using a cursor FOR loop:

1 Type the program code shown in Figure 4-40.

2 Run the program, and debug it if necessary. The LOCATION records are displayed as before.

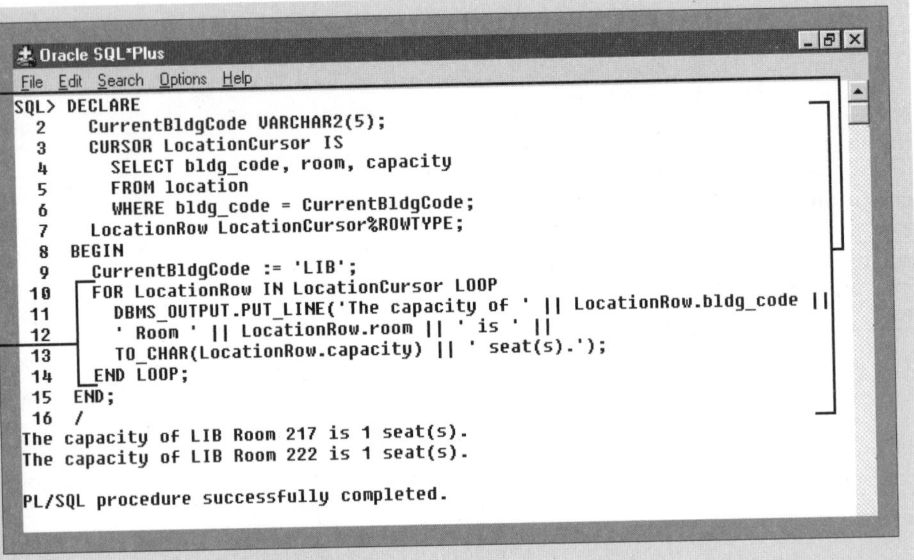

type this code

cursor FOR loop

Oracle SQL*Plus

File Edit Search Options Help

```
SQL> DECLARE
    CurrentBldgCode VARCHAR2(5);
    CURSOR LocationCursor IS
      SELECT bldg_code, room, capacity
      FROM location
      WHERE bldg_code = CurrentBldgCode;
    LocationRow LocationCursor%ROWTYPE;
  BEGIN
    CurrentBldgCode := 'LIB';
    FOR LocationRow IN LocationCursor LOOP
      DBMS_OUTPUT.PUT_LINE('The capacity of ' || LocationRow.bldg_code ||
      ' Room ' || LocationRow.room || ' is ' ||
      TO_CHAR(LocationRow.capacity) || ' seat(s).');
    END LOOP;
  END;
  /
The capacity of LIB Room 217 is 1 seat(s).
The capacity of LIB Room 222 is 1 seat(s).

PL/SQL procedure successfully completed.
```

Figure 4-40: Processing an explicit cursor using a cursor FOR loop

Other Explicit Cursor Attributes

Explicit cursors have **attributes** that describe the cursor's current state, such as whether the cursor is currently open or not, whether records are found or not after a fetch operation, and how many records are fetched. These attributes are appended to the cursor name. Explicit cursor attributes are described in Figure 4-41.

Attribute	Description
%NOTFOUND	Evaluates as TRUE when a cursor has no rows left to fetch, and FALSE when a cursor has remaining rows to fetch
%FOUND	Evaluates as TRUE when a cursor has rows remaining to fetch, and FALSE when a cursor has no rows left to fetch
%ROWCOUNT	Returns the number of rows that a cursor has fetched so far
%ISOPEN	Returns TRUE if the cursor is open and FALSE if the cursor is closed

Figure 4-41: Explicit cursor attributes

The %ISOPEN attribute can be used before or after the cursor is opened. The other attributes can only be used after the cursor is opened.

You used the %NOTFOUND attribute when you used the LOOP ... EXIT WHEN processing structure to exit the loop when all cursor rows were fetched and processed. The %NOTFOUND and %FOUND attributes are also useful for detecting when a cursor query returns no rows—in this case, they allow you to display a customized message to tell the user what happened. Figure 4-42 shows how to use the %NOTFOUND attribute within an IF/THEN structure to notify the user when a cursor returns no rows.

nonexistent search condition will return no rows

error-detection commands

Oracle SQL*Plus

File Edit Search Options Help

```
SQL> DECLARE
  2    CurrentBldgCode VARCHAR2(5);
  3    CURSOR LocationCursor IS
  4      SELECT bldg_code, room, capacity
  5      FROM location
  6      WHERE bldg_code = CurrentBldgCode;
  7    LocationRow LocationCursor%ROWTYPE;
  8  BEGIN
  9    CurrentBldgCode := 'GEOL';
 10    OPEN LocationCursor;
 11    LOOP
 12      FETCH LocationCursor
 13      INTO LocationRow;
 14      EXIT WHEN LocationCursor%NOTFOUND;
 15      DBMS_OUTPUT.PUT_LINE('The capacity of ' || LocationRow.bldg_code ||
 16      ' Room ' || LocationRow.room || ' is ' ||
 17      TO_CHAR(LocationRow.capacity) || ' seat(s).');
 18    END LOOP;
 19    IF LocationCursor%NOTFOUND THEN
 20      DBMS_OUTPUT.PUT_LINE('The building code ' || CurrentBldgCode ||
 21      ' does not return any records.');
 22    END IF;
 23    CLOSE LocationCursor;
 24  END;
 25  /
The building code GEOL does not return any records.

PL/SQL procedure successfully completed.
```

Figure 4-42: Using the %NOTFOUND cursor attribute to generate a custom error message when no rows are returned

Sometimes you use a cursor to determine whether or not a SELECT query will return any records. Figure 4-43 shows how the %FOUND cursor attribute can be used in an IF/THEN structure to determine whether or not a cursor returns any records.

Oracle SQL*Plus

File Edit Search Options Help

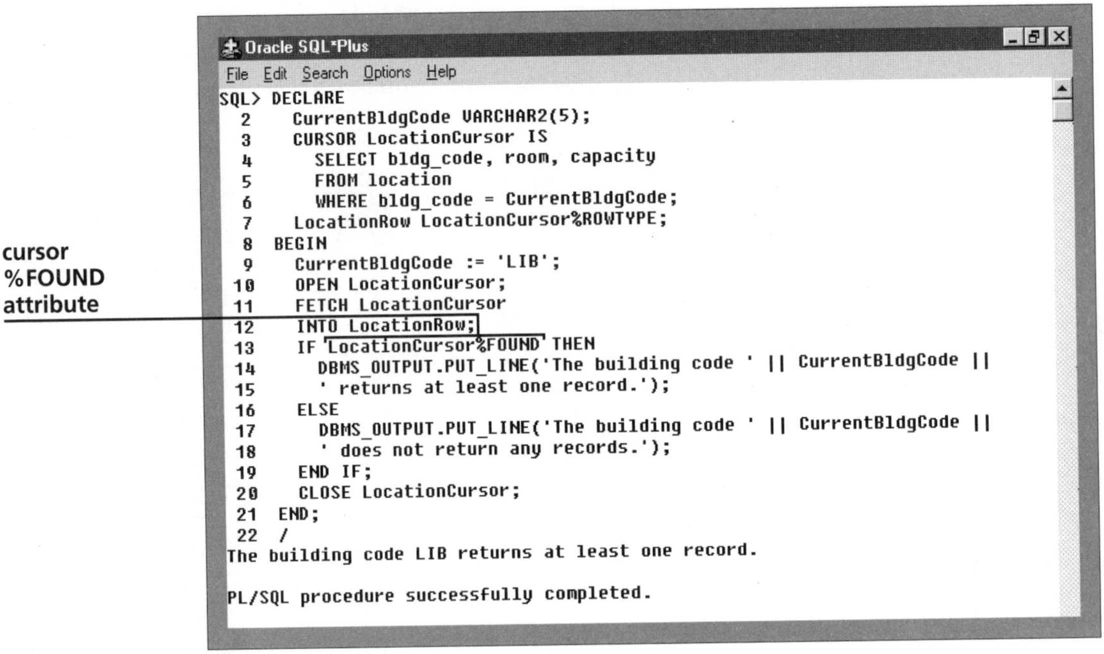

```
SQL> DECLARE
  CurrentBldgCode VARCHAR2(5);
  CURSOR LocationCursor IS
    SELECT bldg_code, room, capacity
    FROM location
    WHERE bldg_code = CurrentBldgCode;
  LocationRow LocationCursor%ROWTYPE;
BEGIN
  CurrentBldgCode := 'LIB';
  OPEN LocationCursor;
  FETCH LocationCursor
  INTO LocationRow;
  IF LocationCursor%FOUND THEN
    DBMS_OUTPUT.PUT_LINE('The building code ' || CurrentBldgCode ||
    ' returns at least one record.');
  ELSE
    DBMS_OUTPUT.PUT_LINE('The building code ' || CurrentBldgCode ||
    ' does not return any records.');
  END IF;
  CLOSE LocationCursor;
END;
/
The building code LIB returns at least one record.

PL/SQL procedure successfully completed.
```

cursor %FOUND attribute

Figure 4-43: Using the %FOUND attribute to signal whether or not a cursor returns any records

The %ROWCOUNT attribute returns the total number of records fetched by the cursor. Figure 4-44 shows how this attribute is used to display the number of records returned by the LocationCursor query.

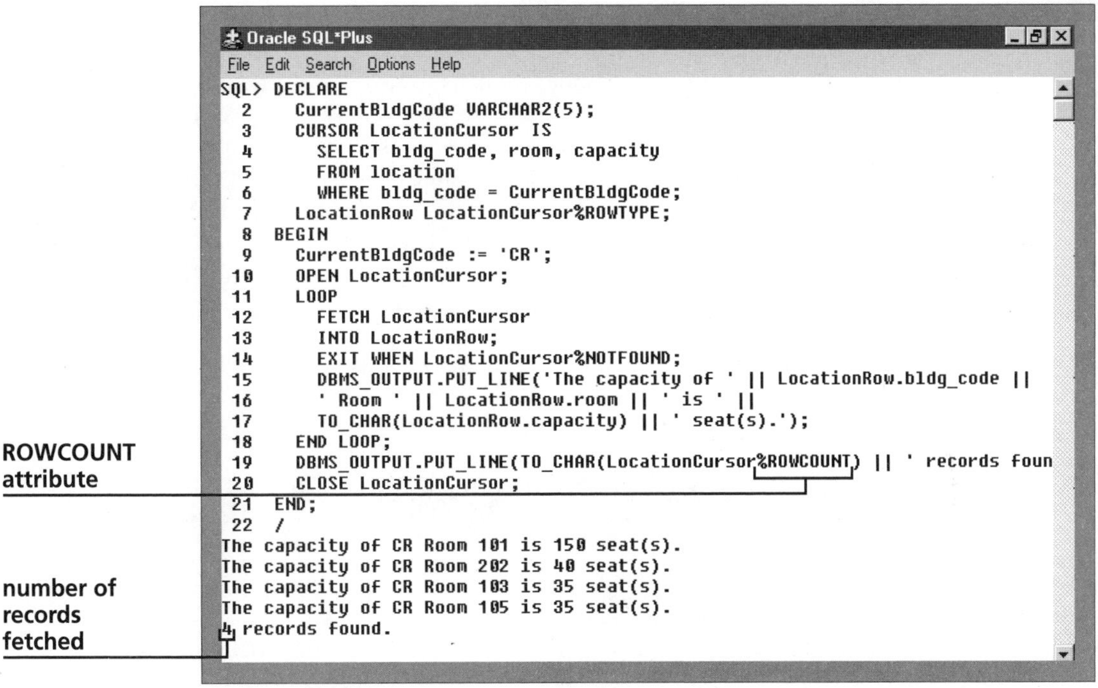

Oracle SQL*Plus

File Edit Search Options Help

```
SQL> DECLARE
    CurrentBldgCode VARCHAR2(5);
    CURSOR LocationCursor IS
      SELECT bldg_code, room, capacity
      FROM location
      WHERE bldg_code = CurrentBldgCode;
    LocationRow LocationCursor%ROWTYPE;
  BEGIN
    CurrentBldgCode := 'CR';
    OPEN LocationCursor;
    LOOP
      FETCH LocationCursor
      INTO LocationRow;
      EXIT WHEN LocationCursor%NOTFOUND;
      DBMS_OUTPUT.PUT_LINE('The capacity of ' || LocationRow.bldg_code ||
      ' Room ' || LocationRow.room || ' is ' ||
      TO_CHAR(LocationRow.capacity) || ' seat(s).');
    END LOOP;
    DBMS_OUTPUT.PUT_LINE(TO_CHAR(LocationCursor%ROWCOUNT) || ' records foun
    CLOSE LocationCursor;
  END;
  /
The capacity of CR Room 101 is 150 seat(s).
The capacity of CR Room 202 is 40 seat(s).
The capacity of CR Room 103 is 35 seat(s).
The capacity of CR Room 105 is 35 seat(s).
4 records found.
```

ROWCOUNT attribute

number of records fetched

Figure 4-44: Using the %ROWCOUNT attribute to display the number of records fetched by an explicit cursor

Handling Exceptions in PL/SQL Programs

Programmers prefer to concentrate on the positive aspects of their programs. Creating programs that generate invoices, process checks, track time usage, and so forth is difficult and time-consuming. As a result, programmers tend to breathe a sigh of relief once the program works and overlook or ignore the consequences of a user pressing the wrong key or a critical network link going down. The reality, however, is that users enter incorrect data and press wrong keys, networks fail, computers crash, and anything else that can go wrong (eventually) will go wrong. Programmers can't do much for the user if the user's computer fails, but programmers can and should do everything possible to prevent incorrect data from being entered and wrong keystrokes from damaging the system, and to inform the user of errors and how to correct them when they occur. PL/SQL offers **exception handling**, whereby all code for displaying error messages and giving users options for fixing errors is placed in the EXCEPTION section of the code. This is PL/SQL's primary method for processing errors and informing users of corrective actions.

Recall that program errors can occur either at compile time or at run time. Compile errors usually involve spelling or syntax problems. Compile errors are reported by the PL/SQL compiler, and have to be corrected by the programmer before the program will compile. Figure 4-45 shows an example of a compile error, where the comparison operator (=) was used instead of the assignment operator (:=). The error code ORA-06550 indicates that it is a compile error, and shows the error's line and column location. The error code PLS-00103 provides specific details about the type of error that occurred.

Error codes with the ORA- prefix are generated by the Oracle DBMS, and usually involve constraint errors, such as trying to insert a record with a NULL value in a field with a NOT NULL constraint. Error codes with the PLS- prefix are generated by the PL/SQL compiler, and involve errors in PL/SQL syntax.

comparison operator used instead of assignment operator

error location

indicates compile error

describes specific compile error

compile error message

Oracle SQL*Plus

File Edit Search Options Help

```
SQL> DECLARE
  2    TermID BINARY_INTEGER;
  3    TermDesc VARCHAR2(20);
  4    TermStatus VARCHAR2(20);
  5  BEGIN
  6    TermID := 1;
  7    TermDesc = 'Fall 2000';
  8    TermStatus := 'CLOSED';
  9    -- insert the records
 10    INSERT INTO term VALUES (TermID, TermDesc, TermStatus);
 11    TermID := TermID + 1;
 12    TermDesc := 'Spring 2001';
 13    COMMIT;
 14  END;
 15  /
 TermDesc = 'Fall 2000';
          *
ERROR at line 7:
ORA-06550: line 7, column 12:
PLS-00103: Encountered the symbol "=" when expecting one of the following:
:= . ( @ % ;
The symbol ":= was inserted before "=" to continue.
```

Figure 4-45: Example of a compile error

Run-time errors are reported by the PL/SQL run-time engine, and must be handled in the exception-handling section of the program. Run-time errors usually involve problems with data values, such as trying to retrieve no rows or several rows using an implicit cursor, trying to divide by zero, or trying to insert an illegal value into a database table. Figure 4-46 shows a run-time error generated when an attempt to insert a record into the TERM table was unsuccessful, because the check constraint ensuring that the value of the STATUS field is either 'OPEN' or

'CLOSED' was violated. The error code ORA-02290 gives details about the nature of the error. The error code ORA-06512 indicates that it is a run-time error rather than a compile error, and shows the line number that generated the error.

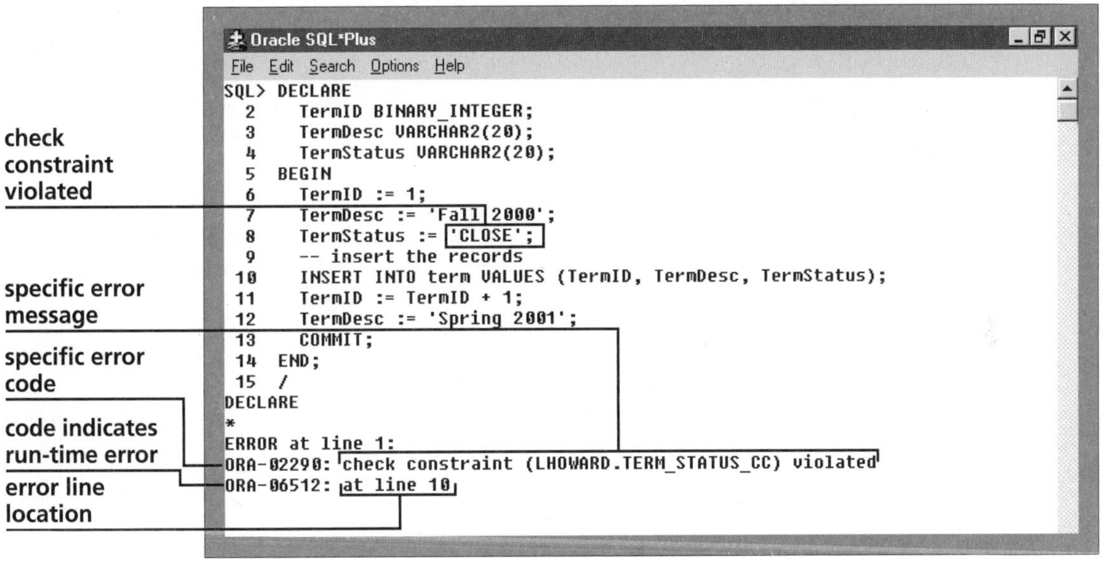

check constraint violated

specific error message

specific error code

code indicates run-time error

error line location

```
Oracle SQL*Plus
File Edit Search Options Help
SQL> DECLARE
  2    TermID BINARY_INTEGER;
  3    TermDesc VARCHAR2(20);
  4    TermStatus VARCHAR2(20);
  5  BEGIN
  6    TermID := 1;
  7    TermDesc := 'Fall 2000';
  8    TermStatus := 'CLOSE';
  9    -- insert the records
 10    INSERT INTO term VALUES (TermID, TermDesc, TermStatus);
 11    TermID := TermID + 1;
 12    TermDesc := 'Spring 2001';
 13    COMMIT;
 14  END;
 15  /
DECLARE
*
ERROR at line 1:
ORA-02290: check constraint (LHOWARD.TERM_STATUS_CC) violated
ORA-06512: at line 10
```

Figure 4-46: Example of a run-time error

When a run-time error occurs, an **exception**, or unwanted event, is **raised** (or occurs). Program control is immediately transferred to the EXCEPTION section of the PL/SQL program. There, the programmer might choose to correct the error without notifying the user of the problem, or the programmer might inform the user of the error without taking corrective action. The programmer also could correct the error and inform the user of the error, or could inform the user of the error and allow the user to decide what action to take.

There are three kinds of exceptions: predefined, undefined, and user-defined. The following paragraphs describe these exceptions and how they are handled.

Predefined Exceptions

Predefined exceptions correspond to common errors that are seen in many programs and that have been given a specific exception name. Figure 4-47 presents a list of the most common predefined exceptions.

Oracle Error Code	Exception Name	Description
ORA-00001	DUP_VAL_ON_INDEX	Unique constraint on primary key violated
ORA-01001	INVALID_CURSOR	Illegal cursor operation
ORA-01403	NO_DATA_FOUND	Query returns no records
ORA-01422	TOO_MANY_ROWS	Query returns more rows than anticipated
ORA-01476	ZERO_DIVIDE	Division by zero
ORA-01722	INVALID_NUMBER	Invalid number conversion (like trying to convert '2B' to a number)
ORA-06502	VALUE_ERROR	Error in truncation, arithmetic, or conversion operation

Figure 4-47: Common Oracle predefined exceptions

Remember that when an exception is raised, program control is immediately transferred to the EXCEPTION block of the program. There, code must be written to explicitly handle every anticipated exception, as well as other exceptions that are not anticipated. The general format for handling predefined and other exceptions is:

```
EXCEPTION
     WHEN <exception1 name> THEN
          <exception1 handling statements>;
     WHEN <exception2 name> THEN
          <exception2 handling statements>;
     ...
     WHEN OTHERS THEN
          <other handling statements>;
END;
```

The `<exception name>` parameter refers to the predefined exception name, as shown in Figure 4-47. The `<exception handling statements>` are the code lines that inform the user of the error, take corrective action, etc. The combination of the WHEN `<exception name>` THEN statement and the associated exception-handling statements is called an **exception handler**. The WHEN OTHERS statement is a catch-all exception handler that allows you to present a general message to describe errors not handled by a specific error-handling statement. After an exception handler finishes processing, program execution terminates. The exception-handling statements are indented to make the exception section easier to read and understand.

Now you will write a program that handles predefined exceptions. In this program, an implicit cursor selects an ITEMDESC from the ITEM table, using an invalid search condition. When an implicit cursor query returns no records, the predefined exception NO_DATA_FOUND is generated. First, you will write the program without an exception handler to view the error message that is generated by the PL/SQL run-time engine. Then, you will add an exception handler that will inform the user of the nature of the error, and suggest a corrective action.

To write the program to handle a predefined exception:

1 Type the code shown in Figure 4-48. Run the program. The Oracle error code ORA-06512 should be displayed, indicating that there is a run-time error. The specific Oracle error code ORA-01403 and associated error message "no data found" also are displayed. (If another error is displayed, debug the program until the output shown in Figure 4-48 is displayed.)

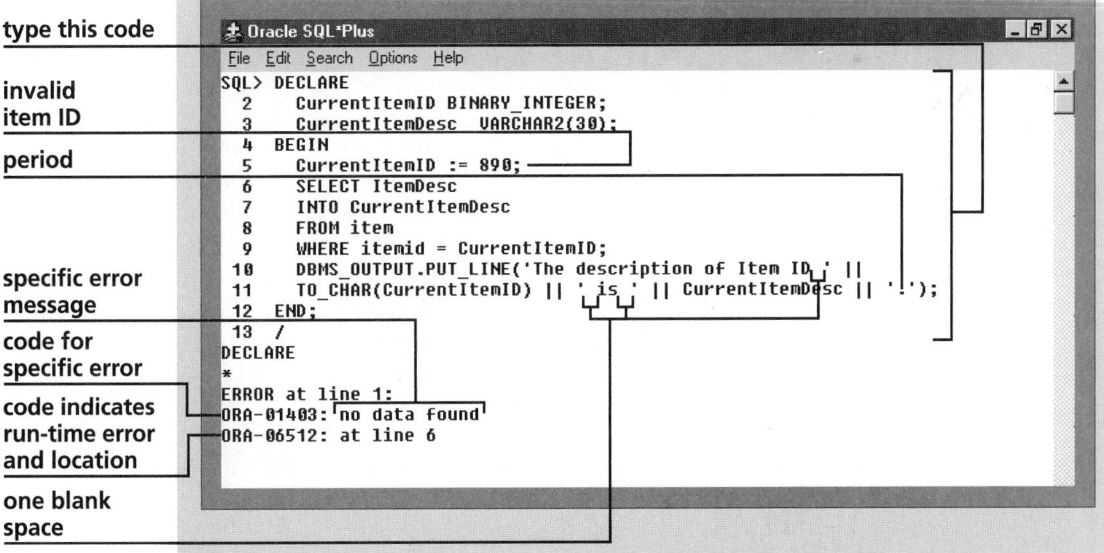
type this code
invalid item ID
period
specific error message
code for specific error
code indicates run-time error and location
one blank space
Oracle SQL*Plus
File Edit Search Options Help
SQL> DECLARE
CurrentItemID BINARY_INTEGER;
CurrentItemDesc VARCHAR2(30);
BEGIN
CurrentItemID := 890;
SELECT ItemDesc
INTO CurrentItemDesc
FROM item
WHERE itemid = CurrentItemID;
DBMS_OUTPUT.PUT_LINE('The description of Item ID ' ||
TO_CHAR(CurrentItemID) || ' is ' || CurrentItemDesc || '.');
END;
/
DECLARE
*
ERROR at line 1:
ORA-01403: no data found
ORA-06512: at line 6

Figure 4-48: PL/SQL program with unhandled NO_DATA_FOUND exception

2 Modify your program code by adding the exception-handling section shown in Figure 4-49. Run the program, and debug it if necessary. The more informative and instructive error messages specified by the exception handler now are displayed.

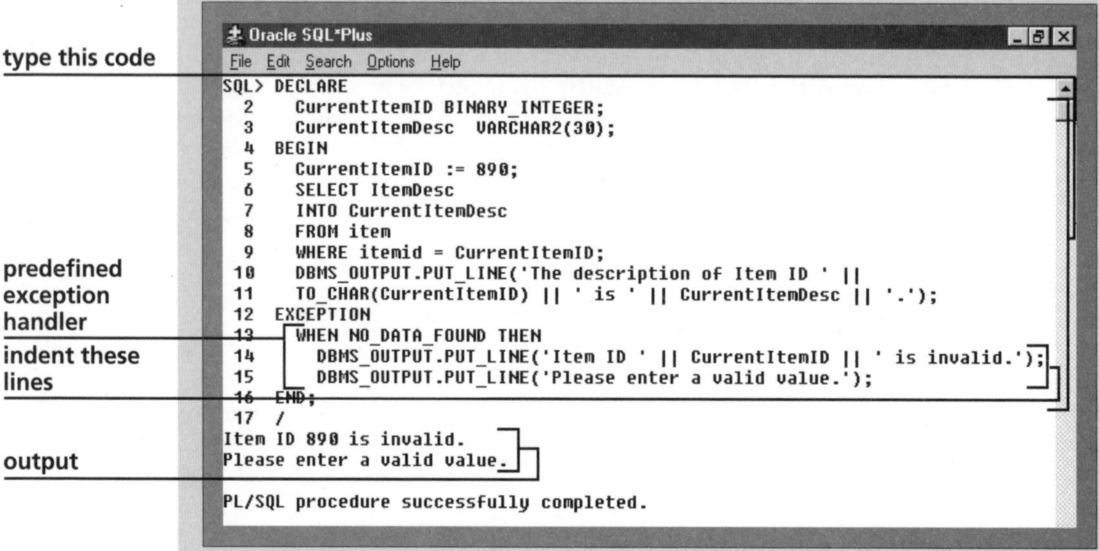

type this code

predefined exception handler

indent these lines

output

```
Oracle SQL*Plus
File Edit Search Options Help
SQL> DECLARE
    CurrentItemID BINARY_INTEGER;
    CurrentItemDesc  VARCHAR2(30);
  BEGIN
    CurrentItemID := 890;
    SELECT ItemDesc
    INTO CurrentItemDesc
    FROM item
    WHERE itemid = CurrentItemID;
    DBMS_OUTPUT.PUT_LINE('The description of Item ID ' ||
    TO_CHAR(CurrentItemID) || ' is ' || CurrentItemDesc || '.');
  EXCEPTION
    WHEN NO_DATA_FOUND THEN
      DBMS_OUTPUT.PUT_LINE('Item ID ' || CurrentItemID || ' is invalid.');
      DBMS_OUTPUT.PUT_LINE('Please enter a valid value.');
  END;
  /
Item ID 890 is invalid.
Please enter a valid value.

PL/SQL procedure successfully completed.
```

Figure 4-49: PL/SQL program that handles a predefined exception

During program development, it is helpful to use the WHEN OTHERS exception handler to display the associated Oracle error code number and error message for unanticipated errors. To do this, you use the SQLERRM function. This function returns a character string that contains the Oracle error code and the text of the error code's error message for the most recent Oracle error generated. To use this function, you must first declare a VARCHAR2 character variable that will be assigned to the text of the error message. The maximum length of the character variable will be 512, because the maximum length of an Oracle error message is 512 characters.

You will now modify your program so that the implicit query returns all of the rows from the INVENTORY table for item ID 894 (Women's Hiking Shorts). Since an implicit query cannot handle more than one row, the query will generate an unhandled exception. You will add the WHEN OTHERS error handler so that it displays the associated Oracle error code and message when an unhandled error is generated.

To modify the WHEN OTHERS error handler to display other error codes and messages:

1 Modify the cursor SELECT command and add the WHEN OTHERS error handler, as shown in Figure 4-50. The error code and message are displayed as shown in the figure.

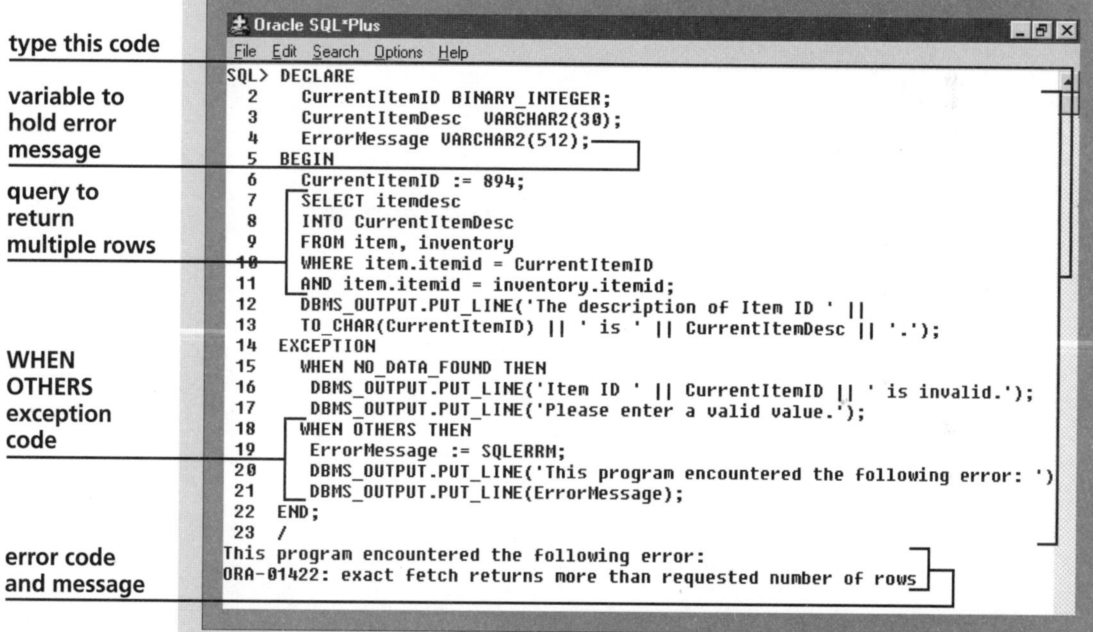

type this code
variable to hold error message
query to return multiple rows
WHEN OTHERS exception code
error code and message
Oracle SQL*Plus
File Edit Search Options Help
SQL> DECLARE
CurrentItemID BINARY_INTEGER;
CurrentItemDesc VARCHAR2(30);
ErrorMessage VARCHAR2(512);
BEGIN
CurrentItemID := 894;
SELECT itemdesc
INTO CurrentItemDesc
FROM item, inventory
WHERE item.itemid = CurrentItemID
AND item.itemid = inventory.itemid;
DBMS_OUTPUT.PUT_LINE('The description of Item ID ' ||
TO_CHAR(CurrentItemID) || ' is ' || CurrentItemDesc || '.');
EXCEPTION
WHEN NO_DATA_FOUND THEN
DBMS_OUTPUT.PUT_LINE('Item ID ' || CurrentItemID || ' is invalid.');
DBMS_OUTPUT.PUT_LINE('Please enter a valid value.');
WHEN OTHERS THEN
ErrorMessage := SQLERRM;
DBMS_OUTPUT.PUT_LINE('This program encountered the following error: ')
DBMS_OUTPUT.PUT_LINE(ErrorMessage);
END;
/
This program encountered the following error:
ORA-01422: exact fetch returns more than requested number of rows

Figure 4-50: Displaying the Oracle error code and message in the WHEN OTHERS exception handler

Undefined Exceptions

Undefined exceptions are the less common errors that have not been given an explicit exception name. (Recall that defined exceptions are the common errors that have been given explicit names, which are listed in Figure 4-47.) Figure 4-51 shows an example of an undefined exception. Here, a program attempts to insert a record into the TERM table where the data value for the third table field (STATUS) is NULL. The TERM table has a NOT NULL constraint for that field, so every record must have a STATUS data value. As a result, the Oracle error code and associated error message are displayed.

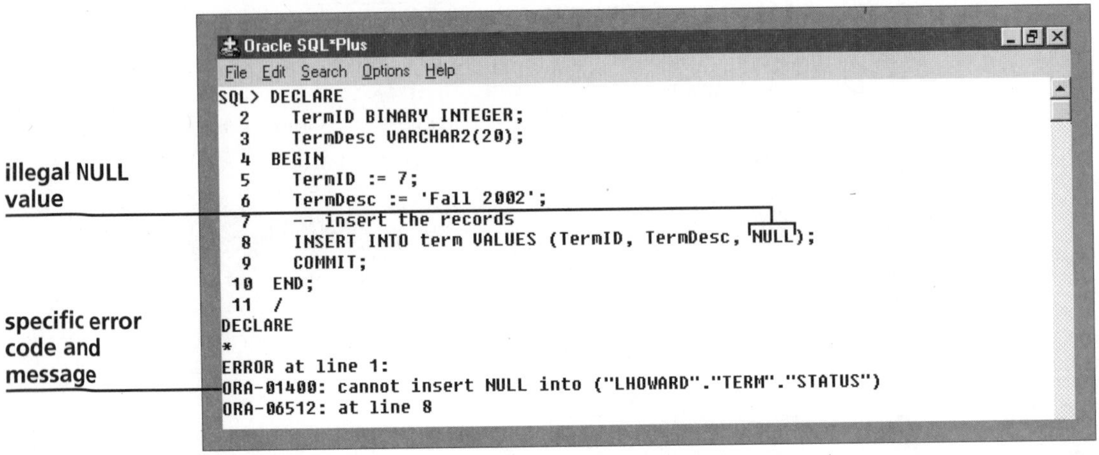

illegal NULL value

specific error code and message

```
Oracle SQL*Plus
File Edit Search Options Help
SQL> DECLARE
  2    TermID BINARY_INTEGER;
  3    TermDesc VARCHAR2(20);
  4  BEGIN
  5    TermID := 7;
  6    TermDesc := 'Fall 2002';
  7    -- insert the records
  8    INSERT INTO term VALUES (TermID, TermDesc, NULL);
  9    COMMIT;
 10  END;
 11  /
DECLARE
*
ERROR at line 1:
ORA-01400: cannot insert NULL into ("LHOWARD"."TERM"."STATUS")
ORA-06512: at line 8
```

Figure 4-51: PL/SQL program with undefined exception

To handle an undefined exception, you must explicitly declare the exception in the DECLARE section of the program and associate it with a specific Oracle error code. Then, you can create an error handler as you did for predefined exceptions. The general format for declaring an exception is:

```
DECLARE
    <e_exception name> EXCEPTION;
    PRAGMA EXCEPTION_INIT(<e_exception name>, <Oracle error code>);
```

The `<e_exception name>` parameter can be any legal variable name. Usually, user-declared exception names are prefixed with `e_` to keep them from being confused with other variables. The `PRAGMA EXCEPTION_INIT` command tells the compiler to associate the given exception name with a specific Oracle error code. The `<Oracle error code>` is the numeric error code that Oracle assigns to run-time errors. The error code number must be preceded by a hyphen (-), and leading zeroes can be omitted. Now, you will write a program that declares an exception named e_NotNullInsert that is associated with Oracle error code ORA-01400. When the user attempts to insert NULL into a field with a NOT NULL constraint, an exception handler will display a message.

Recall that a complete list of Oracle error codes can be found by clicking Start on the Windows taskbar, pointing to Programs, pointing to Oracle for Windows 95 or Windows NT, clicking Oracle8 Error Messages, and then searching for error codes using the prefix ORA-.

To write a program to handle an undefined exception:

1 Type the code shown in Figure 4-52. Run the program and debug it if necessary. The exception handler error message is displayed instead of the Oracle error codes and messages.

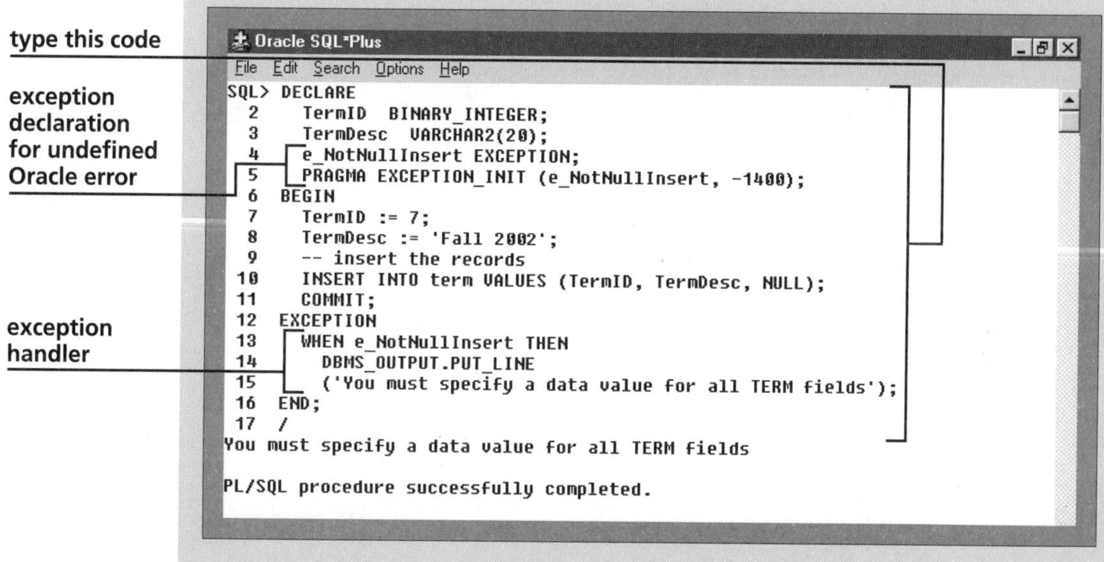

type this code
exception declaration for undefined Oracle error
exception handler
Oracle SQL*Plus
File Edit Search Options Help
SQL> DECLARE
TermID BINARY_INTEGER;
TermDesc VARCHAR2(20);
e_NotNullInsert EXCEPTION;
PRAGMA EXCEPTION_INIT (e_NotNullInsert, -1400);
BEGIN
TermID := 7;
TermDesc := 'Fall 2002';
-- insert the records
INSERT INTO term VALUES (TermID, TermDesc, NULL);
COMMIT;
EXCEPTION
WHEN e_NotNullInsert THEN
DBMS_OUTPUT.PUT_LINE
('You must specify a data value for all TERM fields');
END;
/
You must specify a data value for all TERM fields
PL/SQL procedure successfully completed.

Figure 4-52: PL/SQL program with exception handler for undefined exception

User-Defined Exceptions

User-defined exceptions are used to handle exceptions that will not cause an Oracle run-time error, but that require exception handling to enforce business rules or to ensure the integrity of the database. Suppose that a database program is used to delete records from the ENROLLMENT table. When a student drops a course, the record deleted from the ENROLLMENT table has a NULL value for the GRADE field. However, suppose that a user tries to delete an ENROLLMENT record where GRADE is not NULL. The program should raise an exception to advise the user that the grade field is not NULL, and then not delete the record.

The general format for declaring, raising, and handling a user-defined exception is as follows:

```
DECLARE
     <e_exception name>EXCEPTION;
     <other variable declarations>;
BEGIN
     <other program statements>
     IF <undesirable condition> THEN
          RAISE <e_exception name>
     END IF;
     <other program statements>;
EXCEPTION
     <e_exception name>
          <error-handling statements>;
END;
```

Now you will write a program that uses a user-defined exception to avoid deleting an ENROLLMENT record with an assigned grade.

To write a program with a user-defined exception:

1 Type the code shown in Figure 4-53. Run the program, and debug it if necessary. The user-defined exception message is displayed. This exception-handling message could also be placed directly in the ELSE portion of the IF/THEN statement. However, it is a good programming practice to place all error-handling statements in the EXCEPTION section, because this keeps all of the error-handling program statements in the same place and makes the program easier to understand and maintain.

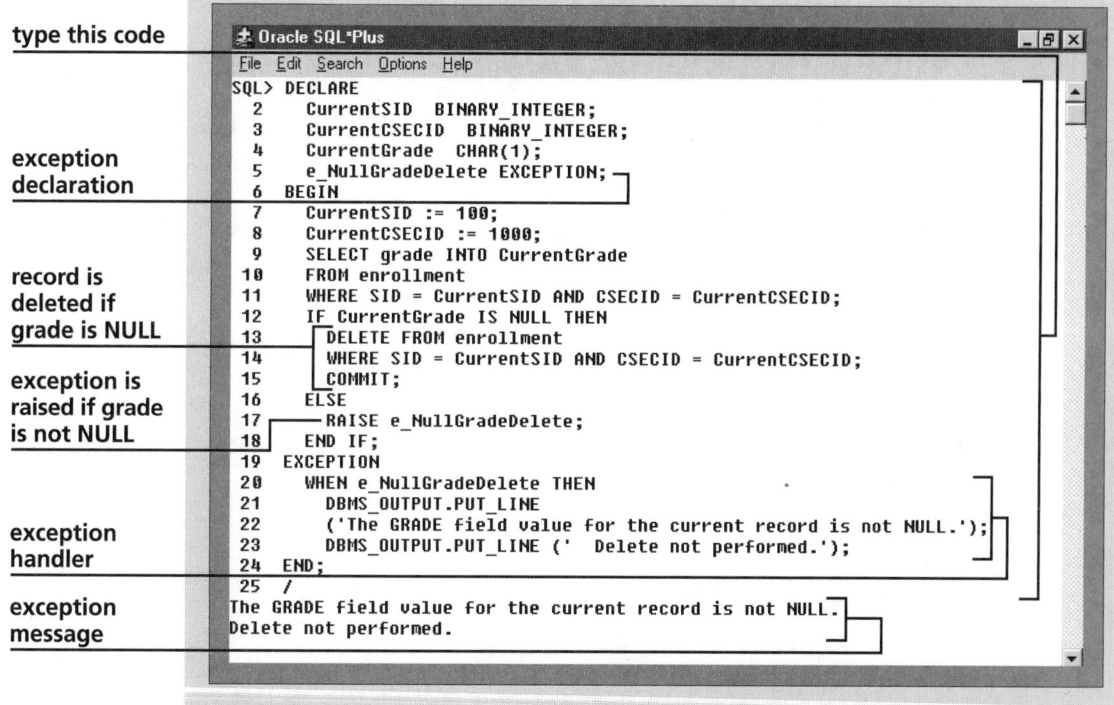
type this code
exception declaration
record is deleted if grade is NULL
exception is raised if grade is not NULL
exception handler
exception message
Oracle SQL*Plus
File Edit Search Options Help
SQL> DECLARE
CurrentSID BINARY_INTEGER;
CurrentCSECID BINARY_INTEGER;
CurrentGrade CHAR(1);
e_NullGradeDelete EXCEPTION;
BEGIN
CurrentSID := 100;
CurrentCSECID := 1000;
SELECT grade INTO CurrentGrade
FROM enrollment
WHERE SID = CurrentSID AND CSECID = CurrentCSECID;
IF CurrentGrade IS NULL THEN
DELETE FROM enrollment
WHERE SID = CurrentSID AND CSECID = CurrentCSECID;
COMMIT;
ELSE
RAISE e_NullGradeDelete;
END IF;
EXCEPTION
WHEN e_NullGradeDelete THEN
DBMS_OUTPUT.PUT_LINE
('The GRADE field value for the current record is not NULL.');
DBMS_OUTPUT.PUT_LINE (' Delete not performed.');
END;
/
The GRADE field value for the current record is not NULL.
Delete not performed.

Figure 4-53: Creating and raising a user-defined exception

2. Exit SQL*Plus.
3. Save your program files in Notepad and close Notepad.

SUMMARY

- SQL commands can be classified as Data Definition Language (DDL) commands, which change the database structure; Data Manipulation Language (DML) commands, which query or manipulate the data in database tables; or transaction control commands, which organize commands into logical transactions and commit them to the database or roll them back.
- DML and transaction control commands can be used in PL/SQL programs, but DDL commands cannot.
- A loop repeats an action multiple times until an ending condition is reached.

- PL/SQL has five different loops: LOOP ... EXIT, LOOP ... EXIT WHEN, WHILE ... LOOP, numeric FOR loops, and cursor FOR loops.
- The LOOP ... EXIT loop can be either a pretest or posttest loop.
- You do not need to declare a counter for a numeric FOR loop. The counter is defined in a series of numbers in the FOR statement, and is automatically incremented each time the loop repeats.
- A cursor is a pointer to a memory area called the context area that Oracle uses to process a query. The context area contains the number of rows processed, the parsed query, and the active set of records that have been fetched.
- Implicit cursors do not need to be formally declared. They are used with SELECT statements that return one and only one record.
- Explicit cursors must be formally declared. They are used with SELECT statements that might retrieve a variable number of records, or that might return no records.
- To use an explicit cursor, you must declare the cursor, open it, fetch the records, and then close the cursor.
- Usually, an explicit cursor fetches records into a variable that is declared using either the %TYPE or %ROWTYPE reference variable data type.
- A cursor should be closed after processing is completed so that its memory area and resources can be released.
- When you process an explicit cursor using a cursor FOR loop, you do not need to explicitly open the cursor, fetch the rows, or close the cursor.
- PL/SQL uses exception handling, whereby all error-handling code is placed in the EXCEPTION section of the code, as its method for processing errors and informing users of corrective actions.
- When a run-time error occurs, an exception (or unwanted event) is raised.
- Run-time errors are reported by the PL/SQL run-time engine, and must be handled in the exception-handling section of the program.
- When an exception is raised, program control is immediately transferred to the EXCEPTION block of the program.
- Predefined exceptions correspond to common errors seen in many programs, and are handled using predefined exception names in error handlers.
- Undefined exceptions are less-common errors that must be given an explicit exception name before they can be handled in the EXCEPTION section.
- User-defined exceptions are used to handle exceptions that will not cause an Oracle run-time error, but that require exception handling to enforce business rules or ensure the integrity of the database.

REVIEW QUESTIONS

1. Specify whether each of the following is a DDL, DML, or transaction control SQL command. Also state whether or not the command can be used in a PL/SQL program.
   a. CREATE SEQUENCE
   b. UPDATE
   c. UNION
   d. ROLLBACK TO SAVEPOINT
   e. GRANT PRIVILEGE
   f. SELECT ... FOR UPDATE
2. Write a loop using the following loop structures that inserts a number variable named Counter into a table named COUNT_TABLE 13 times starting with 0 and ending with 12. COUNT_TABLE has a single numeric field named COUNTID.
   a. LOOP ... EXIT WHEN
   b. WHILE ... LOOP
   c. Numeric FOR loop
3. Of the four basic (noncursor) PL/SQL loop types (LOOP ... EXIT, LOOP ... EXIT WHEN, WHILE ... LOOP, numeric FOR LOOP), which are pretest loops, and which are posttest loops?
4. What is the difference between the data that can be retrieved by an implicit cursor and the data that can be retrieved by an explicit cursor?
5. Write the code for a PL/SQL block that uses an implicit cursor to select LAST from the CUSTOMER table in the Clearwater Traders database into a variable named CurrentLastName. The search condition is CUSTID = 100.
6. Write a PL/SQL block that uses an explicit cursor named CustomerNames to retrieve the first and last names from all of the records in the CUSTOMER table, and then displays each first and last name using the DBMS_OUTPUT command. Use a LOOP ... EXIT WHEN loop to process the cursor. The results should be displayed as follows:

   Paula Harris
   Mitch Edwards
   Maria Garcia
   Lee Miller
   Alissa Chang
7. Repeat Exercise 6 using a cursor FOR loop to process the cursor.
8. What is the difference between a compile error and a run-time error?
9. What is an exception? What happens to program execution when an exception is raised?
10. Do exception handlers handle compile errors or run-time errors?
11. What is the difference between a predefined exception and an undefined exception?
12. When should you create a user-defined exception?

PROBLEM-SOLVING CASES

1. Write a PL/SQL program that uses a loop to calculate the areas of five circles, starting with a circle of radius 1 and ending with a circle of radius 5. The formula to calculate the radius of a circle is πr^2, where $\pi = 3.14159$, and *r* is the radius of the circle. The output of the program should be as follows for each calculated value: The area of a circle with radius = <number> is <calculated area>. Use variable values rather than hard-coded values wherever possible in the DBMS_OUTPUT function. Save your file as Ch4bEx1.txt.

2. Create a database table named YEAR_DATES that has two columns: YEAR_MONTH, which will hold the names of the months of the year (such as 'JANUARY'), and YEAR_DAY, which will hold integers representing each day of the month. Then, write a PL/SQL program that inserts all of the records for every day of the year into the table, using a separate loop for each month. Also, create a counter that counts how many total records are inserted.

   The finished table should look like this:

YEAR_MONTH	YEAR_DAY
JANUARY	1
JANUARY	2
...	
JANUARY	30
JANUARY	31
FEBRUARY	1
FEBRUARY	2
...	
FEBRUARY	27
FEBRUARY	28
...	

   Use the counter variable to display the following output:

   A total of 365 records were inserted.

   Save your file as Ch4bEx2.txt.

3. Write a PL/SQL program that uses an implicit cursor to display the date expected, quantity expected, item description, color, and quantity on hand of SHIPID 212 in the Clearwater Traders database. Include exception handlers for the cases where no data are returned or where multiple records are returned (ORA-01422). Format the program output so that it is displayed as follows:

   Shipment 212 is expected to
   arrive on 11/15/2001
   and will contain 25
   3-Season Tents, Color Forest

Use variable values rather than hard-coded values wherever possible in the DBMS_OUTPUT function. Save your file as Ch4bEx3.txt.

4. Write a PL/SQL program that uses an implicit cursor to display the CALLID, CNAME, TDESC, FLNAME, FFNAME, DAY, TIME, BLDG_CODE, and ROOM for CSECID 1011. Include exception handlers for the cases where no data are returned or where multiple records are returned (ORA-01422). Format the program output so that it displays as follows:

   MIS 301 Systems Analysis
   M-F 08:00 AM
   BUS 404
   Instructor: John Blanchard

   Use variable values rather than hard-coded values wherever possible in the DBMS_OUTPUT function. Save your file as Ch4bEx4.txt.

5. In this exercise you will write a PL/SQL program that uses an explicit cursor to display the item and inventory information for ITEMID 894. The cursor will also calculate the value of each inventory item (CURR_PRICE * QOH), and the total value of all inventory items for a given ITEMID.
   a. Create an implicit cursor that returns the value of ITEMDESC and then displays it as a header row. Format the output as follows:

      Item ID: 894 Item Description: Women's Hiking Shorts

   b. Create an explicit cursor that returns and then displays the size, color, current price, quantity on hand, and total value (price * quantity on hand) for each individual inventory item. Format each individual inventory item return value as follows:

      Size: S
      Color: Khaki
      Price: 29.95
      QOH: 150
      Value: 4492.5

   c. Create a variable that sums the total value of all inventory items and then displays the total value after all rows are processed. Format the total value output as follows:

      TOTAL VALUE: 20605.6

   d. Create a predefined exception handler for the case where no data are returned.
   e. Save the file as Ch4bEx5.txt.

6. Write a PL/SQL program that uses an explicit cursor to display the call ID, course description, course credits, and grade for every course that has been taken by SID 104 where a grade has been assigned. Calculate the total credits and overall grade point average. Grade point average is calculated as follows:

$$\frac{\Sigma(\text{Course Credits} * \text{Course Grade Points})}{\Sigma(\text{Course Credits})}$$

Course grade points are awarded as follows:

Grade	Grade Points
A	4
B	3
C	2
D	1
F	0

Format the output as follows:

MIS 101 Intro. to Info. Systems 3 B
MIS 301 Systems Analysis 3 C
MIS 451 Client/Server Systems 3 C
Total Credits: 9
Overall GPA: 2.33

Hint: Round the overall GPA output using the ROUND function, which has the following general format: ROUND(<number to be rounded>, <number of decimal places>). Save the file as Ch4bEx6.txt.

7. Computer programs often use check digits as an error-checking technique for input data. A check digit is an extra reference number that is appended to the end of a numerical data value, such as a product identification code. This extra digit has a mathematical relationship to the data value. The check digit is input with the data value, and then recomputed by the computer program. The result is compared with the input, and an exception is raised if the input is incorrect. The most common check digit system is the Modulus 11 system. In this system, each digit of a product's INVID is multiplied by the corresponding digit of the product's ITEMID. The result is summed, and then divided by a fixed number called the modulus. The resulting remainder is then subtracted from the modulus, with the result being the check digit. Here is an example of a check digit computation for an INVID in the Clearwater Traders database:

INVID	1	1	6	6	8	
ITEMID	0	0	7	8	6	
Multiply each INVID by Corresponding ITEMID	0	0	42	48	48	
Sum the results		0 + 0 + 42 + 48 + 48 = 138				
Divide the sum by the modulus (in this case, 11)		138/11 = 12 with remainder of 6				
Subtract remainder from modulus to get check digit		11 – 6 = 5				
Append check digit to original INVID to get new INVID		1	1	6	6	8 5

a. Write a PL/SQL program that generates check digits for the INVID values in the Clearwater Traders database, using the ITEMID and a modulus value of 11. Use an explicit cursor to retrieve each INVID and corresponding ITEMID, compute the check digit, and then output the value of the new INVID with the check digit appended to it. The remainder of a division operation can be returned using the MOD function, which has the following general format:

```
MOD(<number being divided>,*<divisor>)
```

For example, MOD(9,2) would return 1, since 9/2 = 4 with a remainder of 1.

Hint: Change the numeric values to strings, and parse them using the SUBSTR function. Assume that all ITEMID values have three digits, and all INVID values have five digits.

Hint: In the case where the remainder is 1 and the check digit is 10, change the check digit to 0. In the case where the remainder is 0 and the check digit is 11, change the check digit to 1.

Format your output as follows:

Original INVID: 11668 INVID with Check Digit: 116685
Original INVID: 11669 INVID with Check Digit: 116690
Original INVID: 11775 INVID with Check Digit: 117754
Original INVID: 11776 INVID with Check Digit: 117761

c. Save the file as Ch4bEx7.txt.

LESSON C

objectives

- Understand PL/SQL program units
- Use Oracle Procedure Builder to write procedures and functions
- Create libraries, stored program units, and packages
- Create database triggers

Procedures and Functions in PL/SQL

A **program unit** is a self-contained group of program statements that exists within a larger program. You should always break complex programs into smaller program units because it is easier to conceptualize, design, and debug a small procedure than a large program. When all of the smaller program units work correctly, you can then link them into the large program. Program units can also be reused, saving valuable programming time. In most programming languages, there are two main kinds of program units: **procedures**, which can return and change several data values, and **functions**, which return a single data value. The following paragraphs discuss how to create procedures and functions in PL/SQL.

Procedures

Procedures act like "black boxes," which are objects that transform an input into an output; you don't really care how it happens, just as long as it happens correctly. All the user needs to know are the required inputs and the expected outputs. In order for a procedure to be a "black box," it must have a way to receive inputs from other programs and a way to send results back. This is done using **parameters**, which are variables used to pass information from one program to another. Recall that PL/SQL programs have three sections: declarations, body, and exceptions. Instead of a declarations section, procedures have a section called a **header**, where the procedure's name and parameters are specified and the procedure variables are declared. A procedure header has the following general format:

```
PROCEDURE <procedure name>  (<parameter1 name> <mode> <data type>,
                             <parameter2 name> <mode> <data type>,
                             ...) IS
     <variable declarations>
```

Note that in the header, the parameter names are indented so that they are aligned vertically. The <mode> describes how the parameter value can be changed in the procedure. Parameters can have three modes: IN, OUT, or IN OUT. The different modes are described in Figure 4-54.

Mode	Description
IN	Parameter is passed to the procedure as a read-only value that cannot be changed within the procedure
OUT	Parameter is a write-only value that can only appear on the left side of an assignment statement in the procedure
IN OUT	Combination of IN and OUT; the parameter is passed to the procedure, and its value can be changed within the procedure

Figure 4-54: Procedure parameter modes

If a parameter is declared without the IN, OUT, or IN OUT specification, it is by default an IN parameter. When you specify the parameter data type, you do not include the precision or scale value for a numerical data type, or the maximum width for a character data type.

The general format of the rest of a procedure is the same as a regular PL/SQL program block, with a body, and an exception section.

The general format for calling a procedure is: `<procedure name>(<variable or value to pass to parameter1>, <variable or value to pass to parameter2>, ...);`. When you are calling a procedure, you must distinguish between formal parameters and actual parameters. **Formal parameters** are the parameters that are declared in the header of the procedure. **Actual parameters** are the values placed in the procedure parameter list when the procedure is called. Actual parameters can be constants or variables that have assigned values, and the actual parameter names do not have to be the same as the formal parameter names. Suppose you have a procedure named CalcGPA that calculates the grade point average of a student for a given term. The procedure header would look like this:

```
PROCEDURE CalcGPA (     StudentID         IN     NUMBER,
                        CurrentTermID     IN     NUMBER,
                        CalculatedGPA     OUT    NUMBER) IS
```

You would call the procedure using the following program statement:

```
CalcGPA (CurrentSID, 4, CurrentGPA);
```

The student ID parameter is passed as a variable value, and the term ID value is passed as a constant. The calculated GPA parameter is returned as a variable value. Notice that the actual parameter variable names in the command calling the procedure are different from the formal parameter names in the procedure declaration.

Figure 4-55 shows the relationship between the formal parameters and the actual parameters. It is important to remember that the variable or constants that are passed to each parameter must be in the same order in which the parameters are declared in the procedure declaration, since that is how the values are associated. The first variable value in the procedure calling statement will be assigned to the first parameter in the procedure declaration; the second variable will be assigned to the second parameter, etc.

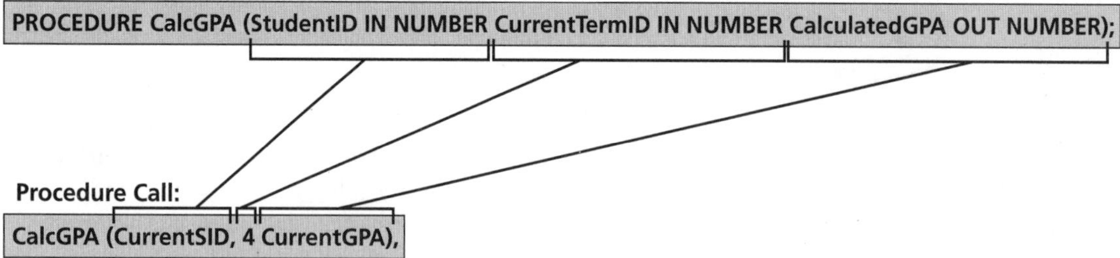

Procedure Header:
PROCEDURE CalcGPA (StudentID IN NUMBER CurrentTermID IN NUMBER CalculatedGPA OUT NUMBER);
Procedure Call:
CalcGPA (CurrentSID, 4 CurrentGPA),

Figure 4-55: Relationship between formal parameters in procedure header and actual parameters in procedure call

Using Oracle Procedure Builder

To create program units, you will use Oracle Procedure Builder, which provides a graphical interactive environment for creating, compiling, testing, and debugging procedures. First, you will run the scripts to refresh your Clearwater Traders and Northwoods University database tables. Then you will start Procedure Builder and become familiar with its environment.

To refresh your database tables and start Procedure Builder:

1 Start SQL*Plus, run the clearwat.sql script from the Chapter4 folder of your Student Disk.

2 Run the northwoo.sql script from the Chapter4 folder of your Student Disk.

3 Click **Start** on the Windows taskbar, point to **Programs,** point to **Oracle Developer 6.0,** then click **Procedure Builder.** The Procedure Builder opens, as shown in Figure 4-56. If necessary, click on □ to maximize the window.

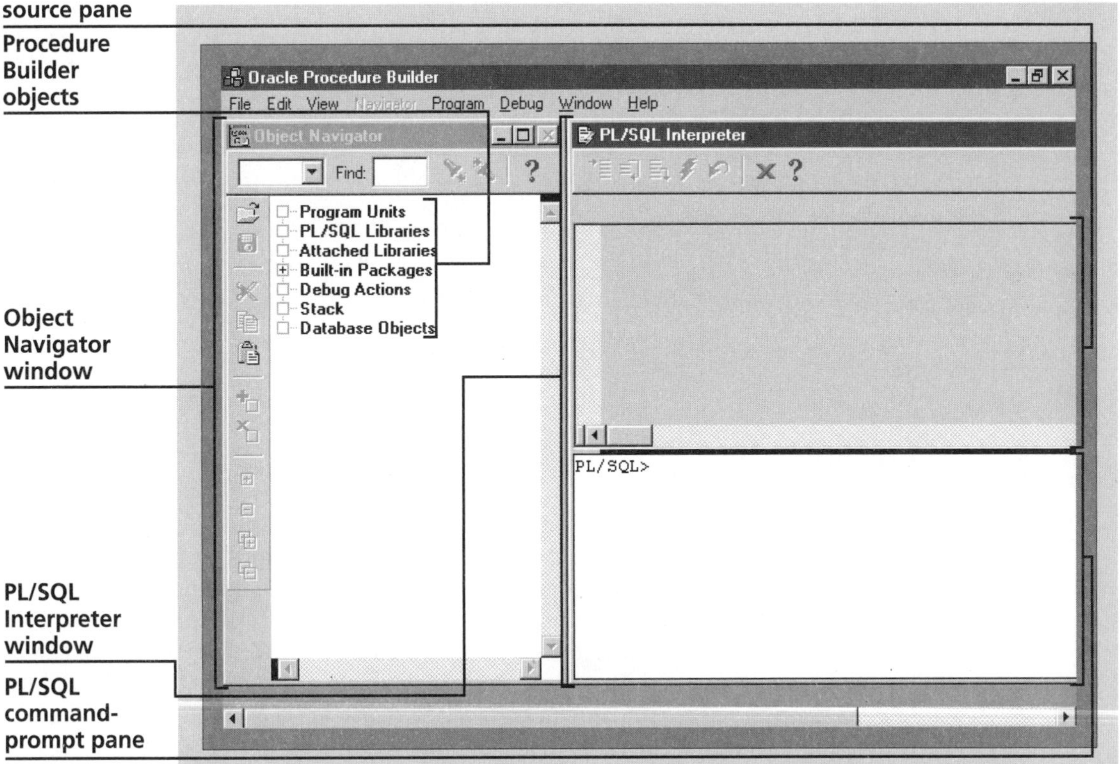

source pane
Procedure Builder objects
Object Navigator window
PL/SQL Interpreter window
PL/SQL command-prompt pane

Oracle Procedure Builder
File Edit View Navigator Program Debug Window Help
Object Navigator
Find:
Program Units
PL/SQL Libraries
Attached Libraries
Built-in Packages
Debug Actions
Stack
Database Objects
PL/SQL Interpreter
PL/SQL>

Figure 4-56: Procedure Builder windows

Another way to start Procedure Builder is to start Windows Explorer, and then double-click the file ORAWIN95\BIN\DE60.EXE.

Procedure Builder has two main windows: the **Object Navigator**, which provides an environment for moving among different Procedure Builder objects, and the **PL/SQL Interpreter**. The PL/SQL Interpreter has two panes. The upper pane, called the **source pane**, displays the source code for existing PL/SQL programs and provides a debugging environment. The lower pane, called the **PL/SQL command prompt**, provides a SQL command-line environment that you can use like SQL*Plus.

The Object Navigator provides a hierarchical view of the different types of objects that can be created and modified in Procedure Builder. To view the different objects that you can access and modify, you click the **plus sign** [+] beside the object type, and a list of current objects appears. If there is no [+] in front of an object type, then there are no objects of that type currently available to edit or run. To close a list of objects, you click the **minus sign** [−] beside the object type. When you first start Procedure Builder, the only available object type is **Built-in Packages.** Built-in packages are special PL/SQL program libraries that are stored in the database when it is created, and contain the code for built-in procedures and functions such as TO_DATE and ROUND. To become familiar with opening and closing objects in the Object Navigator, you will view the different built-in packages that are available.

To open an object in the Object Navigator:

1. Click [+] beside Built-in Packages. A list of different packages appears. Notice that the [+] changes to a [–] to indicate that the object is open.
2. Click [–] beside Built-in Packages to close the object. The list of Built-in Packages disappears, and only the different object types are displayed.

Creating a New Procedure

At Clearwater Traders, you want to modify the INVENTORY table so that it stores the current value of each inventory item, which is equal to CURR_PRICE * QOH. First, you will add a column named INV_VALUE to the INVENTORY table. Then, you will write a procedure to calculate the value for each inventory item and insert the value into the associated record.

When you first start Procedure Builder, you are not automatically prompted to connect to the database, so you have to explicitly make a connection. First, you will connect to the database. Then, you will modify the INVENTORY database table by typing the ALTER TABLE command at the Procedure Builder SQL command prompt.

To connect to the database and modify the INVENTORY table using Procedure Builder:

1. Click **File** on the top menu bar, then click **Connect**. Log on in the usual way.
2. Type the following code at the **PL/SQL>prompt** to modify the INVENTORY table, and then press the **Enter** key. The PL/SQL prompt will be displayed again, indicating that the command executed successfully.

```
ALTER TABLE inventory
ADDInv-Value NUMBER (9.2);
```

Now you will create a procedure named UpdateInvValue that calculates the value for each inventory item and updates the INV_VALUE field.

To create the procedure:

1. Click **Program Units** in the Object Navigator window, then click the **Create** button on the Object Navigator toolbar.
2. Make sure that the Procedure option button is selected, and then type **UpdateInvValue** for the procedure name.
3. Click **OK**. The Program Unit Editor window opens, as shown in Figure 4-57. Click the **Maximize** button to maximize the window.

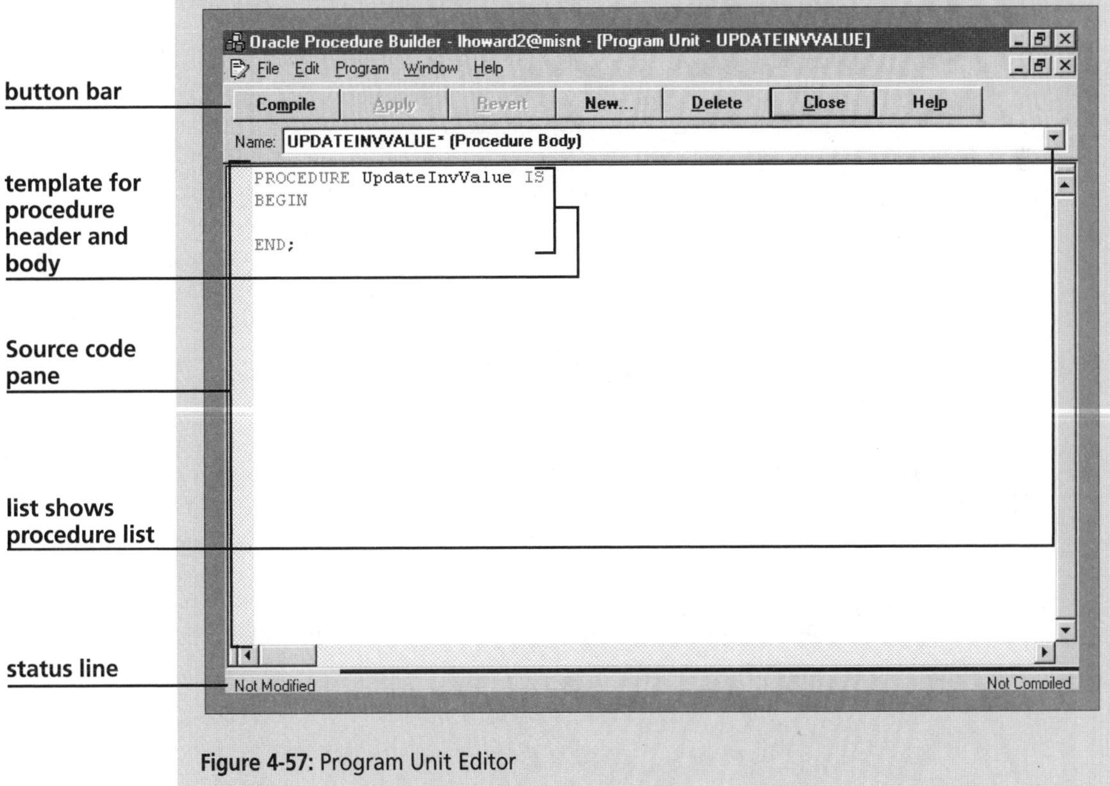

Figure 4-57: Program Unit Editor

The Program Unit Editor is an environment for writing, compiling, and editing PL/SQL programs. The **Procedure list** shows the name of the current program unit, and allows you to access other program units. The **Source code** pane is where you type PL/SQL program statements. The **status line** shows the program unit's current modification status (Modified or Not Modified) and compile status (Not Compiled, Successfully Compiled, or Compiled with Errors). The **Program Unit Editor button bar** has the following buttons:

- **Compile**, which compiles the program statements in the Source code pane. The compiler detects syntax errors and reports run-time errors.
- **Apply**, which saves changes since the Program Unit Editor was opened, or since the last time the Apply button was clicked
- **Revert**, which reverts the source code to its status as of the last time the Apply or Revert button was clicked
- **New...**, which creates a new program unit
- **Delete**, which deletes the current program unit
- **Close**, which closes the Program Unit Editor
- **Help**, which enables the user to get help while working in the Program Unit Editor

When you create a new procedure, a template is automatically inserted into the Program Unit Editor for the procedure header and body. Note that the text in the Source code pane uses color highlighting to define different command elements. Reserved words (such as BEGIN and UPDATE) appear in blue. Command operators for arithmetic, assignment, and comparison operators are red. User-defined variables are black, literal values (such as the number 11) are blue-green, and comment statements are light green.

Now you will enter the procedure code. You will need to create an explicit cursor that retrieves each record in the INVENTORY table. The procedure will then calculate the INV_VALUE for each record, and update the record.

To enter the procedure code:

1 Click the mouse pointer before the B in BEGIN, then press the **Enter** key.

2 Type the code shown in Figure 4-58.

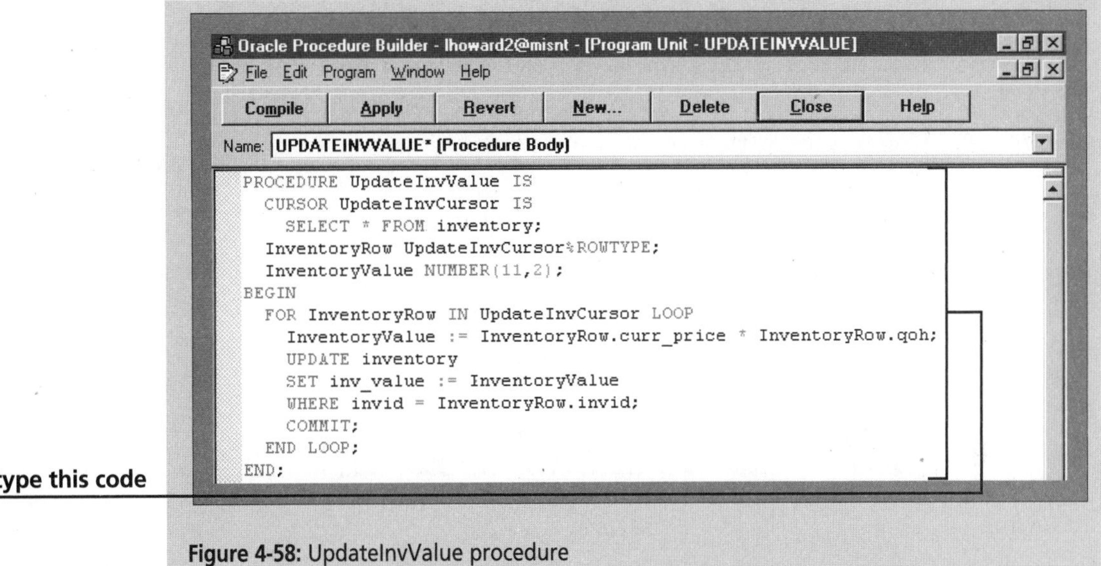
Oracle Procedure Builder - lhoward2@misnt - [Program Unit - UPDATEINVVALUE]
File Edit Program Window Help
Compile
Apply
Revert
New...
Delete
Close
Help
Name: UPDATEINVVALUE* (Procedure Body)
PROCEDURE UpdateInvValue IS
CURSOR UpdateInvCursor IS
SELECT * FROM inventory;
InventoryRow UpdateInvCursor%ROWTYPE;
InventoryValue NUMBER(11,2);
BEGIN
FOR InventoryRow IN UpdateInvCursor LOOP
InventoryValue := InventoryRow.curr_price * InventoryRow.qoh;
UPDATE inventory
SET inv_value := InventoryValue
WHERE invid = InventoryRow.invid;
COMMIT;
END LOOP;
END;
type this code

Figure 4-58: UpdateInvValue procedure

When you ran PL/SQL programs in SQL*Plus, compiling and running the program was performed in a single step. In Procedure Builder, PL/SQL programs must be explicitly compiled before they can be run. When a compile error is detected, the program displays the line number of the statement causing the error and an error code and description. Next, you will compile the procedure.

To compile the procedure:

1. Click **Compile** on the Program Unit Editor button bar. Your procedure should display the compile error shown in Figure 4-59. (Your procedure might display a different compile error if you made other errors.) The following paragraphs discuss how to interpret and correct compile errors in the Program Unit Editor environment.

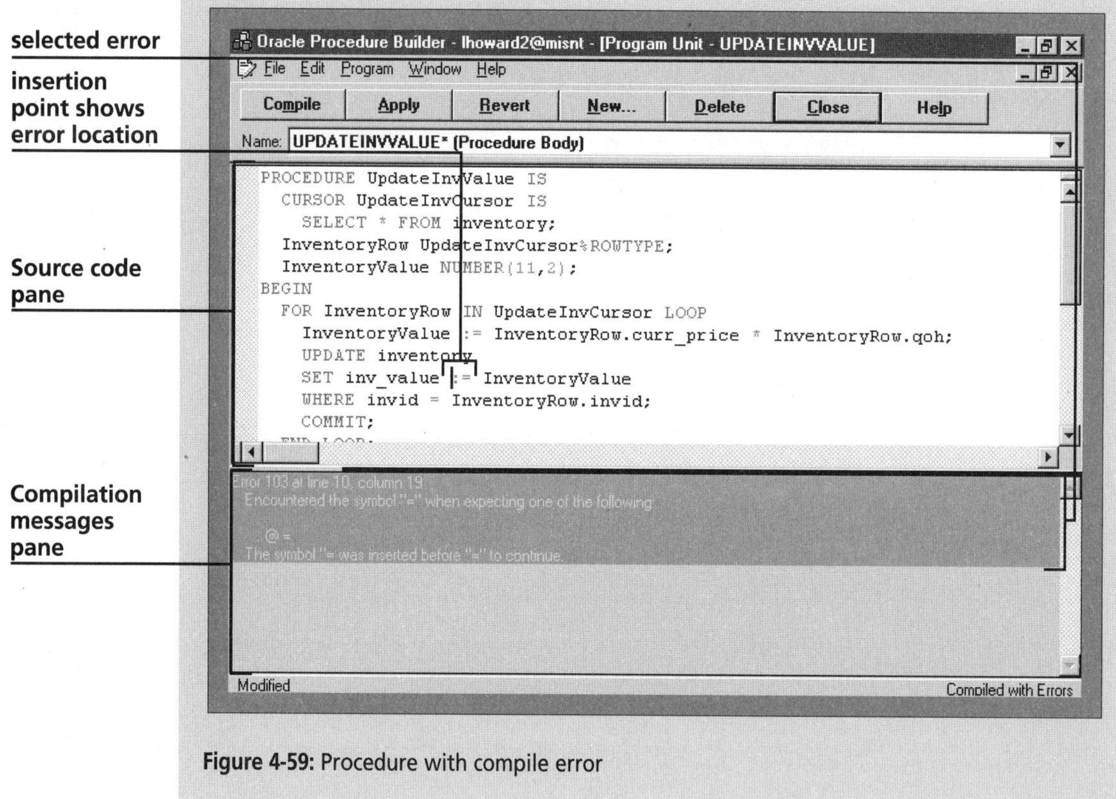

Figure 4-59: Procedure with compile error

Correcting Compile Errors

As you can see in Figure 4-59, the Program Unit Editor's **Compilation messages pane** displays the line number of the error and the error message, as well as correction suggestions. The selected error is the error message that is selected in the Compilation messages pane. The insertion point shows the location for the selected error in the Source code pane. If there are multiple errors, you can select a different error message by clicking the message in the Compilation messages pane.

The error shown in Figure 4-59 occurs because the compiler expected the comparison operator (=) instead of the assignment operator (:=) in the SQL command. This error can be corrected by changing := to = and then clicking Compile again to recompile the procedure. You will now correct this error.

To correct the compile error:

1 Delete the colon (:) after the insertion point to change the assignment operator to a comparison operator, then click **Compile** to recompile the procedure. If the procedure successfully compiles, the message "Successfully Compiled" appears on the status line. If the procedure does not successfully compile, continue to debug your procedure until it does. The following paragraphs provide examples of common compile errors.

Another common compile error is shown in Figure 4-60. The user mistakenly typed INVENTORY as INVENTRY, and the compiler flagged this as an undeclared variable.

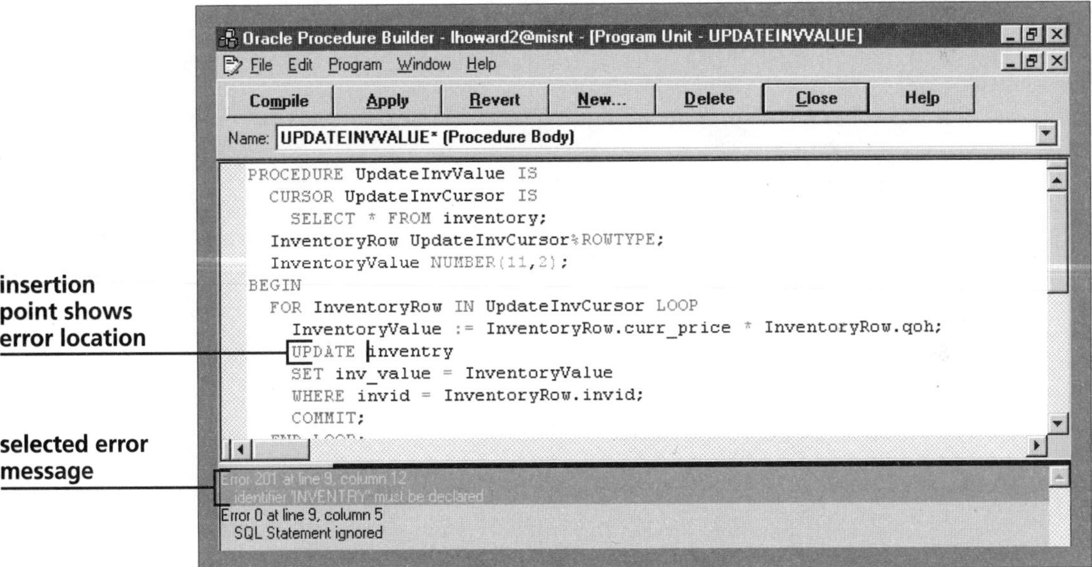

Figure 4-60: Compile error caused by a misspelled database object

Figure 4-61 shows the same error message, but for a different reason. The database table name is spelled correctly, but the user forgot to connect to the database when he started Procedure Builder. To correct this error, click File on the top menu bar, then click Connect, and log on to the database.

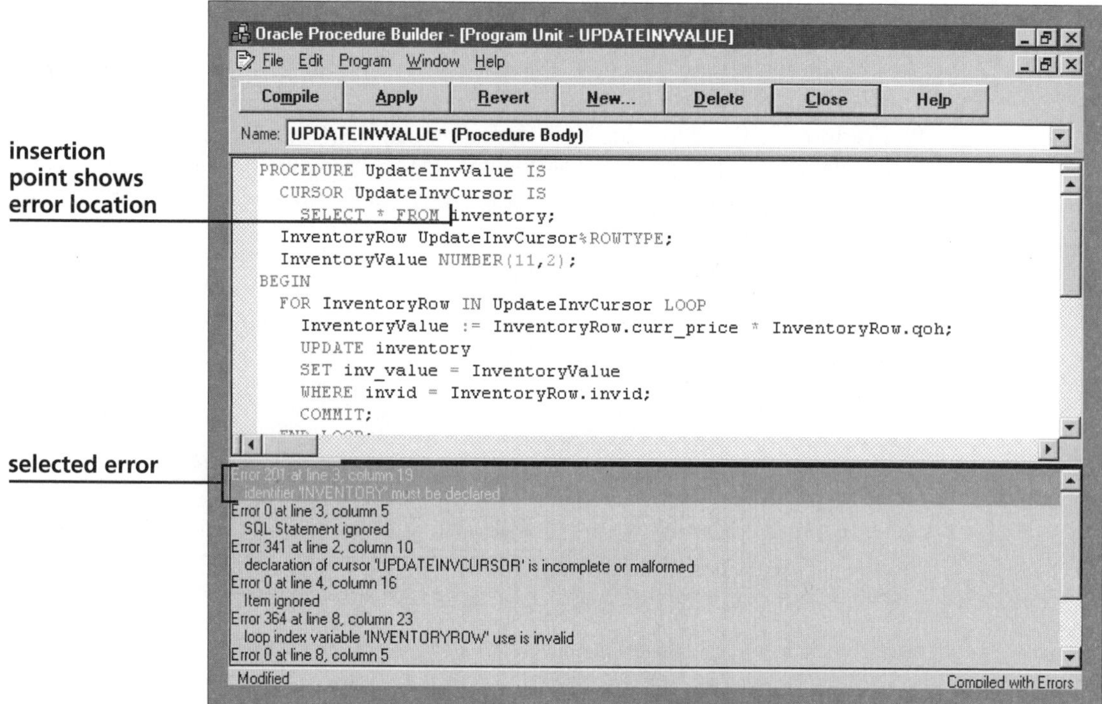

Figure 4-61: Compile error generated when user does not connect to the database before compiling

There are many more possible compile errors, but these are some of the common ones. The best way to locate errors is to examine the code carefully, and simplify it by commenting out program lines. It is also a good idea to copy SQL commands into SQL*Plus and test them by themselves to see if they are generating compile errors.

Running a Procedure

Once you have successfully compiled your procedure, you can run it in Procedure Builder by typing the procedure name at the PL/SQL> prompt. Now you will run the procedure, then view the database records to make sure that they were updated correctly.

To run the procedure and then view the updated records:

1. After you have successfully compiled your procedure, click **Close** on the Program Unit Editor button bar.
2. Click to the right of the PL/SQL> prompt to activate the PL/SQL command prompt pane in the lower pane of the PL/SQL Interpreter window, as shown in Figure 4-62.

help

You might need to click Program on the menu bar, and then click PL/SQL Interpreter to open that window. You might also need to click Window on the menu bar, click Tile, and then rearrange and resize your windows so your screen looks like Figure 4-62.

click here to activate command-line prompt

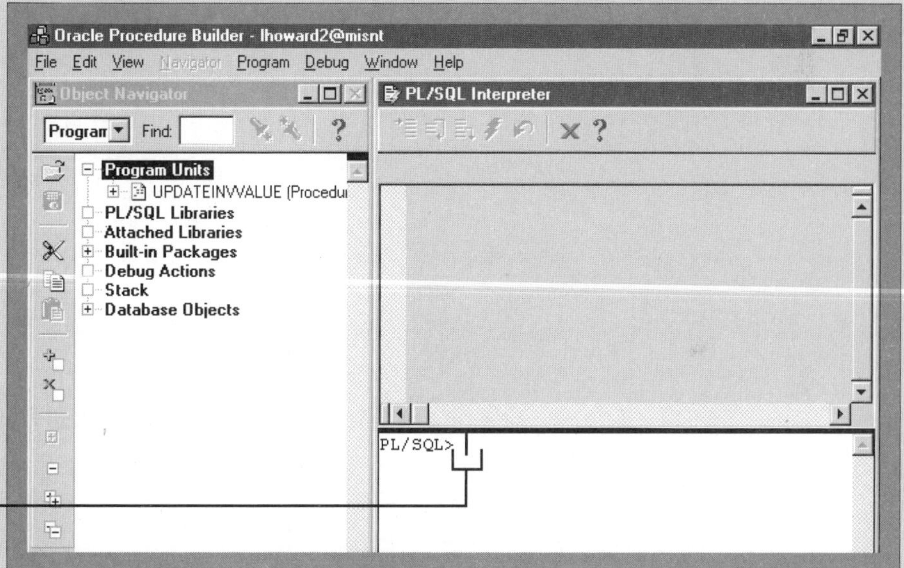

Figure 4-62: Activating the command-line prompt pane

3. Type **UpdateInvValue;** at the PL/SQL> prompt, as shown in Figure 4-63, and then press the **Enter** key to run the procedure. If the procedure runs successfully and no run-time errors are generated, then the PL/SQL> prompt is redisplayed. If the procedure did not run successfully, run-time error messages are displayed. If your procedure did not run correctly, refer to the next section, which discusses how to use the PL/SQL Interpreter to detect run-time errors. If your procedure ran successfully, you should now view the INVENTORY records to make sure that they were updated correctly.

4 Type the SELECT command shown in Figure 4-63 to view the updated records. Use the scrollbar in the PL/SQL command prompt pane to view all of the retrieved records.

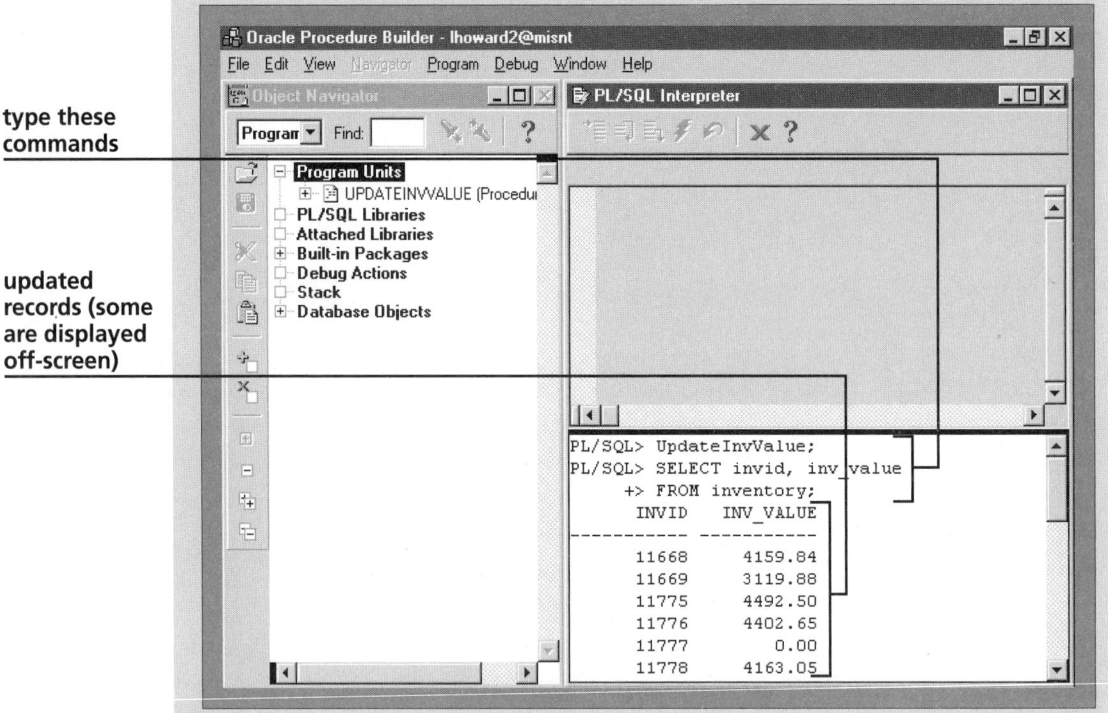

type these commands

updated records (some are displayed off-screen)

```
PL/SQL> UpdateInvValue;
PL/SQL> SELECT invid, inv_value
     +> FROM inventory;
     INVID   INV_VALUE
----------- -----------
     11668     4159.84
     11669     3119.88
     11775     4492.50
     11776     4402.65
     11777        0.00
     11778     4163.05
```

Figure 4-63: Viewing the updated INVENTORY records

Using the PL/SQL Interpreter to Find Run-time Errors

Now you will learn how to use the PL/SQL Interpreter as a debugging environment. You will load the procedure into the PL/SQL Interpreter, and then set a **breakpoint,** which is a place in a program where execution pauses in a debugging environment. Then you will single-step through the procedure to observe execution and examine variable values. You can set breakpoints only on program lines that contain executable program statements. You cannot set breakpoints on comment lines, variable declarations, or SQL statements.

To load the procedure, set a breakpoint, and single-step through the program:

1 In the Object Navigator window, click the **Program Unit** icon beside UPDATEINVVALUE to load the procedure code into the PL/SQL Interpreter window. The procedure code is displayed in the upper pane of the PL/SQL Interpreter window.

2 Place the pointer on the number **00007** in the PL/SQL Interpreter window or the program line directly below the BEGIN command, if your line numbers are different. Double-click to set a breakpoint on that line. The program line number changes to B (01) to indicate that Breakpoint 01 occurs on Line 7, as shown in Figure 4-64.

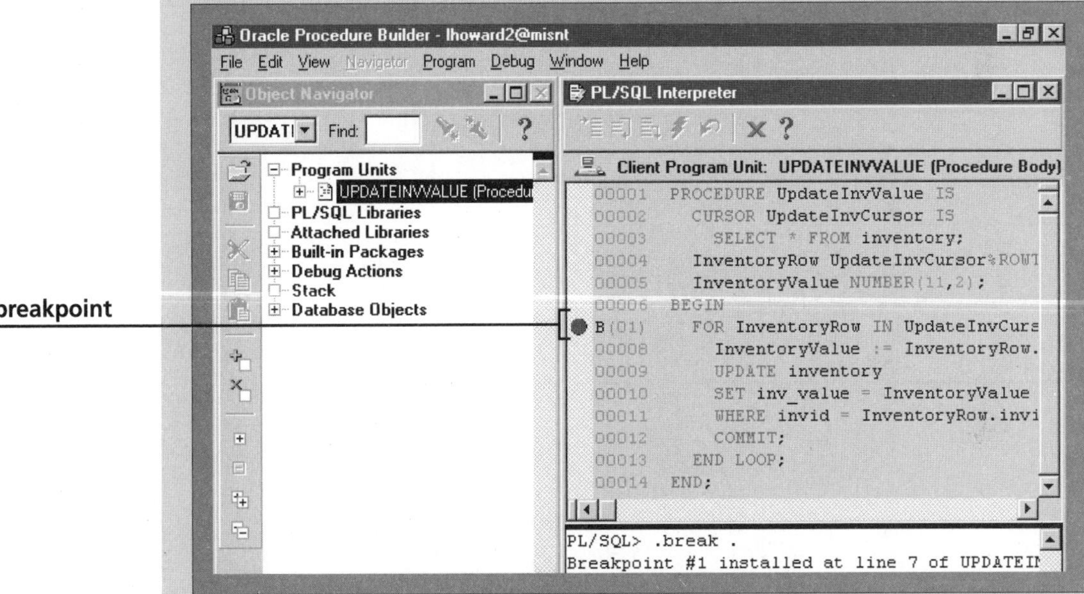

Figure 4-64: Setting a breakpoint on Line 7

3 Click to the right of the PL/SQL> prompt to activate the SQL command prompt pane in the lower pane of the PL/SQL Interpreter window, and then type **UpdateInvValue;** at the PL/SQL> prompt and press the **Enter** key to run the procedure. The program runs, and program execution stops at the breakpoint on Line 7.

Figure 4-65 shows the position of the execution arrow on Line 7 of the procedure. The execution arrow shows which line of the procedure will be executed next. As you step through the program, the execution arrow stops on the PL/SQL statements and skips comment lines and SQL statements.

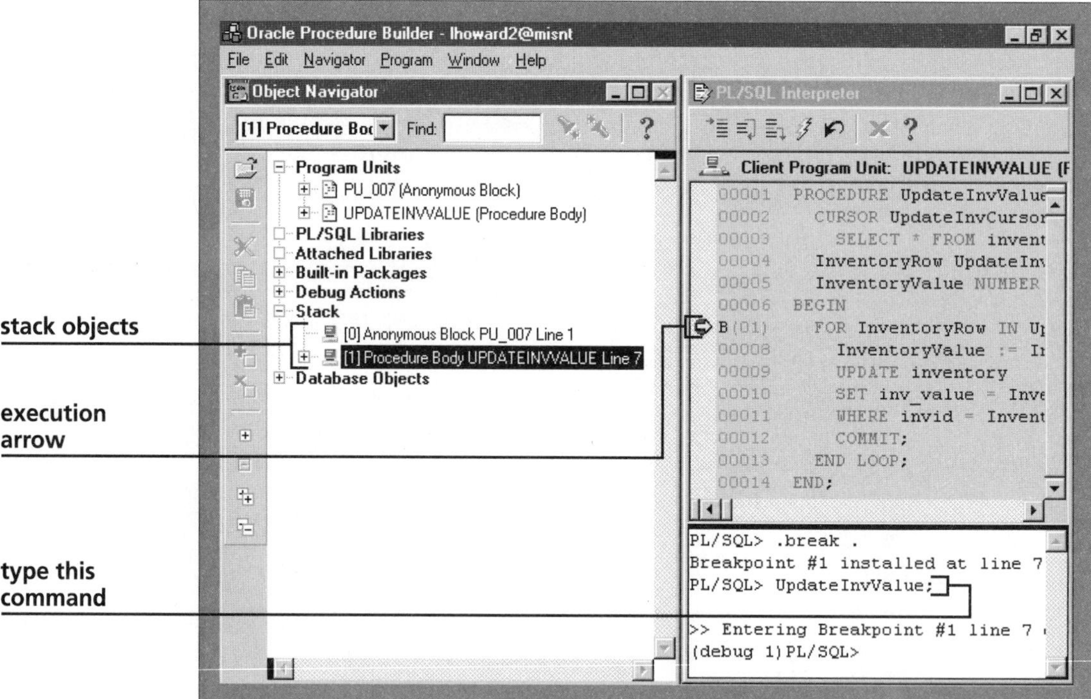

Figure 4-65: PL/SQL Interpreter window during program execution

Notice the objects displayed in the Object Navigator window. The **Stack** now has two objects. The first object, **Anonymous Block**, is an unnamed program unit that is created whenever a program unit runs, and has no objects beneath it. The second object, **Procedure Body UPDATEINVVALUE**, displays the current values of procedure variables. Now, you will examine the current values of the procedure variables.

To examine the current values of procedure variables:

1. Click [+] beside **Procedure Body UPDATEINVVALUE.** The procedure variables are displayed, as shown in Figure 4-66. Notice that there are two INVENTORYROW variables with datatype RECORD. The first INVENTORYROW variable represents the values that were most recently fetched by the cursor. The second INVENTORYROW variable represents the current values stored in the INVENTORYROW variable.

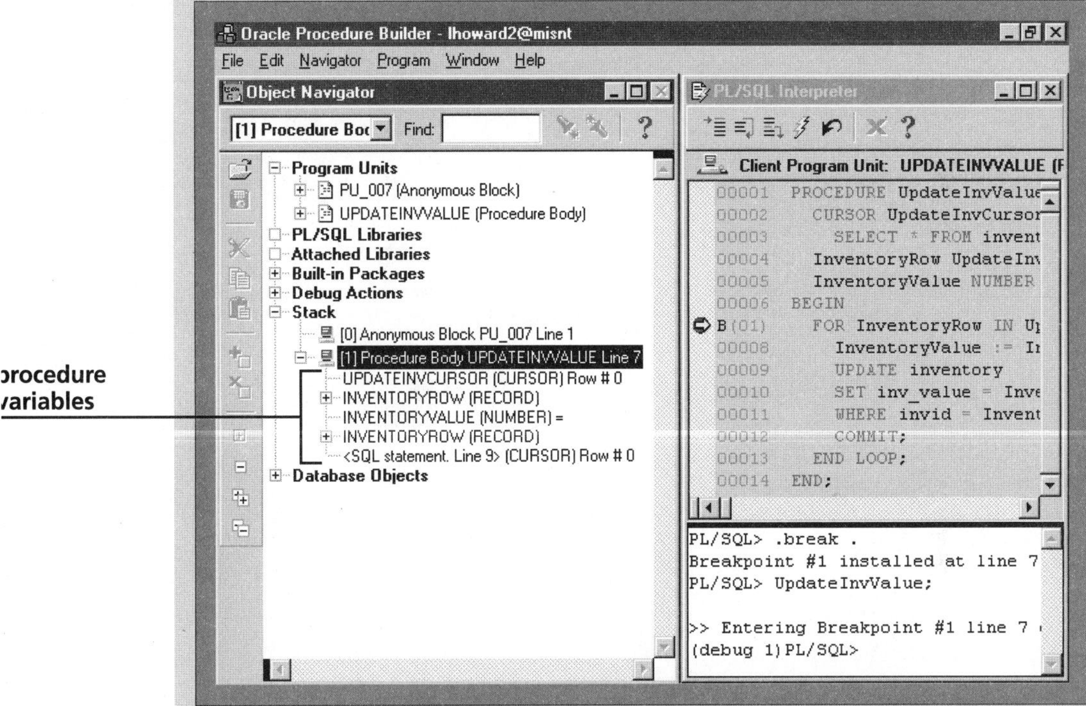

Figure 4-66: Viewing procedure variables during program execution

2. Click [+] beside the first INVENTORYROW variable item. The individual items that are fetched by the cursor are displayed, as shown in Figure 4-67.

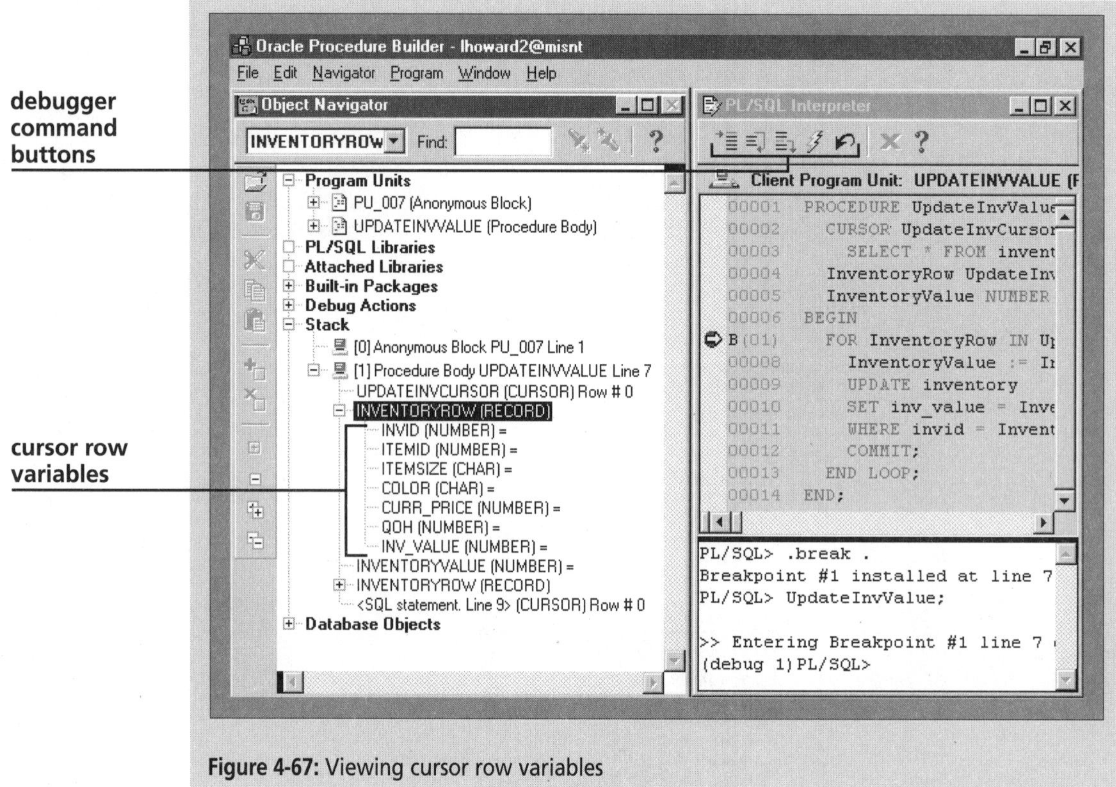

Figure 4-67: Viewing cursor row variables

Also notice the debugger command buttons that are active on the PL/SQL Interpreter window toolbar. The **Step Into** button allows you to step through the program one line at a time. The **Step Over** button allows you to bypass a procedure that is called by the current program unit. The **Step Out** button executes all program lines to the end of the current program unit. The **Go** button allows you to pass over the current breakpoint and run the program until the next breakpoint or until the program terminates. The **Reset** button terminates execution. Now you will single-step through the procedure and watch how the variable values change.

To single-step through the procedure:

1. Click the **Step Into** button. The execution arrow moves to Line 3 to process the cursor and fetch the first record.
2. Click again. The execution arrow moves to Line 8. The cursor record values have been retrieved into the second INVENTORYROW record variable shown in the Object Navigator Stack.

3 Click [+] beside Procedure Body UPDATEINVVALUE, then click [+] beside the second INVENTORYROW(RECORD) variable. The values fetched by the INVENTORYROW cursor are displayed, as shown in Figure 4-68. These data values correspond with the first data row in the INVENTORY table.

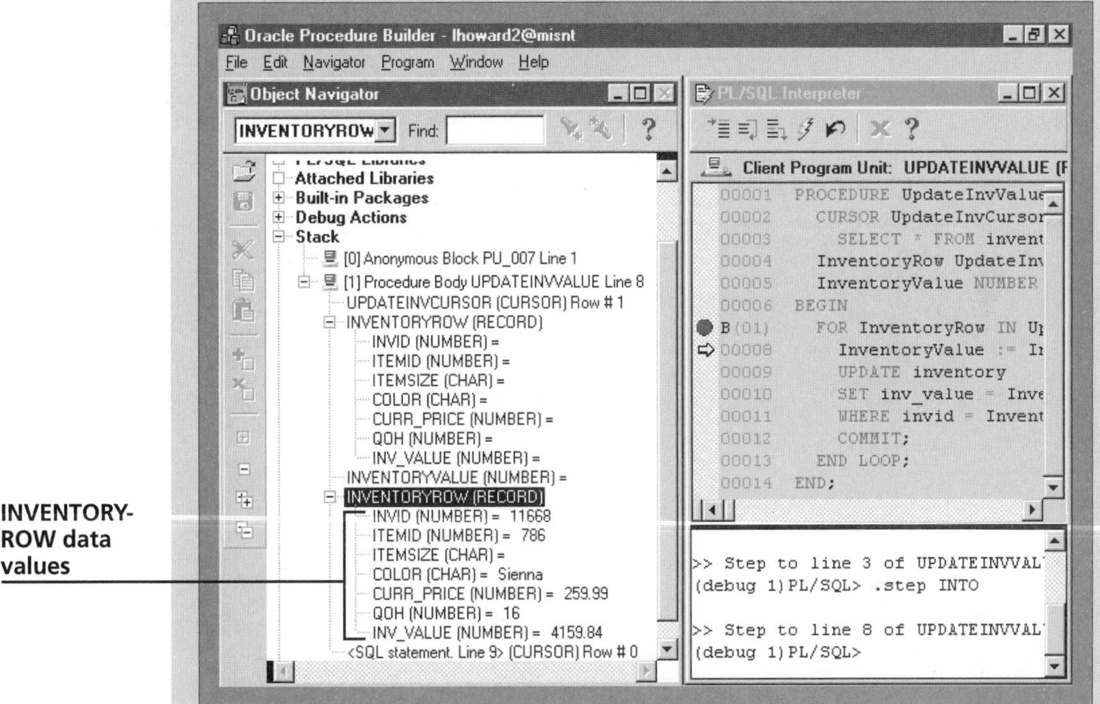

Figure 4-68: Viewing data values fetched into INVENTORYROW

4 Click again to execute Line 8 and calculate the InventoryValue variable. After the line is executed, the INVENTORYVALUE(NUMBER) variable updates to 4159.84 in the Object Navigator window.

5 Click again to execute the UPDATE command.

6 Click again to execute the COMMIT command. The execution arrow moves back to Line 7 to fetch the next record.

7 Click the **Go** button to finish running the program without stopping on the breakpoint again. The PL/SQL> prompt should be displayed when the execution is completed.

Running a program in a debugging environment is useful for locating program lines causing run-time errors, or for identifying the cause of logic errors. For example, Figure 4-69 shows a run-time error that is received when running the UPDATEINVVALUE procedure.

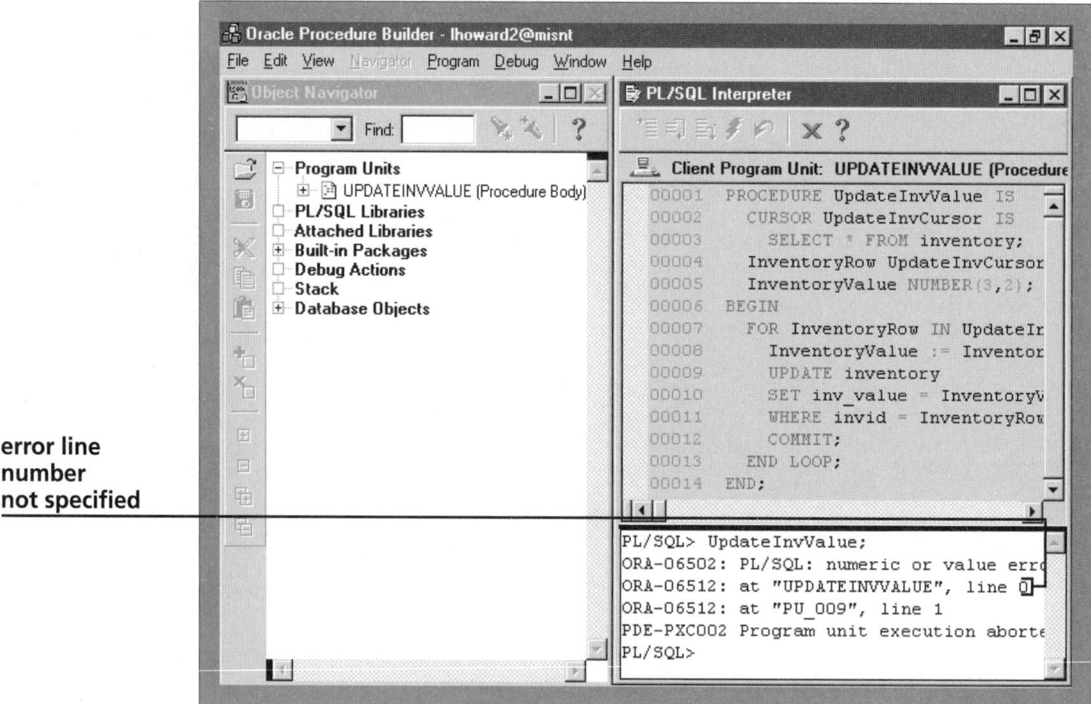

Figure 4-69: Run-time error with line number not specified

The error message states that the error is a numeric or value error, but the error location is given as Line 0. To find the specific program line causing the error, you would set a breakpoint at the beginning of the procedure and single-step through the program lines until you find the line that is generating the error.

If you did this, you would find that the execution error message is displayed when the program is attempting to execute Line 8, as shown in Figure 4-70. You would then restart the program, and examine the variable values just prior to the execution of Line 8.

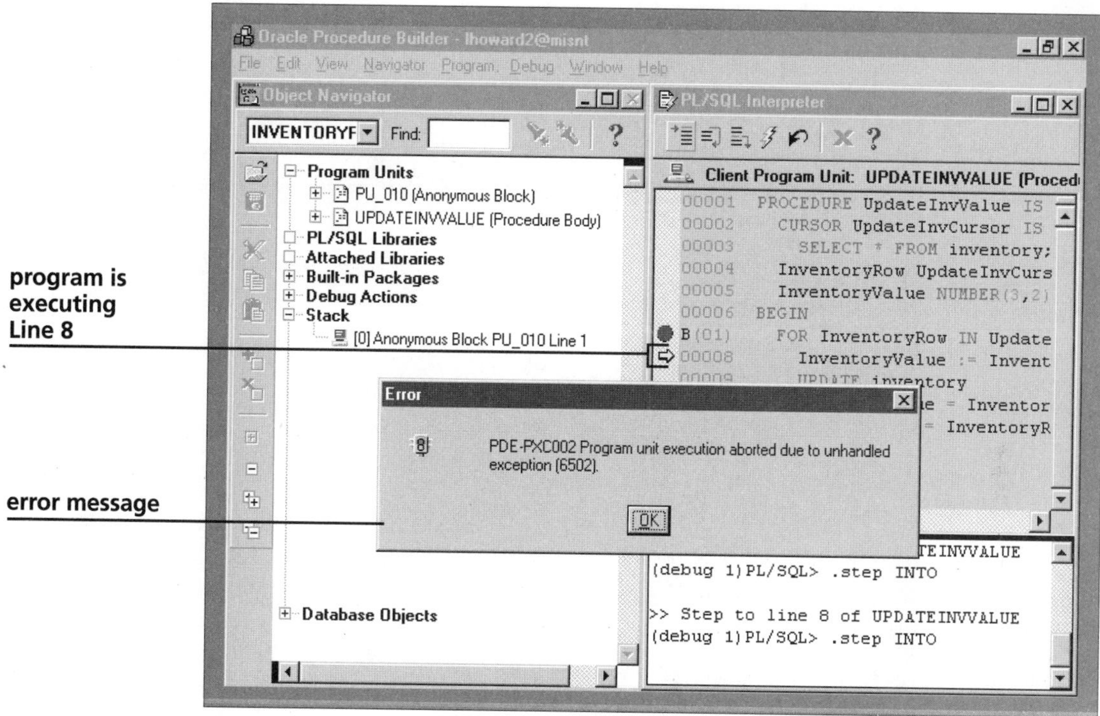

Figure 4-70: Using the PL/SQL Interpreter to find location of run-time error

At this point, you would see that the numeric value to be assigned to the InventoryValue variable is 4159.84 as shown in Figure 4-71. However, when InventoryValue was declared in the procedure, it was specified as a NUMBER with precision 3 and scale 2. Recall that the precision of a number specifies its total number of digits including decimal places. InventoryValue must have a precision of at least 6 for the value 4159.84 to be successfully assigned. To correct the error, you would open the Program Unit Editor, change the variable declaration to NUMBER(6,2), and then recompile and rerun the procedure. In our example, we change the precision to 11 so that very large values will not generate errors.

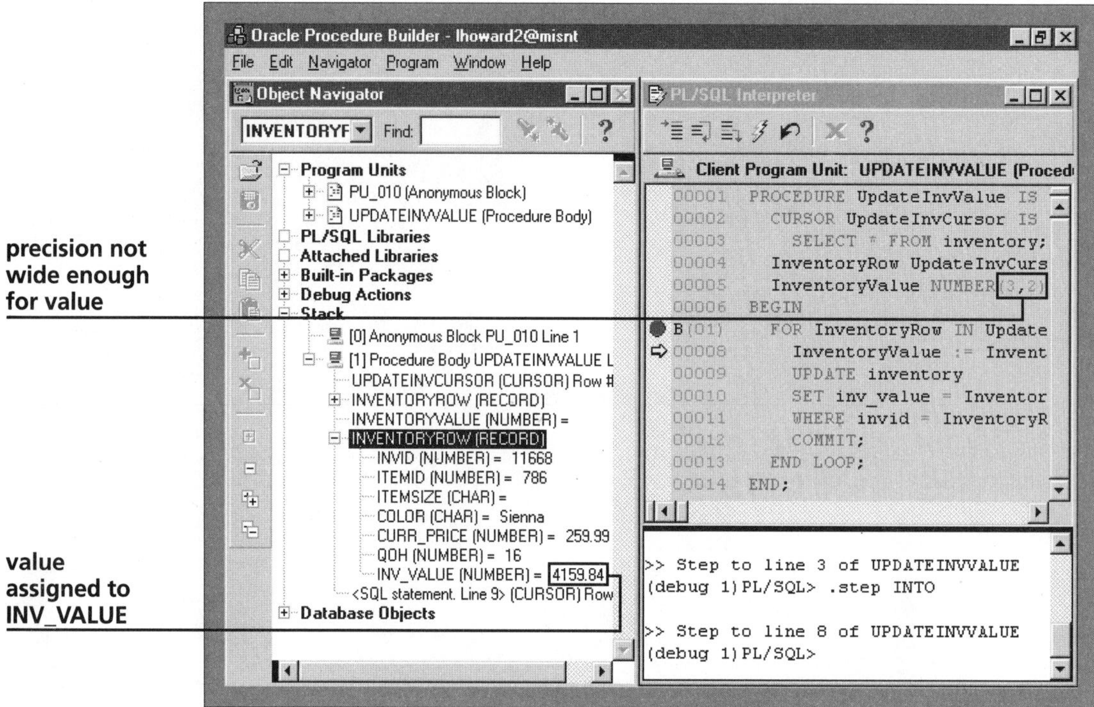

Figure 4-71: Procedure variable values prior to execution of Line 8

If a procedure executes successfully but has data values that are not what you expected, you can use the same process for locating the error. Figure 4-72 shows the result of running another version of the UPDATEINVVALUE procedure. The INV_VALUE for the resulting records is incorrect—the value for INVID 11668 should be 4159.84.

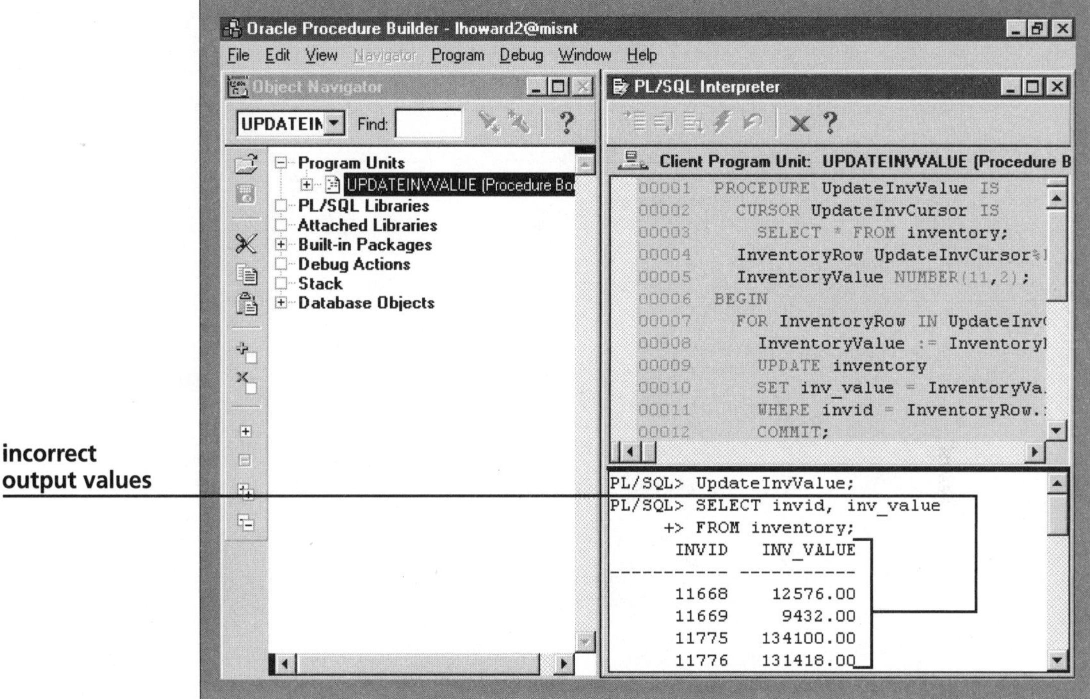

Figure 4-72: Procedure with incorrect output values

You would again set a breakpoint on the first line of the procedure, and examine the values that are being used to calculate the InventoryValue variable that is used to update the table, as shown in Figure 4-73. When you do this, you see that InventoryValue has the incorrect value (12576), but that the variables that are used to calculate InventoryValue (the INVENTORYROW CURR_PRICE and QOH fields) have the correct values.

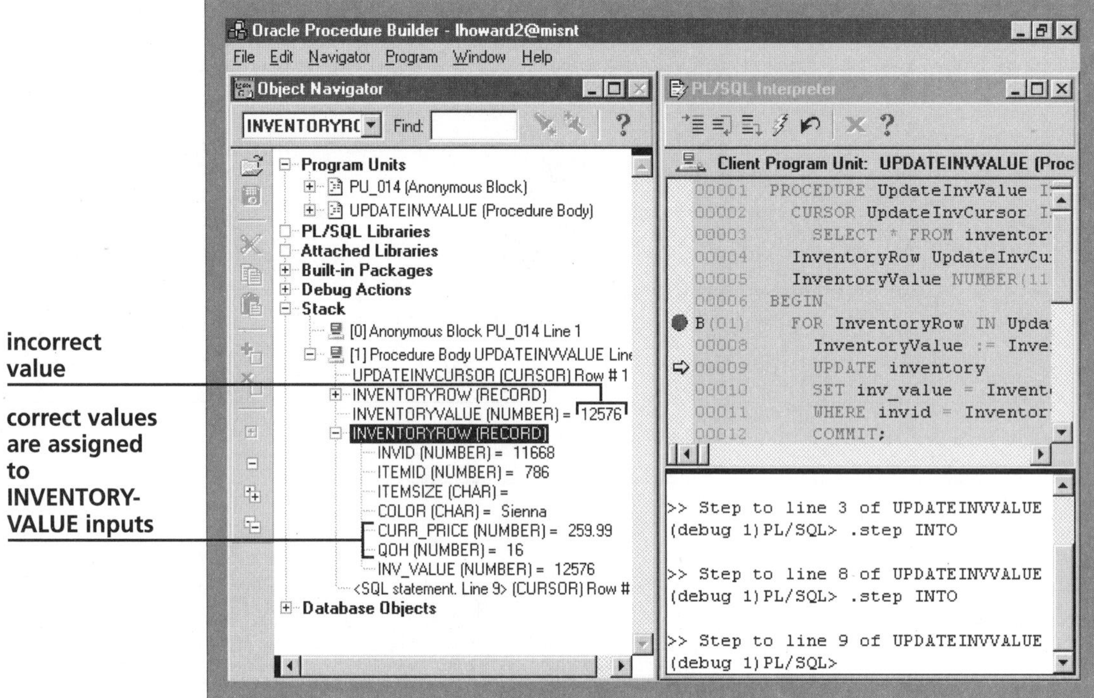

Figure 4-73: Examining variable values to find logic error

This should lead you to examine your program code more carefully, and locate the error: InventoryRow.itemid was used instead of InventoryRow.curr_price to calculate the value of the inventory, as shown in Figure 4-74.

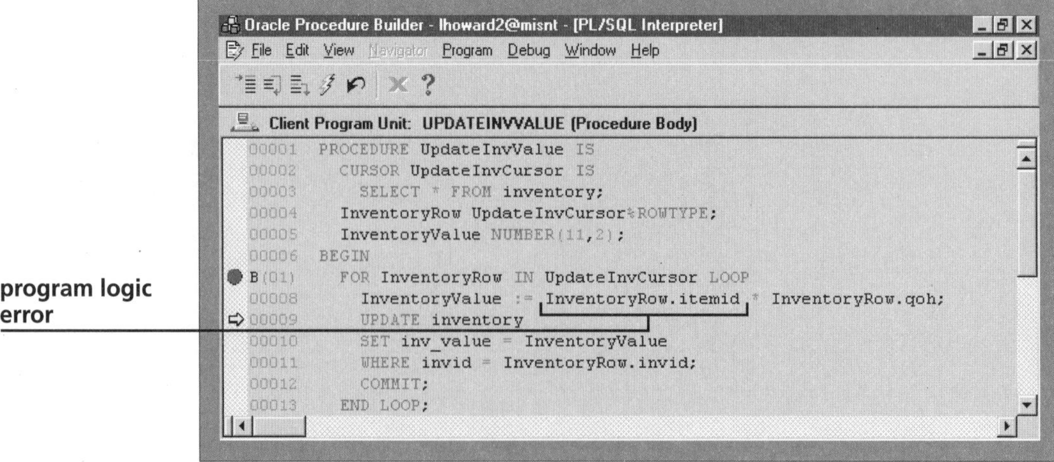

Figure 4-74: Locating the logic error

Saving a Program Unit

Now you will save your procedure to your Student Disk, and then open the saved procedure to confirm that it was saved correctly.

To save and reopen your program unit:

1. Double-click the **Program Unit** icon beside **UPDATEINVVALUE** in the Object Navigator window to open the Program Unit Editor and display the code. If necessary, maximize the Program Unit Editor window.

You can also open the Program Unit Editor by clicking Program on the menu bar, and then clicking Program Unit Editor. Or, you can right-click in the upper pane of the PL/SQL Interpreter window, and then click Edit.

2. Click **Edit** on the menu bar of the Procedure Builder, then click **Select All.**
3. Click **File** on the menu bar, click **Export Text**, and then save the selected text to a file named **updateinvvalue.pls** in the Chapter4 folder on your Student Disk.
4. Click **Close** to close the Program Unit Editor.
5. Unload the UPDATEINVVALUE program unit from Procedure Builder by selecting it in the Object Navigator window, clicking the **Delete** button on the Object Navigator toolbar, and then clicking **Yes** to confirm removing the unit. This unloads the procedure from Procedure Builder, but does not delete it from your file system.
6. Click **File** on the menu bar, click **Load**, click **Browse**, select your saved file from the Chapter4 folder on your Student Disk, click **Open**, and then click **Load**. The stored program unit appears in the Program Units section in the Object Navigator.

Calling a Procedure and Passing Parameters

Now you will learn how to call a procedure from another procedure and pass parameters. Since you have updated the INVENTORY table so that every record has a stored inventory value, it would be useful to have a procedure that is called every time the QOH field is updated that calculates and updates the INV_VALUE field using the updated QOH value. To do this, you will write a procedure that updates the QOH in the INVENTORY table, and then calls a second procedure that receives the INVID as an input parameter, and then calculates the new INV_VALUE and updates it in the database. You will write the procedure that updates the INV_VALUE first. In this procedure, the INVID of the inventory row to be updated is passed as an IN parameter. The procedure retrieves the QOH and CURR_PRICE for this INVID using an implicit cursor, and then calculates the new INV_VALUE by multiplying the QOH by the CURR_PRICE. It then updates the INV_VALUE of the record, using the calculated value.

To write a procedure that updates a single INV_VALUE:

1. Click **Program Units** in the Object Navigator window, and then click the **Create** button to create a new procedure. Type **UpdateInvValueRecord** in the name box, then click **OK**.
2. Type the code shown in Figure 4-75. Compile the procedure, debug it if necessary, and then close the Program Unit Editor.

type this code

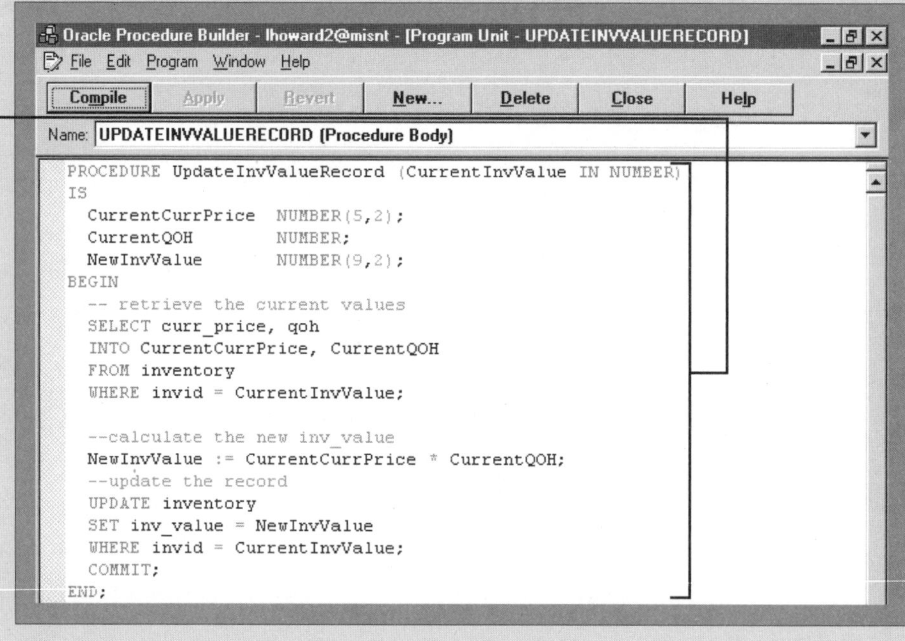
Oracle Procedure Builder - lhoward2@misnt - [Program Unit - UPDATEINVVALUERECORD]
File Edit Program Window Help
Compile Apply Revert New... Delete Close Help
Name: UPDATEINVVALUERECORD (Procedure Body)
PROCEDURE UpdateInvValueRecord (CurrentInvValue IN NUMBER)
IS
CurrentCurrPrice NUMBER(5,2);
CurrentQOH NUMBER;
NewInvValue NUMBER(9,2);
BEGIN
-- retrieve the current values
SELECT curr_price, qoh
INTO CurrentCurrPrice, CurrentQOH
FROM inventory
WHERE invid = CurrentInvValue;
--calculate the new inv_value
NewInvValue := CurrentCurrPrice * CurrentQOH;
--update the record
UPDATE inventory
SET inv_value = NewInvValue
WHERE invid = CurrentInvValue;
COMMIT;
END;

Figure 4-75: Procedure to update INV_VALUE field for a single record

Next, you will test your procedure to make sure it is working correctly. To do this, you will set a breakpoint on the first line of the procedure, and then run the procedure from the PL/SQL> prompt in Procedure Builder. You will pass the CurrentInvValue parameter as a literal value by typing an actual INVID number. Then, you will examine the record in the INVENTORY table to be sure the correct value was updated.

To test the procedure:

1. Click the **Program Unit** icon beside UPDATEINVVALUERECORD in the Object Navigator window to load UPDATEINVVALUERECORD into the PL/SQL Interpreter window.
2. Create a breakpoint on Line 8 of the procedure.

3 Type **UpdateInvValueRecord(11668);** at the PL/SQL> prompt in the SQL command prompt pane, and then press the **Enter** key. This runs the procedure and passes the value 11668 to the procedure as the CurrentInvValue parameter. The program execution arrow appears on Line 8.

4 Single-step through the procedure until the execution arrow is on Line 16, then examine the variable values that will be used to update the INVENTORY record. Confirm that the value for INVID is 11668, and the value for NEWINVVALUE is 4159.84, as shown in Figure 4-76.

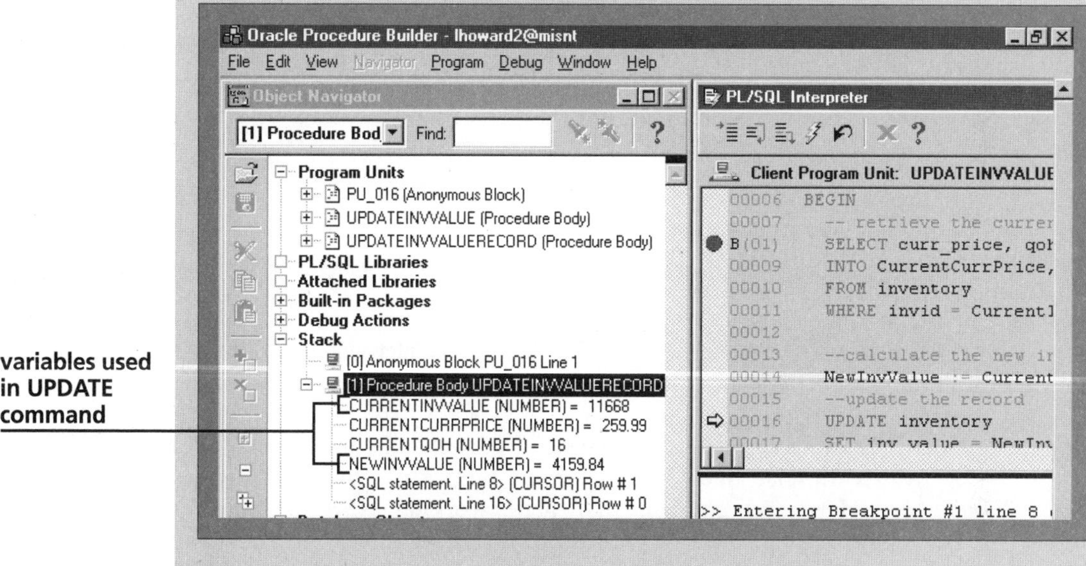

Figure 4-76: Viewing variable values used in UPDATE command

help

If your values are different from the ones shown, examine your program logic and variable values to identify the problem.

5 Click [reset button] to reset the PL/SQL Interpreter.

Next, you will create the calling procedure that calls the UpdateInvValueRecord procedure that you just wrote. This procedure will receive the INVID and new QOH as input parameters, update QOH, and then call the procedure to update INV_VALUE. Then you will test the procedure to make sure it is working correctly.

To write the calling procedure:

1 Create a new procedure named **UpdateQOH.**

2 Type the code shown in Figure 4-77. Compile the code, debug it if necessary, and then close the Program Unit Editor window.

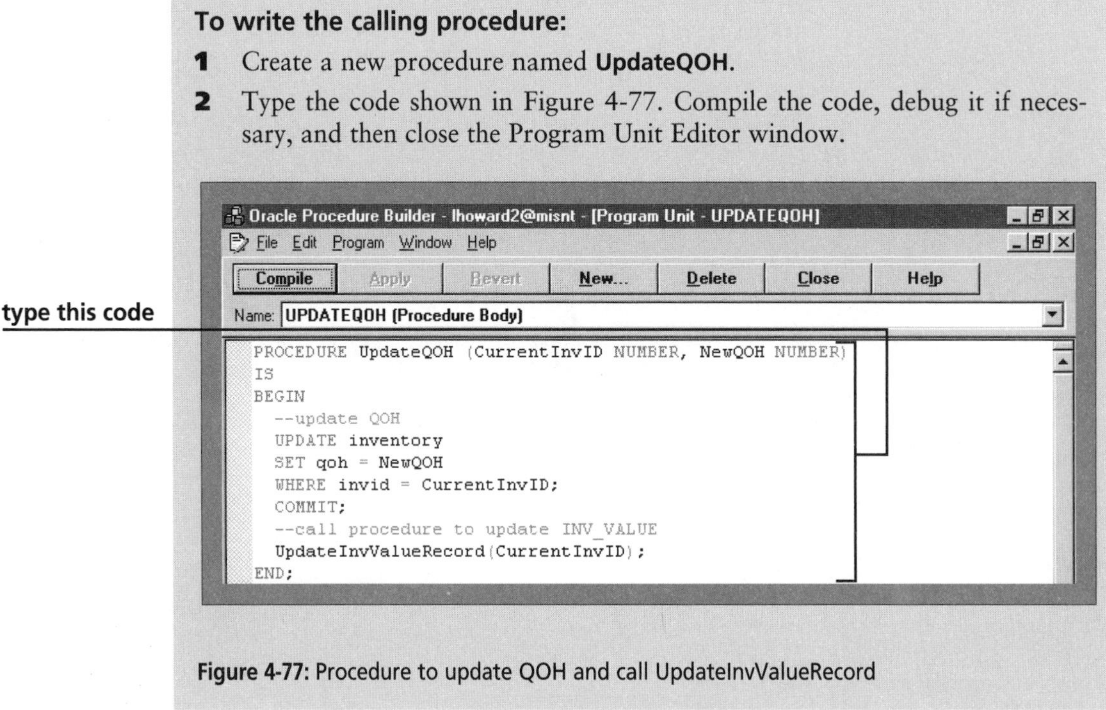
Oracle Procedure Builder - lhoward2@misnt - [Program Unit - UPDATEQOH]
File Edit Program Window Help
Compile
Apply
Revert
New...
Delete
Close
Help
type this code
Name: UPDATEQOH (Procedure Body)
PROCEDURE UpdateQOH (CurrentInvID NUMBER, NewQOH NUMBER)
IS
BEGIN
--update QOH
UPDATE inventory
SET qoh = NewQOH
WHERE invid = CurrentInvID;
COMMIT;
--call procedure to update INV_VALUE
UpdateInvValueRecord(CurrentInvID);
END;

Figure 4-77: Procedure to update QOH and call UpdateInvValueRecord

Now you will test UpdateQOH. You will run it using values of 11668 for CurrentInvID, and 15 for CurrentQOH. Then you will single-step through the procedure and confirm that it is working correctly, and that it calls the UpdateInvValueRecord procedure and passes the correct parameter value.

To test UpdateQOH:

1 If necessary, load UpdateQOH into the PL/SQL Interpreter.

2 Create a breakpoint on Line 5, or on the first line of the UPDATE command, if your line numbers are different.

3 Type **UpdateQoh(11668, 15);** then press the **Enter** key at the PL/SQL> prompt to run the procedure and pass 11668 as the value for CurrentInvID, and 15 for the value of NewQOH. The program execution arrow moves to Line 5.

help

If the program execution arrow does not appear on Line 5, click [reset icon] to reset the PL/SQL Interpreter, and then repeat Steps 1 and 2.

4 Check the variable values that are used in the UPDATE command to confirm that CURRENTINVID = 11668 and NEWQOH = 15.

5 Click the **Step Into** button until the execution arrow stops on Line 10, which is the program line that calls the UpdateInvValueRecord procedure. Click again, and note that the UpdateInvValueRecord procedure appears in the PL/SQL Interpreter pane.

6 Continue to single-step through the procedure. When the execution arrow is on Line 16 of the UpdateInvValueRecord procedure, confirm that the correct CURRENTINVVALUE (11668) and NEWINVVALUE (3899.85) values are used in the UPDATE command in the called procedure.

7 Click to reset the PL/SQL Interpreter.

8 Save the procedures as **updateinvvaluerecord.pls** and **updateqoh.pls** in the Chapter4 folder on your Student Disk.

Functions

A function is similar to a procedure, except that when it is called, it returns a single value that is assigned to a variable. The general format of a function header is:

```
FUNCTION <function name>     (<parameter1 name> <mode> <data type>,
                              <parameter2 name> <mode> <data type>,
                              ...)
                             RETURN <function return value data type> IS
<variable declarations>
```

Functions use parameters the same way procedures use parameters. The RETURN command at the end of the header (before the variable declarations) specifies the data type of the value that the function returns. Note that the function parameters are indented so they are vertically aligned in the function header.

The general format of the body of a function is:

```
BEGIN
    <program statements>
    RETURN <return value>
```

The RETURN command specifies the actual value that the function will return. Usually, the return value is a variable that has been assigned a value resulting from function computations.

The exception section is also slightly different for a function, with the following format:

```
EXCEPTION
    <exception handlers>
    RETURN EXCEPTION_NOTICE;
END;
```

The RETURN EXCEPTION_NOTICE command instructs the function to display the exception notice in the program that calls the function.

Calling a function requires assigning its return value to a variable. The general format for calling a function is <variable name> := <function name>(<parameter list>);.

You will now write a function that receives a Student ID, and returns the value for the student's age based on the difference between the current system date and the student's date of birth that is stored in the SDOB field in the STUDENT table of the Northwoods University database.

To create the function that calculates a student's age:

1 Select **Program Units** in the Object Navigator window, then click the **Create** button.

2 Type **StudentAge** for the function name, select the **Function** option button, then click **OK**. The Program Unit Editor window opens with the function template header that includes the RETURN command.

3 Type the code shown in Figure 4-78 in the Program Unit Editor. Compile the function, debug it if necessary, and then close the Program Unit Editor.

type this code

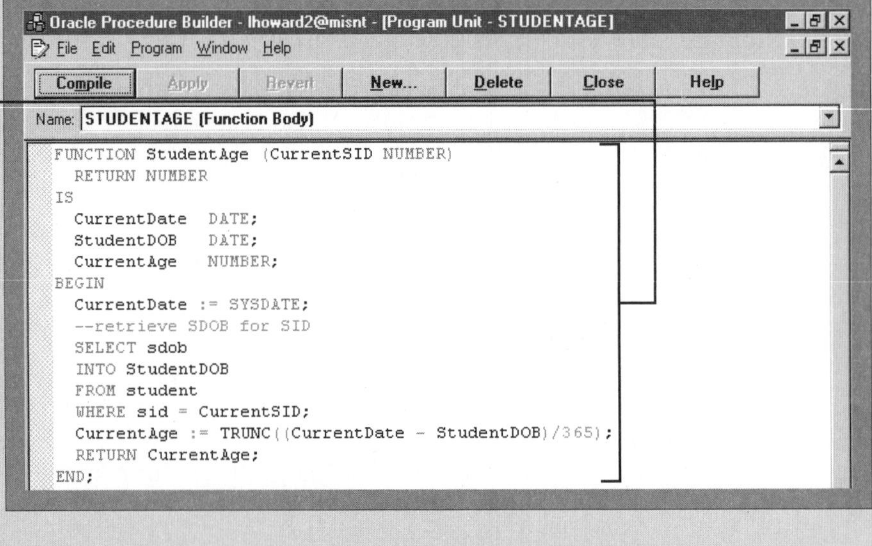

Figure 4-78: Code for StudentAge function

To test the function, you will have to write a procedure that calls the function and passes it a value for student ID. Then, you will test the procedure and function by single-stepping through the program units and examining the variable values. You will do this next.

To create a procedure to call and test the StudentAge function:

1. Create a new procedure named **UpdateStudentData.**
2. Type the code shown in Figure 4-79. Compile the procedure, debug it if necessary, and then close the Program Unit Editor.

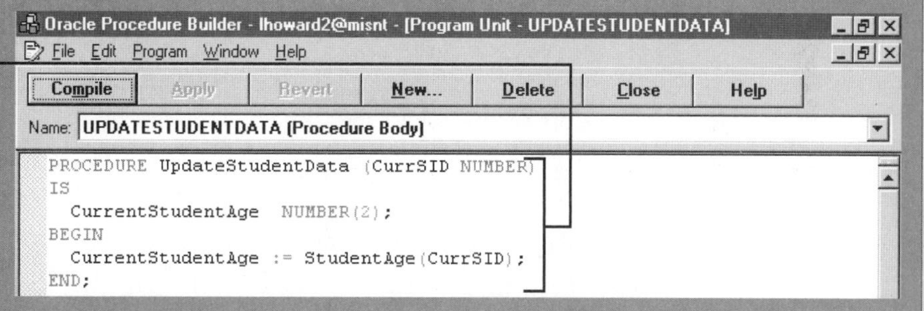

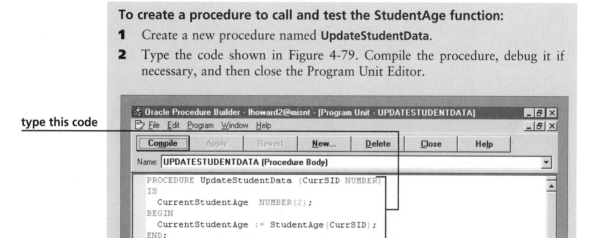

Figure 4-79: UpdateStudentData procedure code

3. If necessary, load UpdateStudentData in the PL/SQL Interpreter.
4. Set a breakpoint on Line 5.
5. Type **UpdateStudentData(100);** at the PL/SQL> prompt, and then press the **Enter** key to test the function using SID 100. The program execution arrow stops at Line 5, which is the line that calls the function.
6. Click the **Step Into** button. The StudentAge function is displayed in the PL/SQL Interpreter window. Continue to single-step through the function until the execution arrow is on Line 15 of the function.
7. Click + beside **Function Body STUDENTAGE** in the Object Navigator window under Stack to examine the current variable values in the function. The values should show the student date of birth, the current date, and the calculated age, as shown in Figure 4-80. (Your current date will be different.) Confirm that the student age value is correct for your current system date.

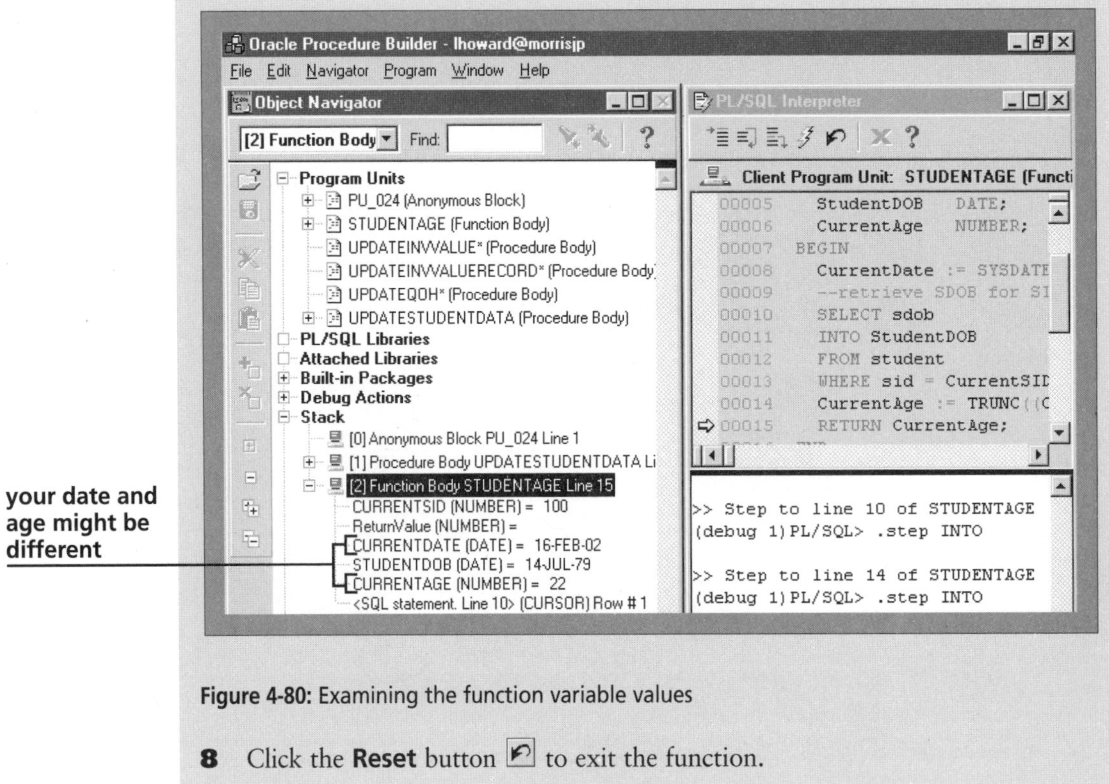

Figure 4-80: Examining the function variable values

8 Click the **Reset** button to exit the function.

You can save functions as .pls text files just as you can save procedures. Next, you will save your function and your calling procedure.

To save the function and calling procedure:

1. Open the StudentAge function in the Program Unit Editor and select all of the text.
2. Click **File** on the menu bar, click **Export Text**, navigate to the Chapter4 folder on your Student Disk, enter **studentage.pls** as the filename, then click **Save.**
3. Save the UpdateStudentData procedure as **updatestudentdata.pls** in the Chapter4 folder on your Student Disk.
4. Unload the StudentAge function from Procedure Builder by selecting it in the Object Navigator window, then clicking .
5. Unload the UpdateStudentData procedure from Procedure Builder by selecting it in the Object Navigator window, then clicking .

PL/SQL Libraries

So far, you have saved your PL/SQL procedures and functions as exported text files with a .pls extension. Another method for saving and restoring program units is to create a **PL/SQL library**, which is a single file that contains one or more related procedures or functions. This provides an efficient way of managing related PL/SQL programs—instead of having to load multiple program units, you can load a single library containing several programs and functions. Program units within a library run on the client workstation rather than on the database server. Now you will create a library that contains the UpdateInvValue, UpdateInvValueRecord, and UpdateQOH procedures that you created earlier.

To create a library:

1. Select **PL/SQL Libraries** in the Object Navigator window, then click the **Create** button. A new library appears under the PL/SQL Libraries category in the Object Navigator window.

help

You might need to click Window on the menu bar, then click Object Navigator to redisplay the Object Navigator window.

When new objects are created in Oracle applications, they are often given default names. To change the library name, you must save the library either in the database or in the file system of your computer. You will save the library in the file system in the Chapter4 folder on your Student Disk.

To save the library:

1. Make sure that the new library is selected in the Object Navigator window, then click the **Save** button on the Object Navigator toolbar. The Save Library dialog box opens.
2. Type **UpdateInventory** as the library name, and make sure that the File System option button is selected.
3. Click **Browse**, and find the Chapter4 folder on your Student Disk. Type **updateinventory.pll** as the filename, click **Open**, then click **OK** to save the library. The new library name is displayed in the Object Navigator window.

The next step is to add the program units to the library. To add a program unit to a library, the program unit must currently be loaded in Procedure Builder. Then, you simply drag the program unit from the Program Units category to the library's program unit subcategory. Now, you will add the program units to the Update_Inventory library.

To add the program units to the library:

1 Make sure that UPDATEINVVALUE, UPDATEINVVALUERECORD, and UPDATEQOH are displayed in the Program Units category in the Object Navigator window.

help If the required program units are not displayed, click File, then click Load, and load the procedures from the Chapter4 folder on your Student Disk.

2 Click [+] beside the UPDATEINVENTORY library if necessary, so its objects are displayed.

3 Click the **UPDATEINVVALUE** program unit, then drag it to the UPDATEINVENTORY library. The pointer changes to .

4 When the tip of the pointer arrow is on the Program Units category under the UPDATEINVENTORY library, drop the program unit. It now appears under the library, as shown in Figure 4-81.

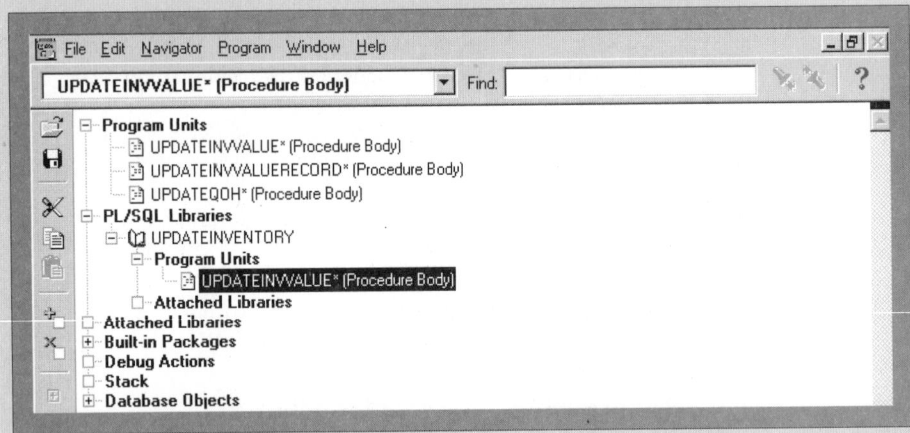

Figure 4-81: Adding a program unit to a library

4 Add the UPDATEINVVALUERECORD and UPDATEQOH program units to the library also. All three program units should be displayed under the Program Units subcategory under the UpdateInventory library.

5 Click the **UPDATEINVENTORY** library to select it, then click the **Save** button to save the changes to the library.

Finally, you will learn how to load and use library files. First, you must unload the current program units and the Update_Inventory library from Procedure Builder.

To unload the program units and library:

1. Click the **UPDATEINVENTORY** library, then click the **Delete** button to unload the library.
2. Select the **UPDATEINVVALUE** program unit, then click to unload the procedure.
3. Unload UPDATEINVVALUERECORD and UPDATEQOH also.

Now you will reload the Update_Inventory library. Then, you will be able to run and edit the individual program units.

To reload the library:

1. Click **PL/SQL Libraries** in the Object Navigator window, then click the **Open** button on the Object Navigator toolbar. The Open Library dialog box is displayed.
2. Make sure the File System option button is selected, then click **Browse**, open the Chapter4 folder on your Student Disk, select **updateinventory.pll**, click **Open**, then click **OK**. The UPDATEINVENTORY library is displayed in the Object Navigator window.
3. Click beside Program Units under UPDATEINVENTORY. The individual program units are displayed.

You can now run the individual program units by typing the program unit name and any necessary parameters at the PL/SQL> command line. If you want to debug a library file using the PL/SQL Interpreter, you must first drag the program unit to the top-level Object Navigator Program Units category so that it is displayed as an individual program unit that is not part of a library. From that point, debugging procedures are identical to those described before. Now, you will unload the library you created.

To unload the library:

1. Unload the UPDATEINVENTORY library by selecting it in the Object Navigator window and then clicking the **Delete** button.

Stored Program Units

Stored program units, also called stored procedures, are program units that are stored in the database, and run on the database server. Creating a stored program unit is desirable, because it makes the program unit available at all times, and you do not have to load it from an external file system or floppy disk. Stored program units are different from libraries, because libraries can be stored either in the database or the file system. Another difference is that library program units always run on the client machine, while stored program units always run on the database server. This is a desirable trait with client/server systems, because by using both libraries and stored program units, processing can be distributed across several machines without overloading the server. However, some programs have to be processed on the server, because the client workstation might not have the required applications or hardware to run the program.

Now, you will convert the StudentAge function that you wrote earlier to a stored program unit that will be stored in the database. First, you will load the function from your Student Disk into Procedure Builder. Then, you will view the objects that are currently stored in your user area on the database.

To load the function and view your database objects:

1. Select **Program Units** in the Object Navigator, click **File** on the menu bar, click **Load**, then click **Browse**.
2. Select **studentage.pls** from the Chapter4 folder on your Student Disk, click **Open**, then click **Load**.
3. Click [+] beside Program Units. The StudentAge function is displayed.
4. In the Object Navigator window, click [+] beside Database Objects. This displays a list of all database users.
5. Find your user name, and click [+] to view your database objects. This lists the objects that are stored in your user area of the database. Objects that can be stored in the database include stored program units, PL/SQL libraries, tables, and views.

Currently, you probably do not have any stored program units or libraries. Now, you will convert the StudentAge function to a stored program unit. Then it will be available for use in later PL/SQL programs.

To convert StudentAge to a stored program unit:

1 Click the **STUDENTAGE** function under the top-level Program Units category, and then drag it to the Stored Program Units subcategory under your user name. After a moment, STUDENTAGE appears as a stored program unit, as shown in Figure 4-82.

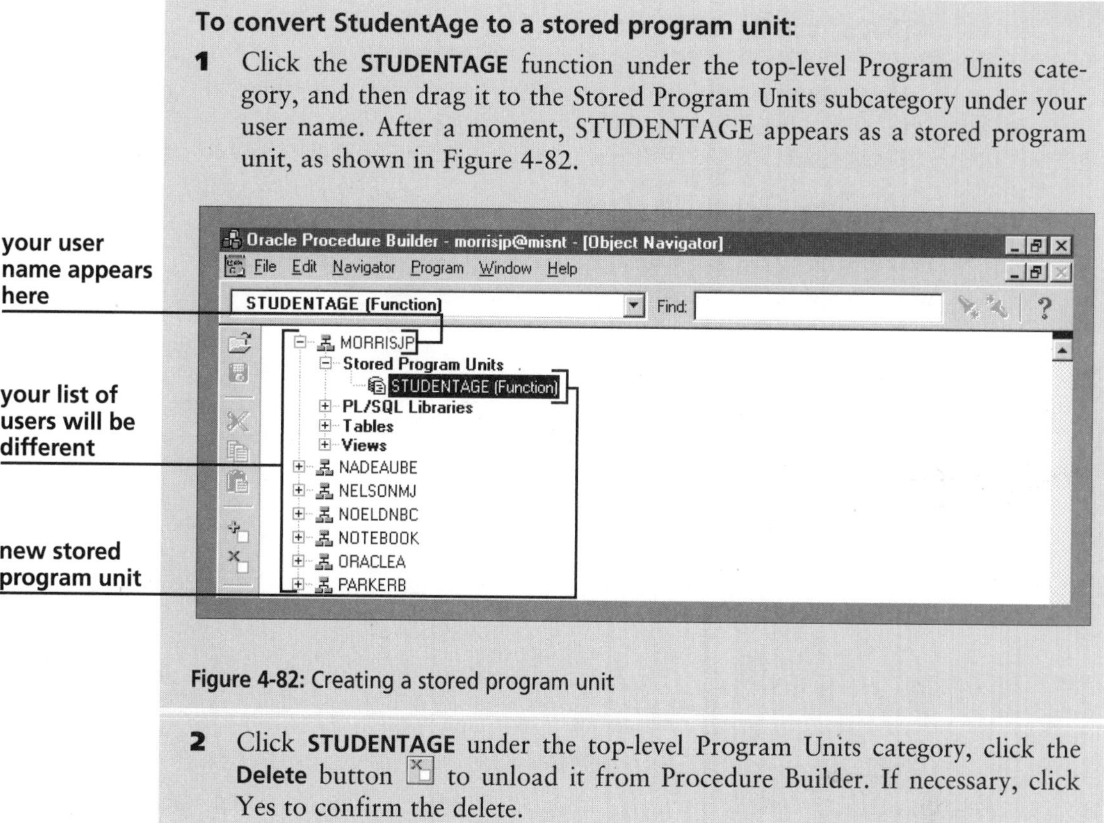

Figure 4-82: Creating a stored program unit

2 Click **STUDENTAGE** under the top-level Program Units category, click the **Delete** button to unload it from Procedure Builder. If necessary, click Yes to confirm the delete.

Packages

A **package** is a code library that provides a way to manage large numbers of related programs and variables. All of the variables that you have used so far are **local** variables, meaning that they are visible only in the program in which they are declared. As soon as the program terminates, the memory used to store the variable is freed, and the variable value cannot be accessed again. Sometimes it is necessary to have **global** variables, which are variables that are visible to many different PL/SQL programs. A global variable is declared using the same syntax as is used for declaring a local variable, and while local variables are declared in the DECLARE section of an individual program, global variables are declared in the DECLARE section of a package. Global variables can be used in programs in the same way that local variables are used. Also, sometimes you need to make PL/SQL procedures and functions available to other database users and other programs. To do this in PL/SQL, you must create a package. A package has two components: the package specification and the package body.

The Package Specification

The **package specification**, also called the package header, declares global variables, cursors, procedures, and functions that can be called or accessed by other program units. The general format for declaring a package specification is:

```
PACKAGE <package name>
IS
     <variable declarations>;
     <cursor declarations>;
     <procedure and function declarations>;
END <package name>;
```

Variables are declared the same way in packages as they are in any declaration. Cursors must be declared using the following format:

```
CURSOR <cursor name>
RETURN <cursor return row variable>
IS <cursor SELECT statement>;
```

To declare a procedure in a package, you must specify the procedure name, followed by the parameters and variable types, using the following format:

```
PROCEDURE <procedure name> (param1 param1datatype, param2 param2datatype, …);
```

To declare a function in a package, you must specify the function name, parameters, and return variable type, as follows:

```
FUNCTION <function name> (param1 param1datatype, param2 param2datatype, …)
RETURN <return data type>;
```

Now you will create a package specification for a package named InventoryPackage that will contain three procedures. The first procedure is UpdateInvValue, which updates every INV_VALUE field in the INVENTORY table. It will also contain UpdateInvValueRecord, which updates the single record in the INVENTORY table. Finally, it will contain UpdateQOH, a procedure that updates the quantity on hand of a single inventory record and then calls the UpdateInvValueRecord procedure to update the INV_VALUE for the record whose QOH was just updated. Instead of passing the INVID value as a parameter, the package will declare a global variable named GlobalInvID that will be assigned to the current INVID in UpdateQOH, and then be referenced by the global variable name in UpdateInvValueRecord.

To create the package specification:

1 Select the Program Units top-level category in the Object Navigator window, then click the **Create** button to create a new program unit. The Program Unit dialog box opens.

2. Type **InventoryPackage** for the program unit name, select the **Package Spec** option button, then click **OK**. The Package Specification template appears in the Program Unit Editor.
3. Type the package specification shown in Figure 4-83. Compile the package specification, debug it if necessary, then close the Program Unit Editor.

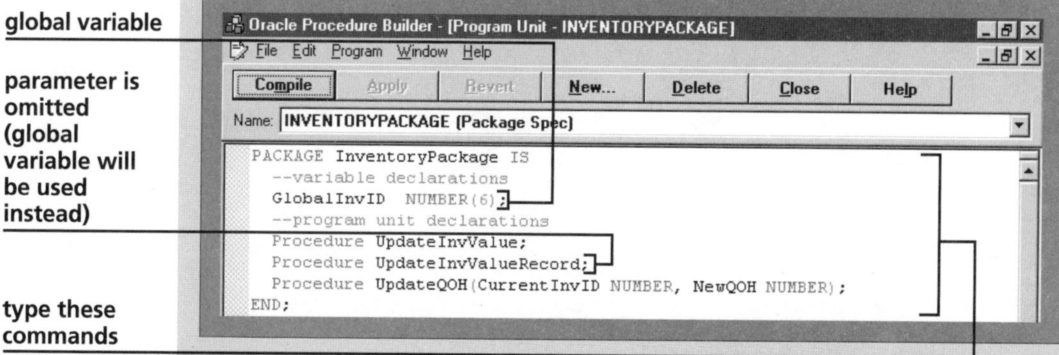
global variable
parameter is omitted (global variable will be used instead)
type these commands
Oracle Procedure Builder - [Program Unit - INVENTORYPACKAGE]
File Edit Program Window Help
Compile
Apply
Revert
New...
Delete
Close
Help
Name: INVENTORYPACKAGE (Package Spec)
PACKAGE InventoryPackage IS
--variable declarations
GlobalInvID NUMBER(6);
--program unit declarations
Procedure UpdateInvValue;
Procedure UpdateInvValueRecord;
Procedure UpdateQOH(CurrentInvID NUMBER, NewQOH NUMBER);
END;

Figure 4-83: Package specification

The Package Body

The **package body** contains the code for the programs declared in the package specification. The package body is optional, because sometimes a package contains only variable declarations and no programs. The general format for the package body is:

```
PACKAGE BODY <package name>
IS
 <variable declarations>
 <cursor specifications>
 <module bodies>
END <package name>;
```

Variables declared at the beginning of the body are visible to all modules in the package body, but are not visible to modules outside of the package body. Variables declared within the individual modules are local to the individual module. Each individual module has its own BEGIN and END statements. Now you will create the package body.

To create the package body:

1. Select the Program Units top-level category in the Object Navigator window, then click the **Create** button to create a new program unit. The Program Unit dialog box opens.
2. Type **InventoryPackage** for the program unit name, select the **Package Body** option button, then click **OK**. The package body template appears in the Program Unit Editor.
3. Type the code shown in Figure 4-84 to create the package body variable declarations. This includes the cursor that is used in the UpdateInvValue program unit, and the other variables that are used in the UpdateInvValue, UpdateInvValueRecord, and UpdateQOH procedures. Do NOT compile the code, because the package body is not yet complete.

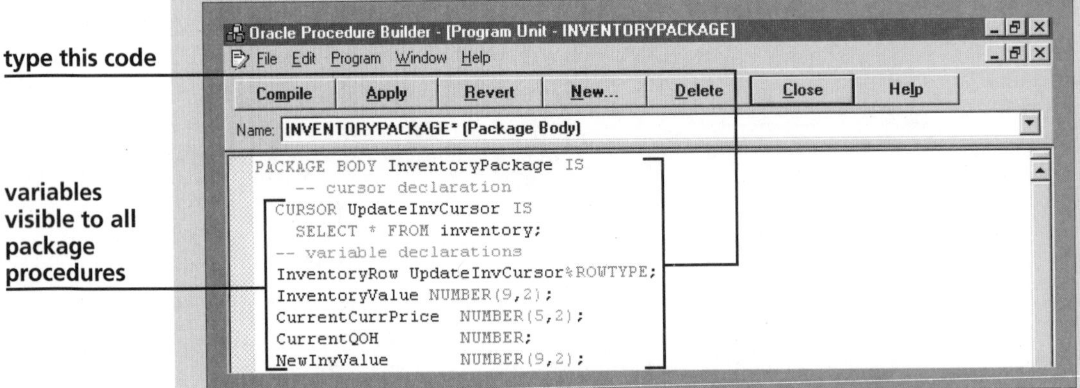

Figure 4-84: Package body cursor and variable declarations

4. Type the code for the body of the UpdateInvValue procedure, as shown in Figure 4-85, in the Program Editor window. This code should be entered directly below the variable declarations shown in Figure 4-84.

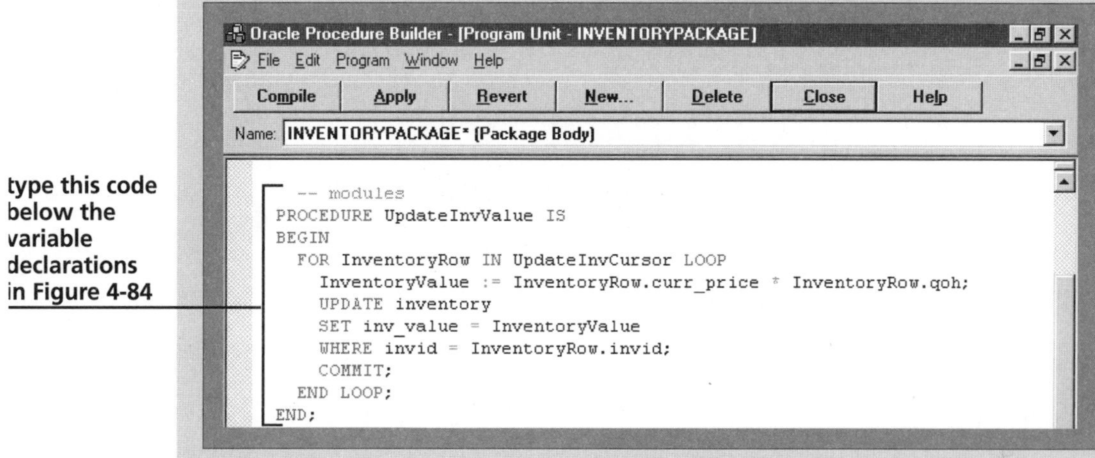

Figure 4-85: UpdateInvValue procedure body

5 Type the code for the body of the UpdateInvValueRecord procedure, as shown in Figure 4-86, directly below the code for the body of the UpdateInvValue procedure.

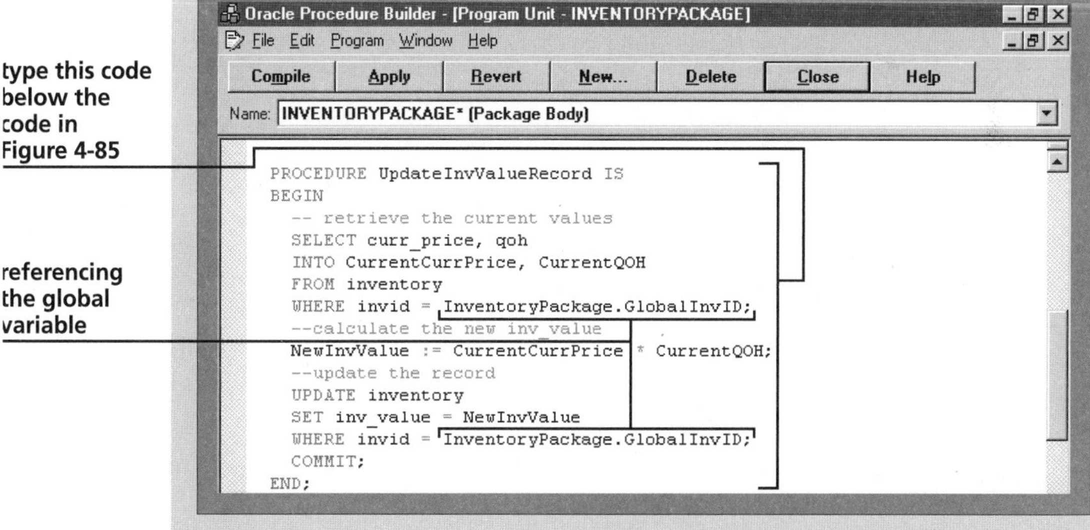

Figure 4-86: UpdateInvValueRecord procedure body

6 Type the code for the body of the UpdateQOH procedure, as shown in Figure 4-87, directly below the code for the body of the UpdateInv-ValueRecord procedure. Note that the UpdateInvValueRecord procedure must be entered before the UpdateQOH procedure, because it is called by the UpdateQOH procedure. If it is not entered first, a compile error will occur.

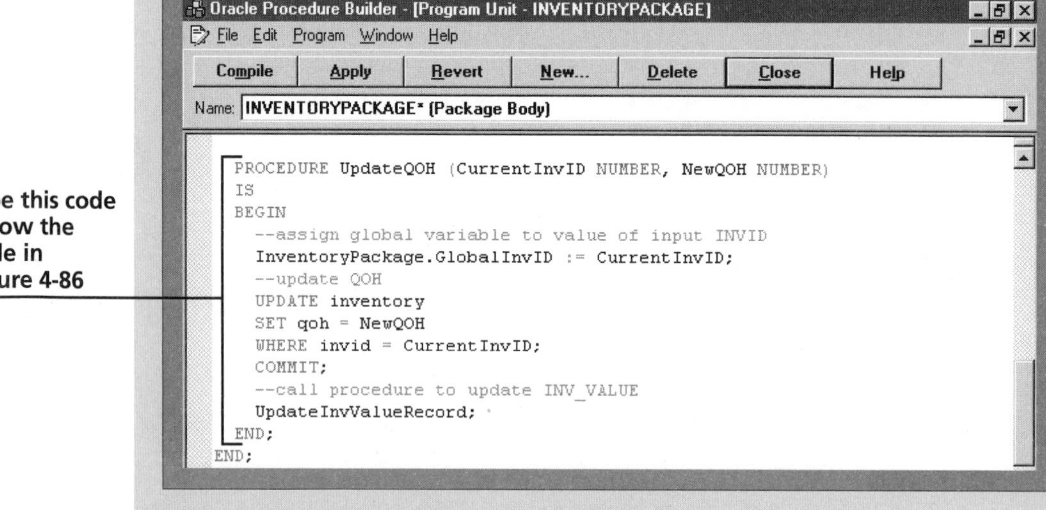
type this code below the code in Figure 4-86
Oracle Procedure Builder - [Program Unit - INVENTORYPACKAGE]
File Edit Program Window Help
Compile Apply Revert New... Delete Close Help
Name: INVENTORYPACKAGE* (Package Body)
PROCEDURE UpdateQOH (CurrentInvID NUMBER, NewQOH NUMBER)
IS
BEGIN
--assign global variable to value of input INVID
InventoryPackage.GlobalInvID := CurrentInvID;
--update QOH
UPDATE inventory
SET qoh = NewQOH
WHERE invid = CurrentInvID;
COMMIT;
--call procedure to update INV_VALUE
UpdateInvValueRecord;
END;
END;

Figure 4-87: UpdateQOH procedure body

7 Compile the package body, correct any compile errors, and close the Program Unit Editor.

To reference an item (such as a variable or program unit) that is declared in a package specification, you must preface the item with the name of the package, using the general format <package name>.<item name>. In Figure 4-87, notice that the value for the CurrentInvID parameter is assigned to the global variable GlobalInvID using the following program statement:

```
InventoryPackage.GlobalInvID := CurrentInvID;
```

When the global variable is referenced in the UpdateQOH procedure in Figure 4-87, the variable name is again prefaced by the package name.

To run a procedure within a package, you must preface the procedure name with the package name, using the general format <package name>.<procedure name> (<parameter list>). Now you will test the package by running it from the PL/SQL> prompt. You will run the UpdateQOH procedure, and pass it parameter values for INVID = 11669, and new QOH = 10. Then, you will verify that the package updated the record correctly.

To run the UpdateQOH procedure in the package and verify that it updated the record:

1 Open the PL/SQL Interpreter pane if necessary, and type the command shown in Figure 4-88 to run the procedure.

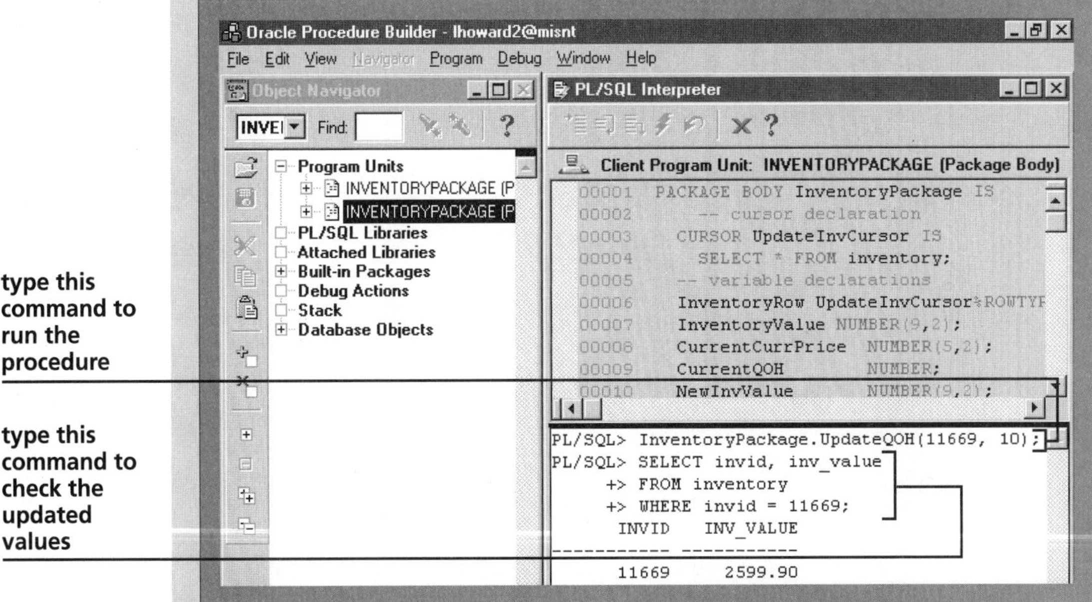

Figure 4-88: Running the procedure and verifying the result

2 Verify that the record was successfully updated by typing the SQL command shown in Figure 4-88. The updated value of INV_VALUE should be displayed as shown.

3 Save the package body and specification as stored program units by dragging and dropping them from the top-level Program Units category to the Program Units category under your user name, as shown in Figure 4-89.

You can select and save the package specification and body together by selecting the package specification, pressing the Shift key, and then selecting the package body while keeping the Shift key pressed. Then, drag them both under your user name at the same time.

4 Select **INVENTORYPACKAGE(Package Spec)**, and click the **Delete** button to unload it from Procedure Builder. Click **Yes** to confirm unloading.

5 Unload INVENTORYPACKAGE(Package Body) from Procedure Builder.

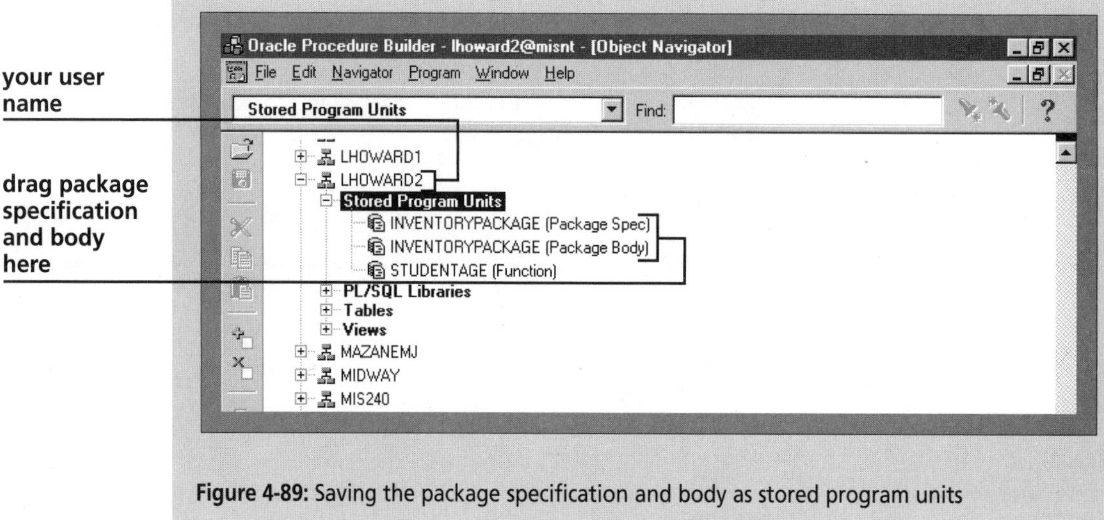

Figure 4-89: Saving the package specification and body as stored program units

Elements in the package body can be public or private. **Public elements** are defined in the package specification, and can be accessed from other PL/SQL programs. **Private elements** are not defined in the package specification, and can only be accessed within the package. For example, if the UpdateInvValueRecord procedure, which is called by the UpdateQOH procedure, was omitted from the package specification but included in the package body, it would be a private procedure.

In addition to user-made packages such as the one you created, Oracle provides many built-in packages that are accessible to both client and server programs. The core functions of the PL/SQL language are contained in the STANDARD and DBMS_STANDARD built-in packages. For the core functions, you do not need to preface the function name with the package name. Other built-in packages extend the core functions. Calling functions in these extension packages requires prefacing the function with the name of the package.

Database Triggers

Database triggers are program units that execute in response to the database events of inserting, updating, or deleting a record. Database triggers are similar to procedures, except that a trigger executes whenever the triggering event occurs, and a trigger does not accept parameters. When a trigger executes, it is said to have *fired*. Triggers are attached to tables, and can be used to maintain complex integrity constraints or to audit changes made in a table, or to signal to other programs that changes were made to a table.

Types of Triggers

A trigger can be categorized by the statement that causes it to fire, the timing of when it fires, and the level at which it fires. Figure 4-90 summarizes the different types of triggers.

Type	Values	Description
Statement	INSERT, DELETE, UPDATE	Defines statement that causes trigger to fire
Timing	BEFORE, AFTER	Defines whether trigger fires before or after statement is executed
Level	ROW, STATEMENT	Defines whether trigger fires once for each triggering statement, or once for each row affected by the triggering statement

Figure 4-90 Types of triggers

SQL INSERT, DELETE, and UPDATE commands can cause triggers to fire. **Statement-level triggers** fire once, either before or after the SQL triggering statement executes. **Row-level triggers** fire once for each row affected by the triggering statement. A table can have many triggers attached to it.

Now you will create a trigger attached to the INVENTORY table in the Clearwater Traders database. This trigger will be a row-level trigger that will fire just before the QOH is updated for a record, and will automatically update the INV_VALUE field using the updated QOH value.

To create the trigger:

1. If necessary, start Procedure Builder and connect to the database. Open the following objects in the Object Navigator: **Database Objects**, **<your account>**, **Tables**, **INVENTORY**, and **Triggers**.
2. Be sure that Triggers is selected, then create a new trigger by clicking the **Create** button on the Object Navigator toolbar. The Database Trigger dialog box opens.
3. Click **New** on the bottom-left area of the dialog box. Delete the default trigger name, and enter **UpdateInventoryTrigger** for the trigger name, as shown in Figure 4-91.

4 Be sure the Before option button is selected to specify the trigger timing, check **UPDATE** to specify the trigger statement, and click **QOH** in the Of Columns list box to specify the update field that will fire the trigger. Check the **Row** option button to specify that this is a row-level trigger. The trigger's types and actions look like Figure 4-91.

The trigger needs to fire *before* the record is updated, because the trigger is fired whenever the record is updated, and the trigger command also updates the record. Therefore, if the trigger were to fire *after* the update, then the record would be updated, the trigger would fire and update the record again, and then the trigger would fire again and update the record, and fire again and update, repeating in an endless loop of fire and update commands.

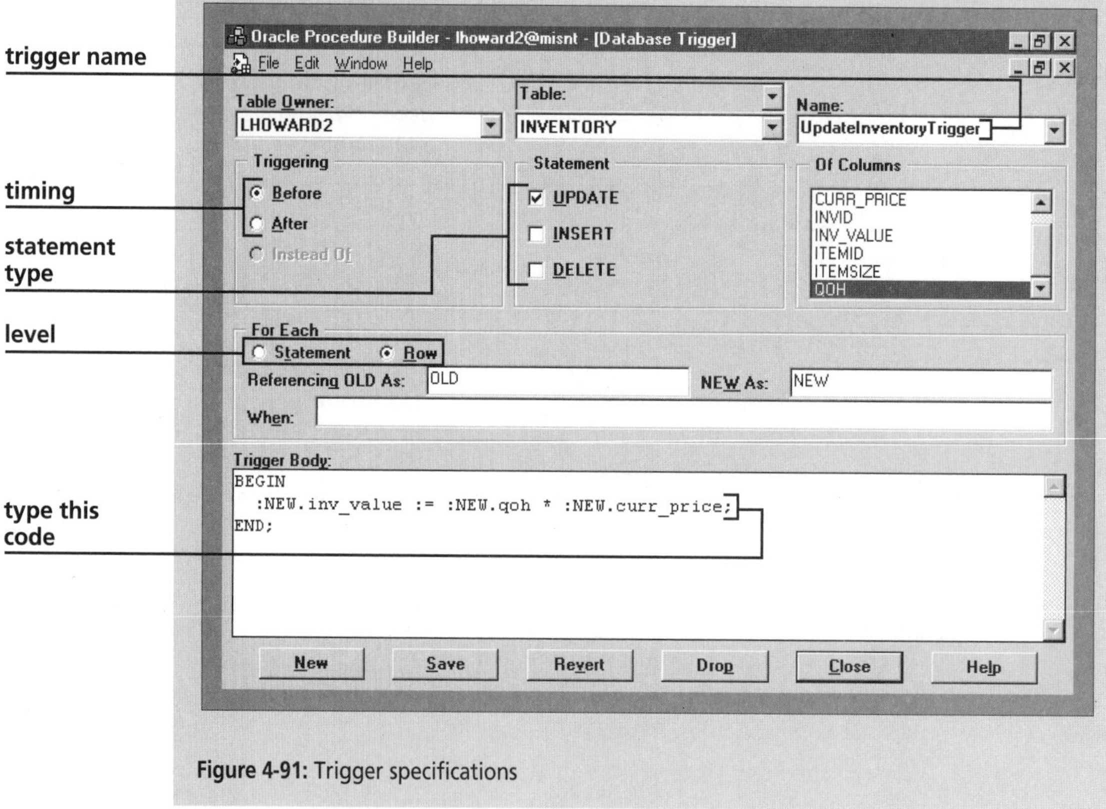

Figure 4-91: Trigger specifications

The final step is to enter the code that is executed when the trigger fires. The code will retrieve the values of CURR_PRICE and QOH from the updated record, calculate the new INV_VALUE, and then update the record using the new INV_VALUE.

You can reference the value of a field in the current record both before and after the triggering statement executes. To reference a value before the triggering statement executes, you use the following general format: `:OLD.<field name>`. To reference the value after the triggering statement executes, you use the following format: `:NEW.<field name>`. For example, to reference the value of QOH after it is updated, you would use the statement `:NEW.qoh`. The Database Trigger dialog box allows you to enter alternate aliases for "before" and "after" values, since you might have database tables named OLD or NEW. Since you don't have tables named OLD or NEW, you will specify the default alias names of OLD and NEW, and enter the trigger code.

To enter the trigger aliases and trigger body code:

1 Type **OLD** in the Referencing OLD As: text box, and type **NEW** in the NEW As: text box, as shown in Figure 4-91.
2 Type the trigger body code shown in Figure 4-91.
3 Click **Save** to create the trigger.
4 Click **Close** to close the Database Trigger dialog box.

Now you will test the trigger to make sure it is working correctly. First, you will update the QOH of INVID 11775 to 100 in the INVENTORY table to fire the trigger. Then you will check the value of INV_VALUE for the record, to be sure it has been updated to reflect the new quantity on hand.

To test the trigger:

1 Open the PL/SQL Interpreter window if necessary.
2 Type the UPDATE command shown in Figure 4-92 at the PL/SQL> prompt, and then press the **Enter** key.
3 Type the SELECT command shown in Figure 4-92 at the PL/SQL> prompt to verify that the trigger updated the INV_VALUE to 2995.00. Now you will update the QOH of this inventory item again to verify that the INV_VALUE has changed.

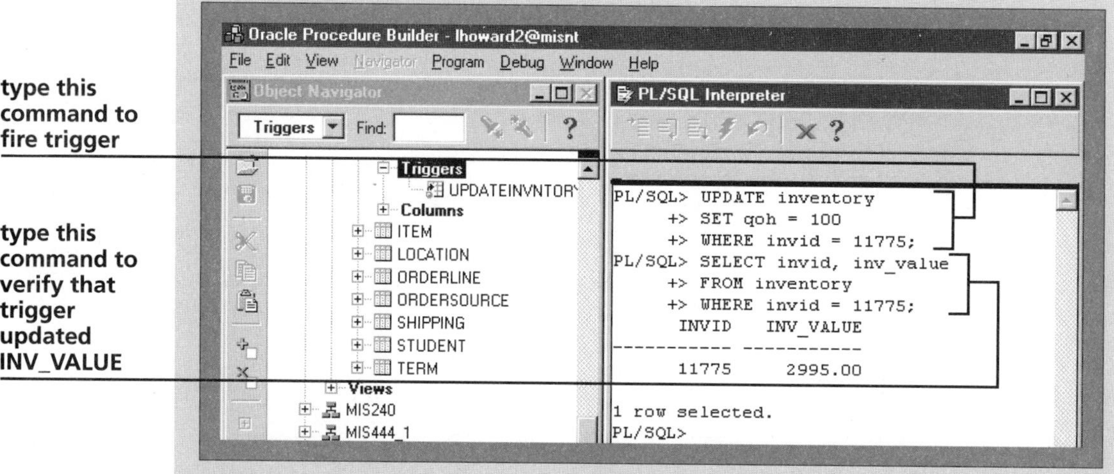

Figure 4-92: Testing the trigger

4 Type the UPDATE query shown in Figure 4-93 at the PL/SQL> prompt.

5 Type the SELECT query shown in Figure 4-93 at the PL/SQL> prompt. Your updated INV_VALUE of 3024.95 should be displayed.

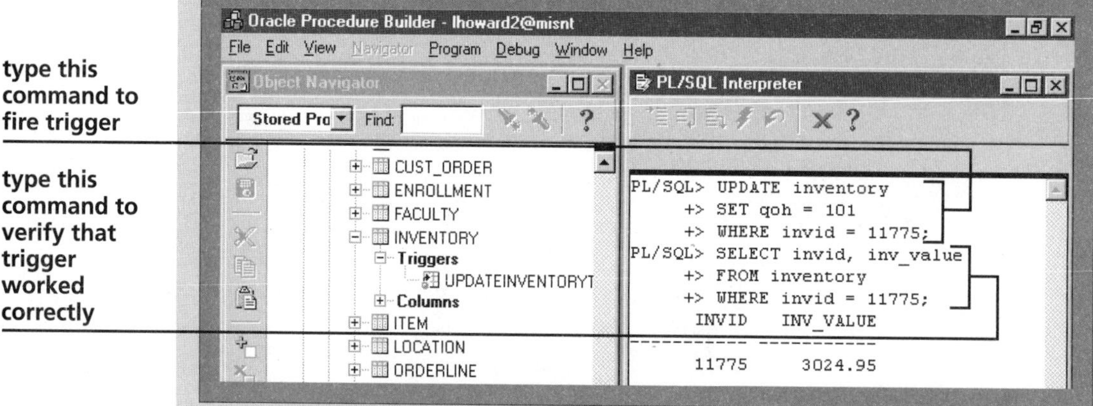

Figure 4-93: Testing the trigger again

Now you will create a trigger that corresponds to multiple SQL statements. Whenever an item is ordered at Clearwater Traders, a record is inserted in the ORDERLINE table specifying the order ID, inventory ID of the item, price of the item being ordered, and quantity ordered. The quantity ordered must then be subtracted from the inventory item's QOH field in the INVENTORY table. If an ORDERLINE record is updated, the existing quantity must be added back to the inventory QOH, and the new quantity must be subtracted from the QOH. And, if an ORDERLINE record is deleted, the quantity must be added back to the inventory QOH. This trigger will correspond to the INSERT, UPDATE, and DELETE statements for the ORDERLINE table. Now you will create the trigger for the ORDERLINE table.

To create the ORDERLINE table trigger:

1 Click [+] beside ORDERLINE under your list of database tables, then click **Triggers**. Click the **Create** button to create a new trigger for the ORDERLINE table. The Database Triggers dialog box opens.

2 Click **New** on the bottom-left of the dialog box. Delete the default trigger name, and type **OrderlineTrigger** for the trigger name.

3 Make sure that the Before option button is selected, and check **UPDATE**, **INSERT**, and **DELETE** to specify the trigger statements.

4 Click **QUANTITY** in the Of Columns: list box to specify the update field that will fire the trigger. Select the **Row** option button to specify that this is a row-level trigger. Type **OLD** in the Referencing OLD As: text box, and type **NEW** in the NEW As: text box.

In the trigger body, you will use an IF/ELSIF structure to specify the appropriate action depending on the current SQL command. For an INSERT operation, the new QUANTITY is subtracted from the inventory QOH. For an UPDATE operation, the old QUANTITY is added back to the inventory QOH, and the new QUANTITY is subtracted from the QOH. For a DELETE operation, the old QUANTITY is subtracted from the inventory QOH. Now, you will create the trigger body.

To create the trigger body:

1 Type the trigger body code shown in Figure 4-94. Click **Save** to create the trigger, then click **Close** to close the Database Trigger dialog window. Now you will test the trigger. First, you will test to confirm that the INSERT part of the trigger works correctly.

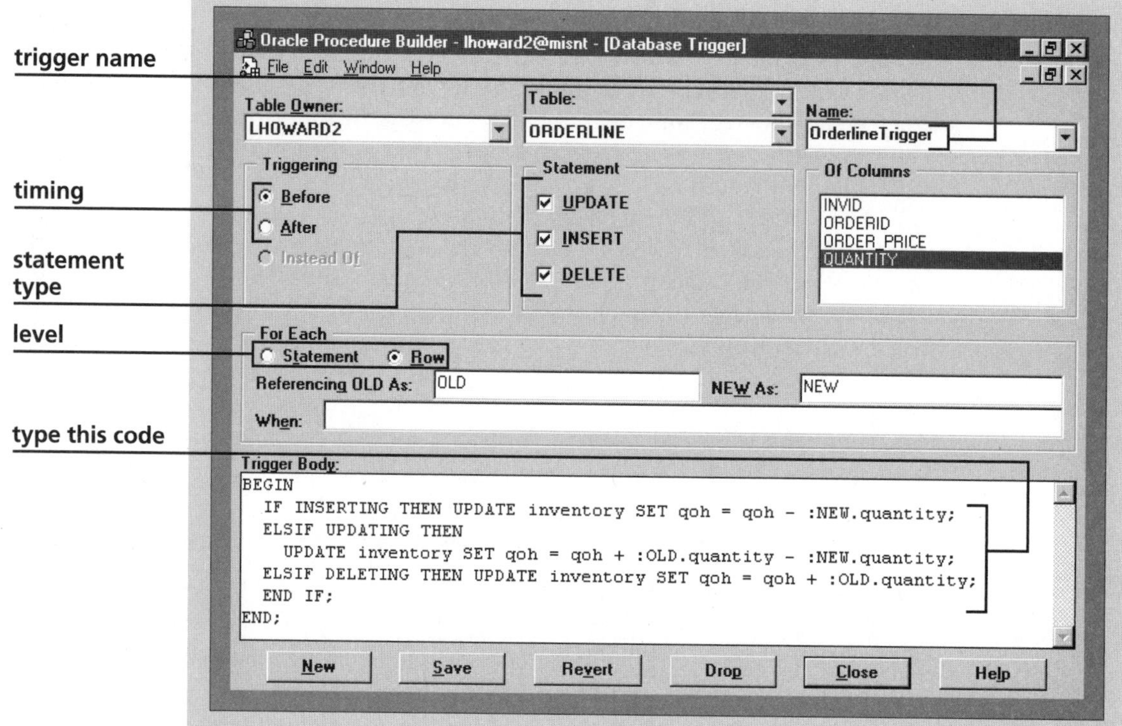

Figure 4-94: OrderlineTrigger specification

2 Open the PL/SQL Interpreter window, and type the SELECT command shown in Figure 4-95 at the PL/SQL> prompt to determine the current QOH of inventory item 11848. The current QOH should be 12.

3 Type the INSERT command shown in Figure 4-95 to insert a new record into ORDERLINE.

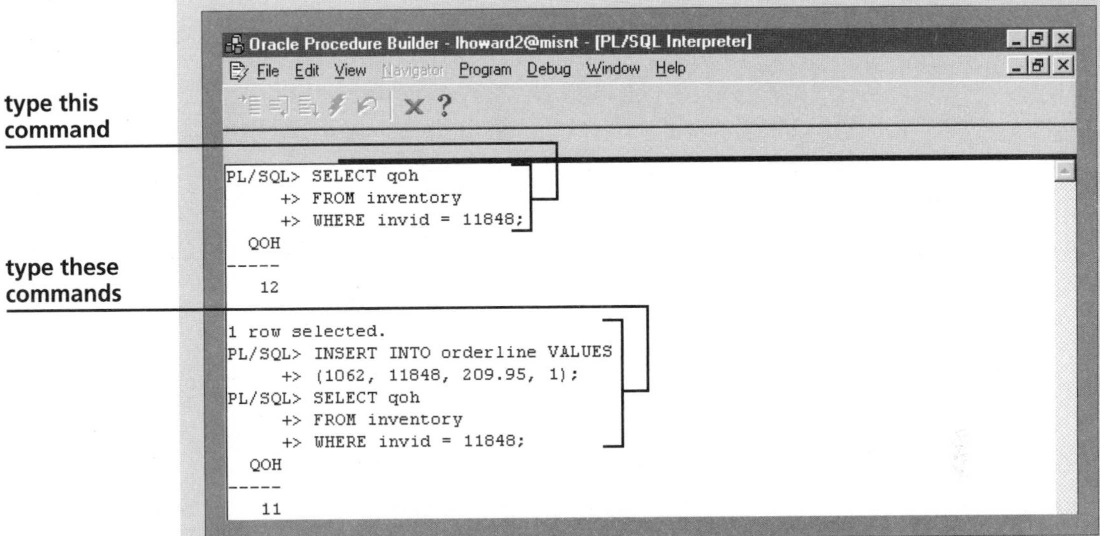
Oracle Procedure Builder - lhoward2@misnt - [PL/SQL Interpreter]
File Edit View Navigator Program Debug Window Help
type this command
type these commands
PL/SQL> SELECT qoh
+> FROM inventory
+> WHERE invid = 11848;
QOH

12
1 row selected.
PL/SQL> INSERT INTO orderline VALUES
+> (1062, 11848, 209.95, 1);
PL/SQL> SELECT qoh
+> FROM inventory
+> WHERE invid = 11848;
QOH

11

Figure 4-95: Testing the INSERT trigger

4 Type the SELECT command again, as shown, to confirm that the QOH of INVID 11848 was decreased by one from 12 to 11 as a result of adding the record to ORDERLINE. Now you will test to make sure the UPDATE and DELETE trigger statements work correctly.

5 Type the UPDATE command shown in Figure 4-96 to update the quantity ordered from one item to two items. Type the SELECT command to confirm that the QOH was decreased by one (from 11 to 10) to reflect the change.

6 Type the DELETE command shown in Figure 4-96 to delete the order from the database.

7 Type the second SELECT command to confirm that the QOH was increased by two (from 10 to 12) to reflect the change caused by deleting the order.

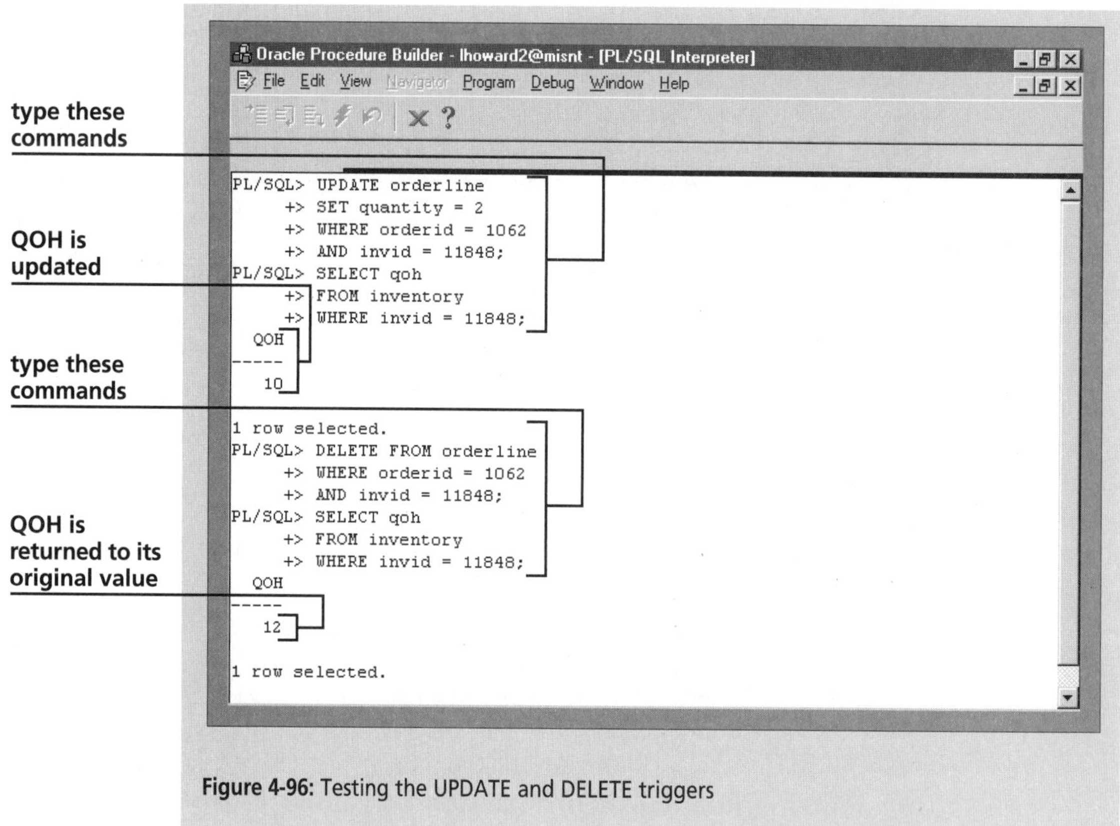
type these commands
QOH is updated
type these commands
QOH is returned to its original value
Oracle Procedure Builder - lhoward2@misnt - [PL/SQL Interpreter]
File Edit View Navigator Program Debug Window Help
PL/SQL> UPDATE orderline
+> SET quantity = 2
+> WHERE orderid = 1062
+> AND invid = 11848;
PL/SQL> SELECT qoh
+> FROM inventory
+> WHERE invid = 11848;
QOH

10
1 row selected.
PL/SQL> DELETE FROM orderline
+> WHERE orderid = 1062
+> AND invid = 11848;
PL/SQL> SELECT qoh
+> FROM inventory
+> WHERE invid = 11848;
QOH

12
1 row selected.

Figure 4-96: Testing the UPDATE and DELETE triggers

Now you will disable your triggers and exit Procedure Builder.

To disable the triggers and exit Procedure Builder:

1. Open the Object Navigator window, click [+] beside the INVENTORY table if necessary to view its objects, click [+] beside Triggers, right-click **UPDATEINVENTORYTRIGGER**, then click **Disable**.
2. Disable **ORDERLINETRIGGER**.
3. Click **File** on the top menu, click **Exit** to exit Procedure Builder, then click **Yes** to confirm exiting.

SUMMARY

- You should always break complex programs into smaller self-contained program units.
- The main types of program units are procedures, which can return several data values, and functions, which return a single data value.
- Procedures receive and deliver output values using parameters. By default, parameters are passed as IN parameters even if IN is omitted, and an IN parameter can be read but not modified. A parameter passed using the OUT mode can be written to, but cannot be read within the procedure. A parameter passed as both IN and OUT can be read and modified.
- Formal parameters are the parameters that are declared in the header of the procedure, and actual parameters are the values placed in the procedure parameter list when the procedure is called.
- Oracle Procedure Builder provides a graphical interactive environment for creating, compiling, testing, and debugging PL/SQL programs.
- The PL/SQL Interpreter enables you to single-step through programs to observe execution and examine variable values at run time.
- PL/SQL libraries are files that contain one or more procedures or functions. A library can be stored either in the database or in the file system, and always runs on the client workstation.
- A stored program unit is a program unit that is stored in the database, and runs on the database server.
- The scope of a variable is the portion of the program that can access the variable. Local variables are only visible in the program where they are declared, while global variables are visible outside the program where they are declared.
- A package is a code library that provides a way to manage large numbers of related programs and variables, and to create global procedures and variables. Oracle provides many built-in packages that are accessible to both client and server programs.
- Database triggers are procedures that execute in response to the database events of inserting, updating, or deleting a record.

REVIEW QUESTIONS

1. Why is it a good practice to break complex programs into self-contained program units?
2. What is the difference between a procedure and a function? When should you use a procedure, and when should you use a function?
3. What is a parameter?

4. What is the difference between formal and actual parameters, and what is the relationship between formal and actual parameters?
5. Write a declaration for a procedure named MyProcedure that has the following parameters:

Name	Sample Data	Description
item_cost	$329.55	Is read and modified by the procedure
cust_name	Paula Harris	Variable-length character string with maximum width of 30
		Read-only; cannot be changed by the procedure
invoice_total	$9,999.99	Is not read by the procedure, but is calculated in the procedure and returned to the calling program

6. How can you run a procedure in Oracle Procedure Builder? How can you run a procedure that expects an input parameter?
7. What is a breakpoint? Where can you put a breakpoint when you are debugging a program?
8. What is the difference between a library and a stored program unit in terms of where each is stored and where each executes? When should you create a library, and when should you create a stored program unit?
9. When should you create a package?
10. What is a database trigger?

PROBLEM-SOLVING CASES

1. Create a procedure named UpdateEnrollment that receives a specific SID and CSECID as input parameters named CurrentSID and CurrentCSECID. This procedure first inserts the SID and CSECID into the ENROLLMENT table, with a grade value of NULL. The procedure then updates the CURRENRL value in the COURSE_SECTION table to reflect that one less seat is available.
   a. Save the procedure as a .pls file named Ch4cEx1.pls in the Chapter4 folder on your Student Disk.
   b. Save the procedure as a stored program unit in the database.
2. Create a procedure named UpdateOrderline that receives information about a customer order that includes a specific ORDERID, INVID, and QUANTITY. These values are input as parameters named CurrentOrderID, CurrentInvID, and CurrentQuantity. The procedure must look up the CURR_PRICE of the inventory item in the INVENTORY table, and then insert the record into the ORDERLINE table. The procedure also must update the QOH field in the INVENTORY table by subtracting the quantity ordered from the quantity on hand.
   a. Save the procedure as a .pls file named Ch4cEx2.pls in the Chapter4 folder on your Student Disk.
   b. Save the procedure as a stored program unit in the database.

3. Create a procedure named UpdateShipping that receives input parameters specifying a SHIPID, INVID, and QUANTITY_RECEIVED. These values are input as parameters named CurrentShipID, CurrentInvID, and CurrentQRec. The procedure will first update the associated record in the SHIPPING table by updating DATE_RECEIVED to the system date, and updating QUANTITY_RECEIVED to the input value. The procedure will also update the QOH field in the INVENTORY table by adding the quantity received to the current QOH value of the inventory item.
   a. Save the procedure as a .pls file named Ch4cEx3.pls in the Chapter4 folder on your Student Disk.
   b. Save the procedure as a stored program unit in the database.
4. In this exercise, you will modify the STUDENT table in the Northwoods University database so that it has two additional fields to store a student's age and grade point average. Then, you will create a series of procedures and functions to insert the appropriate records into these fields.
   a. Add the following columns to the STUDENT table: S_AGE, which will hold integers with a maximum value of 999 (it is not impossible that a student might be 100 years old or older), and S_GPA, which will hold floating-point numbers.
   b. Create a procedure named UpdateSAge that updates the S_AGE field of all records in the STUDENT table based on the current system date. Save the procedure in a file named Ch4cEx4b.pls in the Chapter4 folder on your Student Disk and as a stored program unit in the database.
   c. Create a procedure named UpdateSAgeRecord that receives a specific SID as an input parameter named CurrentSID, and then updates the S_AGE field for this record based on the current system date. Save the procedure as a file named Ch4cEx4c.pls in the Chapter4 folder on your Student Disk and as a stored procedure in the database.
   d. Create a function named CalcGPA that receives a specific SID as an input parameter, and then calculates the student's current grade point average. Save the function as a file named Ch4cEx4d.pls in the Chapter4 folder on your Student Disk and as a stored program unit in the database. Recall that grade point average is calculated as follows:

      $$\frac{\Sigma(\text{Course Credits} * \text{Course Grade Points})}{\Sigma(\text{Course Credits})}$$

      Course grade points are awarded as follows:

Grade	Grade Points
A	4
B	3
C	2
D	1
F	0

   e. Create a procedure named UpdateSRecord that receives a single SID as an input parameter named CurrentSID, and then calls the UpdateSAgeRecord procedure to update the student's age, uses the CalcGPA function to find the student's GPA, and then updates the student's GPA record. Save the procedure as a file named Ch4cEx5e.pls in the Chapter4 folder on your Student Disk and as a stored program unit in the database.

f. Create a library named STUDENTLIBRARY that contains UpdateSAge, UpdateSAgeRecord, CalcGPA, and UpdateSRecord. Save the library as Ch4cEx5f.pll in the Chapter4 folder on your Student Disk.

5. Create a package named STUDENTPACKAGE that contains the UpdateSAge, UpdateSAgeRecord, CalcGPA, and UpdateSRecord procedures that were created in the last exercise. Save the package specification as a file named Ch4cEx6spec.pls in the Chapter4 folder on your Student Disk, and save the package body as a file named Ch4cEx6body.pls in the Chapter4 folder on your Student Disk. Also save the package specification and body as stored program units in the database.
   a. Create a global variable named GlobalSID, and modify the UpdateSRecord procedure so that it assigns the value of the CurrentSID input parameter to the global variable.
   b. Modify the UpdateSAgeRecord procedure and CalcGPA function so that they no longer receive an input parameter when they are called, but they instead use the global variable value for the current SID. (**Note**: the CalcGPA will still receive the CurrentSID input parameter.)

6. Whenever a new shipment is received at Clearwater Traders, the quantity received in the shipment has to be added to the QOH field for the associated inventory item in the INVENTORY table. Create a database trigger named ShippingUpdateTrigger that automatically updates the INVENTORY table's QOH field by the correct amount every time a SHIPPING record is updated and the QUANTITY_RECEIVED field is greater than 0.

7. Create a database trigger named CurrEnrlTrigger that fires whenever a record is inserted into or deleted from the ENROLLMENT table. This trigger should automatically increase the correct course section's CURRENRL value by one whenever an ENROLLMENT record is inserted for the course, and decrease the course section's CURRENRL value by one whenever a record is deleted.

CHAPTER 5

Creating Oracle Data Block Forms

Introduction▶ You have learned to use SQL commands to insert, update, delete, and view database data. However, it is not practical to expect users to regularly create SQL queries to interact with a database. Instead, users use applications called **forms**. A form looks like a paper form, and provides a graphical interface that allow users to easily insert new database records and to modify, delete, or view existing records. Programmers use an Oracle utility named Form Builder to create Windows-based forms. Form Builder is part of the Oracle application development utility called Developer. In this chapter, you will learn about **data block forms**, which are forms that are explicitly connected to specific database tables.

LESSON A

objectives

- View, insert, update, and delete data records using a data block form
- Create a single-table data block form
- Understand how to navigate among form objects using the Object Navigator
- Use the Object Navigator within Form Builder to change form properties

Using a Data Block Form

Before you learn how to create a data block form, you will run an existing form to become familiar with its appearance and functionality. This form is associated with the Clearwater Traders CUSTOMER table. First you will insert a new record into the CUSTOMER table. Using a form to add records is the equivalent of adding records in SQL using the INSERT command.

Before you open the form, you will run the SQL script that drops any existing tables for the Clearwater Traders database and then re-creates the new tables and inserts all of the data values into them.

To run the script:

1. Start SQL*Plus and log on to the database.
2. Run the clearwat.sql script from the Chapter5 folder on your Student Disk and close SQL*Plus.

help If you get error messages when you run the script, be sure you have disabled all the database triggers that you created in Chapter 4. To disable a database trigger, start Procedure Builder, open Database Objects, open your user name, open Database Triggers, and then right-click each trigger and click Disable.

help If you have trouble running the script file, ask your instructor or technical support person for help.

Next you will run the form application file that is stored on your Student Disk.

To run the form application:

1 Start Windows Explorer, navigate to the Chapter5 folder on your Student Disk, and then double-click the **customer_demo.fmx** file. The Developer Forms Runtime Logon window opens. Log on to the database, and then click **Connect**. The Oracle Developer Forms Runtime application opens and displays the Clearwater Traders Customers form, as shown in Figure 5-1.

tip

When you start an FMX file by double-clicking it in Windows Explorer, the filename and pathname cannot contain any spaces.

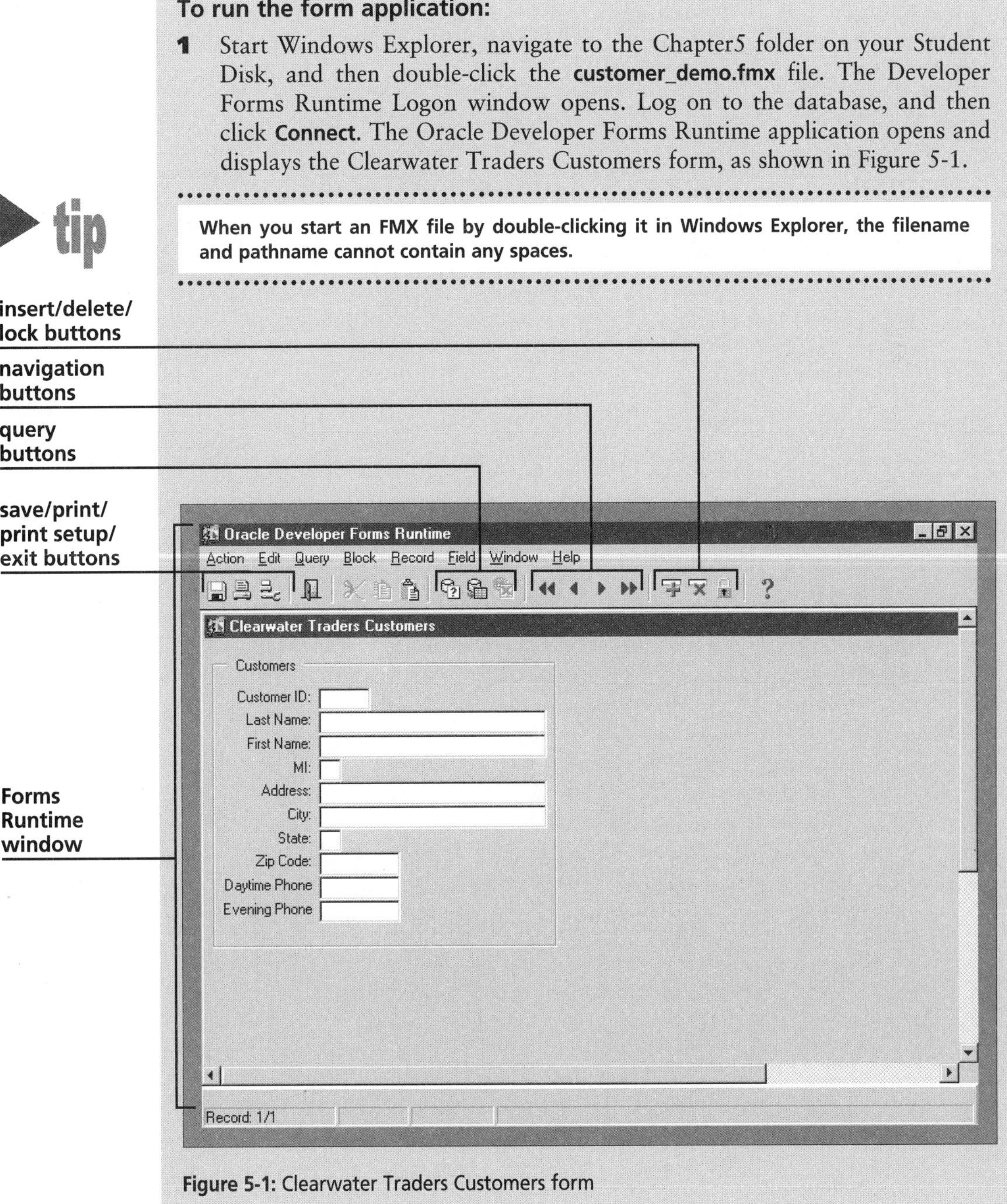

Figure 5-1: Clearwater Traders Customers form

The outer window is the Developer Forms Runtime window, which is where all form applications run. The inner window displays the Clearwater Traders Customer form, which can be used to insert a new record, and to modify, delete, or view existing records in the CUSTOMER table in the Clearwater Traders database. The Forms Runtime window has pull-down menus and a toolbar that can be used to insert, view, modify and delete records. The toolbar has buttons for calling application functions and manipulating data. The first button group includes **Save**, used to save a new record or to save changes made to an existing record; **Print**, to print the current form; and **Print Setup**, to reconfigure the printer setup. The **Exit** button exits the form.

The query button group is used for querying data. When a form is running, it can be in one of two modes: Normal or Enter Query. When a form is in Normal mode, you can view data records and sequentially step through the records. When a form is in Enter Query mode, you can enter a search parameter in the form fields, and then retrieve the associated records. To go into Enter Query mode, you click the **Enter Query** button on the toolbar. This clears the form fields and allows you to enter a search condition. To return to Normal mode, you must either execute the query or cancel the query. To execute the query, you click the **Execute Query** button. This returns the form to Normal mode, and the records associated with the search condition appear in the form fields. To cancel the query, you click the **Cancel Query** button. This also returns the form to Normal mode, but without retrieving any data.

The next button group is for navigating among different records and different blocks. A **block** is a group of related form items, such as text fields and buttons. An Oracle data block form contains one or more **data blocks**. Each data block is a group of related form items that are associated with a single database table. A data block usually contains text fields that correspond to one or more of the table's data fields.

The Clearwater Traders Customers form contains a single block, with text fields to represent all of the fields in the CUSTOMER table. Since this form has data from only one table, it is a single-block form. However, a form can display data from multiple tables using multiple blocks. The **Previous Block** button moves the cursor to the previous data block in a multiple block form, and the **Next Block** button moves the cursor to the next data block.

When the results of a query are displayed, you can sequentially step forward and backward through the records to view them one at a time. The **Previous Record** button moves back to the previous table record, and the **Next Record** button moves to the next record in the table.

The final button group is used to insert, delete, and lock records. The **Insert Record** button clears the form fields and creates a blank record into which the user may enter new data. The user must enter all required fields and then save the record, or else delete the new record by clicking the **Remove Record** button. The button can also be used to remove an existing record. The **Lock Record** button locks the current record so that other users cannot update or delete it.

Now you will use the form to insert a new record into the CUSTOMER table.

To insert a new record using the form:

1. Click in the Customer ID field, and then enter **1000**.
2. Use the **Tab** key to navigate from field to field and enter the data values shown in Figure 5-2.
3. Click the **Save** button on the toolbar. The confirmation message "FRM-40400: Transaction complete: 1 records applied and saved" appears briefly on the status line in the lower-left corner of the screen. (If your mouse pointer is still on the Save button, then this message will be overwritten with the "Save" Tooltip.)

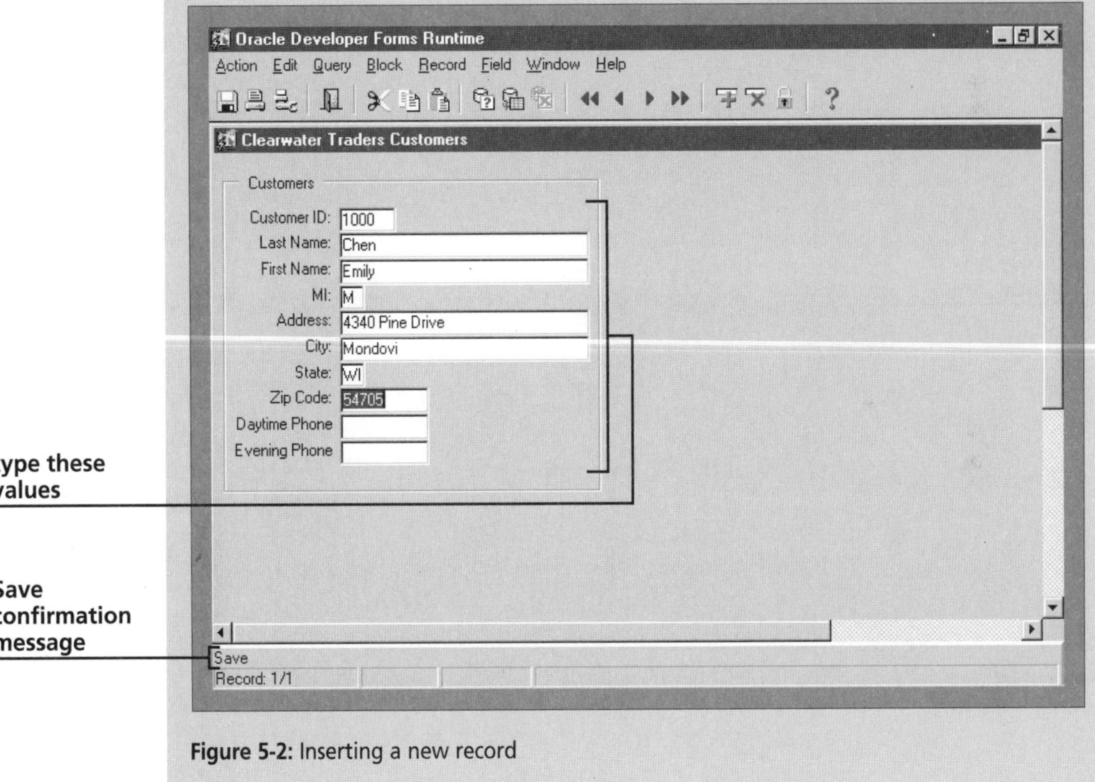

Figure 5-2: Inserting a new record

Next you will modify an existing record in the CUSTOMER table. Using a form to modify records is the equivalent of modifying records using the SQL UPDATE command. You will change customer Paula Harris's evening phone number next.

To update an existing record:

1. Click the **Enter Query** button on the toolbar. The form fields are cleared, and the insertion point appears in the Customer ID field, as shown in Figure 5-3. Notice the message that appears in the status line. Also, notice that the **mode indicator** indicates that the form is in Enter Query mode.

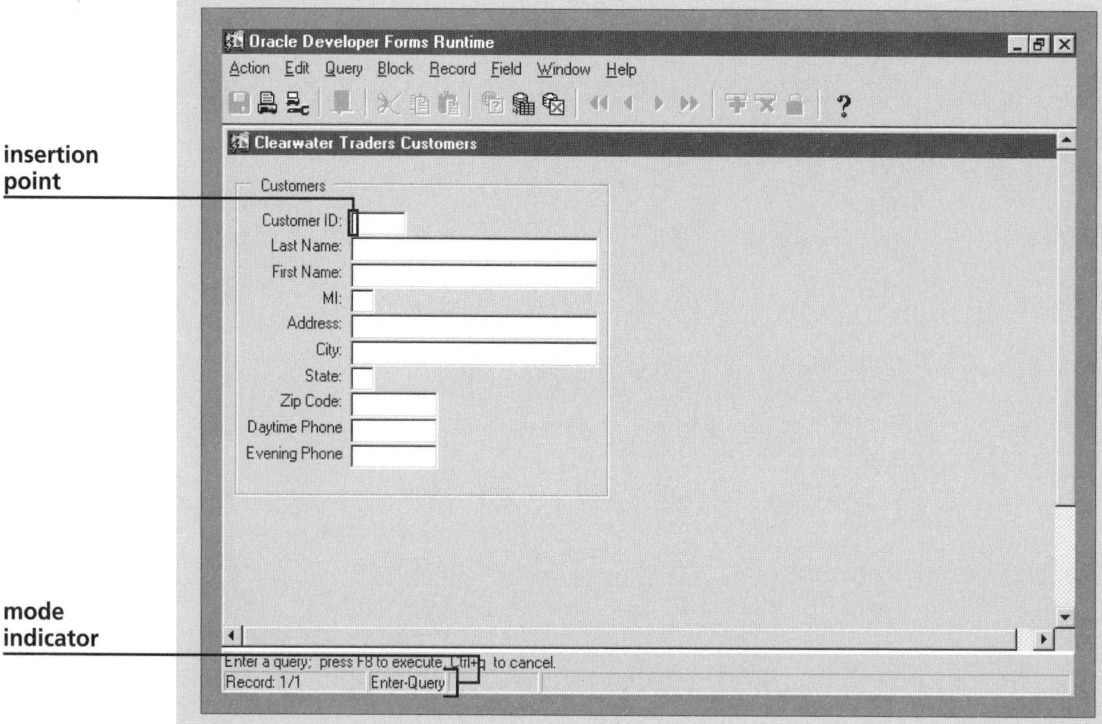

insertion point

mode indicator

Figure 5-3: Form in Enter Query mode

2 Type **107** in the Customer ID field.

3 Click the **Execute Query** button to execute the query. The complete data record for customer Paula Harris appears in the form. Notice that the messages in the status line and in the mode indicator disappear.

Pressing the F8 key is equivalent to clicking the Execute Query button on the toolbar, and pressing Ctrl + Q is equivalent to clicking the Cancel Query button.

4 Click in the Evening Phone field, and then change the phone number to **7155558975.**

5 Click . The confirmation message "FRM-40400: Transaction complete: 1 records applied and saved" appears on the status line.

You also can use this form to view all of the records in the table and stepthrough them sequentially. To view the first record in a table, place the form in Enter Query mode, but do not enter a search condition. Instead, click the Execute Query button. Since no search condition was specified, the first record in the table is displayed. Then you can use the record navigation buttons to step through the records sequentially.

To display the first CUSTOMER record and step through the records sequentially:

1. Click the **Enter Query** button, then click the **Execute Query** button on the toolbar. Paula Harris's data record (which is the first CUSTOMER record) is displayed in the form.
2. Click the **Next Record** button on the toolbar. The record displaying data for the next customer in the database (customer Mitch Edwards) appears.
3. Click the **Previous Record** button on the toolbar. Paula Harris's record appears again because it is the record directly in front of Mitch Edwards's record.
4. Click again. Continue clicking until you scroll through all of the CUSTOMER records.

You can use a form to delete data records. You will do this next.

To delete a record from the CUSTOMER table:

1. Click or on the toolbar until you see customer Emily Chen's record.
2. Click the **Remove Record** button on the toolbar to delete Emily's record from the form. Alissa Chang's record now appears in the form, because her record was in front of Emily's record in the database.

When you close a form, you are given an option to commit or roll back your changes. You will commit your changes now.

To close the form and commit the changes:

1. Click the **Close** button on the Forms Runtime title bar. A message box asking if you want to save your changes appears. (Personal Oracle8 users might not get this message.) You can click Yes to commit your changes, No to roll back your changes, or Cancel to return to the form.
2. Click **Yes** to commit your changes and, if necessary, click **OK** to confirm the commit and close the form.

Creating the CUSTOMER Form

The CUSTOMER form is a data block form. Recall that a data block is a block that is related to a database table. When you create a data block form, the system automatically generates the text fields and labels for data fields in that table and then creates the code for inserting, modifying, deleting, and viewing data records. In this chapter, you will create form applications using data blocks.

Now you will begin to create the CUSTOMER form. First you will start Form Builder.

To start Form Builder:

1 Click **Start** on the Windows taskbar, point to **Programs**, point to **Oracle Developer 6.0**, and then click **Form Builder**. The Welcome to the Form Builder dialog box is displayed.

help

If you cannot find the Form Builder program, ask your instructor or technical support person for help.

tip

Another way to start Form Builder is to start Windows Explorer, change to the ORAWIN95\BIN folder, and then double-click the ifbld60.exe file (Windows NT users will click the ORANT\BIN\ifbld60.exe file).

This dialog box gives you options for creating new forms and opening existing forms. It also allows you to go directly to the Quick Tour and Cue Cards learning features. You will first create a new data block form using the Data Block Wizard, which displays a series of pages that automatically guide you through the form-building process.

help

If the Object Navigator window appears instead of the Welcome to the Form Builder dialog box, click Tools on the menu bar, then click Data Block Wizard, and continue with the next set of steps.

To use the Data Block Wizard to create a new form:

1 Be sure the **Use the Data Block Wizard** radio button is selected, then click **OK**. The Data Block Wizard Welcome page is displayed. This page describes how you can use the Data Block Wizard to create a data block based on a table, view, or set of stored procedures.

2 Click **Next**. The Data Block Wizard Type page is displayed.

3 The Type page allows you to choose whether you will create the data block by using a table or view, or by using a stored procedure. Since this data block will be associated with the CUSTOMER table, confirm that the **Table or View** option is selected, and then click **Next**. The Data Block Wizard Table page is displayed.

The Data Block Wizard Table page allows you to select the database table that you will use to create the data block. It also allows you to select the specific table fields that will be included in the block, and whether you want integrity constraints enforced in the form application.

Before you can select a database table, you must connect to the database using your usual user name, password, and connect string. Now you will connect to the database, select the CUSTOMER table, and include all of its fields in the data block.

To connect to the database and select the CUSTOMER table and all of its fields:

1 Click **Browse.** The Connect dialog box opens. Type your user name, password, and connect string as usual, then click **Connect**. The Tables dialog box opens, as shown in Figure 5-4. If the Current User check box is checked, then every table that you own is listed in the Table list. If the Other Users check box is checked, every table in the database is listed. You would check the Other Users check box only when you want to use another user's database table in your form. (If you use a table that was created by another user, that user must first grant you the required privileges for inserting, updating, deleting, or viewing data.) You also are given the option of displaying different kinds of database objects: Tables, Views, or Synonyms. In this lesson, you will use tables. Views will be discussed in a later lesson, and Synonyms are an advanced topic that will not be covered in this book.

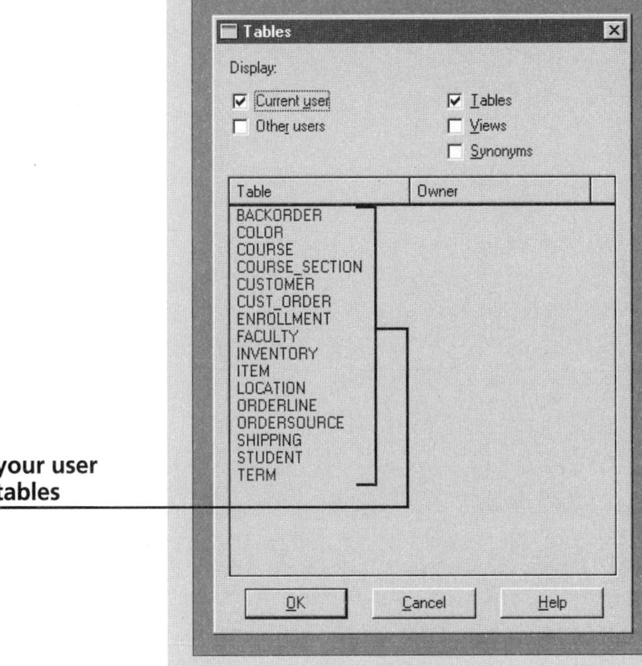

Figure 5-4: Tables dialog box

2 If necessary, make sure that the **Current user** and **Tables** check boxes are checked and that the other check boxes are cleared. Note that the list displays every database table that you own.

3 Click the **CUSTOMER** table, and then click **OK** to return to the Data Block Wizard Table page, as shown in Figure 5-4. CUSTOMER appears as the table, and the customer fields are displayed in the Available Columns list. The icon beside each field indicates whether the field has a number or character data type.

To include a specific column in the data block, you can click on the column name to select it, and then click the Select button [>]. To remove a column from the data block, you can select it in the Database Items list, and click the Deselect button [<]. To select all of the table columns, click the Select All button [>>]. To remove all of the selected columns, click the Deselect All button [<<]. To select several adjacent columns, click the first column, press and hold the Shift key, then click the last column. To select several nonadjacent columns, click the first column, press the Ctrl key, then click the other columns while keeping the Ctrl key pressed. Now, you will select all CUSTOMER column fields to be included in the data block.

To select all of the column fields:

1 Click the **Select All** button [>>].

2 The selected columns appear in the Database Items list.

If you check the Enforce data integrity check box on this dialog box, the form will flag violations to the integrity constraints that you established when you created the table (unique primary keys, foreign key references, etc.). This means that you will see FRM-error codes generated by Form Builder, rather than ORA-error codes generated by the DBMS. The integrity constraints will be enforced by the DBMS even if you do not check the Integrity Constraints check box, and users will receive ORA-error codes directly from the DBMS if they violate a table integrity constraint. In this application, you will not enforce data integrity in the form, so you will not check the Enforce data integrity check box. Now, you will finish creating the block.

To finish creating the block:

1 Click **Next**. The Data Block Wizard Finish page is displayed.

The Finish page presents the options of creating the data block, then starting the Layout Wizard, or just creating the data block. The form **layout** specifies how the form looks to the user. The basic components of a layout include the selection of data block fields that are displayed, the labels of the displayed fields, the number of records displayed at one time, and the form title. The Layout Wizard displays a series of pages to help you specify the form layout. Now you will start the Layout Wizard.

To start the Layout Wizard:

1 Be sure that the **Create the data block, then call the Layout Wizard** radio button is selected, then click **Finish.** The Layout Wizard Welcome page is displayed. Click **Next.** The Layout Wizard Canvas page is displayed.

A **canvas** is the area on a form where you place **graphical user interface (GUI) objects** such as buttons and text fields. The Layout Wizard Canvas page lets you specify which canvas the block will appear on, and the canvas properties. Since you have not created any canvases yet, you will accept the default (New Canvas) selection. Canvases can be of different types: **content** canvases, which fill the entire window; **stacked** canvases, in which two or more canvases are stacked in a window (used for creating a separate working area, such as a toolbar or tool palette, that can be moved to a different location); canvases with **horizontal** or **vertical toolbars**; or **tab** canvases, whereby different related canvases appear on a tab page, and each canvas can be accessed by clicking a tab that is labeled with a description of the particular canvas. Since you want the CUSTOMER data to appear on a single canvas that fills the entire window, you will create a content canvas that will display all of the CUSTOMER fields.

To create a content canvas:

1 Be sure that **(New Canvas)** is selected as the canvas, and that **Content** is selected as the canvas type, then click **Next**. The Data Block page allows you to select the display fields.

2 Click » to select all block fields for display, then click **Next**. The Items page is displayed.

The Items page allows you to specify the column prompts, widths, and heights. A column **prompt** is a label that describes the data value that appears in the text box, such as Customer ID or Evening Phone Number. By default, the prompts are the same as the database column names. The column widths and heights are specified using **points**, which correspond to font sizes, as the default measurement unit. The default column widths correspond to the maximum column data widths specified in the database tables. Now you will modify the column prompts so that they are more user-friendly.

To modify the column prompts:

1 Click in the first Prompt row, and change the text from Custid to **Customer ID:**. Press the **Tab** key three times, then modify the Last field's prompt to **Last Name:**.

2 Modify the rest of the prompts, as shown in Figure 5-5. Click **Next**. The Style page is displayed.

▶ tip

Since the end user of the form is being prompted to enter a value, prompts should end with a colon (:).

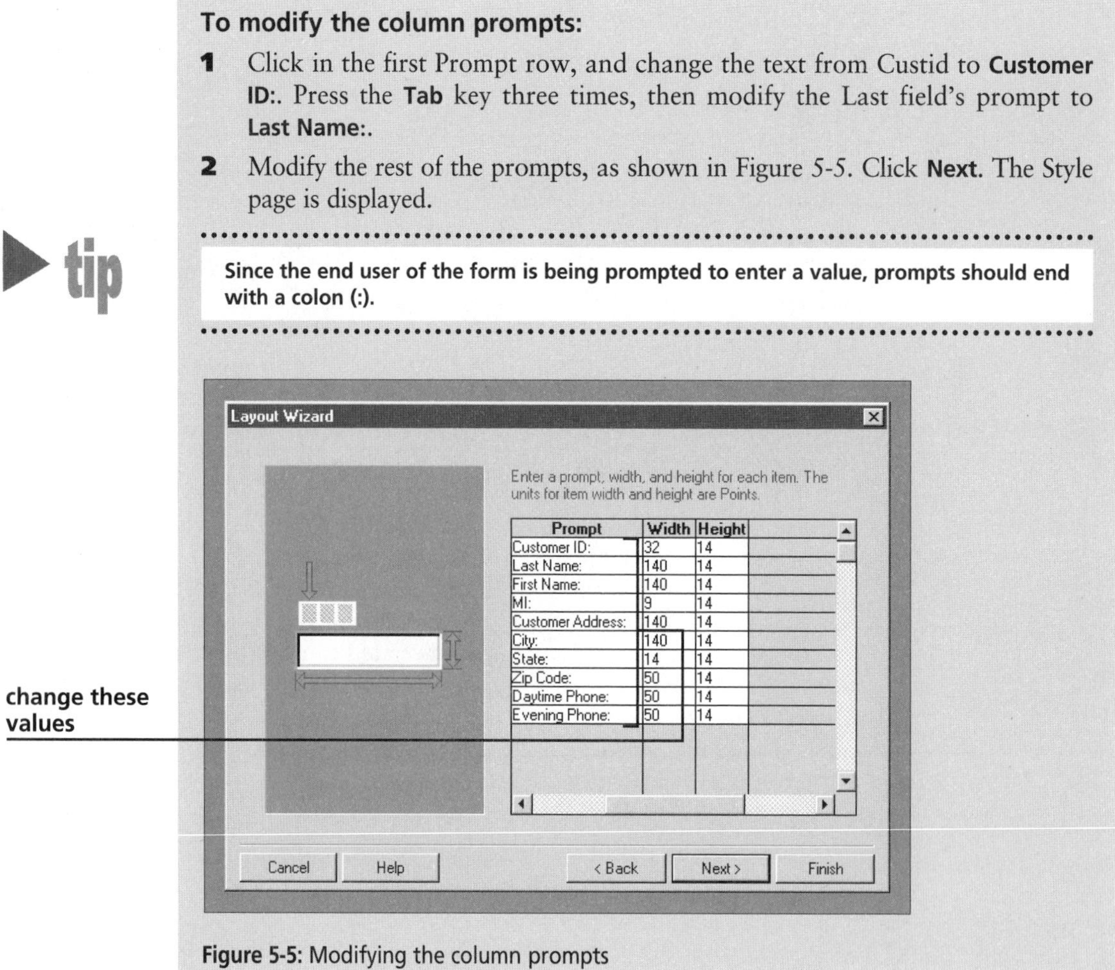

Figure 5-5: Modifying the column prompts

The Style page allows you to specify the layout style and properties, to determine how the data are displayed. In a **form** style, only one record is displayed at a time. In a **tabular** style, multiple records are displayed on the form. On a tabular form, if more records exist than can be displayed at one time, the user can use a scroll bar to view records. Since you will display only one customer record at a time, you will specify form as the layout style. Then, you will enter the title that appears in the frame around the record, specify how many records are displayed on the form at one time, and enter the distance between successive records.

To specify the layout style and frame title:

1. Be sure that the **Form** radio button is selected, then click **Next**. The Rows page is displayed.
2. Type **Customers** in the Frame title box. Since you specified a form layout style, there is always only one record displayed at a time, and you do not need to specify the distance between records. Be sure that Records Displayed is 1, and Distance Between Records is 0. You do not need a scroll bar, so you will leave the scroll bar box unchecked. Click **Next**, then click **Finish** to finish the layout.
3. Click **Tools** on the menu bar, then click **Layout Editor**. The finished layout is displayed in the Form Builder Layout Editor window, as shown in Figure 5-6.

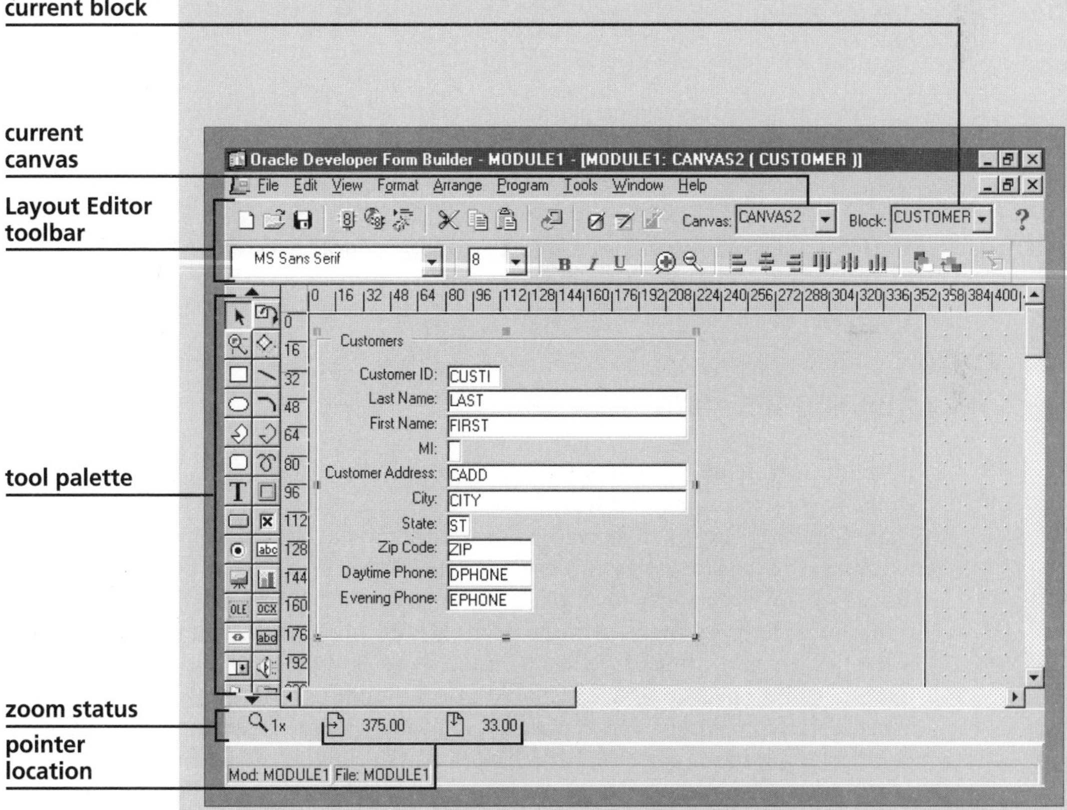

Figure 5-6: Finished form layout in Layout Editor window

The **Layout Editor** provides a graphical display of the form canvas that can be used to draw and position form items, and to add **boilerplate objects** such as labels, titles, and graphic images. The Canvas indicator at the top of the Layout Editor window shows the name of the current canvas. When you created a new canvas using the Layout Wizard, the canvas received a default name. In Figure 5-6, this default name was CANVAS2, but your default name might be different. The current block is the CUSTOMER block.

The **Layout Editor toolbar** allows you to save and edit the form, and modify the form's text properties. The **tool palette** provides tools for creating boilerplate objects and other objects. Rulers appear along the top and left edges of the canvas, and zoom status and pointer location indicators are on the bottom-left edge of the window.

help If you don't see one of the Layout Editor window elements in Figure 5-6 on your screen, maximize the window, click View on the menu bar, and then click the name of the element that you need to turn on.

Now you will save your work.

To save your file:

1 Click [Save] on the toolbar. Open the Chapter5 folder on your Student Disk, and save the file as **ch5acustomer.fmb**.

Running the Form

Form Builder design files are saved as files with .fmb (form module binary) extensions. Before you can run a form to test and see if it works, you must compile the .fmb file to create an executable form file, which has an .fmx (form module executable) extension. As you develop a new form using Form Builder, you periodically "run" the form to test it. When you run a form in Form Builder, two things must happen: first, the form design (.fmb) file must be compiled to generate a new executable (.fmx) file that contains your latest design specifications, then, the .fmx file needs to be executed. Form Builder .fmx files are not directly executable from your workstation's operating system, but must be run using another application. There are three options for running Form Builder .fmx files: in the Forms Runtime application; as a World Wide Web application in a Web browser; and in the Form Builder Debugger environment. In this chapter, you will run forms in the Forms Runtime application. (You will run forms in these other environments in later chapters.)

When you click the **Run Form Client/server** button [Run Form] on the toolbar in either the Object Navigator or Layout Editor window, Form Builder automatically recompiles your form, and then starts Forms Runtime and displays the current form in the Forms Runtime environment. Before you run your form, you need to change some of the setup options to instruct the system to automatically save your current Form Builder design (.fmb) file and regenerate your .fmx file.

To change the setup options:

1. Click **Tools** on the menu bar, and then click **Preferences**. The Preferences dialog box opens. The Save Before Building check box instructs the system to save your Form Builder design (.fmb) file automatically each time you run your form. The Build Before Running check box instructs the system to generate your .fmx file again automatically before running it. It is advisable to check both of these boxes so that your design file will be saved automatically and generated each time you run your form.
2. Make sure that the **Save Before Building**, and **Build Before Running** check boxes are checked, and then click **OK**. Do not check or clear any other check boxes or change any other options.

Now you will run the form. When you run your form, Form Builder creates a new .fmx file, and also creates a text file named <form file name>.err that records compilation messages and errors. This file is generated even when there are no compilation errors. If there are compilation errors, the error messages are automatically displayed on the screen, so you probably will not need to use this file. However, you can open it in any text editor to review its contents.

To run the form:

1. Click the **Run Form Client/Server** button on the Layout Editor toolbar to run your form. After a few moments, your form should appear in the Developer Forms Runtime window.

You can run a form by clicking on the Object Navigator toolbar, or by pressing Ctrl + R.

2. Click the **Close** button to close the form.

When users run your form, they will not run it from the Form Builder environment, but will run it directly from the operating system by double-clicking the form's .fmx file in the Windows Explorer, or by clicking on a shortcut that has been created for the form. For the form to run successfully, the .fmx file type must be **registered** on the user's workstation. Registering a file type involves specifying a file extension, and then associating this extension with a particular application.

When a user double-clicks an .fmx file, the operating system starts Forms Runtime, and then loads the .fmx file that was double-clicked.

The .fmx file type must be registered with the Forms Runtime application, which is an executable file named ifrun60.exe and is located in the ORAWIN95\BIN\ folder on the user's workstation. (For Windows NT users, the file is located in the \ORANT\BIN\ folder.) This file registration is created when Form Builder was installed on your workstation. Now you will start Windows Explorer, and view the file registration information.

To view the .fmx file registration information:

1. Start Windows Explorer, click **View** on the menu bar, then click **Folder Options**. (Windows NT users will click View, then click Options.) The Options dialog box is displayed. Click the **File Types** tab.
2. Scroll down the Registered File Types list box until you find the **FMX Files** entry, then select it (your file entry might be named **Forms Builder Executable**). The File Type details should appear. This shows that files with an .fmx extension open with the ifrun60.exe (Forms Runtime) application.
3. Click **Cancel** to close the Options dialog box.

To confirm that you can run the ch5acustomer.fmx file directly, you will run the form's .fmx file by double-clicking it in Windows Explorer.

To run the form from Windows Explorer:

1. Navigate to the Chapter5 folder on your Student Disk, and double-click **ch5acustomer.fmx.** The Logon dialog box is displayed.
2. Log on to the database in the usual way. The form is displayed in the Forms Runtime window.

Viewing Table Records Sequentially

First you will test to confirm that you can scroll through the CUSTOMER records sequentially.

To scroll through the records:

1. Click the **Enter Query** button on the Forms Runtime toolbar, and then click the **Execute Query** button to execute the query. The first record from the CUSTOMER table (customer Paula Harris) appears in your form.

2 Click the **Next Record** button to display the record for Mitch Edwards. Note that Mitch's middle initial is not displayed. This is because the Middle Initial field is not wide enough to display wide letters, such as M. You will fix this later. Click until you have viewed all of the records.

3 Click the **Previous Record** button and view the records until you return to the first record.

Viewing Specific Records Using Search Conditions

Sequentially viewing all table records works well for the small sample databases you are using within this book, but it won't work when you are looking for a specific record in a database that contains thousands of records. When you are working with large databases, you can use another approach. You can enter search conditions in a form to retrieve specific records, just as when you use the WHERE clause in a SQL command. In Form Builder, as in SQL*Plus, searches involving text strings are always case sensitive.

To use a query to find a specific record:

1 Click the **Enter Query** button on the toolbar. The message in the message area prompts you to enter a query.

2 Click in the Zip Code field, enter **53821**, and then click to execute the query. Lee Miller's customer record appears. Queries can retrieve multiple records, as you will see next.

3 Click again, and note that the fields are cleared. Enter **WI** in the State field, and then click to execute the query.

4 Click to confirm that all five CUSTOMER records are retrieved.

So far you have done exact searches. You can do approximate searches by using the percent sign (%) to indicate that there can be a variable number of wildcard characters either before or after a search string. For example, if you want to retrieve the records for all customers whose Zip codes begin with 54, you enter the search condition as 54%. If you want to retrieve the records of all customers who have the characters "Apt" anywhere in their address, you enter the search condition %Apt%.

To search using a wildcard character:

1 Click , enter **54%** as the Zip search condition, and then click . Four records are retrieved.

2 Click again, enter **%Apt%** as the Customer Address search condition, and then click . The query finds the record for Paula Harris.

Counting Query Hits

Before you execute a query, it is useful to know how many total records will be returned without having to execute the query and download all of the data from the server. You can count the number of records that would be retrieved by using the Count hits command prior to executing the query.

To use the Count hits command:

1. Click the **Enter Query** button and then type the search condition **%Street%** in the Customer Address field to determine which customers have the word "Street" in their address, but *do not click* .

help: If you click by mistake, just repeat Step 1.

2. Click **Query** on the menu bar, and then click **Count Hits**. The message "Query will retrieve 2 records" appears in the message area at the bottom of the window.
3. Click to execute the query and view the two records.

Inserting New Records

Now you will test to confirm that the form you created inserts new CUSTOMER records correctly.

To use your form to add a new record to the CUSTOMER table:

1. Click the **Insert Record** button on the toolbar. A blank record appears, and the insertion point moves to the Customer ID field.
2. Enter the following data: Customer ID: **1001**, Last Name: **Brian**, First Name: **Sarah**, Middle Initial: **A**, Address: **1444 Spring Street**, City: **Elk Mound**, State: **WI**, and ZIP: **54705**. You will not enter the daytime and evening phone numbers.
3. Click on the toolbar. The confirmation message "FRM-40400: Transaction complete: 1 records applied and saved" appears in the message area.

Updating Records

To update an existing database record, you first need to display the record. Then you can modify the field you want to update, and save the change.

To update a record:

1. Execute a new query to retrieve Alissa Chang's record by entering a query using **Chang** as the search condition in the Last Name field.
2. Select the value in the Evening Phone field, then type **7155557644**.
3. Click [Save] on the toolbar. The message "1 records applied and saved" appears in the message area.
4. Change Alissa's evening phone number back to the original value (**7155550087**), and then save your changes.

Deleting Records

To delete a record you must execute a query to display the record, and then click the Remove Record button [X] on the toolbar.

To delete a record:

1. Execute a query to retrieve Sarah Brian's record, using **Brian** as the search condition in the Last Name field.
2. Click the **Remove Record** button [X] on the toolbar.
3. Click [Save] on the toolbar to commit the delete to the database.
4. Execute another query to retrieve Sarah Brian's record, using **Brian** as the search condition for Last Name, to confirm that it has been deleted. The message "FRM-40301: Query caused no records to be retrieved. Re-enter." indicates that the record does not exist in the database.
5. Press **Ctrl + Q** to cancel the query and return to Normal mode.

To close Forms Runtime, click the Close button [X] on the title bar. If you have made any changes (insertions, updates, or deletions) since the last time you clicked the Save button [Save], you will be asked if you want to save your changes. If you click Yes, your changes will be committed to the database. If you click No, your changes will not be saved. Always exit Forms Runtime when you return to Form Builder. If you don't, and then you run the form again, you will have multiple Forms Runtime processes running on your computer and you will eventually run out of memory.

To exit Forms Runtime:

1 Click the **Close** button [X] on the Forms Runtime title bar.

If you cannot exit the program, press the F8 key to return to Normal mode, and then repeat Step 1.

tip

You also can click Action on the menu bar, and then click Exit to close Forms Runtime.

2 If a message box appears and asks if you want to save your changes, click the **Yes** button to commit your changes.

If you want to take a break, you can save your form file, and then close Form Builder. You can reopen your file later and continue the lesson.

To save the file and close Form Builder:

1 Click [Save] on the toolbar to save the file.

2 Click **File** on the menu bar, then click **Exit** to close Form Builder.

tip

You also can click the Close button [X] on the top application bar to close Form Builder.

Form Components and the Object Navigator

Now that you have created a form, you need to become familiar with its components by viewing the components using a Form Builder window called the **Object Navigator**, which is similar to the Object Navigator window you used with Query Builder in Chapter 4. Now you will view the components of the form in the Object Navigator.

To view the form components:

1 If necessary, start Form Builder, select the **Open an existing form** option button, and click **OK**. Select **ch5acustomer.fmb** in the Chapter5 folder on your Student Disk, then click **Open**. The form opens in the Object Navigator window, as shown in Figure 5-7.

2 If you did not restart Form Builder, click **Window** on the menu bar in the Layout Editor, then click **Object Navigator**. The Object Navigator windows opens.

3 Click **CH5ACUSTOMER**, then click the **Collapse All** button beside CH5ACUSTOMER. Click beside CH5ACUSTOMER so that only the top-level form components (which are also called objects) are displayed, as shown in Figure 5-7.

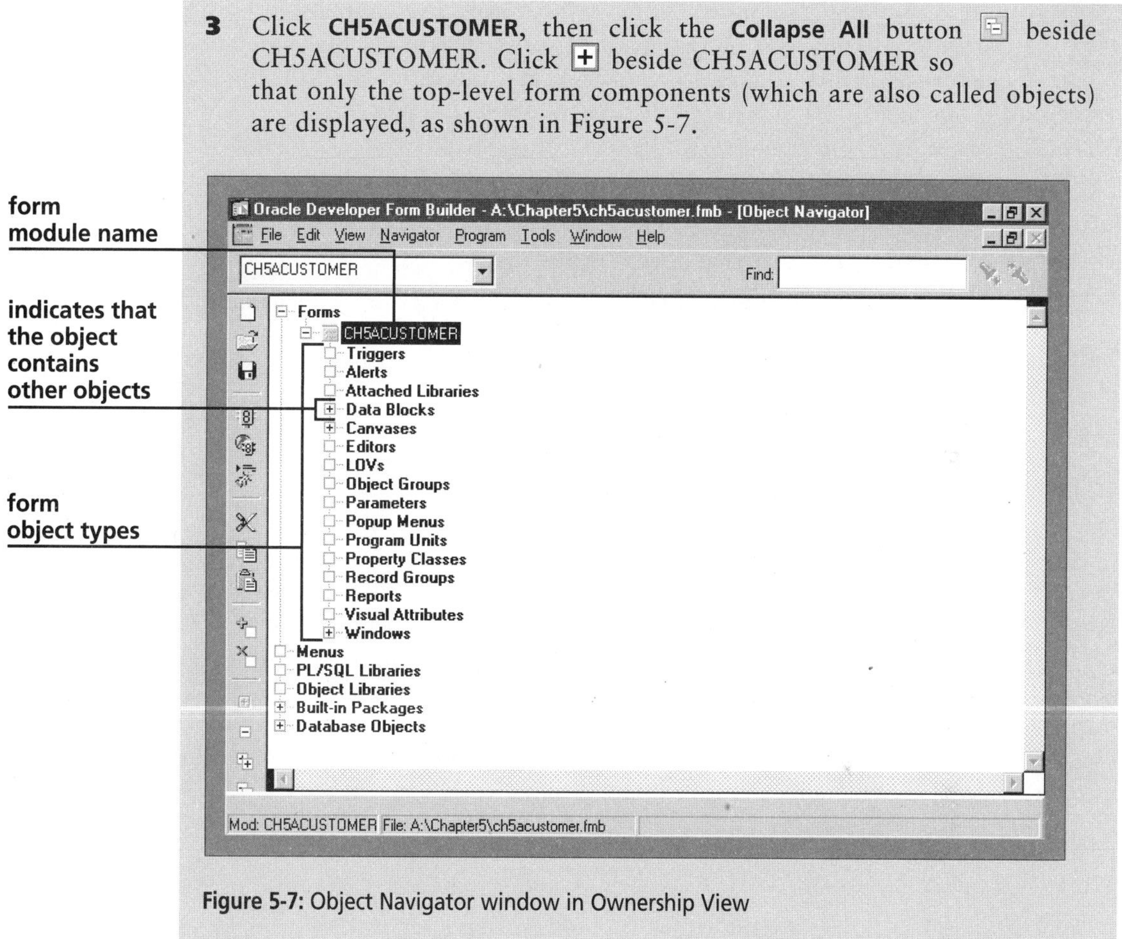

Figure 5-7: Object Navigator window in Ownership View

The Object Navigator window shows the form objects in **Ownership View**, in which all form object types appear directly below the form module. If an object type has a plus box to the left of its name, it means that objects of that type are present in the form. A Form Builder application such as CH5ACUSTOMER is called a **form module,** or just a form. A form can contain all of the form object types listed. The CH5ACUSTOMER form specifically contains data blocks, canvases, and windows, as indicated by the plus box to the left of the names of these object types. These particular objects are the basic building blocks of every form—a form must have a window, canvas, block, and at least one block item.

A **window** is the familiar rectangular area on a computer screen that has a title bar at the top. Windows usually have horizontal and vertical scroll bars, and usually can be resized, maximized, and minimized. In Form Builder forms, you can specify window properties such as title, size, and position on the screen. Recall that a **canvas** is the area in a window where you place **graphical user interface (GUI) objects**, such as buttons and text fields. A form window can have multiple canvases. A block is a structure that contains a group of GUI objects. A canvas might display one or more blocks. A data block is a block that is related to a database table. When you create a data block, the system automatically generates the text fields and labels for data fields in that table and then creates the code for inserting, modifying, deleting, and viewing data records.

A form can contain one or more windows. Simple applications usually have only one window, and more complex applications might have several windows. A window can have multiple canvases. A canvas can have multiple blocks, but individual block items can be displayed on different canvases. It is useful to think of the form as a painting, with the window as the painting's frame, the canvas as the canvas of the painting, and a block as a particular area of the painting.

You can expand each object type in the Object Navigator window by clicking the plus box [+] to the left of the object's name. Now you will expand the form objects.

To expand the form objects:

1 Click [+] beside **Data Blocks**, click [+] beside **Canvases**, and click [+] beside **Windows.**

Figure 5-8 shows the expanded view of the form objects in the Object Navigator window. The form has one block (CUSTOMER), one canvas (CANVAS2), and one window (WINDOW1). CANVAS2 and WINDOW1 are default names given to the canvas and window that were created when you made the form using the Layout Wizard, and your default names might be different.

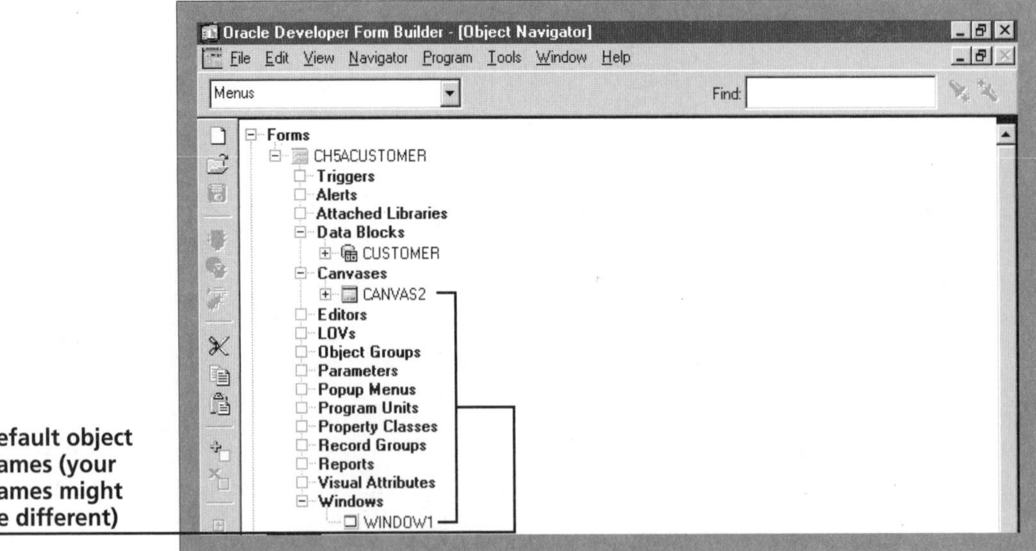

Figure 5-8: Expanding the form objects

The plus boxes to the left of the CUSTOMER block and CANVAS2 indicate that they contain more objects that can be viewed by clicking the plus boxes. Now you will expand the block objects further.

To expand the block objects:

1 Click ⊞ beside **CUSTOMER** under Data Blocks, and click ⊞ beside **Items** under CUSTOMER.

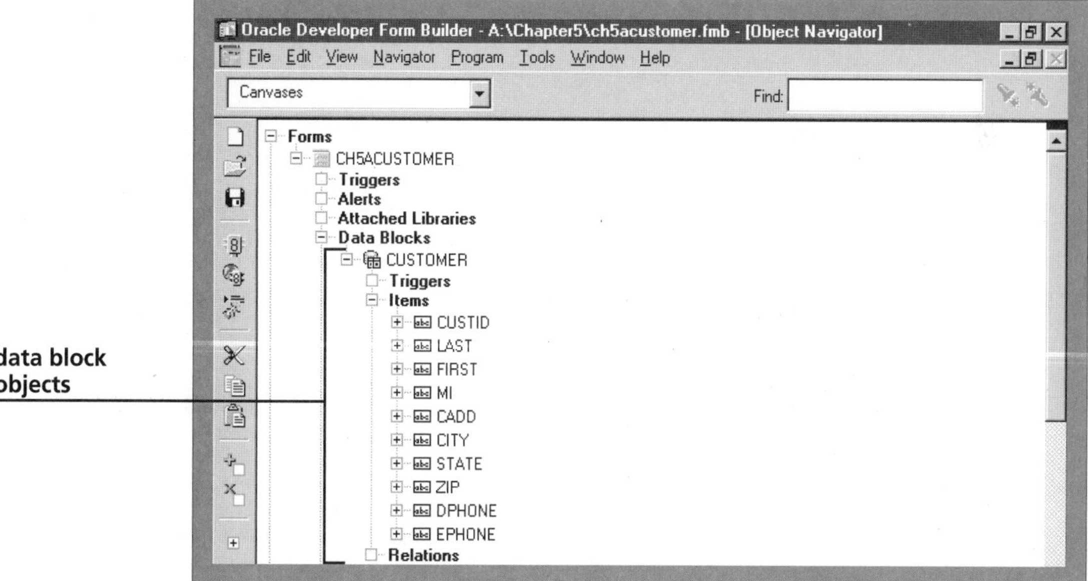

Figure 5-9: Form block objects

Figure 5-9 shows that the CUSTOMER block contains items. Blocks also can contain triggers and relations, although the blank boxes in front of Triggers and Relations for the CUSTOMER block indicate that no triggers or relations are contained in the block.

Windows applications respond to a user action, such as clicking a button, or to a system action, such as loading a form. These actions are called **events**. In Form Builder, events start programs called **triggers** that are written in PL/SQL. Triggers that respond to form events are not the same as the database triggers you created in Chapter 4. Database triggers respond to an event in the database, such as inserting, updating, or deleting a record. Form event triggers respond to user events such as clicking a button. You will create triggers when you create custom forms in Lesson C of this chapter.

A **relation** is a form object that is created when you specify a relationship between two data blocks with a foreign key relationship. To create a relationship, you must specify that the value of a primary key field in one block is equal to the corresponding foreign key field in the second block. Since the customer_form involves only one database table, it has no relations. **Items** are the GUI objects that a user sees and interacts with on the canvas. Figure 5-10 lists some of the common form item types.

Item Type	Description
Check box	Use for selecting options; an option is selected when the box is checked. You can check one or more check boxes in a group.
Button	Use for specifying an action, such as selecting an item from a list or closing an application.
Radio button (also called Option button)	Use for selecting options; an option is selected when the center of the button is black. You can select only one radio button in a group, and choosing a different button deselects the current one.
Text item	Use for entering or modifying text. Text items often are associated with database fields.

Figure 5-10: Common block item types

Now you will expand the form canvas and examine the objects it contains.

To expand the form canvas:

1. Click [+] beside **CANVAS2** (or the default name of your canvas). The canvas contains Graphics, which are boilerplate objects such as frames, lines, graphic images, etc., that enhance the form's appearance but do not contribute directly to its functionality.
2. Click [+] beside **Graphics**. The FRAME3 (your default frame name might be different) object appears, representing the frame around the data block items. There are no other boilerplate objects in the form.

So far you have been using the Object Navigator in Ownership View, which presents the form as the top-level object, and then lists all form objects in the next level. The Object Navigator also has a **Visual View** that shows how form objects "contain" other objects: a form contains windows, a window contains canvases, and a canvas contains data blocks. Figure 5-11 shows the CH5ACUSTOMER form in Ownership View, and Figure 5-12 shows the same form in Visual View.

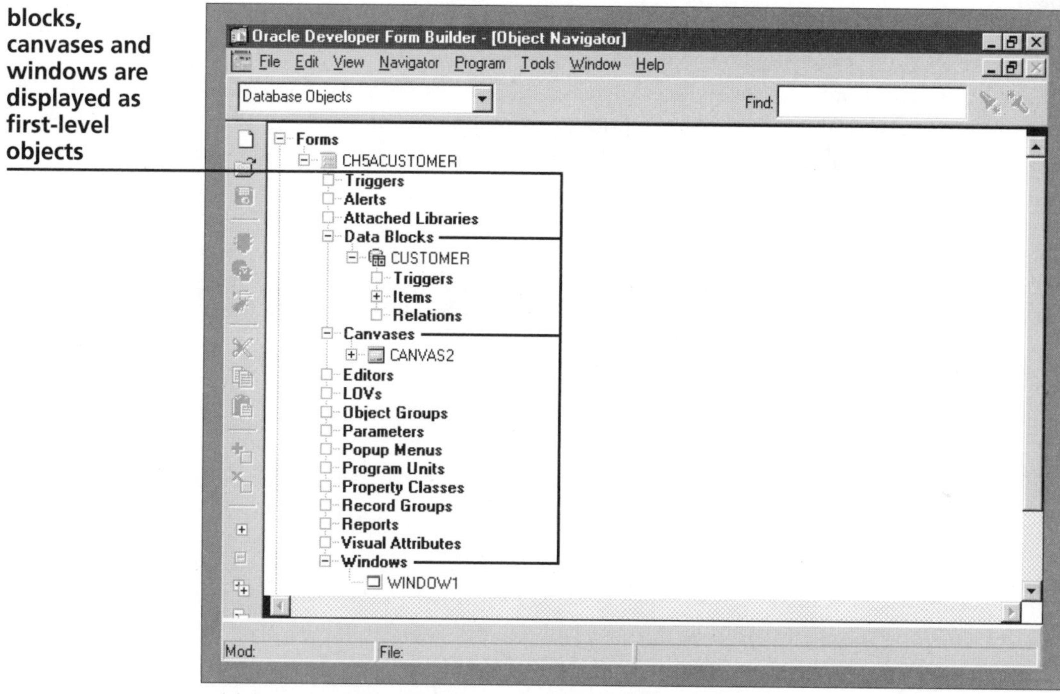

Figure 5-11: CH5ACUSTOMER form in Ownership View

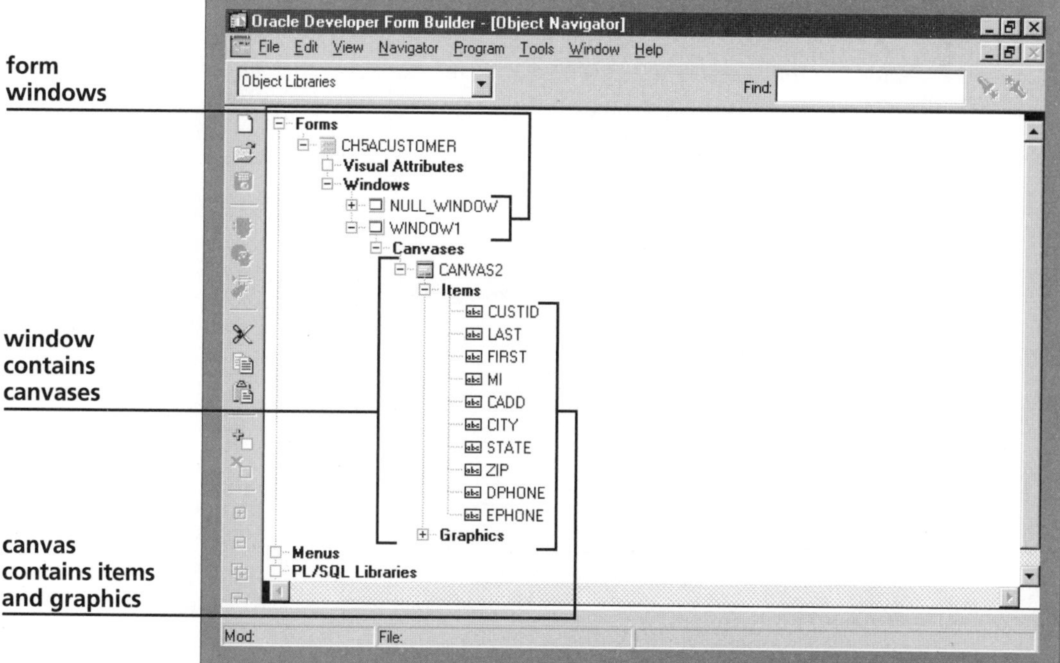

Figure 5-12: CH5ACUSTOMER form in Visual View

Note that in Ownership View, data blocks, canvases, and windows are all displayed as first-level objects, and the hierarchical relationships among windows, canvases, and blocks are not displayed. (Ownership View does provide one hierarchical relationship: it shows which items are contained in a specific data block.) Conversely, Visual View shows that WINDOW1 contains CANVAS2, and that the form items and graphics are on CANVAS2. Note that data blocks are not shown in Visual View. Ownership View is useful for quickly accessing specific objects without having to open all of the objects' parents, while Visual View is useful for viewing and understanding how form objects are related.

Now you will examine the form objects in Visual View.

To examine the form objects in Visual View:

1. Select **CH5ACUSTOMER**, then click the **Collapse All** button to close all form objects.
2. Click **View** on the menu bar, then click **Visual View**. The top-level object—the form—still is displayed.
3. Click [+] beside **CH5ACUSTOMER** to display the next-level object. Windows is displayed, since it is the highest-level object within a form. The plus box to the left of Windows indicates that the form contains Windows objects.
4. Click [+] beside **Windows.** The windows within the CH5ACUSTOMER form are displayed.

The form's Windows objects contain all of the other form objects. Two Windows objects are displayed: the NULL Window, which is created automatically when the form is created but is not displayed on the form, and WINDOW1 (your window name might be different), which was automatically created by the Layout Wizard and contains the rest of the form objects. Now you will expand WINDOW1 to examine the rest of the form objects.

To expand WINDOW1:

1. Click [+] beside **WINDOW1.** The Canvases object appears.
2. Click [+] beside **Canvases.** The default canvas (CANVAS2) appears. (Your canvas name might be different.)
3. Click [+] beside **CANVAS2.** The Object Navigator shows that the canvas contains items and graphics.
4. Click [+] beside **Items.** The form items appear.
5. Click [+] beside **Graphics.** The form frame appears.

   The canvas contains the final level of the Visual View, which contains the form items and graphics. Note that data blocks are not shown in Visual View.

Changing Object Names in the Object Navigator

After you create a form using the Data Block and Layout Wizards, it is a good idea to change the default object names to more descriptive names. When you start creating forms with multiple windows, canvases, and frames, it is hard to visually distinguish between different objects in the Object Navigator unless they have descriptive names. Now you will change the window name to CUSTOMER_WINDOW, the canvas name to CUSTOMER_CANVAS, and the frame name to CUSTOMER_FRAME.

To change the object names:

1. Click on **WINDOW1** (or whatever your default window name is) to select it. Click it again so its background color changes to blue, type **CUSTOMER_WINDOW**, then press the **Enter** key.
2. Click on **CANVAS2** (or whatever your default canvas name is) to select the canvas, click it again so its background color changes to blue, and type **CUSTOMER_CANVAS**. Press the **Enter** key.
3. Click on **FRAME3** (or whatever your default frame name is), and change the name to **CUSTOMER_FRAME**. The new window and canvas names should appear, as shown in Figure 5-13.

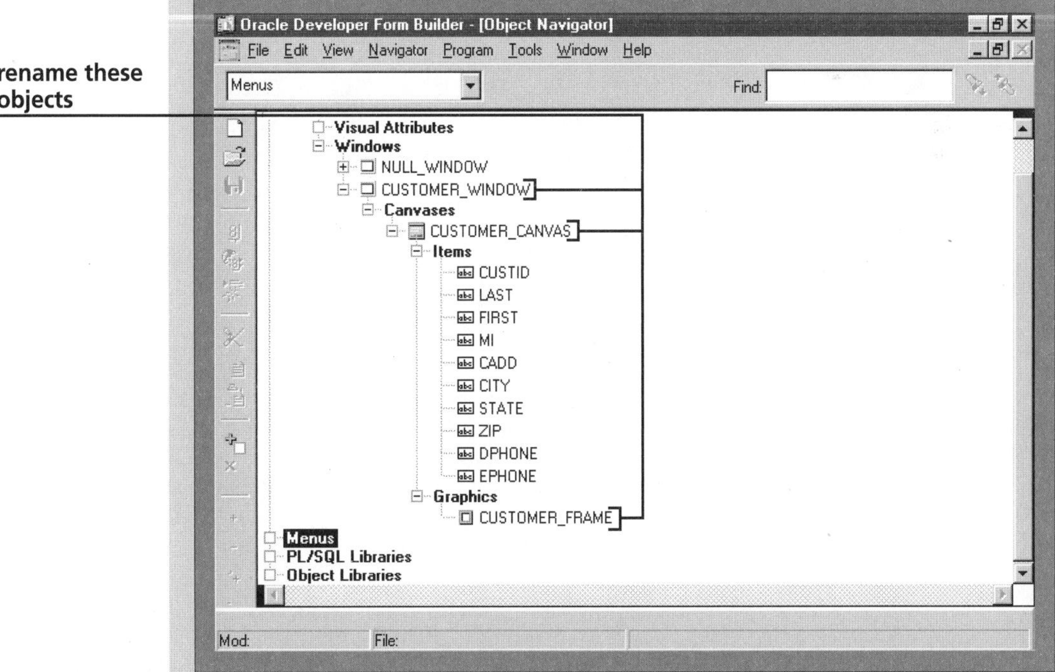

Figure 5-13: Renaming form objects with descriptive names

4. Save the form.

Modifying a Form Using the Data Block Wizard and Layout Wizard

A powerful characteristic of the Data Block Wizard and Layout Wizard is that they are **re-entrant**, which means that they can be used to modify the properties of an existing block or layout. To use a wizard to modify an existing block or layout, it must be in **re-entrant mode**. To start a wizard in re-entrant mode, you must first select the data block or layout frame in the Layout Editor that you wish to modify, and then start the Data Block Wizard or Layout Wizard. (If you do not want to start the Data Block Wizard or Layout Wizard in re-entrant mode, you must be sure that no data block or layout frame is currently selected when you start the wizard.)

You can visually tell when a wizard is in re-entrant mode, because the different wizard pages are displayed as tabs on the top of the wizard dialog box. Now you will modify the CUSTOMER data block so that it enforces database integrity constraints directly in the form. You will do this by selecting a field in the block, and then re-entering the block in the Data Block Wizard.

To modify the block using the Data Block Wizard:

1. Click **Tools** on the menu bar, then click **Layout Editor** to open the Layout Editor window.

help

If the Layout Editor does not open, make sure that a form object is currently selected in the Object Navigator

2. Click the **CUSTID** field to select the block. (You could click any of the other block fields to select the block as well.)
3. Click **Tools** on the menu bar, then click **Data Block Wizard** to open the Data Block Wizard in re-entrant mode. The Data Block Wizard appears with the Type and Table tabs displayed at the top of the window, indicating that it is in re-entrant mode.
4. Click the **Table** tab, then click the **Enforce Data Integrity** check box so it is checked. Log on to the database if necessary. Click **Finish** to save the change and close the Data Block Wizard.

Clicking Apply in the Data Block Wizard saves the change, but does not close the Data Block Wizard. Clicking Finish saves the change and also closes the Data Block Wizard. Clicking Cancel closes the Data Block Wizard without saving any changes.

Next, you will modify the form layout so that the prompt for the Customer Address field is displayed as "Address" instead of "Customer Address." To do this, you must start the Layout Wizard in re-entrant mode by first selecting the layout frame, and then starting the Layout Wizard.

If you delete the frame around a layout, you cannot revise the layout using the Layout Wizard in re-entrant mode.

To modify the form layout using the Layout Wizard:

1. Click the Customers frame around the form fields to select it.
2. Click **Tools** on the menu bar, then click **Layout Wizard**. The Layout Wizard opens with Data Block, Items, Style, and Rows tabs displayed at the top of the page indicating that it is in re-entrant mode.
3. Click the **Items** tab to move to the Items page, then change the prompt for the CADD field to **Address:**.
4. Click **Finish**. The modified prompt is displayed in the Layout Editor.

You can also modify form layout properties manually in the Layout Editor window. Recall that when you scrolled through the CUSTOMER records, some of the customers' middle initials did not appear, because the MI field was not wide enough. Next, you will make the MI field wider, using the Layout Editor.

To make the MI field wider:

1. If necessary, click **Tools** on the menu bar, then click **Layout Editor** to open the Layout Editor window.
2. Click on the MI field in the layout. Six selection handles appear on its left and right edges.
3. Click on the middle selection handle on the right edge, and drag to the right so that the MI field is slightly wider and "MI" is displayed in the field.
4. Run the form, click the **Enter Query** button, and then click the **Execute Query** button to step through all of the records sequentially to verify that the MI for each customer is displayed correctly. (Recall that Lee Miller does not have a middle initial, so no value should be displayed for this record.)
5. Exit Forms Runtime.

Now you will save your Form Builder file, close the form, and exit Form Builder.

To save the file, close the form, and exit Form Builder:

1. Click the **Save** button on the toolbar to save the form.
2. Click **File** on the menu bar, then click **Close** to close the form. Click **Yes** if you are prompted to save your changes.
3. Click **File** on the menu bar, then click **Exit** to exit Form Builder, and all other open Oracle applications.

SUMMARY

- A form is a database application that looks like a paper form, and provides an easy-to-use interface that allows users to add new database records and to modify, delete, or view existing records.
- When a form is running, it can be in either Normal or Enter Query mode. In Normal mode, you can view data records and sequentially step through the records. In Enter Query mode, you can enter one or more search parameters in the form fields, and then retrieve the associated records.
- Using a form to add records is the equivalent of adding records in SQL using the INSERT command. Using a form to modify records is the equivalent of updating records in SQL using the UPDATE command.
- In an Oracle data block form, each data block is associated with a database table.
- When you create a data block form, the system automatically generates the text fields and labels for data fields in that table and then creates the code for inserting, modifying, deleting, and viewing data records.
- If you enforce data integrity constraints in a form, the form will flag violations to the integrity constraints that you established when you created the table (unique primary keys, foreign key references, etc.), and you will see Form Builder error messages.
- If you do not enforce integrity constraints in a form, the integrity constraints still will be enforced by the DBMS, and users will receive ORA- error codes directly from the DBMS if they violate a table integrity constraint.
- The form layout specifies how the form looks to the user.
- A canvas is the area on a form where you place graphical user interface (GUI) objects, such as buttons and text fields.
- In a form layout style, only one record is displayed at one time, while in a tabular layout style, multiple records are displayed on the form at one time.
- The Layout Editor provides a graphical display of the form canvas that can be used to draw and position form items, and to add boilerplate objects such as labels, titles, and graphic images.
- While running a form, you can do approximate searches by using the percent sign (%) to indicate that there can be any number of wildcard characters either before or after a search string.
- The Object Navigator in Form Builder enables you to access all form objects either by object type (Ownership View) or hierarchically by opening objects that contain other objects (Visual View).
- Form modules contain windows, windows contain canvases, and canvases contain items such as data fields and labels.
- Boilerplate objects are customizable text, graphics, and other objects that do not contribute to the functionality of a form but enhance its appearance and ease of use.

- Form Builder files are saved with an .fmb extension, whereas executable Forms Runtime files are saved with an .fmx extension. Whenever you make a change to your .fmb file, you must regenerate it to create an updated .fmx file before you can run it.
- The Data Block Wizard and Layout Wizard can be used to modify existing data blocks and form layouts by selecting the data block or layout frame, and then starting the wizard in re-entrant mode.

REVIEW QUESTIONS

1. Describe the purpose of each of the following buttons on the Forms Runtime toolbar:

   Button **Purpose**

2. When a form is running, it can be in one of two modes: Normal or Enter Query. Which mode must it be in for you to:
   a. enter a search condition?
   b. insert a new record?
   c. step through table records sequentially?
3. Describe two ways to change from Enter Query mode back to Normal mode.
4. True or false: A data block can be associated with multiple database tables.
5. List how the following form objects appear from top to bottom level in Visual View in the Object Navigator: Canvas, Item, Form, Window, Data Block.
6. How do you rename objects in the Object Navigator?
7. What is the difference between the form and tabular layout styles in a data block form?
8. Why is it not necessary to check the Enforce Data Integrity check box when you create a new data block?
9. What is a boilerplate object?
10. What is the difference between the files named customer.fmb and customer.fmx?
11. Specify an approximate search condition that would return the records for all items with the word "Children" in the description. (Specify the search condition so there could be text either before or after the word "Children.")

PROBLEM-SOLVING CASES

Run the northwoo.sql script to refresh the Northwoods University database, and run the clearwat.sql script to refresh the Clearwater Traders database. The script files are stored in the Chapter5 folder on your Student Disk.

1. Create a single-table form for inserting, updating, and viewing records in the LOCATION database table in the Northwoods University database. Name the window LOCATION_WINDOW and the canvas LOCATION_CANVAS. Use a form layout style, create descriptive item prompts for the form fields, and use the frame label "Northwoods Building Locations." Save the Form Builder file as Ch5aEx1.fmb in the Chapter5 folder on your Student Disk. Use the form to execute the following operations:
   a. Add a new LOCATION record with LOCID = 65, BLDG_CODE = BUS, Room = 100, Capacity = 150.
   b. Modify the capacity of LOCID 52 to 75.
   c. Use an approximate search to count the number of records where BLDG_CODE = BUS.
2. Create a single-table form for inserting, updating, and viewing records in the ITEM database table in the Clearwater Traders database. Call the window ITEM_WINDOW, and the canvas ITEM_CANVAS. Use a form layout style, create descriptive item prompts for the form fields, and use the frame label "Clearwater Traders Items." Save the Form Builder file as Ch5aEx2.fmb in the Chapter5 folder on your Student Disk. Use the form to perform the following operations:
   a. Update the ITEMDESC of ITEMID 786 to "4-Season Tent."
   b. Add the following record: ITEMID = 800, ITEMDESC = Fleece Vest, Category = Women's Clothing.
   c. Use an approximate search to count the number of items in the category Women's Clothing.
3. Create a single-table form for inserting, updating, and viewing records in the FACULTY database table in the Northwoods University database. Name the window FACULTY_WINDOW, and the canvas FACULTY_CANVAS. Use a form layout style, create descriptive item prompts for the form fields, and use the frame label "Faculty." Save the Form Builder file as Ch5aEx3.fmb in the Chapter5 folder on your Student Disk. Use the form to execute the following operations:
   a. Insert the following record: FID = 10, FLNAME = Wilson, FFNAME = Ephraim, FMI = V, LOCID = 57, FPHONE = 7155556023, FRANK = ASST, FPIN = 4433.
   b. Update the FPHONE of faculty member Laura Sheng to 7155559878.
   c. Count the number of records where FRANK = ASSO. Use 'ASSO' as the search condition.

4. Create a single-table form for inserting, updating, and viewing records in the COURSE database table in the Northwoods University database. Name the window COURSE_WINDOW, and the canvas COURSE_CANVAS. Use a form layout style, create descriptive item prompts for the form fields, and use the frame label "Northwoods University Courses." Save the Form Builder file as Ch5aEx4.fmb in the Chapter5 folder on your Student Disk. Use the form to execute the following operations:
   a. Insert the following new record: CID = 9, CALLID = MIS 460, CNAME = 'Information Systems Management', CCREDIT = 3.
   b. Modify the Database Management course so CCREDIT = 4.
   c. Use an approximate search to determine the number of courses with MIS in the CALLID.

LESSON B

objectives

- Create a single-table form that displays multiple records on the same form
- Create a data block form using a database view
- Modify form properties to improve form appearance and function
- Create a form with multiple data blocks
- Format form data fields using format masks

Creating a Form to Display Multiple Records

The CH5ACUSTOMER form that you created in the previous lesson displays a single record at a time using a form-style layout. Now you will create a **tabular** data block form that displays multiple records on the same form. In this form, you will display all of the records in the INVENTORY table. First, you will refresh your Clearwater Traders database tables. Then, you will start Form Builder and create the new data block.

To refresh your database tables and create the new data block:

1. Start SQL*Plus, and run the clearwat.sql script in the Chapter5 folder on your Student Disk.
2. Start Form Builder, make sure the **Use the Data Block Wizard** option button is selected, then click **OK.** The Welcome to the Data Block Wizard page is displayed. Click **Next.**

help If Form Builder is already running from the previous lesson, close all open forms, click File on the top menu bar, point to New, then click Form. A new form appears in the Object Navigator window. To start the Data Block Wizard, click Tools on the menu bar, then click Data Block Wizard.

3. Make sure that the **Table or View** option is selected, then click **Next.**
4. Click **Browse** on the Type page, and connect to the database if necessary. Be sure that the **Current User** and **Tables** check boxes are checked, select **INVENTORY** on the table list, then click **OK.**
5. Select all of the columns in the INVENTORY table to be included in the data block. Do not click the Enforce data integrity box. Click **Next.**

6 Make sure that the **Create the Data Block, then call the Layout Wizard** option button is selected, then click **Finish.**

Now that the data block has been created, you must specify the layout properties, using the Layout Wizard. This form layout will be different from the one you made before, because you will specify the tabular layout style and display five records on the form at one time. You will also include scroll bars on the layout so the user can scroll up and down through the data records. You will now create the INVENTORY block layout.

To create the layout:

1 When the Layout Wizard Welcome page is displayed, click **Next.**

2 Accept **(New Canvas)** as the layout canvas, and **Content** as the canvas type, then click **Next.**

3 Select all of the data block items to be displayed in the layout, then click **Next.**

4 Modify the item prompts, as shown in Figure 5-14. Since the data items will be displayed in columns, you will not put a colon after the column headings.

5 Click **Next.**

change these values

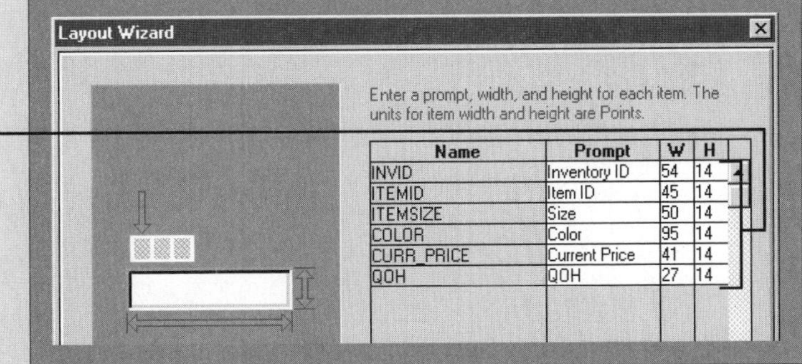

Figure 5-14: Modifying the Inventory form item prompts

6 Click the **Tabular** option button to specify that multiple records are displayed on the form, then click **Next.**

7 Type **Inventory** for the frame title, and **5** for Records Displayed. Leave Distance Between Records as 0. Check the **Display Scrollbar** check box.

8 Click **Next,** then click **Finish.** The completed Inventory form layout appears in the Layout Editor, as shown in Figure 5-15.

If the Object Navigator window opens, click Tools on the menu bar, then click Layout Editor to see the form in the Layout Editor.

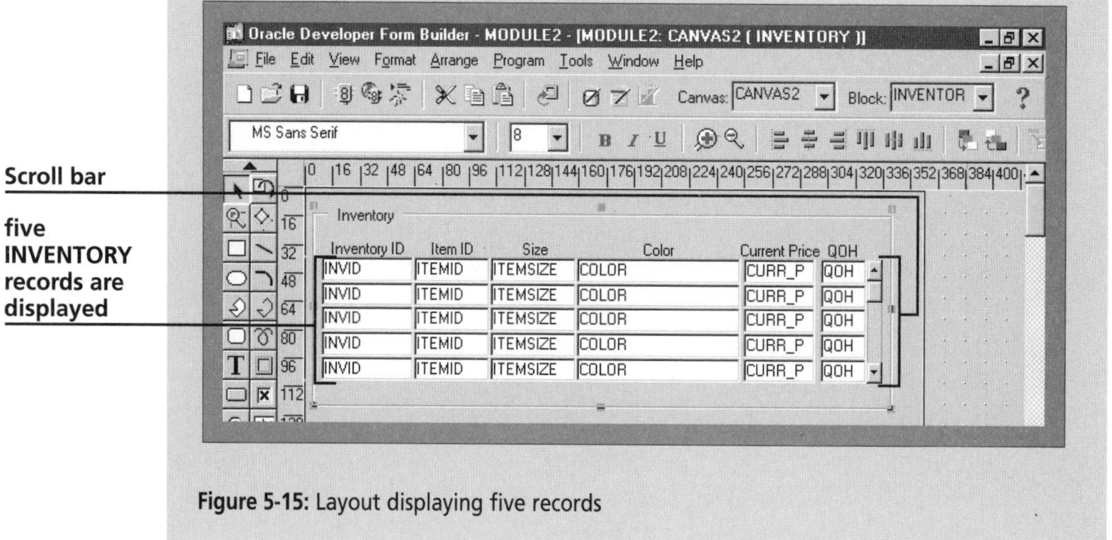

Figure 5-15: Layout displaying five records

The form layout in Figure 5-15 shows that five records from the INVENTORY table will be displayed on the form at one time. The scroll bar at the right edge of the records will allow the user to scroll up and down through the records if more than five records are retrieved into the form. Because you specified the distance between records as 0 in the layout properties, the records are stacked directly on top of one another.

Now you will save the form and then run it.

To save and run the form:

1. Click the **Save** button, and save the file as **ch5binventory.fmb** in the Chapter5 folder on your Student Disk.
2. Click the **Run Form Client/Server** button to compile the form and display it in the Forms Runtime window. Maximize the Forms Runtime window if necessary.

You can perform insert, update, delete, and query operations in a multiple-record form just as you did in a single-record form. First, you will view all of the INVENTORY records, and then add a new record.

To view the INVENTORY records and add a new record:

1. Click the **Enter Query** button, then click the **Execute Query** button. All of the INVENTORY records are retrieved into the form.
2. Scroll down through the records using the scroll bar, and then click Inventory ID **11848** to select it. Now you will insert a new record below Inventory ID 11848.

3 Click the **Insert Record** button. A new blank record appears below Inventory ID 11848.

4 Type the information for the new INVENTORY row, as shown in Figure 5-16, to add Inventory ID 11849 (50 Blue 3-Season tents), then click to save the new record.

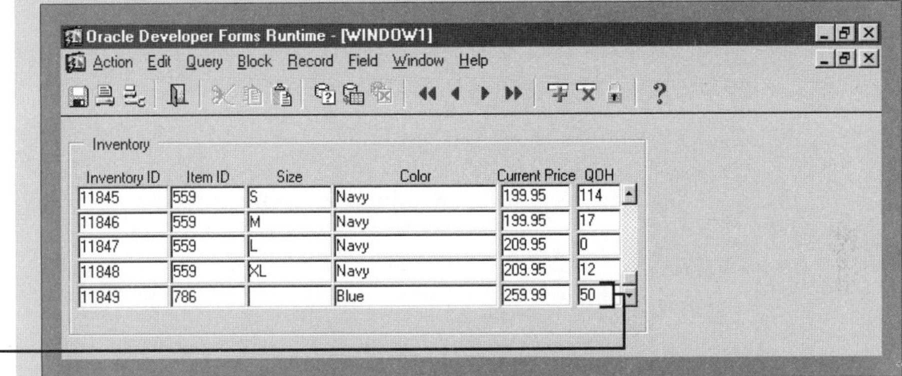

add this record

Figure 5-16: Adding a new record on a tabular form

You can also search for specific records on a tabular form. Next you will retrieve all inventory items corresponding to ITEMID 786. Then you will delete the record you just created.

To search for specific records and delete a record:

1 Click the **Enter Query** button to go to Enter Query mode. The insertion point appears in the first Inventory ID field.

2 Press the **Tab** key, then type **786** in the Item ID field for the query search condition, and then click the **Execute Query** button to execute the query. The three records corresponding to Item ID 786 are displayed.

3 Click Inventory ID **11849** to select it, then click the **Remove Record** button to delete the record. The record disappears.

4 Click the **Save** button to save your changes.

5 Exit Forms Runtime.

6 Open the Object Navigator window by clicking **Window** on the menu bar, then clicking **Object Navigator**.

7 Close the form by selecting **CH5BINVENTORY** in the Object Navigator window, clicking **File** on the menu bar, then clicking **Close**.

Creating a Form Using a Database View

Recall that a database view looks and acts like a database table, but is derived from other database tables. It is useful to create views that retrieve actual data values for records that are related by foreign keys and then display the related data in a form, rather than display the foreign key value. For example, in the Inventory form you just created, it would probably be more informative to show the item description for each inventory ID rather than just the item ID. However, when you create a data block using a view, you can only examine the data. You cannot modify the data by inserting new records or updating or deleting existing records.

Now you are going to create a view that retrieves all of the INVENTORY records, along with their associated item descriptions. Then, you will create a tabular form to display the information. First, you will create the view using SQL*Plus.

To create the view in SQL*Plus:

1 Switch to SQL*Plus, and type the query shown in Figure 5-17 to create the view. The message "View created" is displayed.

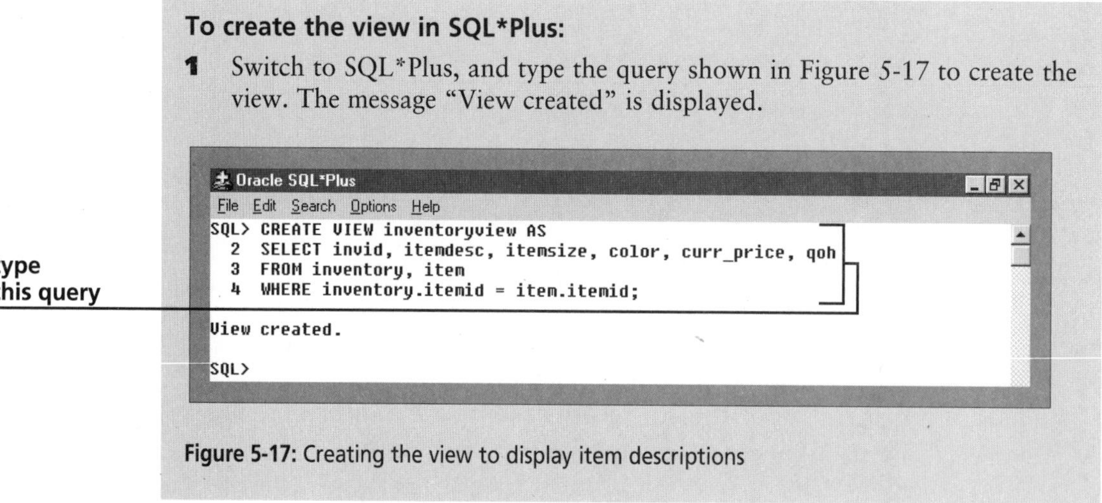

Oracle SQL*Plus
File Edit Search Options Help
SQL> CREATE VIEW inventoryview AS
2 SELECT invid, itemdesc, itemsize, color, curr_price, qoh
3 FROM inventory, item
4 WHERE inventory.itemid = item.itemid;
View created.
SQL>
type this query

Figure 5-17: Creating the view to display item descriptions

Now you will create a new form that uses the view as its data source and displays five records at one time. You will select the view from your list of database objects when you create the data block. First, you will create a new form.

To create a new form:

1 Switch to Form Builder and, if necessary, open the Object Navigator window. Be sure that **Forms** is selected as the top-level object, then click the **Create** button to create a new form module. A new form module appears.

2 Select the new form module, then click the **Save** button and save the form as **ch5binvview.fmb** in the Chapter5 folder on your Student Disk. Notice that the name of the form module changes to the filename in the Object Navigator window.

Now you will use the Data Block Wizard to create the data block, using the INVENTORYVIEW view you just made in SQL*Plus. To do this, you will select the view as the data source on the Data Block Wizard Table page.

To create the data block using a view:

1 Make sure that the **CH5BINVVIEW** form module is selected in the Object Navigator window. Click **Tools** on the menu bar, click **Data Block Wizard,** then click **Next.**

2 Make sure **Table or View** is selected as the data block type in the Tables dialog box, then click **Next.**

3 When the Table page is displayed, click **Browse**. Make sure that **Current User** is checked. If necessary, uncheck **Tables**, then check **Views** so that only your database views are displayed. Select **INVENTORYVIEW**, then click **OK.**

4 Select all of the view columns to be included in the data block, then click **Next.** Make sure the **Create the data block, then call the Layout Wizard** option button is selected, then click **Finish.**

The form layout will display five records and include scroll bars. Now you will create the layout using the Layout Wizard.

To create the form layout:

1 Click **Next** on the Layout Wizard Welcome page.

2 Accept the (New Canvas) and Content defaults, then click **Next.**

3 Select all of the block items for display in the layout, then click **Next.**

4 Modify the item prompts, as shown in Figure 5-18, then click **Next.**

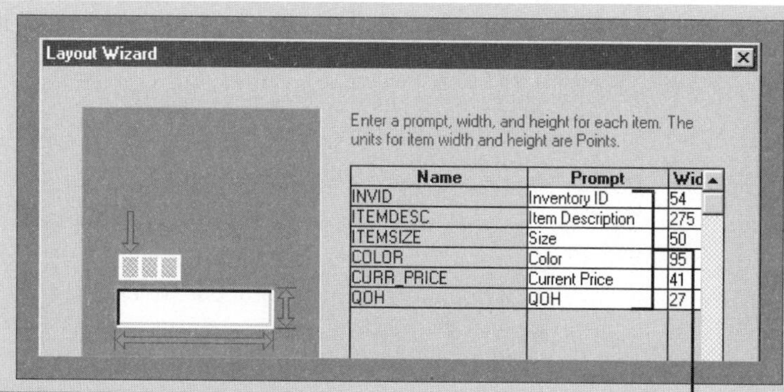

Figure 5-18: Changing the item prompts

5 Select the **Tabular** layout style option button, then click **Next.**

6 Type **Inventory** for the frame title, change Records Displayed to **5**, leave Distance Between Records as 0, and check the **Display Scrollbar** check box. Click **Next**, then click **Finish.**

Now you will save the form, and change the form's window, canvas, and frame names in the Object Navigator.

To save the form and change the form item names:

1 Click the **Save** button [icon] to save the form.

2 Click **Window** on the top menu bar, then click **Object Navigator.**

3 Click **View** on the top menu bar, then click **Ownership View** to make sure the Object Navigator is in Ownership View.

4 Click [+] beside Windows, and change the form window name to **INVVIEW_WINDOW.**

5 Change the form canvas name to **INVVIEW_CANVAS**, and the form frame name to **INVVIEW_FRAME**. Your Object Navigator window should look like Figure 5-19.

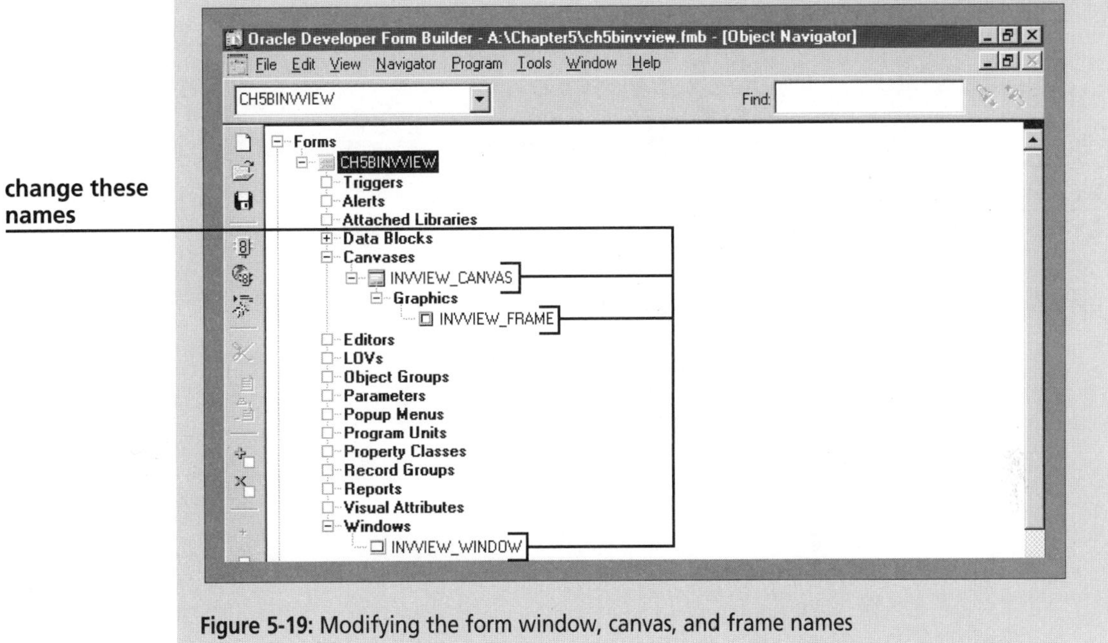

Figure 5-19: Modifying the form window, canvas, and frame names

Now you will run the form and retrieve all of the INVENTORYVIEW records to test the form.

To test the form:

1 Click the **Run Form Client/Server** button to run the form. The Forms Runtime window opens.

2 Click the **Enter Query** button, then click the **Execute Query** button to retrieve all of the table records into the form. The form looks like Figure 5-20. (Your form might look slightly different, depending on the display settings for your monitor.)

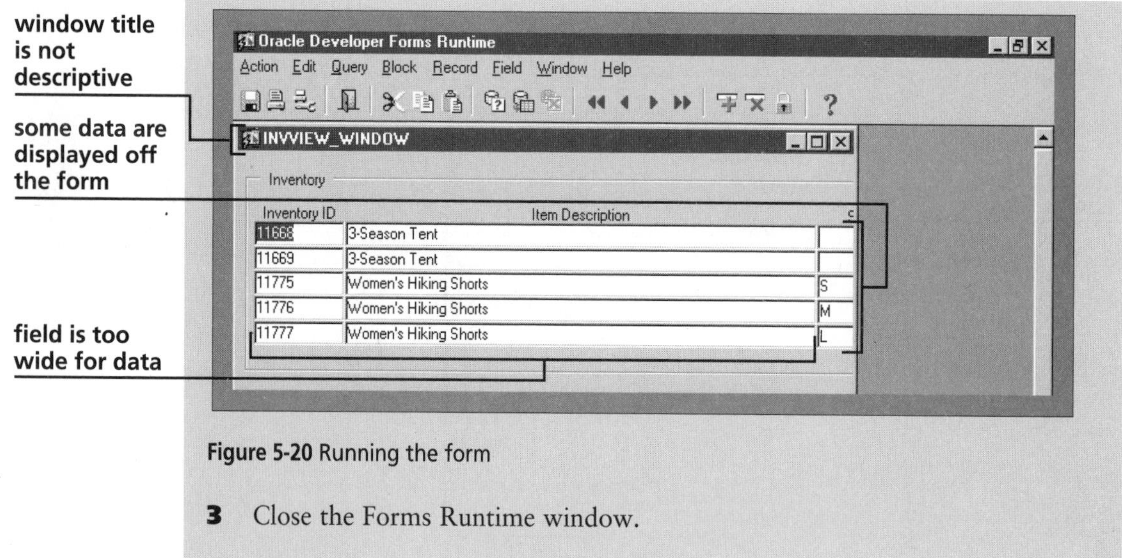

Figure 5-20 Running the form

3 Close the Forms Runtime window.

Modifying Form Properties

Figure 5-20 shows that some of the properties of the form need to be changed to improve the form's appearance. The Item Description field is far wider than the widest data item. Some of the fields are displayed off-screen, and the window title (INVVIEW_WINDOW) should be more descriptive. (By default, the window title is the same as the window name in the Object Navigator.) Visual applications should be attractive, easy to use, and configurable, and this application needs some modifications to meet these standards. First, you will modify the properties of the form window. You should always configure form windows using the following checklist:

1. Change the window title to a descriptive title.
2. Always allow the user to be able to minimize the window.
3. Do not allow the user to maximize the window unless you intend for the window to run maximized, since form objects might appear off-center in a maximized window.
4. Do not allow the user to resize the window, since resizing might cause some canvas objects to be clipped off.
5. Include horizontal and vertical scroll bars if some canvas objects extend beyond the window boundaries.
6. Make the form window large enough to fill the Forms Runtime window.

You will change the window title, disable the window's Maximize button and resize property, make the form window larger, and add horizontal and vertical scroll bars to the window so that you can scroll to view the form areas that are off-screen.

To change these window properties, you will need to modify the properties in the window's Property Palette. Every form object has a Property Palette that allows you to configure the object's properties. Different types of objects have different properties.

To modify the form window properties:

1 Right-click on the **Window** icon beside INVVIEW_WINDOW in the Object Navigator, then click **Property Palette**. The Property Palette for the window opens, as shown in Figure 5-21.

You can also open the Property Palette for most objects by double-clicking the object icon in the Object Navigator window.

Oracle Developer Form Builder - A:\Chapter5\ch5binvview.fmb - [Property Palette]

File Edit Property Program Tools Window Help

Find:

Window: INVVIEW_WINDOW

General	
Name	INVVIEW_WINDOW
Subclass Information	
Comments	
Functional	
Title	
Primary Canvas	<Null>
Horizontal Toolbar Canvas	<Null>
Vertical Toolbar Canvas	<Null>

Figure 5-21: INVVIEW_WINDOW Property Palette

2 Click in the space next to the **Title** property, and type **Clearwater Traders Inventory** for the new window title.

3 Click in the space next to the **Resize Allowed** property. A drop-down list appears. Open the list, and select **No** to disable the Resize property.

4 Scroll down in the Property Palette window if necessary, click in the space next to the **Maximize Allowed** property, open the list, and select **No**.

5 Scroll down in the Property Palette window, and change the Width property to **640**, and the Height property to **480**.

6 Click in the space next to **Show Horizontal Scrollbar**, and select **Yes**.

7 Click in the space next to **Show Vertical Scrollbar**, and select **Yes**.

8 Click [X] on the Property Palette title bar to close the Property Palette window.

help

If you receive the message "Save changes to CH5INVVIEW," when you try to close the Property Palette, you clicked [X] on the Form Builder window and not on the Property Palette window. Click Cancel, then click [X] on the inner Property Palette window.

You also need to change the width of the Item Description field in the form layout. Currently, the field is about twice as long as the longest description. You will change the field width by selecting the layout frame, and then revising the layout using the Layout Wizard in re-entrant mode. You will do this next.

To revise the layout with a shorter Item Description field:

1. In the Object Navigator window, click **Tools** on the menu bar, then click **Layout Editor** to open the Layout Editor. Click the frame around the INVVIEW fields to select the layout.
2. Click **Tools** on the menu bar, click **Layout Wizard** to enter the Layout Wizard in re-entrant mode, then click the **Items** tab.
3. Click in the **Width** space beside Item Description, and change the width to **130.** Click **Finish.** The revised layout is displayed in the Layout Editor.

Recall that the canvas is like the canvas in a painting, and the window is like the painting's frame. Since you changed the window width and height, you will also need to change the canvas width and height so it fills the larger window. Now you will open the form canvas's Property Palette and change the canvas's width and height.

To change the canvas width and height:

1. Click **Window** on the top menu bar, then click **Object Navigator** to display the Object Navigator window.
2. Right-click on the **Canvas** icon beside INVVIEW_CANVAS, then click **Property Palette.**
3. Change the canvas Width to **640** and the Height to **480**, then close the canvas's Property Palette.

The final step is to save the form, run it, and look at the properties you changed. You will do this next.

To run the form and view the changes:

1. Click the **Save** button.
2. Click the **Run Form Client/Server** button to run the form. Maximize the Forms Runtime window if necessary. Now you will retrieve all of the records into the form.
3. Click the **Enter Query** button, then click the **Execute Query** button. The INVENTORYVIEW records are displayed, as shown in Figure 5-22.

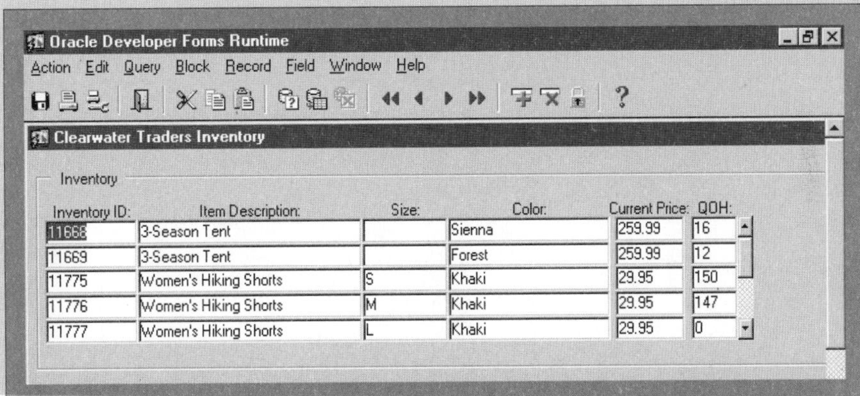

Figure 5-22: Modified form

4. Scroll to the right edge of the Forms Runtime Window to view the right edge of the form.
5. Close the Forms Runtime application.

Figure 5-22 shows that most of the form's appearance problems have been corrected: the new window title is displayed, the Item Description field is the correct width, the window and canvas fill the Forms Runtime window, and the window scroll bars are displayed. However, there is still one problem: the form's Minimize button and Close button might not be displayed unless you scroll to the right edge of the form. (This depends on your display settings; your buttons might be visible without scrolling.) If your buttons are not visible without scrolling, you will need to change the window width property so that it just fits the Forms Runtime window. This width will vary, depending on the display setting of your monitor. Now you will change the window width.

To change the window width:

1 Go to the Object Navigator window, open the **INVVIEW_WINDOW** window's Property Palette, and change the window Width to **500**. Close the Property Palette.

2 Run the form, maximize the Forms Runtime window if necessary, and see if the window width is correct and the [-] and [X] buttons are displayed, as shown in Figure 5-23. If the buttons do not appear, readjust the width until they appear as shown in the figure.

window buttons are displayed

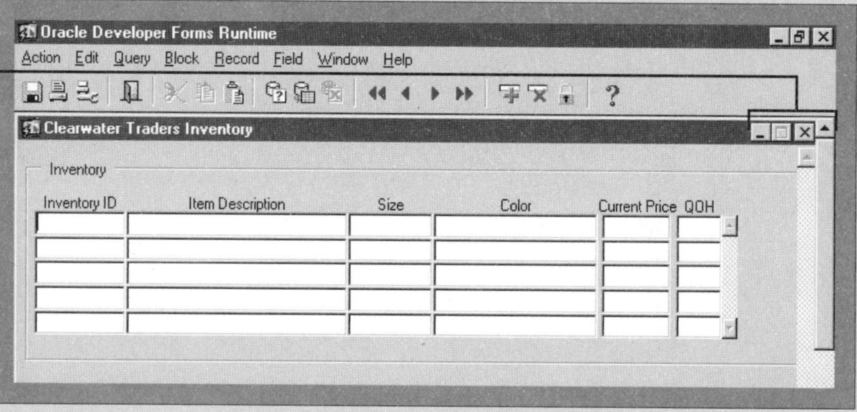

Figure 5-23: Modified form window display

Recall that when you create a data block using a view, you can view database records, but you cannot insert new records or modify or delete existing records. Now you will try to insert a record using the form, to see what happens.

To insert a record in the form:

1 If necessary, run the form.

2 Click the **Enter Query** button, click the **Execute Query** button to display the INVENTORYVIEW records, then select Inventory ID **11669**.

3 Click the **Insert New Record** button to insert a new record. A blank record appears under Inventory ID 11669.

4 Type the following inventory record information: Inventory ID = **11850**, Item Description = **3-Season Tent**, Size = **<Null>**, Color = **Blue**, Current Price = **259.99**, QOH = **50**.

5 Click to save the new record. The message "FRM-40508: ORACLE error: unable to insert record" is displayed, confirming that you cannot insert a record using a form created from a view. Click the **Remove Record** button to remove the record.

Recall that you cannot modify or delete a record in a data block created from a view. Now, you will try to update Inventory ID 11669 to change the color from Forest to Hunter. Then, you will try to delete a record.

To try to modify and delete a record:

1. Select the color field for Inventory ID 11669, change the color value from Forest to **Hunter**, then click [save icon] to save the change. The message "FRM-40509: ORACLE error: unable to UPDATE record" is displayed briefly, confirming that you cannot update a record in a form that is created using a view. Change the color for Inventory ID 11669 back to **Forest**.
2. Select Inventory ID **11669**, then click the **Remove Record** button [remove icon]. The record disappears, but the change has not yet been committed to the database.
3. Click [save icon] to commit the change to the database. The message "FRM-40510: ORACLE error: unable to DELETE record" is displayed briefly, confirming that you cannot delete a record in a form that is created using a view.

Now you will exit Forms Runtime, and close the form in Form Builder.

To close the form:

1. Exit Forms Runtime and close the form in Form Builder. Click **No** when you are asked if you want to save your changes.
2. Close the form by selecting **CH5BINVVIEW** in the Object Navigator window, clicking **File** on the menu bar, then clicking **Close**. Click **Yes** if you are asked if you want to save your changes.

Creating a Multiple-Table Form

Now that you have learned how to create a single-table data block form and display one or more records on a form, the next task is to create a form using multiple tables with master-detail relationships. In a master-detail relationship, one database record (the master record) can have multiple related (detail) records through foreign key relationships. In the Clearwater Traders database, one CUSTOMER record can have multiple associated CUST_ORDER records. For example, customer Alissa Chang (CUSTID 179) has two different CUST_ORDER records (ORDERID 1061 and ORDERID 1062). The CUSTOMER record is the master side of the relationship, since a master (CUSTOMER) can have many detail (CUST_ORDER) records, but a detail record can have only one associated master record. Now you will create a form that allows users to select a customer record, and then display and edit customer order information.

When you create a master-detail form, you always create the master block, which will be the list of customers, first. For the master block, you will use the ch5acustomer.fmb file that you created in the previous lesson, which displays a single customer record on a form. Then, you will create the detail block, which is the list of orders for a specific customer. You specify the master-detail relationship when you create the detail block. First, you will open the ch5acustomer.fmb file, save it using a different filename, and modify some of the form properties to improve its appearance.

To open ch5acustomer.fmb, save it using a different filename, and modify its properties:

1 If necessary, open the Object Navigator window.

2 Click the **Open** button, then open **ch5acustomer.fmb** from the Chapter5 folder on your Student Disk. Save the file as **ch5bcustorder.fmb.**

Your system might not show the .fmb file extensions when you open the ch5acustomer.fmb file. If you try to open an .fmx instead of an .fmb file in Form Builder, the system will display the error message "FRM-10043: Cannot open file."

3 Select **CH5ACUSTOMER** under Forms in the Object Navigator, and change the form module name to **CH5BCUSTORDER.**

4 Open the CUSTOMER_WINDOW Property Palette, and change the window title to **Clearwater Traders Customer Orders.** Change the Maximize Allowed property to **No**, and change the Resize Allowed property to **No.** Resize the window so that it fits the Forms Runtime window correctly. Change the Show Horizontal Scrollbar property to **Yes**, and change the Show Vertical property to **Yes.**

5 Change the canvas's width and height so that it is the same width and height as the window.

6 Run the form to check your changes. Your form should look like Figure 5-24.

7 Exit Forms Runtime.

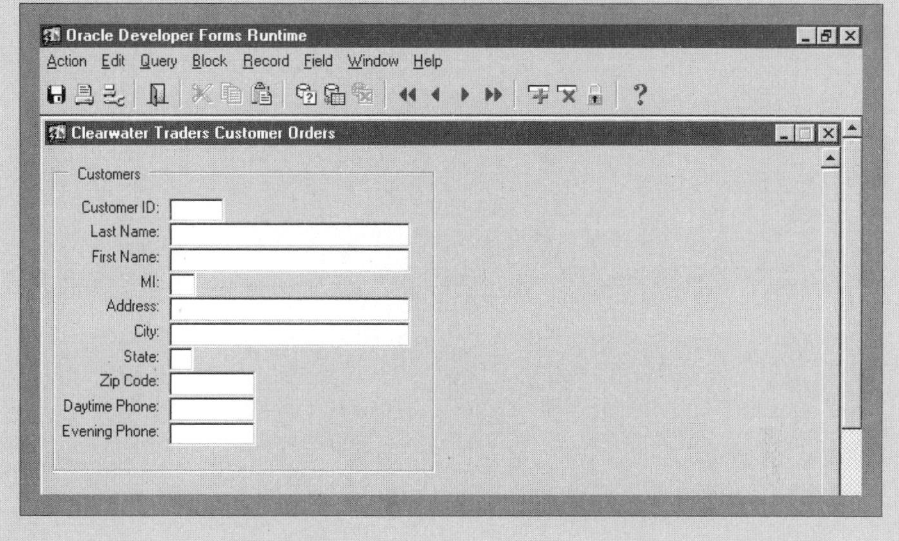

Figure 5-24: Modified CUSTOMER form

Creating the CUST_ORDER Detail Data Block

When the user selects a customer in the CUSTOMER block, the form will show the selected customer's order information. Next, you will create a detail block called CUST_ORDER that is associated with the Clearwater Traders CUST_ORDER table to show the related order information. To create a new block on a form using the Data Block Wizard, you must select the form in the Object Navigator. This block will be the detail block, so you will specify the master-detail relationship when you create it.

To create the CUST_ORDER data block and specify the master-detail relationship:

1. Make sure the **CH5BCUSTORDER** form module is selected in the Object Navigator, click **Tools** on the menu bar, click **Data Block Wizard**, then click **Next.**
2. Make sure the **Table or View** option button is selected, then click **Next.**
3. Click **Browse**, make sure the Tables check box is checked, select **CUST_ORDER** as the database table, then click **OK.**
4. Select all of the block fields to be included in the data block, then click **Next.** The Wizard master-detail page is displayed, as shown in Figure 5-25.

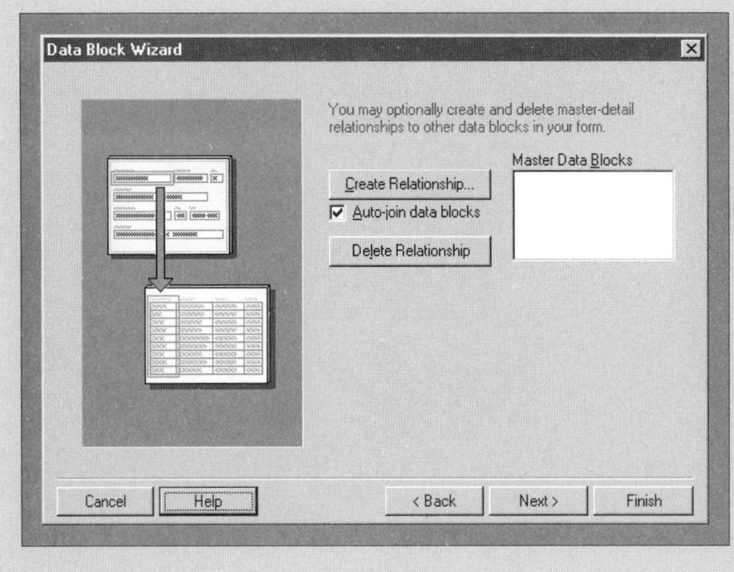

Figure 5-25: Data Block Wizard master-detail page

This page allows you to specify master-detail relationships among data blocks. The Data Block Wizard displays this page only when you create a new data block in a form that already has at least one other data block. First, you will select the name of the master block, which will appear in the Master Data Block list. If the **Auto-join data blocks** check box is checked, then the system will automatically create the join condition between the field in the selected master block that has a foreign key relationship with a field in the current block, which is the CUSTID data field. If the Auto-join data blocks box is not checked, then you will need to manually select the field in the master block and the field in the detail block that will join the two blocks. First, you will create the relationship using the Auto-join data blocks feature. You will click the Create Relationship button, and be presented with a list of the current form blocks. You will select the CUSTOMER block to be the master block. The join condition will automatically be created based on the foreign key relationship between the CUSTID field in the CUSTOMER table and the CUSTID field in the CUST_ORDER table.

To create the relationship using the Auto-join data blocks feature:

1. Make sure the Auto-join data blocks check box is checked, then click **Create Relationship.** The CH5BCUSTORDER: Data Blocks dialog box opens, which shows all of the form blocks.
2. Select **CUSTOMER** (which is the only choice), then click **OK.** CUSTOMER is displayed in the Master Data Blocks list, and the join condition for the two blocks is displayed in the Join Condition box, as shown in Figure 5-26.

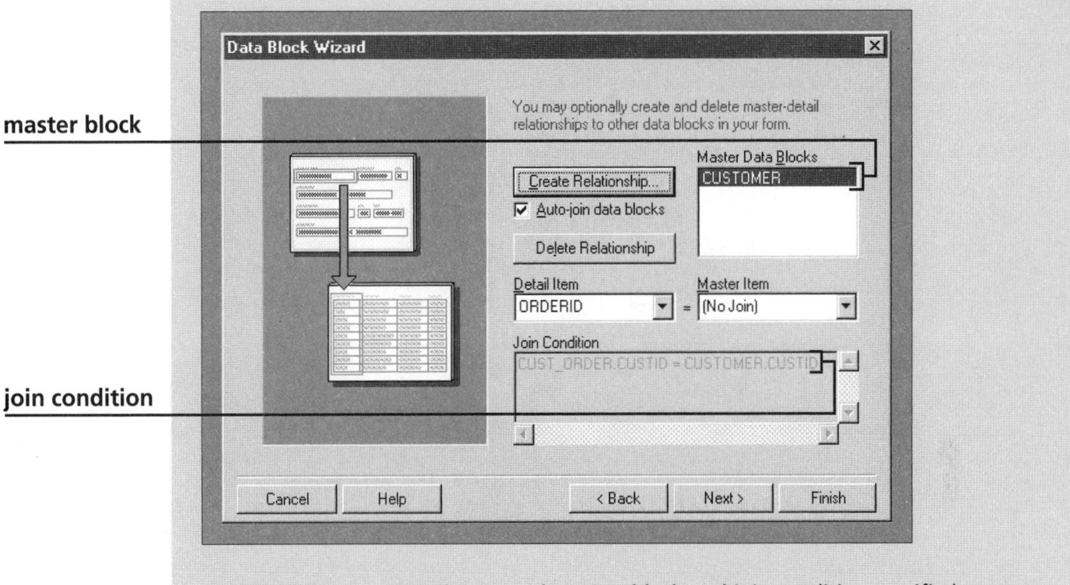

Figure 5-26: Master-Detail page with master block and join condition specified

The join condition is listed in the Join Condition box, and uses the general format <detail block name>.<join item name> = <master block name>.<join item name> to specify the block and item names on which you are joining the two blocks. For this form, the join condition is CUST_ORDER.CUSTID = CUSTOMER.CUSTID.

Now, you will create a master-detail relationship manually. First you will delete the relationship you just created, and uncheck the Auto-join data blocks check box. Then you will create the join condition manually. You do not need to select the join fields, since the join condition was created automatically. Always double-check to make sure that the join condition is correct when you use the Auto-join feature.

To delete the relationship and create the join condition manually:

1 Click **Delete Relationship.** The master data block and join condition disappear.

2 Clear the **Auto-join data blocks** check box.

3 Click **Create Relationship.** The Relation Type dialog box is displayed. You can create the relation based on a join condition, or on a REF item. (A REF item is a new Oracle8 data type that contains pointers, or physical addresses of related data objects.) Since your tables do not contain any REFs, you will base the relation on a join condition.

4 Confirm that the **Based on a join condition** radio button is selected, then click **OK**. The CH5BCUSTORDER: Data Blocks dialog box is displayed, which shows all of the form blocks.

5 Select **CUSTOMER**, then click **OK**. Notice that no join condition appears in the Join Condition box.

6 Open the Detail Item list box. This lists all of the fields in the detail (CUST_ORDER) block. Select **CUSTID**.

7 Open the Master Item list box. This lists all of the fields in the master (CUSTOMER) block. Select **CUSTID**. (You might need to scroll up to display CUSTID.) The join condition appears in the Join Condition box.

8 Click **Next**, then click **Finish**.

help

Sometimes when you click Create Relationship when the Auto-join data blocks check box is checked, the message "FRM-10757: No master blocks are available" might display. If it does, click OK, uncheck the Auto-join data blocks check box, and then create the join condition manually.

Next, you need to create the layout for the CUST_ORDER block. You will use a tabular layout, and display five order records and use a scroll bar.

To create the CUST_ORDER layout:

1 Click **Next** on the Welcome to the Layout Wizard dialog box.

2 Accept CUSTOMER_CANVAS as the layout canvas, then click **Next**.

3 Select all block fields to be displayed in the layout, then click **Next**.

4 Modify the item prompts, as shown in Figure 5-27, then click **Next**.

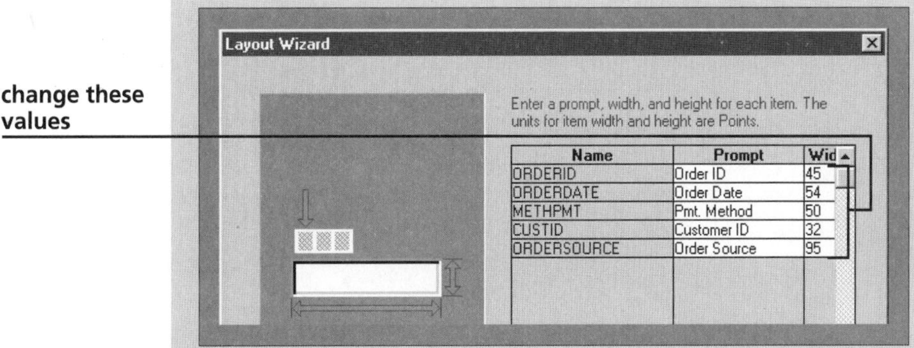

Figure 5-27: Item prompts for CUST_ORDER layout

5 Select the **Tabular** layout style option, then click **Next.**

6 Type **Customer Orders** for the frame title, change Records Displayed to **5,** then click the **Display Scrollbar** check box. Click **Next**, then click **Finish.**

7 Open the Object Navigator window, and change the name of the new frame to **CUST_ORDER_FRAME.**

8 Save the form.

Running the Master-Detail Form

You can add, update, delete, and query database records in a form that is based on multiple tables, just as you did in the single-table form. The only difference is that you need to activate the data block that contains the record that you want to add, modify, or query. Now you will run the form to see how a master-detail form works. You will retrieve all of the customer orders for Alissa Chang, and then enter a new order.

To run the form:

1 Click the **Run Form Client/Server** button to run the form. Maximize the Forms Runtime window if necessary.

2 Click the **Enter Query** button, press the **Tab** key to move to the Last Name field, then type **Chang** as the search condition. Click the **Execute Query** button to execute the query. Alissa Chang's CUSTOMER record appears in the Customers block fields, and her associated customer order records are displayed in the Customer Orders block, as shown in Figure 5-28.

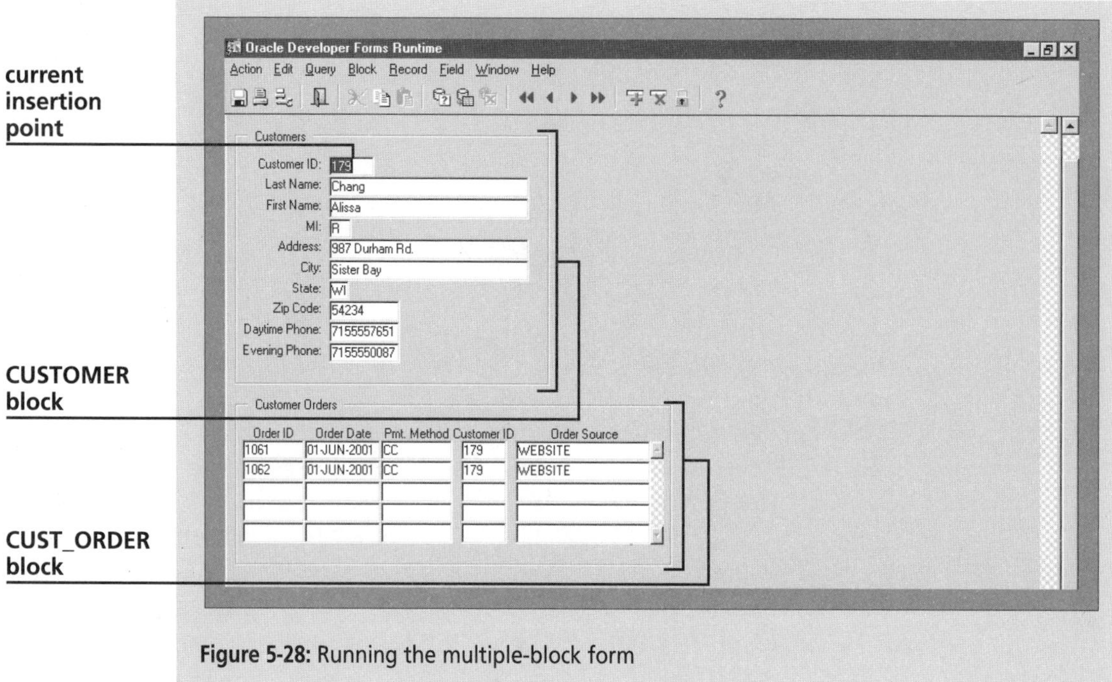

Figure 5-28: Running the multiple-block form

When a form displays multiple data blocks, you can only insert, update, delete, or query records in the block that is currently selected and displays the insertion point. Currently, the insertion point is in the CUSTOMER block. Now you will enter a new customer order record. To do this, you must place the form insertion point in the CUST_ORDER block. There are two ways to do this: click the Next Block button on the toolbar to move from the CUSTOMER block to the CUST_ORDER block, or click the mouse pointer on any field in the CUST_ORDER block. Now you will place the insertion point in the CUST_ORDER block, and insert a new order record.

To enter a new customer order record:

1. Click the **Next Block** button to move to the CUST_ORDER block. The insertion point moves to the first record in the CUST_ORDER block.
2. Click the **Insert Record** button to insert a new record. A blank record appears below Order ID 1061.
3. Enter the following new record: Order ID **1063**, Order Date **10-JUN-2001**, Pmt. Method **CHECK**, Customer ID **179**, Order Source **152.**
4. Click the **Save** button to save the record. The message "FRM-40400: Transaction complete: 1 records applied and saved" appears on the message line to confirm that the new record is committed to the database.

Now you will move back to the CUSTOMER block, and sequentially step through all of the customer records in the database. When you display a new customer record, the customer order information changes in the CUST_ORDER block of the form.

To move to the CUSTOMER block and step through the customer records:

1. Click the **Previous Block** button to move back to the CUSTOMER block.
2. Click the **Enter Query** button, then click the **Execute Query** button to retrieve all of the customer records. The record for customer Paula Harris is displayed in the CUSTOMER block. Her associated order information is displayed in the CUST_ORDER block.
3. Click the **Next Record** button to move to the next CUSTOMER record.The associated CUST_ORDER records are displayed. Continue to click until you have displayed every customer record.
4. Click to close Forms Runtime.

Adding Another Detail Data Block to the Form

A block can be the master block in multiple master-detail relationships. For example, an inventory item can have many associated shipments, and an inventory item can also have many associated back orders. The inventory block is the master block in the inventory/shipping relationship, and it is also the master block in the inventory/back order relationship.

A block can also be the detail block in one master-detail relationship, and the master block in a second master-detail relationship. For example, a customer can have many orders, and an order can have many order lines. In this scenario, the order is the detail block in the customer/order relationship, and it is the master block in the order/order line relationship. Currently in your form, CUSTOMER is the master block, and CUST_ORDER is the detail block. Next you will create a third block that displays data from the ORDERLINE table to show the order line details that correspond to a particular customer order. CUST_ORDER will be the master block in this new relationship, and ORDERLINE will be the detail block.

To create the second master-detail data block relationship:

1. Select the **CH5BCUSTORDER** form in the Object Navigator window, click **Tools** on the menu bar, then click **Data Block Wizard.**
2. Create a new data block using all of the columns in the ORDERLINE table.
3. For the master-detail relationship, select **CUST_ORDER** as the master block, and **ORDERID** as the join field. Your master-detail specification page should look like Figure 5-29

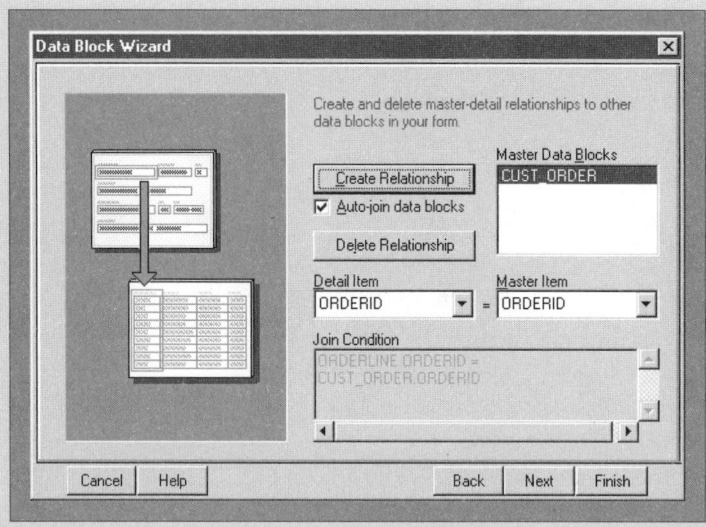

Figure 5-29: Specifying the CUST_ORDER/ORDERLINE master-detail relationship

4 Select all of the ORDERLINE data block fields for the layout, and modify the field prompts, as shown in Figure 5-30.

change these values

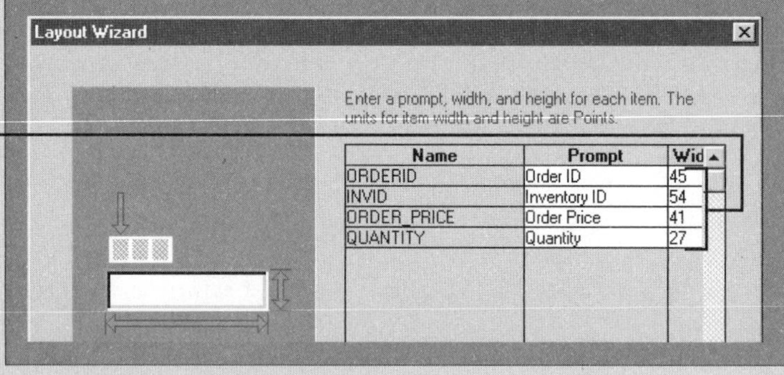

Figure 5-30: Specifying the ORDERLINE field prompts

5 Use a tabular layout, enter **Order Line Detail** for the frame title, show 5 records, and display a scroll bar.

Currently, the new ORDERLINE block is displayed at the bottom of the form. Although the user can use the vertical scroll bar to view the ORDERLINE information, it is best to consolidate all of a form's information onto a single visible area. You can move the ORDERLINE block so that it is beside the CUSTOMER block, and all blocks are simultaneously visible. You will do this next.

To move the ORDERLINE block:

1 Scroll down in the Layout Editor window until the ORDERLINE block is visible, and then select it by clicking its frame. Selection handles appear around all of the block fields.

help If selection handles only appear around one set of fields, click anywhere outside the block, then select the block again by clicking directly on the frame around the block.

2 Drag and drop the block from its current position until it is beside the CUSTOMER block, as shown in Figure 5-31.

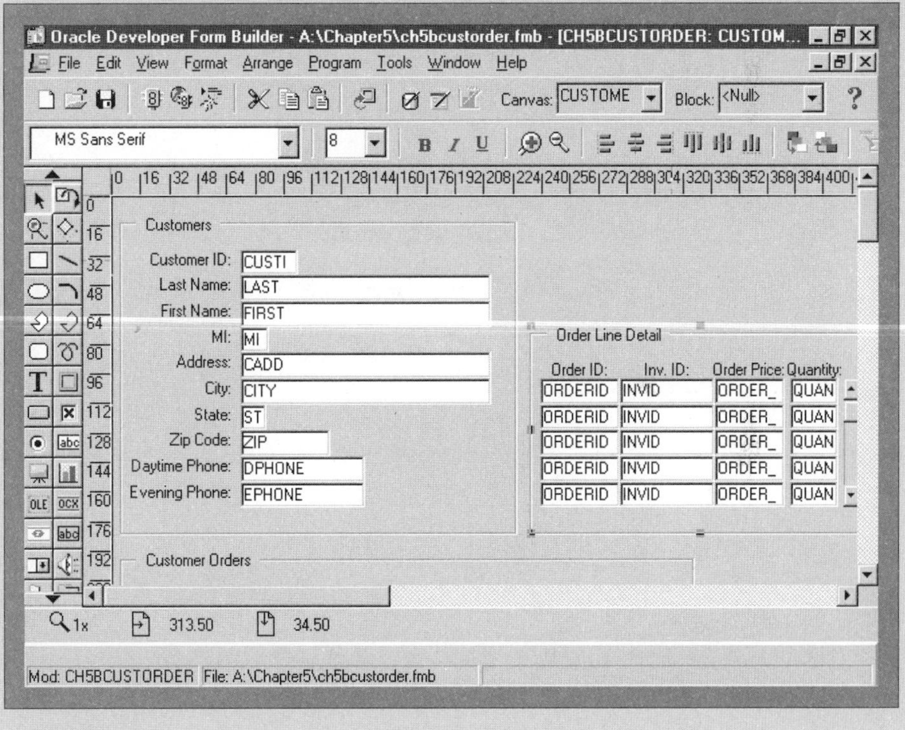

Figure 5-31: Repositioning the Order Line Detail frame

3 Save the form.

When you select a particular CUSTOMER record, the associated customer order records are displayed in the CUST_ORDER block, and the associated ORDERLINE records are displayed for the record that is currently selected in the CUST_ORDER block. The insertion point shows the currently selected record in a block. Now you will run the form, and view Alissa Chang's orders and their associated order lines.

To run the form and view the customer orders and order lines:

1. Run the form.
2. Click the **Enter Query** button, press the **Tab** key, type **Chang** as the search condition in the Last Name field, and then click the **Execute Query** button.

Alissa Chang's CUSTOMER record is displayed in the CUSTOMER block, and her associated CUST_ORDER records are displayed in the CUST_ORDER block. The ORDERLINE block displays the ORDERLINE records associated with CUST_ORDER 1061, because it is the first record in the table. Now you will display the ORDERLINE records for CUST_ORDER 1062 by moving to the CUST_ORDER block and selecting Order ID 1062.

To view the ORDERLINE records for Order 1062:

1. Click the **Next Block** button to move to the CUST_ORDER block.
2. Click the **Next Record** button to select Order ID 1062. Notice that the ORDERLINE records are updated and show information for Order ID 1062.
3. Click to select Order ID 1063. This is the new order that you entered before, and no ORDERLINE records have been inserted, so none are displayed in the ORDERLINE block.

You can insert, update, and delete data records just as before. Next, you will add an order line to Order ID 1063.

To add an order line to an order:

1. Click the mouse pointer in the Inventory ID field next to Order ID 1063 in the Order Line Detail frame and enter the following data in the form: Inventory ID = **11846**, Order Price = **199.95**, and Quantity = **1**.
2. Click the **Save** button. Note the confirmation message at the bottom left of the screen.
3. Close Forms Runtime.

Using Format Masks to Format Data Fields

Recall that a format mask can be used to specify data output formatting. Figure 5-32 shows that some of the data fields in the CH5BCUSTORDER form need to be formatted. The daytime phone number for customer Paula Harris is displayed as 7155558943, but it would be more understandable if the digits were separated as (715) 555-8943. The order price should be formatted as currency. And, you will change the date format from the default DD-MON-YYYY format to MM/DD/YYYY.

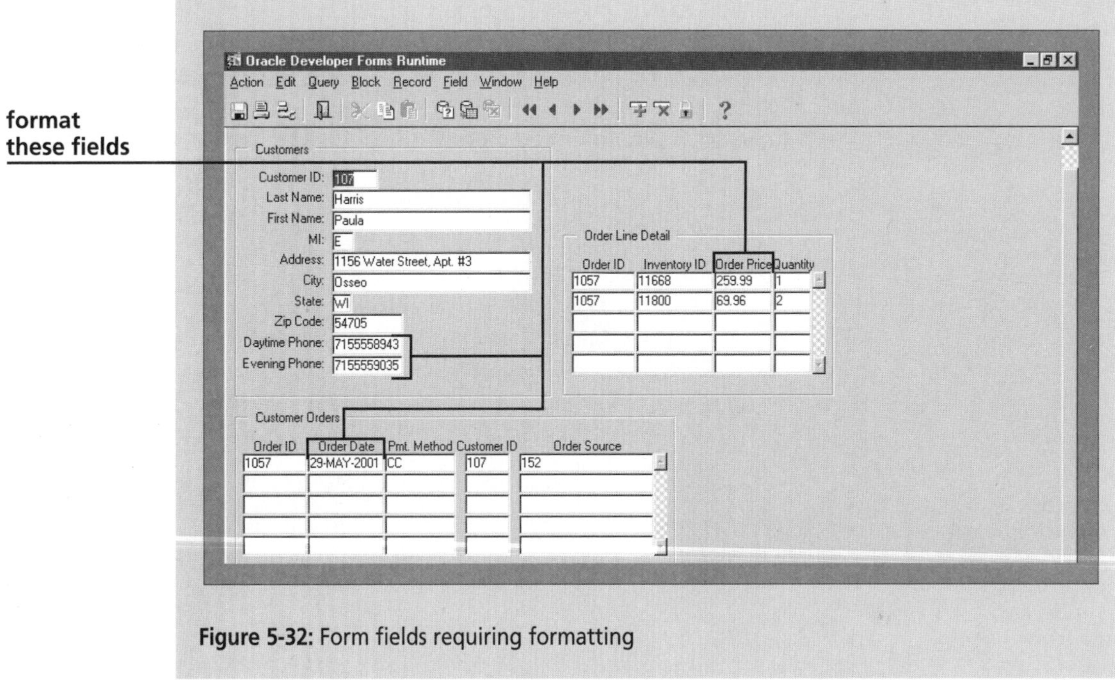

Figure 5-32: Form fields requiring formatting

You can specify format masks for a form data field by entering the desired format mask in the field's Property Palette.

A list of common Oracle format masks is provided in Chapter 2.

Now you will change the format masks for the customer daytime and evening phone numbers, the order price, and the order date.

To change the format masks:

1. Click **Window** on the menu bar, then click **CH5BCUSTORDER: CUSTOMER_CANVAS (CUSTOMER)** to view the form in the Layout Editor, if necessary. (The name in parentheses indicates the currently selected block, and yours might be different.)

You can also view the form in the Layout Editor by clicking Tools on the menu bar, then clicking Layout Editor.

2 Right-click the **DPHONE** data field, then click **Property Palette**. The item's Property Palette opens.

Recall that you can embed characters in character string data fields by preceding the format mask with the characters FM, and placing embedded characters in quotation marks. For the phone number, you will embed an opening parenthesis ("("), list the first three numbers, embed a closing parenthesis and a blank space (") "), list the next three numbers, embed a hyphen ("-"), then list the last four numbers. Now you will enter the format mask for the phone number.

To enter the phone number format mask:

1 Scroll down to the Data section of the Property Palette, then click in the space next to **Format Mask**. Type the format mask exactly as shown in Figure 5-33. There are three embedded character strings in the format mask: an opening parenthesis, a closing parenthesis followed by a space, and a hyphen. Close the DPHONE Property Palette.

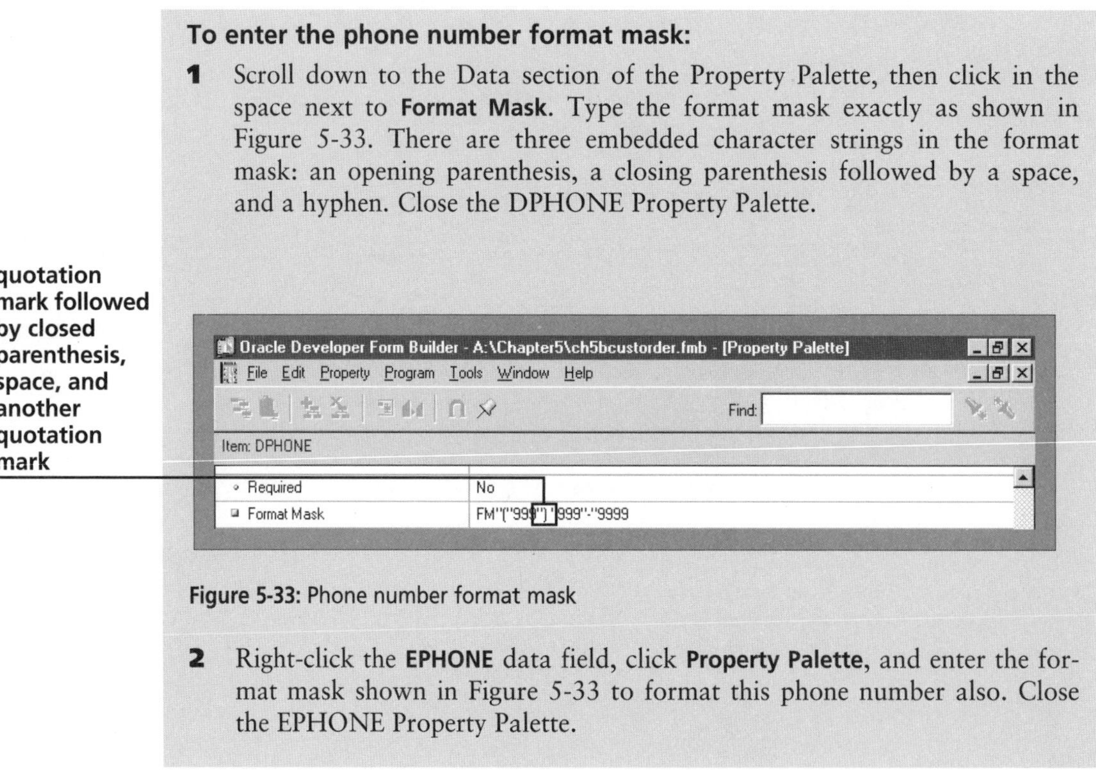

Figure 5-33: Phone number format mask

2 Right-click the **EPHONE** data field, click **Property Palette**, and enter the format mask shown in Figure 5-33 to format this phone number also. Close the EPHONE Property Palette.

Now you will format the ORDER_PRICE field so it displays as currency. To do this, you will preface the format mask with a dollar sign ($), and then enter the digits and embed commas and the decimal point as desired. Now you will format the ORDER_PRICE field.

To format a field as currency:

1 Right-click **ORDER_PRICE** in the Layout Editor. Selection handles appear around all of the ORDER_PRICE fields. Each displayed record in a tabular form shares a common Property Palette. Right-click, then click **Property Palette.**

2 Click in the space beside **Format Mask**, and enter the following format mask: **$99,999.99.** Close the Property Palette.

Now you will format the ORDERDATE field. Recall that front slashes, hyphens, and colons can be used as separators between different date and time elements in DATE data fields. For this format mask, you will embed front slashes between the month and the day fields.

To enter the date format mask:

1 Right-click the **ORDERDATE** fields, and click **Property Palette.**

2 Enter the following format mask: **MM/DD/YYYY.** Close the Property Palette.

After you have entered format masks for data fields, you must always make sure that the displayed fields are wide enough to display the data as well as extra formatting characters, such as slashes and hyphens. Currently, the DPHONE and EPHONE fields are just wide enough to display the phone numbers' digits. When the embedded formatting characters are included, the fields will not be wide enough. You will need to change two properties in the phone number fields' Property Palettes: the Data Width, which determines how much data can be displayed, and the Visible Width, which determines how wide the displayed field is on the form. You can change these properties for both fields in one step, by selecting both fields at the same time and formatting them using a group Property Palette. Group Property Palettes only work on items of the same type. When all items in the group have the same value for a property, the property value is displayed. When the items have different values, the property value is displayed as "*****" in the group Property Palette.

If a field's Data Width is not wide enough for a format mask, the field will be displayed as pound signs (####). If a field's Visible Width is not wide enough, the data will be clipped.

To change the phone number field widths using a group Property Palette:

1 Select **DPHONE**, press the **Shift** key, then select **EPHONE**. Both phone number fields are selected.

2 Right-click, then click **Property Palette**. Note that the name field is displayed as *****, which indicates that this is a group Property Palette. Scroll down to the Data section, and find the Maximum Length, which currently has a value of 10. You are adding four additional formatting characters, so the maximum length must be at least 14. Since it is a good idea to always add extra characters so you are sure your data will be displayed, you will change the width to 16.

3 Change the Maximum Length value to **16**.

4 Scroll down to the Physical section, and change the Width to **70**. Close the Property Palette.

Now you will save the form and run it to view the formatted fields. Then you will exit Form Builder.

To save and run the form:

1 Click the **Save** button.

2 Click the **Run Form Client/Server** button, click the **Enter Query** button, then click the **Execute Query** button to retrieve all customer records. Your formatted fields should look like Figure 5-34.

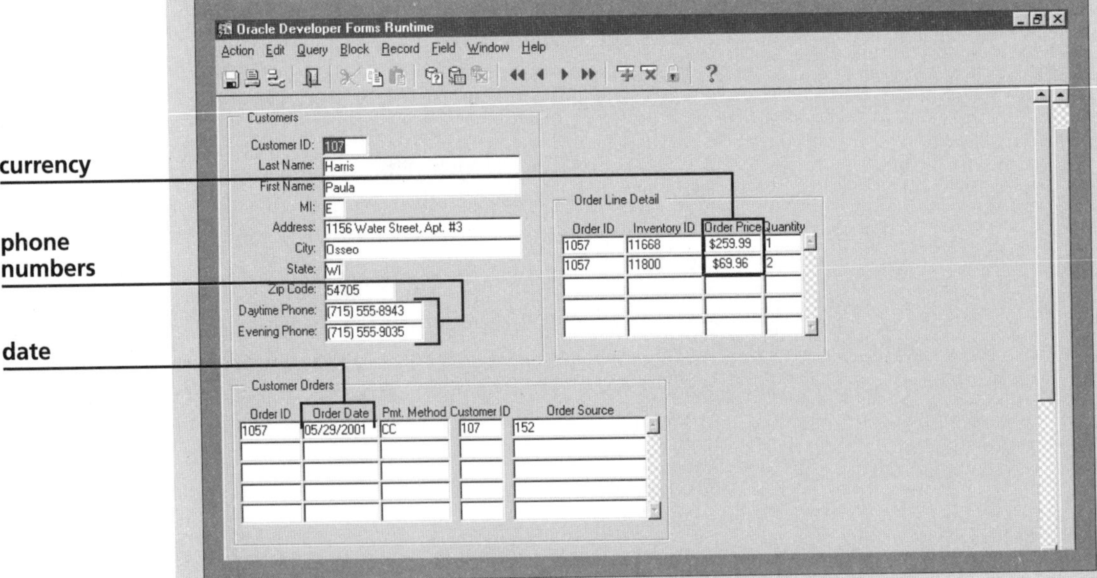

Figure 5-34: Formatted form fields

3 Exit Forms Runtime.

4 Exit Form Builder and all other open Oracle applications.

SUMMARY

- A tabular data block form displays multiple records on the same form.
- When you create a data block using a view, you can view data, but you cannot modify the database by inserting new records or updating or deleting existing records.
- You should not allow users to maximize or resize a form window, because the form objects might appear off-center or clipped off.
- Every form object has a Property Palette that lists properties specific to the object type.
- The form window width will vary depending on the display setting of your computer.
- In a master-detail relationship, one database record can have multiple related records through foreign key relationships. A master record can have many detail records, but a detail record always has only one associated master record.
- A block can be the master block in multiple master-detail relationships.
- A block can be the detail block in one master-detail relationship, and the master block in a second master-detail relationship with a different block.
- When you create a master-detail data block relationship, you always create the master block first, and then specify the master-detail relationship when you create the detail block.
- When you create a master-detail relationship, checking the Auto-join data blocks check box instructs the system to automatically create the join condition between the field in the selected master block that has a foreign key constraint and a field in the current block. You also can create the join condition manually.
- You can add, update, delete, and query database records in master-detail blocks just as in single-table data blocks, except that you need to first place the insertion point in the data block before performing the insert, update, or query.
- You can specify format masks for a form data field by entering the desired format mask in the field's Property Palette.
- After you have entered a format mask for a data field, you must always make sure that the field is wide enough to display the data as well as extra formatting characters. If it is not wide enough, you need to adjust the field's Data Width and Visible Width properties.

REVIEW QUESTIONS

1. What is the difference between a form layout style and a tabular layout style in a form layout?
2. When do you need to include a scroll bar in a tabular layout style?

3. Why would you create a form using a database view?
4. True or false: You can delete records from the database when using a form created from a view.
5. Why is it important to change default window and canvas names to descriptive names in a form?
6. When do you need to include horizontal and vertical scroll bars in a Form window?
7. How do you determine the width property of a Form window so that the window fills the Forms Runtime window?
8. When creating a master-detail relationship, how do you decide which block is the master block and which block is the detail block? Which block do you create first?
9. For the following data fields, indicate the required values for format mask and data width to get the desired formatted data:

Variable Name	Data Type	Data Value	Desired Format
CUSTSSN	VARCHAR2(9)	888993452	888-99-3452
APPTTIME	DATE	12-DEC-2001 10:00 AM	10:00 AM
ACCTBALANCE	NUMBER(6,2)	-1,221.56	<$1,221.56>

PROBLEM-SOLVING CASES

Run the northwoo.sql script to refresh the Northwoods University database, and run the clearwat.sql script to refresh the Clearwater Traders database. The script files are stored in the Chapter5 folder on your Student Disk.

1. Create a form that displays all of the fields in the LOCATION table in the Northwoods University database. The form should display five records at a time and use a scroll bar to display additional records. Create descriptive field prompts, and title the frame "Northwoods University Locations." Rename the window, canvas, and frame using descriptive names, and change the window and canvas properties, using the guidelines described in the lesson. Save the file as Ch5bEx1.fmb.
2. In this exercise you will create a master-detail form that displays records from the FACULTY and COURSE_SECTION tables.
   a. Create a master block that displays all of the fields from the FACULTY table, one record at a time. Create descriptive field prompts, and use "Faculty" as the frame title. Format the phone number field with parentheses and hyphens so that it is easier to read; for example, Kim Cox's phone number is displayed as (715) 555-1234.
   b. Create a related detail block that displays all of the related fields from the COURSE_SECTION table for the current faculty member. Show five records at one time, and use a scroll bar to display additional records. Create descriptive field prompts. Format the COURSE_SECTION TIME field so it displays as "10:00 AM."
   c. Rename the window, canvas, and frame using descriptive names, and change the window and canvas properties, using the guidelines described in the lesson.
   d. Save the file as Ch5bEx2.fmb.

3. In this exercise, you will create a master-detail form where the detail block is created from a database view.

   a. Create a master block that displays all of the fields from the STUDENT table, one record at a time. Create descriptive field prompts. Format the phone number field so Sarah Miller's phone number is displayed as 715-555-9876.

   b. Create a view named COURSEENRLVIEW that displays CSECID, SID, CALLID, CNAME, CCREDIT, and GRADE for all course section records.

   c. Create a detail block that displays five COURSEENRLVIEW records for the current student at one time, and use a scroll bar to display additional records. Create descriptive field prompts. (*Hint*: To display the view records, you will need to make it the current block, and then click .)

   d. Rename the window, canvas, and frame using descriptive names, and change the window and canvas properties, using the guidelines described in the lesson.

   e. Save the file as Ch5bEx3.fmb.

4. In this exercise, you will create a multiple-table form that shows courses, course sections, and their associated students.

   a. Create a master block that shows all records in the COURSE table. Show five records at one time, and use a scroll bar to display additional records. Create descriptive field prompts.

   b. Create a detail block that shows all records in the COURSE_SECTION table for the selected course. Show three records at one time, and use a scroll bar to display additional records. Create descriptive field prompts.

   c. Create a view named STUDENTENRLVIEW that displays SID, SLNAME, SFNAME, CSECID, and GRADE for all students.

   d. Create a detail block that displays all fields from STUDENTENRLVIEW for a selected course section. Show three records at one time, and use a scroll bar to display additional records. Create descriptive field prompts. (*Hint*: COURSE_SECTION will be the master block.)

   e. Rename the window, canvas, and frame using descriptive names, and change the window and canvas properties, using the guidelines described in the lesson.

   f. Save the file as Ch5bEx4.fmb.

LESSON C

objectives

- Use sequences to automatically generate primary key values in a form
- Create single-table and multiple-table lists of values (LOVs) to provide lists for foreign key values
- Create radio button groups to display data values
- Add a graphic image to a form

Using Sequences in Forms to Generate Primary Key Values

Whenever you insert a new record into the database using a form, you have to type in the value for the record's primary key. This could be a source of errors if two users use the same primary key value for different records, and it is inconvenient for the user to always have to query the database or look up the next primary key value on a printed report. Recall that in Chapter 3, you learned how to create sequences to generate surrogate key values for primary key fields automatically. In this lesson, you are going to create a form for inserting and viewing records in the INVENTORY table in the Clearwater Traders database that automatically retrieves the next value in a sequence and inserts it into the Inventory ID field on the form. First, you will run the script that refreshes the Clearwater Traders database tables. Then, you will create the sequence to generate the INVID values that will be used in the form. Since the highest current INVID value is 11848, you will make the sequence start at value 11900.

To run the script and create the sequence:

1 Start SQL*Plus and log on to the database.

2 Run the clearwat.sql script that is in the Chapter5 folder on your Student Disk.

3 Type the code shown in Figure 5-35 to create the INVID_SEQUENCE.

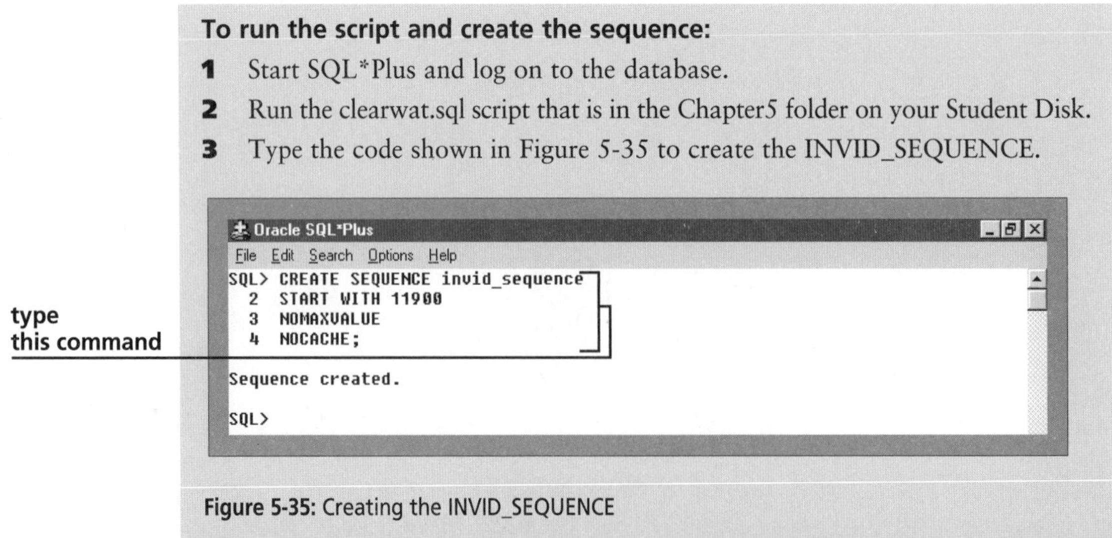

Oracle SQL*Plus
File Edit Search Options Help
SQL> CREATE SEQUENCE invid_sequence
2 START WITH 11900
3 NOMAXVALUE
4 NOCACHE;
Sequence created.
SQL>
type this command

Figure 5-35: Creating the INVID_SEQUENCE

Now you will start Form Builder, and create the form's data block and layout. The data block will use all of the INVENTORY table fields, and will use a form layout style and display one record at a time.

To create the form:

1. Start Form Builder.
2. Use the Data Block Wizard to create a new data block using all of the INVENTORY table fields. Change the field prompts so your finished form will look like Figure 5-36.
3. Use a form-style layout, and change the frame title property to **Clearwater Traders Inventory**.
4. Change the canvas name to **INVENTORY_CANVAS**, the window name to **INVENTORY_WINDOW**, and the frame name to **INVENTORY_FRAME**.
5. Change the window and canvas widths so they are the correct size to fit the Forms Runtime window. Change the window title to **Clearwater Traders Inventory**, and set the window's Resize Allowed and Maximum Allowed properties to **No**.
6. Format the CURR_PRICE field as currency using the format mask **$9,999.99**.
7. Save the file as **ch5cinventory.fmb** in the Chapter5 folder on your Student Disk, then run the form. Your formatted form should look like Figure 5-36. Close Forms Runtime.

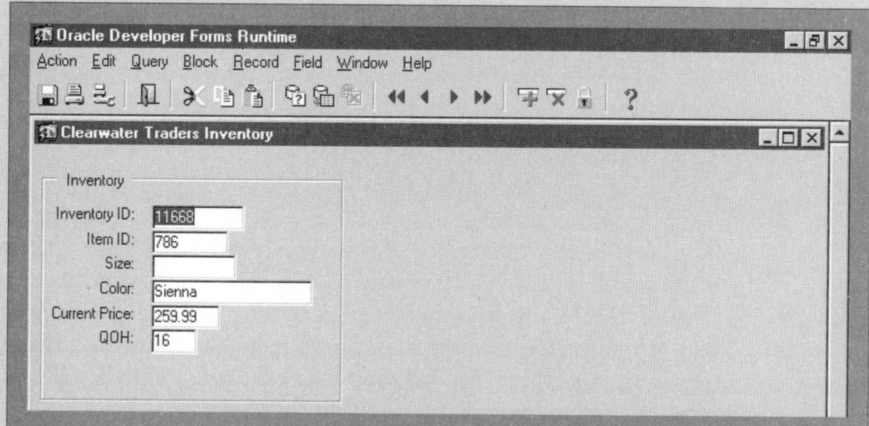

Figure 5-36: Formatted Inventory form

help

If your form fields are not stacked on top of each other, and are displayed in two columns on the form, select the frame and make it narrower. The form fields will automatically adjust to the positions shown in Figure 5-36.

Creating Form Triggers

To automatically insert the next INVID_SEQUENCE value into the Inventory ID form field, you will create a PL/SQL program unit called a trigger. **Triggers** are program units in forms that run when a user performs an action such as clicking a button. Triggers also can be associated with specific actions that are performed by the system, such as loading the form or exiting the form. Triggers are associated with specific form objects, such as data fields, buttons, data blocks, or the form itself. For example, when you click the Save button on the Forms Runtime toolbar, a trigger containing a SQL INSERT command runs. The trigger's form object was the Save button, and the action was clicking the button. To write a trigger, you must specify the object that the trigger is associated with, and the action, or **event**, that starts the trigger. Every form object has specific events that can be used to start triggers. For example, buttons have an event called WHEN-BUTTON-PRESSED that starts a trigger just before a user clicks the button. Forms have an event called PRE-FORM that starts a trigger just before the form is first started.

Now you will create a trigger that automatically inserts the next sequence value into the Inventory ID field on the INVENTORY form whenever the user inserts a new record. This trigger is associated with the INVENTORY block, and is started by a block event called WHEN-CREATE-RECORD. This block event occurs whenever a new blank record is created in a data block.

To create the trigger:

1 Open the Object Navigator window in Ownership View, click + beside DataBlocks, click + beside INVENTORY, and then select **Triggers** under the INVENTORY data block.

2 Click the **Create** button to create a new trigger. The CH5CINVENTORY: Triggers dialog box opens.

The Triggers dialog box shows all of the block events that can have associated triggers. An object can have several associated events, so the Find text box allows you to enter an approximate search string, helping you to quickly find the event that you want to associate with the trigger. You want to attach the trigger code to the WHEN-CREATE-RECORD event, so you will do an approximate search using the string "when" to help find this event.

To find the trigger event:

1. Click the mouse pointer just before the **%** in the Find text box, type **when**, and then press the **Enter** key. This automatically scrolls the list box down to the beginning of the events that start with the word *when*.
2. Select the **WHEN-CREATE-RECORD** event, then click **OK**. The PL/SQL Editor opens, as shown in Figure 5-37. If necessary, click ☐ to maximize the window.

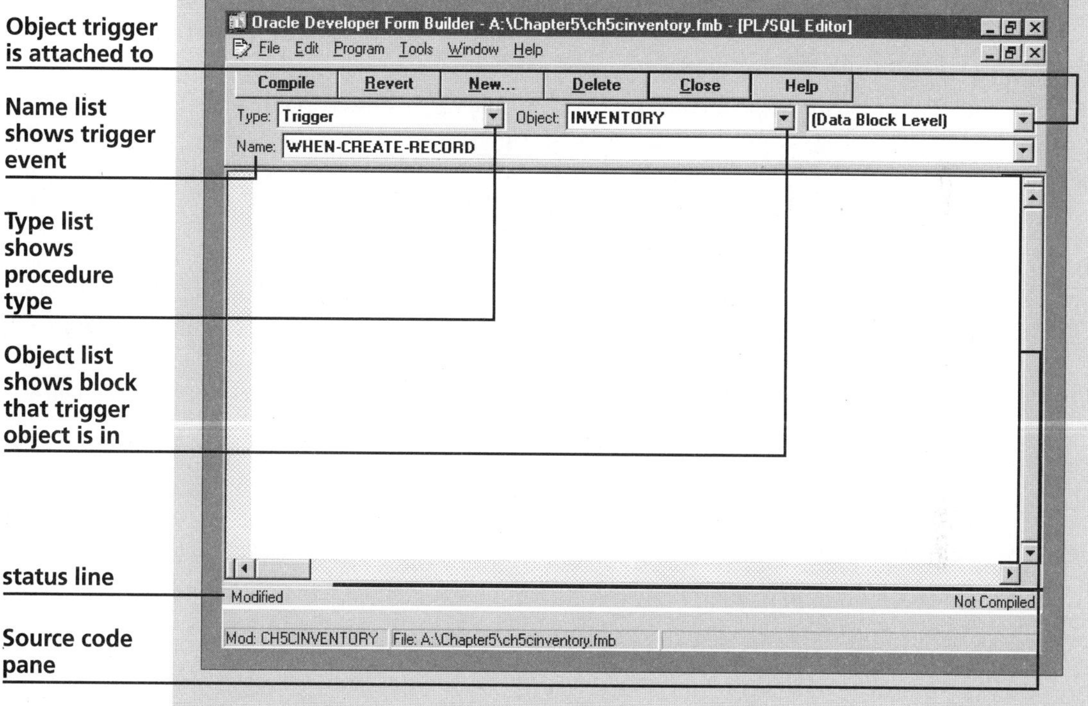

Figure 5-37: PL/SQL Editor

The PL/SQL Editor is similar to the editor you used in the Procedure Builder in Chapter 4, except that it has been customized to run in the Form Builder envronment. The **Type list** shows the procedure type, which is a trigger. (Procedures also can be **program units**, which are subprograms that are called by triggers.) The **Object list** displays the name of the form data block that the trigger is in, and the list beside the Object list shows the name of the form object the trigger is attached to. The **Name list** shows the event associated with the trigger. The **Source code pane** is where you type the PL/SQL program statements. The **status line** shows the trigger's current modification status (Modified or Not Modified) and compile status (Not Compiled, Successfully Compiled, or Compiled with Errors).

The PL/SQL Editor button bar has the following buttons:

- **Compile**, which compiles the program statements in the Source code pane. The compiler detects syntax errors and references to database tables, fields, and form objects that do not exist or are referenced incorrectly.
- **Revert**, which reverts the source code to its status prior to the last compile or since the last time the Revert button was clicked
- **New**, which creates a new procedure
- **Delete**, which deletes the current procedure
- **Close**, which closes the PL/SQL Editor
- **Help**, which allows the user to access the Help System while working in the PL/SQL Editor

Now you will enter the code for the WHEN-CREATE-RECORD trigger. This trigger needs to retrieve the next value in the INVID_SEQUENCE, and insert it into the INVID item on the form. To do this, you will use an implicit cursor, which can be used to retrieve a single query value and assign the result to a variable. You will select the next value of the sequence and display it in the form inventory ID field.

Form data items can be referenced in a PL/SQL program using the following general format **:<block name>.<item name>**. The block name and item name must be the same as the names that are shown in the Object Navigator, and the block name is always preceded with a colon (:). In this form, the block name is INVENTORY, and the item name is INVID. Therefore, this item will be referenced as **:inventory.invid**. Now you will enter the PL/SQL code for the WHEN-CREATE-RECORD trigger.

To enter the WHEN-CREATE-RECORD trigger code:

1 Click anywhere in the Source code pane in the PL/SQL Editor, and then type the code shown in Figure 5-38. Note that PL/SQL and SQL commands are displayed in blue text, and user-entered variables are displayed in black text. This makes the code easier to understand and debug.

type this code

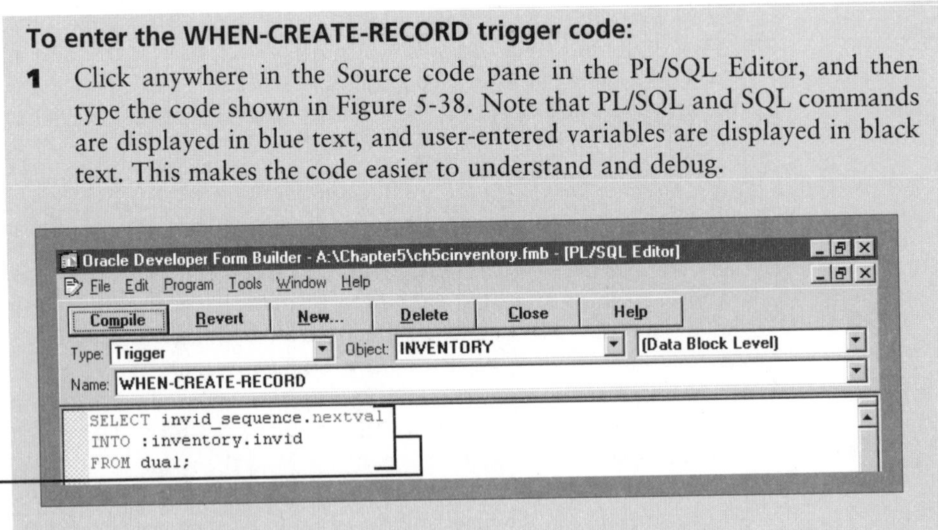

Figure 5-38: Trigger code

Like other PL/SQL programs, triggers must be **compiled**, or converted to executable code. When a trigger is compiled, it is checked for syntax errors and for references to database tables, or fields, or form objects that are not specified correctly or do not exist. Form Builder automatically compiles all form triggers when you generate a form to create the executable (.fmx) file, but it is better to compile each trigger just after you enter the source code, so you can find and correct errors immediately. When an error is detected, the program displays the line number of the statement causing the error, with an error description. Next you will compile the WHEN-CREATE-RECORD trigger.

To compile the trigger:

1. Click the **Compile** button on the PL/SQL Editor button bar. If the trigger compiles successfully, the "Successfully Compiled" message appears on the status bar.
2. If your trigger compiles successfully, click **Close** on the PL/SQL button bar to close the PL/SQL Editor. If it does not compile successfully, you will need to debug it. Debugging form triggers is similar to debugging other PL/SQL programs, except that there are some different kinds of errors that you might encounter. The next section will discuss common errors seen in form triggers.
3. Save the form.

Debugging Form Triggers

Figure 5-39 shows an example of a compile error generated when compiling the WHEN-CREATE-RECORD trigger. As in the Procedure Builder, the **Compilation messages pane** displays the line numbers and error messages. The selected error is the error message that is selected in the Compilation messages pane. The insertion point shows the location for the selected error in the Source code pane.

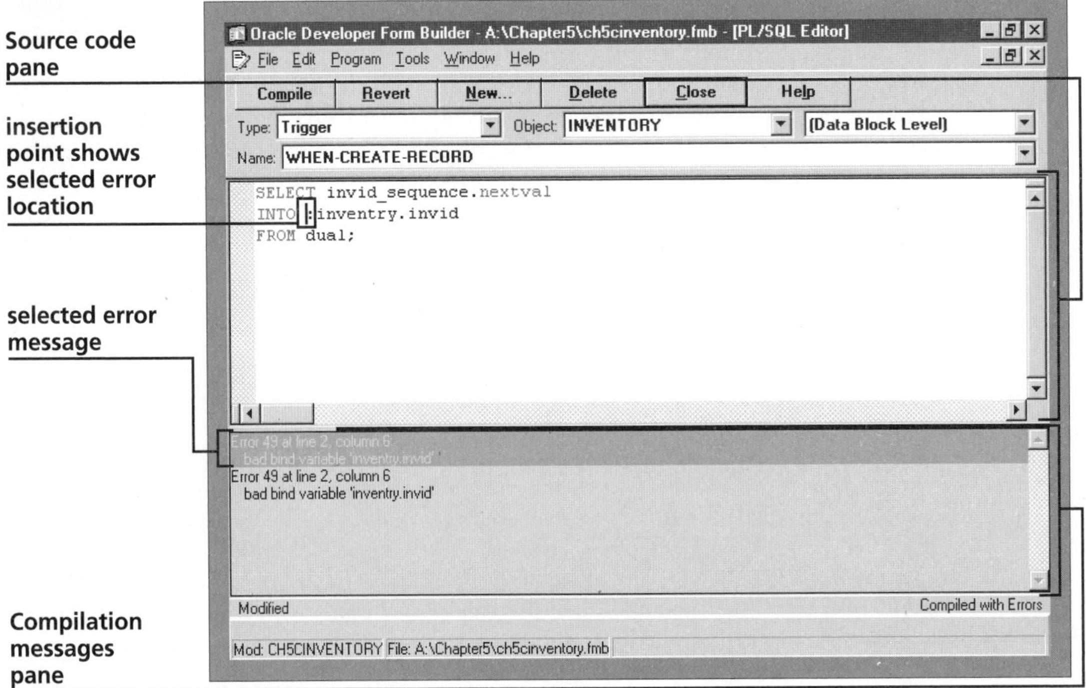

Figure 5-39: Trigger compile error generated by referring to a nonexistent form object

The error shown in Figure 5-39 is a common compile error. The error message "bad bind variable 'inventry.invid'" indicates that the indicated form item does not exist. The probable cause is that the block name or text item name was entered incorrectly. In this case, the block name is INVENTORY, not INVENTRY. To correct the error, you would correct the block name, and then recompile the trigger.

Another common compile error message, shown in Figure 5-40, is "identifier 'DUAL' (or some other table name used in the trigger) must be declared." This error (or one naming a database table other than DUAL) is generated when you are not connected to the database at the time you compile the trigger. This happens when you start Form Builder, open an existing form, revise a trigger, and then try to compile it. Since you are not explicitly prompted to connect to the database, you might forget to connect before compiling. The solution is to click Connect on the File menu, connect to the database in the usual way, and then recompile the trigger.

The "user must have access" error message will be generated if you are not connected to the database when you compile a trigger with a SELECT command for *any* database table, not just DUAL.

A common compile error message, shown in Figure 5-40, is "identifier 'DUAL' (or some other table name used in the trigger) must be declared." This error (or one naming a database table other than DUAL) is generated when you are not connected to the database at the time you compile the trigger. This happens when you start Form Builder, open an existin form, revise a trigger, and then try to compile it. Since you are not explicitly prompted to connect to the database, you might forgat to connect before compiling. The solution is to click Connect on the File menu, connect to the database in the usual way, and then recomplie the trigger.

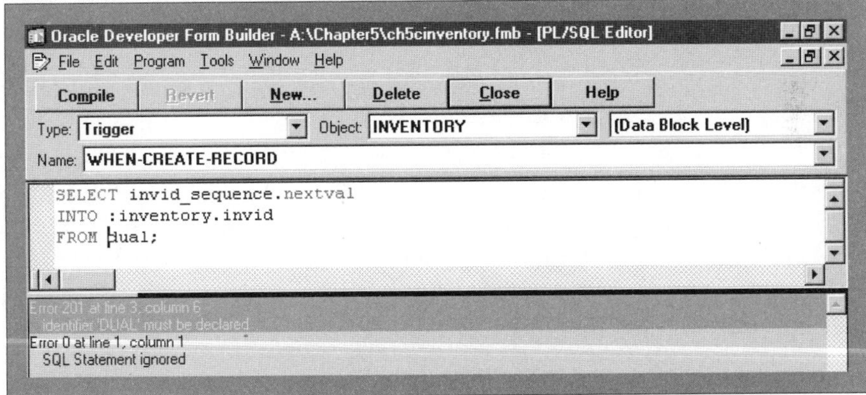

Figure 5-40: Compile error generated when the user is not connected to the database

Figure 5-41 shows another common error message. The error message "identifier 'INV_ID_SEQUENCE.NEXTVAL' must be declared" indicates that the compiler cannot identify the given variable name. The variable is a database object, and the compiler cannot find it in any of the user's database objects, which include tables, fields, and sequences. If the user checked the names of his sequences using SQL*Plus, he would find that the name of the sequence is INVID_SEQUENCE.NEXTVAL, not INV_ID_SEQUENCE.NEXTVAL. Entering the sequence name correctly and then recompiling the trigger corrects this error.

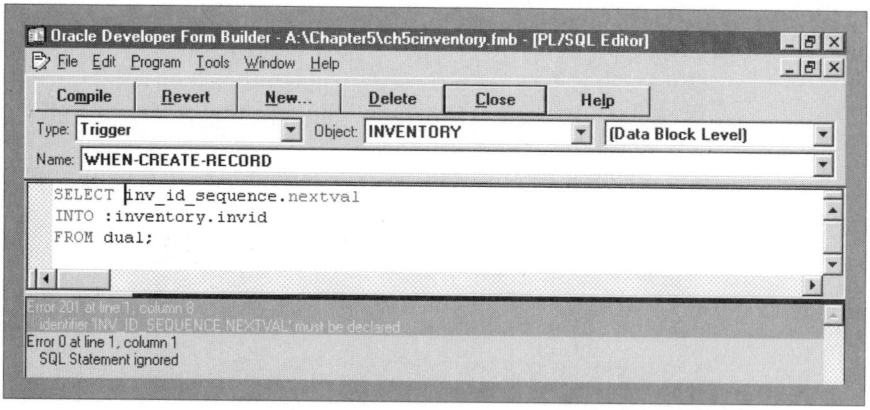

Oracle Developer Form Builder - A:\Chapter5\ch5cinventory.fmb - [PL/SQL Editor]

File Edit Program Tools Window Help

Compile Revert New... Delete Close Help

Type: Trigger Object: INVENTORY (Data Block Level)

Name: WHEN-CREATE-RECORD

```
SELECT inv_id_sequence.nextval
INTO :inventory.invid
FROM dual;
```

Error 201 at line 1, column 8
identifier 'INV_ID_SEQUENCE.NEXTVAL' must be declared
Error 0 at line 1, column 1
SQL Statement ignored

Figure 5-41: Compile error generated by referring to a nonexistent database object

Errors often occur when a SQL command is not correct. If you suspect that this is the case, copy the SQL command into a text editor, and then remove the INTO command used in the trigger PL/SQL statement to place the query output into the form fields. Paste the edited SQL query into SQL*Plus and see if the query output is correct. Figure 5-42 shows the SELECT command from Figure 5-41 running in SQL*Plus with its INTO command commented out. The SQL*Plus error message immediately identifies that the error occurred because the sequence name does not exist.

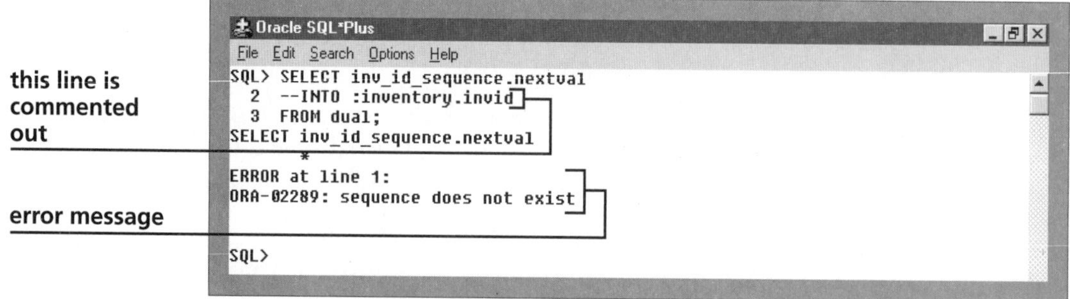

Figure 5-42: Testing a trigger SQL command in SQL*Plus

There are many more possible compile errors, but these are some of the most common ones. The best way to locate errors is to carefully examine the source code line that contains the error and make sure that the form objects and database objects are specified correctly and that the PL/SQL syntax is correct. Test the SQL commands in SQL*Plus to make sure that they are correct.

Testing the Trigger

After you successfully compile the trigger, the next step is to test it to be sure that it is working correctly. It is a good idea to run the form each time you create a new trigger or add new code to an existing trigger, so that if there are errors, you know exactly what code caused them. Now you will run the form to test the trigger.

To test the trigger:

1 If necessary, correct any syntax errors in your trigger, and then recompile the code until it compiles successfully.

2 If necessary, click **Close** on the PL/SQL Editor button bar to close the PL/SQL Editor. The WHEN-RECORD-CREATED trigger is displayed under the INVENTORY block.

3 Click the **Run Form Client/Server** button to run the form. Maximize the Forms Runtime window if necessary. The form is displayed with the next sequence number inserted into the Inventory ID field.

Now you will enter the rest of the data fields and insert a new record to confirm that the form still inserts new records correctly.346

To insert a new record:

1 Type the new INVENTORY item data fields as follows:

Inv. ID: **11900**

Item ID: **559**

Size: **XXL**

Color: **Navy**

Price: **$209.95**

QOH: **50**

2 Click the **Save** button. The confirmation message is displayed, confirming that the record is successfully inserted.

3 Click the **Insert Record** button to insert another record. A new blank record appears, with the next sequence value inserted into the Inventory ID field.

4 Click the **Remove Record** button to remove the blank record.

5 Close Forms Runtime.

Creating a List of Values (LOV)

Currently when you insert a new record into the INVENTORY table using the INVENTORY form, you must type data values for foreign keys such as ITEMID and COLOR. To make data entry easier and avoid errors, it is a good practice to guide the user to select from a list of allowable values for data fields that contain foreign key values or restricted data values. (For example, allowable values for ITEMID are 894, 897, 995, 559, and 786.) To do this, you will create a **list of values (LOV)** (pronounced lahv). A LOV displays a list of data values when the user clicks a button, called the **LOV command button**, that is beside the data field that has restricted values. The user can select one of the data records, and the selected record values will be inserted into one or more form data fields. An example of a LOV display is shown on the Inventory form in Figure 5-43. When the user clicks the LOV command button, the **LOV display** shows a list of records returned by a query. The user can do approximate searches using the Find text box. When the user selects a record and clicks OK, one or more values from the selected record are returned to specified form fields.

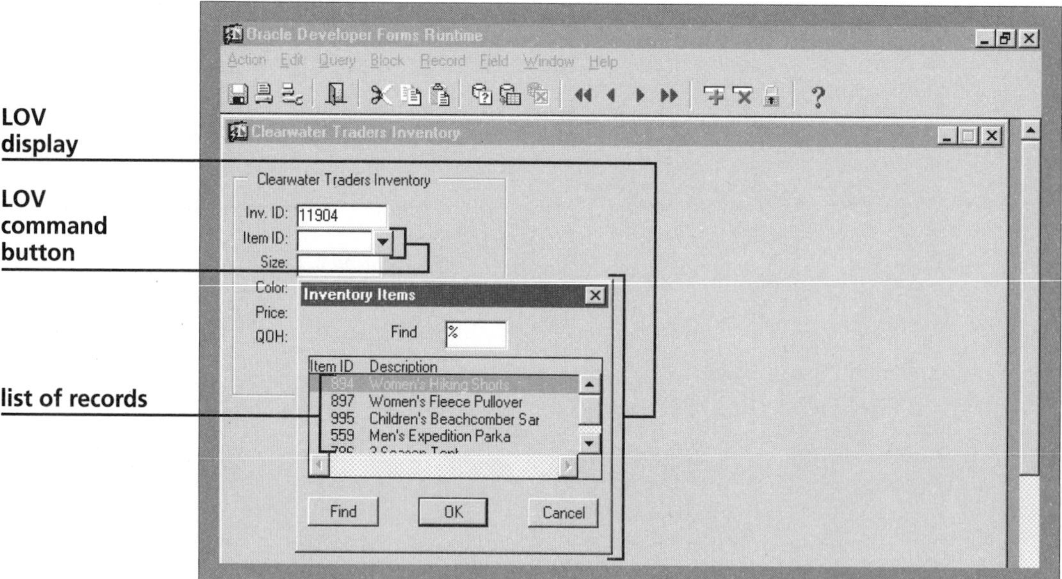

Figure 5-43: LOV display on form

Creating a LOV is a five-step process:

1. Specify the LOV display values. When you create a new LOV, you enter a SQL query that returns the data values that will be displayed in the list. This query creates a new **record group**, which is a group of records returned by a SQL query. You can also specify for the LOV to use an existing record group.

2. Format the LOV display. This involves changing the column titles and widths, specifying which display fields are returned to form data fields, modifying the LOV display title, and changing the position of the LOV display on the screen.
3. Attach the LOV to a data field. When the form insertion point is in this field and the LOV command button is clicked, the LOV display appears. Usually, this is the data field that is beside the LOV command button.
4. Change the names of the new LOV and record group to something more descriptive than the default names.
5. Create a command button to activate the LOV display.

Using the LOV Wizard

Now you will create the LOV shown in Figure 5-43. The first three steps can be done using the LOV Wizard. Now you will use the LOV Wizard to specify the LOV display values, format the display, and attach the LOV to a data field.

To use the LOV Wizard:

1 Click **Tools** on the menu bar, then click **LOV Wizard**. The LOV Source page is displayed, which allows you to create a new record group or use an existing record group. Since there are no existing record groups in the form, this option is not available. Click **Next**. The SQL query page is displayed.

2 The SQL query page allows you to type a SQL query in the SQL Query statement box, build a SQL query button using Query Builder, or import a query from a text file. Type the query shown in Figure 5-44 to retrieve the ITEMID and ITEMDESC for the records in the ITEM table. *Do not* type a semicolon(;) at the end of the query. Click **Next**. The Column Selection page is displayed.

type this query

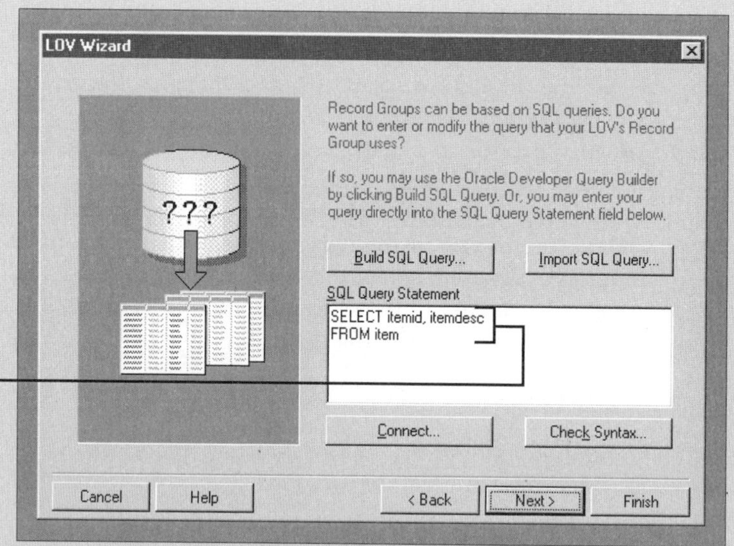

Figure 5-44: LOV Wizard SQL query page

help

If you received an error message, it means you did not enter the SQL query correctly. Here are some debugging tips:

- Immediately copy the query text into a text editor by clicking and dragging over the query text to select it, and then pressing Ctrl + C.
- Errors usually are caused by writing the query incorrectly. If you cannot spot the error visually, copy the query from the text editor and paste it into SQL*Plus and run and debug it.
- Do not type a semicolon (;) at the end of the query.

3 The Column Selection page specifies which record group columns will be displayed in the LOV display. All of the retrieved record group fields will be displayed in the LOV display, so click the **Select All** button ›› so all of the record group columns are displayed as LOV columns.

4 Click **Next.** The Column Display page appears. Click the scroll bar to view all of the display column properties.

The Column Display page enables you to format the LOV display. The Title column shows the column title that will appear for each column in the LOV display and the Width column shows the width of each column in the LOV display. By default, the LOV display fields have the same column widths and titles as the corresponding database fields, which might result in column titles that are hard to understand and column widths that are too wide for the display screen.

The Return Value column specifies the form field (preceded by its block name and a period) where the user's LOV selection will be transferred for the current column. For example, if the user selects Women's Hiking Shorts in the LOV display and clicks OK, the Item ID for Women's Hiking Shorts (894) will be transferred to the ITEMID field on the form. On the column display page, the form block and field name (INVENTORY.ITEMID) will be entered as the return value for the ITEMID column. In this LOV, the selected ITEMDESC value will not be returned to the form. Now you will specify the LOV display column titles, and specify the Return value.

To specify the LOV Wizard Display Column page:

1. Modify the column titles, as follows:

Title	Width
Item ID:	**108**
Description:	**455**

2. Click the mouse pointer in the Return value space next to Item ID, then click **Look up return item**. The Items and Parameters dialog box is displayed, which is a list of all the items in the INVENTORY block. Each item is preceded by the block name and a period (INVENTORY.). These are the form fields where the selected item ID values can be returned.
3. Click **INVENTORY.ITEMID**. This is the field where the selected item ID will be displayed after the user makes a selection from the LOV display.
4. Click **OK**. INVENTORY.ITEMID is displayed as the Return value for the Item ID field on the Column Display page. Click **Next**. The LOV Display page is displayed.
5. The LOV Display page is used to specify the title, size, and position of the LOV display. Type **Clearwater Traders Items** in the Title box, and accept the default values for Width and Height. Be sure the Yes, Let Forms position the LOV automatically radio button is selected. Your completed LOV Display page should like Figure 5-45.

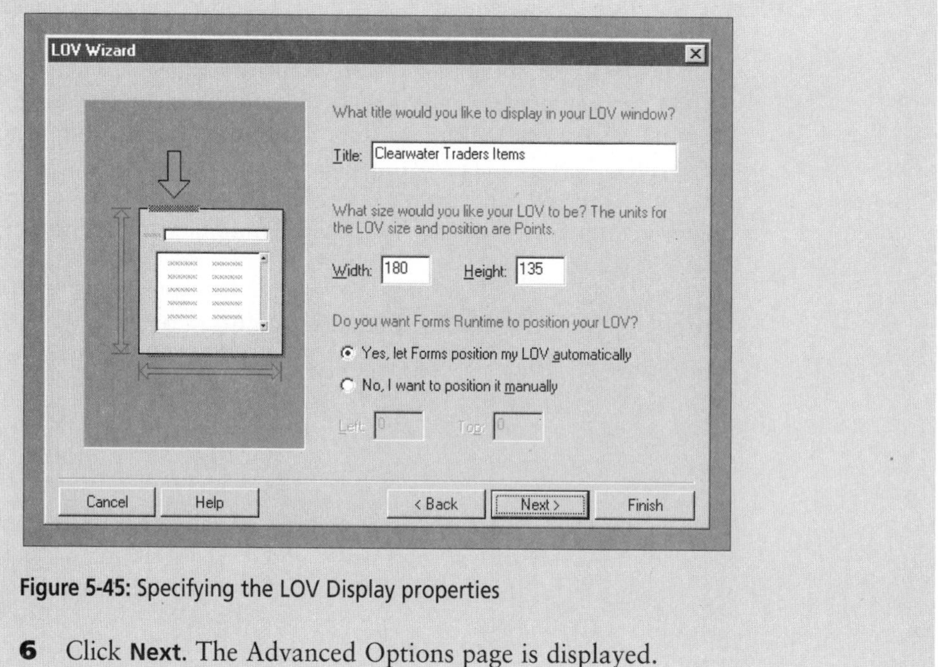

Figure 5-45: Specifying the LOV Display properties

6 Click **Next**. The Advanced Options page is displayed.

The Advanced Options page is used to specify how many records are retrieved by the LOV. For a LOV that might return hundreds or thousands of records, it is desirable to limit the number of records that are retrieved to improve system response time, so the default value for number of records to retrieve is 20. The first check box on the page (Refresh record group data before displaying LOV) allows the developer to specify whether the record values are refreshed each time the LOV is displayed. If this check box is cleared, the data are retrieved only once from the database when the user first displays the LOV, and the data values are stored. When the user displays the LOV subsequent times, she is presented with the stored data values. This would be a desirable option for a LOV that retrieves many data values, and the values don't change very often. The second check box (Let the user filter records before displaying them) allows the user to type a search condition in the Find text box on the LOV display before any records are retrieved. This will reduce the number of records that are retrieved in the LOV display, and improve system response time for a LOV that retrieves many records. Since your LOV display will retrieve only the five records in the ITEM table, you will accept the LOV defaults (retrieve 20 records, refresh the data each time the LOV is displayed, and do not let the user filter the data). Now you will complete the LOV Advanced Options page, and finish creating the LOV.

To complete the LOV Advanced Options page, and finish the LOV:

1. Accept the default selections on the Advanced Options page, then click **Next**. The Items page is displayed. Recall that a LOV is attached to a form data field, and when the form insertion point is in this field and the LOV command button is clicked, then the LOV display appears. The field the LOV is attached to is always one of the LOV return values.
2. Select **INVENTORY.ITEMID** in the Return Items list if necessary, then click the **Select** button [>] to place it in the Assigned Items list.
3. Click **Next**, then click **Finish** to finish the LOV.
4. Click **Window** on the menu bar, then click **Object Navigator** if necessary, to display the new LOV and record group in the Object Navigator.
5. Save the form.

LOVs are only displayed in the Object Navigator Ownership View. They are not displayed in the Object Navigator in Visual View.

After you create the LOV using the LOV Wizard, two new items appear in the Object Navigator Screen under LOVs and Record groups. The LOV item specifies the LOV display properties, and the record group item specifies the data that the LOV displays. Now you need to give these new items descriptive names.

To change the names of the LOV and record group:

1. Rename both the LOV and record group to ITEM_LOV, as shown in Figure 5-46.

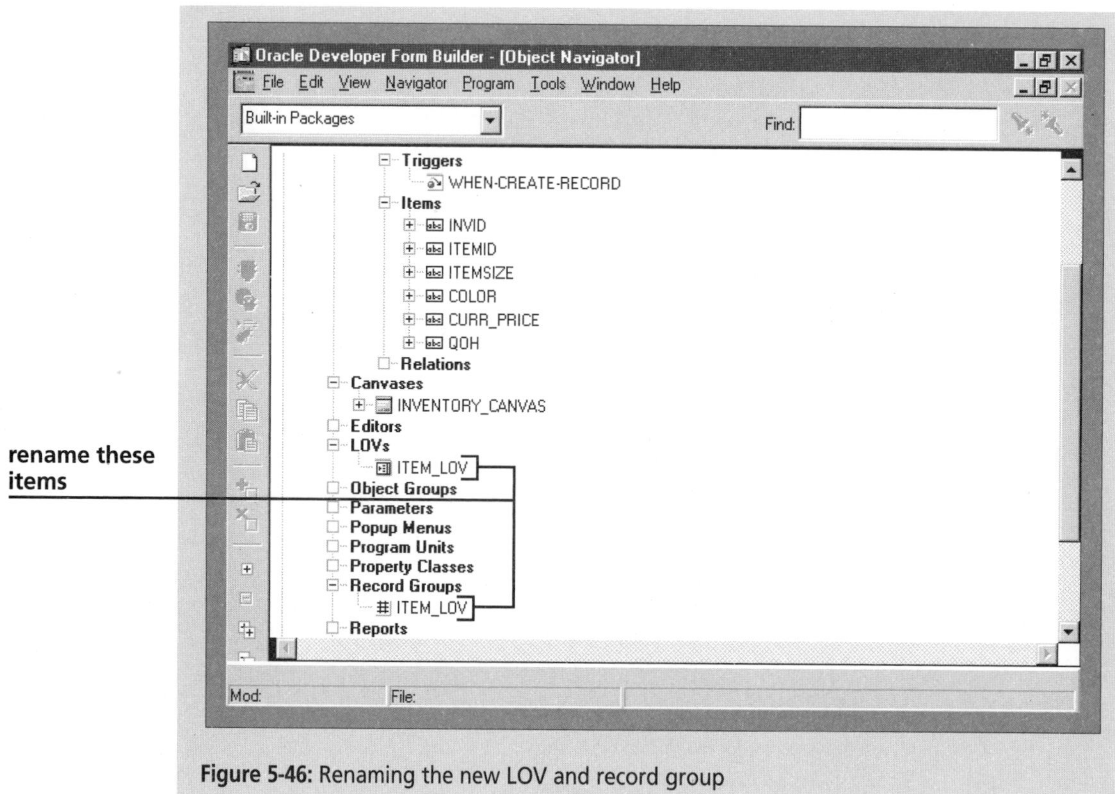

Figure 5-46: Renaming the new LOV and record group

Creating the LOV Command Button

The next step is to create the **LOV command button**, which is a button that the user clicks to open the LOV display. This will be a small square button positioned on the right edge of the ITEMID data field that displays a down arrow symbol (like a Windows list arrow). Buttons that display pictures are called **iconic buttons,** because they display bitmapped images called **icons.** To create the LOV command button, you will open the form in the Layout Editor, and draw a command button on the canvas using the Button tool in the Layout Editor tool palette. Then you will open the button's Property Palette, resize the button, change the button's Iconic property to True, and change its Icon Name property to the name of the icon file that you want to display on the button.

To specify an icon file, you must enter the complete path (including drive letter and folder path) to the icon file. Or, you can store the icon file in the default icon folder on your client workstation and omit the path information. Since you might want to run the form on many different workstations, it is a good idea to store the icon in the default icon folder and omit the absolute path information. The default icon folder is ORAWIN95\TOOLS\DEVDEMO60\BIN\ICON (or ORANT\TOOLS\DEVDEMO60\BIN\ICON). If you reference an icon that is stored in the default icon folder, you do not need to include the path or extension on the filename in the Icon Name property.

An icon file is a file with an .ico extension that contains a special kind of bitmap image file that is sized so it can be displayed as an icon. Icon images are usually 16 pixels wide by 16 pixels long, or 32 pixels wide by 32 pixels long. To create a new icon image or convert a regular bitmap image to an .ico image, you must have a special graphics application for creating .ico files. This book does not cover creating or editing icons.

To create the LOV command button:

1 Click **Tools** on the menu bar, then click **Layout Editor** to open the Layout Editor window.

2 Click the **Button** tool on the Tool palette, and draw a small square button positioned just to the right of the ITEMID data field, as shown in Figure 5-47.

draw this button

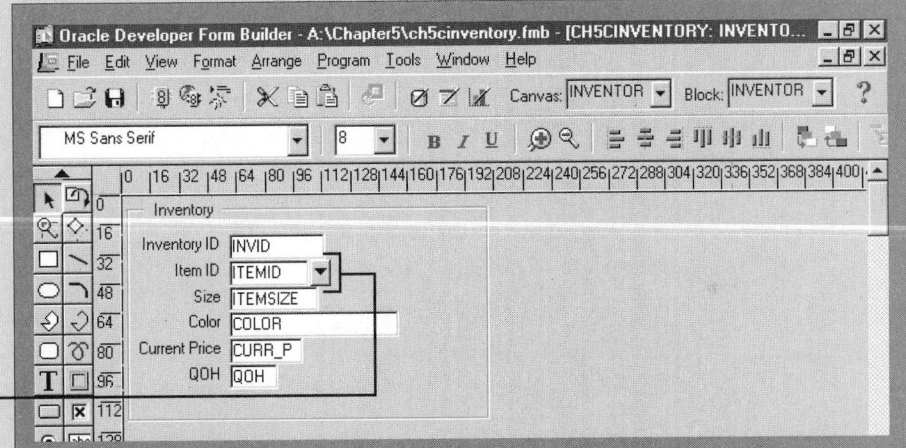

Figure 5-47: Drawing the LOV command button

help

Sometimes when you try to draw a small button, Form Builder will automatically transform it to a default-sized larger button. If this happens, resize the button to match Figure 5-47.

3 Select the button, right-click, then click **Property Palette.** The new command button's Property Palette opens. Change the button's properties as follows:

Name	**ITEM_LOV_BUTTON**
Label	(Delete the default label—it will have an icon instead)
Iconic	**Yes**

Icon Filename **a:\Chapter5\down3**
Width **12**
Height **14**

4. Close the ITEM_LOV_BUTTON Property Palette. The icon should appear on your button.

help If the icon does not appear, verify that you have changed the button's Iconic property to Yes, and entered a:\Chapter5\down3 for the Icon Filename property. Verify that the down3.ico file is in the Chapter5 folder on your Student Disk.

To complete the LOV command button, you must create the trigger, or program code, that displays the LOV when the user clicks the LOV command button. This trigger will be associated with the button's WHEN-BUTTON-PRESSED event. The code will first put the form insertion point into the field the LOV is attached to (ITEMID), using the GO_ITEM command. The general format of this command is: `GO_ITEM ('<form data field name>')`. The form data field name is in single quotes. The trigger will then issue the `LIST_VALUES` command, which instructs the system to display the LOV that is attached to the text field that contains the form insertion point.

To create the LOV command button trigger:

1. Right-click the LOV command button you just created, point to **Smart Triggers**, then click **WHEN-BUTTON-PRESSED** to create a new trigger.
2. Type the following code in the Source code pane:

```
GO_ITEM ('itemid');
LIST_VALUES;
```

3. Compile the trigger, correct any errors, then close the PL/SQL Editor.
4. Save the form.

Testing the LOV

Next, you will test the form to verify that the LOV display appears and that the selected record's item ID is placed in the ITEMID form field.

To test the LOV:

1 Run the form.

2 Click the **LOV command button** to display the LOV. The LOV display should appear, as shown in Figure 5-48. The column widths of the LOV display will need to be reformatted, but you will do that later. First, you will confirm that the selected data value is returned to the form field correctly.

columns widths need to be reformatted

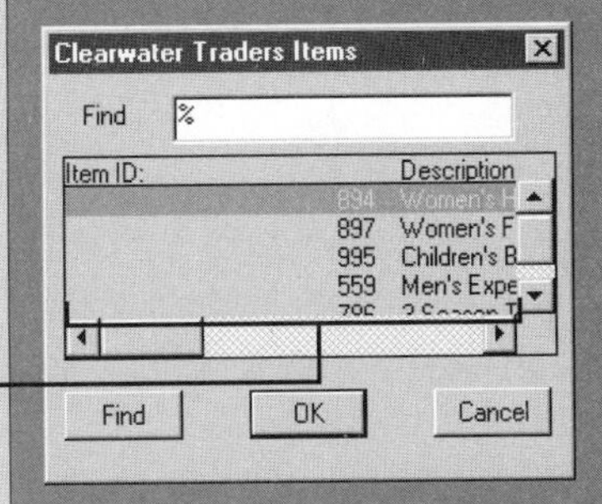

Figure 5-48: Viewing the LOV display

3 Make sure Item ID **894** (Women's Hiking Shorts) is selected, then click **OK**. Item ID 894 is displayed in the Item ID field on the form.

4 Close Forms Runtime.

help

If you click the Cancel button on the LOV display, no value is returned for Item ID, and the message "FRM-40202: Field must be entered" appears on the status line. This is because you are inserting a new record, and the Item ID field is a required (NULL not allowed) field in the INVENTORY table. To display the LOV again, you must click [X] to remove the current record and insert a new blank record.

Reformatting the LOV Display

There are a few final formatting details that you need to adjust so that the LOV display is more attractive and functional. Currently, the Item ID field is too wide for the data values, and the Description field is clipped off, because the overall LOV display is not wide enough to display both fields. Also, some of the records do not appear, because the LOV display is not quite tall enough. Now, you will reformat the LOV display to shorten the width of the Item ID field, widen the Description field, and make the LOV display taller. To do this, you will select the LOV in the Object Navigator, and then enter the LOV Wizard in re-entrant mode.

To modify the format of the LOV display:

1. Open the Object Navigator window, and select the LOV named ITEM_LOV.
2. Click **Tools** on the menu bar, then click the **LOV Wizard**. The LOV Wizard opens in re-entrant mode, with tabs across the top of the page to allow you to access each of the Wizard pages.
3. Click the **Column Display** tab, and change the width of the Item ID column to **15**. Click **Apply** to save the change, but leave the LOV Wizard open.
4. Click the **LOV DISPLAY** tab, and change the LOV display Width to **200**, and the Height to **150**, then click **Finish** to save your changes and close the LOV Wizard.
5. Save the form, run the form, and display the LOV. The formatted display should look like Figure 5-49.

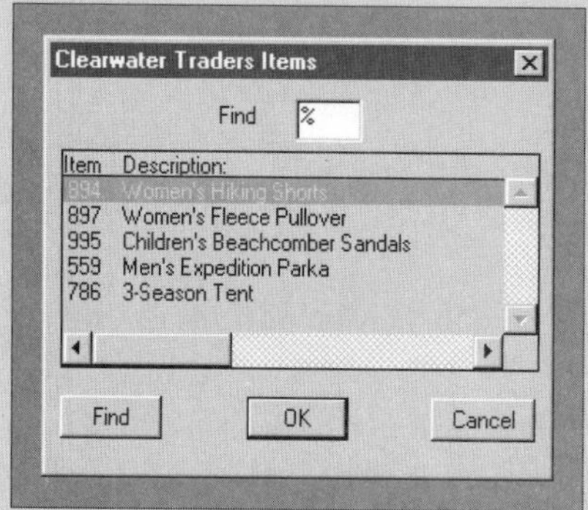

Figure 5-49: Formatted LOV display

6. Close Forms Runtime.
7. Close the ch5cinventory.fmb form in Form Builder.

Displaying Data Using a Radio Button Group

When a data field has a limited number of possible values, you can use radio buttons to enter and display data. **Radio buttons**, also called option buttons, limit the user to only one of two or more related choices. In Form Builder, individual radio buttons exist within a **radio button group**. Each individual radio button has an associated data value, and the radio button group has the data value of the currently selected radio button.

When you use the Layout Wizard to create a form layout, you can select the type of form item that is used to represent each data field. So far, you have used text items, which display data values as text. Now you will create a form that displays a data field using a radio button group. You will make a form for inserting, updating, viewing, and deleting records in the ITEM table of the Clearwater Traders database. In this table, the CATEGORY field currently is limited to four values: Women's Clothing, Men's Clothing, Children's Clothing, and Outdoor Gear. You will display the CATEGORY field using a radio button group. First, you will create a new form and data block associated with the ITEM table. Then, you will create the form layout and specify that the CATEGORY field will be displayed in a radio button group.

To create the new form, data block, and layout:

1. Create a new form, and save the form as **ch5citem.fmb** in the Chapter5 folder on your Student Disk.
2. Use the Data Block Wizard to create a new data block associated with the ITEM table that includes all ITEM data fields.
3. Use the Layout Wizard to create a form layout. Accept the (New Canvas) and Content canvas default options, then click **Next**.
4. On the Layout Wizard Data Block page, select all data block items for display.
5. Click the mouse pointer on the **CATEGORY** item in the Displayed Items list to select it by itself. The Item Type list box is activated, as shown in Figure 5-50. Open the list to view the list choices.

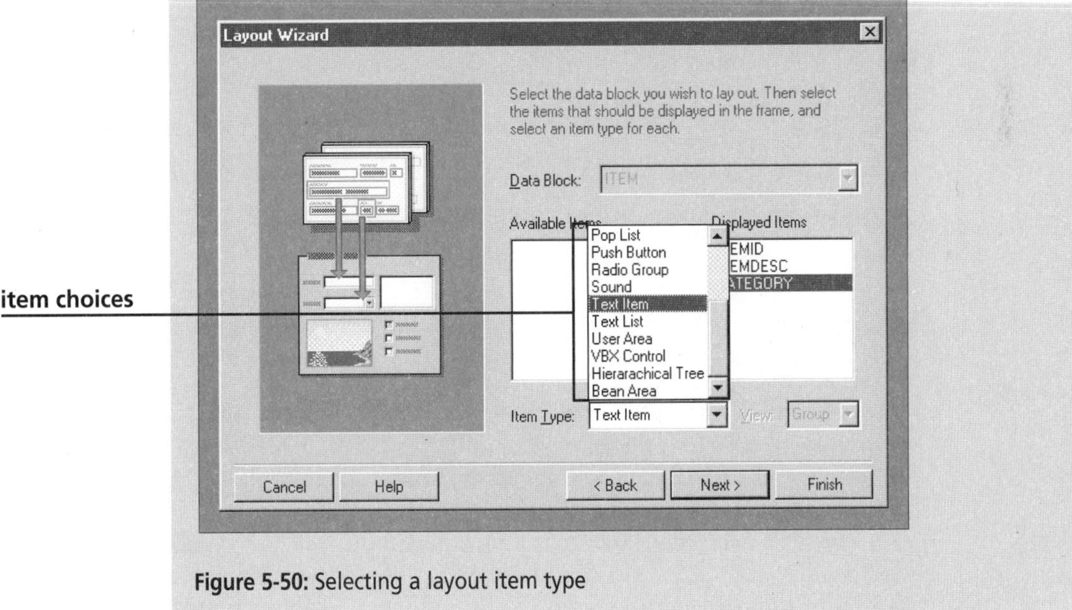

Figure 5-50: Selecting a layout item type

The default item type is Text Item, as shown in Figure 5-50. Data fields can be displayed using a variety of item types, including charts, check boxes (for data that have only two possible values), images, and lists. Now you will select Radio Group for the CATEGORY item type, and finish creating the layout.

To finish the layout:

1. Select **Radio Group** as the item type, then click **Next**.
2. Modify the layout prompts and widths as follows, then click **Next**.

Name	Prompt	Width
ITEMID	**Item ID:**	**45**
ITEMDESC	**Description:**	**150**
CATEGORY	**Category:**	**117**

3. Make sure that the **Form** option is selected for the layout style, then click **Next**.
4. Enter **Clearwater Traders Items** for the frame title, leave Records Displayed as 1, click **Next**, then click **Finish**.

   The layout is displayed in the Layout Editor, as shown in Figure 5-51. However, no radio buttons are displayed, and nothing is displayed for the CATEGORY data field.

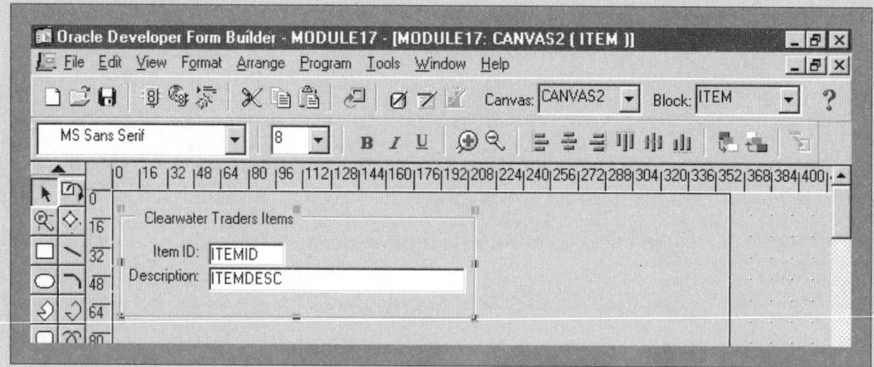

Figure 5-51: Current form layout (radio buttons are not displayed)

5. Open the Object Navigator window, and click [+] beside Items under the ITEM data block.

The Object Navigator display for the data block items shows an icon in front of each item. The **Text item** icon indicates that ITEMID and ITEMDESC data are displayed as text items, and the **Radio button** icon shows that CATEGORY data are displayed as a radio button group.

You will need to create four individual radio buttons to represent the four data values that can be used in the CATEGORY data field. To do this, you will draw the radio buttons on the canvas using the **Radio Button** tool on the Layout Editor tool palette. First, you will format the form window and canvas. Then you will make the data block frame larger to accommodate the radio buttons. You will also change the frame's Update Layout property. Currently, the frame automatically positions the block items within the frame. You will change the Update Layout property to manual so that you can better control the item positions. And, you will create the first radio button.

To modify the window, canvas, and frame, and create the first radio button:

1. Format the form window so it cannot be maximized or resized, and so it displays horizontal and vertical scroll bars. Resize the window so it fills the Forms Runtime window. Change the window name to **ITEM_WINDOW**, and the window title to **Clearwater Traders.**
2. Resize the canvas so it is the same size as the window. Change the canvas name to **ITEM_CANVAS.**
3. Select the frame, open its Property Palette, click the Update Layout property, and change it to **Manually**. Close the Property Palette.
4. Open Layout Editor window, select the Clearwater Traders Items frame, and drag the lower-right selection handle to the right to make the frame wider and longer, as shown in Figure 5-52.

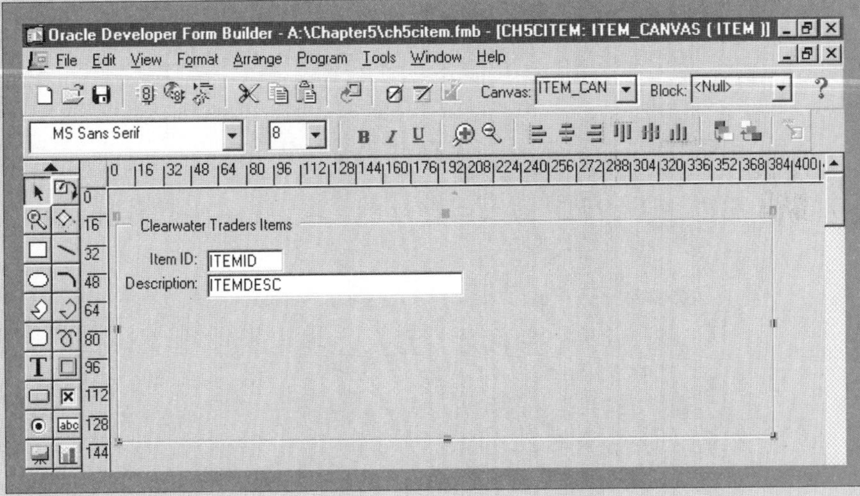

Figure 5-52: Enlarging the layout frame

5. Click the **Radio Button** tool on the Tool palette, and draw a rectangle on the canvas to correspond with the first radio button, as shown in Figure 5-53. After you release the mouse button, the Radio Groups dialog box is displayed, asking which radio group to place the radio button in. This dialog box allows you to place a new radio button in an existing radio group, or to create a new radio group. You will place the new radio button in the CATEGORY radio button group that was created when you made the form layout.

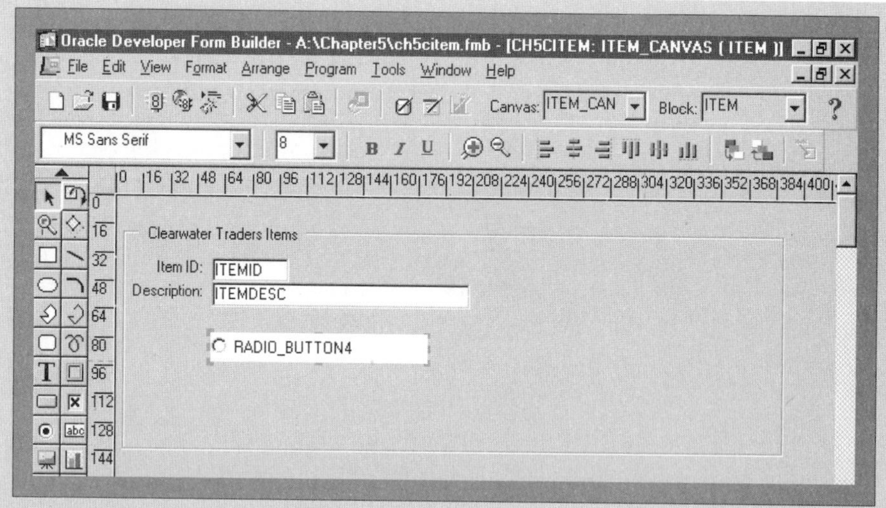

Figure 5-53: Drawing the first radio button

6 Make sure the **CATEGORY** radio group is selected, then click **OK**. The radio button and its corresponding label are displayed, as shown in Figure 5-53.

Now you will change some of the radio button's properties. You will specify the button's **Name**, which is how it is referenced within the form, and its **Label**, which is the description that appears beside the button on the canvas. And, you must specify the button's **Radio Button Value**, which corresponds to the actual data value stored in the CATEGORY field that the radio button represents.

To change the radio button properties:

1 Double-click the new radio button to open its Property Palette, and then change the following properties. When you are finished, close the Property Palette and maximize the Layout Editor if necessary.

Property	New Value
Name	**WC_RADIO_BUTTON**
Label	**Women's Clothing**
Radio Button Value	**Women's Clothing**

The Value property is the actual database data value that is associated with the radio button—in this case it happens to be the same as the label, but this is not always the case. For example, if you create a radio group to display the METHPMT values in the CUST_ORDER table, the actual data values stored in the database are CC (for credit card) and CHECK. The corresponding radio button labels are "Credit Card" and "Check."

Now you need to change the background color of the radio button so that it is the same color as the form. To do this, you will use the **Fill Color** tool. This tool allows you to change the fill color of any form item. The tool is displayed at the bottom of the tool palette. Depending on your screen display setting, it might not currently be displayed. If it is not displayed, you will need to scroll down on the Tool palette to find it.

To change the radio button fill color:

1. If the Fill Color tool is not displayed, click the down arrow on the Tool palette to scroll down.
2. Select **WC_RADIO_BUTTON**, then click. The Fill Color palette opens. Select a color to match the form background.

You cannot change the Fill Color property of a radio button to No Fill. Instead, you need to match its fill color to the background color of the canvas.

Now that the Women's Clothing radio button is finished, you can easily create the rest of the buttons in the radio group by copying this button, pasting the copy onto the canvas, and changing the individual properties of each button to correspond to each of the CATEGORY data values. First, you will copy the radio button and paste it three times, so that you have a total of four radio buttons.

To copy and paste the radio button:

1. Select **WC_RADIO_BUTTON**, and click the **Copy** button on the Layout Editor toolbar to make a copy of the radio button.

You also can press Ctrl + C to copy and Ctrl + V to paste, or click Copy or Paste on the Edit menu on the menu bar.

2. Click the **Paste** button on the Layout Editor toolbar, make sure **CATEGORY** is the selected radio group, then click **OK** to paste the copied radio button to the canvas. The new radio button is pasted on top of the first radio button.

3 Select the newly pasted radio button, and drag it down so it is positioned below the first radio button.

4 Click [paste icon], select the **CATEGORY** radio group and click **OK,** then drag the new button down so it is below the current radio buttons. Repeat this process one more time to create one more radio button. Your pasted radio buttons should look something like Figure 5-54.

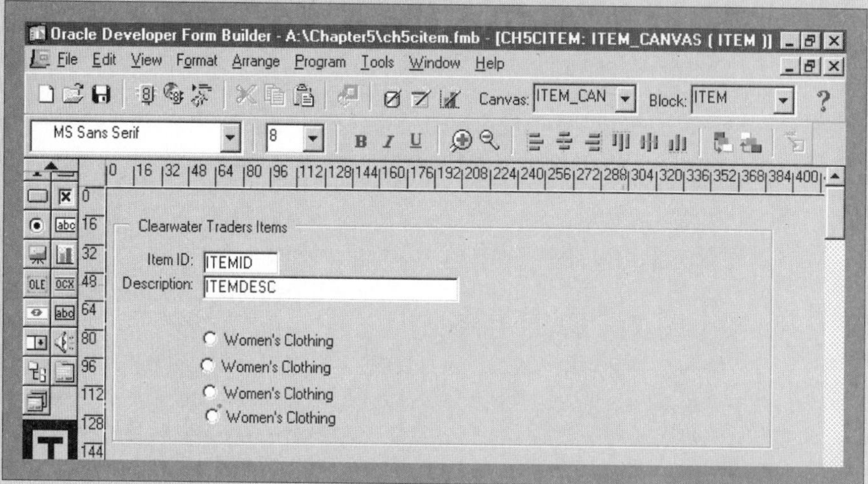

Figure 5-54: Pasting the new radio buttons

Now you will need to reposition the radio buttons so that they are aligned and spaced evenly. To do this, you will select all of the radio buttons into an object group, and then align the left edges and stack them vertically.

To reposition the radio buttons:

1 Select the radio buttons as an object group by clicking the first radio button, pressing and holding down the **Shift** key, clicking the remaining radio buttons, and then releasing the Shift key. Selection handles appear around the object group.

2 With the buttons still selected as an object group, click **Arrange** on the menu bar, click **Align Objects**, and then click the **Horizontally Align Left** and **Vertically Stack** option buttons. Click **OK**. The four radio buttons should now appear with their left edges aligned and stacked on top of each other.

3 Click anywhere on the canvas to deselect the object group.

4 Change the properties of the second, third, and fourth buttons in the group as follows:

Property	Button 2	Button 3	Button 4
Name	**MC_RADIO_BUTTON**	**CC_RADIO_BUTTON**	**OG_RADIO_BUTTON**
Label	**Men's Clothing**	**Children's Clothing**	**Outdoor Gear**
Radio Button Value	**Men's Clothing**	**Children's Clothing**	**Outdoor Gear**

5 When you are finished changing the radio button properties, close the Property Palette. Reposition the radio buttons as necessary so that your canvas looks like Figure 5-55.

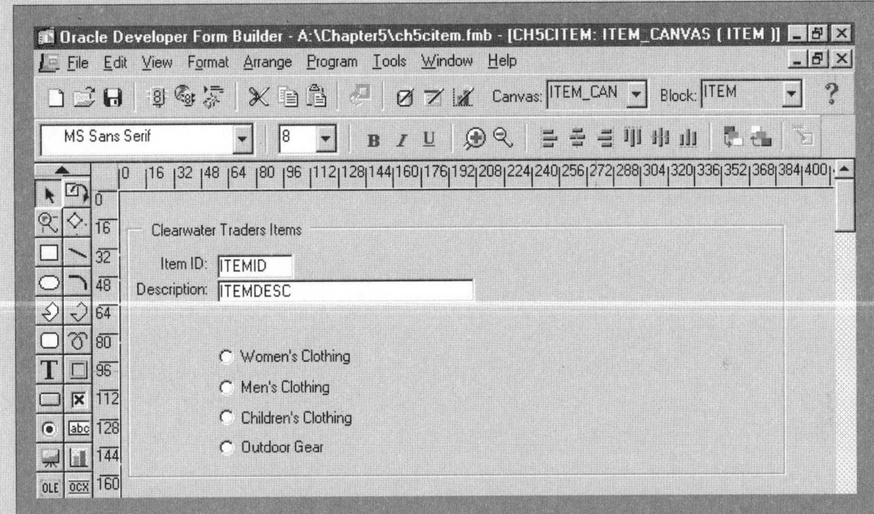

Figure 5-55: Finished radio buttons

6 Save the form.

Next, you will specify the radio group's initial value. The **initial value** of a radio button group is the radio button that is selected when the form is first displayed. To specify the initial value, you must change the Initial Value property on the radio button group's Property Palette so it has the same value of the Radio Button Value of the radio button that you want selected when the form is first displayed. (Recall that the Radio Button Value is the database value that corresponds to the radio button.) For example, in the CATEGORY radio button group, the Radio Button Value for the first radio button is "Women's Clothing." If you want the first radio button to be selected when the form is first displayed, you will specify the Initial Value property of the CATEGORY radio group to be "Women's Clothing" also. Usually, the initial value is the most common choice or the first choice in the radio button group. You will change the initial value of the CATEGORY radio button group to the value of the first button, which is "Women's Clothing."

To change the radio group's initial value:

1. Open the Object Navigator window, right-click the **CATEGORY** radio group, and then click **Property Palette.**
2. Scroll down the Property Palette, and type **Women's Clothing** for the Initial Value property. Close the Property Palette.
3. Save the form.

Next you will format the radio button group to enhance its appearance. You will create a frame around the radio button group.

To create a frame around the radio button group:

1. Open the Layout Editor and click the **Frame** tool on the Tool palette, then draw a frame around the radio buttons.
2. If the background color of the new frame is not the same as the form's background color, select the new frame, click the **Fill Color** tool on the Tool palette, and change the frame's fill color to **No Fill**. Select the frame, click **Format** on the menu bar, select **Bevel**, then click **Inset.**
3. Right-click the frame to open its Property Palette, and change the frame's Name property to **ITEMDESC_FRAME**, and change the Frame Title property to **Category.** Close the Property Palette.
4. Reposition the radio buttons, and adjust the frame size and form objects so your form looks like Figure 5-56. You might need to make your radio buttons narrower so they do not cover the right edge of the frame.
5. Save the form.

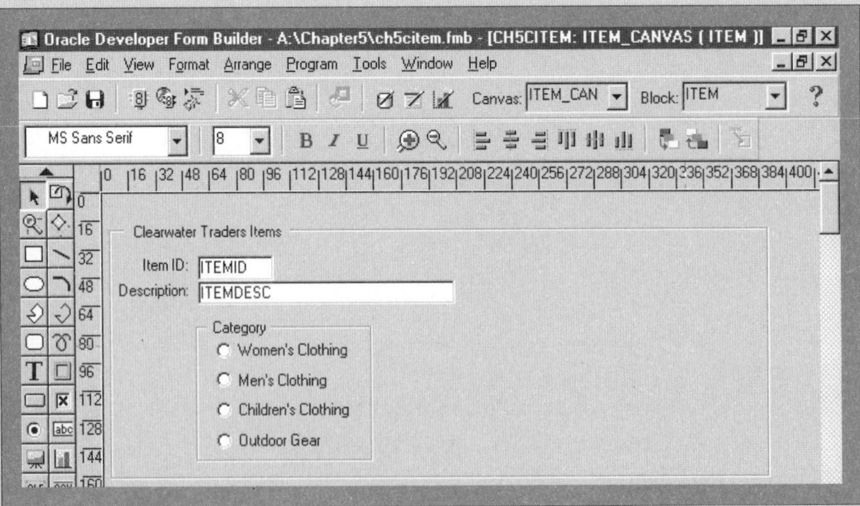

Figure 5-56: Formatted radio button group

Now you will test the form to confirm that the radio buttons display the ITEM data correctly. First you will run the form, and step through all of the table records.

To retrieve all of the ITEM records:

1. Run the form.
2. Click the **Enter Query** button, then click the **Execute Query** button. The first ITEM record (Women's Hiking Shorts) is displayed, with the Women's Clothing radio button selected. Click the **Next Record** button.
3. Click until you have stepped through all of the ITEM records and confirmed that the radio button group displays the correct data value.

Now you will test the form to confirm that you can use the radio button group to insert a new data value. You will do this next.

To insert a new ITEM record using the radio button group:

1. Click the **Insert Record** button to insert a new blank record. Note that the Women's Clothing radio button is selected, because it is the radio button group's initial value.
2. Enter the following data values, then click the **Save** button to save the new record. The save confirmation message is displayed to confirm that the record is inserted correctly.

   Item ID: **900**

   Description: **Denim Overalls**

   Category: **Children's Clothing**

Finally, you will use the radio button group to create a query. You will enter a query with a search condition that returns all items where the CATEGORY value is "Women's Clothing."

To create a query using the radio button group:

1. Click the **Enter Query** button, then click the **Women's Clothing** radio button.
2. Click the **Execute Query** button. The first Women's Clothing item (Women's Hiking Shorts) is displayed.
3. Click the **Next Record** button to display the next returned record (Women's Fleece Pullover).
4. Close Forms Runtime.

Enhancing the Form's Appearance

When you create applications using a graphical user interface, you should always use subtle colors. Intense, saturated colors look less professional than subtle shades, and can cause eyestrain. Always ensure that there is sufficient contrast between the text in data fields and the field background color. So far, you have accepted the default colors for the forms you have created, and in general, this is a good practice. Color should be used sparingly, to give extra visual cues or to enhance the form's appearance in an understated way. Often in database applications, developers use clip art images to enhance the appearance of forms and reports. Now, you will import a graphic image of the Clearwater Traders logo into the ch5citem.fmb form to enhance its appearance.

In a form, there are two ways to display a graphic image: as a **graphic object**, which is incorporated into the form design (.fmb) file and compiled into the .fmx file, or as an **image item**, which is loaded into the form when the user runs the form. The trade-off is that graphic objects make the .fmb and .fmx files larger, while image items make the form run more slowly, and force you to always ensure that the graphic file is available either in the file system or the database when the user runs the form. Since this is a fairly small graphics file (69 KB), you will import the image as a graphic object. (In Chapter 9, you will use the image item approach to create a splash screen.)

The Import Image dialog box, as shown in Figure 5-57, allows you to import an image that is stored either in the file system or in the database. You can specify the **image format**, which specifies the type of image file to be used. The different image formats depend on the graphics application that is used to create the file, and on how the image is compressed to make it occupy less file space. Popular image types include bitmaps (files with a .bmp extension), PC Paintbrush (.pcx), .gif, and .tif files. Bitmap files are usually uncompressed, while .pcx, .gif, and .tif files use different compression methods. Most graphics applications support one or more of these file types. If you select the "Any" option, the Form Builder application will automatically import the image using the correct image type.

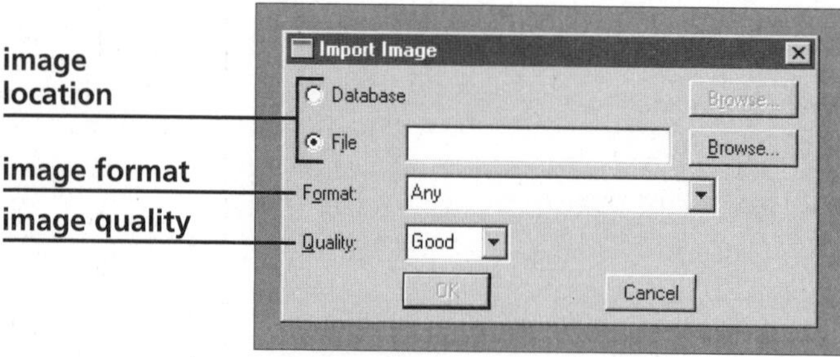

Figure 5-57: Import Image dialog box

The **image quality** property determines how the image is stored in terms of image resolution and number of colors. Possible image quality choices include

Excellent, Very Good, Good, Fair, and Poor. If you import an image using the Excellent quality selection, the image will be saved using the maximum number of colors and highest possible resolution. It will also use the maximum amount of file space, and take the longest time to load. The default image quality is Good, which is satisfactory for most graphic images. You can experiment with the different quality levels to determine the minimum acceptable quality for each image.

Now you will import the graphic image of the Clearwater Traders logo into the form as a graphic object. The image is stored as clearlogo.tif in the Chapter5 folder on your Student Disk. The image will be imported using the default (Good) image quality selection.

To import the image as a graphic object:

1 Click **File** on the menu bar, point to **Import**, then click **Image**. The Import Image dialog box opens, as shown in Figure 5-57.

2 Click **Browse**, select the **clearlogo.tif** file in the Chapter5 folder on your Student Disk, then click **Open**. Make sure that **Any** is the selected image format, select **Good** for the image quality, then click **OK**. The image appears in the upper-left corner of the form.

3 Select the image, and drag it so that it is positioned in the upper-right corner of the frame, as shown in Figure 5-58.

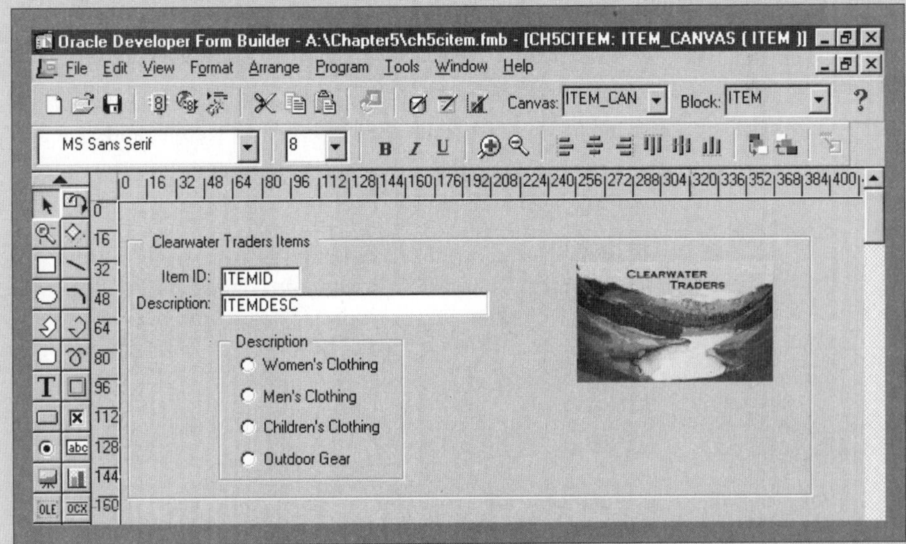

Figure 5-58: Positioning the imported image

4 Run the form to view the graphic image in Forms Runtime.

5 Exit Forms Runtime, then exit Form Builder and all other open Oracle applications.

SUMMARY

- Triggers are program units that run when a user performs an action, such as clicking a button, or when the system performs an action, such as loading a form.
- Triggers are associated with specific form objects, such as data fields, buttons, data blocks, or the form itself.
- To write a trigger, you must specify the object that the trigger is associated with, and the action, or event, that starts the trigger.
- To reference a form item in a PL/SQL trigger, use the format `:<block name>.<item name>`.
- It is a good idea to run the form each time you create a new trigger or add new code to an existing trigger, so that if there are errors, you know exactly what code caused them.
- A list of values (LOV) displays a list of data values that can be used to populate foreign key fields or fields with restricted data values.
- To create a LOV, you create a record group using a SQL query. The LOV display shows records returned from the SQL query.
- The LOV Wizard is used to create the LOV display and record group, specify the display properties, and attach the LOV to a form field.
- When you create an iconic button, you must specify the complete path to the .ico file. If you do not specify the complete path to the file, then the file containing the button icon must be in the default icon folder on your workstation.
- To display a LOV, you must put the insertion point in the text field that the LOV is attached to, and then issue the LIST_VALUES command.
- Radio buttons, also called option buttons, are used to limit the user to a few related choices.
- In Form Builder, individual radio buttons exist within a radio button group. Each individual radio button has an associated data value, and the radio button group has the data value of the currently selected radio button.
- You must specify a radio button's Name, which is how it is referenced within the form, its Label, which is the description that appears beside the button on the canvas, and its Value, which corresponds to the actual data value stored in the table field that radio button represents.
- The initial value of a radio button group is the value of the radio button that is selected when the form is first displayed.
- If a frame around a data block is long enough, the layout items are automatically positioned directly below each other. If the frame is not long enough, the items appear beside each other.

- You can display a graphic image as a graphic object, which is incorporated into the form .fmb file and compiled into the .fmx file, or as an image item, which is loaded into the form when the user runs the form.

REVIEW QUESTIONS

1. Why is it a good idea to use a sequence to automatically generate primary key values in a form?
2. What is a trigger? How is it different from other PL/SQL program units?
3. True or false: All form items have the same trigger events.
4. A form contains an item named CUSTID that is in the CUSTOMER_BLOCK. Write a PL/SQL assignment statement that assigns the current value displayed in CUSTID to a variable named CurrentCustID.
5. When you compile a trigger, you receive the following error message: "'DUAL' must name a table to which the user has access." What is the cause of the error?
6. What is the difference between a LOV and a record group?
7. How do you create an iconic button?
8. List the five steps for creating a LOV.
9. What is a radio group? What is the Initial Value property of a radio group?
10. What does the Radio Button Value property for a radio button specify?
11. What is the relationship between the form insertion point and the LIST_VALUES command?
12. What is the difference between a graphic object and an image item? Which should you use when you want to minimize the size of the form executable (.fmx) file?

PROBLEM-SOLVING CASES

Before you begin these exercises, run the northwoo.sql SQL script from the Chapter5 folder on your Student Disk to refresh your Northwoods University database tables, and the clearwat.sql SQL Script to refresh your Clearwater Traders database tables. Save all forms created in the exercises to the Chapter5 folder on your Student Disk.

1. Modify the ch5cinventory.fmb file that you created in the tutorial exercises so that it also includes a LOV that lists all of the possible color choices from the COLOR table, and returns the selected value to the form's COLOR field. Format the LOV display using the guidelines in the lesson. Save the form as Ch5cEx1.fmb.

2. Create a form that displays all of the fields in the CUSTOMER table in the Clearwater Traders database. Save the file as Ch5cEx2.fmb.
   a. Create a new data block and form-style layout using descriptive field prompts. Rename the window, canvas, and frame using descriptive names, and change the window and canvas formatting properties, using the guidelines described earlier in the chapter.
   b. Create a sequence named CUSTID_SEQUENCE that starts at 250 and has no maximum value, and modify the form so that it automatically inserts the new sequence value every time the user creates a new record.
3. Create a form that displays all of the fields in the CUST_ORDER table in the Clearwater Traders database. Save the file as Ch5cEx3.fmb.
   a. Create a new data block and form-style layout using descriptive field prompts. Create a radio button group to display the METHPMT field, with choices for Credit Card (value CC) and Check (value CHECK). Create a frame around the radio button group, with the frame title "Payment Method."
   b. Create a sequence named ORDERID_SEQUENCE that starts at 1070 and has no maximum value. Create a form trigger to automatically insert the next sequence value when the user creates a new record.
   c. Create a LOV named CUST_LOV that displays current customer last names, first names, and middle initials, and automatically inserts the selected customer ID value into the CUSTID field on the form. Format the LOV display using the guidelines in the lesson.
   d. Import the clearlogo.tif graphic image onto the form, and position the data fields so they are arranged symmetrically with respect to the graphic image.
   e. Rename the window, canvas, and frame using descriptive names, and change the window and canvas formatting properties, using the guidelines described earlier in the chapter.
   f. Format the ORDERDATE field so the order date is displayed as "May 29, 2001."
4. Create a form that displays all of the fields in the STUDENT table in the Northwoods University database. Save the file as Ch5cEx4.fmb.
   a. Create a new data block and form-style layout using descriptive field prompts. Create a radio button group to display the SCLASS field, with labels (and associated data values) for Freshman (FR), Sophomore (SO), Junior (JR), and Senior (SR). Create a frame around the radio button group with the frame title "Class."
   b. Create a sequence named SID_SEQUENCE that starts at 110 and has no maximum value. Create a form trigger to automatically insert the next sequence value when the user inserts a new record.
   c. Create a LOV named FAC_LOV that displays the first and last names of current faculty members, and automatically inserts the FID of the selected record into the FID field on the form. Format the LOV display using the guidelines in the lesson.
   d. Import the nwlogo.jpg graphic image from the Chapter5 folder on your Student Disk onto the form, and position the data fields so they are arranged symmetrically with respect to the graphic image.

e. Rename the window, canvas, and frame using descriptive names, and format the window and canvas properties, using the guidelines described earlier in the chapter.

f. Modify the SDOB field's format mask so the date is displayed as 07/14/1979.

5. Create a form that displays all of the fields in the ENROLLMENT table in the Northwoods University database. Save the file as Ch5cEx5.fmb.

a. Create a new data block and form-style layout using descriptive field prompts. Create a radio button group to display the GRADE field, with choices for A, B, C, D, or F.

b. Create a LOV named SID_LOV that displays current student last and first names, and automatically inserts the selected value into the SID field on the form. Format the LOV display using the guidelines in the lesson. (*Hint*: You cannot use the SID_LOV immediately after starting the form. You must do a query first.)

c. Create a LOV named CSECID_LOV that displays current course CSECIDs, course call IDs, and section numbers, and automatically inserts the selected value into the CSECID field on the form. Format the LOV display using the guidelines in the lesson. (*Hint*: You can create a LOV record group using a SQL query that joins multiple tables. *Hint*: You cannot use the CSECID_LOV immediately after starting the form. You must do a query first.)

d. Import the nwlogo.jpg graphic image from the Chapter5 folder on your Student Disk onto the form, and position the data fields so they are arranged symmetrically with respect to the graphic image.

e. Rename the window, canvas, and frame using descriptive names, and format the window and canvas properties, using the guidelines described earlier in the chapter.

6. Create a form that displays all of the fields in the TERM table in the Northwoods University database. Save the file as Ch5cEx6.fmb.

a. Create a new data block and form-style layout using descriptive field prompts. Use a check box to display the value of the STATUS field. The label of the check box should be Closed. If the box is checked, then the value of STATUS is "CLOSED." If the box is not checked, then the value of STATUS is "OPEN."

b. Create a sequence named TERMID_SEQUENCE that starts at 7 and has no maximum value. Create a form trigger to automatically insert the next sequence value when the user inserts a new record.

c. Import the nwlogo.jpg graphic image from the Chapter5 folder on your Student Disk onto the form, and position the data fields so they are arranged symmetrically with respect to the graphic image.

d. Rename the window, canvas, and frame using descriptive names, and format the window and canvas properties, using the guidelines described earlier in the chapter.

7. Create a form that displays all of the fields in the FACULTY table in the Northwoods University database. Save the file as Ch5cEx7.fmb.

a. Create a new data block and form-style layout using descriptive field prompts. Create a radio button group to display the FRANK field, with choices (and corresponding data values) for Full (FULL), Associate (ASSO), Assistant (ASST), and Instructor (INST). Draw a frame around the radio button group, and use "Rank" for the frame title.

b. Create a sequence named FID_SEQUENCE that starts at 10 and has no maximum value. Create a form trigger to automatically insert the next sequence value when the user inserts a new record.
c. Create a LOV named LOC_LOV that displays current location building codes and rooms, and automatically inserts the selected value into the LOCID field on the form. Format the LOV display using the guidelines in the lesson.
d. Rename the window, canvas, and frame using descriptive names, and format the window and canvas properties, using the guidelines described earlier in the chapter.
e. Format the format mask of the FPHONE field so that the data is displayed as (715) 555-1234.

CHAPTER

6

Creating Custom Forms to Support Business Applications

Introduction▶ The forms that you have created so far were based on a single database table. Some of your forms retrieved data from other tables to use as foreign key values, but the basic INSERT, UPDATE, and DELETE operations involved a single table. However, business operations often involve inserting, updating, and deleting records from many related tables. In this chapter you will learn how to create forms that integrate several database tables and support realistic business processes. You will create two custom business forms: one for supporting the Clearwater Traders merchandise-receiving process, and one for handling the customer order process. While developing these forms, you will learn how to use the Form Builder Debugger and explore how to intercept system-level messages and replace them with custom messages.

LESSON A

objectives

- Design and create a custom form
- Write PL/SQL triggers to process database records
- Create program units that are called by a trigger
- Use the Form Builder Debugger

The Clearwater Traders Merchandise-Receiving Process

To create a custom database application, first you must identify the necessary processes you wish to undertake and the corresponding database tables. The best way to start is to describe the process. The first form you will develop supports the merchandise-receiving function at Clearwater Traders, which is described next.

When Clearwater Traders purchases new merchandise from its suppliers, it keeps track of information about anticipated shipments in the SHIPPING table. This information includes the inventory ID of the purchased item, the quantity expected, and the date the shipment is expected. When a new shipment arrives, the receiving clerk first has to determine if the incoming shipment is a pending or back-ordered shipment. The records for a pending shipment are stored in the SHIPPING table, and the records for a backordered shipment are stored in the BACKORDER table. After finding the records associated with an incoming shipment, the clerk updates the records by entering the following information for each inventory item in the shipment from the incoming shipment's packing list: the quantity received, quantity back ordered (if any), and expected date that the backordered items will arrive. If the record came from the SHIPPING table (i.e., it was not a back order), then the system updates the DATE_RECEIVED field in the SHIPPING table to the current date and updates the QUANTITY_RECEIVED field to the quantity received. If the record came from the BACKORDER table, then these same updates are made in the BACKORDER table. If the incoming shipment reports a new back-ordered item, then a new record must be created in the BACKORDER table, and values are inserted for the item inventory ID, expected backorder arrival date, and the expected quantity. Finally, the system updates the QOH field in the INVENTORY table to include the quantity of goods actually received.

Figure 6-1 illustrates these processing steps using a flowchart to show more detailed design information. The clerk retrieves all information about pending new and backordered shipments from the SHIPPING and BACKORDER tables, using a Shipment LOV that displays all regular and backordered shipments that have not yet been received. After finding the correct shipment, the clerk enters the actual quantity received, quantity back ordered, and expected backorder receipt date. The system determines whether the selected shipment came from the SHIPPING or BACKORDER table by executing a SELECT query on the SHIPPING table first. If no record is found in the SHIPPING table, the system executes a SELECT query on the BACKORDER table. The appropriate SHIPPING or BACKORDER record is updated with the date and quantity received. If there is a new back order, the system creates a new backorder record. Regardless of whether there is a back order or not, the system updates the current QOH in the INVENTORY table by adding the quantity received to the current QOH.

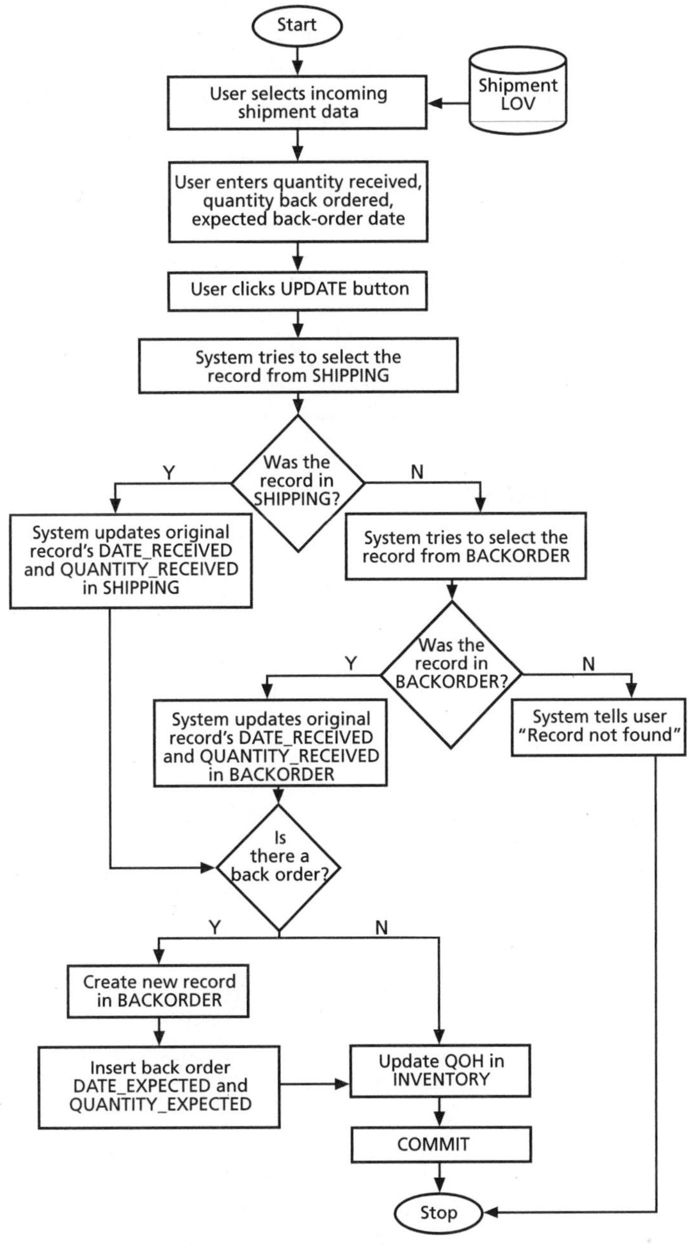

Figure 6-1: Clearwater Traders' receiving process flowchart

Creating the Interface Design

The next step is to visualize how the user interface will look. Figure 6-2 shows a design sketch of the interface design. The form will display the shipment ID, inventory ID, and inventory description, size, and color information. It will also display the date the shipment was expected, and the quantity expected. The date received field will automatically display the current system date, which the user can change if necessary (for example, if the entry of shipment data is delayed beyond the receipt date). The form will have fields for inserting the quantity received, quantity back ordered, and the expected back order arrival date. Buttons will be included to display the SHIPPING LOV, to update the shipment and inventory tables, to cancel the current operation and clear the screen, and to exit the form.

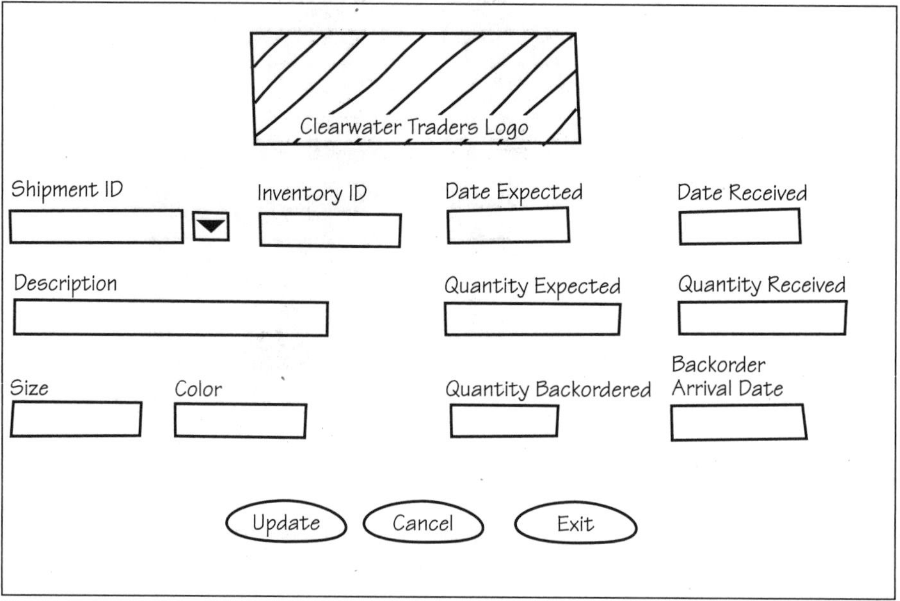

Figure 6-2: Design of the receiving form

Before working on the form, you will refresh your Clearwater Traders database tables by running the script that re-creates the tables and inserts all of the sample data records. You will also create a sequence to automatically generate primary keys for the BACKORDER table.

To refresh the Clearwater Traders database and create the sequence:

1. Start SQL*Plus, log on to the database, and then run the clearwat.sql script from the Chapter6 folder of your Student Disk to refresh your database tables.
2. Type the command shown in Figure 6-3 to create the new sequence. If you already created this sequence in a previous lesson or exercise, skip this step.

```
SQL> CREATE SEQUENCE backorderid_sequence
  2  START WITH 10
  3  NOMAXVALUE;

Sequence created.
```

type this command

Figure 6-3: Creating the BACKORDERID_SEQUENCE

Creating a Custom Form

To create a custom form, you have to create the form manually rather than use the Data Block and Layout Wizards, because the form displays fields from multiple tables. When you create a form manually, you create the form window and canvas in the Object Navigator. Then, you create the form items manually by "painting" them on the canvas, using buttons and tools on the Tool palette within Form Builder. Finally, you write the code that controls the form functions. Now you will start Form Builder, and select the option for building a form manually. Then you will rename the default form window, create a new canvas, and modify the form window properties.

To create a form manually:

1. Start Form Builder, select the **Build a New Form Manually** option button, and click **OK**. The Object Navigator window opens.

If Form Builder is already running, open the Object Navigator window, click Forms, then click the Create button to create a new Form module.

2. Click **File** on the menu bar, then click **Connect**, and connect to the database in the usual way.
3. Click **View** on the menu bar, then click **Visual View**.
4. Click [+] beside **Windows**, click **WINDOW1** twice, change the window name to **RECEIVING_WINDOW**, and then press the **Enter** key.

5 Click [+] beside **RECEIVING_WINDOW**, select **Canvases**, and then click the **Create** button on the Oracle Navigator toolbar to create a new canvas. Change the new canvas's name to **RECEIVING_CANVAS**.

6 Open the RECEIVING_WINDOW Property Palette, and change the window title to **Clearwater Traders Receiving**. Change the Resize Allowed and Maximize Allowed properties to **No**.

7 Change the window width to **460** (or whatever width is necessary to completely display the window in the Forms Runtime window without scrolling), change the Show Horizontal Scrollbar property to **Yes**, and change the Show Vertical Scrollbar property to **Yes**.

8 Open the canvas Property Palette and change its width and height properties so that they are the same as the window's width and height.

9 Save the form as **ch6areceiving.fmb** in the Chapter6 folder on your Student Disk.

Note: If you are storing your Student Disk files on a floppy disk, you will need to use two floppy disks for this chapter. When your first disk becomes full, save your solution files to the second disk.

Control blocks are data blocks that are not connected to a particular database table. Instead, they contain text items, command buttons, radio buttons, and other form items that you manually draw on the form canvas, and then control through triggers that you write using PL/SQL. Now you will create a control block.

To create the control block:

1 Click **View** on the menu bar, then click **Ownership View**. (Recall that Ownership View shows all of the form items as a flat list, while Visual View shows hierarchical relationships showing how form objects contain other form objects. Data blocks are only shown in Ownership View.)

2 Click **Data Blocks**, then click the **Create** button to create a new data block. Click the **Build a new data block manually** option button, then click **OK**. The new data block appears in the Object Navigator window. Change the block name to **RECEIVING_BLOCK**.

Next you will open the Layout Editor, and create the text items and labels for the fields that will appear on the form. You will create a text item for each data field shown in Figure 6-2, and a label for each text item. The text item is the box that will contain the actual data retrieved from the database, and the label is a boilerplate object that indicates the type of data that are displayed in the corresponding text item (such as "Last Name"). First you will create the text item and label for the Shipment ID data field.

To create the Shipment ID text item and label:

1 Click **Tools** on the menu bar, then click **Layout Editor**. The Layout Editor displays a blank canvas and shows the current canvas and block. Be sure that RECEIVING_CANVAS is displayed as the current canvas, and RECEIVING_BLOCK is displayed as the current block.

Note: If you are not working in RECEIVING_BLOCK, or if it is a data block rather than a control block, your form will not work correctly. To move to a different block or canvas in the Layout Editor window, click the drop-down arrow in the Block or Canvas list, and select the desired block or canvas.

2 Click the **Text Item** tool on the Tool palette, and then draw a rectangular box for the Shipment ID field, as shown in Figure 6-4.

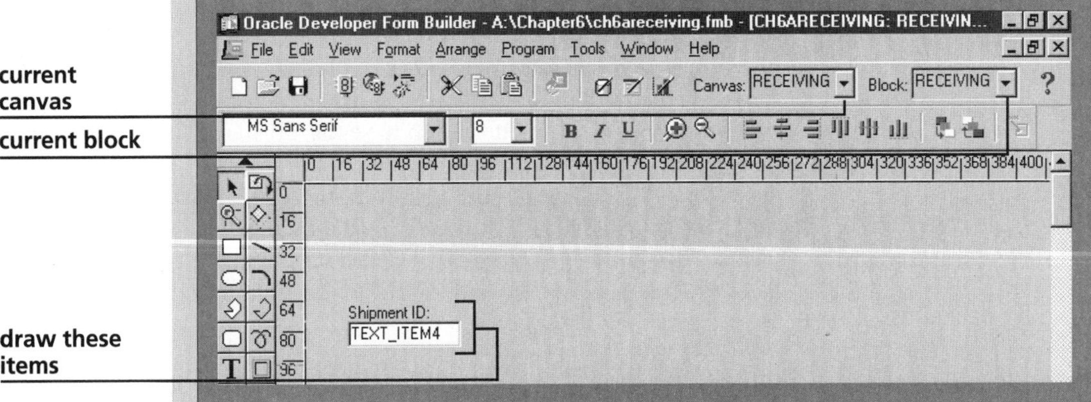

Figure 6-4: Drawing the first text item and label

help

The default name of your text item might be different. This is a system-generated name that you will change.

3 Click the **Text** tool, click directly above the left corner of the new data field, and then type **Shipment ID:** for the field label.

4 Select the new text item and field label as an object group, click **Arrange** on the menu bar, click **Align Objects**, select the **Horizontally Align Left** and **Vertically Stack** options, then click **OK**.

The next step is to specify the properties of the new text item. You will change the text item's Name property, which is used whenever you refer to the text item in a trigger. You will also change the text item's data type and maximum length. Text items that display database records must have the same data type as the corresponding data field in the database. The maximum length must be large enough to accommodate the maximum length of the data, plus any characters included in a format mask. When the SHIPPING table was created, SHIPID was specified as a NUMBER field of size 10, so you will change the text item's data type to NUMBER, and its maximum length to 10.

To specify the text item's properties:

1. Right-click the new text item, and then click **Property Palette.** Confirm that the Item Type property is Text Item, then change the Name property to **SHIPID_TEXT.**
2. Scroll down to the Data section of the Property Palette, click in the space next to the **Data Type** property, and select **Number.**
3. Change the Maximum Length property to **10.**
4. Close the Property Palette. The new name appears in the text item.

Now you will create the rest of the text items for the form. After you create each item, you will change the item's name and data type to correspond with the item's associated database field. You will change the Format Mask property for the date fields to MM/DD/YYYY. For text items with a format mask, you must add extra characters to the maximum length to accommodate the embedded format mask. The date fields with format mask MM/DD/YYYY must have a length of at least 10 characters for each of the data values plus the embedded slashes. It is a good idea to make the length a little longer to ensure that all of the data will be displayed.

To create the rest of the form text items:

1. Create text items with the following properties. Then position the text items, as shown in Figure 6-5, and create a field label for each text item, as shown in the figure.

Name	Data Type	Maximum Length	Initial Value	Format Mask
INVID_TEXT	Number	10		
DESC_TEXT	Char	100		
SIZE_TEXT	Char	10		
COLOR_TEXT	Char	20		
DATE_EXP_TEXT	Date	20		MM/DD/YYYY
DATE_REC_TEXT	Date	20		MM/DD/YYYY
QUANT_EXP_TEXT	Number	4		
QUANT_REC_TEXT	Number	4		
QUANT_BO_TEXT	Number	4	0	
BO_ARRIVAL_TEXT	Date	20		MM/DD/YYYY

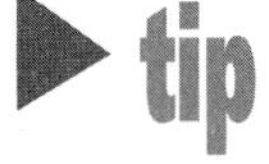

To help position the text items, select a group of text items that need to be aligned. Click Arrange on the menu bar, click Align Objects, then select the proper option button combination to align the objects horizontally or vertically.

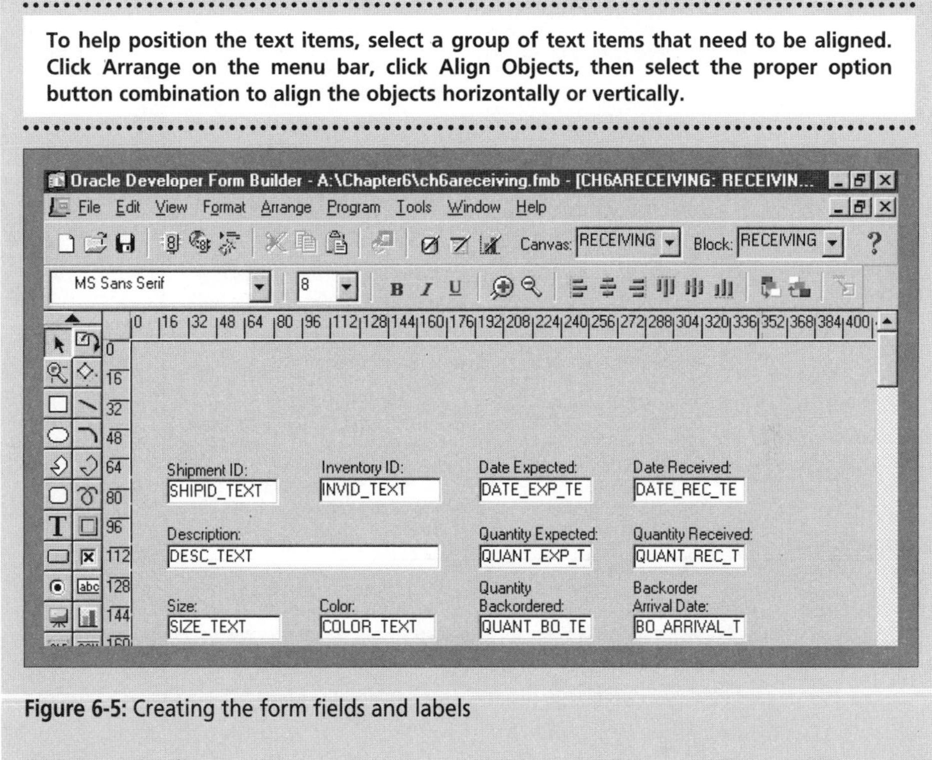

Figure 6-5: Creating the form fields and labels

2 Save the form.

Creating the LOV

Next you will create the LOV to retrieve data from the SHIPPING and BACKORDER tables for items that have been ordered but have not yet arrived at Clearwater Traders (their DATE_RECEIVED fields are NULL). This LOV will display the SHIPID, INVID, and QUANTITY_EXPECTED fields from SHIPPING or BACKORDER, and the corresponding ITEMDESC, ITEMSIZE, and COLOR data for each inventory item. This query requires a UNION, because it needs to return all records from both the SHIPPING and BACKORDER tables where the DATE_RECEIVED field is NULL. However, a LOV in a form cannot be derived from a UNION—it can be derived only from a conventional SELECT query. To get around this limitation, you will create a database view. This view will be derived from a query that retrieves all SHIPPING records and their associated inventory information where the date received is NULL, and then use a UNION to join these records with a second query that retrieves all BACKORDER records and their associated inventory information where the date received is NULL.

To create the database view:

1. Click the **Oracle SQL*Plus** button on the taskbar to switch to SQL*Plus.
2. Type the command shown in Figure 6-6 to create the view named SHIPVIEW.

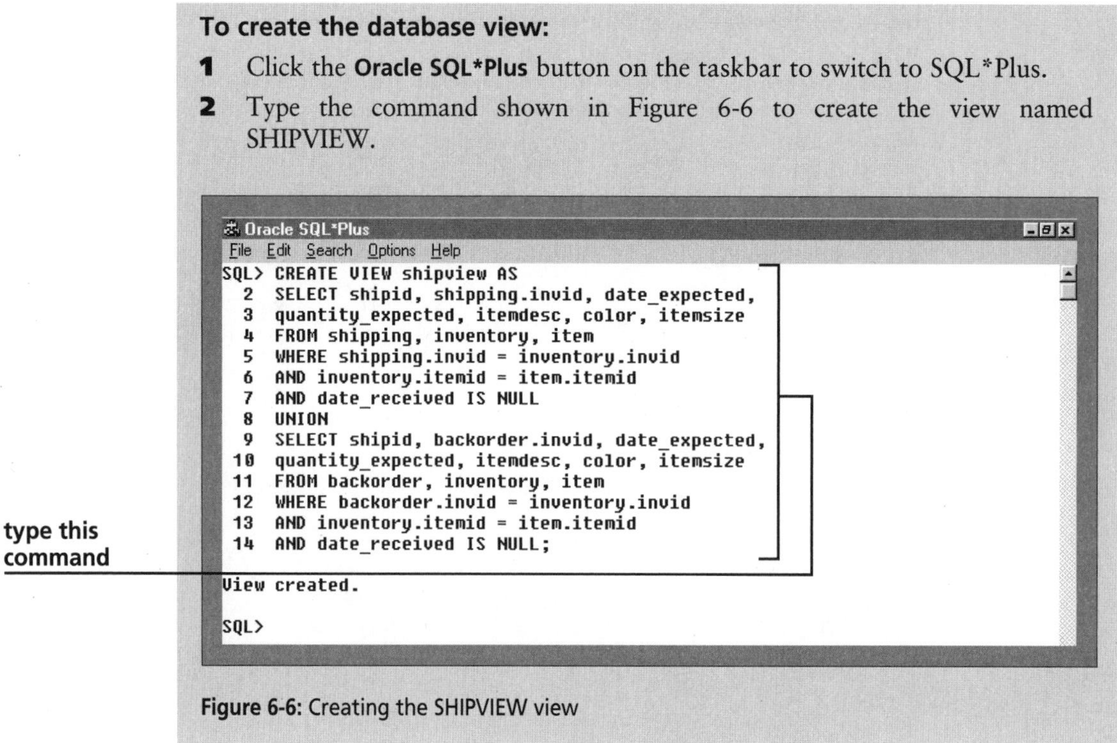

Figure 6-6: Creating the SHIPVIEW view

Next you will create the LOV to display the SHIPVIEW data.

To create the LOV:

1. Click the **Form Builder** button on the taskbar, click **Tools** on the menu bar, then click **LOV Wizard**. The LOV source page opens. Accept the default values, then click **Next**.
2. Type the SQL command shown in Figure 6-7 to select all of the view fields for the LOV record group, then click **Next**. The Column Selection page is displayed.

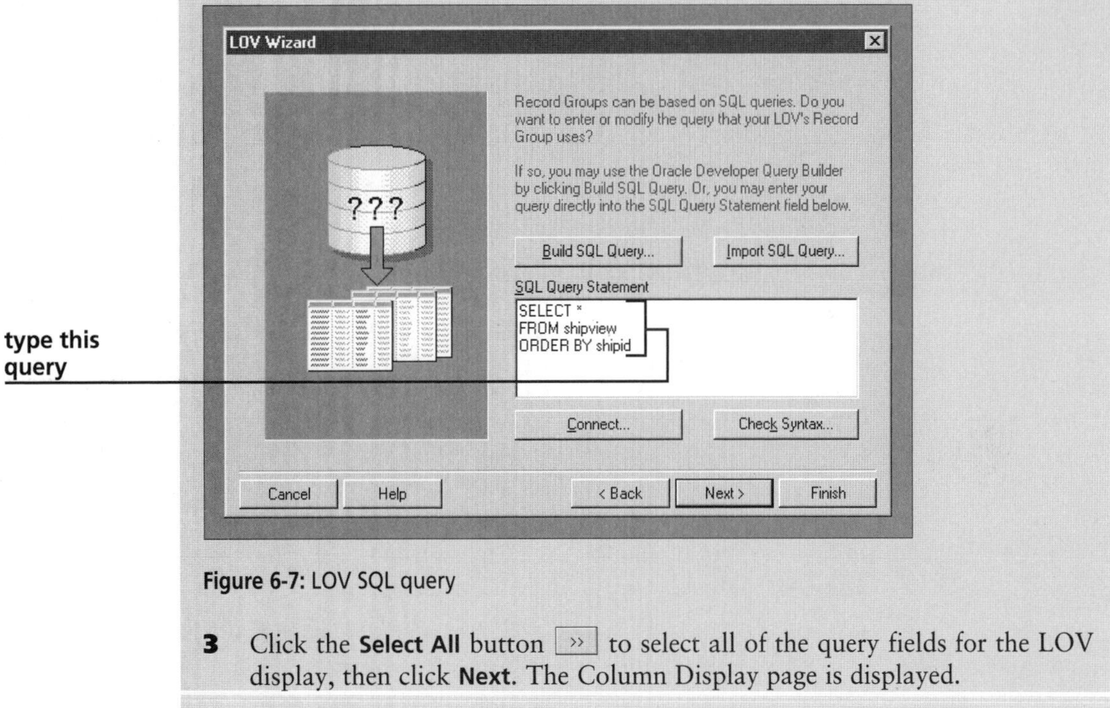

Figure 6-7: LOV SQL query

3 Click the **Select All** button [>>] to select all of the query fields for the LOV display, then click **Next.** The Column Display page is displayed.

Now you will specify the column titles and widths for the LOV display, and the column return items. When the user makes a selection from the LOV display, each display column value will be returned to a form field. Now you will format the LOV display and specify the return values.

To format the LOV display and specify the return values:

1 Modify the column titles and widths as shown in Figure 6-8.

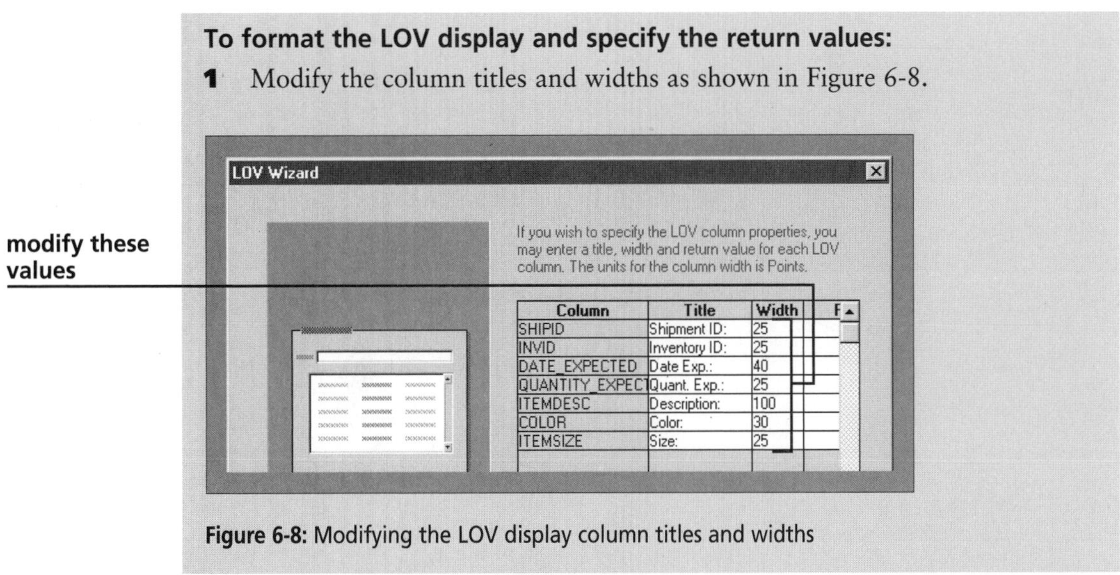

Figure 6-8: Modifying the LOV display column titles and widths

2 To specify the return value for the Shipment ID column, click the mouse pointer in the Return value column beside Shipment ID, click the **Look up return item** button, select **RECEIVING_BLOCK.SHIPID_TEXT**, then click **OK.** When the user selects a record in the LOV display, the selected value's shipment ID will be displayed in the Shipment ID field on the form.

3 Specify the Return values for the rest of the LOV display fields as shown below.

Column	ReturnValue
INVID	RECEIVING_BLOCK.INVID_TEXT
DATE_EXPECTED	RECEIVING_BLOCK.DATE_EXP_TEXT
QUANTITY_EXPECTED	RECEIVING_BLOCK.QUANT_EXP_TEXT
ITEMDESC	RECEIVING_BLOCK.DESC_TEXT
COLOR	RECEIVING _BLOCK.COLOR_TEXT
SIZE	RECEIVING _BLOCK.SIZE_TEXT

4 Click **Next.** The LOV Display page is displayed. Type **Expected Shipments** for the LOV title, change the Width to **380**, and the Height to **200**, then click **Next.** The Advanced Options page displays.

5 You will accept the default options, so click **Next.** The Items page is displayed.

This page enables you to specify the form field that the LOV is assigned to. (Recall that to display the LOV, you must place the form insertion point in the field the LOV is assigned to, then issue the LIST_VALUES command.) Since the LOV command button will be next to the SHIPID_TEXT form field, you will assign the LOV to SHIPID_TEXT. Now you will assign the LOV to SHIPID_TEXT, finish the LOV, and then rename the LOV and record group in the Object Navigator.

To assign the LOV to the form field, finish the LOV, and rename the LOV and record group:

1 Click **RECEIVING_BLOCK.SHIPID_TEXT** in the Return Items list, then click the **Select** button [>] to assign the LOV to SHIPID_TEXT. Click **Next**, then click **Finish** to finish the LOV.

2 Open the Object Navigator window, and change the name of the new LOV item and record group to **SHIPPING_LOV**. Your Object Navigator window should look like Figure 6-9.

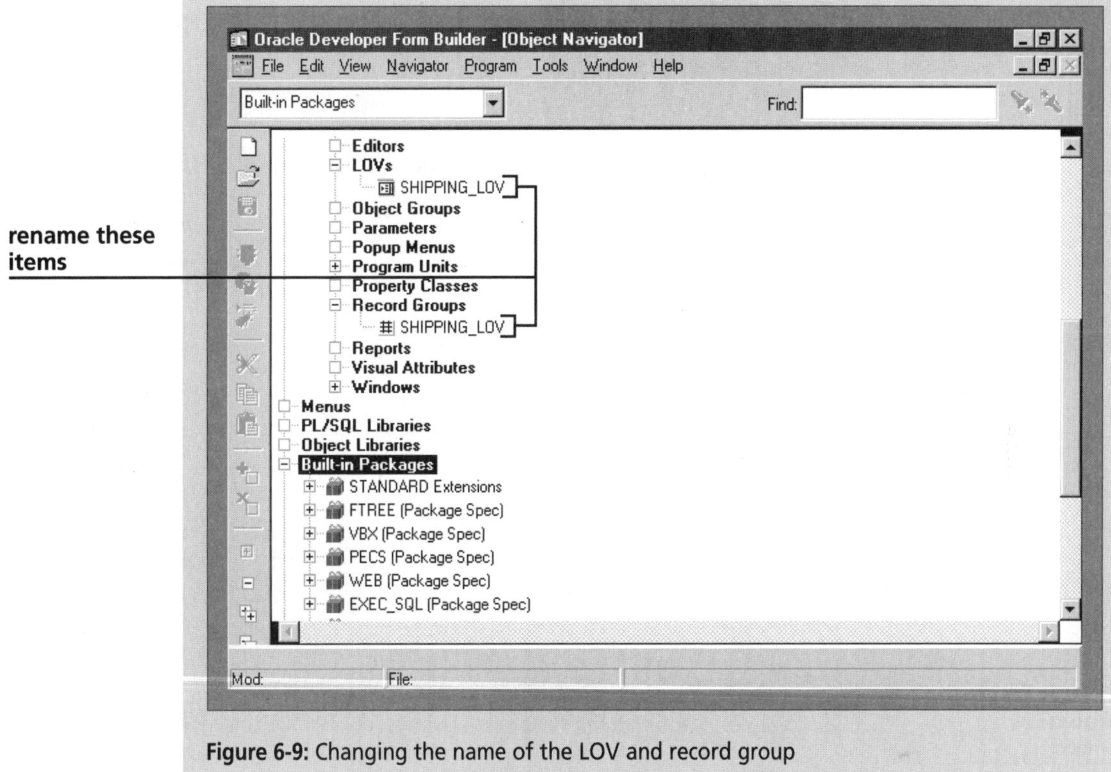

Figure 6-9: Changing the name of the LOV and record group

Next you will create the LOV command button to activate the LOV. You will create the command button and reset its properties, and then create the trigger to place the insertion point in the SHIPID_TEXT field and activate the LOV display. This command button will also insert the current system date in the DATE_RECEIVED_TEXT field, and reset the quantity back ordered to zero for each new shipment that is selected.

To create the LOV command button:

1. Switch to the Layout Editor, draw a command button, and position it to the right of the SHIPID_TEXT text item.
2. Open the button's Property Palette, and change the button's properties as follows:

Name	**SHIPPING_LOV_BUTTON**
Label	<deleted>
Iconic	**Yes**
Icon Filename	**a:\Chapter6\down3**
Width	**12**
Height	**14**

3 Create a WHEN-BUTTON-PRESSED trigger for the SHIPPING_LOV_BUTTON, and type the following command text:

```
GO_ITEM('shipid_text');
LIST_VALUES;
:receiving_block.date_rec_text := SYSDATE;
:receiving_block.quant_bo_text := 0;
```

4 Compile the trigger and close the PL/SQL Editor.

Now you will test the form. You will run the form, and click the LOV command button to confirm that the LOV is displayed correctly, and that the values of the selected shipment are displayed correctly in the form fields.

To test the LOV:

1 Run the form, and click the **LOV command button**. The LOV display opens, as shown in Figure 6-10. Note that the Shipment ID and Inventory.ID column titles are clipped off, so the display fields need to be wider. You will adjust these later.

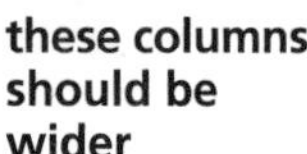

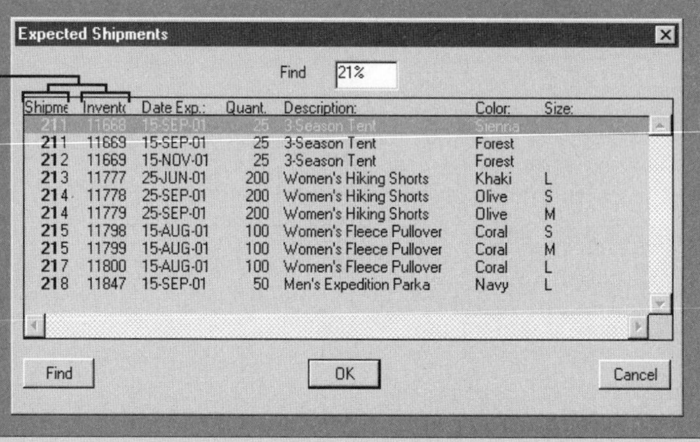

Figure 6-10: Expected Shipments LOV display

The first two digits of the Shipping ID are displayed as bold because the first two digits are the same value (21) for every record returned in the LOV display. The bolded text indicates that you must refer to the rest of the record data to differentiate between the data records.

2 Select **Shipment 212** and **Inventory ID 11669**, then click **OK.** The selected data fields are displayed in the form fields, as shown in Figure 6-11. Note that the initial value for Quantity Backordered is automatically entered as 0.

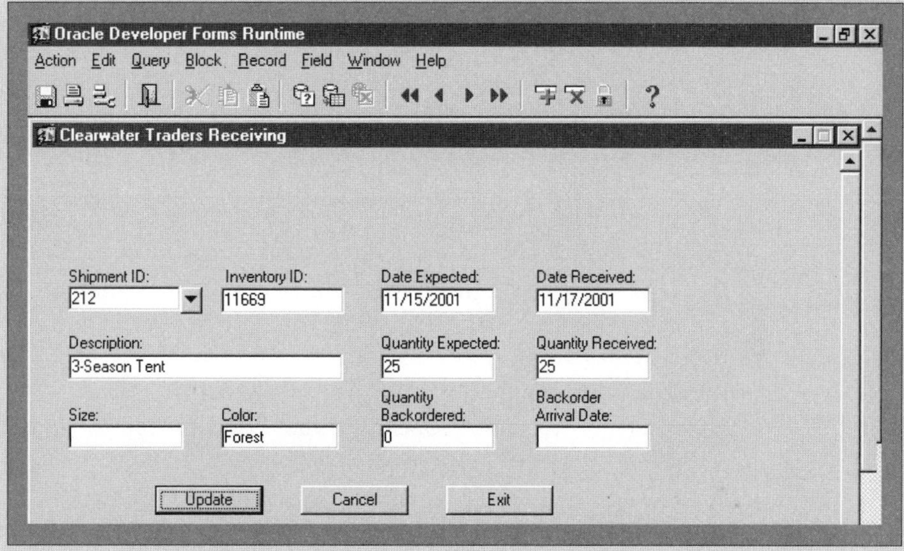

Figure 6-11: Viewing the selected fields in the form text items

help

If the returned values are not displayed correctly, close the Forms Runtime application, select the SHIPPING_LOV (LOV) in the Object Navigator, open the LOV Wizard in re-entrant mode, and then examine the return values on the Column Display page to confirm that the column names are listed correctly and that each column corresponds with the correct return value. Also, make sure that the data type and maximum width for each form text item are the same as for the text item's corresponding database field.

3 Close Forms Runtime.

4 Open the Object Navigator window, select the LOV named SHIPPING_LOV, open the LOV Wizard in re-entrant mode, and adjust the column widths of the Shipment ID and Inventory ID fields so that the column titles are displayed in full.

5 Save the form.

The next step is to create the Update, Cancel, and Exit buttons. First, you will draw the buttons on the canvas. Whenever you create a button group in a graphical user interface, the buttons should all be the same size, and should be wide enough to accommodate the longest button label, which is the text displayed on the button. You will draw the first button and change its name and label properties. Then you will copy the button and paste it on the canvas twice to create the other two buttons.

To create the Update, Cancel, and Exit buttons:

1. In the Layout Editor, use the **Button** tool to draw a command button in the bottom-center area of the form below the data fields, as shown in Figure 6-12.
2. Open the new button's Property Palette, change its Name property to **UPDATE_BUTTON,** Label to **Update**, Width to **60**, and Height to **16**, and then close the Property Palette.
3. In the Layout Editor, make sure the button is selected, click the **Copy** button on the toolbar to copy the button, then click the **Paste** button two times to paste the button twice. The new buttons are pasted directly on top of the first button.
4. Select the top button and drag it to the right side of the form. Select the next button, and drag it to the middle of the form.
5. Select all three buttons as an object group, click **Arrange** on the menu bar, click **Align Objects**, click the **Horizontally Distribute** and **Vertically Align Bottom** option buttons, then click **OK.** Center the object group under the text items on the form.
6. Change the Name property of the center button to **CANCEL_BUTTON** and change its Label property to **Cancel.**
7. Change the Name property of the right button to **EXIT_BUTTON** and change its Label property to **Exit.** The buttons should appear on the canvas as shown in Figure 6-12.

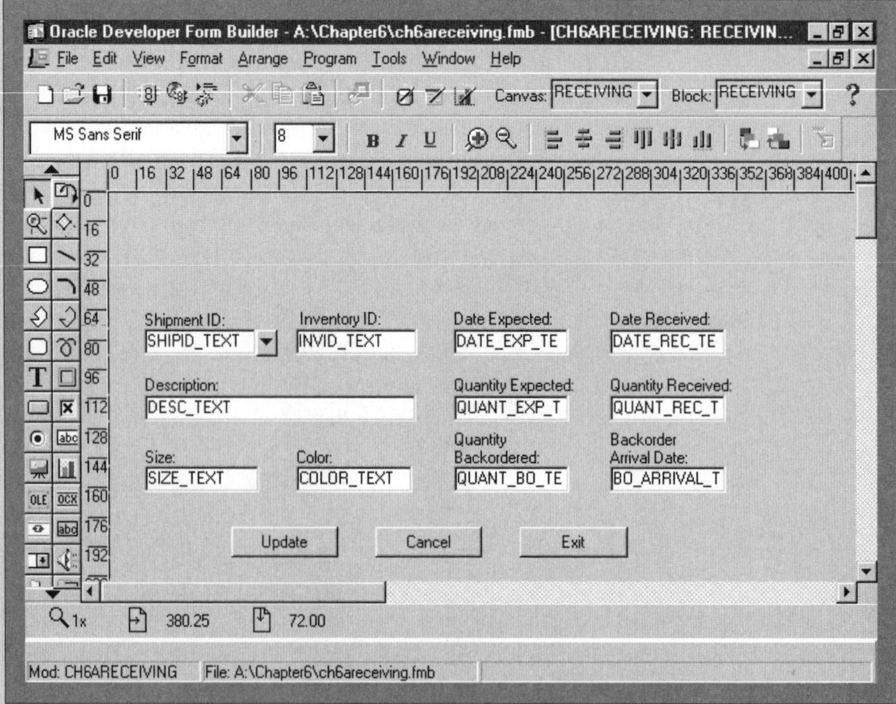

Figure 6-12: Positioning the form buttons

For data block forms like the ones you created in Chapter 5, the Forms Runtime environment automatically provides triggers for retrieving, inserting, updating, and deleting data. For the custom forms you will create in this chapter, you will create these triggers manually. When you create a custom form, you create a control block rather than a data block.

First, you will write the trigger for the Update button. Review the process design in Figure 6-1. The user first retrieves the shipment information, enters the quantity received, quantity back ordered, and expected backorder date, and then clicks the Update button. The UPDATE_BUTTON trigger first must determine if the selected record came from the SHIPPING table, by trying to select the record. If the record is not found in the SHIPPING table, then the trigger must try to select the record from the BACKORDER table. To create these queries, you will have to create an explicit cursor.

Recall that an explicit cursor must be used in a PL/SQL SELECT command that might retrieve a variable number of records, or no records at all. To determine whether the shipment record selected through the LOV came from the SHIPPING or BACKORDER table, the trigger first will query the SHIPPING table. There is a possibility that no records will be returned. If no records are returned from the SHIPPING table, then the trigger will query the BACKORDER table. Again, there is a possibility that no records will be returned. Therefore this operation requires explicit cursors for the SELECT commands for both the SHIPPING and BACKORDER tables. Recall that explicit cursors are declared formally in the DECLARE section of a PL/SQL procedure or trigger. You will type the DECLARE section for the Update button trigger that declares the explicit cursors next.

To create the DECLARE section of the UPDATE_BUTTON trigger:

1. Create a WHEN-BUTTON-PRESSED trigger for the UPDATE_BUTTON.
2. Type the code shown in Figure 6-13 in the PL/SQL Editor Source code pane. *Do not* compile the trigger yet. You entered only the DECLARE section, so you will receive compile errors if you attempt to compile now.

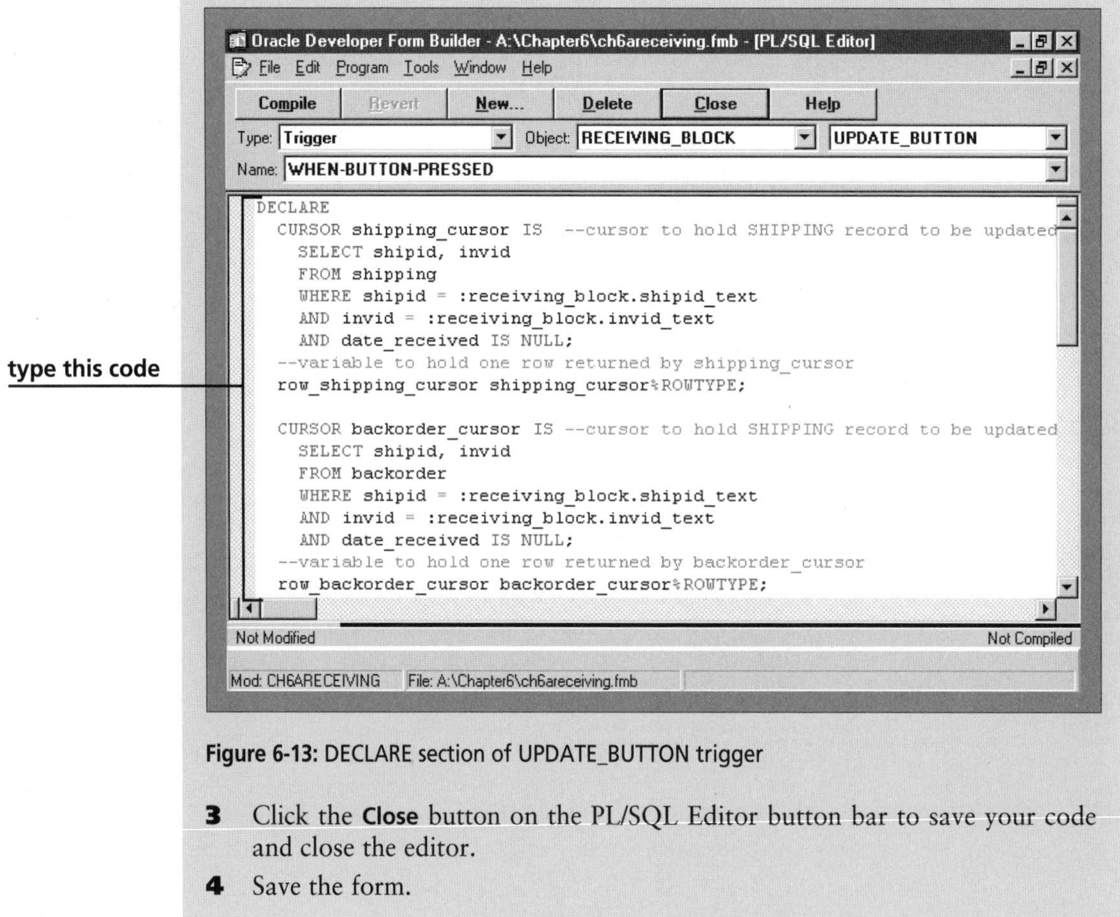

Figure 6-13: DECLARE section of UPDATE_BUTTON trigger

3 Click the **Close** button on the PL/SQL Editor button bar to save your code and close the editor.

4 Save the form.

A good coding practice when creating a complex trigger like the one in the UPDATE_BUTTON is to divide it into smaller program units that are easier to understand and debug. Recall that a program unit is a series of executable statements that corresponds to a single process, such as what is shown in one or more related flowchart steps. A program unit can be a procedure, which is simply a series of statements that change variable values, or a function, which returns a single value. You can pass variables to and from program units.

Figure 6-14 shows the program units you will create for the UPDATE_BUTTON trigger. First you will create a program unit named UPDATE_SHIPPING, which will update the selected record in the SHIPPING table.

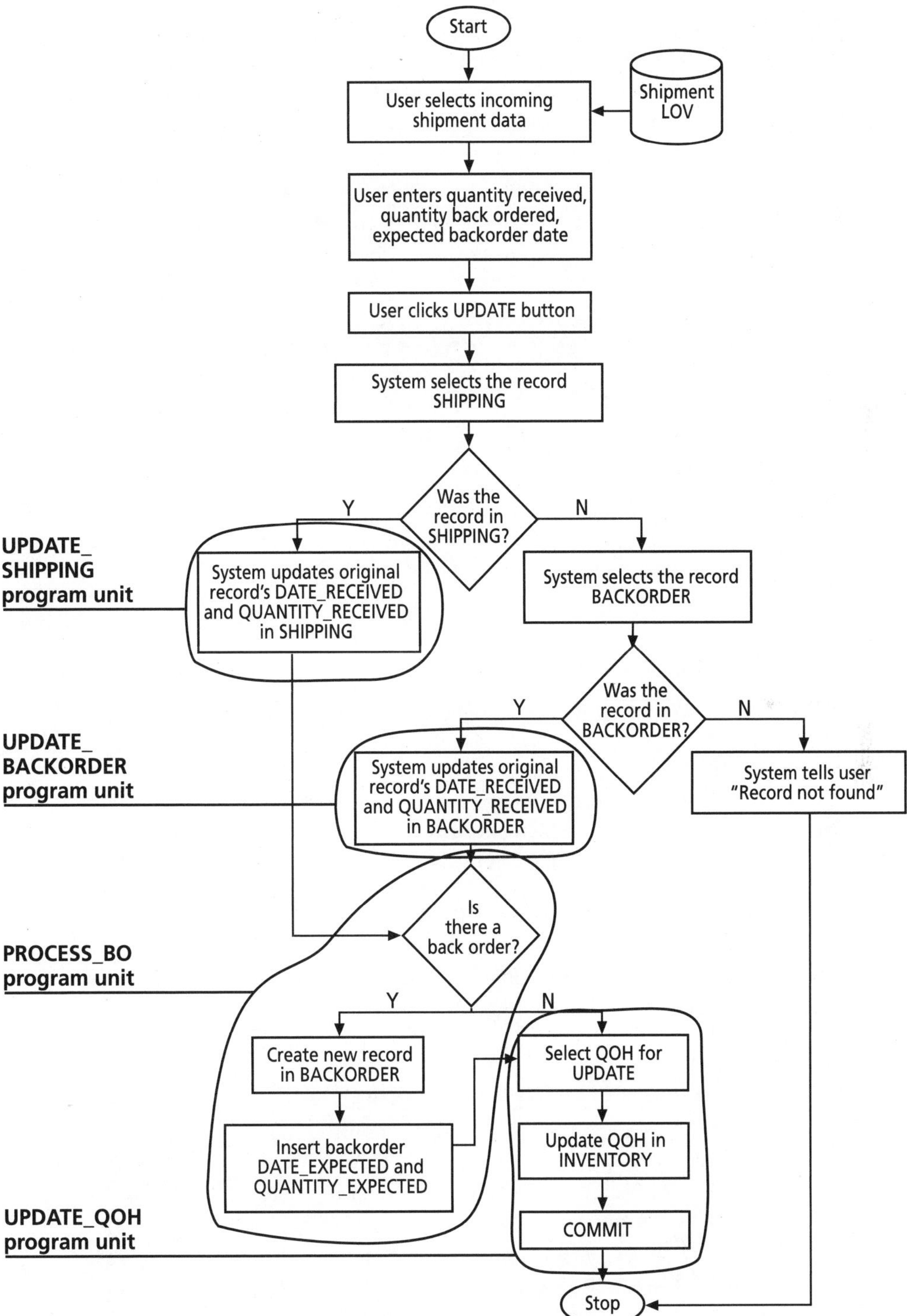

Figure 6-14: UPDATE_BUTTON program units

To create the UPDATE_SHIPPING program unit:

1. Switch to the Object Navigator window in Ownership View, click **Program Units**, and then click the **Create** button on the Object Navigator toolbar. The New Program Unit dialog box opens.
2. Type **UPDATE_SHIPPING** in the Name text box, make sure that the **Procedure** option button is selected, and then click **OK**. The PL/SQL Editor opens.
3. Type the code shown in Figure 6-15, and then compile your code and correct any syntax errors. *Do not* close the PL/SQL Editor.

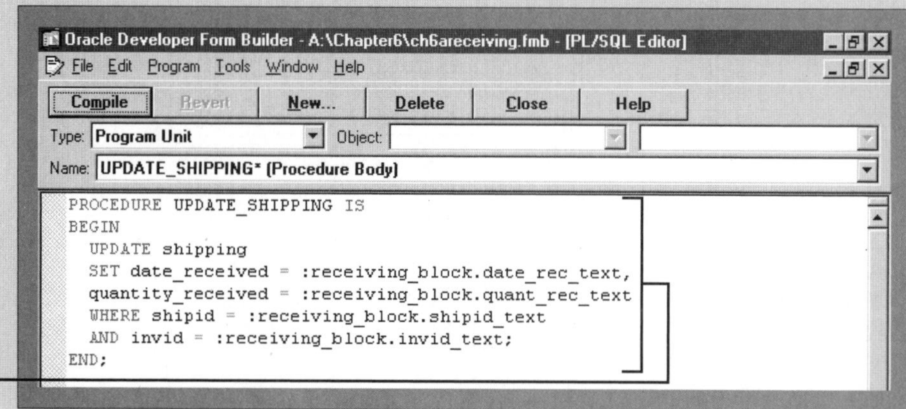
Oracle Developer Form Builder - A:\Chapter6\ch6areceiving.fmb - [PL/SQL Editor]
File Edit Program Tools Window Help
Compile Revert New... Delete Close Help
Type: Program Unit Object:
Name: UPDATE_SHIPPING* (Procedure Body)
PROCEDURE UPDATE_SHIPPING IS
BEGIN
UPDATE shipping
SET date_received = :receiving_block.date_rec_text,
quantity_received = :receiving_block.quant_rec_text
WHERE shipid = :receiving_block.shipid_text
AND invid = :receiving_block.invid_text;
END;
type this code

Figure 6-15: UPDATE_SHIPPING program unit

Next, you will create a program unit named UPDATE_BACKORDER to update the selected record in the BACKORDER table. Its code is exactly the same as the code in UPDATE_SHIPPING, except that it updates the BACKORDER table instead of the SHIPPING table. You will copy the code for the UPDATE_SHIPPING program unit, and then create the UPDATE_BACKORDER program unit directly within the PL/SQL Editor by clicking the New button on the PL/SQL Editor button bar. Then you will paste the copied code into the new program unit source code pane, and modify the UPDATE command.

To create the UPDATE_BACKORDER program unit:

1. With the UPDATE_SHIPPING program unit still displayed in the PL/SQL Editor, drag the mouse pointer over the text of the UPDATE command between the BEGIN and END statements to select it, click **Edit** on the menu bar, then click **Copy**.
2. Click the **New** button on the PL/SQL Editor button bar, type **UPDATE_BACKORDER** in the Name text box, make sure that the **Procedure** option button is selected, and then click **OK**.

3 With the insertion point between the BEGIN and END statements of the new program unit, click **Edit** on the menu bar, then click **Paste** to paste the copied UPDATE command. Change the table name from SHIPPING to BACKORDER in the first line of the UPDATE command, then compile your code and correct any syntax errors. Your completed UPDATE_BACKORDER program unit looks like Figure 6-16.

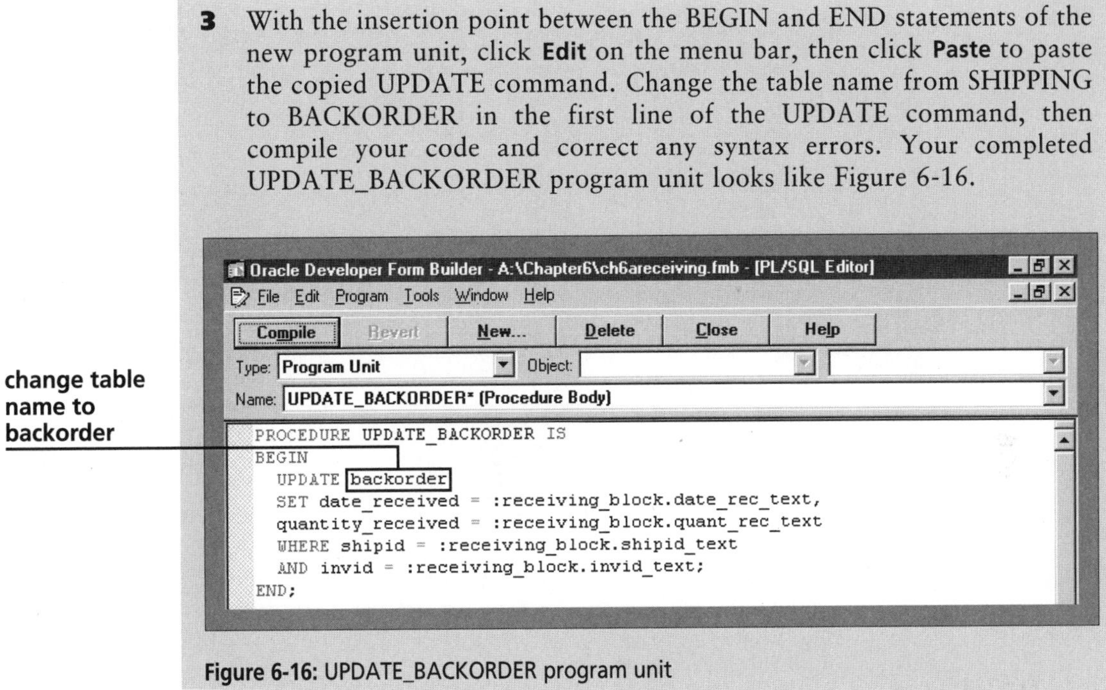
change table name to backorder
Oracle Developer Form Builder - A:\Chapter6\ch6areceiving.fmb - [PL/SQL Editor]
File Edit Program Tools Window Help
Compile
Revert
New...
Delete
Close
Help
Type: Program Unit
Object:
Name: UPDATE_BACKORDER* (Procedure Body)
PROCEDURE UPDATE_BACKORDER IS
BEGIN
UPDATE backorder
SET date_received = :receiving_block.date_rec_text,
quantity_received = :receiving_block.quant_rec_text
WHERE shipid = :receiving_block.shipid_text
AND invid = :receiving_block.invid_text;
END;

Figure 6-16: UPDATE_BACKORDER program unit

Next, you will create the PROCESS_BO program unit. This program unit will determine if the user has entered a backordered quantity, and if so, it will create a new BACKORDER record. To do this, it must use a value for BACKORDERID that is generated by the sequence you created earlier.

To create the PROCESS_BO program unit:

1 Click **New** on the PL/SQL Editor button bar, type **PROCESS_BO** in the Name text box, make sure that the **Procedure** option button is selected, and then click **OK**.

2 Type the code shown in Figure 6-17, then compile your code and correct any syntax errors.

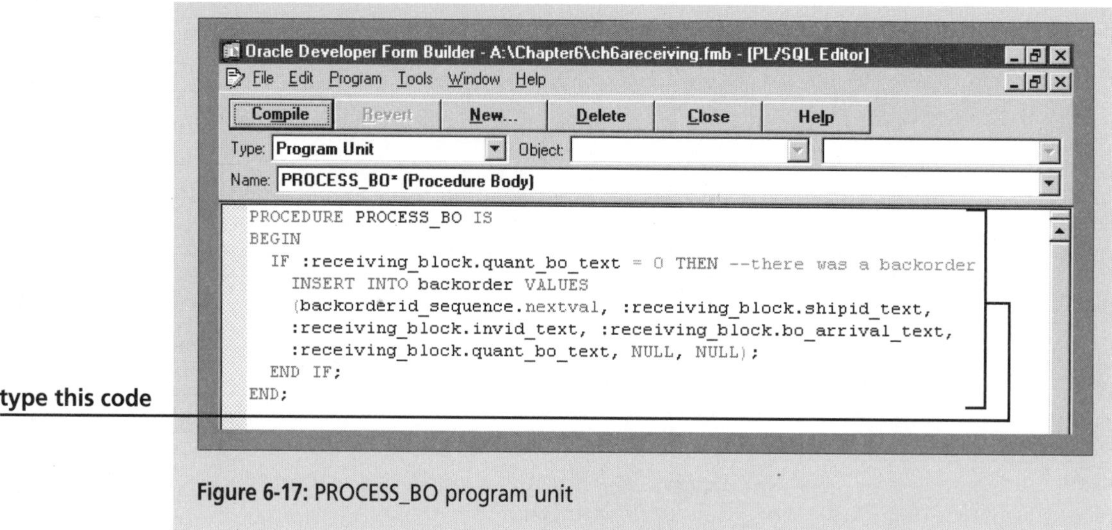

Figure 6-17: PROCESS_BO program unit

Finally, you will create the UPDATE_QOH program unit. This program unit will select the current QOH for update, update the quantity received, and commit the queries. You will create this program unit next.

To create the UPDATE_QOH program unit:

1 Click **New** on the PL/SQL Editor button bar, type **UPDATE_QOH** in the Name text box, make sure that the **Procedure** option button is selected, and then click **OK**.

2 Type the code shown in Figure 6-18, then compile your code and correct any syntax errors. *Do not* close the PL/SQL Editor.

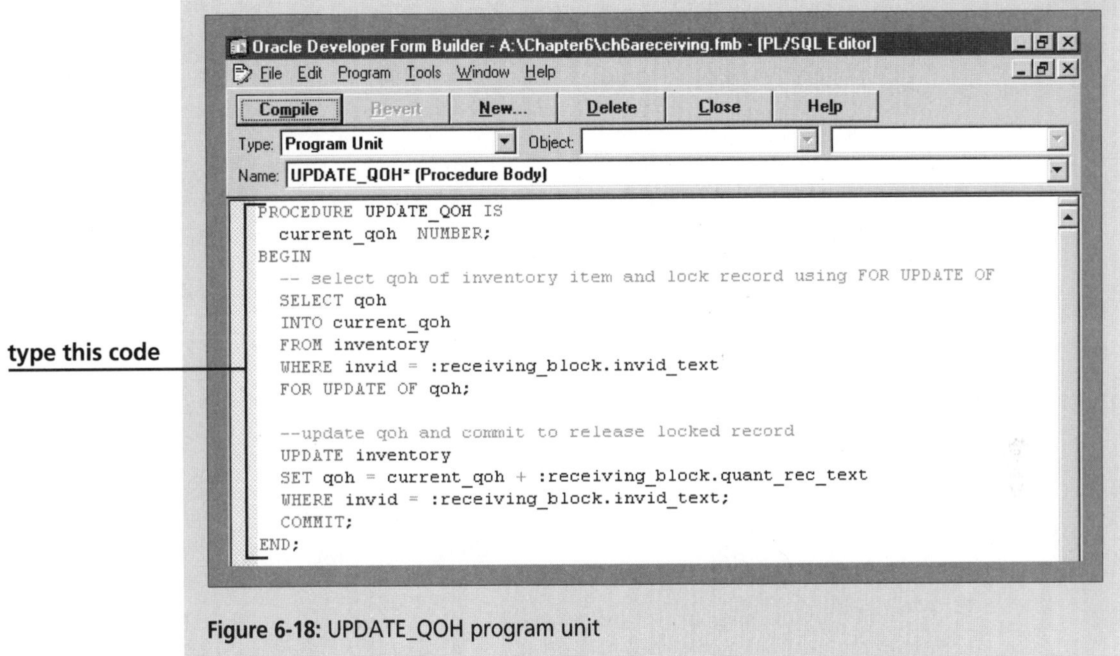
Oracle Developer Form Builder - A:\Chapter6\ch6areceiving.fmb - [PL/SQL Editor]
File Edit Program Tools Window Help
Compile Revert New... Delete Close Help
Type: Program Unit Object:
Name: UPDATE_QOH* (Procedure Body)
type this code
PROCEDURE UPDATE_QOH IS
current_qoh NUMBER;
BEGIN
-- select qoh of inventory item and lock record using FOR UPDATE OF
SELECT qoh
INTO current_qoh
FROM inventory
WHERE invid = :receiving_block.invid_text
FOR UPDATE OF qoh;
--update qoh and commit to release locked record
UPDATE inventory
SET qoh = current_qoh + :receiving_block.quant_rec_text
WHERE invid = :receiving_block.invid_text;
COMMIT;
END;

Figure 6-18: UPDATE_QOH program unit

Now you will enter the body of the trigger for the Update button. Recall that earlier you created the UPDATE_BUTTON trigger, and entered the commands that declared the SHIPPING_CURSOR, which attempts to retrieve the shipment from the SHIPPING table, and the BACKORDER_CURSOR, which attempts to retrieve the shipment from the BACKORDER table. You can easily move to the UPDATE_BUTTON trigger without closing the PL/SQL Editor. The following steps will show you how.

To move to the UPDATE_BUTTON trigger:

1. In the PL/SQL Editor, click the **Type** list arrow, and then click **Trigger**. Next you will select the object the trigger code is attached to, which can be either a form or a specific block in a form.
2. If RECEIVING_BLOCK is not the current Object, click the first **Object** list arrow, and then click **RECEIVING_BLOCK**. Finally, you will select the specific item (button, text field, etc.) associated with the trigger.
3. Click the second **Object** list arrow, and then click **UPDATE_BUTTON** to display the code for the UPDATE_BUTTON trigger.

Now you will enter the body of the UPDATE_BUTTON trigger. The trigger body will open the SHIPPING_CURSOR, and will try to retrieve the shipment record. If the shipment record is found, it will call the UPDATE_SHIPPING, PROCESS_BO, and UPDATE_QOH program units, and then clear the form fields using the CLEAR_FORM procedure, which is a built-in procedure that clears all the form fields. If no record is found, the trigger will open the BACKORDER_CURSOR, and if the shipment is found in the BACKORDER table, the trigger will call the UPDATE_BACKORDER, PROCESS_BO, and UPDATE_QOH program units and clear the form fields.

If the record is not found in BACKORDER, the MESSAGE procedure will be used to display a message in the form message area stating that the record was not found. MESSAGE is a Form Builder built-in procedure that is used to display messages to users in the message area at the bottom of the form. It has the following general format:

```
MESSAGE('<message text to be displayed>');
```

The message text can be text that is typed directly, such as `'No records found.'`, or text plus variable values, such as `'No records found for ' || :student_block.slname_text;`.

To enter the body of the UPDATE_BUTTON trigger:

1 Type the code shown in Figure 6-19 at the end of the DECLARE section to complete the UPDATE_BUTTON trigger.

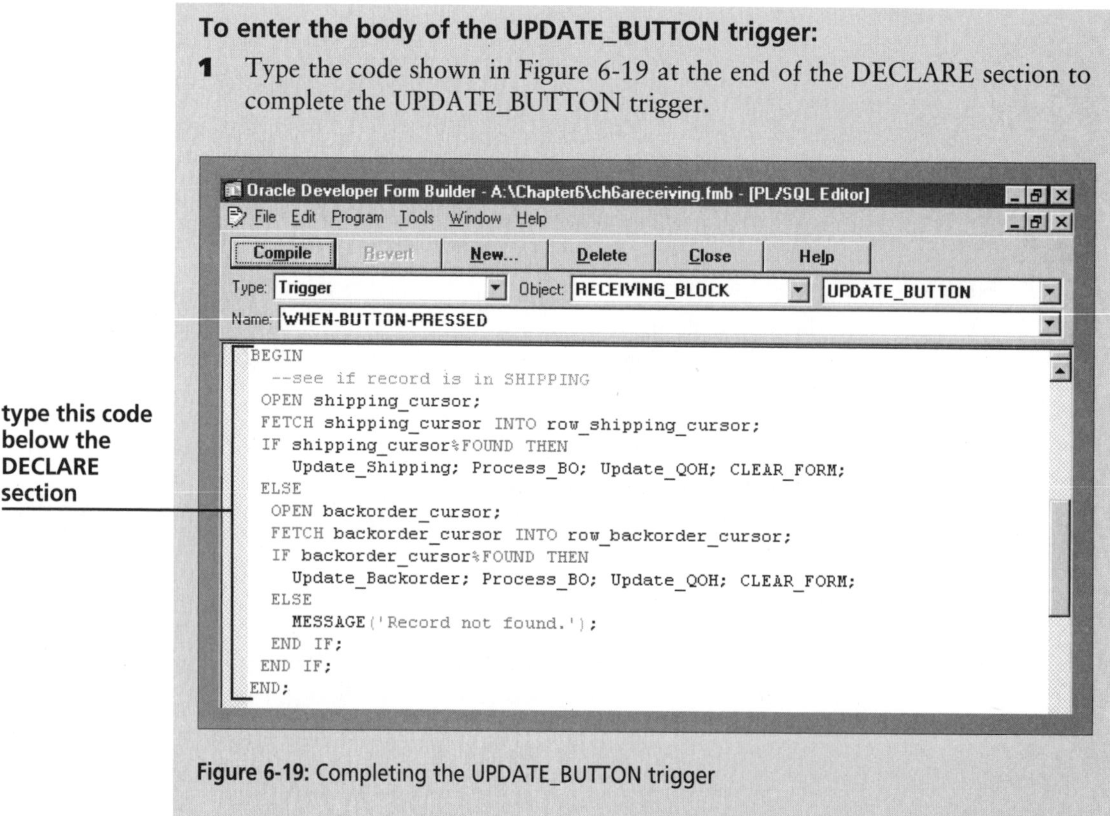

Figure 6-19: Completing the UPDATE_BUTTON trigger

2 Compile the trigger, correct any syntax errors, and then click **Close** on the PL/SQL Editor button bar.

help

If you receive an error message stating that one of your program units is "not declared in this scope" (such as "PROCESS_BACKORDER not declared in this scope"), it means that the program unit has not been created, was not referenced by the correct name, or has not been successfully compiled. Confirm that the specified program unit exists by opening the Object Navigator window, clicking [+] beside Program Units, and making sure the program unit is there and is referenced by the right name. If the program unit exists and is referenced correctly, open the program unit in the PL/SQL Editor window by double-clicking the Program Unit icon beside its name, then clicking the Compile button on the PL/SQL Editor button bar.

Now you will create the triggers for the Cancel and Exit buttons. The Cancel button will clear the form text items using the CLEAR_FORM procedure. The Exit button exits the form using the EXIT_FORM procedure, which is a built-in procedure that exits Forms Runtime.

To create the Cancel and Exit buttons:

1. Create a WHEN-BUTTON-PRESSED trigger for the Cancel button with the following code: **`CLEAR_FORM;`**.
2. Create a WHEN-BUTTON-PRESSED trigger for the Exit button with the following code: **`EXIT_FORM;`**.
3. Close the PL/SQL Editor.
4. Save the form.

Testing the Clearwater Traders Receiving Form

To test the Clearwater Traders Receiving form and confirm that it works correctly, you need to examine the following possible cases:

1. The selected shipment is from the SHIPPING table, and there is no back order.
2. The selected shipment is from the BACKORDER table, and there is no new back order.
3. The selected shipment is from the SHIPPING table, and there is a new back order.
4. The selected shipment is from the BACKORDER table, and there is another back order.

You will test the first case now. You will record receipt of Shipment ID 212 (3-Season Tent, Color Forest, and Quantity 25) from the SHIPPING table, and assume that all 25 tents were received.

To test the first case in the Clearwater Traders Receiving form:

1. Run the form.
2. Click the **Shipment ID LOV** command button, select shipment ID **212**, and then click **OK**.
3. Type **25** in the Quantity Received field.
4. Make sure that the **Quantity BackOrdered** field displays 0.
5. Click the **Update** button. The following error message should appear at the bottom of the screen: "FRM-40735: WHEN-BUTTON-PRESSED trigger raised unhandled exception ORA-01400."

help If you did not receive the error message shown, or if you received a different error message, continue with the lesson anyway to learn how to use the Form Builder Debugger.

This is a good time to review how to interpret error codes and learn how to use the Form Builder Debugger.

Using the Form Builder Debugger to Find and Correct Run-Time Errors

The first step in correcting an error is to find out what the FRM- and ORA- error codes mean. Error codes prefaced with FRM- are Form Builder error codes, while ORA- error codes are generated by the database. First you will look up the description of the FRM- error code.

To find the description of the error message:

1. Close Forms Runtime.
2. In Form Builder, click **Help** on the menu bar, then click **Form Builder Help Topics.** Click the **Index** tab, type **FRM-40735** in the search text box, click **Display**, make sure that FRM-40735 is selected in the Topics Found list, then click **Display** again.

This error message indicates that the generated error needed to be handled in the EXCEPTION section of the trigger. However, you still need to find what caused the error so you can write the code to handle it. Next, you will look up the description of the ORA-01400 error code that was displayed on the form. To do this, you will look in the database error code Help system that you used in the earlier chapters on SQL*Plus. You will do this next.

To look up the ORA- error description:

1. Click the **Start** button on the Windows taskbar, point to **Programs,** point to **Oracle for Windows 95** (or **Windows NT**), and then click **Oracle8 Error Messages.**

help

If Oracle8 Error Messages is not on your Start menu, start Windows Explorer, navigate to the ORAWIN95\MSHELP\ folder and double-click the ora.hlp file.

2. Click the **Index** tab, type **ORA-01400** in the search text box, and then click the **Display** button to see the error explanation. The ORA-01400 message indicates that you did not specify a value for a column that cannot be NULL. Close the Oracle Messages and Codes window. Now you will use the Form Builder Debugger to locate where the error is occurring.

The Form Builder Debugger, like the debugger you used in Procedure Builder, allows you to step through triggers and other PL/SQL programs one line at a time to examine the values of variables during execution. To use the Form Builder Debugger, you must run the form by clicking the Run Form Debug button and specify **breakpoints** that temporarily halt execution so you can single-step through specific program lines to examine program flow and variable values.

To use the Form Builder Debugger:

1. Open the Object Navigator window, and then click the **Run Form Debug** button on the Object Navigator toolbar.

You also can click Program on the menu bar, point to Run Form, and then click Debug.

2. Maximize the Debugger window so it looks like Figure 6-20.

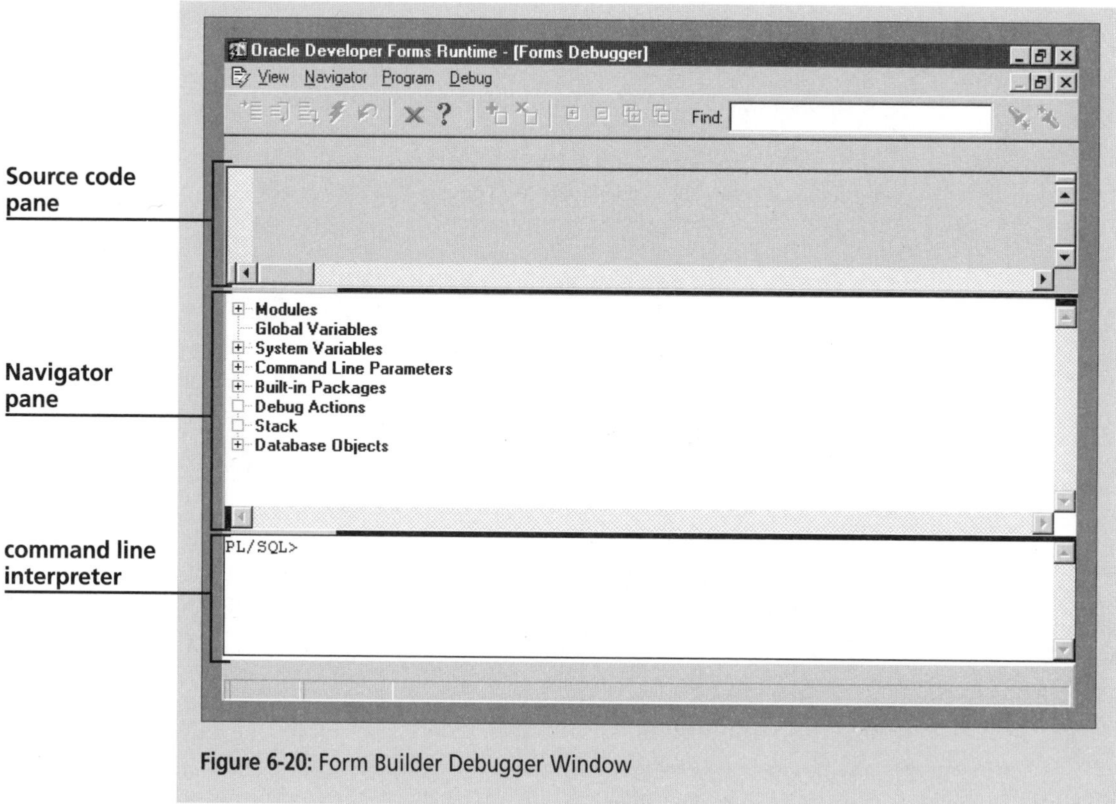

Figure 6-20: Form Builder Debugger Window

The Form Builder Debugger window (see Figure 6-20) has a toolbar with buttons that perform debugger commands. It also has a Source code pane to show the code you are debugging, a Navigator pane similar to the Object Navigator that allows you to navigate among form and database objects, and a command line interpreter for entering command-line debugger and SQL*Plus commands.

The Navigator pane lists the system objects that you might want to access while you are debugging. An object with a plus box [+] to the left of its name indicates that the specific object might exist in the code you currently are running. To access the trigger code for the Update button, you will first open the Modules object, which lists the forms and form objects in your current application. Then you will navigate to the RECEIVING_BLOCK, since this is the block that contains the Update button. Then you will select the UPDATE_BUTTON item, and display its WHEN-BUTTON-PRESSED trigger, since it contains the code that you want to debug.

To access the Update button code:

1 Click [+] beside Modules in the Navigator pane, and then open the following objects: **CH6ARECEIVING, Blocks, RECEIVING_BLOCK, Items, UPDATE_BUTTON, Triggers,** and **WHEN-BUTTON-PRESSED.** The source code for the UPDATE_BUTTON trigger appears in the Source code pane.

Next you need to reconfigure the Debugger window. You will enlarge the Source code and Navigator panes so that you can see most of the trigger code in the Source code pane. Since you will control the Debugger by clicking buttons rather than typing commands, you can close the Command line interpreter pane.

To reconfigure the Form Builder Debugger window:

1 Move the pointer over the line between the Navigator and Command line interpreter panes so the pointer changes to +. Click and drag down so the Command line interpreter pane is no longer visible.

2 Move the pointer over the line between the Source code and Navigator panes so the pointer changes to + again. Click and drag down so each pane occupies about one-half of the screen.

Setting Breakpoints

The first step in using the Form Builder Debugger is to set breakpoints to halt program execution. Now you will set a breakpoint on the first executable line of the UPDATE_BUTTON trigger so you can single-step through it and examine how the program (and the Debugger) works. As in the Procedure Builder debugger, breakpoints can only occur on executable PL/SQL statements—they cannot be placed on variable declaration statements, comment statements, or other nonexecutable commands.

To set a breakpoint:

1 Double-click the program line **OPEN shipping_cursor** in the Source code pane. A breakpoint icon appears before the program line number and the program line number changes to B (01), as shown in Figure 6-21, to indicate that the first breakpoint occurs on this line. (Your line number for this program line might be different than the one shown in the figure).

You can set breakpoints only in PL/SQL code lines. SQL statements, blank lines, or comment lines do not support breakpoints.

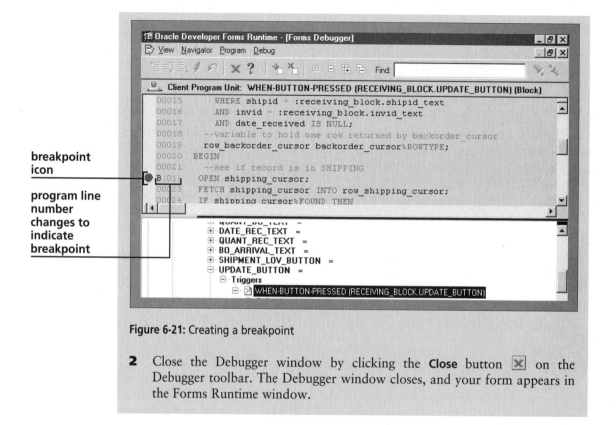

breakpoint icon

program line number changes to indicate breakpoint

Figure 6-21: Creating a breakpoint

2 Close the Debugger window by clicking the **Close** button on the Debugger toolbar. The Debugger window closes, and your form appears in the Forms Runtime window.

Monitoring Program Execution and Variable Values

Now you will select the shipment ID, enter the quantity received, and click the Update button. As the trigger runs, you will be able to single-step through the source code one line at a time to see how the program executes, and examine the values of form variables.

To single-step through the source code and monitor program execution:

1 Click the **Shipment ID LOV** command button, click **212**, and then click **OK**. The shipment information is displayed in the form text items.

If Shipment 212 does not appear in your list of shipments, run the clearwat.sql script in SQL*Plus to refresh your database tables, then click the Shipment ID LOV command button again.

2 Type **25** in the Quantity Received field, and then click **Update**. The Debugger window opens, as shown in Figure 6-22.

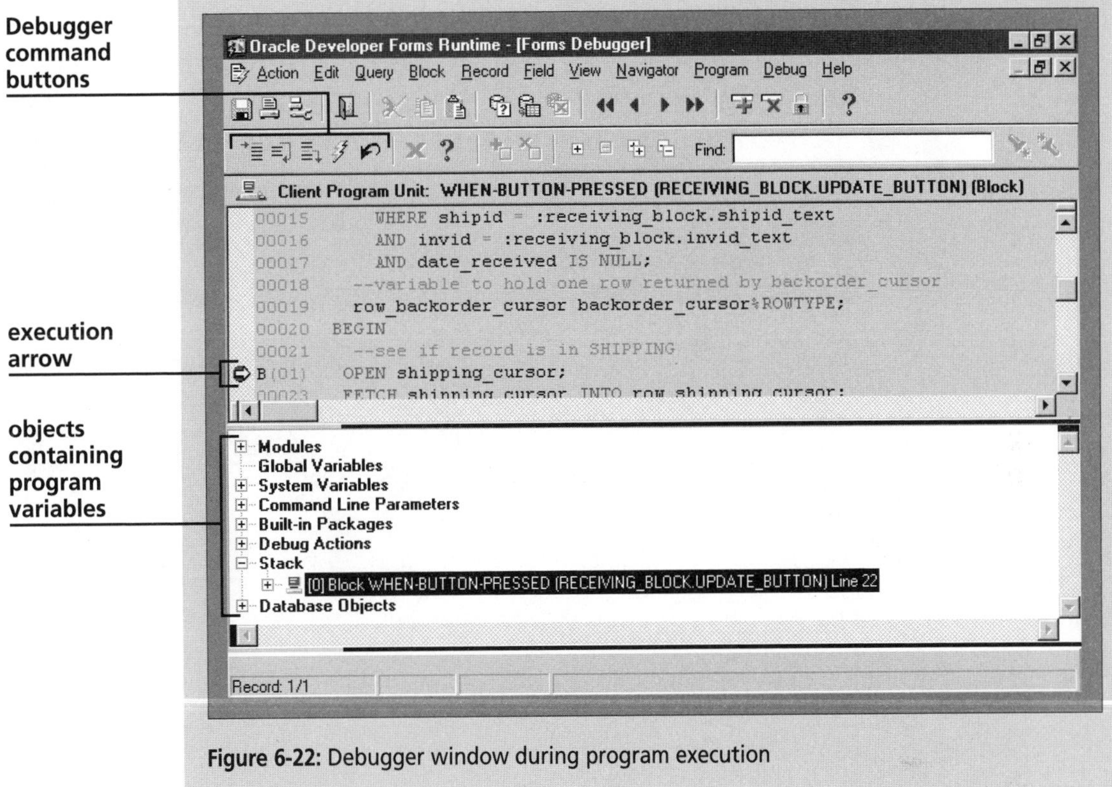

Figure 6-22: Debugger window during program execution

Notice that the Debugger command buttons that are active on the toolbar are the same as the ones used in the Procedure Builder Debugger. The **Step Into** button allows you to step through the program one line at a time. The **Step Over** button allows you to bypass a call to a program unit. The **Step Out** button executes all program lines to the end of the current trigger. The **Go** button allows you to pass over the current breakpoint and run the program until the next breakpoint is called or the program is terminated. The **Reset** button terminates execution and returns the form to the point it was at before you clicked the Update button and started the current trigger.

Also notice the position of the execution arrow on Line 22. The **execution arrow** shows which line of the procedure the Forms Debugger will execute next. As you step through the program, the execution arrow stops on PL/SQL statements and skips comment lines and SQL statements.

Finally, notice the objects displayed in the Navigator pane. You can open these objects during program execution to examine the current values of program variables. The three most common object types that you will examine are Stack, Modules, and Global Variables. The **Stack** object contains **stack variables**, which are the variables that are initialized in the DECLARE statements of individual triggers and procedures. Stack variables sometimes are called **local variables**, because they are only used by the procedure or trigger where they are declared.

To examine the current stack variable values:

1 Click ⊞ beside **[0] Block WHEN-BUTTON-PRESSED (RECEIVING_BLOCK. UPDATE_BUTTON) Line 22** under Stack to view the current values of the stack variables, as shown in Figure 6-23.

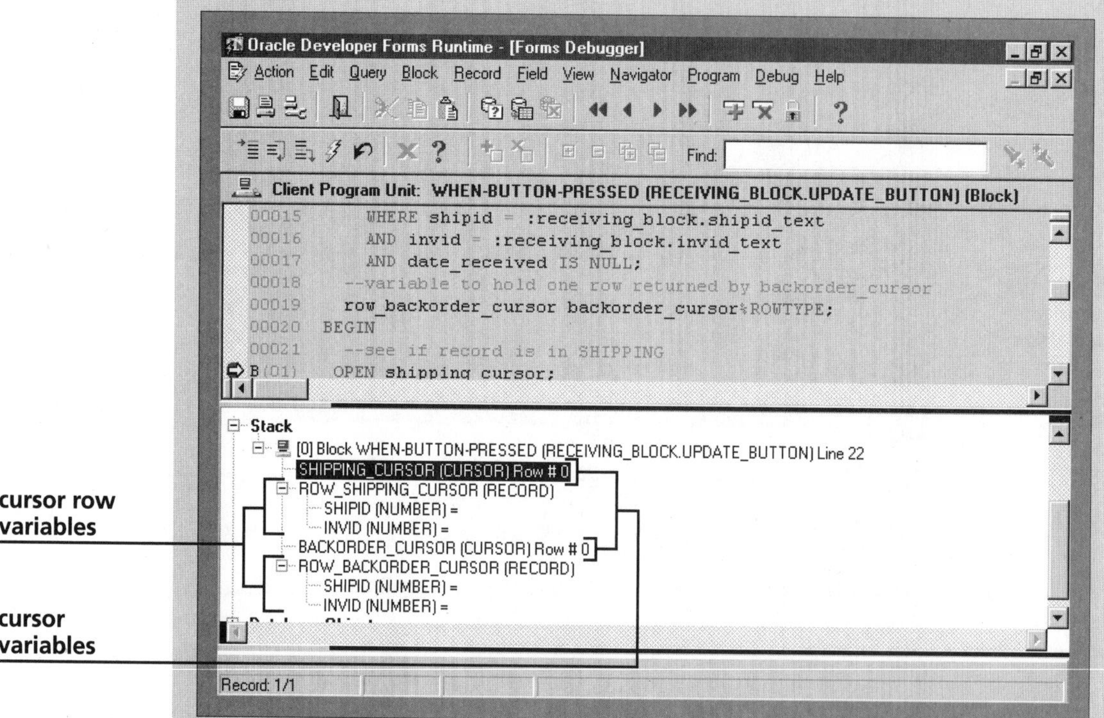

Figure 6-23: Current Stack variable values

2 Note the first variable, named SHIPPING_CURSOR(Cursor) Row #0. This variable represents the values in the cursor SELECT statement that will be retrieved by the cursor FETCH operation. Since no FETCH operation has been performed yet, the cursor's current position is Row #0.

3 Click ⊞ beside ROW_SHIPPING_CURSOR (RECORD). This line represents the values that will be stored in the variable that was declared to hold one row returned by the SHIPPING_CURSOR. Again, since no FETCH operation has been performed yet, the values are not yet defined.

4 Click ⊞ beside ROW_BACKORDER_CURSOR (RECORD) to confirm that the BACKORDER_CURSOR variables are not defined yet either.

Another important variable type is **Modules**, which allows you to select a specific block and then examine the **block variables**, which are the current values of block items in the form. Since you have already selected a shipment, many of the form block variables have values. You will view these next.

To view the form module block variables:

1 Scroll up in the Navigator pane to display Modules, then click ⊞ beside **Modules.** Open the following objects: **CH6ARECEIVING, Blocks, RECEIVING_BLOCK, Items.** The current form block variable values are displayed, as shown in Figure 6-24.

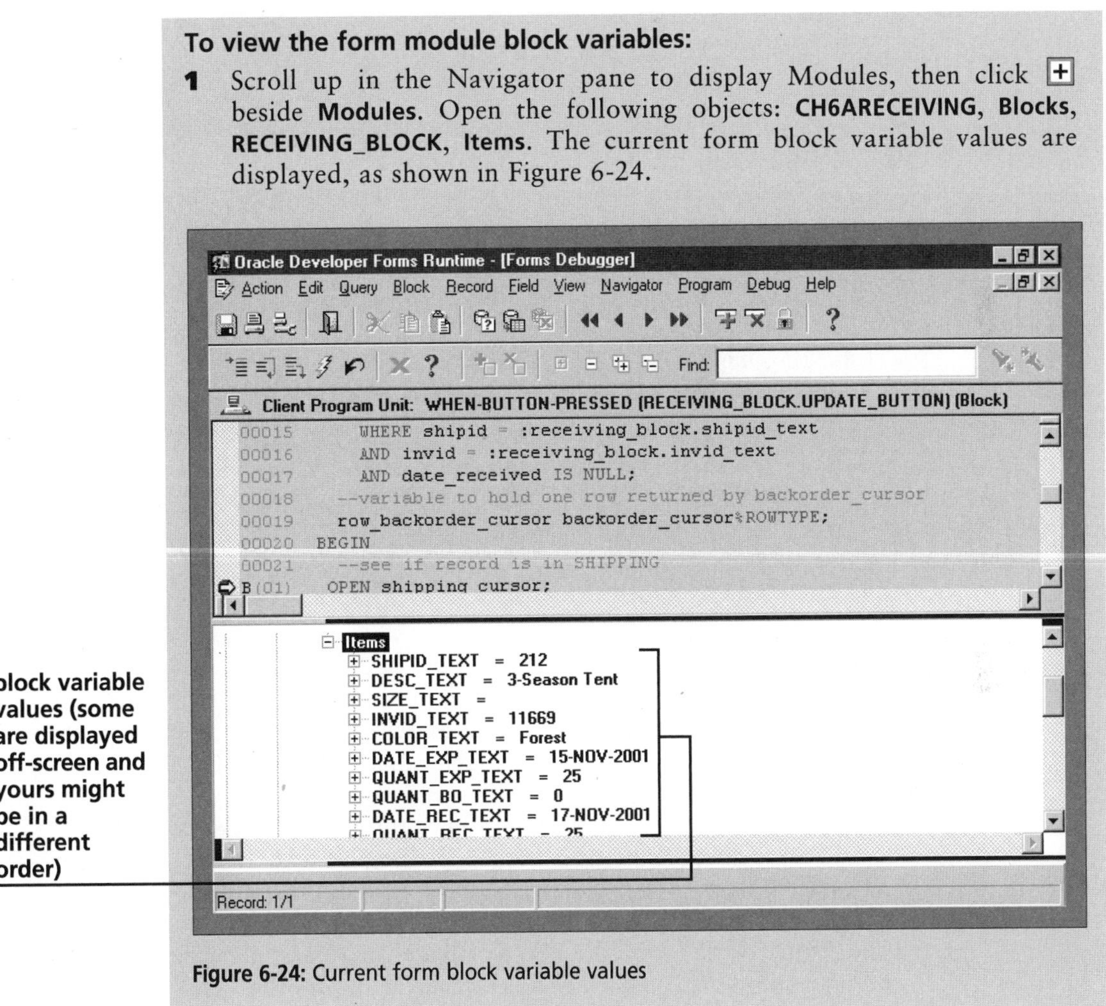

Figure 6-24: Current form block variable values

Now you will single-step through the trigger code and watch how the form variable values change. You will also find the program line that is causing the ORA-01400 error, and determine the cause of the error by examining the values of the variables.

To single-step through the trigger:

1. Click the **Step Into** button on the Debugger toolbar. The execution arrow stops on the SQL command in the first variable declaration (SELECT shipid, invid) as the cursor is opened.
2. Click again. If necessary, scroll down in the Source pane to find the execution arrow, which is on the command to fetch the first cursor row (FETCH shipping_cursor INTO row_shipping_cursor;).
3. Click again to fetch the row. Now you will examine the fetched values.
4. In the Navigator pane, click beside [0] block WHEN-BUTTON-PRESSED (RECEIVING_BLOCK.UPDATE_BUTTON) Line 22, and then click beside ROW_SHIPPING_CURSOR (RECORD). SHIPID(NUMBER) = 212 and INVID(NUMBER) = 11669 are displayed, indicating that the cursor has retrieved these data values.
5. Note the position of the execution arrow, which is on Line 24 (IF shipping_cursor%FOUND THEN). Since a record was found by shipping_cursor, the program statements after the IF command should execute.
6. Click again. The execution arrow stops on the command to call the program units. The UpdateShipping program unit is called first.
7. Click again. The execution arrow moves to the first line of the UpdateShipping program unit.

At this point, you need to examine the block variable values to confirm that legal values will be used in the UPDATE command in the UpdateShipping program unit. It is very important to anticipate where errors might occur in SQL statements, and examine the variable values *before* the ORA- error is generated. Once the error is generated, you cannot go back and look at the variable values. You will need to look at the values for the variables that will be used in the UPDATE command: :receiving_block.date_rec_text, :receiving_block.quant_rec_text, :receiving_block.shipid_text, and :receiving_block.invid_text. It is a good idea to write these variable values down. If the error is generated by this command, then you can type the SQL command in SQL*Plus, substitute the actual values for the variable names, and determine what problem in the query is causing the error message to be displayed. You cannot correct the error in the Debugger, but you can locate the code that is causing the error.

To examine the SQL statement variable values:

1. Scroll down in the Navigator pane if necessary, and note the current values of the variables that will be used in the SQL command, as shown in Figure 6-25. All of these values look like they should work, so you will execute the next program line that updates the SHIPPING table.

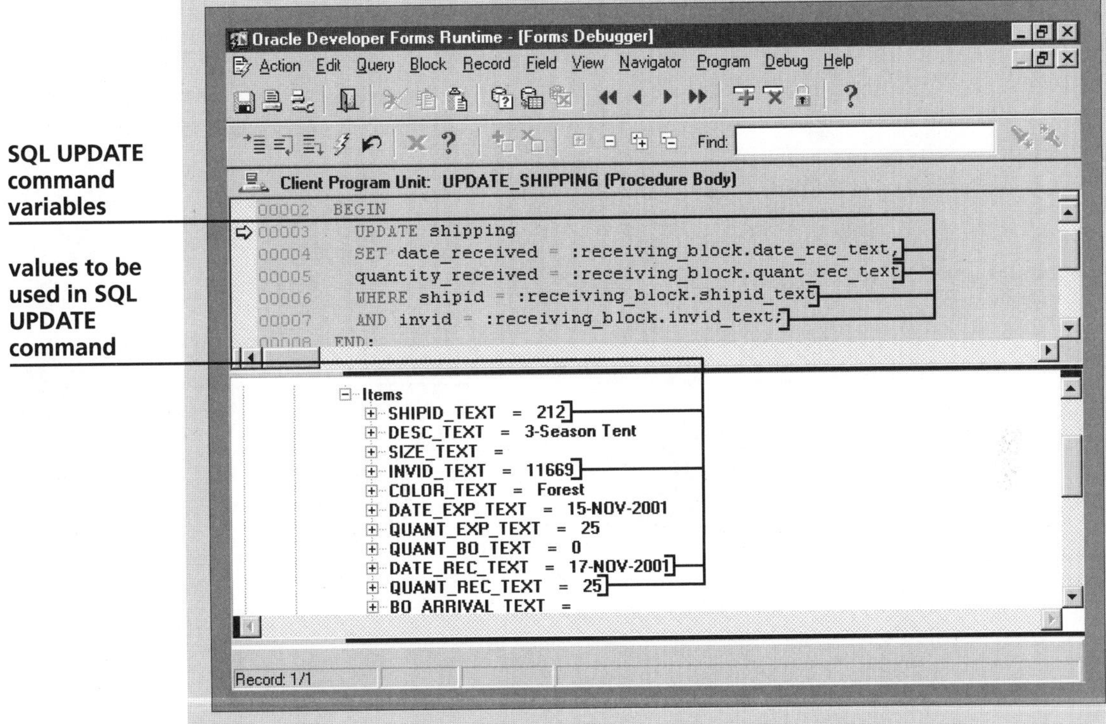

Figure 6-25: Block data values just before SQL UPDATE command is executed

2 Click the **Step Into** button. The execution arrow moves to Line 8 (END) of the UpdateShipping program unit. No error message is displayed, so this UPDATE command did not generate the ORA-01400 error.

Since no error occurred in the UPDATE_SHIPPING program unit, you know that the error is in one of the other parts of the code. Next you will step through the next program unit, which is PROCESS_BO. This is the program unit that processes potential back orders. Your test case does not involve a back order, so no new back-order record should be inserted.

To continue stepping through the program:

1 Click the **Step Into** button again. The execution arrow returns to the line of the UPDATE_BUTTON trigger that calls the PROCESS_BO program unit.

2 Click again. The execution arrow is displayed on Line 3 of the PROCESS_BO program unit. If necessary, scroll in the Navigator pane so the block data values are displayed, as shown in Figure 6-26.

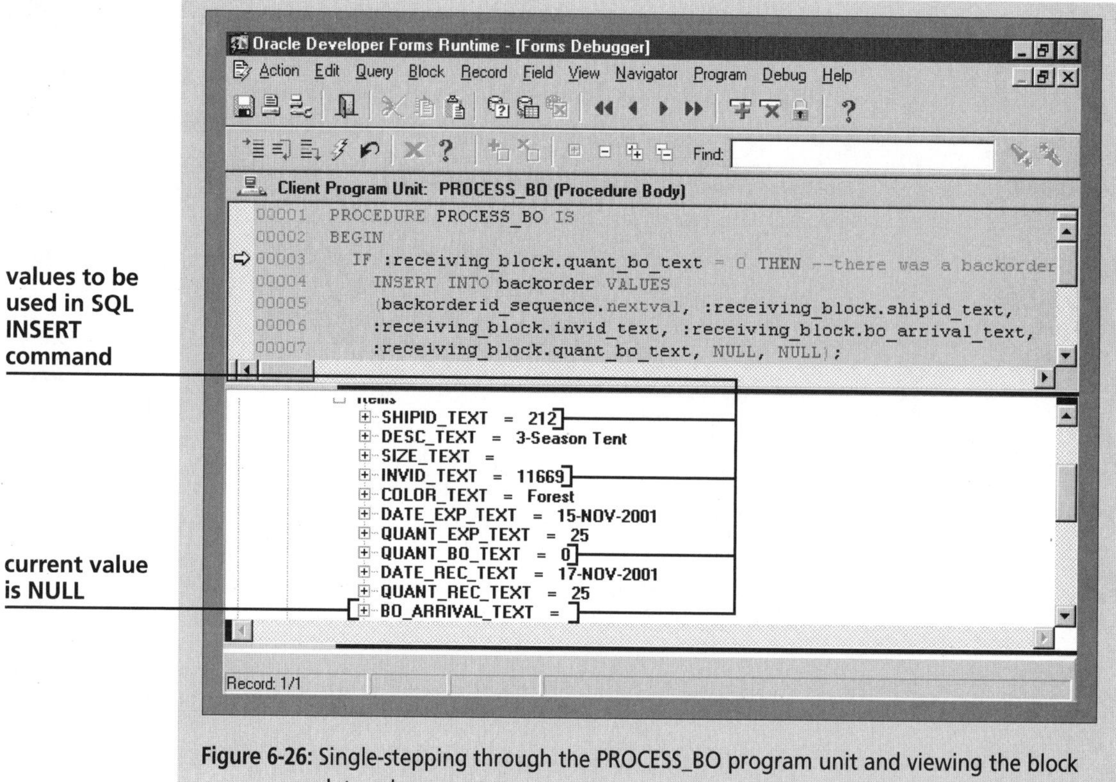

Figure 6-26: Single-stepping through the PROCESS_BO program unit and viewing the block data values

The PROCESS_BO program unit will execute the SQL command to insert a record into the BACKORDER table only if a back-order quantity was entered. You need to examine and note the variable values before the SQL INSERT command is executed, to anticipate if this is the query that is causing the error message.

To examine the block variable values:

1 View your current block data values, and note that the value of QUANT_BO_TEXT is 0, which is correct, since the current order does not have a back order.

2 Click the **Step Into** button. Since the IF statement condition was true, the execution arrow moves to the code line directly below the IF statement.

Note the values of the variables in the INSERT command as shown in Figure 6-26. One of the values, BO_ARRIVAL_TEXT, is NULL. This could be the source of the error—if this value is required in the BACKORDER table, then inserting a NULL value would generate the error message that you saw earlier.

Next you will switch to SQL*Plus, view the properties of the BACKORDER table, and determine if the field in the table corresponding to the back order arrival date has a NOT NULL constraint.

To view the BACKORDER table properties:

1. Click SQL*Plus on the Windows taskbar to switch to SQL*Plus.
2. Type the command shown in Figure 6-27. The result shows that the first five data fields cannot be NULL. This INSERT statement is generating the error.

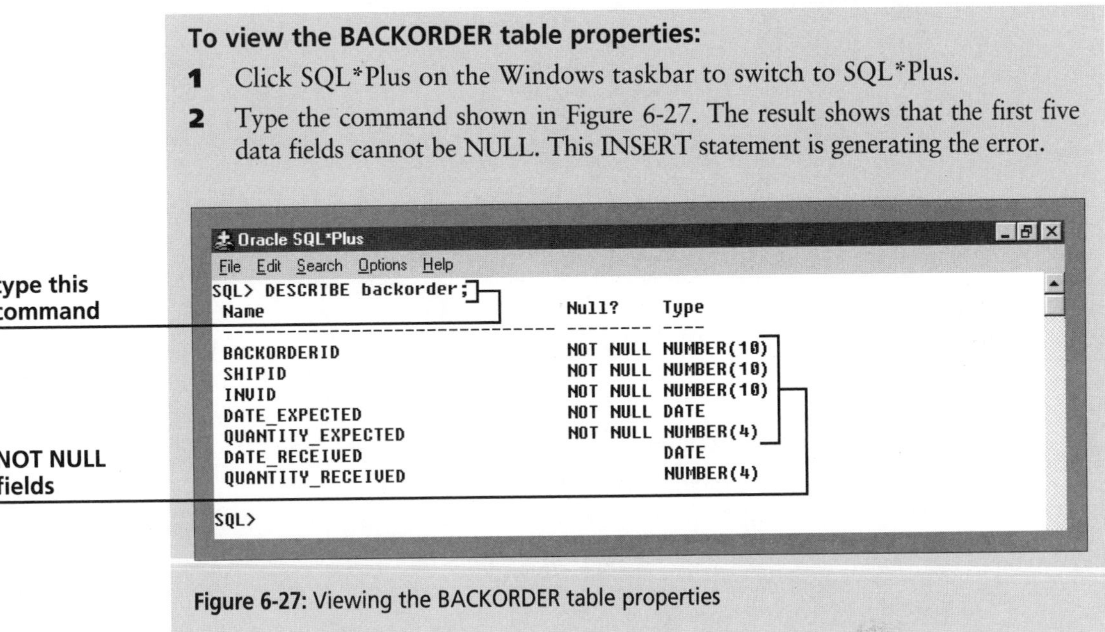

Figure 6-27: Viewing the BACKORDER table properties

Now you will switch back to the Debugger window, and execute the INSERT command to confirm that it is the source of the program error.

To execute the INSERT command:

1. Click . An error message stating that the program unit execution was aborted due to an unhandled exception is displayed.
2. Click **OK**, then click **Exit** to exit the form.

Now you need to figure out what caused the error. No value was entered for the back order arrival date, because the order did not have a back order. Look at the source code for the PROCESS_BO trigger in Figure 6-26: the IF statement show that if the QUANT_BO_TEXT equals zero, then a new record will be inserted into the BACKORDER table. That doesn't make sense—you should insert a new backorder record only if QUANT_BO_TEXT is *greater* than zero, not equal to zero. If there is no back order, then the user does not need to enter a backorder arrival date. Currently, the database requires a value for the back order arrival date, and generates an error when an INSERT command with no value for backorder arrival date is executed. Next you will correct the PROCESS_BO program unit code.

To correct the program unit code:

1. If necessary, click the **Form Builder** button on the Windows taskbar to activate the program window.
2. If necessary, open the Object Navigator window in Ownership View, open **Program Units**, and then double-click the **Program unit** icon beside the PROCESS_BO program unit.
3. Change the program line IF :receiving_block.quant_bo_text = 0 THEN to **IF :receiving_block.quant_bo_text > 0 THEN.**
4. Compile the program unit, correct any syntax errors, and then close the PL/SQL Editor.
5. Save the form.

Now you need to test the form again to see if this change corrects the error.

To test the form again:

1. Click the **Run Form Client/Server** button on the Object Navigator toolbar to run the form.
2. Click the **Shipping ID LOV** command button, select Shipment ID **212** and Inventory ID **11669**, and then click **OK.**
3. Type **25** in the Quantity Received field, and then click **Update.** The form fields are cleared, and the confirmation message "FRM-40401: No changes to save." is displayed.

help If an error message appears instead of the confirmation message, repeat the debugging exercise until you find the location of the next error. Analyze the values of the stack and block variables just before the error occurs so you can determine the cause of the error.

Although the control block confirmation message ("FRM-40401: No changes to save") seems counterintuitive, it is the confirmation message issued by the Form Builder when a COMMIT command is successfully executed by the database for a control block form. Before you can be totally confident that this operation worked correctly, you should double-check the values in the SHIPPING and INVENTORY tables. First, you will confirm that the SHIPPING record has been updated so the DATE_RECEIVED field is the current date, and the QUANTITY_RECEIVED field is 25. You will do that next in SQL*Plus.

To confirm that the SHIPPING record has been updated:

1 Click the **Oracle SQL*Plus** button on the taskbar.

2 Type the query shown in Figure 6-28 to confirm that the DATE_RECEIVED and QUANTITY_EXPECTED fields were updated correctly.

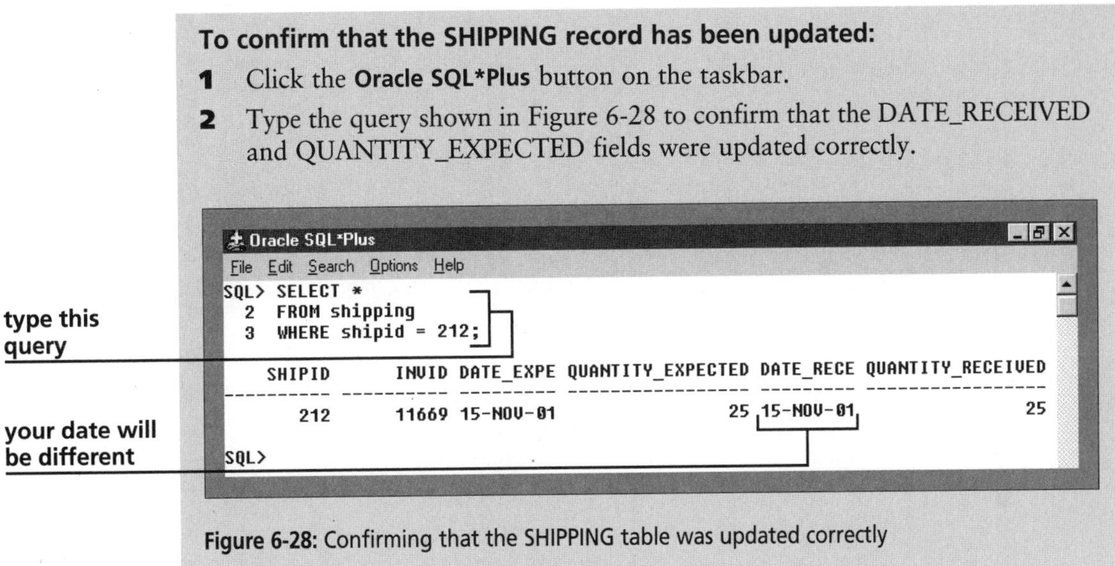
Oracle SQL*Plus
File Edit Search Options Help
SQL> SELECT *
2 FROM shipping
3 WHERE shipid = 212;
SHIPID INVID DATE_EXPE QUANTITY_EXPECTED DATE_RECE QUANTITY_RECEIVED
212 11669 15-NOV-01 25 15-NOV-01 25
SQL>
type this query
your date will be different

Figure 6-28: Confirming that the SHIPPING table was updated correctly

The query output confirms that DATE_RECEIVED is the current date and that the QUANTITY_RECEIVED is 25. Next, you need to determine if QOH was updated properly in the INVENTORY table. The original QOH for item 11669 in the INVENTORY table was 12. You just received 25 additional items, so the new QOH should be 37.

To verify that the INVENTORY table was updated correctly:

1 Type the query shown in Figure 6-29. The query output indicates that QOH was updated correctly.

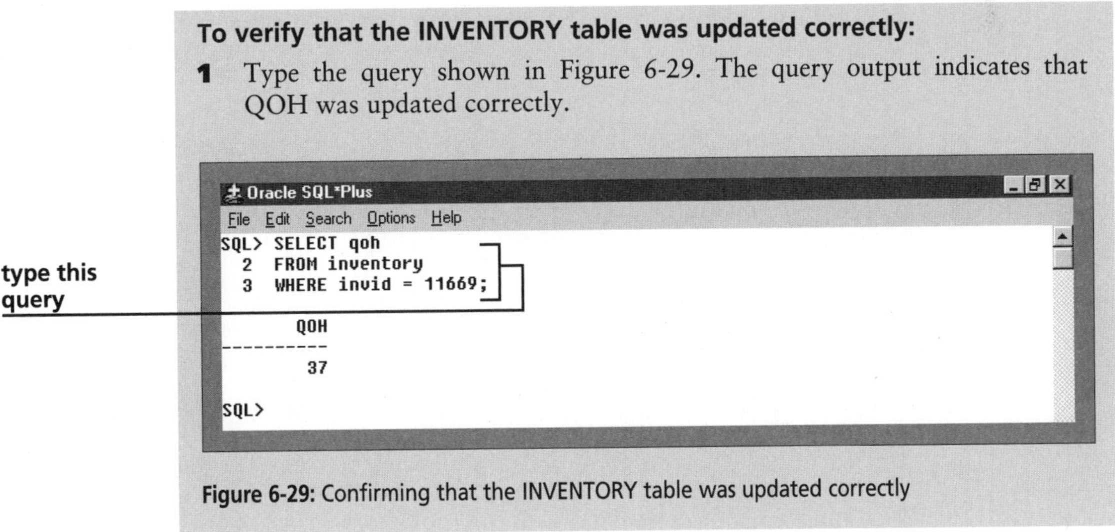
Oracle SQL*Plus
File Edit Search Options Help
SQL> SELECT qoh
2 FROM inventory
3 WHERE invid = 11669;
QOH
37
SQL>
type this query

Figure 6-29: Confirming that the INVENTORY table was updated correctly

Next, you need to test whether the form works correctly when a shipment with a back order is recorded. Assume that shipment 215 was received today, and all 100 of the Women's Fleece Pullovers, color Coral, size S were received, but none of the size M were received. One hundred items with size M are back ordered and are expected to arrive on December 1, 2001.

To test the form when a shipment with a back order is recorded:

1. Click the **Forms Runtime** button on the taskbar.
2. Click the **LOV** command button, select the size **S** items for shipment **215**, and then click **OK**. Enter that all **100** items were received, and then click **Update**. The "No changes to save" confirmation message is displayed.
3. Click the **LOV** command button, and select the size **M** items for shipment **215**, then click **OK**. Enter that **0** items were received, and that **100** items are back ordered and expected on **12/01/2001**.
4. Click **Update**. The "No changes to save" confirmation message indicates that the update works correctly.

help

If an error message appears instead of the confirmation message, repeat the debugging steps to locate the error and determine its cause. Fix your code, and then test it again.

5. When you successfully insert the back order, you will confirm that the information was updated correctly in the database. Click the **Oracle SQL*Plus** button on the Windows taskbar.
6. Type the query shown in Figure 6-30 to check the status of the SHIPPING table. The query output confirms that the DATE_RECEIVED was updated to today's date and that QUANTITY_RECEIVED is 0.

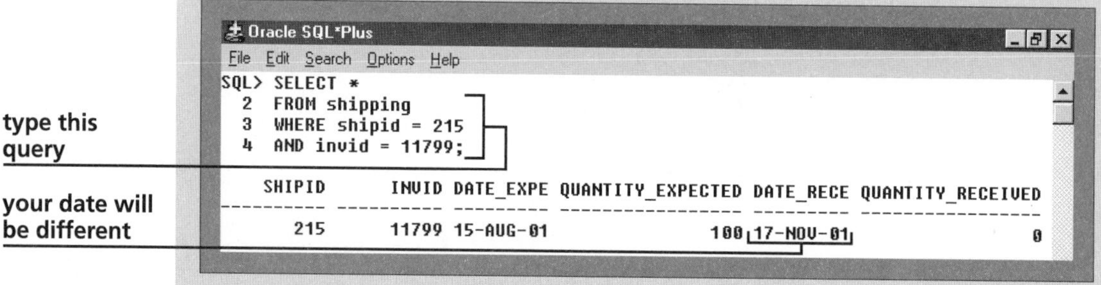

type this query

your date will be different

```
Oracle SQL*Plus
File Edit Search Options Help
SQL> SELECT *
  2  FROM shipping
  3  WHERE shipid = 215
  4  AND invid = 11799;

    SHIPID      INVID DATE_EXPE QUANTITY_EXPECTED DATE_RECE QUANTITY_RECEIVED
---------- ---------- --------- ----------------- --------- -----------------
       215      11799 15-AUG-01               100 17-NOV-01                 0
```

Figure 6-30: Confirming that the SHIPPING table was updated correctly

7. Type the query shown in Figure 6-31 to confirm that a new BACKORDER record was added. The query output confirms that a new BACKORDER record was created with the DATE_EXPECTED field as 01-DEC-2001 and the QUANTITY_EXPECTED equal to 100. (Your BACKORDERID value might be different, depending on the current value of your sequence.)

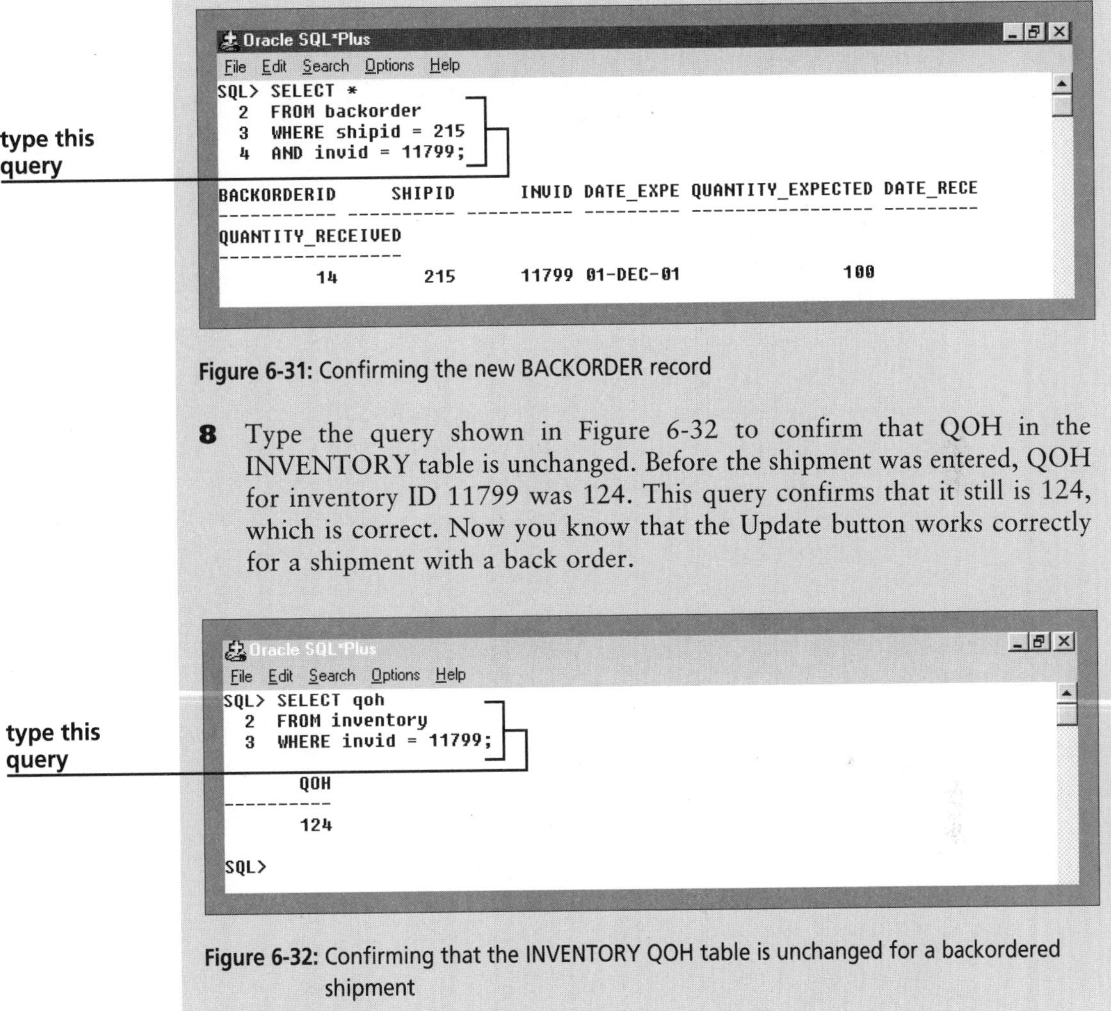
type this query
Oracle SQL*Plus
File Edit Search Options Help
SQL> SELECT *
2 FROM backorder
3 WHERE shipid = 215
4 AND invid = 11799;
BACKORDERID SHIPID INVID DATE_EXPE QUANTITY_EXPECTED DATE_RECE
QUANTITY_RECEIVED
14 215 11799 01-DEC-01 100

Figure 6-31: Confirming the new BACKORDER record

8 Type the query shown in Figure 6-32 to confirm that QOH in the INVENTORY table is unchanged. Before the shipment was entered, QOH for inventory ID 11799 was 124. This query confirms that it still is 124, which is correct. Now you know that the Update button works correctly for a shipment with a back order.

type this query
Oracle SQL*Plus
File Edit Search Options Help
SQL> SELECT qoh
2 FROM inventory
3 WHERE invid = 11799;
QOH
124
SQL>

Figure 6-32: Confirming that the INVENTORY QOH table is unchanged for a backordered shipment

Sometimes when a company like Clearwater Traders orders items, the supplier does not have the items in stock. The supplier must notify Clearwater Traders that the items are back ordered. Sometimes the supplier cannot get the backordered items from their supplier, or they cannot manufacture the items due to some unforeseen circumstances. This forces a backordered item to be back ordered again. When this happens, the supplier sends a notification to Clearwater Traders to let them know when the shipment of the backordered items can be expected. Your next task is to ensure that the database is updated correctly when Clearwater Traders receives a backordered shipment that does not require another back order. You will need to enter the receiving information using the receiving form and then confirm the updates made on the BACKORDER and INVENTORY tables.

To check the updates when Clearwater Traders receives a back-ordered shipment with no additional back order:

1. Click the **Forms Runtime** button on the taskbar.
2. Click the **Shipping ID LOV** button, and then select Shipment ID **215** and Inventory ID **11799**. This shipment is the one that you back ordered in the previous set of steps. Click **OK**.
3. Type **100** in the Quantity Received field.
4. Click **Update**. The "No changes to save" confirmation message indicates that the change was made. (Recall that this is the counterintuitive error message that Oracle always generates whenever a COMMIT command is successfully executed in a form trigger.) Now you need to check the BACKORDER and INVENTORY tables to confirm that they were updated correctly.
5. Click the **Oracle SQL*Plus** button on the taskbar.
6. Type the query shown in Figure 6-33. The query output shows that the DATE_RECEIVED is today's date and the QUANTITY_RECEIVED is 100 items.

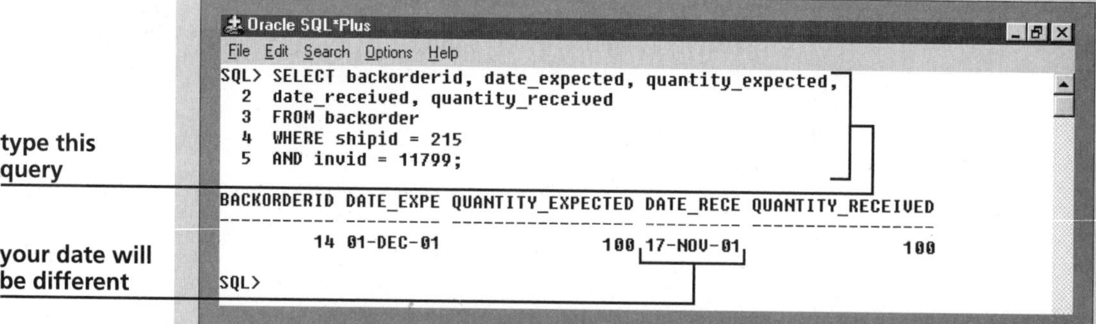

Figure 6-33: Confirming that the BACKORDER record was updated correctly

7. Type the query shown in Figure 6-34. As noted before, the quantity on hand for Inventory ID 11799 was 124. Now 100 new items have been received, and the updated quantity on hand should be 224, as confirmed in the query output.

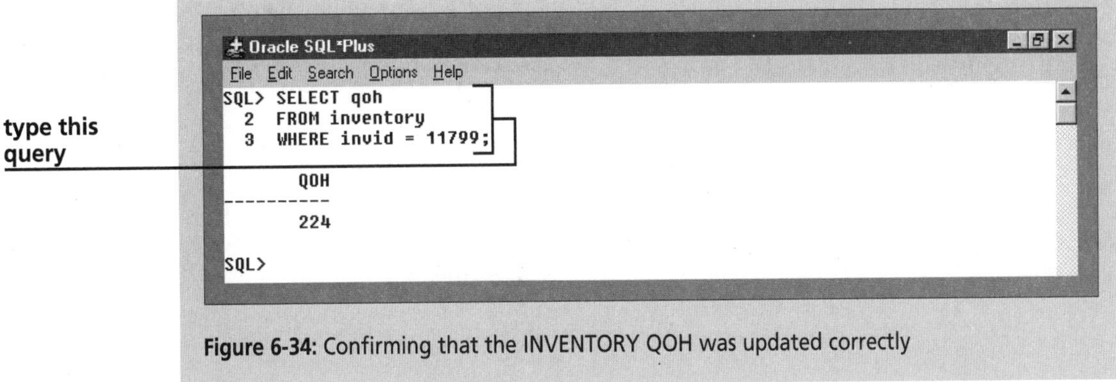
type this query
Oracle SQL*Plus
File Edit Search Options Help
SQL> SELECT qoh
2 FROM inventory
3 WHERE invid = 11799;
QOH

224
SQL>

Figure 6-34: Confirming that the INVENTORY QOH was updated correctly

In the last test case, Clearwater Traders receives notification of a backordered shipment that is back ordered again. Initially, you received a notification for shipment 213 (200 pairs of Women's Hiking Shorts, color Khaki, size L) stating that all 200 pairs are back ordered with an expected shipment date of 11/15/2001. Then, a few days later, you receive another notification about the shipment, stating the expected shipment date has been changed to 12/15/2001. You will create the first back-order record, and then you will update it with the second back-order notification.

To check whether the update is correct for a backordered record that is backordered again:

1. Click the **Forms Runtime** button on the taskbar.
2. Click the **Shipment ID LOV** command button, and then select Shipment ID **213** and Inventory ID **11777**. Click **OK**.
3. Enter **0** for Quantity Received, **200** for Quantity Backordered, and **11/15/2001** for Backorder Arrival Date, and then click **Update**. The "No changes to save" confirmation message is displayed. Now enter the record for the second back-order notification with the later shipment date.
4. Click the **Shipment ID LOV** command button, and select the same shipment record again (Shipment ID **213** and Inventory ID **11777**). Note that the date expected field is now 11/15/2001. Click **OK**.
5. Enter **0** for Quantity Received, **200** for Quantity Backordered, and **12/15/2001** for Backorder Arrival Date, and then click the **Update** button. The "No changes to save" confirmation message is displayed.

Next you need to use SQL*Plus to check how the records were updated in the BACKORDER table, and confirm that QOH was not updated by the first back-order entry.

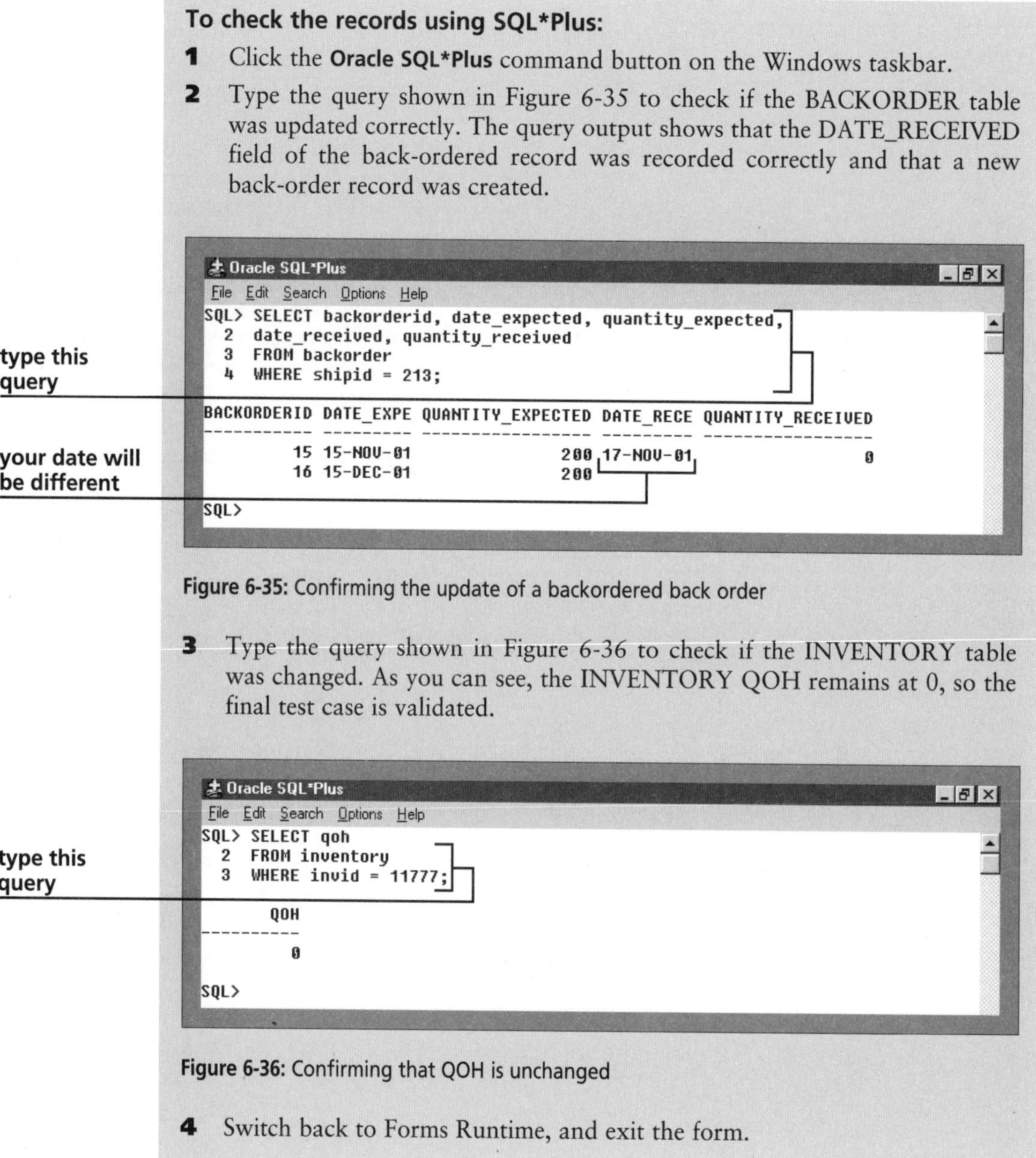

Oracle SQL*Plus
File Edit Search Options Help
SQL> SELECT backorderid, date_expected, quantity_expected,
2 date_received, quantity_received
3 FROM backorder
4 WHERE shipid = 213;
BACKORDERID DATE_EXPE QUANTITY_EXPECTED DATE_RECE QUANTITY_RECEIVED
15 15-NOV-01 200 17-NOV-01 0
16 15-DEC-01 200
SQL>
type this query
your date will be different
Oracle SQL*Plus
File Edit Search Options Help
SQL> SELECT qoh
2 FROM inventory
3 WHERE invid = 11777;
QOH
0
SQL>
type this query

To check the records using SQL*Plus:

1. Click the **Oracle SQL*Plus** command button on the Windows taskbar.
2. Type the query shown in Figure 6-35 to check if the BACKORDER table was updated correctly. The query output shows that the DATE_RECEIVED field of the back-ordered record was recorded correctly and that a new back-order record was created.

Figure 6-35: Confirming the update of a backordered back order

3. Type the query shown in Figure 6-36 to check if the INVENTORY table was changed. As you can see, the INVENTORY QOH remains at 0, so the final test case is validated.

Figure 6-36: Confirming that QOH is unchanged

4. Switch back to Forms Runtime, and exit the form.

Finalizing the Clearwater Traders Receiving Form

The final step in finishing the Clearwater Traders Receiving form is to import the Clearwater Traders logo, create a trigger to automatically maximize the Forms Runtime window when the form is loaded, and update the tab order for the form fields. First you will import the Clearwater Traders logo, and create a PRE-FORM trigger that has a command to automatically maximize the Forms Runtime window when the form runs. Within the Forms application, the Forms Runtime window is called the **MDI (multiple-document interface)** window. **MDI** is the Microsoft Windows window management system that allows programs to display an outer "parent" window (called the application window) and multiple inner windows. Multiple form windows can be active within the Forms Runtime MDI window.

To import the Clearwater Traders logo and create the PRE-FORM trigger for maximizing the Forms Runtime MDI window:

1. Import the Clearwater Traders logo as a graphic image using the clearlogo.tif file from the Chapter6 folder of your Student Disk. Center the logo at the top of the form.
2. Open the Object Navigator window in Ownership View, select **Triggers** (directly under the CH6ARECEIVING form module), right-click, point to **Smart Triggers**, then click **PRE-FORM**.
3. Type the following code in the PL/SQL Editor to maximize the application window (which is named FORMS_MDI_WINDOW) at runtime:

```
SET_Window_Property(FORMS_MDI_WINDOW, WINDOW_STATE, MAXIMIZE);
```

4. Compile the code, correct any syntax errors, close the PL/SQL Editor, and save the form.
5. Run the form to confirm that the Forms Runtime window maximizes automatically, then close Forms Runtime.

In most Windows applications, the user can press the Tab key to navigate between text fields and other form items. Usually the navigation order is top to bottom, right to left. To set the tab order of items in a Forms Builder application, you need to place the items in the correct order under the Data Blocks object in the Object Navigator window. The text item where the insertion point will appear at form startup should be listed first, the field or item where the insertion point will go when the user first presses the Tab key should be listed next, and so on. Next, you will adjust the tab order of the form items.

To adjust the tab order of the form items:

1. Select SHIPID_TEXT in the Object Navigator window, and drag and drop it so it is the first field listed under Items in the RECEIVING_BLOCK.
2. Select SHIPPING_LOV_BUTTON, and drag and drop it so it is the second item listed under the RECEIVING_BLOCK.
3. Continue to move the form fields until they are in the order shown in Figure 6-37.

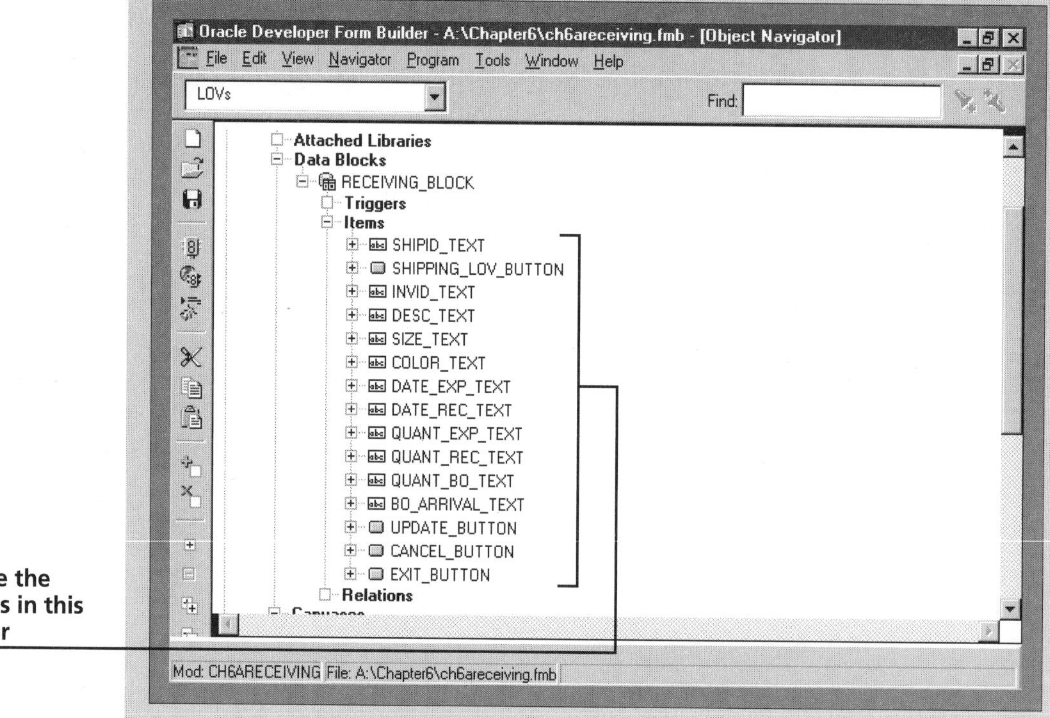

Figure 6-37: Arranging the tab order of the form items

4. Run the form to check the tab order. The order that you tab through the fields should be the same as the order in which the fields are listed in Figure 6-37.
5. Close Forms Runtime.
6. Save the form, and close Form Builder and SQL*Plus.

SUMMARY

- Process text descriptions, flowcharts, and sketches of the user interface for a form are important tools to use when designing custom forms.
- To create a custom form, you have to create the form manually, "paint" the form items on the canvas, and write the code that controls the form functions.
- Custom forms have control blocks that are not connected to a particular database table.
- Text items that display database records must have the same data type as the corresponding data field in the database, and their maximum length must be large enough to accommodate the maximum length of the data, plus any characters included in a format mask.
- A LOV cannot be derived from a UNION, but can be derived only from a conventional SELECT query.
- Complex triggers should be divided into smaller program units that are easier to understand and debug.
- The Form Builder Debugger allows you to step through PL/SQL programs one line at a time to examine values of variables during execution.
- Breakpoints are places in PL/SQL programs where you can pause program execution and single-step through the code that follows and/or examine current variable values.
- The Stack object in the Form Builder Debugger allows you to examine the values of local variables. The Modules object in the Form Builder Debugger allows you to examine the values of form text items.
- MDI (multiple-document interface) is the Microsoft Windows window management system in which applications display an outer "parent" window (called the application window) and multiple inner windows. You can automatically maximize the MDI run-time window using a PRE-FORM trigger.
- To set the tab order of items in a Forms application, you need to place the items in the correct order under the block object in the Object Navigator window.

REVIEW QUESTIONS

1. What three tools can be used to design a Form Builder application to support a complex business process?
2. When can you use a data block form, and when do you need to create a custom form?
3. What is the difference between a custom form and a data block form?

4. List the text item data types and minimum lengths you would be required to use to display the following data fields:

Table	Field	Format Mask
CUST_ORDER	ORDERDATE	29-MAY-2001
STUDENT	SADD	none
FACULTY	FPHONE	(715) 555-1234
LOCATION	CAPACITY	none
TERM	STATUS	none

5. Write a UNION query to create a view named NorthwoodsView that contains all student first and last names and phone numbers, and all faculty first and last names and phone numbers from the Northwoods University database. Then, write a query for a LOV that will display all of the view data fields.
6. What is a breakpoint?
7. On what kinds of program lines can you set breakpoints? On what kinds of lines can you not set breakpoints?
8. How do you set a breakpoint? How can you visually tell that a breakpoint has been set on a program line?
9. What is the purpose of the Step Into button in the Debugger window?
10. What is the execution arrow?
11. What are stack variables in the Form Builder Debugger?
12. What are block variables in the Form Builder Debugger?
13. Describe the path that you would follow in the Form Builder Debugger Navigation pane to find the value of a variable named :custom_block.cust_name_text that is in a form named TEST_FORM.
14. What is the MDI window? How can you maximize it at runtime?

PROBLEM-SOLVING CASES

1. Create a custom form to insert new records in the FACULTY table in the Northwoods University database with a user interface like the one shown in Figure 6-38. When the user clicks the Create button, a new faculty ID appears. The user then enters the faculty member's last and first name, middle initial, and phone number, and selects the faculty member's location ID and rank from a list of values. When the user clicks the Save button, the record is saved in the FACULTY table.

   a. Create a new form and save it as Ch6aEx1.fmb in the Chapter6 folder on your Student Disk.
   b. Create a PRE-FORM trigger to automatically maximize the Forms Runtime window at startup.

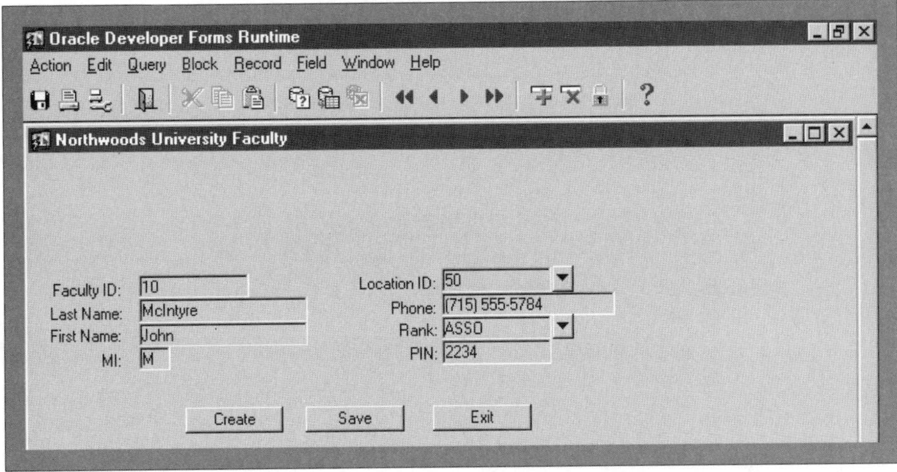

Figure 6-38

c. Change the form window name to FACULTY_WINDOW. Change the window title to Northwoods University Faculty, and change the window size so it fits correctly within the Forms Runtime window.

d. Create a new canvas named FACULTY_CANVAS.

e. Create a control block named FACULTY_BLOCK.

f. Create text items and labels as shown in Figure 6-38. Give each text item an appropriate and descriptive name, and change each text item's data type and maximum length to correspond to the associated database field.

g. Change the format mask of the FPHONE text item so phone numbers are displayed as (715) 555-1234. Change the field length as necessary.

h. Create a LOV associated with the LOCID text item that displays LOCID, BLDG_CODE, and ROOM from the LOCATION table, and returns the selected LOCID to the Location ID field on the form.

i. Create a LOV associated with the FRANK text item that displays all possible FRANK values from the FACULTY table, and returns the selected FRANK to the Rank field on the form. (*Hint*: Use the DISTINCT qualifier in the LOV query to suppress duplicates.)

j. Create three buttons named CREATE_BUTTON, SAVE_BUTTON, and EXIT_BUTTON. Modify their labels as shown, and align the button group and center it at the bottom of the form.

k. Create a new sequence named FID_SEQUENCE that starts with 10 and has no maximum value.

l. Create a WHEN-BUTTON-PRESSED trigger for the CREATE_BUTTON that retrieves the next value of the FID_SEQUENCE, and displays it in the Faculty ID text field. (*Hint*: You will use the following command: SELECT fid_sequence.nextval INTO :faculty_block.fid_text FROM dual;.)

m. Create a WHEN-BUTTON-PRESSED trigger for the SAVE_BUTTON that inserts the new data into the FACULTY table and COMMITS it to the database.

n. Create a WHEN-BUTTON-PRESSED trigger for the EXIT_BUTTON that exits the form.

2. Create a custom form to update records in the COURSE_SECTION table in the Northwoods University database with a user interface like the one shown in Figure 6-39. When the user clicks the LOV command button next to the Section ID, a list of values for all course sections is displayed. The user can select a specific course section, and its associated data fields appear on the form. The user can modify the fields as necessary, and then click the Update button to save the changes to the database.

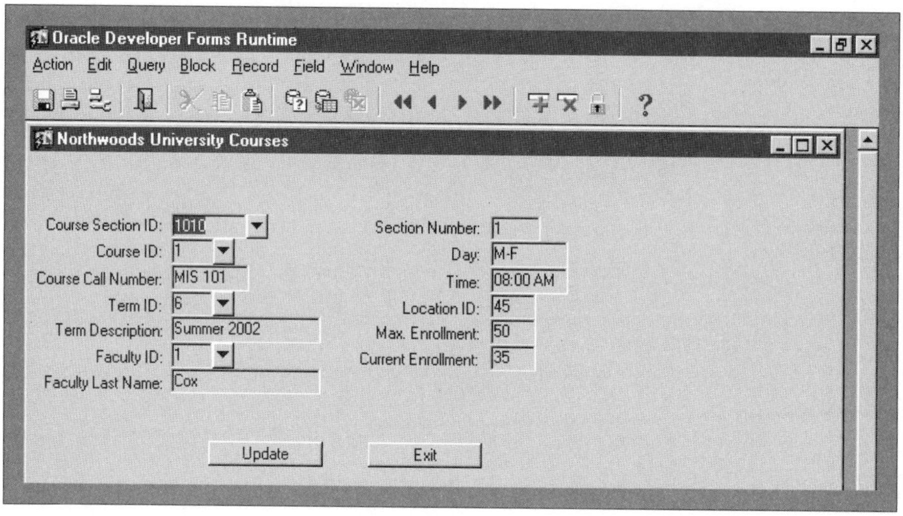

Figure 6-39

a. Create a new form and save it as Ch6aEx2.fmb in the Chapter6 folder on your Student Disk.

b. Create a PRE-FORM trigger to automatically maximize the Forms Runtime window at startup.

c. Change the form window name to CSEC_WINDOW. Change the window title to Northwoods University Courses, and change the window size so it fits correctly within the Forms Runtime window.

d. Create a new canvas named CSEC_CANVAS.

e. Create a control block named CSEC_BLOCK.

f. Create text items and labels as shown in Figure 6-39. Give each text item an appropriate and descriptive name, and change each text item's data type and maximum length to correspond to the associated database field.

g. Create a LOV associated with the CSECID text item that displays the course section ID, course call ID, term ID, term description, section number, day, and time. The LOV should return the selected record's CSECID, CID, CALLID, TERMID, TDESC, FID, FLNAME, SECNUM, DAY, TIME, LOCID, MAXENRL, and CURRENRL to the form fields. (*Hint*: Your LOV query will have some fields in the SELECT statement that are not displayed in the LOV display. To hide a column in the LOV display, set its display width equal to zero.)(*Hint*: To make the Time data field display as a time value, convert the returned data value to a character data type using the following query to creat the LOV: SELECT csecid, callid, term.termid, tdesc, secnum, day, TO_CHAR(time, 'HH:MI AM'), course.cid, faculty.fid, flname, course_section.locid, maxernrl, currenrl FROM course_section, course, term, faculty WHERE course_section.cid=course.did AND course_section.termid=term.termed AND course_section.fid=faculty.fid;. Make sure that the TIME field on the form so it has a character data type.)

h. Create a LOV associated with the Course ID text item that displays all possible values for CID, CALLID, and CNAME, and returns the selected CID and CALLID to the Course ID and Call ID fields on the form.

i. Create a LOV associated with the Term ID text item that displays all possible values for TERMID, TDESC, and STATUS, and returns the selected TERMID and TDESC to the Term ID and Term Description fields on the form.

j. Create a LOV associated with the Faculty ID text item that displays all possible values for FID, FLNAME, and FFNAME, and returns the selected FID and FLNAME to the Faculty ID and Faculty Last Name fields on the form.

k. Create two buttons named UPDATE_BUTTON and EXIT_BUTTON. Modify their labels as shown in Figure 6-39, and align the button group and center it at the bottom of the form.

l. Create a WHEN-BUTTON-PRESSED trigger for the UPDATE_BUTTON that updates the current CSECID record in the COURSE_SECTION table using the current values displayed in the form for CID, TERMID, SECNUM, FID, DAY, TIME, LOCID, MAXENRL, and CURRENRL, and COMMITS the change to the database.

m. Create a WHEN-BUTTON-PRESSED trigger for the EXIT_BUTTON that exits the form.

3. Create a custom form to add new records to the ENROLLMENT table in the Northwoods University database with a user interface like the one shown in Figure 6-40. When the user clicks the LOV command button next to the Student ID, a list of values that shows student information is displayed. When the user clicks the LOV command button next to the Section ID, a list of values is displayed for all course sections that have open enrollment (term status = 'OPEN') and are not filled to capacity. The user can select a specific course section, and its associated data fields appear on the form. When the user clicks the Add Record button, the SID and CSECID values are inserted into the ENROLLMENT table, and the CURRENRL of the course section is incremented by one.

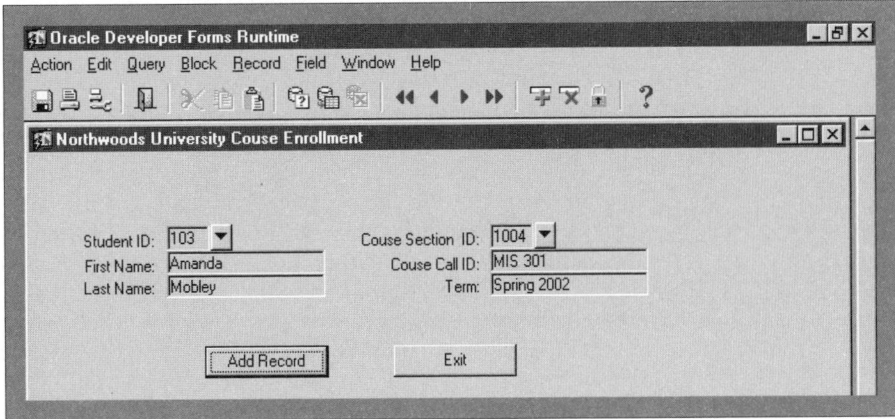

Figure 6-40

a. Create a new form and save it as Ch6aEx3.fmb in the Chapter6 folder on your Student Disk.
b. Create a PRE-FORM trigger to automatically maximize the Forms Runtime window at startup.
c. Change the form window name to ENRL_WINDOW. Change the window title to Northwoods University Course Enrollment, and change the window size so it fits correctly within the Forms Runtime window.
d. Create a new canvas named ENRL_CANVAS.
e. Create a control block named ENRL_BLOCK.
f. Create text items and labels as shown in Figure 6-40. Give each text item an appropriate and descriptive name, and change each text item's data type and maximum length to correspond to the associated database field.
g. Create a LOV associated with the SID text item that displays the student ID, last name, and first name. The LOV should return the selected record's SID, SFNAME, and SLNAME to the form fields.
h. Create a LOV associated with the Course Section ID text item that displays all possible values for CSECID, CALLID, CNAME, SECNUM, DAY, TIME, and TERMDESC where TERM STATUS = 'OPEN' and MAXENRL is greater than CURRENRL. The LOV should return the selected CSECID, CALLID, and TDESC to the fields on the form. (*Hint*: To make the TIME data field display as a time value, convert the returned data value to a character data type using the following query to create the LOV: SELECT course_section.csecid, callid, cname, secnum, day, TO CHAR(time, 'HH:MI AM'), tdesc FROM course_section, course, termWHERE course_section.cid = course.cid AND course_section.termid=term.termed AND term.status = 'OPEN' and MAXENRL>CURRENRL;.)
i. Create two buttons named ADD_BUTTON and EXIT_BUTTON. Modify their labels as shown, and align the button group and center it at the bottom of the form.
j. Create a WHEN-BUTTON-PRESSED trigger for the ADD_BUTTON that adds the current SID and CSECID to the ENROLLMENT table, with a value of NULL for the grade. The trigger must also update the COURSE_SECTION table so that when the record is added to the ENROLLMENT table, the CURRENRL of the associated CSECID is incremented by one. Be sure to COMMIT the changes.
k. Create a WHEN-BUTTON-PRESSED trigger for the EXIT_BUTTON that exits the form.

4. Create a custom form to update records in the ENROLLMENT table in the Northwoods University database with a user interface like the one shown in Figure 6-41. When the user clicks the LOV command button next to the Student ID, a list of values for all students is displayed. When the user clicks the LOV command button next to the Section ID, a list of course sections for that student is displayed from the ENROLLMENT table. The user can select a specific course section, and its associated data fields appear on the form. When the user clicks the Update button, the grade value for the SID and CSECID combination is updated in the ENROLLMENT table. When the user clicks the Delete button, the SID and CSECID combination is deleted from the ENROLLMENT table. The record can only be deleted if the GRADE value for the SID/CSECID combination is NULL.

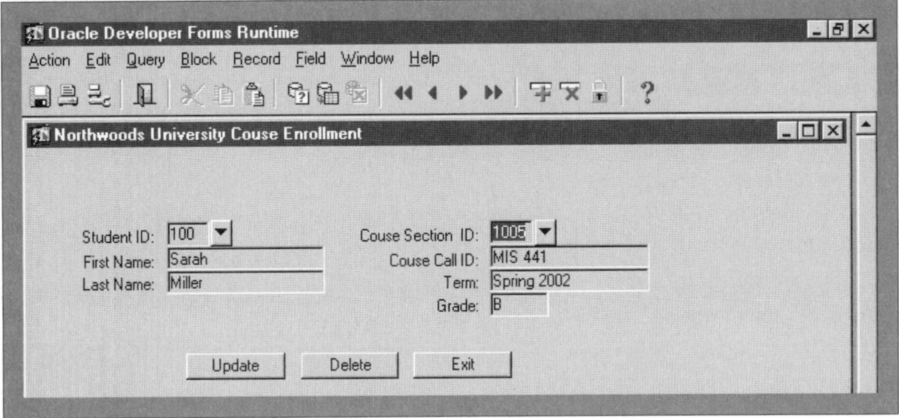

Figure 6-41

a. Create a new form and save it as Ch6aEx4.fmb in the Chapter6 folder on your Student Disk.

b. Create a PRE-FORM trigger to automatically maximize the Forms Runtime window at startup.

c. Change the form window name to ENRL_WINDOW. Change the window title to Northwoods University Course Enrollment, and change the window size so it fits correctly within the Forms Runtime window.

d. Create a new canvas named ENRL_CANVAS.

e. Create a control block named ENRL_BLOCK.

f. Create text items and labels as shown in Figure 6-41. Give each text item an appropriate and descriptive name, and change each text item's data type and maximum length to correspond to the associated database field.

g. Create a LOV associated with the SID text item that displays the student ID, last name, and first name. The LOV should return the selected record's SID, SFNAME, and SLNAME to the form fields.

h. Create a LOV associated with the Course Section ID text item that displays CSECID, CALLID, CNAME, TERMDESC, and GRADE for the student whose SID is currently displayed in the Student ID field. The LOV should return the selected CSECID, CALLID, TDESC, and GRADE to the fields on the form.

i. Create three buttons named UPDATE_BUTTON, DELETE_BUTTON, and EXIT_BUTTON. Modify their labels as shown, and align the button group and center it at the bottom of the form.

j. Create a WHEN-BUTTON-PRESSED trigger for the UPDATE_BUTTON that updates the GRADE field in the ENROLLMENT table for the current SID and CSECID displayed on the form, and then commits the change to the database.

k. Create a WHEN-BUTTON-PRESSED trigger for the DELETE_BUTTON that checks to see if the GRADE field for the SID/CSECID combination that is currently selected is NULL. If it is, the trigger should delete the SID/CSECID combination from the ENROLLMENT table, and then commit the change to the database.

l. Create a WHEN-BUTTON-PRESSED trigger for the EXIT_BUTTON that exits the form.

LESSON B

objectives

- Suppress system messages
- Create alerts and messages to provide system feedback
- Add error traps to an application
- Handle common form errors
- Display system error messages in alerts

Suppressing System Messages

By now you are familiar with the FRM- and ORA- messages that appear in the message area when you execute your forms. These messages are useful for determining if a record is successfully inserted, updated, or deleted, as well as determing what errors are occurring while you are running your forms. Sometimes you might want to suppress these messages and provide your own message descriptions for your forms' users using the MESSAGE procedure. Or, you might want to provide a message that poses a question that a user can respond to, such as whether to commit a change to the database or roll it back. In this lesson, you will learn how to suppress the system-generated messages, generate customized messages, and implement triggers for avoiding common user errors.

Oracle system messages are categorized by severity level and message type. **Informative messages** provide information about what is happening and usually do not require any user action. Informative messages have low severity levels (usually 5). **Error messages** require some user action and their severity levels are higher, with levels of 15, 20, or >25. These error levels are assigned by Oracle, and the scale simply compares relative severity levels, with 5 being of low severity, 15 somewhat higher, 20 somewhat higher still, and >25 of the highest severity. Form Builder determines which message severity level to display in the message area, using a system variable named **:SYSTEM.MESSAGE_LEVEL**. You can set the level of this variable in a PRE-FORM trigger to control which messages are displayed when the form is used. You will now refresh your database tables, open the Clearwater Traders Receiving form that you created in the previous lesson, and modify its PRE-FORM trigger to set the :SYSTEM.MESSAGE_LEVEL variable to 25 so that only the most severe messages (severity level >25) are displayed.

To modify the PRE-FORM trigger:

1. Start SQL*Plus, and run the clearwat.sql script from the Chapter6 folder of your Student Disk.
2. Start Form Builder, open **ch6areceivingdone.fmb** from the Chapter6 folder on your Student Disk, and save the file as **ch6breceiving.fmb.**

3 If necessary, open the Object Navigator window in Ownership View, and change the form name to CH6BRECEIVING.

4 In the Object Navigator window, click [+] beside **Triggers**, then double-click the **trigger** icon next to the PRE-FORM trigger. The PL/SQL Editor opens with the current PRE-FORM trigger displayed.

5 Type the following code below the current code to modify the PRE-FORM trigger:

```
:SYSTEM.MESSAGE_LEVEL := 25;
```

6 Compile the trigger, correct any syntax errors, and then close the PL/SQL Editor. Now run a test to see the effect of this action. Assume that the shipment of 50 Men's Expedition Parkas, color Navy, size S, has arrived.

7 Run the form.

8 Click the **LOV** command button next to Shipment ID, select the record for Shipment ID **212**, then click **OK**.

9 Enter **25** for Quantity Received, and then click **Update**. Notice that the usual confirmation message ("FRM 40401: No changes to save.") does not appear.

10 Close Forms Runtime.

Your form fields were cleared when you clicked the Update button, but you did not receive a confirmation message in the message area. Since you just modified the PRE-FORM trigger to suppress system messages, it is likely that the update was successful and the new PRE-FORM trigger suppressed the confirmation message. However, you need to provide users with explicit system feedback stating that the update was processed correctly, and give them a chance to undo the change. You can do this using form messages called alerts.

Creating Alerts to Provide System Feedback

An important principle of graphic user interface design is to give users feedback as to what is happening and to make applications "forgiving," so that users can undo unintended operations. Right now, this application does not provide any confirmation that the SHIPPING and INVENTORY update operations were successful, nor does it provide an opportunity for canceling the updates. An **alert** is a message box that provides information to users and allows them to choose from different options. You will create an alert named UPDATE_ALERT that informs users that the SHIPPING and INVENTORY tables were updated and gives them the option of continuing or canceling the operation. If a user chooses to continue, he will see a message in the message area confirming that the records were updated successfully. If the user chooses to cancel, he will see a message confirming that the change has been canceled. You could make alerts to inform the user that the operation was confirmed or canceled, but this would force him to click an OK button to acknowledge the alert. If you want to provide a short, informative message that does not involve any user choices or actions, it is probably better to display the message in the message area, using the MESSAGE procedure. If you want to include more than one line of text in the message, give the user different action options, or alert the user to a serious error or problem, you should create an alert. Now you will create the UPDATE_ALERT.

To create the alert:

1 Open the Object Navigator window in Ownership View.
2 Click the **Alerts** object.
3 Click the **Create** button on the Object Navigator toolbar to create a new alert.
4 Rename the alert **UPDATE_ALERT**.

Setting Alert Properties

The alert properties you will use are Title, Style, Button, and Message. The **Title** property determines the title that appears in the alert window title bar. The **Style** property determines the icon that appears on the alert—possible styles are Note, Caution, and Stop. Note alerts contain an "i" for information, Caution alerts contain an exclamation point (!), and Stop alerts contain a red "X" or a red stop light. Note alerts are used to convey information to the user, such as confirming that a record has been inserted. Caution alerts are used to inform the user that he has to make a choice that cannot be undone and could lead to a potentially damaging situation, such as deleting a record. Stop alerts are used to inform the user that he has instructed the system to perform an action that is not possible, such as trying to delete a record that is referenced as a foreign key in another table. An example of a Note alert is shown in Figure 6-42.

Figure 6-42: Example of a Note alert with one button

The **Button** property determines which labels will appear on the alert buttons. If you delete the label for a given button, that button will not be displayed. The **Message** property is the text that is displayed in the alert. Figure 6-42 shows an example of a Note alert with one alert button.

To change the properties of the new alert:

1. Double-click the **Alert** icon beside UPDATE_ALERT to open its Property Palette, and then change its properties as follows:

Title	**Update Alert**
Message	**You are about to update the database.**
Alert Style	**Caution**
Button 1	**OK**
Button 2	**Cancel**

2. Close the Property Palette, then save the form.

Displaying Alerts

To display an alert on the screen, you use the SHOW_ALERT function. In programming, a function always returns a value. The SHOW_ALERT function returns a numeric value. To display an alert during the execution of a trigger, you need to declare a numeric variable and then assign this variable to the value returned by the SHOW_ALERT function. This value corresponds to the alert button that the user clicks. If the user clicks the first button on the alert, this value is assigned to a variable named ALERT_BUTTON1. If the user clicks the second button on the alert, this value is assigned to ALERT_BUTTON2, and if the user clicks the third button, this value is assigned to ALERT_BUTTON3. You can specify different program actions (such as COMMIT or ROLLBACK) depending on which button the user clicks and what associated value is returned. The following example shows the general format of the commands used within a trigger to call and manipulate an alert:

```
DECLARE
    -- declare the variable that will be assigned to SHOW_ALERT
    alert_button number;
BEGIN
-- show the alert
alert_button := SHOW_ALERT ('UPDATE_ALERT');
IF alert_button = ALERT_BUTTON1 THEN
   <program statements corresponding to actions for the
   first button>
ELSE
   IF alert_button = ALERT_BUTTON2 THEN
      <program statements corresponding to actions
      for the second button>
   ELSE
      <program statements corresponding to actions
      for the third button>
   END IF;
END IF;
END;
```

Next, you will create a new program unit named DISPLAY_ALERT that will call the SHOW_ALERT function to display the UPDATE_ALERT, which has two buttons—OK and Cancel. If the user clicks the OK button, the transaction is committed and a confirmation message is displayed. If the user clicks the Cancel button, the transaction is rolled back and a message stating that the transaction was rolled back is displayed.

To create the program unit:

1 In the Object Navigator window, click **Program Units**, and then click the **Create** button on the Object Navigator toolbar.

2 Type **DISPLAY_ALERT** for the new program unit name, and make sure that the **Procedure** option button is selected, then click **OK**.

3 Type the code shown in Figure 6-43 in the PL/SQL Editor.

type this code

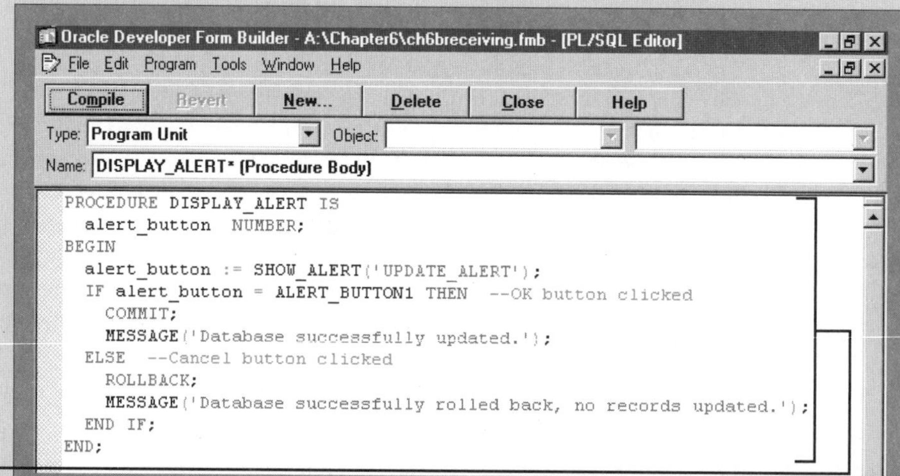

Oracle Developer Form Builder - A:\Chapter6\ch6breceiving.fmb - [PL/SQL Editor]
File Edit Program Tools Window Help
Compile
Revert
New...
Delete
Close
Help
Type: Program Unit
Object:
Name: DISPLAY_ALERT* (Procedure Body)
PROCEDURE DISPLAY_ALERT IS
alert_button NUMBER;
BEGIN
alert_button := SHOW_ALERT('UPDATE_ALERT');
IF alert_button = ALERT_BUTTON1 THEN --OK button clicked
COMMIT;
MESSAGE('Database successfully updated.');
ELSE --Cancel button clicked
ROLLBACK;
MESSAGE('Database successfully rolled back, no records updated.');
END IF;
END;

Figure 6-43: DISPLAY_ALERT program unit

4 Compile the DISPLAY_ALERT program unit code, and correct any syntax errors if necessary. Do not close the PL/SQL Editor.

Next, you need to modify the UPDATE_QOH program unit so that it calls the DISPLAY_ALERT program unit instead of committing the transaction to the database.

To modify the UPDATE_QOH program unit:

1. In the PL/SQL Editor, click the **Name** list arrow to display the list of program units, and then click **UPDATE_QOH**. The UPDATE_QOH program unit code is displayed.
2. Replace the COMMIT command with **DISPLAY_ALERT**. Your modified UPDATE_QOH program unit should look like Figure 6-44.

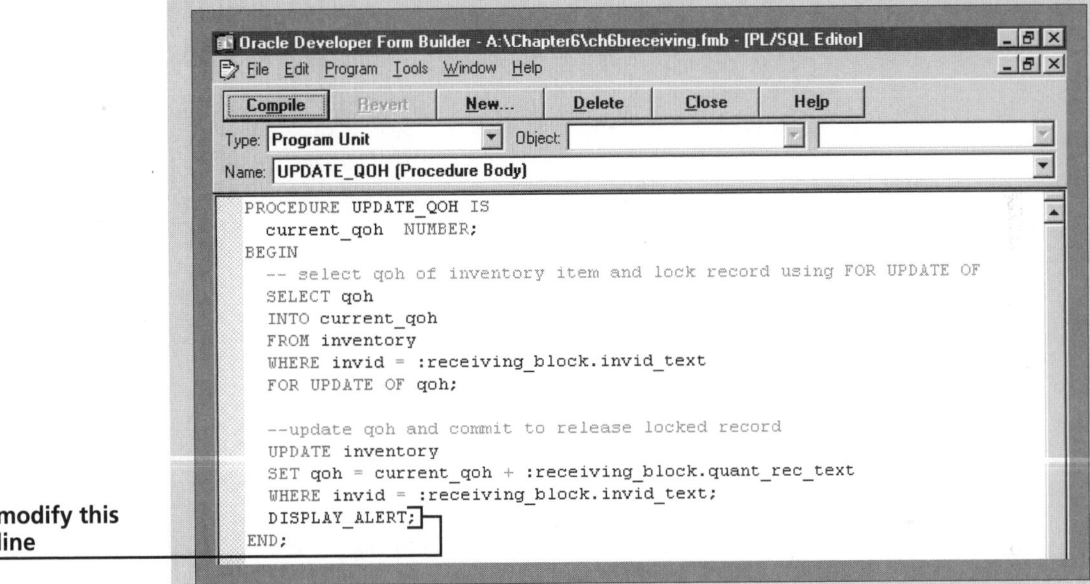

Oracle Developer Form Builder - A:\Chapter6\ch6breceiving.fmb - [PL/SQL Editor]
File Edit Program Tools Window Help
Compile
Revert
New...
Delete
Close
Help
Type: Program Unit
Object:
Name: UPDATE_QOH (Procedure Body)
PROCEDURE UPDATE_QOH IS
current_qoh NUMBER;
BEGIN
-- select qoh of inventory item and lock record using FOR UPDATE OF
SELECT qoh
INTO current_qoh
FROM inventory
WHERE invid = :receiving_block.invid_text
FOR UPDATE OF qoh;
--update qoh and commit to release locked record
UPDATE inventory
SET qoh = current_qoh + :receiving_block.quant_rec_text
WHERE invid = :receiving_block.invid_text;
DISPLAY_ALERT;
END;
modify this line

Figure 6-44: Modifying the UPDATE_QOH program unit code

3. Compile the program unit, correct any syntax errors, and then close the PL/SQL Editor.
4. Save the form.

Next you need to test the UPDATE_ALERT to see if it correctly informs the user that the database is about to be updated and then gives the user the option to save or cancel the changes.

To test the UPDATE_ALERT:

1. Run the form, click the **Shipment ID LOV** command button, click **Shipment 218, Inventory ID 11847** (Men's Expedition Parka, Navy, L), and then click **OK**. All 50 parkas in the order were received, so enter **50** for Quantity Received, and then click the **Update** button. The UPDATE_ALERT is displayed, as shown in Figure 6-45.

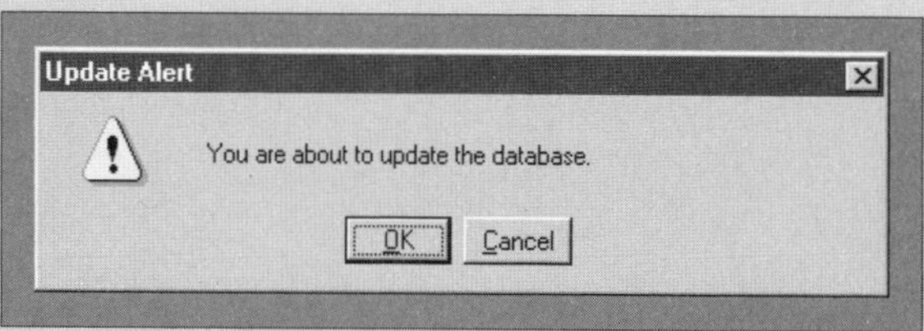

Figure 6-45: Update Alert

2 Click **OK**. The confirmation message is displayed in the message area.

3 Select Shipment ID **214** and Inventory ID **11778** (Women's Hiking Shorts, Olive, S). Enter **200** for Quantity Received, and then click **Update**. You decide not to update the record, so you will cancel your changes.

4 Click **Cancel**. The rollback confirmation message is displayed.

Adding Error Traps

Your forms should include ways to keep users from making mistakes and ways to help them correct their mistakes. To do this, you will create error traps. An **error trap** is program code that does not allow a user to perform an illegal action, or detects an illegal action after it is performed and then provides suggestions to the user for correction.

Next, you will intentionally perform an illegal action in the RECEIVING_FORM. You will change the value of the shipping ID in the Shipment ID text item to a value that does not exist in either the SHIPPING or BACKORDER table. Then you will click the Update button and try to process an incoming shipment for a shipping ID value that does not exist in the database.

To try to process a Shipping ID that does not exist:

1 Click the **LOV** command button on the Clearwater Traders Receiving form. Select Shipment ID **214** and Inventory ID **11778**, and then click **OK**.

2 Select the current shipment ID number (**214**) if necessary, and then change it to **220**. Note that this shipping ID value does not exist in either the SHIPPING or BACKORDER table.

3 Click **Update**. The "Record not found" message appears in the message area. This is the error message you put in the UPDATE_BUTTON trigger that is displayed if no record is found in either the SHIPPING or BACKORDER table that matches the user's LOV selection.

The code in the UPDATE_BUTTON trigger that generated this message is shown in Figure 6-46.

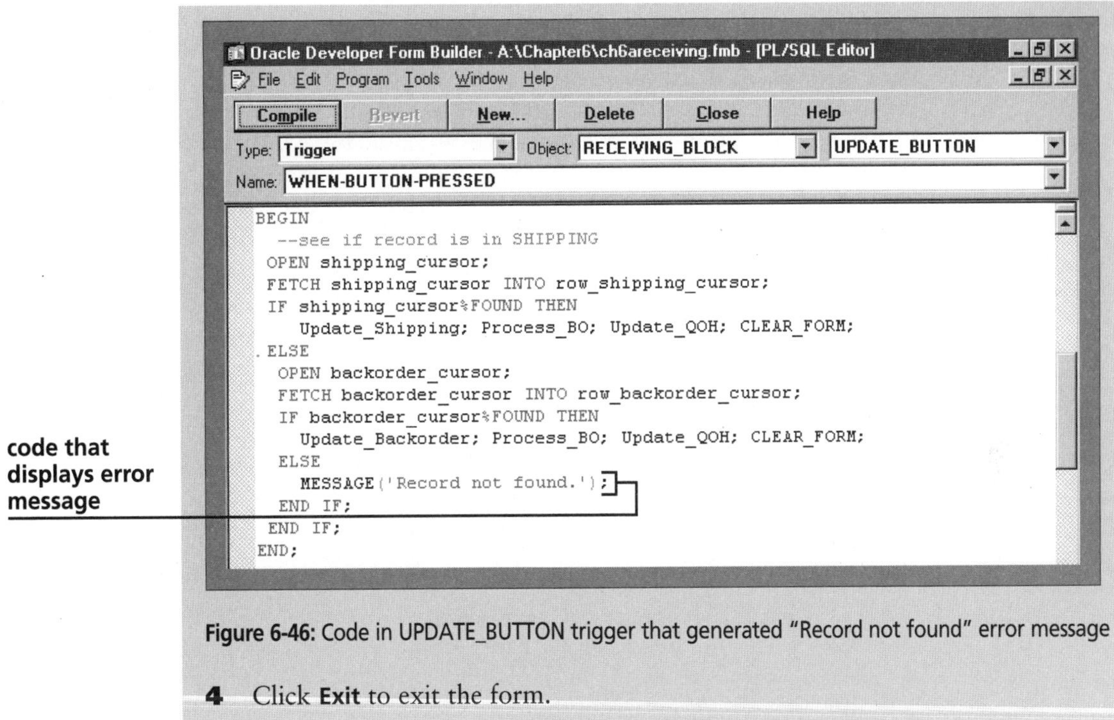

Figure 6-46: Code in UPDATE_BUTTON trigger that generated "Record not found" error message

4 Click **Exit** to exit the form.

When the user makes this error, the trigger successfully traps it, but it would be better to prevent the user from making this error in the first place. Obviously, you do not want to give users the opportunity to directly update or delete primary and foreign key fields, or to delete fields that are for display purposes only. One strategy is to make these fields **non-navigable**, which means the user cannot click the Tab key to place the insertion point on the field. To do this, you need to set the field's Keyboard Navigable property to No, so that the only way a user can get to the field is by using the pointer. Then you need to create a trigger that moves the insertion point, or form focus, to somewhere else on the form whenever a user clicks the field using the mouse. When a form item such as a command button has the **form focus,** it is highlighted. If the user presses the Enter key when a button has the form focus, it is equivalent to clicking the button.

To make the Shipment ID field non-navigable:

1 Open the Layout Editor, right-click **SHIPID_TEXT**, click **Property Palette**, and then change the Keyboard Navigable property to **No**. Close the Property Palette.

2 In the Layout Editor, right-click **SHIPID_TEXT**, and then click **PL/SQL Editor**. You will search for triggers that include mouse-related events.

3 Type **%mouse%** in the Find box of the RECEIVING_FORM: Triggers dialog box, and then press the **Enter** key. The mouse-related triggers appear in the list.

You could use the WHEN-MOUSE-CLICK event, because users usually click the mouse button to place the form insertion point in a text item. However, you would have to create an identical trigger for the WHEN-MOUSE-DOUBLECLICK event, because some users might double-click the field to place the form insertion point in a text item. The best approach is to use the WHEN-MOUSE-UP event, which is called whenever any mouse button is clicked in the item and the mouse button is released. Next you will create a WHEN-MOUSE-UP trigger that places the insertion point in a different form field whenever the user clicks the SHIPID_TEXT field, using the GO_ITEM procedure. Because this is a form where users enter data about orders that have been received, the Quantity Received field seems like a good place to move the insertion point. You could also use the GO_ITEM procedure to switch the form focus to a command button on the form.

To create the WHEN-MOUSE-UP trigger to move the insertion point:

1. Click **WHEN-MOUSE-UP** in the list, and then click **OK**.
2. Type **GO_ITEM('quant_rec_text');** in the PL/SQL Editor to place the form's insertion point in the QUANT_REC_TEXT field.
3. Compile the trigger, correct any syntax errors, and then close the PL/SQL Editor.
4. Save the form and then run it.
5. Click the **Shipment ID LOV** command button, and then select any shipment ID.
6. Try to place the form insertion point in the Shipment ID data field by clicking on the shipment ID data field using the mouse pointer. The insertion point moves to the Quantity Received data field.
7. Click **Exit** to close the form.

Next, you will change the other read-only text fields on the form (Inventory ID, Item Description, Size, Color, Date Expected, and Quantity Expected) to non-navigable by changing their Keyboard Navigate properties to No and creating WHEN-MOUSE-UP triggers that move the form insertion point to QUANT_REC_TEXT.

To change the other read-only text fields to non-navigable:

1. Open the Layout Editor, click **INVID_TEXT**, press and hold down the **Shift** key, and then click **DESC_TEXT, SIZE_TEXT, COLOR_TEXT, DATE_EXP_TEXT,** and **QUANT_EXP_TEXT** to select all of the read-only fields as a group.
2. Right-click while keeping the Shift key pressed, and then click **Property Palette** to open the group Property Palette.
3. Change the Keyboard Navigate property to **No**, and then close the group Property Palette sheet.

4 Open the Object Navigator window, click the **WHEN-MOUSE-UP** trigger below the SHIP_ID_TEXT object, and then click the **Copy** button on the Object Navigator toolbar.
5 Click the **DESC_TEXT** item, and then click the **Paste** button on the Object Navigator toolbar.
6 Paste the trigger into INVID_TEXT, SIZE_TEXT, COLOR_TEXT, DATE_EXP_TEXT, and QUANT_EXP_TEXT.
7 Save the form.
8 Run the form and confirm that the triggers work correctly. You should not be able to move the insertion point into the fields, and the trigger should move the insertion point to Quantity Received field.
9 Click the **Exit** button to close the form.
10 Close ch6breceiving.fmb in Form Builder.

General Error Handling

There are common errors that apply to a variety of forms, such as trying to delete a record that is referenced as a foreign key by other records or trying to insert a new record when fields with NOT NULL constraints are left blank. To investigate how to handle these errors, you will use the Clearwater Traders Inventory Items form, which is a custom form created to retrieve, update, and delete records from the ITEM and INVENTORY tables in the Clearwater Traders database. First you will open this form and run it to understand how it works.

To open and run the Clearwater Traders Inventory Items form:

1 Open **ch6bcustinv.fmb** from the Chapter6 folder on your Student Disk, save it as **ch6bcustinvdone.fmb**, and then run the form.
2 Click the **LOV** command button beside the Item ID field, then select Item ID **894** (Women's Hiking Shorts), then click **OK**. The data appear in the Item fields. You could edit the Item data and click the Update button within the Items frame to edit the record, or click the Delete button within the Items frame to delete the record.
3 Click the **LOV** command button beside the Inventory ID field. The inventory items associated with the selected Item ID appear in the LOV display. Select Inventory ID **11776** (size M, color Khaki), then click **OK**. The selected record's data values appear in the text items in the Inventory Items frame.
4 Delete the current price, type **29.50** in the Price text item, then click the **Update** button within the Inventory Items frame. The Updating Database alert is displayed. Click **OK**. A customized confirmation message ("Database change successfully committed") is displayed in the message area.
5 Click **Clear** in the Inventory Items frame to clear the inventory data.

6 Click **Clear** in the Items frame to clear the item data.

7 Click the **LOV** command button beside Inventory ID again. This time, a Stop alert is displayed stating that you must select an Item ID before you can select an Inventory ID.

8 Click **OK**.

Now you will deliberately make some errors in updating and deleting records to see what happens. First, you will try to delete an ITEM record that has associated INVENTORY records in the INVENTORY table.

To try to delete the ITEM record:

1 Click the **LOV** command button beside Item ID, select Item ID **786** (3-Season Tent, Sienna), then click **OK**. The selected record's data are displayed in the form's text items.

2 Click the **Delete** button in the Items frame. The error message shown in Figure 6-47 appears in the message area.

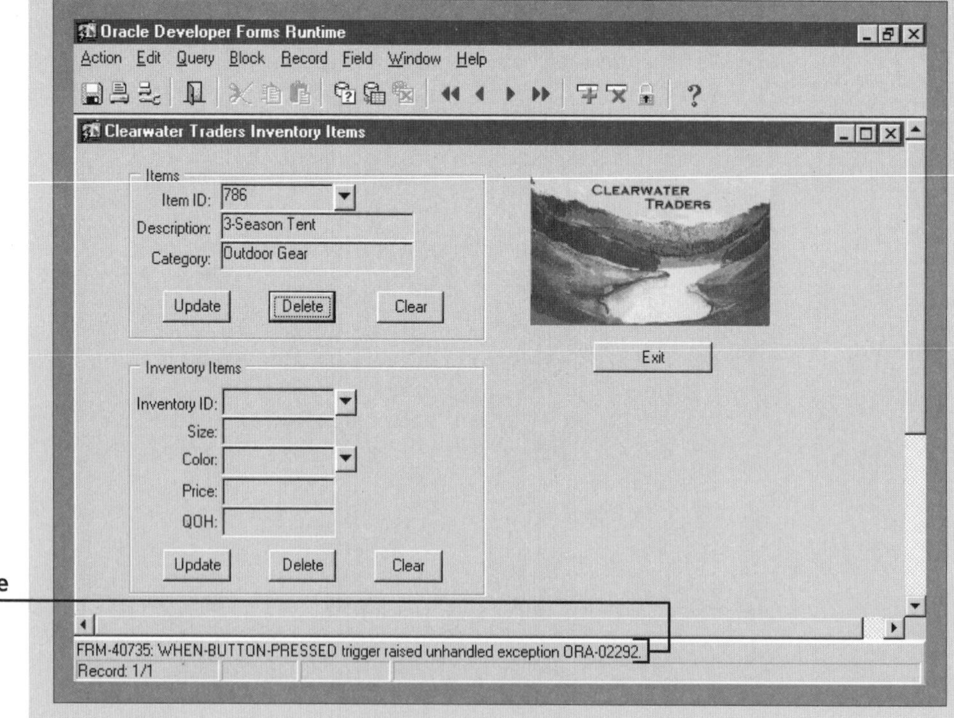

Figure 6-47: Error generated when trying to delete an item referenced in another table

Forms Runtime has error-handling routines that intercept DBMS (ORA-) error messages and then generate the associated Forms (FRM-) error messages. If you look up FRM-40735 in the Form Builder Help System, the explanation states that the trigger raised an "unhandled exception," which means that a database (ORA-) error was generated, and there is no associated FRM- error message for this error. Therefore you need to use the ORA- error message code to determine the cause of the error. If you look up the ORA-02292 error code in the Oracle database Help system, it warns that this record is referenced in other tables and cannot be deleted.

This record is referenced in the INVENTORY table, and the user cannot delete it. But it probably is not reasonable to expect users to look up ORA- error code messages to find out why they cannot complete a task, so you need to intercept the DBMS error message and handle the error in your form.

Next, you will generate an FRM- error message. FRM- errors are generated when the user violates constraints specified in the form. Usually, these are violations of properties set in a text item's Property Palette, such as not entering the correct data type or format mask. Now you will generate an FRM- error by entering an incorrect data type in a form field.

To generate an FRM- error:

1. Click the **LOV** command button beside the Inventory ID text item in the Inventory Items frame, select Inventory ID **11668**, then click **OK**.
2. Delete the current QOH, then type the letter **a** for the new QOH. This is a data type error, since the QOH_TEXT form field is specified as a NUMBER data type.
3. Click **Update** in the Inventory Items frame. The Forms error message "FRM-50016: Legal Characters are 0-9-+E." is displayed.
4. Click **Exit**, and note that the error message is still displayed, and the form insertion point moves back to the QOH field. Since this error is generated within the form, you cannot change the form focus to a different item until you correct the condition that is generating the error.
5. Change the QOH back to **16**, then click **Exit** to exit the form.

You can intercept both FRM- and ORA- error messages by creating a form-level trigger that corresponds to the ON-ERROR event. Whenever an ORA- or FRM- error is generated while a form is running, the ON-ERROR event occurs, and this trigger will execute. If an FRM- error is generated, the corresponding FRM- error code is stored in a system variable named ERROR_CODE. If an ORA- error is generated, the corresponding error code is stored in a system variable named DBMS_ERROR_CODE. By testing to find the values of ERROR_CODE and DBMS_ERROR_CODE, you can anticipate common errors and display custom alerts to provide informative messages and alternatives.

Next, you will create an ON-ERROR trigger that traps FRM- error 50016 (entering an illegal data type in a form field) and ORA- error 02292 (trying to delete a record that is referenced in another table) and displays customized alerts informing the user of the cause of the error. When an error occurs, the trigger first tests to see if the value of ERROR_CODE is 50016. If it is, an appropriate alert is displayed. If ERROR_CODE 40735 (unhandled exception) occurs, then the trigger tests for the value of DBMS_ERROR_CODE, which corresponds to an ORA- error code. If DBMS_ERROR_CODE is 02292, then a different alert is displayed. FRM- error code values are treated as positive values, while ORA- error code values are treated as negative values.

To create the ON-ERROR trigger:

1. If necessary, open the Object Navigator window in Ownership View, and then click **Triggers** directly below the form module to create a form-level trigger.
2. Click the **Create** button on the Object Navigator toolbar, and create a new trigger that corresponds to the ON-ERROR event.
3. Type the code shown in Figure 6-48 in the PL/SQL Editor.

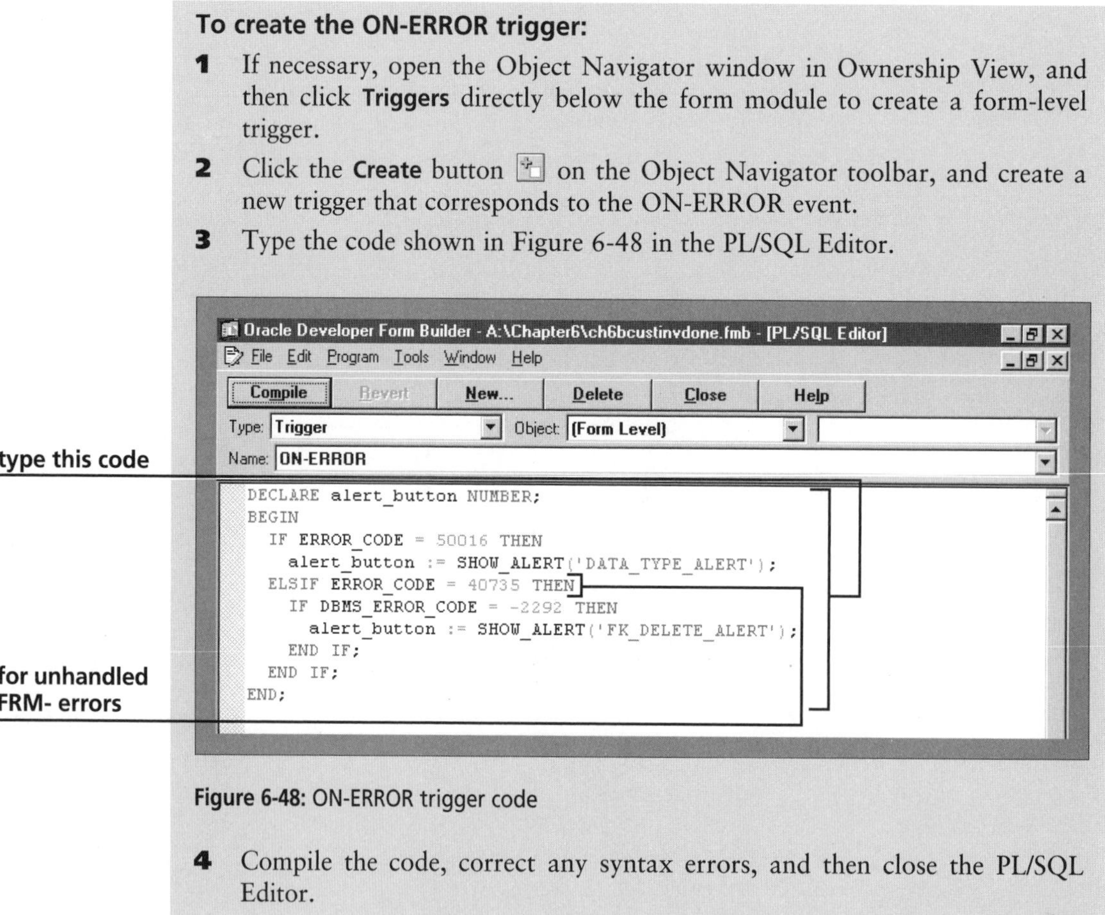
type this code
for unhandled FRM- errors
Oracle Developer Form Builder - A:\Chapter6\ch6bcustinvdone.fmb - [PL/SQL Editor]
File Edit Program Tools Window Help
Compile
Revert
New...
Delete
Close
Help
Type: Trigger
Object: (Form Level)
Name: ON-ERROR
DECLARE alert_button NUMBER;
BEGIN
IF ERROR_CODE = 50016 THEN
alert_button := SHOW_ALERT('DATA_TYPE_ALERT');
ELSIF ERROR_CODE = 40735 THEN
IF DBMS_ERROR_CODE = -2292 THEN
alert_button := SHOW_ALERT('FK_DELETE_ALERT');
END IF;
END IF;
END;

Figure 6-48: ON-ERROR trigger code

4. Compile the code, correct any syntax errors, and then close the PL/SQL Editor.

Next you need to create new alerts named FK_DELETE_ALERT and DATA_TYPE_ALERT that will appear instead of the ORA-02292 and FRM-50016 error codes and messages.

To create the alerts:

1. Return to the Object Navigator window, click **Alerts**, and click the **Create** button on the Object Navigator toolbar.
2. Change the name of the new alert to **FK_DELETE_ALERT**.
3. Click again, and name the new alert **DATA_TYPE_ALERT**.
4. Open the Property Palettes of the new alerts and change their properties as follows, then close the Property Palettes.

Alert	**FK_DELETE_ ALERT**	**DATA_TYPE_ALERT**
Title	**Delete Error**	**Data Type Error**
Alert Style	**Stop**	**Stop**
Button1	**OK**	**OK**
Button2	**<deleted>**	**<deleted>**
Button3	**<deleted>**	**<deleted>**
Message	**You cannot delete this record because it is referenced by other database tables.**	**You have entered a value that is the wrong data type in the selected form field.**

5. Run the form, click the **LOV** command button beside Item ID, select Item ID **786** (3-Season Tent), and then click the **Delete** button in the Items frame. The DELETE_ERROR_ALERT is displayed. Click **OK** to close the alert message box.
6. Click the **LOV** command button beside Inventory ID, and select Inventory ID **11668**. Change the QOH to the letter **a**, and click **Update** in the Inventory Items frame. The DATA_TYPE_ALERT is displayed. Click **OK** to close the alert message box, change the QOH back to **16**, then click **Exit** to close the form.

Next, you need to modify the ON-ERROR trigger to display an alert for unhandled exceptions, which are errors that you don't anticipate. Currently, if an error other than FRM-50016 or ORA-02292 occurs, the ORA- or FRM- error code is displayed in the message area. Now you will modify the ON-ERROR trigger to display an alert whose message contains the error description associated with an unhandled FRM- or ORA- code. To do this, you will first create an alert named UNHANDLED_ERROR_ALERT. Instead of specifying the alert's Message property in the Property Palette, the Message will be set at run time in the ON-ERROR trigger, and the alert's Message property will be left blank in the Property Palette.

To create the UNHANDLED_ERROR_ALERT:

1. Return to the Object Navigator window, click **Alerts**, and click the **Create** button on the Object Navigator toolbar.
2. Change the name of the new alert to **UNHANDLED_ERROR_ALERT**.
3. Open the Property Palette sheet of the new alert and change its properties as follows:

Alert	**UNHANDLED_ERROR_ALERT**
Title	**Error**
Message	**<leave blank>**
Alert Style	**Stop**
Button1	**OK**
Button2	**<deleted>**
Button3	**<deleted>**

4. Close the Property Palette.

When an FRM- error is generated, the Oracle system stores the description of the error message in a variable named ERROR_CODE_TEXT. When an ORA- error is generated, the system stores the text description of the error message in a variable named DBMS_ERROR_TEXT. Now you will modify the ON-ERROR trigger so that when an unhandled error occurs, the system will display the UNHANDLED_ERROR_ALERT showing the appropriate system error description. To show the correct system error description, you will use the SET_ALERT_PROPERTY procedure, which can be used to set alert properties at run time. It has the following general format: `SET_ALERT_PROPERTY(<alert name>, <alert property name>, <property value>);`. The alert name must be placed in single quotation marks. The alert property name is ALERT_MESSAGE_TEXT, and the property value will be the system error description stored in ERROR_TEXT or DBMS_ERROR_TEXT.

To modify the ON-ERROR trigger to display a customized alert for unhandled exceptions:

1. Double-click the **Trigger** icon beside the ON-ERROR trigger. The PL/SQL Editor opens.
2. Modify the ON-ERROR trigger as shown in Figure 6-49. Compile the code, correct any syntax errors, then close the PL/SQL editor window.

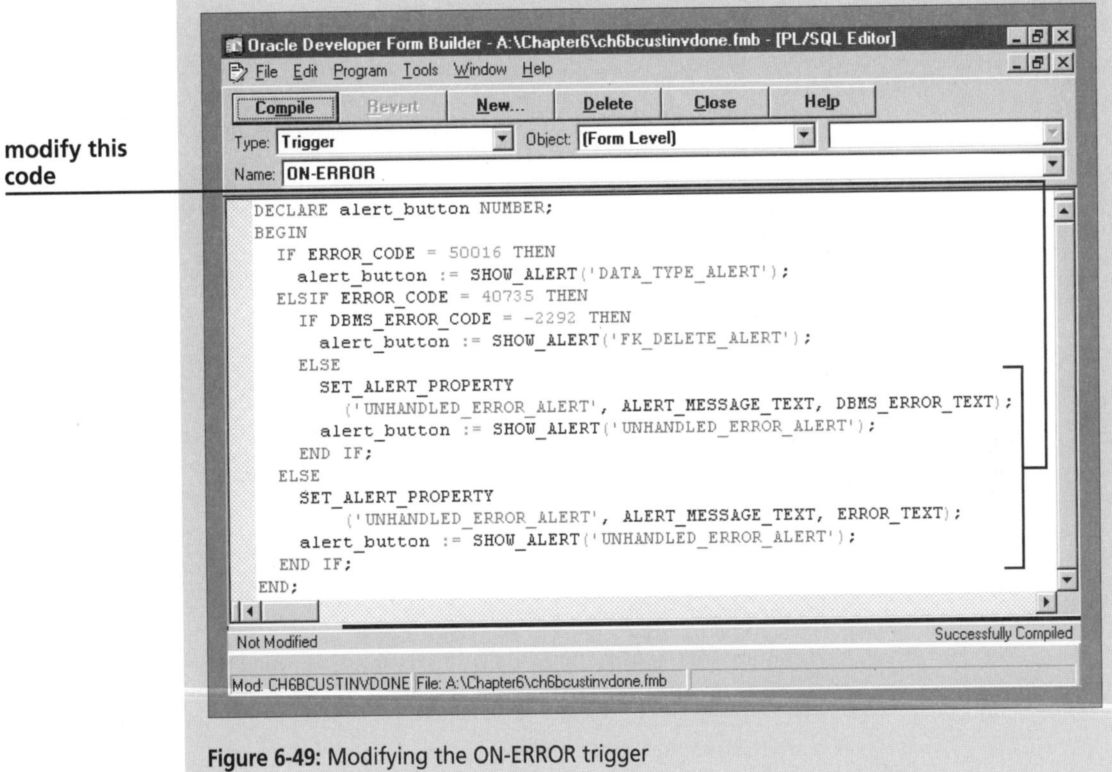

modify this code
Oracle Developer Form Builder - A:\Chapter6\ch6bcustinvdone.fmb - [PL/SQL Editor]
File Edit Program Tools Window Help
Compile Revert New... Delete Close Help
Type: Trigger Object: (Form Level)
Name: ON-ERROR
DECLARE alert_button NUMBER;
BEGIN
IF ERROR_CODE = 50016 THEN
alert_button := SHOW_ALERT('DATA_TYPE_ALERT');
ELSIF ERROR_CODE = 40735 THEN
IF DBMS_ERROR_CODE = -2292 THEN
alert_button := SHOW_ALERT('FK_DELETE_ALERT');
ELSE
SET_ALERT_PROPERTY
('UNHANDLED_ERROR_ALERT', ALERT_MESSAGE_TEXT, DBMS_ERROR_TEXT);
alert_button := SHOW_ALERT('UNHANDLED_ERROR_ALERT');
END IF;
ELSE
SET_ALERT_PROPERTY
('UNHANDLED_ERROR_ALERT', ALERT_MESSAGE_TEXT, ERROR_TEXT);
alert_button := SHOW_ALERT('UNHANDLED_ERROR_ALERT');
END IF;
END;
Not Modified
Successfully Compiled
Mod: CH6BCUSTINVDONE File: A:\Chapter6\ch6bcustinvdone.fmb

Figure 6-49: Modifying the ON-ERROR trigger

Now you will run the form and test to see what happens when an unhandled error that is not explicitly tested for in the ON-ERROR trigger occurs. You will delete the description field for Item 786, and then attempt to update the record. Since the description field (ITEMDESC) is a required (NULL NOT ALLOWED) field in the ITEM table, an ORA- error should be generated.

To generate an unhandled ORA- error:

1. Run the form.
2. Click the **LOV** command button beside the Item ID field, and select Item **786**.
3. Delete the Description field for Item 786.
4. Click the **Update** button to generate an error by not entering data in the NOT NULL fields. The Oracle error message associated with the ORA-01407 database error is displayed by the UNHANDLED_ERROR_ALERT. Click **OK** to close the alert.
5. Click **Clear** in the Items frame to clear the frame fields, then click **Exit** to close the form.

6 Save the form, and close the ch6bcustinvdone.fmb form file.
7 Exit Form Builder.

SUMMARY

- Oracle system messages are categorized by severity level and message type. Informative messages provide information about what is happening, usually do not require any user action, and have low severity levels (usually 5). Error messages require some user action, and their severity levels are higher, and have values of 15, 20, or >25.
- You can suppress lower-level Form Builder messages by setting the :SYSTEM.MESSAGE_LEVEL variable to a higher value in a PRE-FORM trigger.
- An important principle of graphic user interface design is to provide feedback to users about form operations while they are working and to make applications "forgiving," so users can cancel unintended operations.
- Alerts are form messages that provide information to users and allow them to choose from different options. The alert Style property determines which icon is displayed on the alert. Note alerts include an "i" for information, caution alerts include an exclamation point, and stop alerts include a red stoplight.
- If you want to give the user a short, informative message that does not involve any choices, display the message in the message area using the MESSAGE procedure. If you want to include more than one line of text in the message, give the user different action options, or alert the user to a serious error or problem, create an alert.
- To display an alert on the screen, you use the SHOW_ALERT function, which returns a numeric value corresponding to the alert button that the user clicks.
- Forms should include error traps, which are program units that either do not allow a user to perform an illegal action, or that detect an illegal action after it is performed and then provide suggestions to the user for correction.
- System run-time errors correspond to the form-level ON-ERROR event, so you can use a trigger to catch system error messages and substitute your own messages.
- ORA- errors represent errors that violate database rules, such as trying to delete a record that is referenced by another table. FRM- errors violate form properties, such as trying to enter an incorrect data type into a form field.
- To display error messages for unhandled system messages, you can display the Oracle system error message description in an alert by assigning the system message as the alert message at run time.

REVIEW QUESTIONS

1. What is message severity level?
2. Are high-severity level messages more or less serious than low-severity messages?
3. How can you suppress some system-level messages? What messages cannot be suppressed?
4. What is an alert? Why should you make alerts in forms applications?
5. What are the three alert styles, and how are they different?
6. What event causes the ON-ERROR trigger to execute?
7. What is the difference between handling errors by writing an error trap and handling errors in the ON-ERROR trigger?
8. What is the difference between an FRM- and an ORA- error?
9. True or false: In a form, all ORA- errors are associated with the FRM-40735 error code.
10. You have created a form that has four alerts. The first alert is named MASTER_ALERT, and it has an OK button (Button 1), Cancel button (Button 2), and Help button (Button 3). The other three alerts are named ALERT_1, ALERT_2, and ALERT_3, and these three alerts have only an OK button. Write a trigger to display MASTER_ALERT. If the user clicks the OK button, ALERT_1 is displayed. If the user clicks the Cancel button, ALERT_2 is displayed, and if the user clicks the Help button, ALERT_3 is displayed.

PROBLEM-SOLVING CASES

1. Modify ch6breceiving.fmb to include an error trap that displays an alert if the user clicks the Update button without entering a value for Quantity Received. Create a Stop-style alert named NULL_QUANTITY_ALERT with a single OK button and the message "You must enter the quantity received before you can update the database." Place the title "No Quantity Entered" on the alert message box title bar. Save the file as Ch6bEx1.fmb. (*Hint*: Call the alert from the UPDATE_BUTTON trigger.)
2. A custom form named Ch6bEx2.fmb has been created to update records from the STUDENT table in the Northwoods University database. In this exercise, you will add alerts, error trapping, and error handling to make the form easier to use. Open Ch6bEx2.fmb from the Chapter6 folder on your Student Disk, save it as Ch6bEx2done.fmb, and make the following modifications:
   a. Add a Caution alert to the Update button before the UPDATE command is committed with the message "You are about to change the database." If the user clicks the OK button, commit the update and show a message in the message area stating "Record successfully changed." If the user clicks the Cancel button, do not commit the change, and show a message stating "Change rolled back." (*Hint*: Do not forget to modify the :SYSTEM.MESSAGE_LEVEL value.)

b. Add a Caution alert to the Delete button before the DELETE command is committed that displays the message "You are about to delete the selected record." If the user clicks the OK button, commit the delete, clear the form, and display a message stating "Record successfully deleted." If the user clicks the Cancel button, roll back the delete, and show a message stating "Delete rolled back." (*Hint*: to test this alert, you will have to first insert a new record into the STUDENT table with an SID value that is not referenced as a foreign key in any other tables and then delete the record you inserted.)

c. Add an error trap to prevent the user from changing the value of SID, SCLASS, or FID, except through the LOV command button by directing the form's focus to the STUDENT_LOV_BUTTON.

d. Add an ON-ERROR trigger that displays a Note alert with only an OK button to provide information on the following errors: ORA-02292 (trying to delete a record referenced in another table), FRM-50016 (entering an incorrect data type in a form field), ORA-01407 (trying to insert NULL into a field with a NOT NULL constraint). Create customized alerts with informative messages for each of these errors.

e. Modify the ON-ERROR trigger so it displays the associated Oracle system error description for all unhandled FRM- and ORA- errors.

3. Figure 6-50 shows the interface for a form named Ch6bEx3.fmb that is used to insert new records into the FACULTY table in the Northwoods University database. When the user clicks the Create button, a new Faculty ID is retrieved from a sequence. The user enters the data values for the new faculty member and clicks the Save button, and the new record is saved in the database.

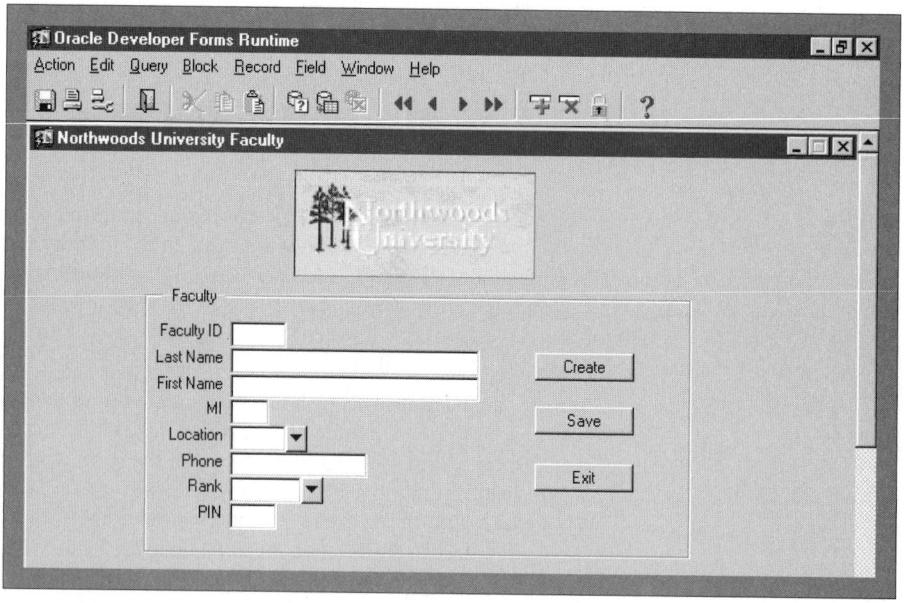

Figure 6-50

a. Open Ch6bEx3.fmb from the Chapter6 folder of your Student Disk, and save it as Ch6bEx3done.fmb.
b. If necessary, create a new sequence named FID_SEQUENCE, starting with 10 and with no maximum value. (You might have already created this sequence in the exercises in a previous lesson.)
c. Create a Caution alert that is displayed when the user clicks the Save button, but before the INSERT command is committed to the database. The alert message should be "Add the new record to the database?" If the user clicks the alert's OK button, the trigger will save the record, clear the form, and show a message in the message area stating "Record successfully inserted." If the user clicks the alert's Cancel button, do not commit the change, clear the form fields and show a message in the message area stating "Changes not saved." (*Hint*: Do not forget to modify the value for :SYSTEM.MESSAGE_LEVEL.)
d. Add an error trap to prevent the user from changing the values of LOCID or FRANK except through the LOV command buttons by redirecting the form focus to the Create button.
e. Add an ON-ERROR trigger that displays a Note alert with an OK button to provide information on the following errors: ORA-00001 (trying to insert a record with a primary key value that already exists in the table), FRM-50016 (entering an incorrect data type in a form field), and ORA-01407 (trying to insert a NULL data value into a field with a NOT NULL constraint). Create customized alerts with informative messages for each of these errors.
f. Modify the ON-ERROR trigger so it displays the associated Oracle system error description for all unhandled FRM- and ORA- errors.

4. Figure 6-51 shows a form named Ch6bEx4.fmb that is designed to allow users to insert new records, update existing records, and delete records from the COURSE_SECTION table of the Northwoods University database. When the user clicks the Create New Section ID button, a new Course Section ID is displayed. The user enters the data values for the new faculty member and clicks the Save New Record button, and the new record is saved in the database. The user can click the LOV command button next to the Course Section ID field and select one of the existing records, edit the data fields, and then click the Update Existing Record button to update a record, or click the Delete Existing Record button to delete a record.

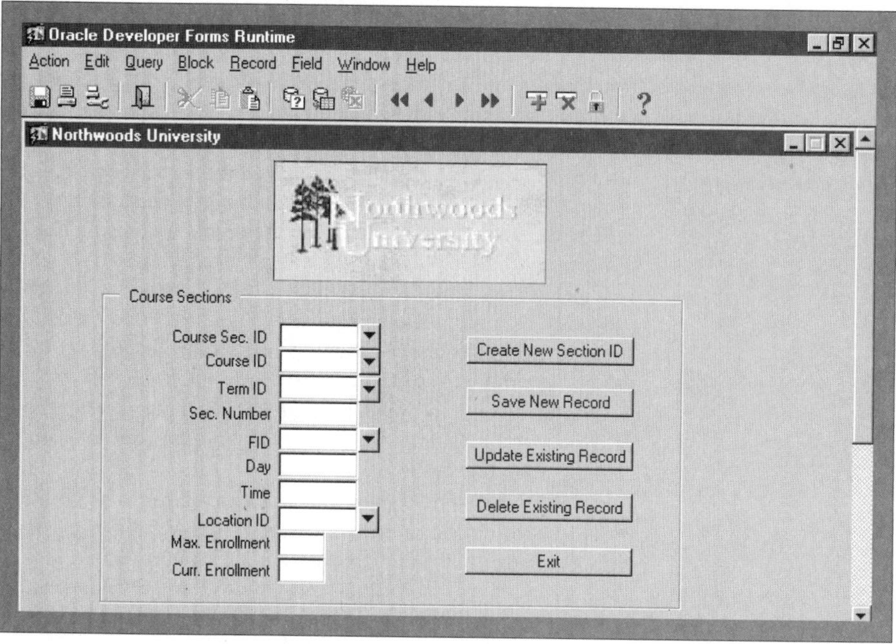

Figure 6-51

a. Open Ch6bEx4.fmb from the Chapter6 folder on your Student Disk, and save it as Ch6bEx4done.fmb.
b. Create a new sequence named CSECID_SEQUENCE that starts with 2000 and has no maximum value.
c. Add a Caution alert that is displayed when the user clicks the Save New Record button before the INSERT command is committed. The alert should have an OK and Cancel button, and display the message "Add the new record to the database?" If the user clicks the OK button, commit the insert and show a message stating "Record successfully inserted." If the user clicks the Cancel button, do not commit the change, and show a message stating "Changes not saved." (*Hint*: Do not forget to modify the value for :SYSTEM.MESSAGE_LEVEL.)
d. Add a Caution alert to the Update Existing Record button before the UPDATE command is committed with the message "Update the current record?" If the user clicks the OK button, commit the update and show a message stating "Record successfully updated." If the user clicks the Cancel button, do not commit the change, and show a message stating "Changes not saved."
e. Add a Caution alert to the Delete Existing Record button before the DELETE command is committed with the message "Delete the current record?" If the user clicks the OK button, commit the delete, clear the form, and show a message stating "Record successfully deleted." If the user clicks the Cancel button, do not commit the change, and show a message stating "Changes not saved."
f. Add an error trap to prevent the user from changing the value of CID, TERMID, CSECID FID, or LOCID except through the LOV command buttons, by redirecting the form focus to the Create New Section ID button.

g. Add an ON-ERROR trigger that displays a Note alert to provide additional information about the following errors: ORA-00001 (trying to insert a record with a primary key value that already exists in the table), ORA-02292 (trying to delete a record that is referenced in another table), ORA-01407 (trying to insert NULL into a field with a NOT NULL constraint). Create customized alerts with informative messages for each of these errors.

h. Modify the ON-ERROR trigger so it displays the associated Oracle system error description for all unhandled FRM- and ORA- errors.

i. Create an error trap and display an associated Stop alert with an OK button only in the Save New Record and Update Existing Record buttons, to ensure that the value entered for current enrollment is never greater than the value entered for maximum enrollment. Place the code to display the alert in the triggers for the Save New Record and Update Existing Record buttons.

5. Figure 6-52 shows a form named Ch6bEx5.fmb that allows users to insert new records, update existing records, and delete records from the CUSTOMER, CUST_ORDER, and ORDERLINE tables of the Clearwater Traders database. When the user clicks the Create New ID button in the Customers frame, a new Customer ID is retrieved from a sequence. The user enters the data values for the new customer and clicks the Save New Customer button, and the new customer record is saved in the database. The user can click the LOV command button next to the Customer ID field to select an existing CUSTOMER record, edit the data fields, and then click the Update Customer button to update the customer record or click the Delete Customer button to delete the customer record.

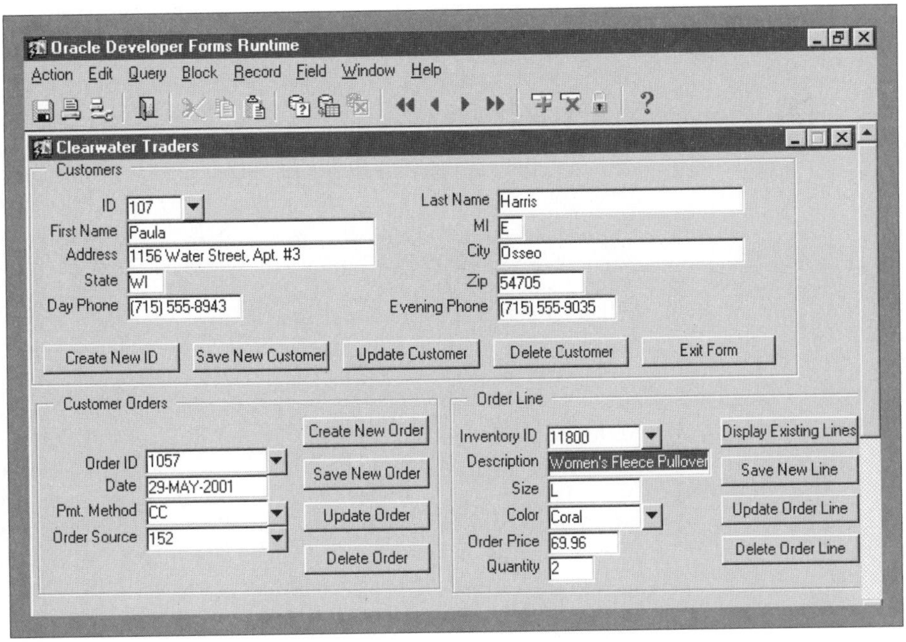

Figure 6-52

After the user creates a new customer record or selects an existing customer record, she can click the Create New Order button to create a new customer order, enter the order data in the data fields, and then click the Save New Order button to save the order. If the user wants to edit an existing order, she clicks the LOV command button next to the Order ID field to view the selected customer's orders. She selects an order, then edits the order information, and clicks the Update Order button to update the order, or the Delete Order button to delete the order.

After the user creates or selects a customer order, she can create new order lines for the order by clicking the LOV command button next to Inventory ID, selecting an Inventory ID and entering a quantity, and then clicking the Save New Line button. (The primary key of the ORDERLINE table is the combination of ORDERID and INVID, so no new primary key needs to be generated.) If the user clicks the Display Existing Lines button, the order line information for the Customer Order ID that is currently selected is displayed. The user can select an existing order line and edit or delete it as needed.

a. Open Ch6bEx5.fmb from the Chapter6 folder on your Student Disk, and save it as Ch6bEx5done.fmb.
b. Create a new sequence named CUSTID_SEQUENCE that starts with 200 and has no maximum value.
c. Create a new sequence named ORDERID_SEQUENCE that starts with 1100 and has no maximum value.
d. Add a Caution alert to notify the user that a record is about to be inserted. The alert should be displayed before the INSERT command is committed, and show the message "Add the new record to the database?" If the user clicks the OK button, commit the insert and show a message stating "Record successfully inserted." If the user clicks the Cancel button, do not commit the change, and show a message stating "Changes not saved." Display the alert when the user clicks the Save New Customer, Save New Order, or Save New Line button.
e. Add a Caution alert to notify the user when a record is going to be updated. The alert should be displayed before the UPDATE command is committed and show the message "Update the current record?" If the user clicks the OK button, commit the update and show a message stating "Record successfully updated." If the user clicks the Cancel button, do not commit the change, and show a message stating "Changes not saved." Display the alert when the user clicks the Update Customer, Update Order, or Update Order Line button.
f. Add a Caution alert to notify the user when a record is about to be deleted. The alert should be displayed before the DELETE command is committed and show the message "Delete the current record?" If the user clicks the OK button, commit the delete and show a message stating "Record successfully deleted." If the user clicks the Cancel button, do not commit the change, and show a message stating "Changes not saved." Display the alert when the user clicks the Delete Customer, Delete Order, or Delete Order Line button. After you delete an order record, the fields in the Customer Orders and Order Line frames should be cleared, but the fields in the Customer frame should still be displayed. After you delete an order line record, the fields in the Order Line frame should be cleared, but the fields in the Customers and Customer Orders frames should still be displayed. (*Hint*: Program units named CLEAR_CUSTORDER_FIELDS and CLEAR_ORDERLINE_FIELDS are included in the file to clear these fields correctly.)

g. Add an error trap to prevent the user from changing the value of CUSTID, ORDERID, METHPMT, INVID or COLOR except through the LOV command buttons, by redirecting the form focus to the Create New ID button.
h. Create a Stop alert with an OK button only that appears if the user clicks the Create New Order button or the LOV command button beside Order ID in the Customer Orders frame when no value has been selected for customer ID. The alert should remind the user that he must select a customer before he can create or select a customer order.
i. Create a Stop alert with an OK button only that appears if the user clicks the Display Existing Lines button or the LOV command button beside Inventory ID in the Order Lines frame when no value has been selected for Order ID. The alert should remind the user that he must select a customer order ID before he can create or select an order line.
j. Create a Stop alert with an OK button only that appears if the user clicks the LOV command button beside the Color field in the Order Lines frame when no value has been selected for INVID. The alert should remind the user that she must select an inventory item before she can select a color for that item.
k. Add an ON-ERROR trigger that displays a customized alert to handle the following errors: ORA-00001 (trying to insert a record with a primary key value that already exists in the table), ORA-02292 (trying to delete a record that is referenced in another table), and ORA-01407 (trying to insert NULL into a field with a NOT NULL constraint). Create customized alerts with informative messages for each of these errors.
l. Modify the ON-ERROR trigger so it displays the associated Oracle system error description for all unhandled FRM- and ORA- errors.
m. Create an error trap and display a Stop alert to notify the user that he must enter a value greater than 0 in the Quantity field in the Order Line Frame. The alert should be displayed when the user clicks the Save New Line or Update Order Line button.

LESSON C

objectives

- Work with a form with multiple canvases
- Link data blocks to control blocks in a custom form
- Pass global variable values between different forms

Working with a Form with Multiple Canvases

Many database applications require multiple canvases to enter and display information. (Recall that a canvas contains the text items, buttons, and other form items that are shown on your computer monitor.) For example, at Clearwater Traders, new customer orders are entered using an on-screen form that the customer service representative completes when the customer calls the company. The first canvas displays customer order information, and calls a second canvas that allows the customer service representative to enter a new customer record or modify customer information if necessary. The application then switches to a third canvas that displays information about the individual order items.

Form Builder provides two approaches for creating applications with multiple screens. The **single-form** approach involves creating one form with multiple canvases. This allows data to easily be shared among the different canvases, but makes it impossible for multiple programmers to program different canvases of the same application at the same time, since the .fmb file can only be opened by one programmer at a time.

The **multiple-form** approach involves creating multiple forms with a different .fmb files for each canvas in the application. This approach works well when multiple programmers are collaborating to create a complex application. It also enables a form to be used in many different applications, thus avoiding redundant programming efforts. Data sharing among forms is more difficult in the multiple-form approach than in the single-form approach, and must be done using global variables, which are variables whose assigned values are visible to different forms. In this lesson, you will create an application to support the Clearwater Traders sales process that uses both approaches. This process is described in detail next.

The Clearwater Customer Order Process

When the customer service representative at Clearwater Traders takes a customer order, she first must either create a new CUSTOMER record for a new customer, or retrieve the CUSTID for an existing customer. Then she must enter the payment method and order source (catalog number or Web site) and create a new

CUST_ORDER record. Next, the customer service representative must enter the item number and quantity desired for each order item, and insert a corresponding record in the ORDERLINE table. After all order items are recorded, she reports the total order cost (including sales tax and shipping and handling) to the customer.

Figure 6-53 illustrates the customer order process in a flowchart. If the customer is a new customer, the application activates a different form for adding new customer information, and then returns the new CUSTID to the current application. If the customer is an existing customer, a LOV is used to retrieve the CUSTID. Then the method of payment and order source are entered, and the CUST_ORDER record is inserted.

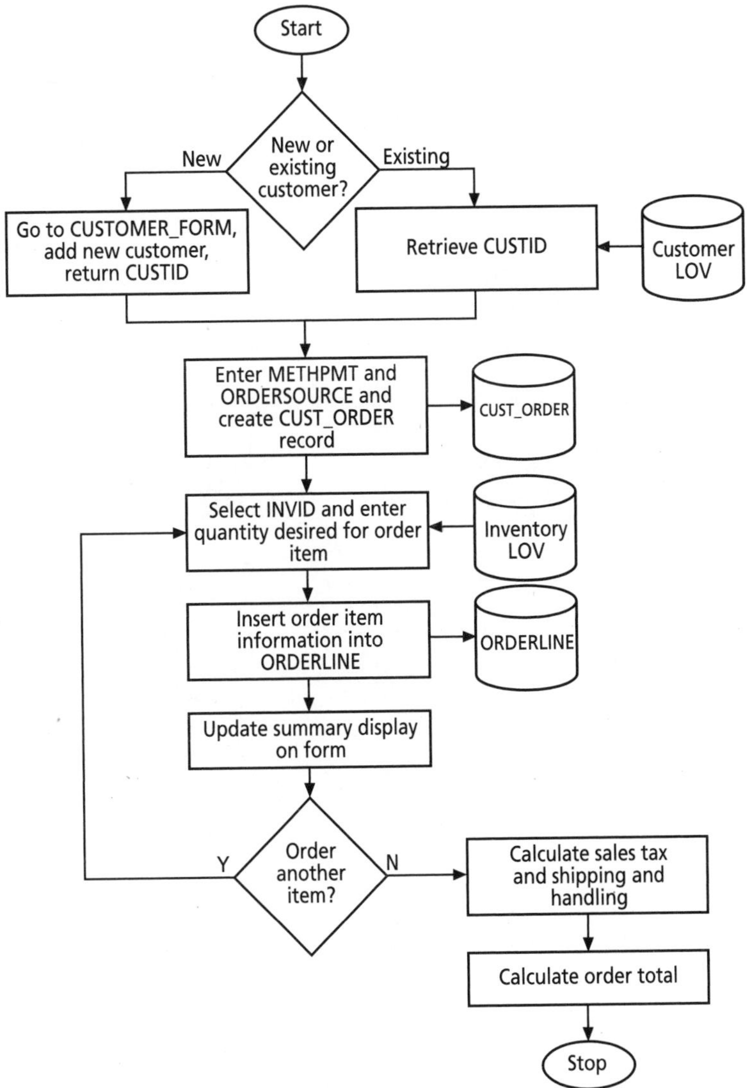

Figure 6-53: Clearwater Traders customer sales process flowchart

The first item's inventory ID is retrieved using the Inventory LOV. The item order specification is inserted into the ORDERLINE table, and a summary of the ordered items is displayed on the form. When the customer is finished ordering, the form calculates the sales tax, shipping and handling charges, and the order total.

Viewing the Interface Design

This application will involve two forms: the Customer form, for creating new customers and editing existing customer information, and the Sales form, for processing customer order and order-line information. Figure 6-54 shows the Customer form interface. When the user clicks the Create New Customer ID button, a new customer ID appears on the form. The user can enter the new customer data and click the Save New Record button to insert a new customer record into the CUSTOMER table. The user can click the LOV command button next to the ID field to select from a list of all customer records, and then edit or delete the selected record. The Return to Customer Order button exits the Customer form.

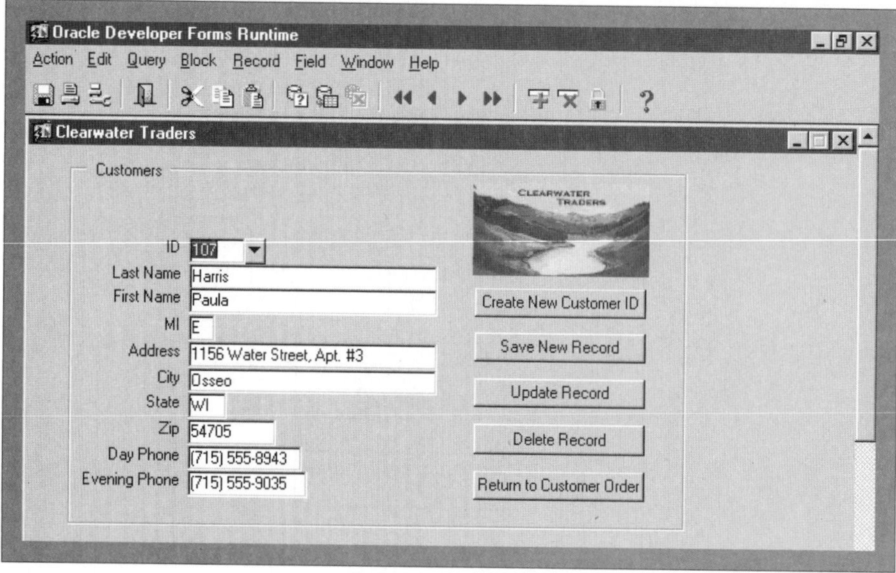

Figure 6-54: Customer form

The interface for the Sales form involves two separate canvases—the Customer Orders canvas displays the customer and order information, and the Order Line canvas will contain the information about the specific order items. The interface for the Customer Order canvas is shown in Figure 6-55. When the form is first displayed, a new order ID is automatically displayed in the Order ID field. The user can select an existing customer ID from the CUSTOMER table by clicking the LOV command button next to the Customer ID field, or the user can click the Create/Edit Customer command button, and display the Customer form and create a new customer or edit a current customer's data. After completing the customer selection, the user can select the payment method (check or credit card type), using the option buttons, and select the order source using a LOV that will show valid order sources from the ORDERSOURCE table. After entering this information, the user clicks the Save Order button to insert the new order record into the CUST_ORDER table. Then the user clicks the Enter/Edit Items button to display the Order Line canvas used to enter the order items. When the user is finished with the current order, the form fields are cleared and a new order ID value is inserted in the Order ID field.

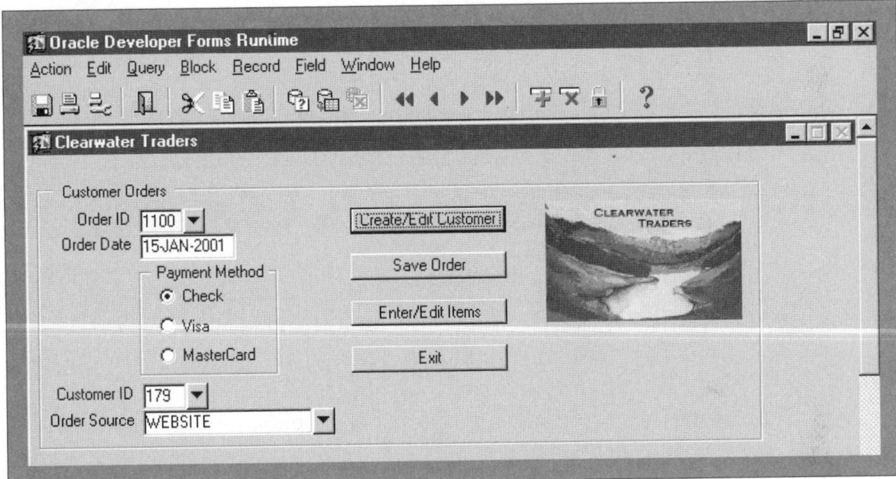

Figure 6-55: Customer Orders canvas

The Order Line canvas (see Figure 6-56) has fields that enable the user to specify the data for a specific order line (inventory ID, description, size, color, price, and desired quantity). When the user clicks the Add Order Line button, the order line information is inserted into the ORDERLINE table. As items are added to the order, the Order Summary block dynamically displays the inventory ID, quantity, description, size, color, and price for each order line in the customer order. The user can click the Select Order Line button to display a LOV that shows all of the order lines for the current order and allows the user to select a specific order line and retrieve its data into the Order Items fields. The user can then modify order line data and update it using the Update Line button, or delete it using the Delete Line button. When all order items are entered, the user clicks the Finish Order button to calculate the subtotal, sales tax, shipping and handling, and final total. The user can then click the Return button to clear the form fields, switch back to the Customer Orders canvas, and begin taking information for a new order.

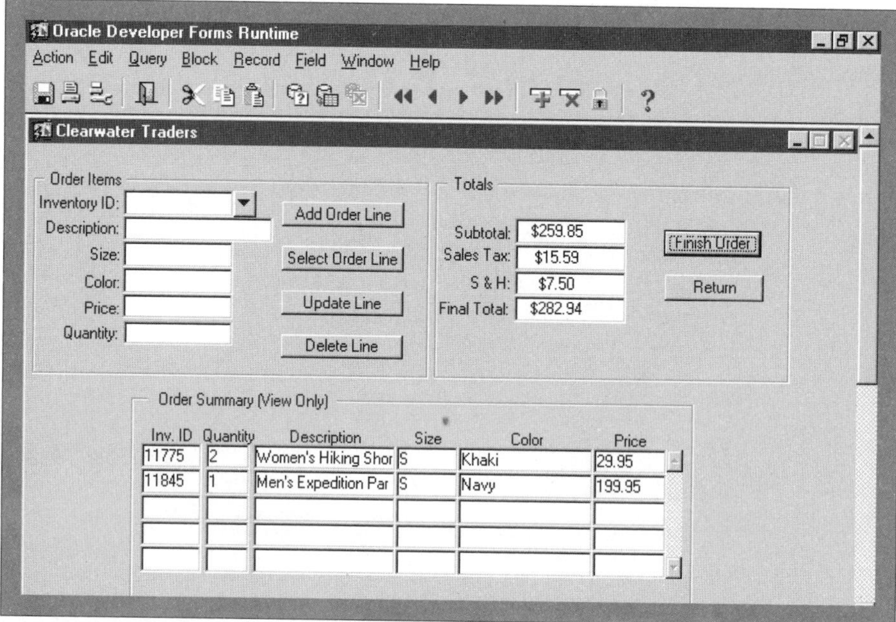

Figure 6-56: Order Line canvas

The Customer and Sales forms have already been created, and are stored in the Chapter6 folder on your Student Disk. You will modify these forms so they work together to make a finished application. First, you will start SQL*Plus and run the clearwat.sql script to refresh your database tables. Then you will open the Sales form.

To refresh your database tables and open the form:

1. Start SQL*Plus, and run the clearwat.sql script from the Chapter6 folder on your Student Disk.
2. Start Form Builder.
3. Open **ch6csales.fmb** from the Chapter6 folder of your Student Disk, and save it as **ch6salesdone.fmb**.
4. Change the form module name to CH6CSALESDONE.

Using a Data Block in a Custom Form

The items on the Customer Order canvas were created in a control block named CUST_ORDER. You will create the triggers for the form's command buttons. First, you will modify the form's PRE-FORM trigger so that when the form starts, a new order ID is inserted into the ORDERID field, and the current system date is displayed in the ORDERDATE field. This trigger will retrieve the next sequence number from a sequence named ORDERID_SEQUENCE, and then insert it into ORDERID. First, you will create the sequence in SQL*Plus. Then, you will modify the PRE-FORM trigger.

To create the sequence and modify the PRE-FORM trigger:

1 Click **SQL*Plus** on the Windows taskbar, and create a new sequence named ORDERID_SEQUENCE that starts with 1100 and has no maximum value.

help If you receive the error message "name is already used by an existing object," then you already created this sequence in a previous exercise, and you can skip this step.

2 Click **Form Builder** on the taskbar, if necessary, open the Object Navigator window in Ownership View, click [+] beside **Triggers** under CH6CSALES-DONE, and double-click the **Trigger** icon beside PRE-FORM to open the PRE-FORM trigger.

3 Modify the code as shown in Figure 6-57.

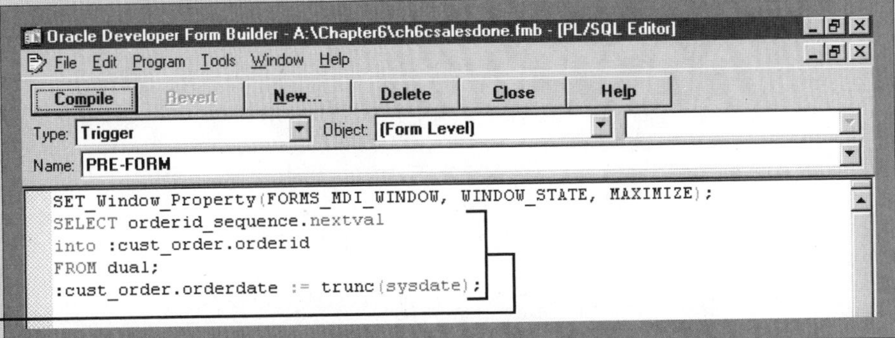

Figure 6-57: Modifying the PRE-FORM trigger

4 Compile the trigger, correct any syntax errors, and close the PL/SQL Editor.

5 Save the form, then run the form to make sure the ORDERID value inserts automatically, and the system date appears in the ORDERDATE field when the form starts.

6 Click **Exit** to close the form. Click **Yes** to confirm closing.

Creating the Form Button Triggers

The first form button trigger you will create is for the Save Order button, which inserts the customer order into the CUST_ORDER table. After you create the trigger, you will run the form to confirm that the trigger works correctly.

To create and test the Save Order button trigger:

1 Create a WHEN-BUTTON-PRESSED trigger for the Save Order button on the CUST_ORDER canvas.

2 Type the code shown in Figure 6-58. Compile the trigger, correct any syntax errors, then close the PL/SQL Editor.

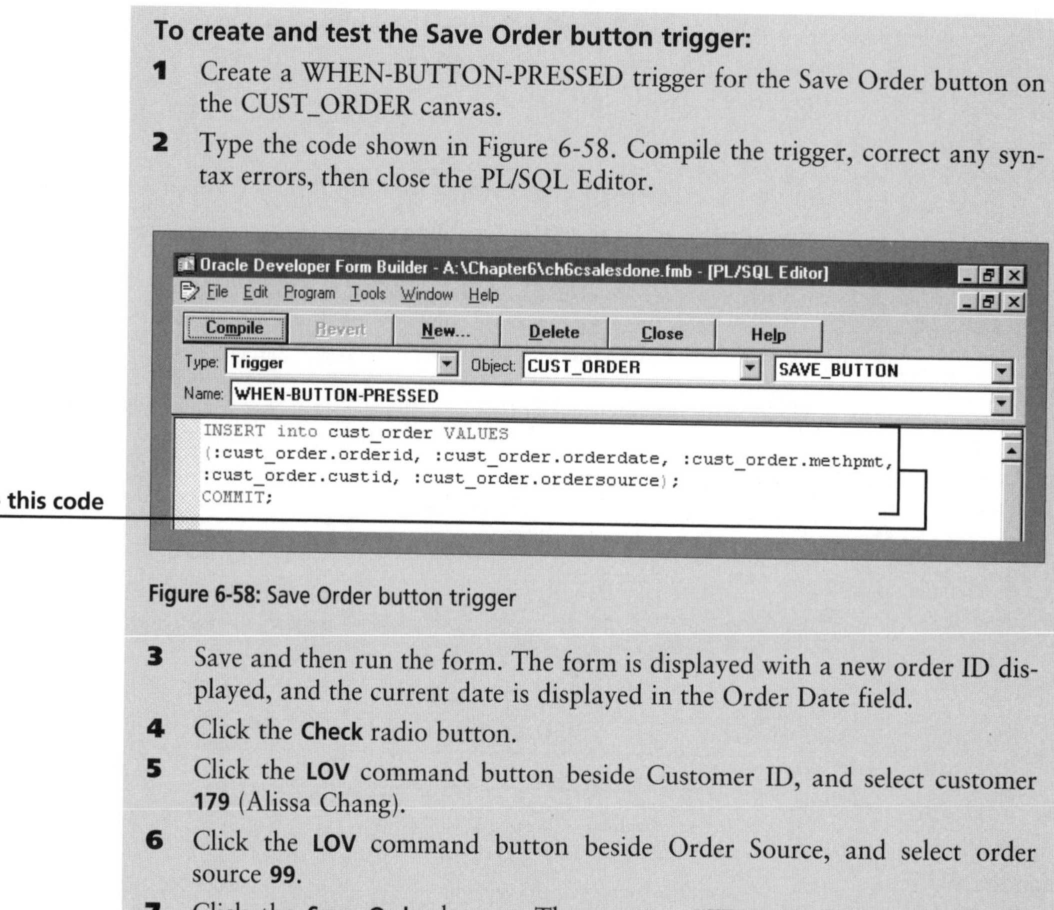

Figure 6-58: Save Order button trigger

3 Save and then run the form. The form is displayed with a new order ID displayed, and the current date is displayed in the Order Date field.

4 Click the **Check** radio button.

5 Click the **LOV** command button beside Customer ID, and select customer **179** (Alissa Chang).

6 Click the **LOV** command button beside Order Source, and select order source **99**.

7 Click the **Save Order** button. The message "FRM-40401: No changes to save." confirms that the record was inserted into the database.

8 Click the **Exit** button to exit the form.

Passing Variable Values Between Forms

The next trigger you need to create is for the Create/Edit Customer button, which will display the Customer form and allow the user to create a new customer record or edit the current customer record. If the user selects an existing customer ID using the LOV command button beside the Customer ID field on the Sales form, then the value of the selected customer ID must be passed from the Sales form to the Customer form. If the user creates a new customer record in the Customer form, then the value of the new customer ID must be passed from the Customer form back to the Sales form. To pass a variable value between two different forms, you must use a **global variable**, which is a variable whose value is visible to multiple forms. Once a global variable is assigned a value in a form, it remains in memory and its value is available to all forms running in the Forms Runtime application on that workstation. The general format of a global variable is `:GLOBAL.<variable name>`. The :GLOBAL qualifier alerts the system that the variable is a global variable. Global variables do not need to be explicitly declared, but are simply assigned a value in a PL/SQL command. The first time a global variable is used in a form or combination of forms, it must always appear on the *left* side of an assignment operator (:=) and be assigned a data value. Placing the global variable on the right side of the assignment operator (assigning another variable to a global variable that does not yet have a value) will generate a run-time error. Using the global variable in a comparison operation (such as `IF :GLOBAL.name = 'Alissa'`) before it has been assigned a value will also generate a run-time error.

When a form is running, you can use the CALL_FORM command to start and display a second form. The first form will still be running, but program control is temporarily transferred to the second form. The CALL_FORM command has the following general format: `CALL_FORM('<full path to form's .fmx file>')`. Note that you must use the executable (.fmx) file in the CALL_FORM statement. When you execute a CALL_FORM command, program control immediately goes to the called form, and stays there until the called form is exited. When the called form is exited, program control returns to the program line immediately after the CALL_FORM command in the first form.

Figure 6-59 shows a flowchart illustrating the interaction between the user and the two forms in this application. When the user clicks the Create/Edit Customer button on the Sales form, the associated trigger assigns :GLOBAL.custid to the current value of CUSTID in the Sales form. (If the user selected a customer in the Sales form, then this customer's CUSTID is now assigned to :GLOBAL.custid. If the user did not select a customer but opted to add a new customer using the Customer form, then the current value of :GLOBAL.custid is NULL.) Then the trigger calls the Customer form.

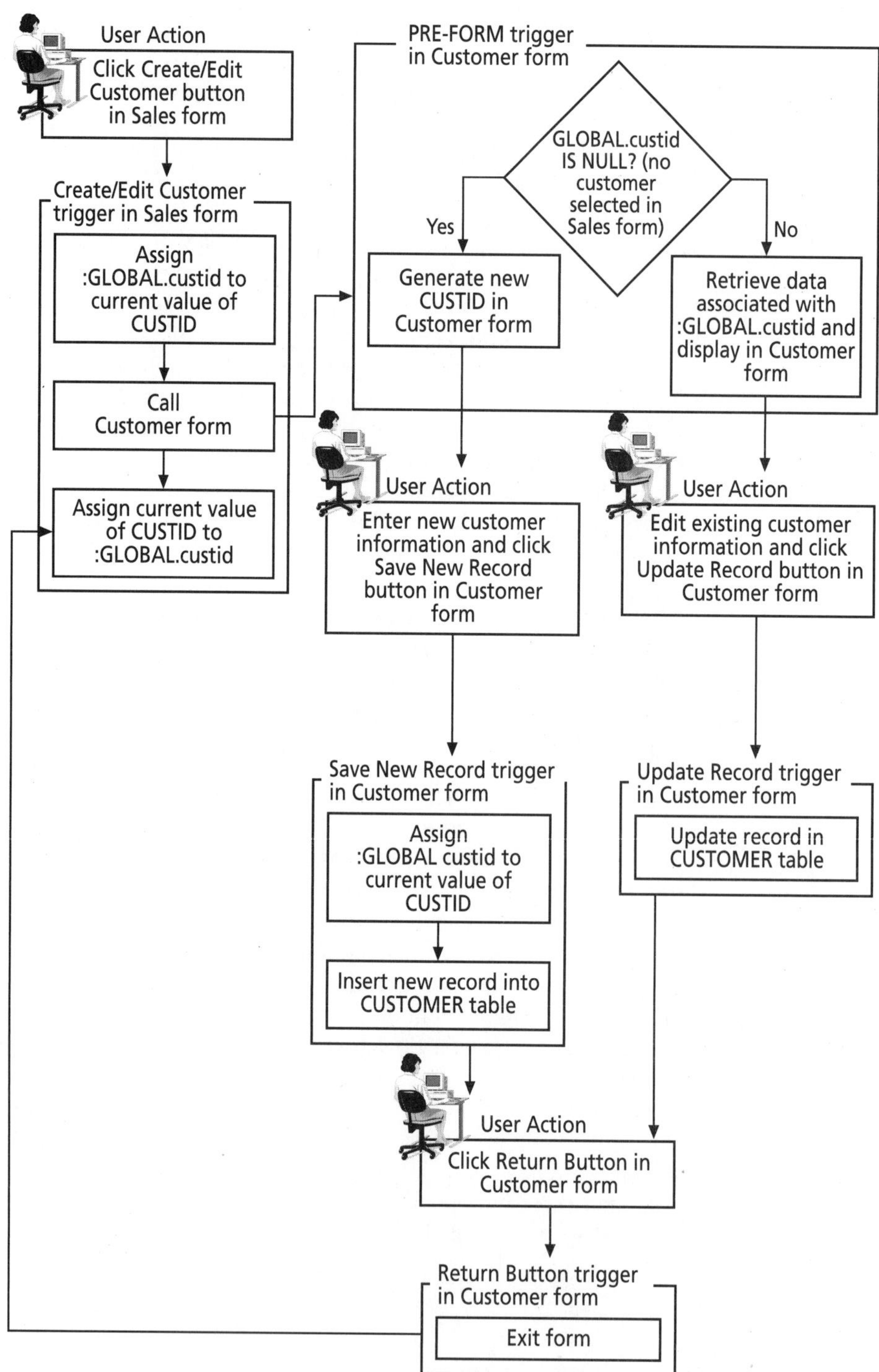

Figure 6-59: Interaction between user and the Sales form and Customer form triggers

When control is transferred to the Customer form, its PRE-FORM trigger immediately checks to see if the value of :GLOBAL.custid is NULL. If it is, then a new CUSTID is retrieved from a sequence and is displayed in the Customer ID field. If :GLOBAL.custid is not NULL, meaning that the user selected a customer ID in the Sales form, then the corresponding customer data fields are displayed in the Customer form.

If a new customer record is being created, then the user enters the customer data in the Customer form data fields, and clicks the Save New Record button. This trigger assigns the newly generated CUSTID value to :GLOBAL.custid, and inserts the new record into the CUSTOMER table. If the customer data are being edited, then the user edits the data and clicks the Update Record button, which updates the record in the CUSTOMER table. When the user clicks the Return button, the Customer form exits, and control is returned to the Sales form at the program line immediately following the CALL_FORM command. The new value of :GLOBAL.custid is then displayed in the Sales form's Customer ID field.

Now you will create the Create/Edit Customer button trigger in the Sales form to assign the global variable value to the current value of CUSTID, and call the Sales form.

To make the Create/Edit Customer button trigger:

1 Make a new WHEN-BUTTON-PRESSED trigger for the Create/Edit Customer button, and then type the code shown in Figure 6-60 in the PL/SQL Editor. The CALL_FORM command assumes that the executable file for the Customer form is stored in the Chapter6 folder on your Student Disk and is named ch6ccustomerdone.fmx.

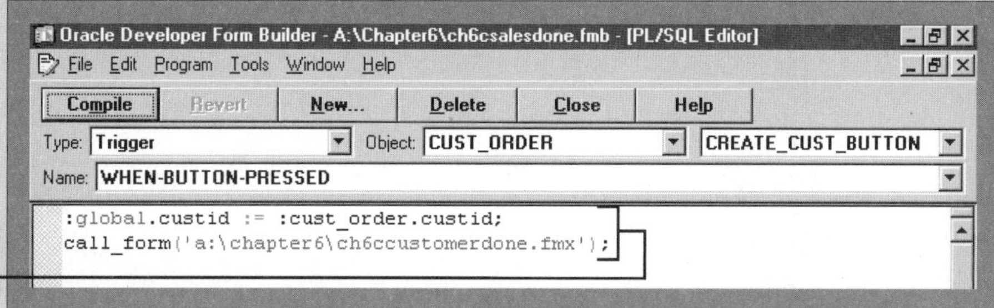

Figure 6-60: Create/Edit Customer button trigger

If your Customer form file is stored on a different drive or in a different folder, you can use it as long as you specify the full pathname. The pathname cannot contain any blank spaces.

2 Compile the trigger, correct any syntax errors, and then close the PL/SQL Editor.

Next, you need to create a new sequence named CUSTID_SEQUENCE that will automatically generate customer ID values. Then, you will modify the PRE-FORM trigger for the Customer form to test that the value of :GLOBAL.custid is NOT NULL. If it is NOT NULL, then the other CUSTOMER data fields associated with this ID value are retrieved from the database and inserted into the form fields using an implicit cursor. If :GLOBAL.custid is NULL, then the next value of CUSTID_SEQUENCE is placed in the Customer ID field.

To create the new sequence and modify the PRE-FORM trigger:

1 Switch to SQL*Plus, and create a sequence named CUSTID_SEQUENCE that starts with 200 and has no maximum value.

help

If you get the message "name is already used by an existing object," it means you already created this sequence in a previous exercise, so you can skip this step.

2 Switch back to Form Builder and open the Object Navigator window, click **CH6CSALESDONE**, then click the **Collapse All** button on the Object Navigator toolbar to close all of the open items in the Sales form.

3 Open the **ch6ccustomer.fmb** file from the Chapter6 folder on your Student Disk, and save it as **ch6ccustomerdone.fmb** in the same folder. Change the name of the form module to **CH6CCUSTOMERDONE** in the Object Navigator Window.

4 Open the PRE-FORM trigger for the Customer form, and modify the code as shown in Figure 6-61. Compile the code, correct any syntax errors, and close the PL/SQL editor.

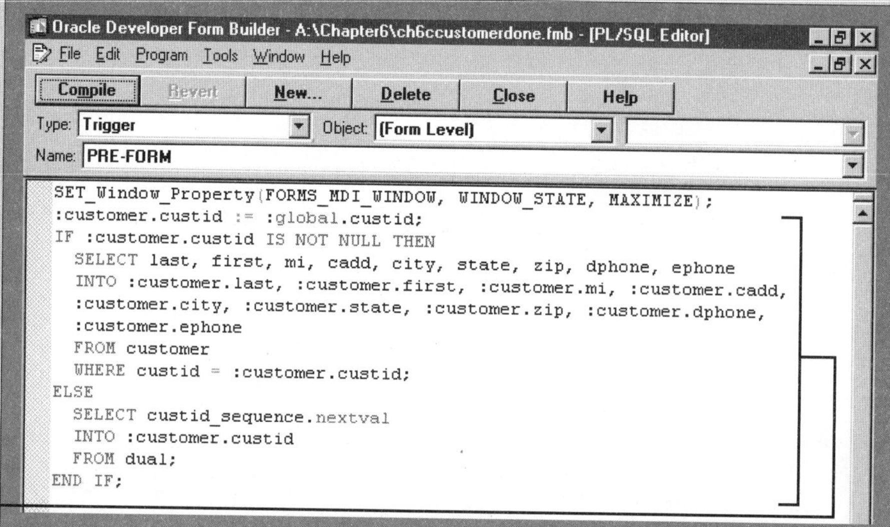

Oracle Developer Form Builder - A:\Chapter6\ch6ccustomerdone.fmb - [PL/SQL Editor]
File Edit Program Tools Window Help
Compile Revert New... Delete Close Help
Type: Trigger
Object: (Form Level)
Name: PRE-FORM
SET_Window_Property(FORMS_MDI_WINDOW, WINDOW_STATE, MAXIMIZE);
:customer.custid := :global.custid;
IF :customer.custid IS NOT NULL THEN
SELECT last, first, mi, cadd, city, state, zip, dphone, ephone
INTO :customer.last, :customer.first, :customer.mi, :customer.cadd,
:customer.city, :customer.state, :customer.zip, :customer.dphone,
:customer.ephone
FROM customer
WHERE custid = :customer.custid;
ELSE
SELECT custid_sequence.nextval
INTO :customer.custid
FROM dual;
END IF;
add these lines

Figure 6-61: Modified Customer form PRE-FORM trigger

5 Save the form.

Next, you need to modify the Save New Record button in the Customer form. When a new customer record is inserted into the CUSTOMER table, the value of the newly inserted customer ID must be assigned to :GLOBAL.custid so the new customer ID value is returned to the Sales form.

To modify the Save New Record button trigger:

1 Select the Customer form (**CH6CCUSTOMERDONE**) in the Object Navigator, click **Tools** on the top menu bar, then click **Layout Editor** to open the Customer form in the Layout Editor window. Select the **Save New Record** button, right-click, then click **PL/SQL Editor**. The button's WHEN-BUTTON-PRESSED trigger is displayed.

2 Modify the trigger code as shown in Figure 6-62. Compile the trigger, correct any syntax errors, then close the PL/SQL Editor window.

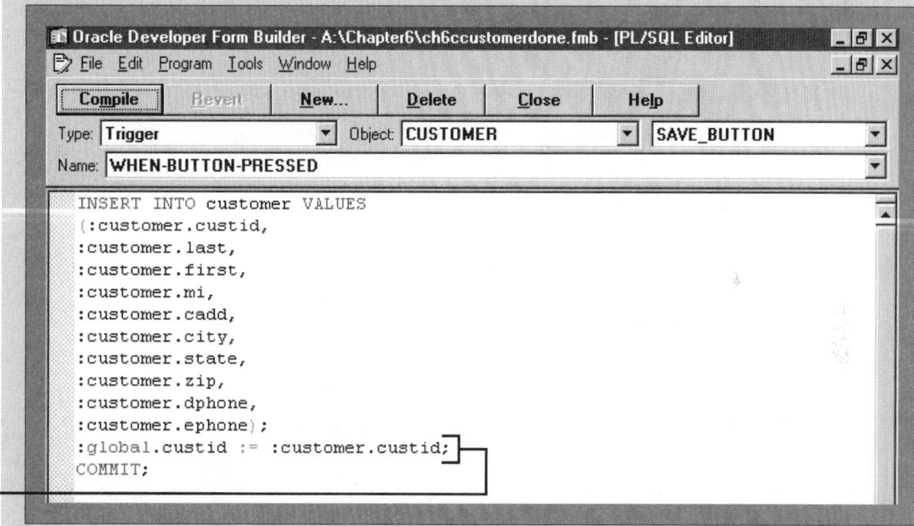

Figure 6-62: Modifying the Save New Record button trigger in the Customer form

3 Save the Customer form.

Before you can test to see if you can successfully display the Customer form from the Sales form, you need to compile your modified Customer form design file (ch6ccustomerdone.fmb) to create a new Customer form executable (ch6ccustomerdone.fmx) file.

When you call a form from another form, remember that the .fmx file (and not the .fmb file) is called. Whenever you make changes to the .fmb file of a form that is being called, don't forget to generate a new .fmx file.

To create the new Customer form executable file:

1. Open the Object Navigator window, select the Customer form (**CH6CCUSTOMERDONE**) in the Object Navigator window.
2. Click **File** on the menu bar, point to **Administration**, and then click **Compile File.**

help

If you run the CH6CCUSTOMERDONE form rather than compile it, you will get an error message stating "variable :GLOBAL.CUSTID does not exist." This is because in this form the global variable has not yet been assigned a value, and it is first used in the right-hand side of an assignment statement. Don't worry—the form will work correctly when it is called from the Sales form, because the global variable will have been assigned a value when the Customer form is displayed.

One last thing needs to be done before the modifications are complete. If the user creates a new customer in the Customer form, the new customer ID is now set to :GLOBAL.custid. When control returns to the Sales form, this value needs to appear in the Customer ID field of the Sales form. Recall that after a CALL_FORM command is executed, program control goes to the called form, and remains there until the called form is exited. Then, program control returns to the program line immediately following the CALL_FORM command. You will set the value of the CUSTID form field equal to :GLOBAL.custid in the Create/Edit Customer button immediately after the CALL_FORM command, so the new customer ID value is displayed on the Customer Orders canvas when you return to it.

To set the value of the customer ID to :GLOBAL.custid:

1. Click **Tools** on the menu bar, click **Layout Editor**, then select the CUST_ORDER_CANVAS from the canvas list. The Customer Order canvas is displayed in the Layout Editor window.
2. Right-click the **Create/Edit Customer** button, then click **PL/SQL Editor.** The button's WHEN-BUTTON-PRESSED trigger code is displayed.
3. Modify the code by adding the assignment statement shown in Figure 6-63.

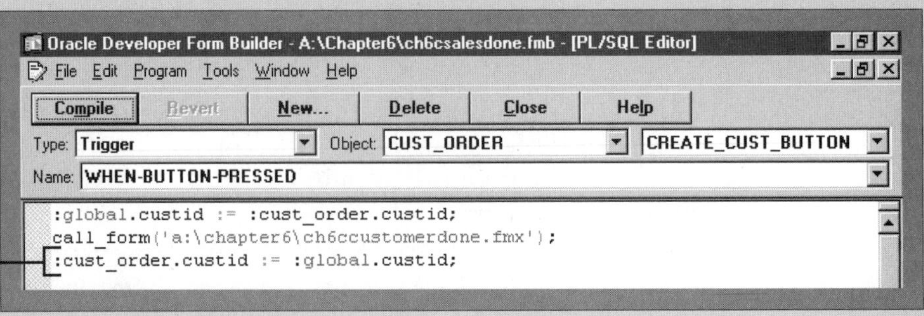

Figure 6-63: Modifying the Create/Edit Customer button trigger

4 Compile the trigger, correct any syntax errors, and close the PL/SQL Editor.

5 Save the Customer Order form.

When you have multiple forms open in Form Builder, clicking the Save button only saves the form that is currently selected in the Object Navigator window.

Now you need to test to make sure that the trigger that calls the Customer form from the Sales form is working correctly. You will run the Sales form, click the Create/Edit Customer button to call the Customer form, insert a new customer record, and then see if the value for the customer ID appears on the Sales form when control is returned to it.

To test if the trigger calling the Customer form from the Sales form is working:

1 Open the Object Navigator window, select **CH6CSALESDONE**, then click the **Run FormClient/Server** button to run the Sales form.

When more than one form is open in Form Builder, the form that is currently selected in the Object Navigator runs when you click the Run Form Client/Server button.

2 Click the **Create/Edit Customer** button. The Customer form is displayed with a new value inserted into the Customer ID field.

If the Customer form is not displayed, check the Create/Edit Customer button trigger in the Sales form and make sure you entered the path to a:\Chapter6\ch6ccustomerdone.fmx correctly. If you did, verify that the ch6ccustomerdone.fmx file exists in the Chapter6 folder on your Student Disk.

help

If the new customer ID does not appear in the Customer form, verify that you made the correct modifications to the ch6ccustomerdone.fmb file, and compile the file again to generate a new .fmx file.

3 Enter the information for the customer record for **Terry L. Harris, 3879 Edgewater Drive, Superior, WI 54880**, daytime phone **7155552008**, and evening phone **7155552388**, then click **Save New Record**. The message "FRM-40401: No changes to save." confirms that the new record was inserted.

4 Click the **Return to Customer Order** button. The Customer form exits, and the new customer's ID should appear in the Customer ID field in the Sales form.

5 Click **Exit** to close the form. Click **Yes** to confirm closing.

6 Select the Customer form (**CH6CCUSTOMERDONE**) in the Object Navigator window, click **File** on the top menu bar, then click **Close** to close the Customer form in Form Builder. If necessary, click **Yes** to save your changes. The Sales form (CH6CSALESDONE) remains open.

Navigating Between Different Form Canvases

To move to a different canvas within a form, you use the GO_ITEM command to switch the form focus to an item (such as a text item or button) on the new canvas. The canvas that contains the item that has the form's focus is then displayed.

When the user clicks the Enter/Edit Order Items button on the Customer Orders canvas, the system should display the Order Line canvas. Now you will create a trigger for the Enter/Edit Items button. This trigger will display the Order line canvas by using the GO_ITEM command to switch the form's focus to the LOV command button next to the Inventory ID field on the Order Line canvas.

To create the Enter/Edit Order Items button trigger:

1 Create a WHEN-BUTTON-PRESSED trigger for the Enter/Edit Items button, and type the following command: `GO_ITEM('inv_lov_button');`.

2 Compile the trigger and correct any syntax errors, then close the PL/SQL Editor.

3 Save the form.

Before you can test the form, you need to ensure that the Customer Orders canvas will appear as the first canvas when the form runs. When you create a form with multiple canvases, the canvas that is displayed first is the canvas whose block items appear first in the Data Blocks section of the Object Navigator window. Next you will open the Object Navigator window, examine the order of the blocks, and confirm that the CUST_ORDER block is the first block displayed.

To check the order of the form blocks:

1 Open the Object Navigator window in Ownership View.

2 Select **Data Blocks**, then click the **Collapse All** button on the toolbar.

3 Click beside Data Blocks to examine the block order.

4 Confirm that CUST_ORDER is the first block displayed, as shown in Figure 6-64.

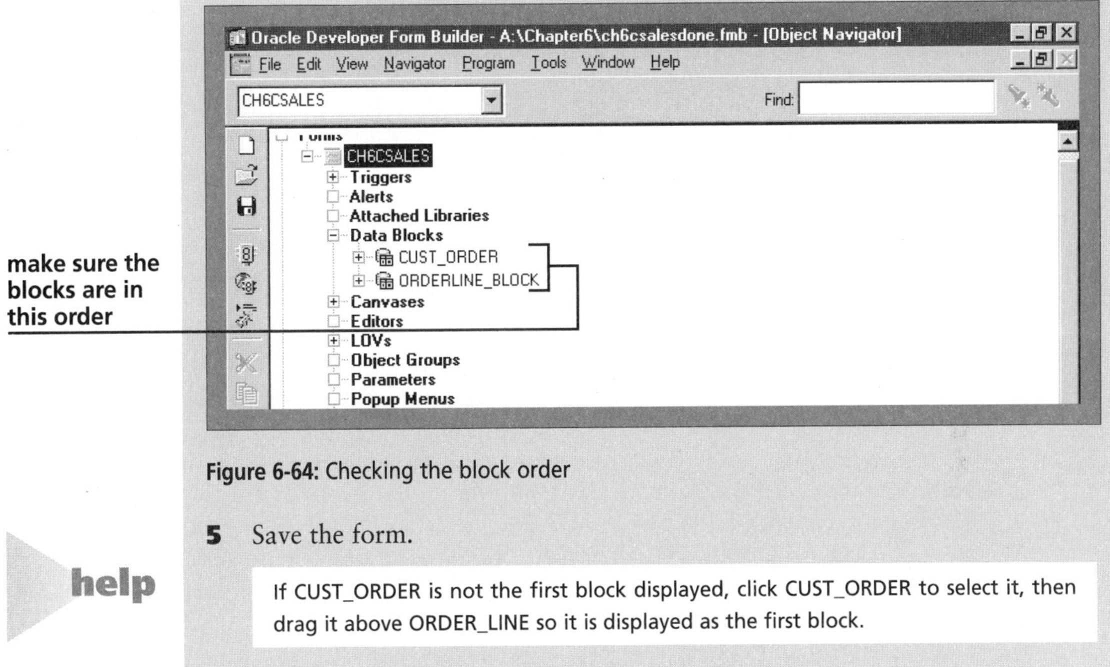

Figure 6-64: Checking the block order

5 Save the form.

help

If CUST_ORDER is not the first block displayed, click CUST_ORDER to select it, then drag it above ORDER_LINE so it is displayed as the first block.

Now you will test the form. Customer Mitch Edwards is placing an order. He wants to pay by check, and he has indicated that the order source is Catalog 151. You will run the form, enter the customer order information on the Customer Order canvas, and then click the Enter Items button to move to the Order Line canvas.

To test the form:

1 Run the form. The Customer Order canvas is displayed.

2 Click the **LOV** command button beside Customer ID, and select customer **Mitch Edwards.**

3 Make sure the **Check** option button is selected for payment method.

4 Click the **LOV** command button beside Order Source, and select Order Source **151**.

5 Click **Save Order.** The order data are inserted into the CUST_ORDER table, and the confirmation message "FRM-40401: No changes to save." is displayed.

6 Click **Enter/Edit Items.** The Order Line canvas is displayed with the form focus on the LOV command button beside the Inventory ID field.

7 Click [X] to close the Forms Runtime application.

Creating a Relationship Between a Data Block and a Control Block

Recall that a data block is directly associated with a database table, and contains text items that correspond to the table fields. Data blocks can be made quickly and easily using the Data Block and Layout Wizards, and the Forms Runtime application has built-in programs for viewing, inserting, updating, and deleting data in data block forms. However, data block forms do not have the flexibility that is sometimes needed for custom form applications, because they cannot easily insert, update, and delete data fields in multiple tables. When you need to work with fields from multiple tables, you must create a control block, which is not associated with a particular database table, but instead has fields that can contain data values that can be referenced in SQL queries that are written by the form's programmer. Sometimes it is useful to combine data blocks and control blocks on the same canvas to take advantage of the strengths of both types of blocks.

Next, you will create the order summary that is displayed at the bottom of the Order line canvas and shows the information about each ordered item, including the inventory ID, quantity ordered, description, size, color, and price, as shown on the Order Line canvas design in Figure 6-56. Since this information is for display only and does not involve any table insert, update, or delete operations, it can be displayed in a data block with a tabular layout. Since this display involves information from the ITEM, INVENTORY, and ORDERLINE database tables, you will have to create a database view in SQL*Plus as the data block source. The view must also contain the ORDERID, because the data block will have a master/detail relationship with the ORDERID field in the CUST_ORDER block so that it displays the order line information only for the current customer order, rather than for all order lines in the database. Before you create the data block, you will display the Order Line canvas and examine its current data fields.

To display the Order Line canvas:

1 In the Object Navigator window, click [+] beside ORDERLINE_BLOCK, then click [+] beside Items to display the Items in the order line block.

Notice that the name of each of the data items in the ORDERLINE_BLOCK is prefaced by the word CONTROL (CONTROL_INVID, CONTROL_ DESCRIPTION, etc.). If these fields had the default database field names (INVID, DESCRIPTION, etc.), duplicate fields would be created when you create the data block to display the order line information. If all form fields do not have unique names, Form Builder has trouble distinguishing between different fields, even if they are in different blocks. It is a good practice to always give all form fields unique names. Since data block fields automatically assume the names of the corresponding database fields, it is easiest to give unique names to control block fields if there is a conflict. Since ORDERLINE_BLOCK is a control block, each field name that might be duplicated in the data block has been changed in this block so that its name is preceded by the word "CONTROL."

Now you will create the view and corresponding data block to display the order summary information.

To create the view and data block:

1 Switch to SQL*Plus, and type the query shown in Figure 6-65 to create the ORDER_SUMMARY_VIEW.

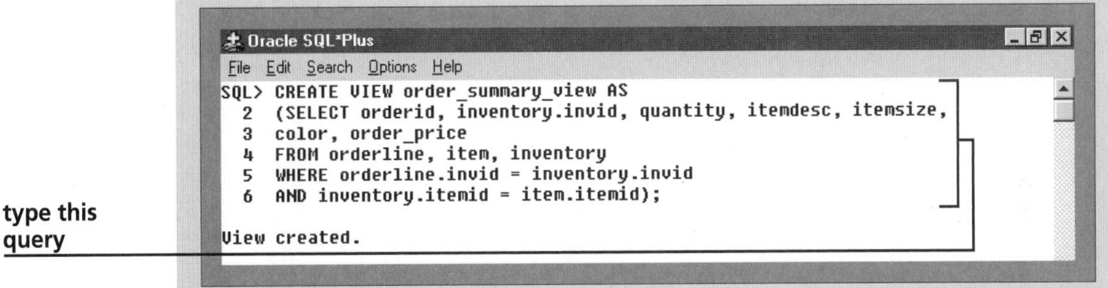

```
Oracle SQL*Plus
File Edit Search Options Help
SQL> CREATE VIEW order_summary_view AS
  2  (SELECT orderid, inventory.invid, quantity, itemdesc, itemsize,
  3  color, order_price
  4  FROM orderline, item, inventory
  5  WHERE orderline.invid = inventory.invid
  6  AND inventory.itemid = item.itemid);

View created.
```

type this query

Figure 6-65: Creating the ORDER_SUMMARY_VIEW

2 Switch back to Form Builder. Open the Object Navigator window if necessary, and select **CH6CSALESDONE**. Since you are going to create a new data block, you do not want to have any data block currently selected. If a data block is selected, the system will open the selected data block in re-entrant mode for editing, rather than creating a new data block.

3 Click **Tools** on the menu bar, click **Data Block Wizard**, then click **Next** if the Data Block Wizard Welcome page is displayed. When the Type page is displayed, make sure the **Table or View** option is selected, then click **Next.**

4 On the Table page, click **Browse**, check the **Views** check box, select **ORDER_SUMMARY_VIEW**, then click **OK.**

5 Select all of the view data fields to be included in the block, then click **Next.** The Master/Detail page is displayed.

When you display a control block and a data block on the same canvas, you cannot directly create a master/detail relationship between the data block and the control block. Instead, you must create the data block without a master/detail relationship, and then later modify the data block's WHERE property on its Property Palette so that the matching key field on the data block is equal to the value of the master key field on the control block. For now, you will omit the master/detail relationship, and finish creating the data block.

To omit the master/detail relationship and finish the data block:

1. Click **Next** to pass the Master/Detail page, then click **Finish**.

Next, you will create the block layout using the Layout Wizard. The block will be displayed on the Order Line canvas, show five records at a time, and have a scroll bar. You will display all of the block items except the ORDERID—it was included in the block to create the master/detail relationship, but it does not need to be displayed on the form, because you are only displaying order items for the current customer order.

To create the block layout:

1. Click **Next** if the Layout Wizard Welcome page is displayed.
2. On the Canvas page, select **ORDERLINE_CANVAS** from the Canvas list if necessary, then click **Next**.
3. On the Data Block page, select all of the block items for the layout except ORDERID. Check to make sure that the items are in the order shown in Figure 6-66, then click **Next**.

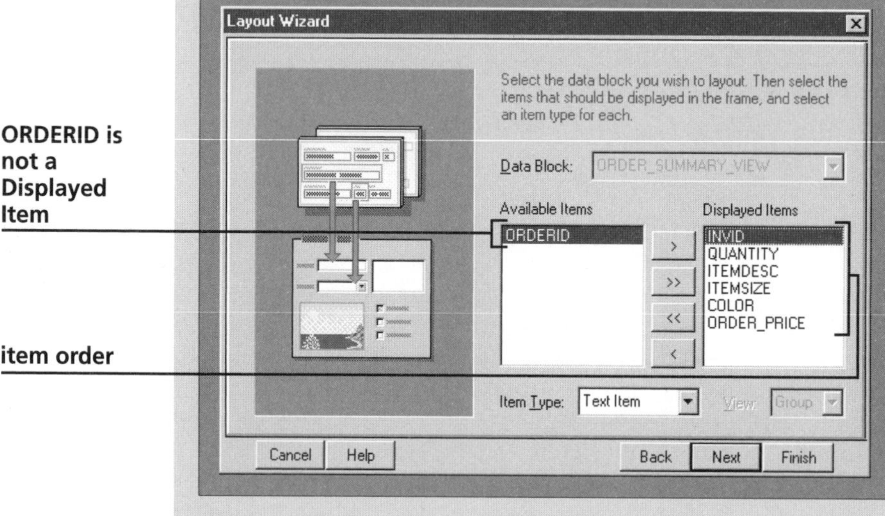

Figure 6-66: Specifying the layout items

4 On the Items page, modify the column prompts and widths as shown in Figure 6-67. Click **Next.**

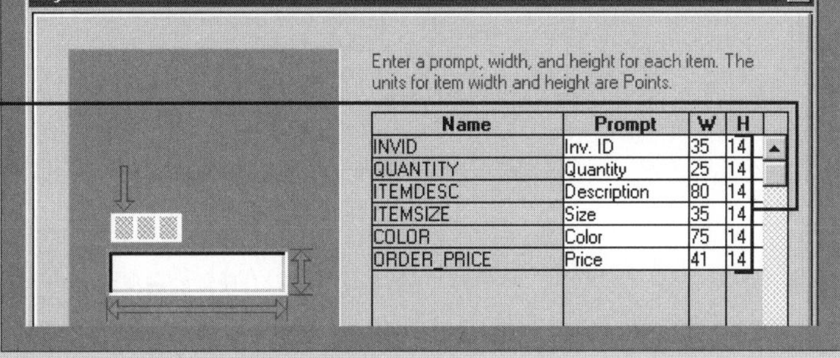

Figure 6-67: Modifying the column prompts and widths

5 On the Style page, click the **Tabular** option style, then click **Next.**

6 On the Rows page, type **Order Summary (View Only)** for the Frame Title. Type **5** for Records Displayed, and check the **Display Scrollbar** check box. Click **Next**, then click **Finish.** The ORDER_DISPLAY_VIEW layout is displayed on the ORDERLINE_CANVAS.

7 Center the Order Summary display under the other form frames.

8 Save the form.

Creating the Relationship Between the Control Block and Data Block

Recall that when you create a relationship between a control block and a data block, you do not specify a master/detail relationship when you create the data block. Instead, you modify the data block's WHERE property on its Property Palette so that the matching key field on the data block is equal to the value of the master key field on the control block. The general format of the WHERE property is as follows: WHERE <data block field name> = :<control block name>.<control block field name>. Next you will create the relationship between the CUST_ORDER control block and the ORDER_SUMMARY_VIEW data block. You will open the Property Palette for the ORDER_SUMMARY_VIEW data block, and specify that the value for ORDERID in the data block is equal to the value of ORDERID in the CUST_ORDER control block.

To create the relationship between the data block and control block:

1. Open the Object Navigator window in Ownership View, and double-click the **Data Block** icon beside ORDER_SUMMARY_VIEW to open the data block's Property Palette.
2. Type the following for the WHERE Clause property:

```
ORDERID = :CUST_ORDER.ORDERID
```

3. Close the Property Palette.

Creating the Order Line Canvas Order Items Frame Triggers

You will create the triggers for the buttons in the Order Items frame first. When the user clicks the Add Order Line button, the form inserts a new record into the ORDERLINE table, and then displays the updated order lines for the order in the Order Summary (View Only) frame. The INSERT command in the Add Order Line button trigger uses the current order ID, which is saved in the CUST_ORDER block's Order ID field. It uses the inventory ID displayed in the Order Items frame's Inventory ID field, which the user selects by clicking the LOV command button next to the Inventory ID field. It also uses the values that the user enters for Price and Quantity. After the INSERT command is executed, the trigger needs to move the form focus to the ORDERLINE_SUMMARY_VIEW block and update its display. This is done using the EXECUTE_QUERY command, which automatically **flushes** the block, or makes its information consistent with the block's corresponding database data. Finally, the trigger clears the form fields in the Order Items frame so the user can select another item to be inserted into the order.

If you use the CLEAR_FORM command to clear the Order Items frame fields, all of the form fields will be cleared, and you don't want to do that because then you will lose the order ID information that is stored on the Customer Order canvas. Therefore, you will call a program unit named CLEAR_ORDERLINE_FIELDS that is included in the Sales form that only clears the fields in the Order Items frame.

To create the Add Order Line button trigger:

1. Create a new WHEN-BUTTON-PRESSED trigger for the Add Order Line button on the Order Line canvas, and type the commands shown in Figure 6-68 to insert a new record into the ORDERLINE table, clear the frame fields, move to the ORDER_SUMMARY_VIEW block, and update the block values.

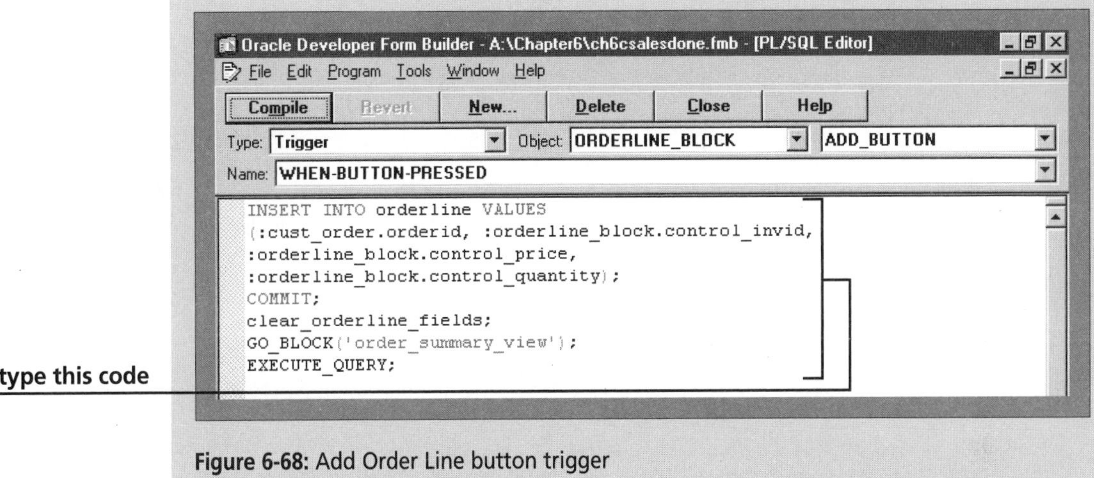

Figure 6-68: Add Order Line button trigger

2 Compile the trigger, correct any syntax errors, and then close the PL/SQL Editor.

3 Save the form.

Next you will create the trigger for the Select Order Line button. This trigger will activate a LOV that will display the inventory ID, quantity, description, size, color, and price for every order line in the current order. To do this, the LOV query will only return order lines where the ORDERID field matches the current order ID displayed in the Order ID field on the Customer Orders canvas. You will assign the LOV to the Description data field in the Order Items frame, and place the GO_ITEM command (to redirect the form's focus to the Description field) and LIST_VALUES command (to display the LOV) in the Select Order Line button's WHEN-BUTTON-PRESSED trigger.

To create the LOV to display the order line information for the current order:

1 Click **Tools** on the menu bar, click **LOV Wizard** to start the LOV Wizard, make sure the **New Record Group based on query** option button is selected, then click **Next.**

2 Enter the following SQL statement in the SQL Query Statement box to display the order lines for the current order:

```
SELECT inventory.invid, quantity, itemdesc, itemsize,
    color, order_price
FROM orderline, item, inventory
WHERE orderline.invid = inventory.invid
AND inventory.itemid = item.itemid
AND orderid = :cust_order.orderid
```

3 Click **Next**, then click the **Select All** button [>>] to display all of the record group columns in the LOV display, then click **Next.**

4 Modify the Display Columns as follows, then click **Next.**

Column	Title	Width	Return Value
INVID	ID	35	ORDERLINE_BLOCK.CONTROL_INVID
QUANTITY	Quantity	35	ORDERLINE_BLOCK.CONTROL_QUANTITY
ITEMDESC	Description	140	ORDERLINE_BLOCK.CONTROL_DESCRIPTION
ITEMSIZE	Size	25	ORDERLINE_BLOCK.CONTROL_SIZE
COLOR	Color	40	ORDERLINE_BLOCK.CONTROL_COLOR
ORDER_PRICE	Price	30	ORDERLINE_BLOCK.CONTROL_PRICE

help

If an error message is displayed when you click Next, it is probably because one of the return values is not specified correctly. Check each return value to make sure it displays the complete block name and item name. Sometimes these values are truncated when they are transferred from the Items and parameters list box to the Return Value Column, and you have to type the Return value manually.

5 Change the LOV title to **Current Order Lines**, and change the display Width to **300**, and the Height to **200**, then click **Next.**

6 On the Advanced Options page, click **Next.**

7 On the Items page, select **ORDERLINE_BLOCK_CONTROL_DESCRIPTION** on the Return Items list, click the **Select** button so this field is displayed in the Assigned Items list, click **Next**, then click **Finish.**

8 In the Object Navigator, change the name of the new LOV and record group to ORDERLINE_LOV.

9 Open the Property Palette of the CONTROL_DESCRIPTION text item, and change its LOV property to **ORDERLINE_LOV.**

10 Create a WHEN-BUTTON-PRESSED trigger for the Select Order Lines button, and enter the following code to display the LOV:

```
GO_ITEM('control_description');
LIST_VALUES;
```

11 Compile the trigger, correct any syntax errors, and close the PL/SQL Editor.

12 Save the form.

Now you need to create the triggers for the Update Line and Delete Line buttons. The Update Line button will update the current order line record with the data displayed in the form, and the Delete Line button will delete the current order line record. Both triggers will clear the frame fields, and update the Order Summary by moving to the ORDER_SUMMARY_VIEW block and then issuing the EXECUTE_QUERY command to update the block display.

To create the Update Line and Delete Line triggers:

1 Create a new WHEN-BUTTON-PRESSED trigger for the Update Line button, and enter the code shown in Figure 6-69.

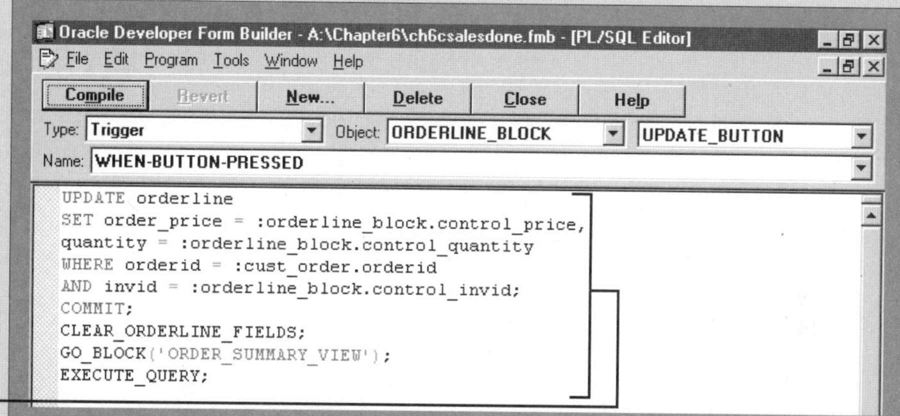

Figure 6-69: Update Line button trigger

2 Compile the trigger and correct any syntax errors.

3 Create a new WHEN-BUTTON-PRESSED trigger for the Delete Line button, and enter the code shown in Figure 6-70.

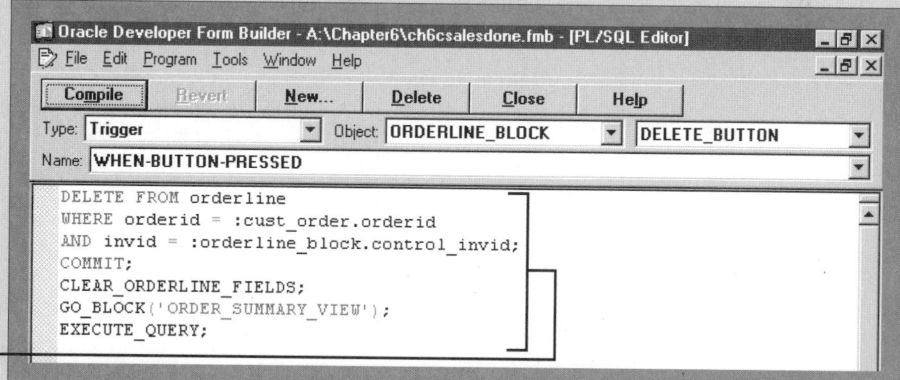

Figure 6-70: Delete Line button trigger

4 Compile the trigger, correct any syntax errors, and close the PL/SQL Editor.

5 Save the form.

Now you will test the form to see if the triggers for the buttons in the Order Items frame work correctly and if the Order Summary display updates correctly. You will create an order for customer Mitch Edwards, who is paying with a Visa credit card and using catalog 22 as the order source, and then select some order items.

To test the form:

1. Run the form.
2. Select customer **Mitch Edwards**, payment method **Visa**, and Order Source **122.** Click **Save Order**. The confirmation message "FRM-40401: No changes to save." is displayed, confirming the CUST_ORDER was successfully inserted.
3. Click the **Enter/Edit Items** button. The Order-Line canvas is displayed.
4. Click the LOV command button beside the Inventory ID field in the Order Items frame, select Inventory ID **11845** (Men's Expedition Parka, S, color Navy), then click **OK**. The selected item's data fields are displayed on the form.
5. Type **1** in the Quantity field.
6. Click **Add Order Line**. If the trigger works correctly, your Order Summary should update, and your screen should look like Figure 6-71.

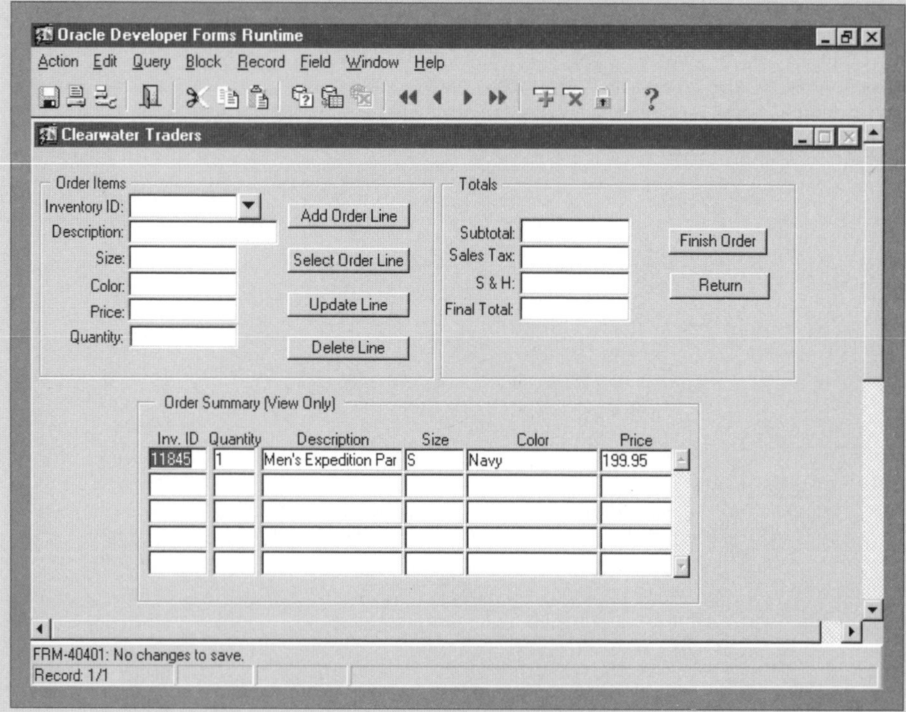

Figure 6-71: Ordered item in the Order Summary display

help

If your trigger does not work correctly, exit the form, and examine the code in the Add Order Line button's WHEN-BUTTON-PRESSED trigger and make sure it looks exactly like the code in Figure 6-68. If it looks the same, look up the explanations for any ORA- or FRM- error codes that occur, and then click the Exit button to exit the form. Run the form using the Form Builder Debugger, place a breakpoint on the first line of the ADD_BUTTON trigger, and single-step through the trigger to determine what part of the code is causing the error.

help

If you receive the error message "FRM-40735: WHEN-BUTTON-PRESSED trigger raised unhandled exception ORA-02291," you probably forgot to click the Save Order button on the Customer Orders canvas before you clicked the Enter/Edit Items button. Close the form, run it again, and be sure to click the Save Order button before you click the Enter/Edit Items button.

7 Click the LOV command button beside Inventory ID, and select inventory ID **11777** (Women's Hiking Shorts, color Khaki, size L). Enter **2** for the Quantity, then click **Add Order Line.** The new item is displayed in the Order Summary.

Next, you will test the Update and Delete buttons. You will change the quantity of the hiking shorts from 2 to 3, and then delete the parka from the order.

To test the Update and Delete buttons:

1 Click the **Select Order Line** button, select inventory ID **11777**, and click **OK**. The order line information is displayed in the Order Items fields.

2 Change the quantity to **3**, then click **Update Line**. The corrected order information is displayed in the Order Summary.

3 Click the **Select Order Line** button, select inventory ID **11845**, then click **OK**. The order line information is displayed in the Order Items fields.

4 Click **Delete Line**. The item is removed from the Order Summary.

5 Close Forms Runtime.

To finish the form, you need to create the triggers for the Finish Order and Return buttons. The Finish Order button calculates the order subtotal, which is the sum of each order line's quantity ordered times the price of each item. It also determines the sales tax amount (which will be 6 percent for Wisconsin residents and zero for all others), and the shipping and handling. The final order cost is the total of the subtotal, tax, and shipping and handling fields. You will create program units to calculate the sales tax and the shipping and handling. You will create the program unit to calculate the sales tax first.

To create the program unit to calculate the sales tax:

1. Open the Object Navigator window in Ownership View, click **Program Units**, and then click the **Create** button on the Object Navigator toolbar.
2. Name the new program unit **CALC_TAX**, be sure the **Procedure** option button is selected, and then click **OK**.
3. Type the code shown in Figure 6-72.

type this code

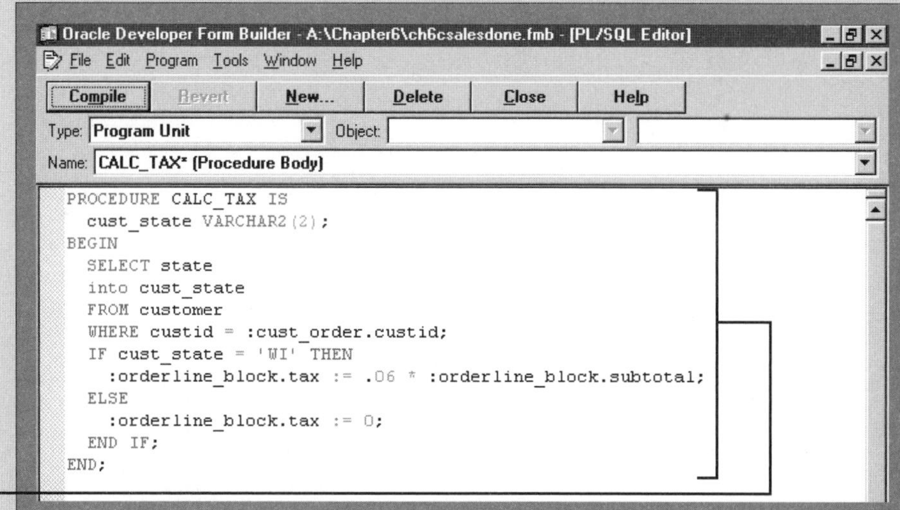
Oracle Developer Form Builder - A:\Chapter6\ch6csalesdone.fmb - [PL/SQL Editor]
File Edit Program Tools Window Help
Compile
Revert
New...
Delete
Close
Help
Type: Program Unit
Object:
Name: CALC_TAX* (Procedure Body)
PROCEDURE CALC_TAX IS
cust_state VARCHAR2(2);
BEGIN
SELECT state
into cust_state
FROM customer
WHERE custid = :cust_order.custid;
IF cust_state = 'WI' THEN
:orderline_block.tax := .06 * :orderline_block.subtotal;
ELSE
:orderline_block.tax := 0;
END IF;
END;

Figure 6-72: CALC_TAX program unit

4. Compile the program unit and correct any syntax errors.
5. Save the form.

Next, you need to create a program unit to calculate shipping and handling. Shipping and handling are calculated as a flat rate, depending on the amount of the order subtotal. The shipping and handling rate for orders for less than or equal to $25 is $3.50; the rate for orders between $25 and $75 is $5; and for orders over $75 the rate is $7.50.

To create the program unit to calculate shipping and handling:

1. In the PL/SQL Editor, click the **New** button on the PL/SQL Editor button bar, type **CALC_SHIPPING** for the program unit name, be sure the **Procedure** option button is selected, and then click **OK**.
2. Type the code shown in Figure 6-73.

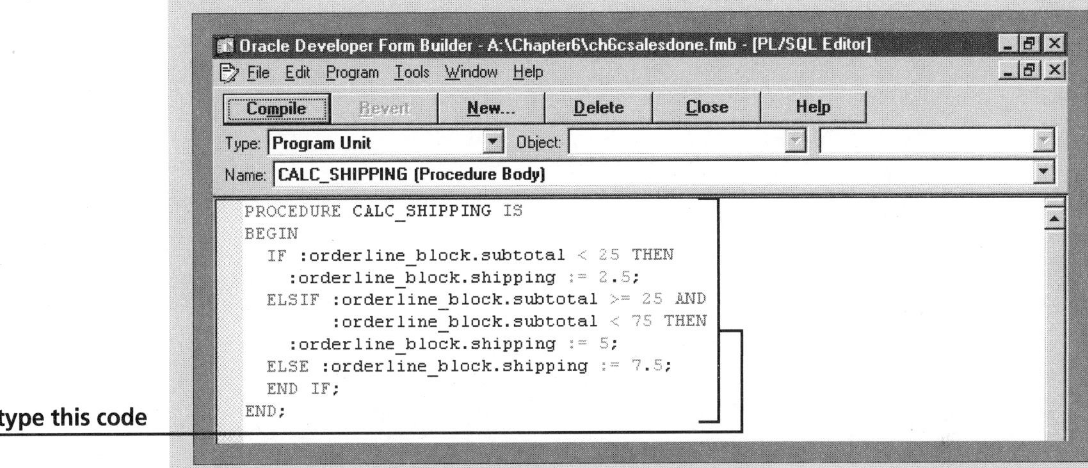
Oracle Developer Form Builder - A:\Chapter6\ch6csalesdone.fmb - [PL/SQL Editor]
File Edit Program Tools Window Help
Compile
Revert
New...
Delete
Close
Help
Type: Program Unit
Object:
Name: CALC_SHIPPING (Procedure Body)
PROCEDURE CALC_SHIPPING IS
BEGIN
IF :orderline_block.subtotal < 25 THEN
:orderline_block.shipping := 2.5;
ELSIF :orderline_block.subtotal >= 25 AND
:orderline_block.subtotal < 75 THEN
:orderline_block.shipping := 5;
ELSE :orderline_block.shipping := 7.5;
END IF;
END;
type this code

Figure 6-73: CALC_SHIPPING program unit

3 Compile the program unit, correct any syntax errors, and then close the PL/SQL Editor.

4 Save the form.

Now you will create the trigger for the FINISH_BUTTON, which will calculate the order subtotal, using a SQL query that multiplies the QUANTITY times the ORDER_PRICE for each record in the order, and then sums the total. The trigger then calls the program units to calculate tax and shipping and handling, and then calculates the order total.

To create the FINISH_BUTTON trigger:

1 Create a new WHEN-BUTTON-PRESSED trigger for the Finish Order button.

2 Type the code shown in Figure 6-74 in the PL/SQL Editor.

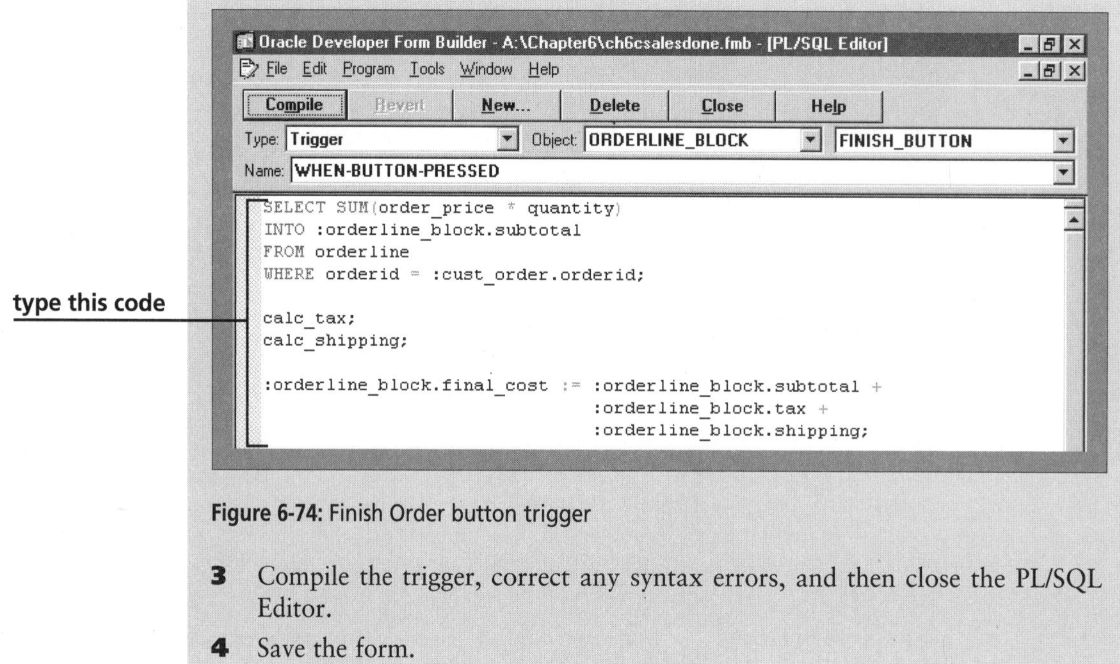

Figure 6-74: Finish Order button trigger

3 Compile the trigger, correct any syntax errors, and then close the PL/SQL Editor.

4 Save the form.

Now you will run the form to test if the Finish Button calculates the order totals correctly.

To test the Finish button:

1 Run the form.

2 Create a new order for customer **Mitch Edwards**, payment type **Check**, order source **122**, and then click the **Save Order** button. The confirmation message "FRM-40401: No changes to save." is displayed.

3 Click the **Enter/Edit Items** button.

4 Select Inventory ID **11845** (Men's Expedition Parka, S, Navy). Enter **1** for Desired Quantity, and then click the **Add Order Line** button. The new item is displayed in the Order Summary (View Only) frame.

5 Select Item ID **11775** (Women's Hiking Shorts, Khaki, S). Enter **2** for Desired Quantity, and then click the **Add Order Line** button. The new item is displayed in the Order Summary (View Only) frame.

6 Click the **Finish Order** button. The subtotal, sales tax, shipping and handling, and final cost are displayed in the form, as shown in Figure 6-75.

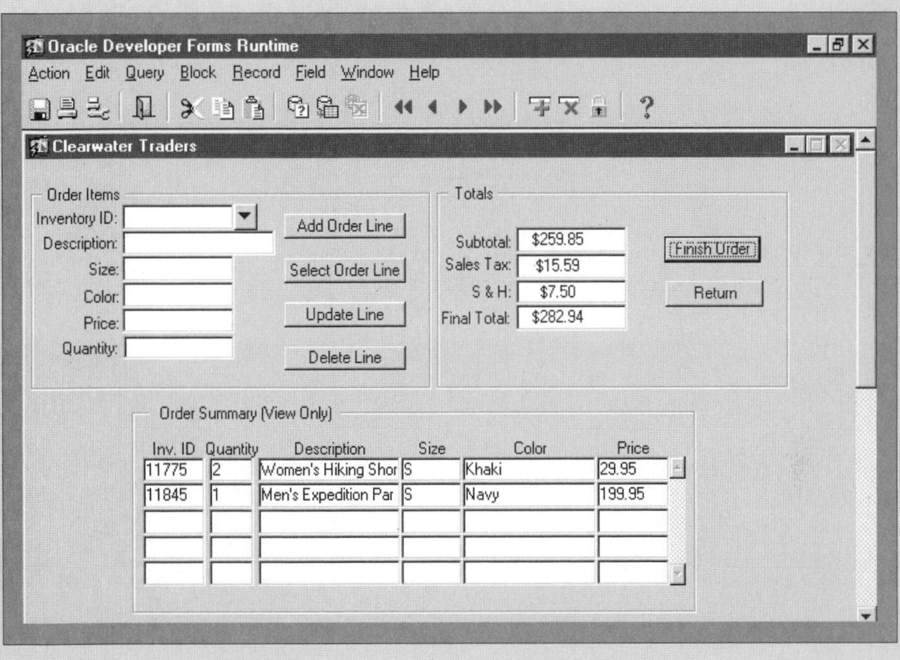

Figure 6-75: Final order totals

7 Close the form.

To complete the form, you need to create the trigger for the Return button. This button needs to clear the form fields, switch back to the CUST_ORDER canvas, and insert a new value into ORDERID so the customer service representative can begin taking another order. To do this, the trigger will first issue the CLEAR_FORM command. Then, the trigger will issue the GO_ITEM command and move to the ORDER_LOV_BUTTON on the CUST_ORDER canvas. Finally, it will insert the next value of the ORDERID_SEQUENCE into the ORDERID field.

To create the RETURN_BUTTON trigger:

1 Create a new WHEN-BUTTON-PRESSED trigger for the Return button.

2 Type the code shown in Figure 6-76. Compile the trigger, correct any syntax errors, then close the PL/SQL Editor.

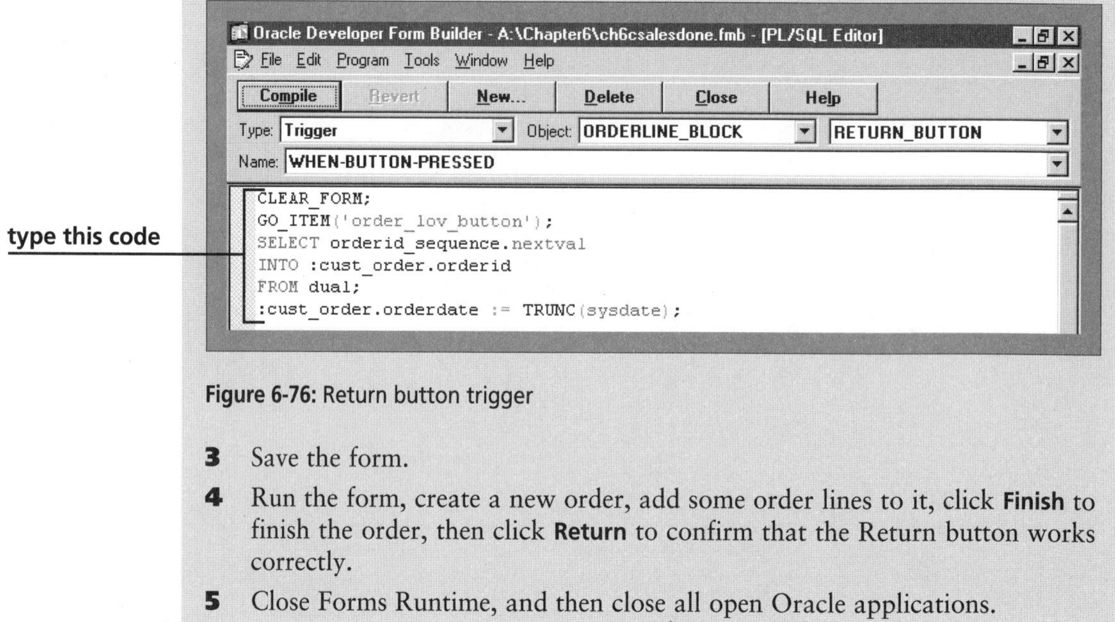

Figure 6-76: Return button trigger

3 Save the form.

4 Run the form, create a new order, add some order lines to it, click **Finish** to finish the order, then click **Return** to confirm that the Return button works correctly.

5 Close Forms Runtime, and then close all open Oracle applications.

SUMMARY

- To create an application with multiple canvases, you can use either a mutiple-form approach or a multiple-canvas approach. With a multiple-canvas approach, you create each separate canvas and save them all in the same form .fmb file. With a multiple-form approach, each canvas is saved in a different form .fmb file.
- In the multiple-canvas approach, data can be easily shared among the different canvases. However, this approach makes it impossible for different programmers to program different screens of the same application at the same time, since the .fmb file can only be modified by one programmer at a time.
- The multiple-form approach is best used when several people are collaborating to create a complex application, and enables the code for a form to be used in different applications. However, data sharing among forms is more difficult, and must be done using global variables.

- A global variable is a variable whose value is visible to multiple forms.
- To display one form from another form, you use the CALL_FORM command in a trigger in the calling form, and specify the executable (.fmx) file for the form being called.
- When a CALL_FORM command executes, program control is immediately transferred to the called form. When the called form exits, program control returns to the program code line immediately following the CALL_FORM command in the called form.
- When you create a form with multiple canvases, the canvas that is displayed first is the canvas whose block items appear first in the Data Blocks section of the Object Navigator window.
- To create a relationship between a control block and a data block, create the data block without a master/detail relationship, and then modify the data block's WHERE property so the matching key field on the data block is equal to the value of the master key field on the control block.
- To programatically update a data block in a form, you must go to the block using the GO_BLOCK command, and then issue the EXECUTE_QUERY command, which automatically flushes the block, or makes its information consistent with the block's corresponding database data.

REVIEW QUESTIONS

1. How do you share data among different form canvases when you call one form from another using the CALL_FORM command?
2. In a form with multiple canvases, how do you move from one form canvas to another?
3. In a form with multiple canvases, how do you share a data value among triggers attached to items on different form canvases?
4. True or false: In a form with multiple canvases, the canvas that is displayed first when the form starts is the canvas that is listed first in the Object Navigator window.
5. When a form is being called by another form using the CALL_FORM command, why is it important to recompile the called form's .fmb file every time you change it?
6. How can you create a relationship between a control block and a data block?

PROBLEM-SOLVING CASES

Run the northwoo.sql script to refresh the Northwoods University database, and run the clearwat.sql script to refresh the Clearwater Traders database. The script files are stored in the Chapter6 folder on your Student Disk.

1. Open the ch6csalesdone.fmb file that you completed in the exercises in this chapter, and save it as Ch6cEx1.fmb. Then, make the following modifications:
   a. Create an error trap in the Enter/Edit Items button on the Customer Orders canvas that checks to make sure that the user has saved the CUST_ORDER record before allowing the program to switch to the Order Line canvas. If the user forgot to save the record, display a message (using either the MESSAGE procedure or an alert) reminding him to save it. (*Hint*: Check to see if the record has been inserted using an explicit cursor and the %NOTFOUND attribute.)
   b. Modify the Add Order Line button on the ORDERLINE_CANVAS so that before the ORDERLINE record is inserted, the system first checks to see if the quantity to be ordered is available in the INVENTORY table. If there is not sufficient inventory quantity on hand, then the system should display a Stop alert with an OK button advising the user that the item is out of stock. When the user clicks OK, the order item fields should be cleared. If there is sufficient quantity, then the INVENTORY table should be updated to show that the QOH has been decreased by the quantity ordered.
2. In this exercise, you will create a form with multiple canvases that will allow a Northwoods University faculty member to log on to the database using a logon screen. If the faculty member logs on successfully, she is presented with a menu that allows her to access forms to view and update personal information, advisee information, and class list information.
   a. Create a new form and save it as Ch6cEx2.fmb.
   b. Create a canvas and associated control block named LOGON_CANVAS and LOGON_BLOCK containing the items shown in Figure 6-77. When the user clicks the Log On button, the system should check to see if the FID and FPIN are valid for any of the current FACULTY records. If they are, a canvas with the menu selections shown in Figure 6-78 is displayed. The user should have three chances to successfully enter her ID and PIN. After the third unsuccessful attempt, the form should exit.

      Hint: Use a global variable to count the number of times the user has attempted to log on.

      Hint: To conceal the PIN data, change the text item's Conceal Data property to Yes.

      Hint: Create an explicit cursor and use the %FOUND attribute to check to see if the faculty ID and PIN entered by the user are included in the FACULTY table.

      Hint: To exit the form without having the system ask for user confirmation, use the following command: EXIT_FORM(NO_VALIDATE);.

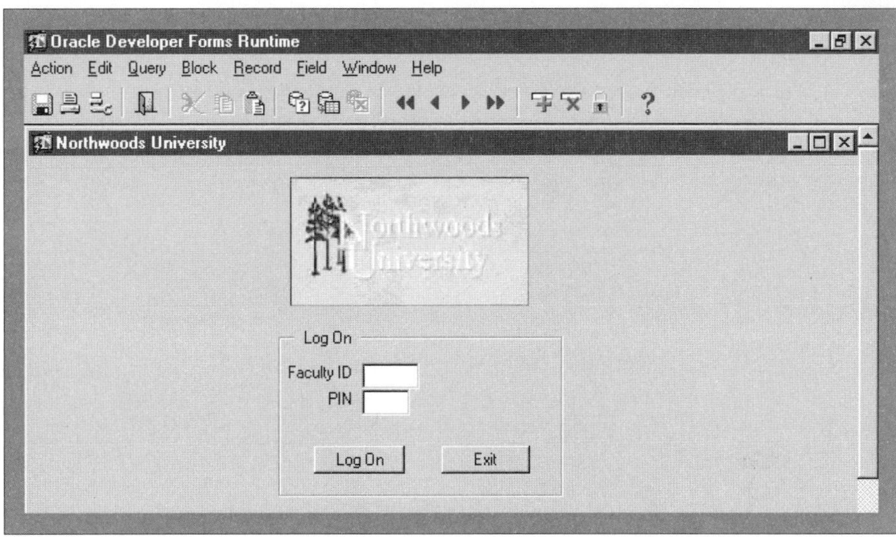

Figure 6-77

c. Create a Menu canvas named MENU_CANVAS and an associated control block containing the command buttons and layout items shown in Figure 6-78.

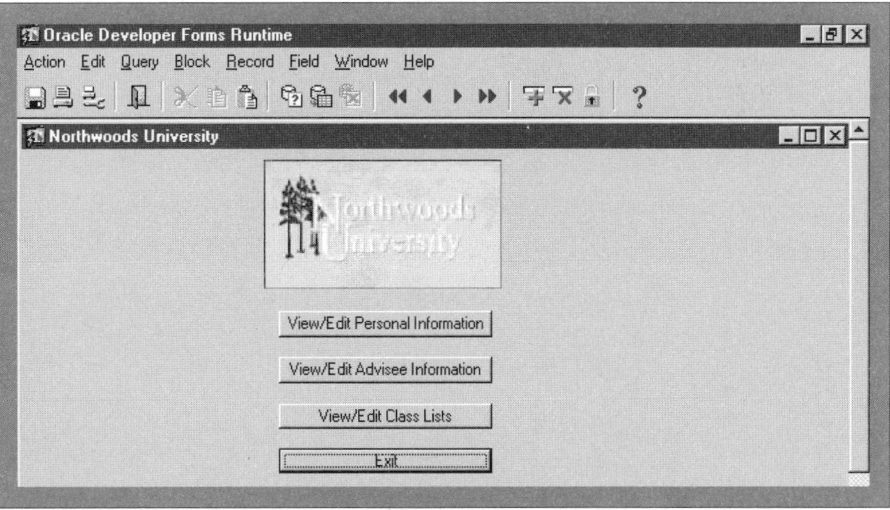

Figure 6-78

d. When the user clicks the View/Edit Personal Information button on the Menu canvas, the canvas shown in Figure 6-79 appears with the user's personal information displayed. Create a control block containing the fields shown. Create a trigger for the Update button to save updated information to the FACULTY table. Create a LOV to display all LOCATION building codes and room numbers, and another LOV to display all choices for FRANK. When the user clicks the Return button, the Menu canvas is displayed again.

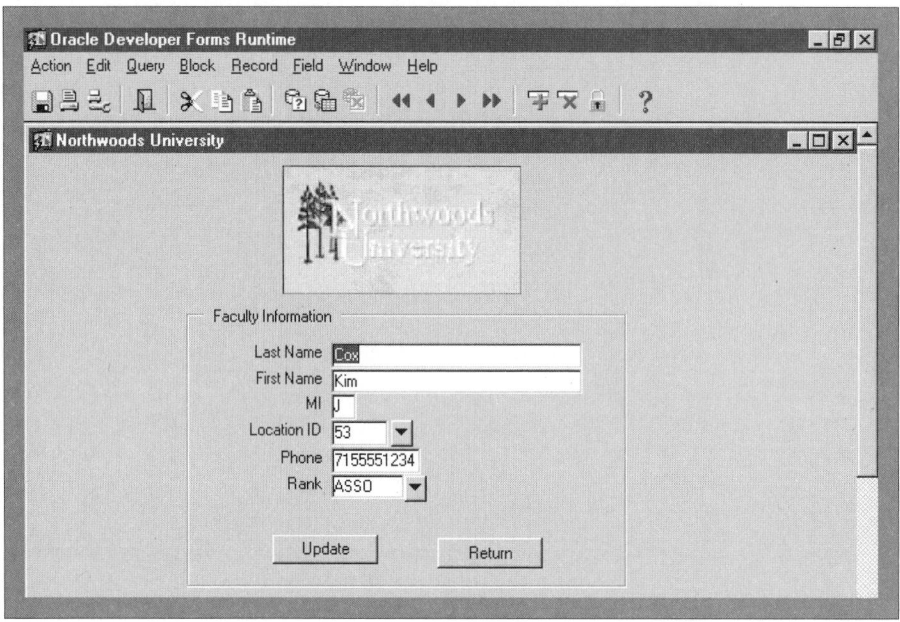

Figure 6-79

e. When the user clicks the View/Edit Advisee Information button on the Menu canvas, the canvas shown in Figure 6-80 is displayed, and the faculty member can view information about her advisees. Create a data block and associated layout that displays the student information for the current faculty member's advisees. When the user clicks the Return button, the Menu canvas should be displayed again.

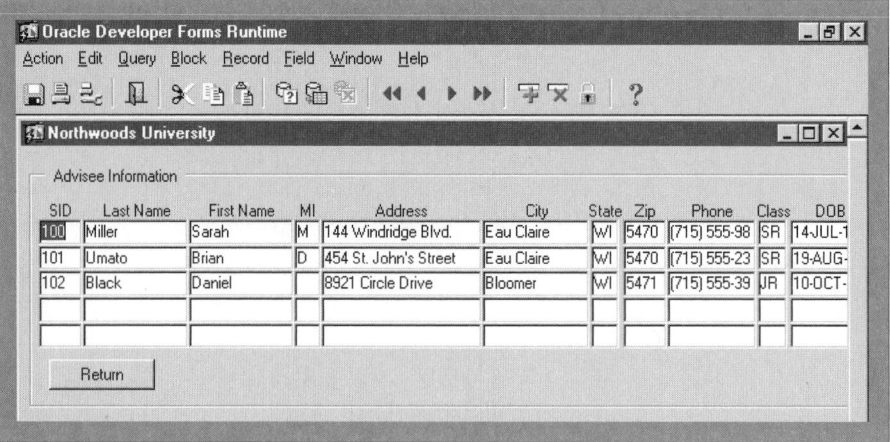

Figure 6-80

f. When the user clicks the View/Edit Class Lists button, the canvas shown in Figure 6-81 is displayed. The Course List frame lists information about all course sections ever taught by the current faculty member. Create a data block based on a database view named COURSE_VIEW that contains faculty ID, course section ID, term description, course call ID, and section number. Do not display the faculty ID in the layout. Make sure that the data block retrieves course information only for the current faculty member. Create a second data block based on a database view named CLASS_LIST that contains the course section ID, student ID, student last and first name and middle initial, student class, and course grade. Do not display the course section ID in the layout. Create a master/detail relationship between the section ID displayed in the Course List data block and the course section ID in the Class List data block. When the user clicks the Return button, the Menu canvas is displayed again.

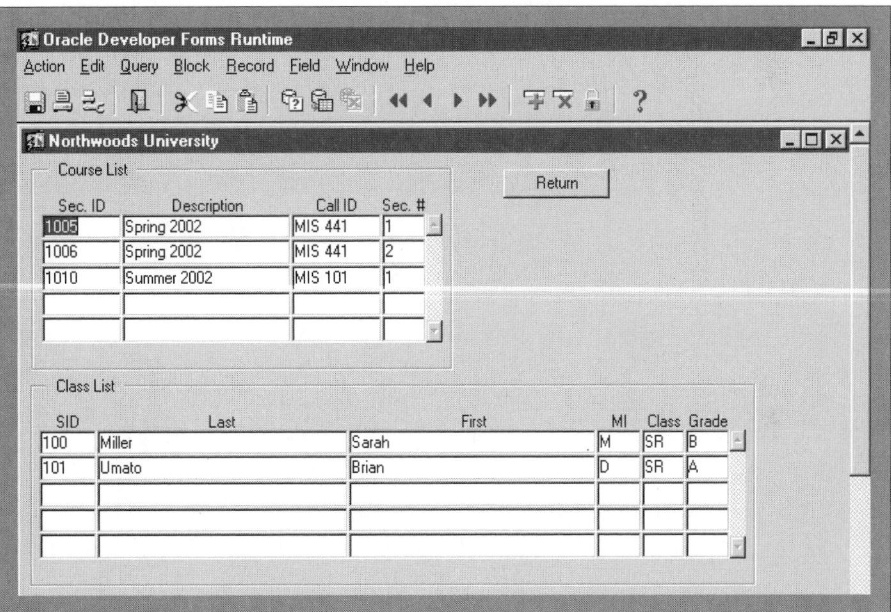

Figure 6-81

3. In this exercise, you will create a form with multiple canvases that will allow a Northwoods University student to use the campus registration system. When the student starts the system, he is presented with a menu that allows him to access forms for viewing and updating personal information, viewing grades for courses he has completed, and enrolling in new courses or dropping courses he is currently enrolled in.

   a. Create a new form and save it as Ch6cEx3.fmb.

b. Create a canvas and associated control block named LOGON_CANVAS and LOGON_BLOCK containing the items shown in Figure 6-82. When a student clicks the Log On button, the system should check to see if the student ID and PIN values are valid for any of the current STUDENT records. If they are, a canvas with the menu selections shown in Figure 6-83 is displayed. The student should have three chances to successfully enter his ID and PIN. After the third unsuccessful attempt, the form should exit.

Hint: Use a global variable to count the number of times the user has attempted to log on.

Hint: To conceal the PIN data, change the text item's Conceal Data property to Yes.

Hint: Create an explicit cursor and use the %FOUND attribute to check to see if the student ID and PIN entered by the user are included in the STUDENT table.

Hint: To exit the form without having the system automatically ask for user confirmation, use the following command: EXIT_FORM (NO_VALIDATE);.

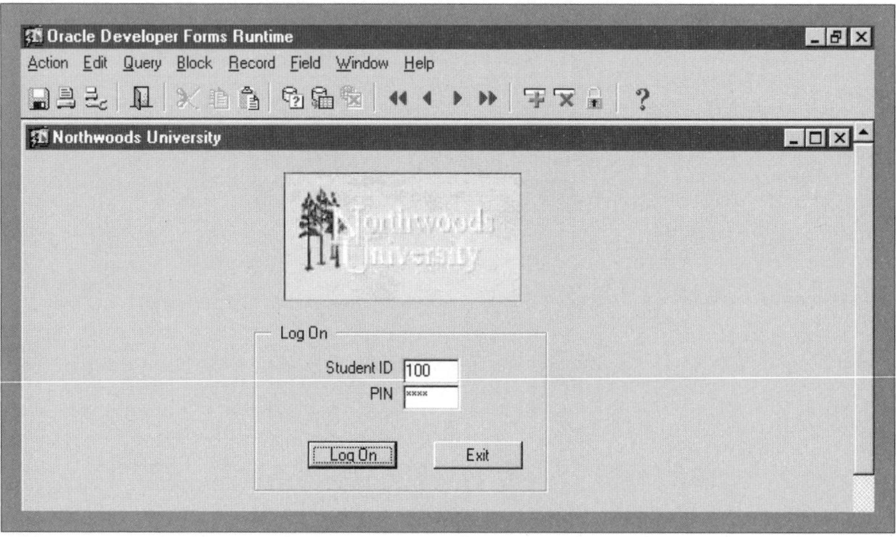

Figure 6-82

c. Create a canvas named MENU_CANVAS and an associated control block containing the command buttons and layout items shown in Figure 6-82.

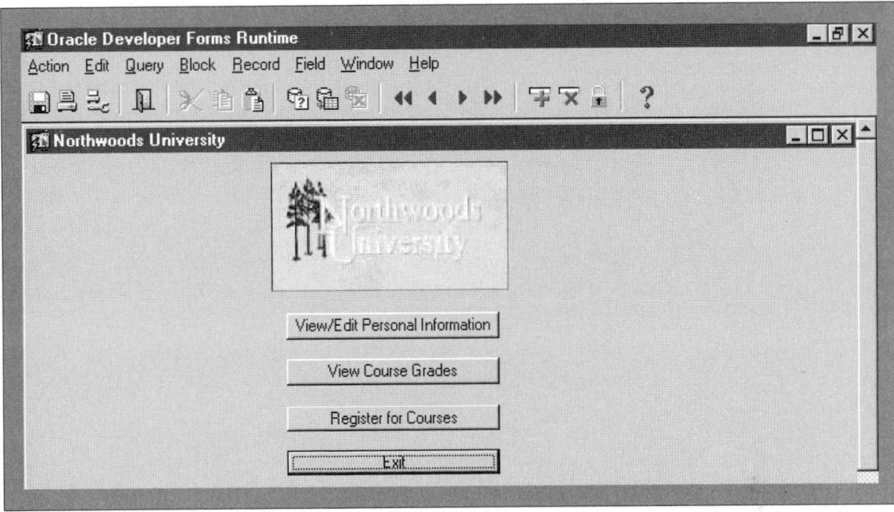

Figure 6-83

d. When the user clicks the View/Edit Personal Information button on the Menu canvas, the canvas shown in Figure 6-84 appears with the student's personal information displayed in the form fields. Create a data block that contains all of the STUDENT data fields and that displays the fields as shown. Create error traps to prevent the user from editing the Class or Advisor ID data fields. When the user clicks the Return button, the Menu canvas is displayed again.

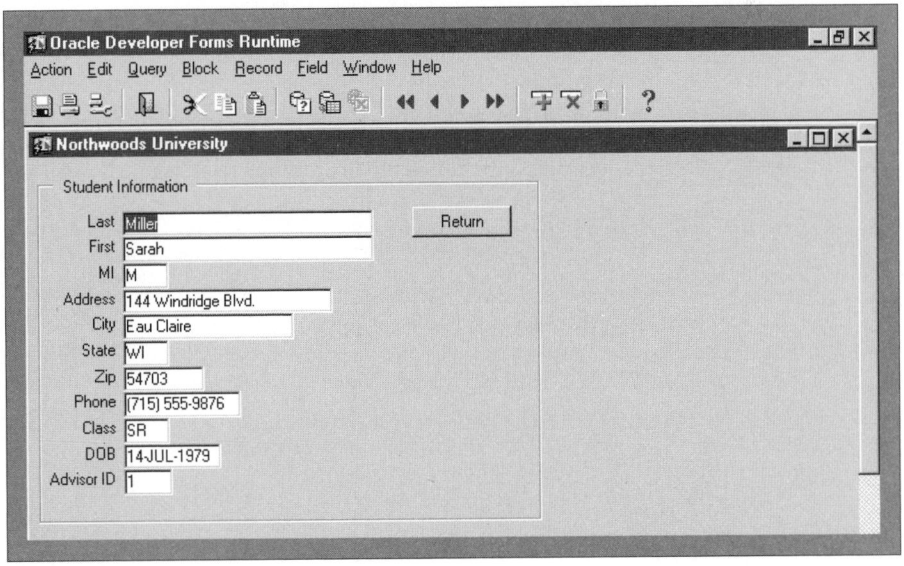

Figure 6-84

e. When the student clicks the View Course Grades button on the Menu canvas, the canvas shown in Figure 6-85 is displayed, enabling the student to view information for all his past course grades. Create a data block based on a database view named GRADE_VIEW that contains the student ID, term ID, term description, course call ID, and grade. Do not retrieve records for course enrollment records that have not yet been assigned a grade. Sort the retrieved course grade information by term ID, and do not display grade information for any other students except the one who has logged on to the system. Do not display the student ID in the layout. When the user clicks the Return button, the Menu canvas is displayed again.

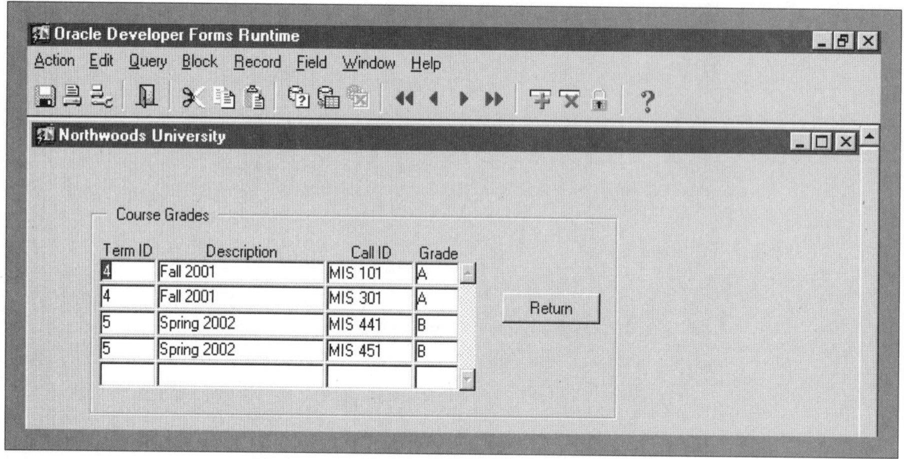

Figure 6-85

f. When the student clicks the Register for Courses button on the Menu canvas, the canvas shown in Figure 6-86 is displayed. Courses that the student is currently enrolled in but has not yet received a grade for are displayed in the Current Schedule frame fields when the form is first displayed. The Current Registration frame is used to add new courses to the student's registration. Create a control block to display the Current Registration frame fields, and create a data block based on a database view named ENROLLMENT_VIEW to display the Current Schedule frame fields. ENROLLMENT_VIEW should contain student ID, section number, faculty last name, day, time, building code, and room number, and only include records for which a grade value has not yet been assigned. Do not display the student ID in the layout. This Current Schedule frame will display only the schedule information for the current student.

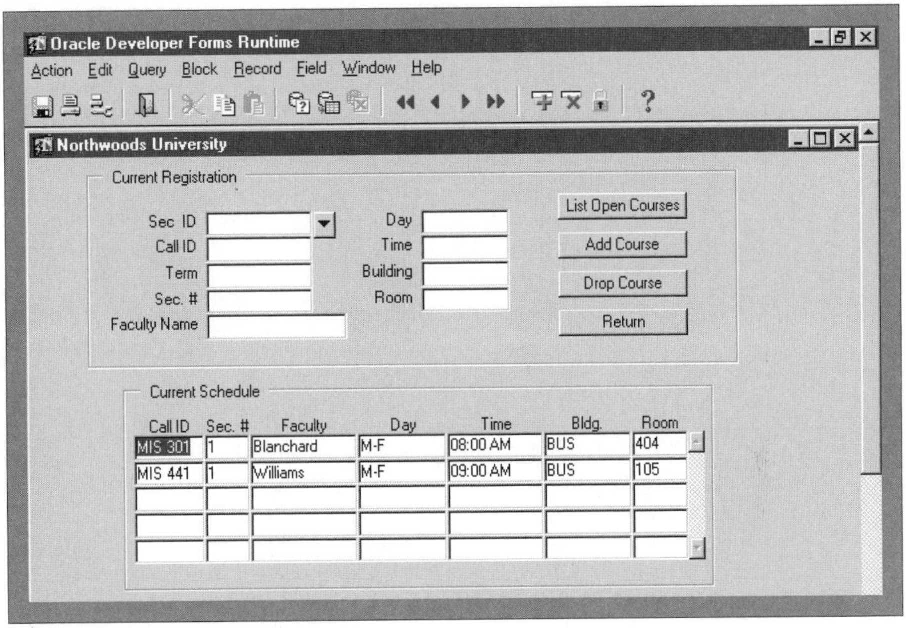

Figure 6-86

When the user clicks the List Open Courses button, a LOV display (see Figure 6-87) shows a listing of all courses from the COURSE_SECTION table for which the associated term registration status is 'OPEN', and for which the course's current enrollment does not exceed the maximum enrollment. When the student selects one of these courses, the associated data fields appear in the Current Registration frame fields. If the student clicks the Add Course button, the course is added to the student's ENROLLMENT records, and the Current Schedule display is automatically updated. The associated COURSE_SECTION record is updated to show that current enrollment has increased by one student. Note that if you try to add a course that is already on the student's schedule, an error message will be generated and the new course will not appear in the Current Schedule. (*Hint:* To view records in the Current Schedule frame, you must have logged on as a student who currently is enrolled in classes for which no grade has been assigned yet.)

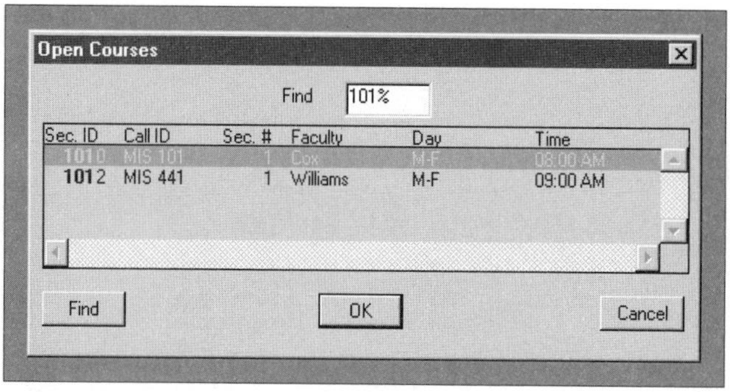

Figure 6-87

When the student clicks the LOV command button next to Section ID, the LOV display shows the courses that the student is enrolled in but has not yet received a grade for. When the student selects one of these courses from the LOV display, the selected course fields appear in the Current Registration frame fields. If the student clicks the Drop Course button, the course is dropped from the student's enrollment in the ENROLLMENT table, the course's current enrollment is decreased by 1, and the Current Schedule display is automatically updated. When the student clicks the Return button, the Menu canvas is displayed again.

CHAPTER

7

Using Report Builder

Introduction▶ Report Builder is the Oracle Developer application used for creating reports, based on database data, that can be viewed on the screen, printed to a file, or printed on paper. Reports can retrieve database data using SQL queries, perform mathematical or summary calculations on the retrieved data, and format the output to look like invoices, form letters, or other business documents.

LESSON A

objectives

- Understand Report Builder report styles
- Use the Report Wizard to create a single-table report
- Examine the Report Builder Object Navigator window
- Create a report that shows master-detail relationships
- Create a report template

Introduction to Report Builder

A form allows a user to interact with the database by inserting, updating, viewing, and deleting data. Conversely, a report is a static, page-oriented view of the database data at a specific point in time that can be viewed or printed. Reports can be structured using the following layout styles:

- **Tabular**, which presents data in a table format with columns and rows. An example of a tabular report is shown in Figure 7-1, which shows a partial listing of the inventory ID, item description, size, color, price, and quantity on hand for the inventory items in the Clearwater Traders database.

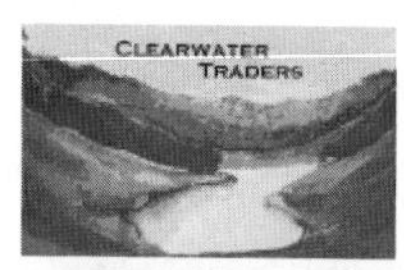

Clearwater Traders Inventory

Report run on: January 19, 2001 4:40 PM

Inv. ID	*Description*	*Size*	*Color*	*Price*	*QOH*
11668	3-Season Tent		Sienna	259.99	14
11669	3-Season Tent		Forest	259.99	12
11775	Women's Hiking Shorts	S	Khaki	29.95	150
11776	Women's Hiking Shorts	M	Khaki	29.95	147
11777	Women's Hiking Shorts	L	Khaki	29.95	0
11778	Women's Hiking Shorts	S	Olive	29.95	139
11779	Women's Hiking Shorts	M	Olive	29.95	137
11780	Women's Hiking Shorts	L	Olive	29.95	115
11795	Women's Fleece Pullover	S	Teal	59.95	135
11796	Women's Fleece Pullover	M	Teal	59.95	168
11797	Women's Fleece Pullover	L	Teal	59.95	187
11798	Women's Fleece Pullover	S	Coral	59.95	0

Figure 7-1: Tabular report

- **Form-like**, which resembles a data block form-style form. This report style displays one record per page and shows data values to the right of field labels.
- **Mailing label**, which prints mailing labels in multiple columns on each page.
- **Form letter**, which includes the recipient's name and address (which are stored in the database) as well as other data values embedded in the text of a letter.
- **Group left** and **group above**, which display master-detail relationships. In a **master-detail relationship**, one master record might have several associated detail records through a foreign key relationship. The relationship between the ITEM and INVENTORY tables in the Clearwater Traders database is an example of a master-detail relationship: Each ITEM record can have many associated INVENTORY records. An example of a **group left** report for this relationship is shown in Figure 7-2.

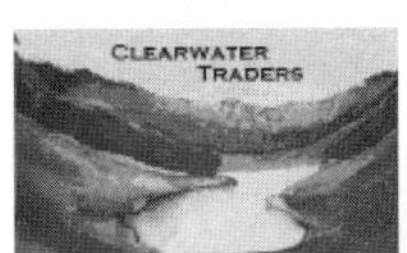

Clearwater Traders Inventory Items

Report run on: January 19, 2001 6:39 PM

Item ID	*Description*	*Inv. ID*	*Price*	*Color*	*Size*	*QOH*
559	Men's Expedition Parka	11845	199.95	Navy	S	114
		11846	199.95	Navy	M	17
		11847	209.95	Navy	L	0
		11848	209.95	Navy	XL	12
786	3-Season Tent	11668	259.99	Sienna		16
		11669	259.99	Forest		12
894	Women's Hiking Shorts	11775	29.95	Khaki	S	150
		11776	29.95	Khaki	M	147
		11777	29.95	Khaki	L	0
		11778	29.95	Olive	S	139
		11779	29.95	Olive	M	137
		11780	29.95	Olive	L	115

Figure 7-2: Group left report

Each item ID and description are listed on the left side of the report, and a detailed listing of the corresponding INVENTORY records is shown on the right side of the master item. The same data are shown in a **group above** report in Figure 7-3, in which each master item appears above the detail lines.

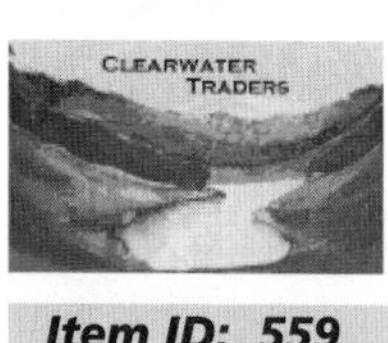

Clearwater Traders Inventory Items

Report run on: January 29, 2002 9:50 AM

Item ID: 559 ***Description: Men's Expedition Parka***

INV. ID	*Size*	*Color*	*Price*	*QOH*
11845	S	Navy	199.95	114
11846	M	Navy	199.95	17
11847	L	Navy	209.95	0
11848	XL	Navy	209.95	12

Item ID: 786 ***Description: 3-Season Tent***

INV. ID	*Size*	*Color*	*Price*	*QOH*
11668		Sienna	259.99	16
11669		Forest	259.99	12

Figure 7-3: Group above report

- **Matrix**, which displays field headings across the top and down the left side of the page. A matrix layout displays data at the intersection point of two data values. Figure 7-4 shows an example of a matrix report in which student names appear in the row headings, course call IDs appear in the column headings, and the specific student's grade for the associated course appears at the intersection.

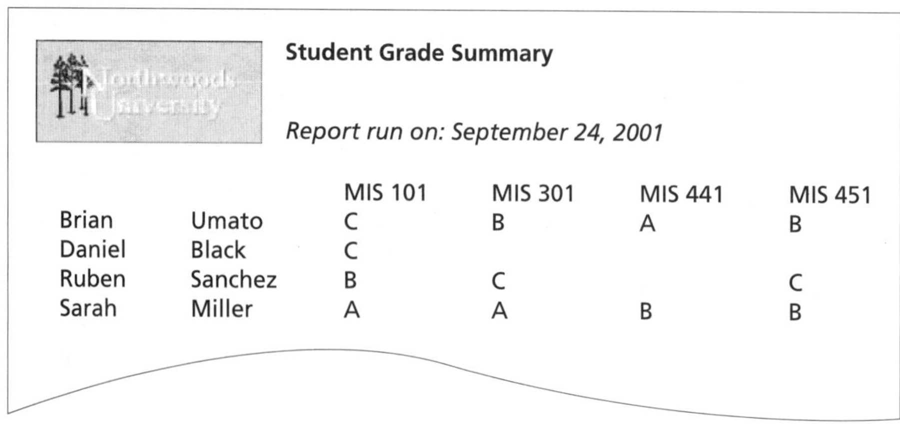

Student Grade Summary

Report run on: September 24, 2001

		MIS 101	MIS 301	MIS 441	MIS 451
Brian	Umato	C	B	A	B
Daniel	Black	C			
Ruben	Sanchez	B	C		C
Sarah	Miller	A	A	B	B

Figure 7-4: Matrix report

- **Matrix with group**, which displays a detail matrix for master data records. Figure 7-5 shows an example of this kind of report, where the master record is the student name, and a detail matrix shows course call IDs in the rows, term descriptions in the columns, and the student's grade at the intersection.

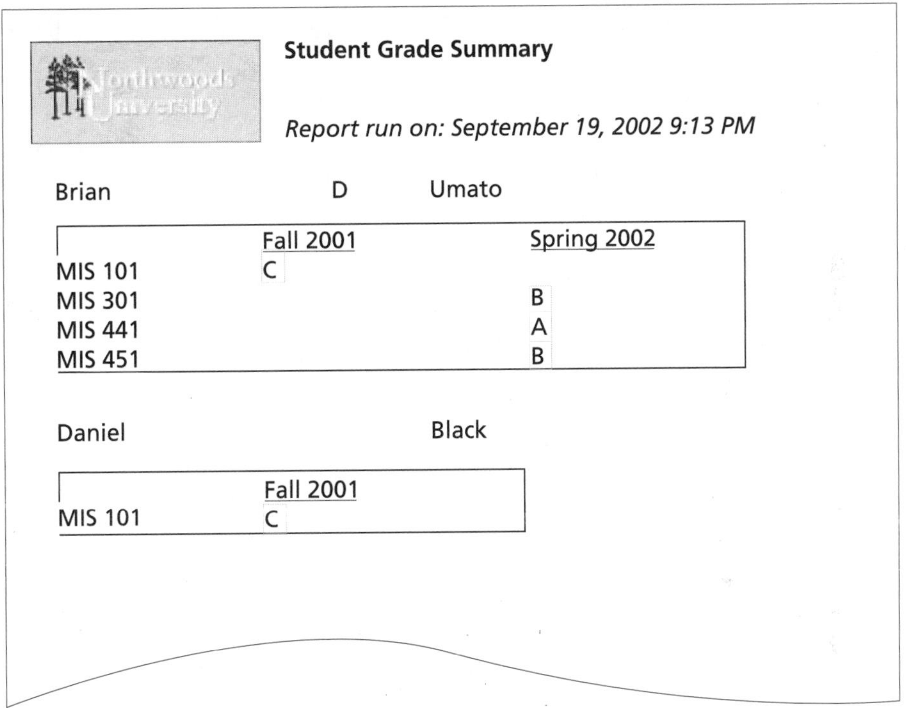

Figure 7-5: Matrix with group report

Creating a Single-Table Report Using the Report Wizard

To create a report, you must use a SQL query to specify the data that will be displayed in the report, and then define the report layout style. Now you will create a tabular report that shows each student's last name, first name, middle initial, address, city, state, ZIP code, phone number, class, date of birth, and advisor's last name from the Northwoods University database. To automate the report creation process, you will use the Report Wizard, which leads you through a series of pages to specify the query and layout. To be sure that you are starting with a complete database, you will first run the northwoo.sql script. Then you will start Report Builder.

To run the script and start Report Builder:

1 Start SQL*Plus and log on to the database.

2 Run the northwoo.sql script from the Chapter7 folder on your Student Disk to refresh your database tables.

Note: If you are storing your Student Disk files on a floppy disk, you will need to use two floppy disks for this chapter. When your first disk becomes full, save your solutions files to the second disk.

3 Exit SQL*Plus.

4 Click **Start** on the taskbar, point to **Programs**, point to **Oracle Developer 6.0**, and then click **Report Builder**. The Welcome to Report Builder dialog box is displayed.

help

Your system configuration might be different, but Report Builder should be in the same list as Form Builder.

tip

You also can start Report Builder by starting Windows Explorer, changing to the ORAWIN95\BIN directory, and then double-clicking the Rwbld60.exe file. (Windows NT users can change to the ORANT\BIN directory, and then double-click the Rwbld60.exe file.)

This dialog box enables you to build a new report using the Report Wizard, build a new report manually, or open an existing report. You will use the Report Wizard to build the student report. Next, you will start the Report Wizard and use it to specify the report style and enter the SQL query to return the fields from the STUDENT and FACULTY tables that will be displayed in the report.

To start the Report Wizard, specify the report style, and enter the SQL query:

1 Make sure the **Use the Report Wizard** option button is selected, then click **OK**.

2 When the Welcome to the Report Wizard page is displayed, click **Next**. The Report Wizard Style page is displayed.

3 Type **Northwoods University Students** in the Title box, make sure the **Tabular** report style option button is selected, then click **Next**. The Type page is displayed. Make sure the SQL Statement option button is selected, then click **Next**. The Data page is displayed.

4 Type the query shown in Figure 7-6 to display the desired data fields, then click **Next**. Since you have not yet connected to the database, the Connect dialog box is displayed.

type this query

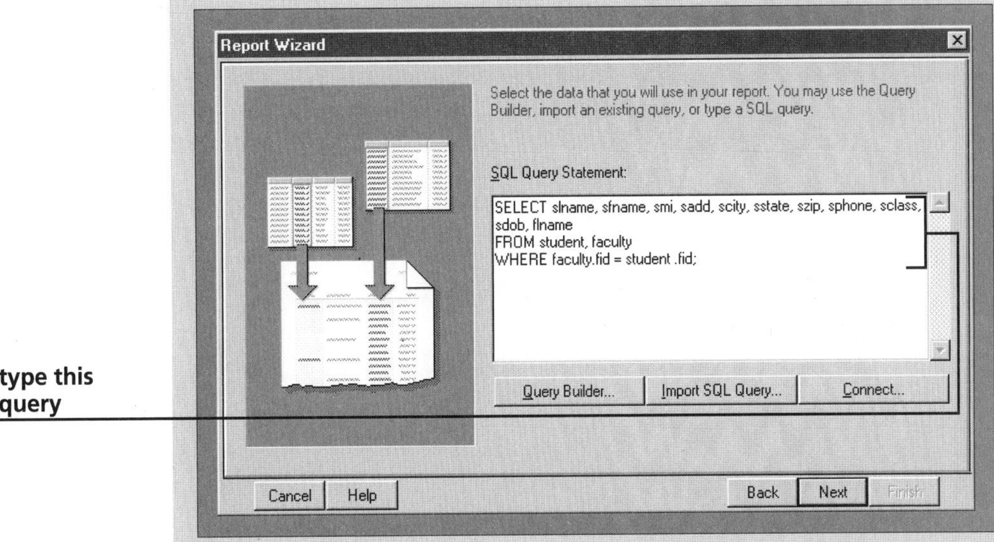

Figure 7-6: Report Wizard Data page

5 Log on to the database in the usual way, then click **Connect**. If your query is entered correctly, the Fields page showing the fields the query returns is displayed.

help

If the Fields page does not appear, debug your query until it works correctly.

In the Available Fields list, the Report Wizard Fields page displays the data fields returned by the query that are available to display in the report. Note that the field data types are indicated by the icon to the left of the field names. The next step is to select the fields that will be displayed in the report, and specify the field labels.

To select the report fields and specify the labels:

1 Click the **Select All** button [>>] to select all query fields for the report. The Available Fields list clears, and all of the query fields are now displayed in the Displayed Fields list.

2 Click **Next**. The Totals page is displayed to allow you to specify one or more fields for which you might want to calculate a total. None of the fields in this report requires a total, so don't select any fields. Click **Next**. The Labels page is displayed.

3 Modify the field labels and widths as follows:

Labels	Width
Last	10
First	10
MI	1
Address	7
City	7
State	2
Zip	9
Phone	10
Class	2
DOB	9
Advisor	10

Since these labels will be displayed as column headings in a table, the label names do not end with a colon. Click **Next**. The Templates page is displayed.

The final step in creating the report is to select a report **template**, which defines the report appearance in terms of fonts, graphics, and color highlights in selected report areas. You can use one of the predefined templates, select a template that you have created, or opt to format the report manually by not selecting a template. You will use one of the predefined templates.

To use a predefined template:

1 Make sure the **Corporate1** predefined template is selected, click **Next,** then click **Finish**. The report is displayed in the Report Editor – Live Previewer window.

2 Maximize the Report Builder window and the Live Previewer window. Your screen should look like Figure 7-7.

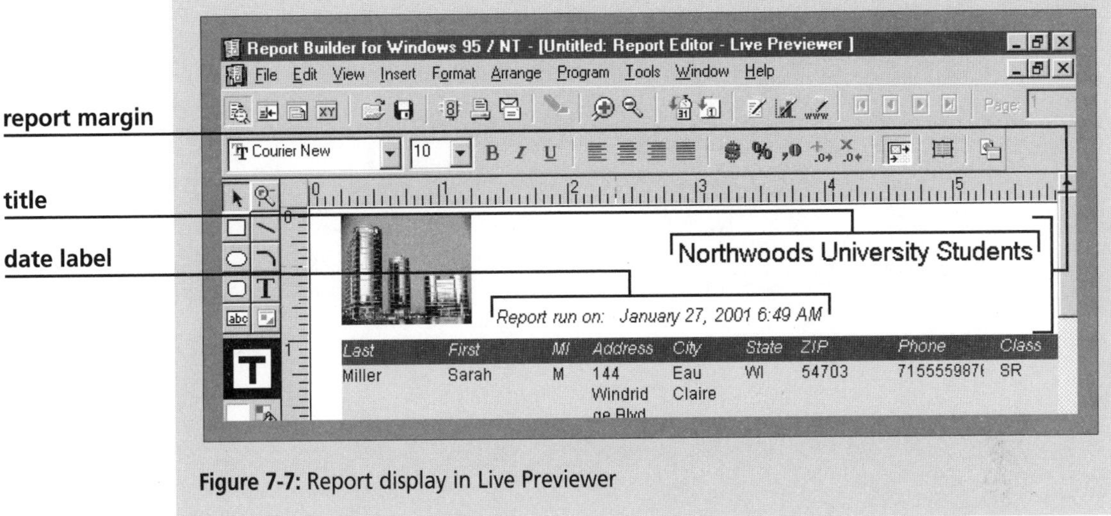

Figure 7-7: Report display in Live Previewer

The **Live Previewer** window shows how the report will look when it is printed. It also provides an environment for refining the report's appearance. This window has toolbars for working in the Report Builder environment, a tool palette for altering the report's appearance, and a status bar for showing the current zoom status and pointer position.

Next you will save your report. Report Builder design files are saved as **report definition** files with an .rdf extension.

To save the report:

1 Click the **Save** button on the toolbar, and save the report as **ch7astudent.rdf** in the Chapter7 folder on your Student Disk.

Next you will edit the report in the Live Previewer window by changing the position of the report margin labels. The **margin** is the area on the page beyond the location where the report data are displayed. It can contain boilerplate text and graphics, and data such as the date the report was created or the page number of the report. Along with the logo, the student report top margin contains the report title (**Northwoods University Students**) and the date label (**Report run on: <today's date>**). You will move these fields so that their left edges are aligned, and position them on the right side of the logo. The date label actually consists of two items: **Report run on** is a label, and the actual date is a data field that displays the current system date. You will move the date label and data field together as a group.

To change the position of the report top margin items:

1. Select the **Northwoods University Students** title label in the Live Previewer window. Currently, the field is center-justified in its text box. Click the **Start Justify** button on the toolbar to make the label left-justified.
2. Drag the **Northwoods University Students** title label so it starts about 0.25 inches from the right edge of the logo.

help

You might not be able to drag the Northwoods University Students label to the right edge of the screen because the label was originally center-justified and is very wide. When you drag it to the right, it extends beyond the right page edge. If necessary, scroll to the right edge of the report, select the label, click on the lower-right selection handle, and drag it to the left, to make the label narrower so you can more easily reposition it on the report.

3. Click the **Report run on:** label, press the **Shift** key, then click the date and time to select the date label and field as a group. Selection handles should appear on both items.
4. Release the Shift key, and drag the date fields up so they are positioned under the report title, as shown in Figure 7-8.

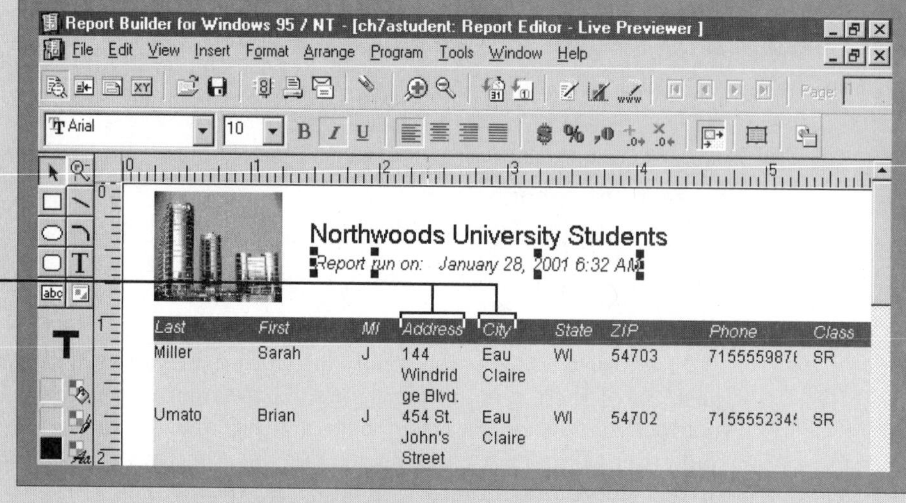

Figure 7-8: Repositioning the report date labels

tip

You can make fine adjustments to an object's position by selecting the object, then moving it using the arrow keys on the keyboard.

Next, you will adjust the data field widths. Figure 7-8 shows that the Address and City data fields wrap to multiple lines, and the report would be easier to read if these fields were wider. Now you will examine the rest of the report data fields, and determine if any other field width adjustments are necessary.

To examine the widths of the report data fields:

1 Click the right horizontal scroll-bar arrow to view the right margin of the report.

Looking at the report's right margin shows that the Phone field is not wide enough for the displayed data, and the DOB field is too wide. Note that the right margin of the report is approximately 8 inches from the left edge of the page. Since standard printer paper is 8½ inches wide, the report can be a little wider than it currently is and still fit on a standard sheet of paper, so you can make some of the fields wider. Now you will adjust the column widths by changing them in the Report Wizard. The Report Wizard is **re-entrant**, which means that after you have completed specifying a report, you can go back into the Wizard and edit the report properties, if necessary. Now you will modify the field column widths using the Report Wizard.

To modify the field column widths:

1 Click **Tools** on the menu bar, then click **Report Wizard**. The Report Wizard opens in re-entrant mode with different tabs showing each of the Report Wizard pages, allowing you to go directly to the page that contains the feature you want to edit.

2 Click the **Labels** tab to view the page for specifying field labels and widths. Change the following field widths:

Address	**15**
City	**10**
Phone	**12**
DOB	**5**

3 Click **Finish** to save your changes and close the Report Wizard. The report is displayed in the Live Previewer window with the field widths modified, as shown in Figure 7-9.

tip

Clicking Apply in the Report Wizard saves your changes without closing the Report Wizard.

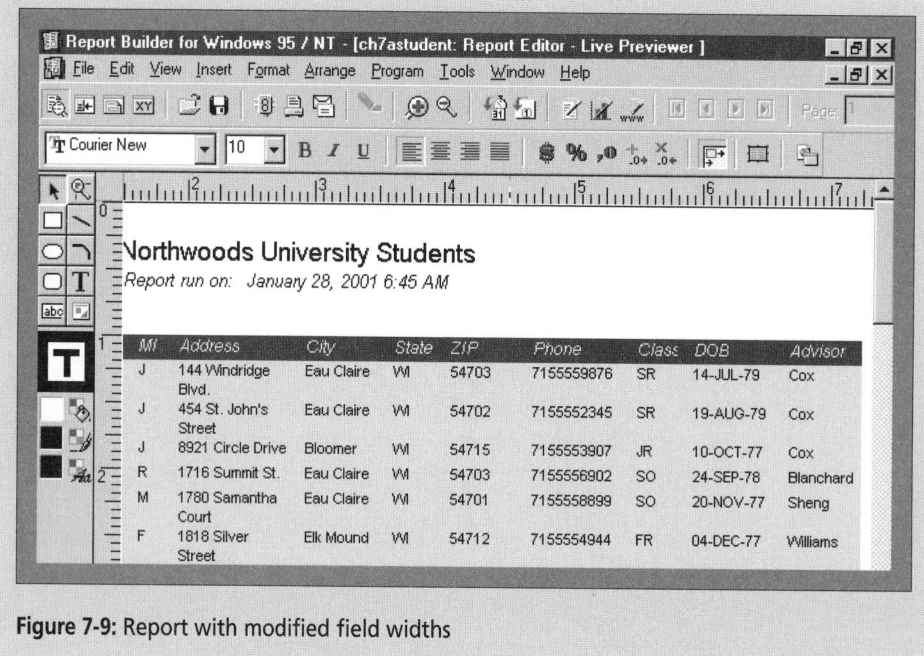

Figure 7-9: Report with modified field widths

The final step is to print your report to see how it looks on paper, and then close the report file. You will do this next.

To print the report:

1. Click the **Print** button on the toolbar. The Print dialog window opens. Click **OK**. After your report has printed, check it to make sure it looks like Figure 7-9.
2. Click **File** on the menu bar, then click **Close**. Click **Yes** to save your changes.

The Report Builder Object Navigator Window

After you close your report, the Report Builder Object Navigator window is displayed, as shown in Figure 7-10. Like the Object Navigator window in Form Builder, it is used to access different components in the Report Builder environment, as well as the components of an individual report. The top-level object is **Reports.** Currently no reports are open in Report Builder, so there is no **plus box** icon next to Reports.

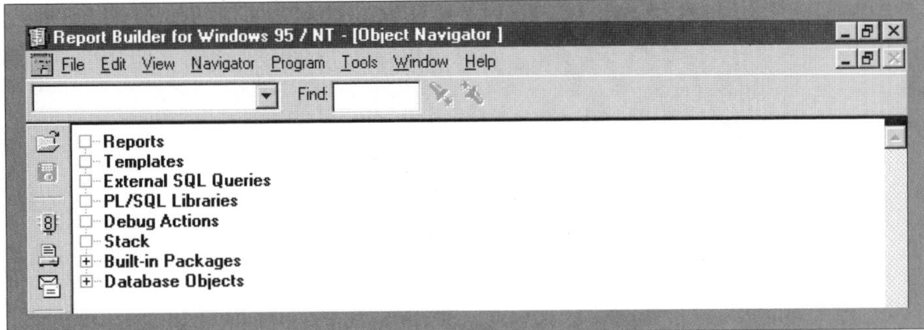

Figure 7-10: Report Builder Object Navigator window

Other environment objects include: **templates**, used to specify report formatting; **external SQL queries**, which allow the use of queries created in another environment such as Query Builder; **PL/SQL libraries**, which are collections of related PL/SQL functions or procedures; **debug actions**, which are actions you can enable or disable while using the Oracle Debugger; **stack**, which shows current values of local variables during a debugging session; **built-in packages**, which are code libraries provided by Oracle to simplify common tasks; and **database objects**, which are all the objects within the database, including users, tables, sequences, triggers, etc.

Creating a Master-Detail Report

Next you will create a master-detail report that shows records that have multiple related detail records. This report (see Figure 7-11) lists term IDs and term descriptions, and the call IDs and course names of courses taught during each term. The report also lists detailed information (section number, instructor, day, time, and location) about each course.

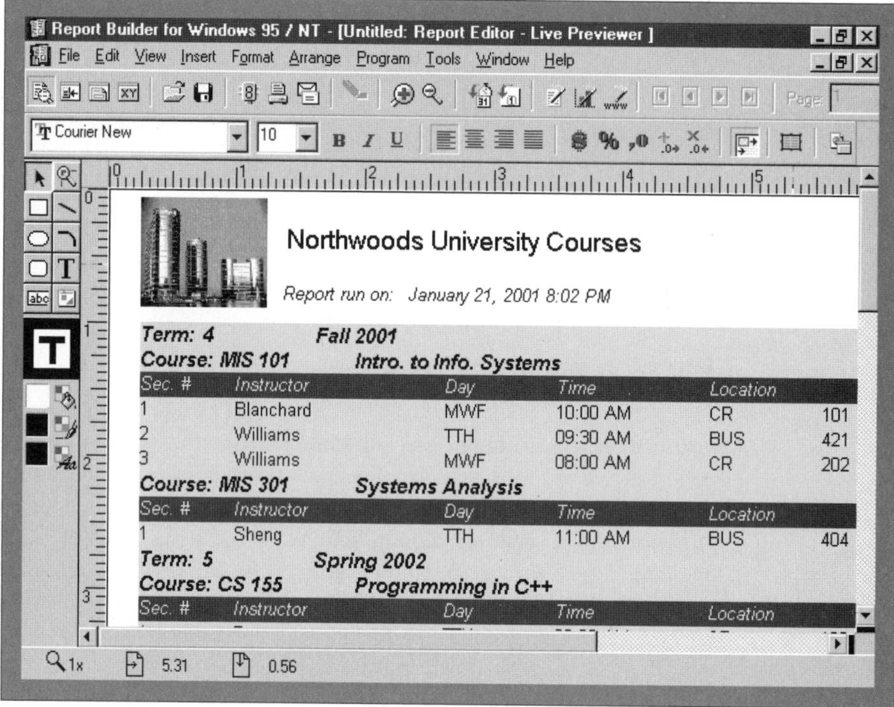

Figure 7-11: Courses report

This report involves two master-detail relationships. A term can have multiple courses, so in this relationship, the term data is the master record, and the course data is the detail record. Each course can have multiple sections, so in this relationship, the course data is the master record, and the course section data is the detail record. The term master list will show term IDs and descriptions, and the course detail list will show the course call IDs and course names. The section number detail list will show the section number, instructor, day, time, building code, and room.

To create a new report after you have already worked with other reports, you must create the new report in the Object Navigator window by clicking the **Create** button. Now you will create the master-detail report.

To create the master-detail report:

1 Click the **Reports** top-level object, then click the **Create** button to create a new report. The New Report dialog window is displayed.

2 Make sure the **Use the Report Wizard** option is selected, then click **OK**.

3 When the Welcome page is displayed, click **Next**. The Style page is displayed.

For this report, you will use the group above style. Now you will specify the report title and style.

To specify the report title and style:

1 Type **Northwoods University Courses** in the Title text box.

2 Click the **Group Above** option button, then click **Next**. The Type page is displayed. Make sure the SQL Statement option button is selected, then click **Next**.The Data page is displayed.

The report's SQL query must return all master and detail records that appear in the report. Therefore, the SQL query will include the term, course, and course section data. Next, you will enter the report's SQL query.

To enter the report's SQL query:

1 Type the SQL query shown in Figure 7-12, then click **Next.** Since this query returns data with master-detail relationships (one term might have multiple courses, and one course might have multiple course sections), the Groups page is displayed.

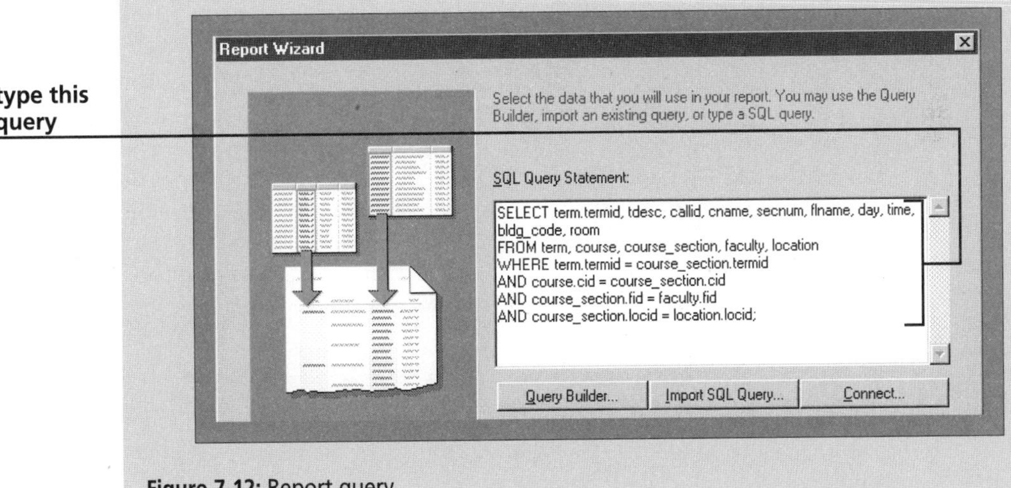

Figure 7-12: Report query

On the Groups page, you specify which fields are in the master group, and which are in the detail groups. A report can have multiple group levels. This report shows term information on the top level, information about courses offered during each term in the next level, and information about individual course sections corresponding to each offered course in the highest (most detailed) level. The top-level fields are designated as Level 1, the next level is called Level 2, the next level is called Level 3, etc.

The first field you select from the Available Fields into the Group Fields list is automatically put in Level 1, and a **Level 1** heading is displayed above the selected field's name in the Group Fields list. To add additional fields to the Level 1 group, you must select one of the fields that is currently in the Level 1 group, select the new field from the Available Fields list, and then add the new field. To create a Level 2 heading, you select the Level 1 heading in the Group Fields list, then add the field from the Available Fields list that will be in Level 2. The Level 2 heading is displayed with the selected field under it. You cannot place all of the report fields into the Group Fields list, so the highest (most detailed) data fields are left in the Available Fields list. Now you will specify the group fields for the report.

To specify the group fields:

1 Click **termid** in the Available Fields list, then click the **Select** button [>]. The Level 1 heading is displayed in the Group Fields list with termid below it, as shown in Figure 7-13.

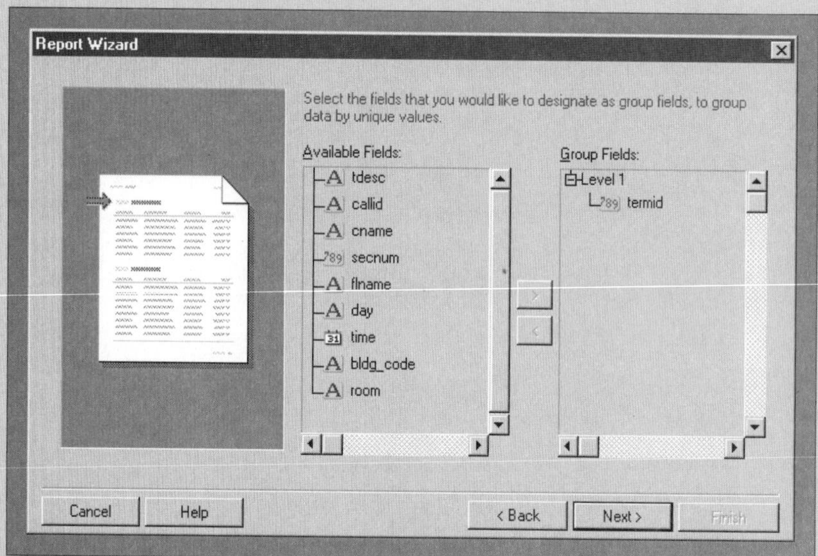

Figure 7-13: Selecting the first Level 1 group field

2 Select **termid** in the Group Fields list, click **tdesc** in the Available Fields list, then click [>]. The tdesc field appears under termid in the Level 1 group definition.

help If tdesc appears under a Level 2 heading, click it, click the Deselect button [<<] to remove it from the Group Fields list, then repeat Steps 1 and 2.

Now you will create the Level 2 group. Recall that to create a new data group, you must select its parent group level in the Group Fields list, select the first field of the new data group from the Available Fields list, then click the Select button [>]. For the Level 2 group in this report, you will select Level 1 in the Group Fields list, and then select the call ID field in the Available Fields list.

To create the Level 2 group:

1 Click **Level 1** in the Group Fields list, click **callid** in the Available fields list, then click [>]. The Level 2 heading appears in the Group Fields list with callid under it.

2 Click **callid** in the Group Fields list, click **cname** in the Available fields list, then click [>]. The completed Groups page looks like Figure 7-14.

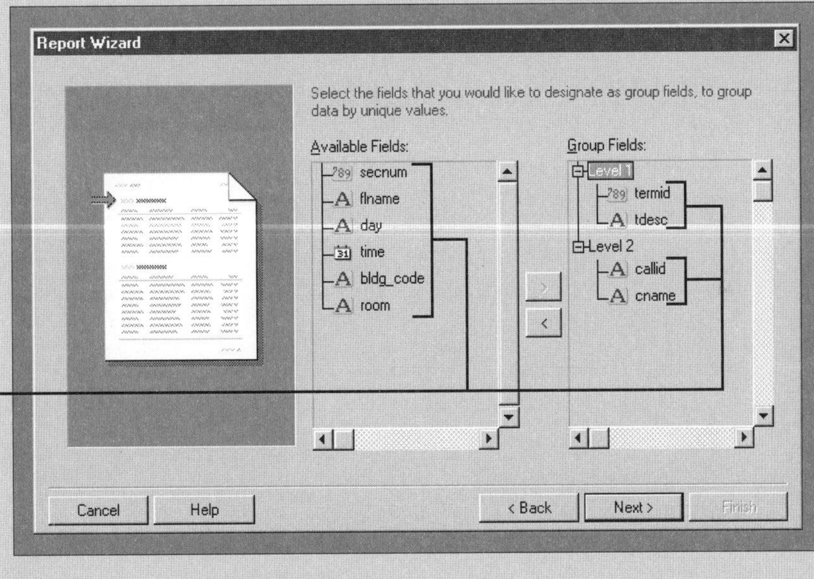

Figure 7-14: Completed Groups page

Recall that the highest-level (most detailed) data fields must remain in the Available Fields list. Since all the remaining fields in the Available Fields list reference a specific course section, they are in the Level 3 group, and the group specifications are complete.

You must leave at least one detail field in the Available Fields list, or a system error will occur.

The order in which the fields appear in the Group Fields list and Available Fields list is the same order that the fields will appear in on the report. If you want to change the order of the fields, you can select a field that is higher in the list, and drag it down to a lower position. To practice doing this, you will now change the order of the Level 1 group fields.

To change the order of the Level 1 group fields:

1. Select **termid**, drag the mouse pointer so that it is positioned below tdesc, then release the mouse button. The termid field should now appear below tdesc.
2. Return the fields to their original order by selecting **tdesc**, and then dragging the mouse pointer so it is below termid. The fields should now be displayed in the order shown in Figure 7-14, with termid first and tdesc second in the Level 1 group list.
3. Click **Next**. The Fields page is displayed.

To finish the report, you must specify the display fields, totals fields, field labels, and template. Then, you will modify the report's appearance in the Live Previewer by repositioning the margin labels.

To finish the report:

1. Click [>>] so that all of the report fields are in the Displayed Fields list. Click **Next**. The Totals page is displayed.
2. Since none of the data fields will be totaled, do not select any fields. Click **Next**. The Labels page is displayed.
3. Modify the field labels and widths as follows:

Labels	**Width**
Term:	**4**
<deleted>	**10**
Course:	**5**
<deleted>	**10**
Sec#	**4**
Instructor	**10**
Day	**5**
Time	**7**
Location	**5**
<deleted>	**6**

Note that the labels for the Level 1 and Level 2 data fields have a colon after the label, because they will be displayed horizontally on the form (see Figure 7-11). The data labels in the highest-level (Level 3) data group do not have a colon, because the data are displayed in a tabular format.

4 Click **Next.** The Templates page is displayed.

5 Make sure the **Corporate1** predefined template is selected, click **Next**, then click **Finish.** The report is displayed in the Live Previewer window.

6 If necessary, reposition the report title so that your report looks like Figure 7-15. Note the relative positions of the group fields, with the Level 2 fields grouped under the Level 1 fields, and the Level 3 fields grouped under each Level 2 field.

time fields appear as dates

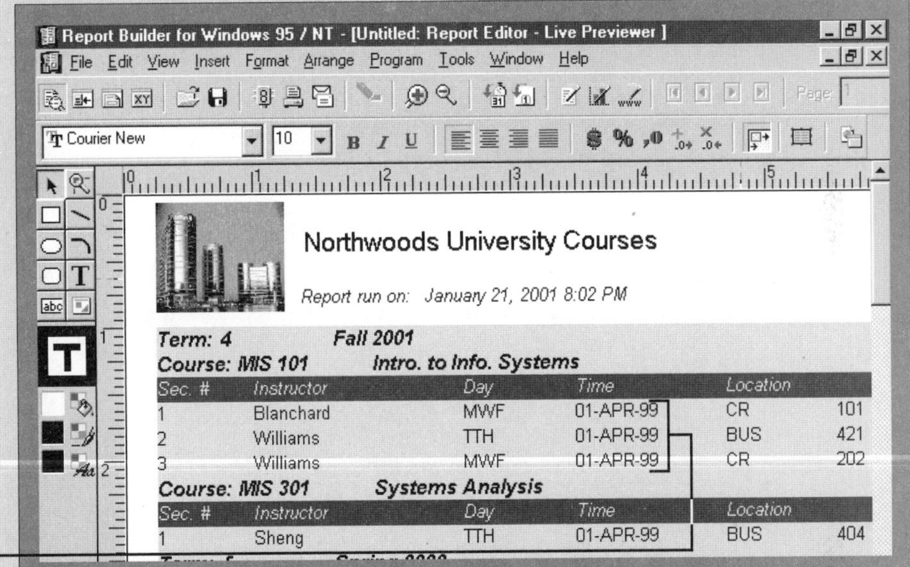

Figure 7-15: Finished Courses report

help

The Time data field currently displays dates because does not use a format mask that displays the time components of the dates. You will learn how to change the format mask of a report field later in this lesson.

7 Save the report as **ch7acourses.rdf** in the Chapter7 folder on your Student Disk.

8 Close the report.

Creating a Report Template

In both of the reports you have created so far, you used one of the predefined templates, and modified it by repositioning the report margin labels. Now you will create a custom template that displays the Northwoods University logo. When you apply a template to a report, objects in the report (such as the report title, date and time the report was generated, and page numbers) are always imported into the same locations in the report. Data field and label fonts and colors, and boilerplate objects such as graphics are also applied to the report. If you apply a template to a report and then later apply another template, the original template objects will be deleted and/or replaced by the ones in the most recent template.

Template definitions are stored in template definition files that have a .tdf extension. When you apply a template to a report, you specify the complete path and filename of the template .tdf file. Now you are going to modify the existing Corporate1 template by deleting the current logo and adding the Northwoods University logo. You will move the report title and date labels so that they appear on the right side of the logo, and you will modify the field that displays the date the report was generated by omitting the time component of the date. You will also change some of the text properties in the report body, and you will add page numbers to each report page. First, you will open the existing Corporate1 template definition (.tdf) file in the Template Editor in Report Builder.

To open the Corporate1.tdf file:

1 Click the **Open** button on the toolbar, navigate to the Chapter7 folder on your Student Disk, and open **corporate1.tdf.** CORPORATE1 is displayed under the Templates object in the Object Navigator.

2 Save the template as **northwoods.tdf** in the Chapter7 folder on your Student Disk.

From the Object Navigator window, you can see that a template has five possible components: Data Model, Layout Model, Report Triggers, Program Units, and Attached Libraries. In this book, the main template component you will work with is the Layout Model, which enables you to define and modify the layout model objects such as text, graphics, and data fields. In the Live Previewer, the report appears on the screen the same way as it will appear printed. In the Layout Model, as in the Layout Editor in Form Builder, objects are represented symbolically to highlight their types and relationships. You cannot view or edit a template in the Live Previewer window, because the template always runs with a report, and cannot run as a standalone object. Now you will open the template in the Layout Model Template Editor, which is used for editing templates.

To open the template in the Layout Model Template Editor:

1 Double-click the **Layout Model** icon under the NORTHWOODS template in the Object Navigator. The report template is displayed in the Layout Model Template Editor, as shown in Figure 7-16.

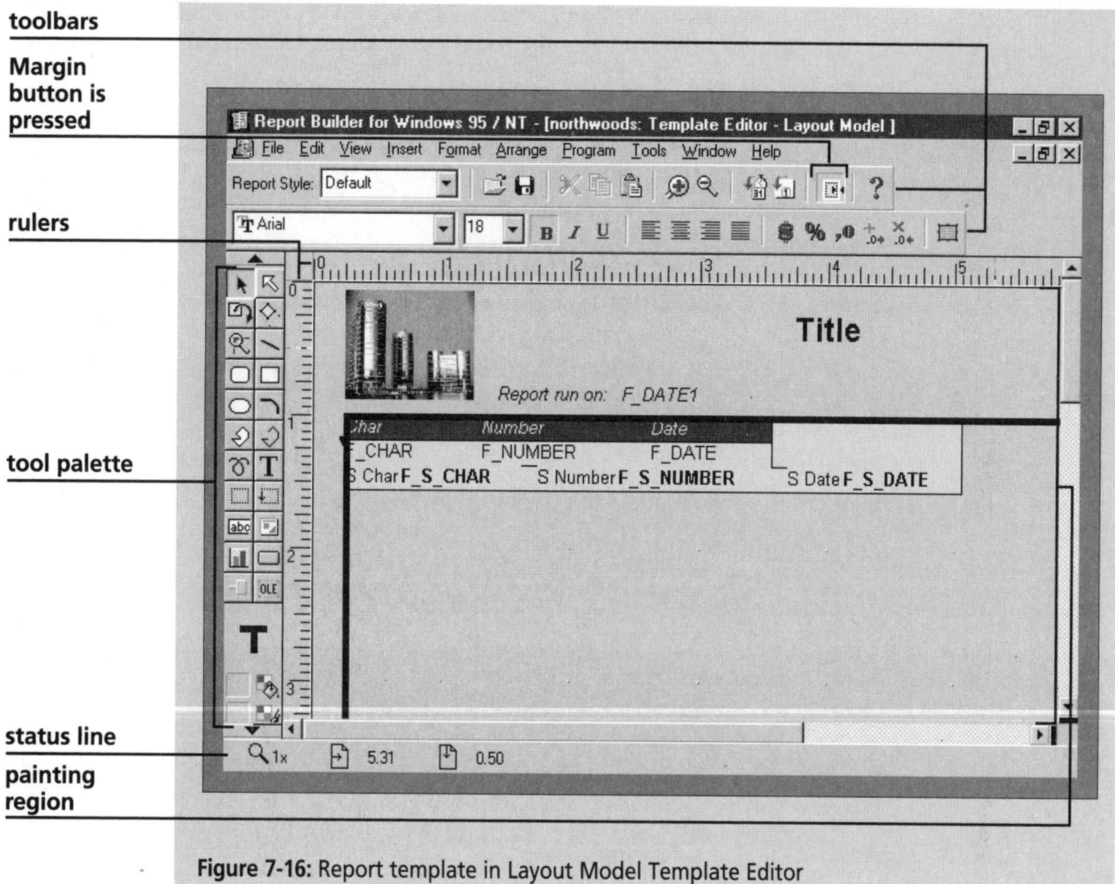

Figure 7-16: Report template in Layout Model Template Editor

The Template Editor, like the Layout Editor in Form Builder, has a tool palette, toolbars, painting region, and status line. The painting region has rulers to help you position report components. There are two regions in a report template: the margins, where the report title, the date the report was generated, and the page number are displayed; and the body, which contains the report data. When you are editing a template in the Template Editor, you can only edit the objects in one region at a time. The Template Editor toolbar contains a Margin button that enables you to toggle between the margin and the report body for editing. Currently, the Margin button is pressed (see Figure 7-16), and the report margins are visible and available for editing.

Now you will edit the objects in the template's top margin. You will delete the current logo and replace it with the Northwoods University logo, and reposition the report title and date label. Then you will change the format of the date data field so that only the date information is displayed, and the time information is omitted.

To edit the top margin objects:

1. Click the **Report run on** label in the top margin, press the **Shift** key, then click the **F_DATE1** field to select the date label and data field as an object group. Drag the label and date field to the right edge of the report, so they are about 2.5 inches from the left edge of the report, as shown in Figure 7-17.

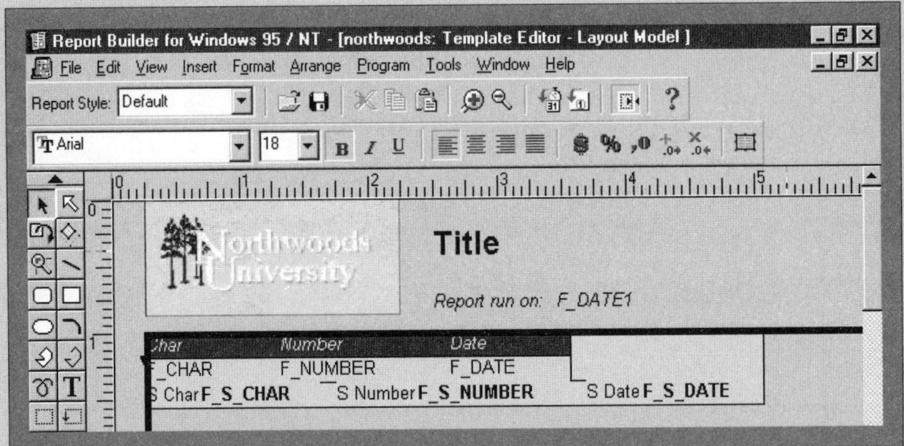

Figure 7-17: Editing the template's top margin

2. Click the current logo to select it, then press the **Delete** key to delete the current logo.
3. Click **File** on the menu bar, point to **Import**, then click **Image**. Click **Browse**, navigate to the Chapter7 folder on your Student Disk, select the **nwlogo.jpg** file, click **Open**, then click **OK**. The Northwoods University logo is displayed on the template.
4. Resize the logo so that it fits in the top margin area, as shown in Figure 7-17.
5. Select the **Title** text object, then click the **Start Justify** button so that the title is left-justified. Drag the Title text object so that it starts about 0.25 inches from the right edge of the logo, as shown in Figure 7-17.
6. Save the template file.

Now you will change the format mask of the date data field. In Report Builder, you can specify format masks for number and date fields in a report, but you cannot specify format masks for character fields. To change the format mask of the date field, you will open the field's Property Palette, and then enter the desired format mask. The format mask property allows you to choose from a list of predefined format masks for a particular data field, or to enter a customized format mask.

To change the date field's format mask:

1. Click **F_DATE1** in the top margin, right-click, then click **Property Palette**. The Property Palette for the field is displayed.
2. Select the **Format Mask** property, then open the list box. A list of possible format masks for the date field is displayed. You can select from the predefined format masks, or enter a custom format mask. You will edit the current format mask to create a custom format mask, so don't select a new format mask from the list.
3. Close the list, then edit the current format mask by deleting the time portion of the mask, so that it appears as shown in Figure 7-18. Close the Property Palette.

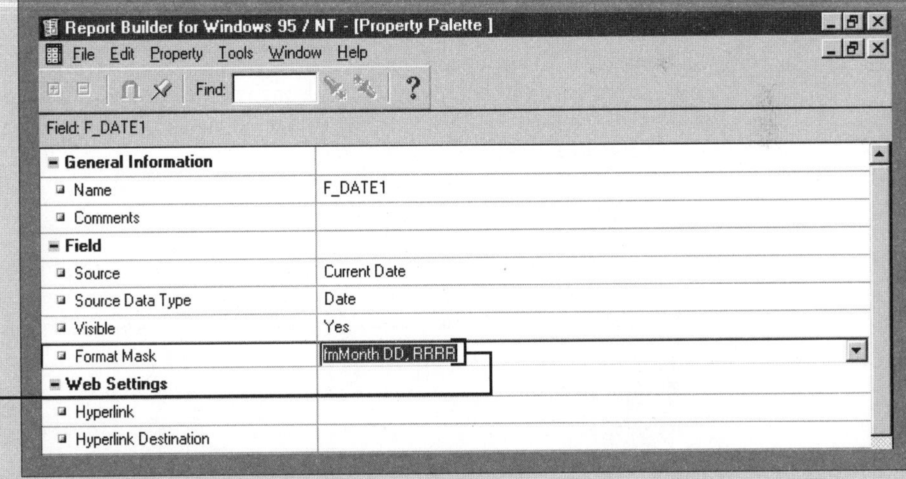

Figure 7-18: Editing the date field's format mask

4. Save the template file.

On the Layout Model toolbar, the **Insert Date and Time** button enables you to insert the current date and time in the report, and the **Insert Page Number** button enables you to insert the current page number. Next you will edit the template's bottom margin so that the report page number is displayed in the center of the bottom margin of each page. You will display the current page and total number of pages.

To display the page number and total number of pages in the bottom margin:

1 Click the **Insert Page Number** button on the toolbar. The Insert Page Number dialog window is displayed.

2 Make sure that **Bottom Center** is selected for the page number placement.

3 Click the **Page Number and Total Pages** option button, then click **OK**. The page number is displayed in the center of the bottom margin in the current system font. You will change the font to 10-point Arial regular.

4 If necessary, select the new page number field, change the font to **10 point Arial**. (If this font is not available on your system, substitute a different font.)

5 Save the template file.

Next, you will modify the colors in the template body. You will change the background color of the column headings from dark blue to dark green. To do this, you must click the Margin button to toggle to the template body and make it available for editing. Then you will select the column headings, and change their fill color to dark green.

To change the column heading fill color:

1 Click the **Margin** button to deselect it and move to the template body. The Template Editor window changes so that only the body is visible, and the toolbar buttons that can be used only for editing the template margins disappear.

2 Move the mouse pointer so that it is just before the word Number in the column headings, then click to select the column heading background. Selection handles should appear around the columns as a group, as shown in Figure 7-19, and the fill color should appear as dark blue on the tool palette.

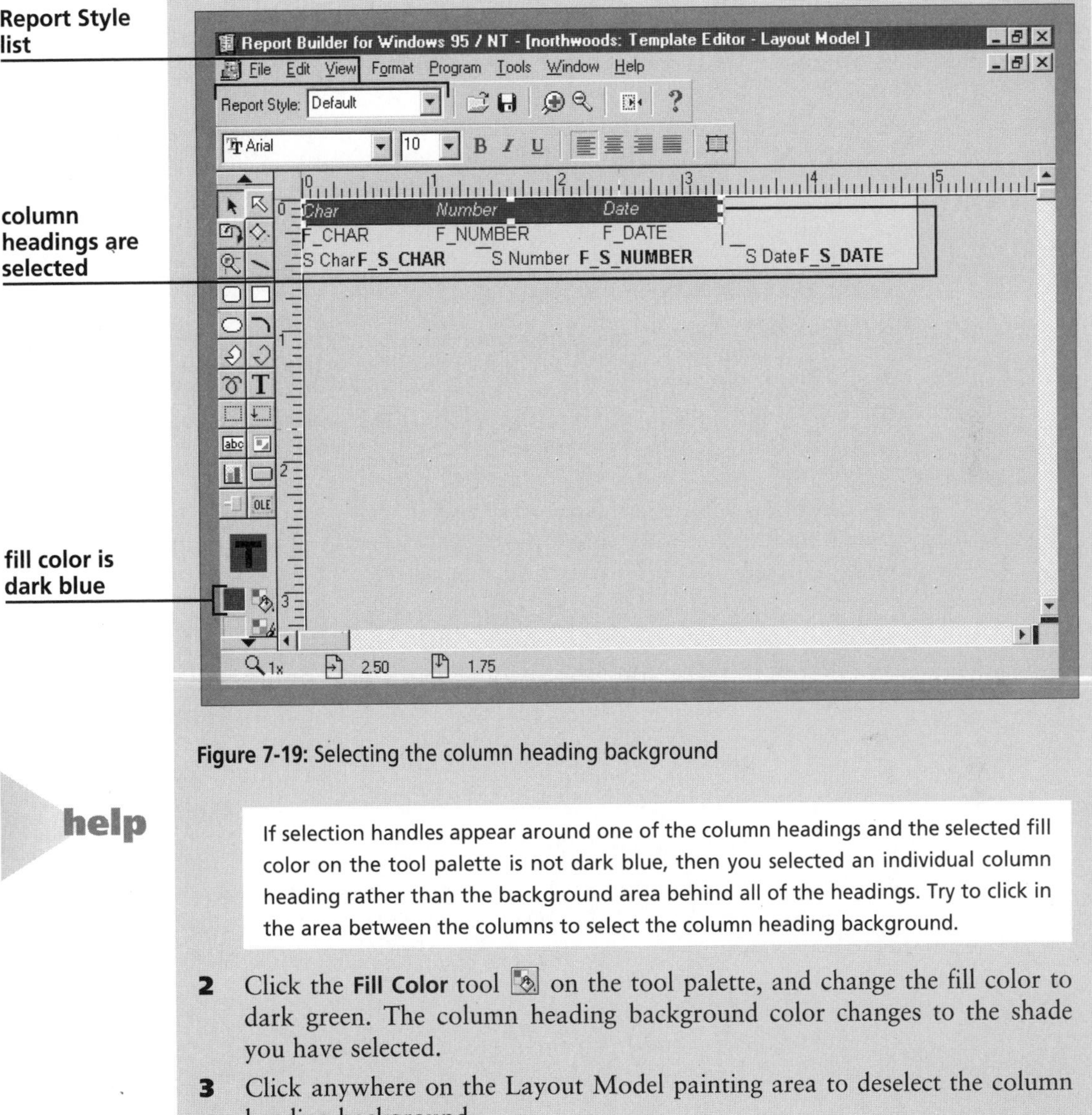

Figure 7-19: Selecting the column heading background

help

If selection handles appear around one of the column headings and the selected fill color on the tool palette is not dark blue, then you selected an individual column heading rather than the background area behind all of the headings. Try to click in the area between the columns to select the column heading background.

2 Click the **Fill Color** tool on the tool palette, and change the fill color to dark green. The column heading background color changes to the shade you have selected.

3 Click anywhere on the Layout Model painting area to deselect the column heading background.

4 Save the template file.

The **Report Style** drop-down list (see Figure 7-19) is displayed in the Layout Model window on the top-left edge of the toolbar. This list specifies the report style that is currently displayed. To be able to apply this template to different report styles, you must modify the template body for each style in the list. So far, you have modified the template for the Default, or Tabular, report style. Whenever you create or modify a template, you must modify the template for every report style that the template might be applied to. Next, you will select a different report style (the group above style), and examine how the template looks. Then you will modify the template so that the column heading background is dark green for the group above style.

To modify the template for the group above report style:

1. Click [▾] on the right edge of the Report Style list. The different report styles are displayed.
2. Select the **Group Above** style. Note that for this report style, the column heading background color has not been changed.
3. Select the column heading background, and change the fill color to dark green.
4. Save the template file.
5. Click **File** on the menu bar, then click **Close** to close the template file.

Now you will open the ch7acourses.rdf report that you created earlier in the lesson, and apply the template file to the report to make sure that the template works correctly.

To test the report template:

1. Click the **Open** button on the toolbar, and open the **ch7acourses.rdf** report from the Chapter7 folder of your Student Disk.
2. Click **Tools** on the menu bar, then click **Report Wizard** to open the Report Wizard in re-entrant mode so you can modify the report template.
3. Click the **Template tab** to open the Template page.
4. Select the **Template file** option button, click **Browse**, then select the **northwoods.tdf** file from the Chapter7 folder on your Student Disk. Click **Finish** to save your changes and close the Report Wizard. The report is displayed using the formatting specified in the new template, as shown in Figure 7-20. Note that the report now displays the Northwoods logo, and has page numbers at the bottom of the page.

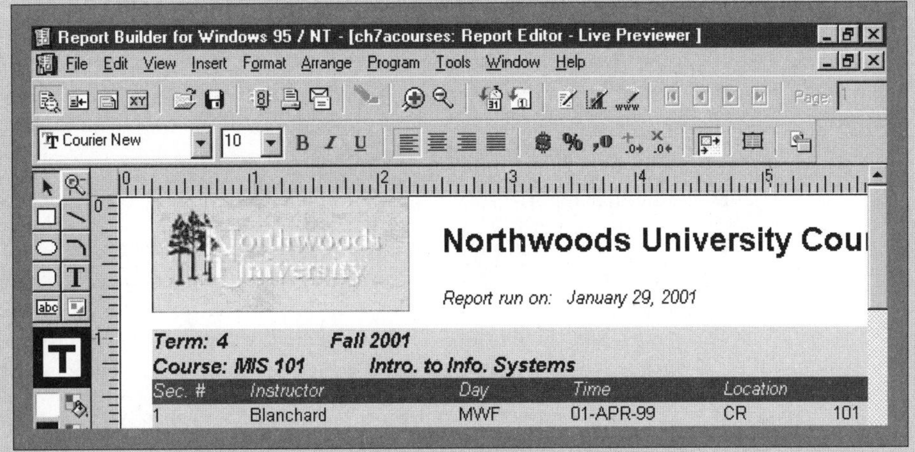

Figure 7-20: Report with formatting from new template

help

If any of the text in the margins on your report display does not appear, select the field that is truncated, and drag the selection handle of the selected object to make it larger, so that all of the text is displayed.

help

Sometimes when you apply a new template to a report, the report title from the previous template is not replaced. If this happens, select the title from the previous template, press the Delete key to delete it, then click the Run button to refresh the report display in the Live Previewer.

5 Save the report as **ch7acoursestemplate.rdf** in the Chapter7 folder on your Student Disk.

6 Close the report.

7 Exit Report Builder.

SUMMARY

- Reports retrieve database data using SQL queries, perform mathematical or summary calculations on the retrieved data, and format the output to look like invoices, form letters, or other business documents.
- A tabular report presents data in a table format with columns and rows.
- Group left and group above reports display master-detail relationships, whereby one master record might have several associated detail records through a foreign key relationship. In a group left report, each master item is listed on the left side of the report, and the multiple detail items are displayed on the right side of the master item. In a group above report, each master item appears above the multiple detail lines.
- To create a report, you must use a SQL query to specify the data that will be viewed in the report, and then define the report layout style.
- A report template defines the report appearance in terms of fonts, graphics, and fill colors in selected report areas.
- The Report Builder Live Previewer window shows how the report will look when it is printed, and provides an environment for refining the report's appearance.
- Report Builder design files are saved as report definition files with an .rdf extension.
- The Report Wizard is re-entrant, which means that after you have completed specifying a report, you can go back into the Wizard and edit the report's properties if necessary.
- When you create a report with a master-detail relationship, the report's SQL query must return all master and detail records that appear on the report.
- When you create a master-detail report, you use the Report-Wizard Groups page to specify which fields are in the master group, and which are in the detail group. A report can have multiple master-detail relationships.
- A report template has a body and a margin region, and you can only edit one region at a time.
- Template definitions are stored in template definition files, which have a .tdf extension.
- In the Report Builder Layout Model, objects are represented symbolically to highlight their types and relationships.
- In Report Builder, you can specify format masks for number and date fields in a report, but you cannot specify format masks for character fields.
- To be able to apply a predefined template to different report styles, you must specify the template properties for each report style in the template body.

REVIEW QUESTIONS

1. What is the difference between a form and a report?
2. Select the appropriate Report Builder report style (tabular, form-like, mailing label, form letter, group left, group above, matrix, matrix with group) for:
   a. a report that is a letter telling Northwoods University students their grades for the past term, and current grade point average
   b. a list of Northwoods University student names, with their associated course call IDs and grades listed below each student name
   c. a report that shows Clearwater Traders order source names on the column headings, inventory IDs on the row headings, and the number of each inventory item ordered from a specific order source at the intersection of the column and row
   d. a list of courses offered during the Summer 2002 term at Northwoods University
3. What is the difference between .rdf and .tdf files?
4. True or false: when you create a master-detail report using the Report Wizard, you retrieve the master records using one SQL query, and the detail records using a second SQL query.
5. Define the term "re-entrant" as it applies to the Report Wizard.
6. State the template region (body, margins) where you would place the following item:
   a. page numbers
   b. report title
   c. date the report was run
   d. data values
   e. field headings
7. True or false: you can view actual data values in a report in the Live Previewer window.
8. You need to create a report for Clearwater Traders that lists item descriptions, associated inventory IDs, sizes, and colors, and the shipment ID, expected date, and expected quantity for incoming shipments for each inventory ID. You have created the report in the Report Wizard as a group left report, and you are now ready to assign the field groups, as shown in Figure 7-21. How would you assign the field group levels?

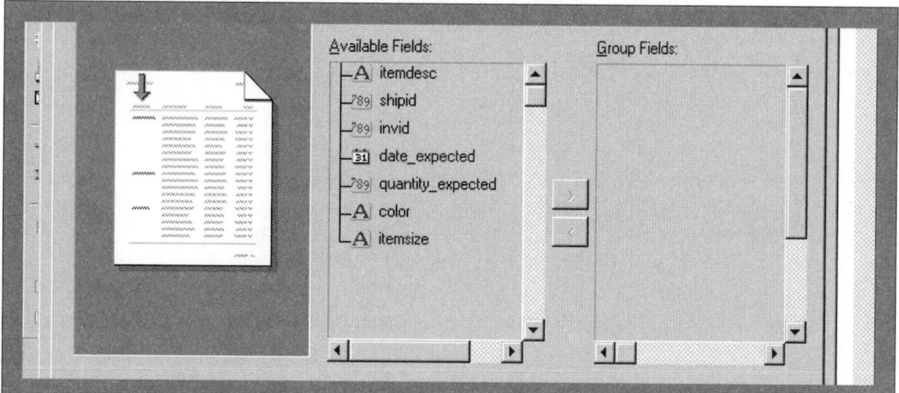

Figure 7-21

9. True or false: when creating a report template, you can edit it either in the Layout Model or in the Live Previewer window.
10. Which data types can you specify format masks for in Report Builder?

PROBLEM-SOLVING CASES

Run the northwoo.sql script to refresh the Northwoods University database, and run the clearwat.sql script to refresh the Clearwater Traders database. The script files are stored in the Chapter7 folder on your Student Disk.

1. In this exercise, you will first create a predefined template for Clearwater Traders. Then you will create a report showing a listing of Clearwater Traders customers.
   a. Open the casual1.tdf template file, which is stored in the Chapter7 folder on your Student Disk, and save it as clearwater.tdf.
   b. Replace the current logo with the Clearwater Traders (clearlogo.tif) logo from the Chapter7 folder on your Student Disk.
   c. Change the title position and justification so that it is centered at the top of the page. Change the title font color to dark blue.
   d. Insert the current date at the top-right corner of the report. Format the date as DD MONTH YYYY. Use a 10-point Italic Comic Sans MS font, and make sure the font color is dark blue. (If this font is not available on your system, substitute a different font.)
   e. Insert the current page number in the top-right corner of each report page, just under the date, using the format "Page <current page number>." Do not include the total number of pages. Format the page number using a 10-point Italic Comic Sans MS font. Make sure the font color is dark blue.
   f. Change the background of the column headings to a light pink, and the text color to black for the tabular, group left, and group above report styles. (*Hint*: To change the text color, you will have to select each column individually.)
   g. Create a new report named Ch7aEx1.rdf that displays all of the fields from the Clearwater Traders CUSTOMER table.
   h. Title the report "Customers," use descriptive column headings, apply the clearwater.tdf template file from the Chapter7 folder on your Student Disk, and format the report attractively.
2. Create a report that lists Clearwater Traders customer IDs and last and first names, and then lists each customer's associated order information (order ID, order date, payment method, and order source). Title the report "Customer Orders," use a group left report style, and use descriptive column labels. Change the order date format to MM/DD/YY. Apply the clearwater.tdf template file you created in the first exercise. (If you did not create the template, use the Corporate1 predefined template.) Format the report attractively, and save it as Ch7aEx2.rdf.

3. Create a report that lists the student IDs and last and first names for all Northwoods University students, and then lists the call IDs of every course a student has taken and the associated grade. Do not list course call IDs for courses that have not been assigned a grade. Title the report "Student Grades," use a group above report style, and include descriptive column labels. Delete column labels if you think it improves the report's appearance. Apply the northwoods.tdf template file you created in the tutorial. Format the report attractively, and save it as Ch7aEx3.rdf.
4. Create a report that lists every term description in the Northwoods University database. For each term, list the call ID of each course offered during the term, and for each call ID, list the section number, faculty member's last name, day, time, location building and room, and current enrollment. Title the report "Course Sections," use a group above report style, and use descriptive column labels. Apply the northwoods.tdf template file you created in the tutorial. Format the report attractively, and save it as Ch7aEx4.rdf.
5. Create an incoming shipment report for Clearwater Traders that lists inventory ID, item description, item size, and color, and then lists the shipment ID, date expected, and quantity expected for all incoming shipments that have not yet been received. Title the report "Pending Shipments," use a group left report style, and include descriptive column labels. Change the format mask of the date expected field to MM/DD/YY. Apply the clearwater.tdf template you created in the first exercise. (If you did not create the template, apply the Corporate1 predefined template, and delete the current logo and replace it with the Clearwater Traders logo.) Format the report attractively, and save it as Ch7aEx5.tdf.
6. Create a report that lists the schedule for each student who is enrolled in classes for the Summer 2002 term at Northwoods University. The report should list the term description once, then list each student's ID, last name, first name, and middle initial, then the student's schedule information (course call ID, section number, day, time, building, and room number). Use tdesc = 'Summer 2002' for the query search condition. Title the report "Student Schedules," and use the group left report style. Format the TIME field to show the hours and minutes and whether it is morning or afternoon. Apply the northwoods.tdf template file you created in the tutorial. Format the report attractively, and save it as Ch7aEx6.rdf.
7. Create a report that summarizes the customer orders for Clearwater Traders. The report should list the customer ID and last and first name, then the order ID and order date for each order placed by that customer. For each order, list the inventory ID, item description, color, size, order price, and quantity ordered. Title the report "Customer Order Details," and use the group above report style. Apply the clearwater.tdf template file you created in Exercise 1. (If you did not create the template, apply the Corporate1 predefined template, and delete the current logo and replace it with the Clearwater Traders logo.) Format the fields so that the report is displayed as shown in Figure 7-22, and save it as Ch7aEx7.tdf.

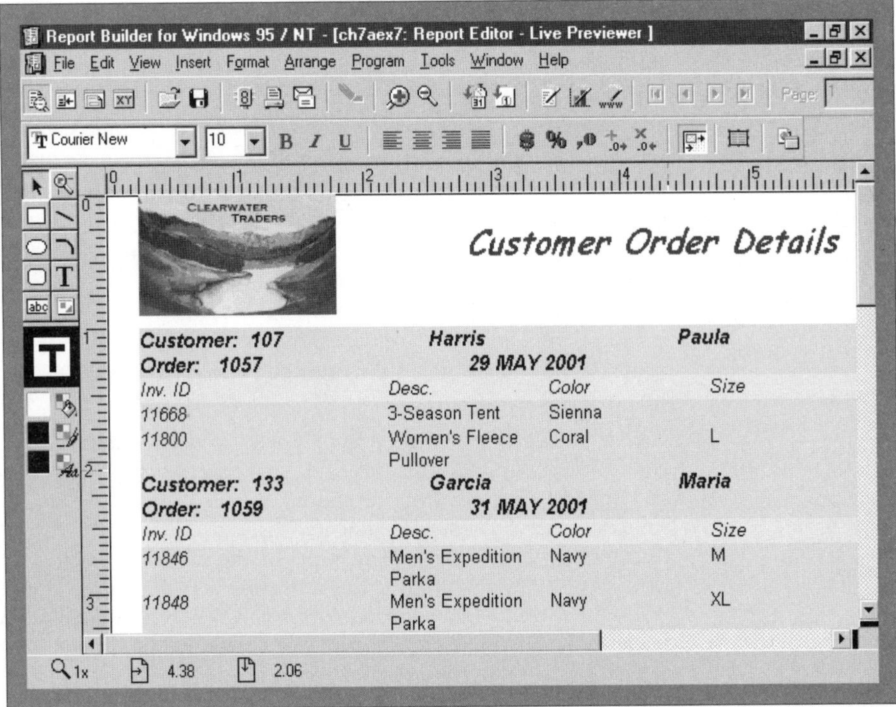

Figure 7-22

8. Create a report that makes mailing labels for the customers in the Clearwater Traders database. Format the mailing labels as follows:

   Paula E Harris
   1156 Water Street, Apt. #3
   Osseo, WI 54705
   Hello, Paula!

   Format the labels using an 8-point Comic Sans MS font. Save the report as Ch7aEx8.rdf.

9. Create a report that lists course call IDs as column headings, student first and last names as row headings, and the grade that the student received in a particular course at the intersection point. Do not include courses for which grades have not yet been assigned. Title the report "Student Grade Summary," delete all column headings and apply the northwoods.tdf template you created in the tutorial. Save the report as Ch7aEx9.rdf. Your completed report should look like Figure 7-4. (*Hint*: You will need to modify the northwoods.tdf template file so that the report is formatted correctly for the report style).

LESSON B

objectives

- Understand the components of a report
- Work with master-detail reports
- Modify the report format
- Create a user parameter to allow the user to customize report data

Report Components

In Lesson A, you learned how to create reports quickly and easily using the Report Wizard. In this lesson, you will become familiar with the report components that are automatically created by the Report Wizard and learn how to modify these components to customize the report's output and appearance. You will be working with a tabular report created with the Report Wizard that shows the location ID, building code, room, and capacity values from the LOCATION table in the Northwoods University database. First you will start Report Builder and open the report file.

To start Report Builder and open the file:

1. Start Report Builder. Click the **Open an existing report** option button, then click **OK**.
2. Open **location.rdf** from the Chapter7 folder on your Student Disk. The Object Navigator window is displayed. Save the file as **ch7blocation.rdf**.
3. Double-click the **Live Previewer** icon and connect to the database in the usual way. The report is displayed in the Live Previewer window, as shown in Figure 7-23. Maximize the Live Previewer window if necessary.

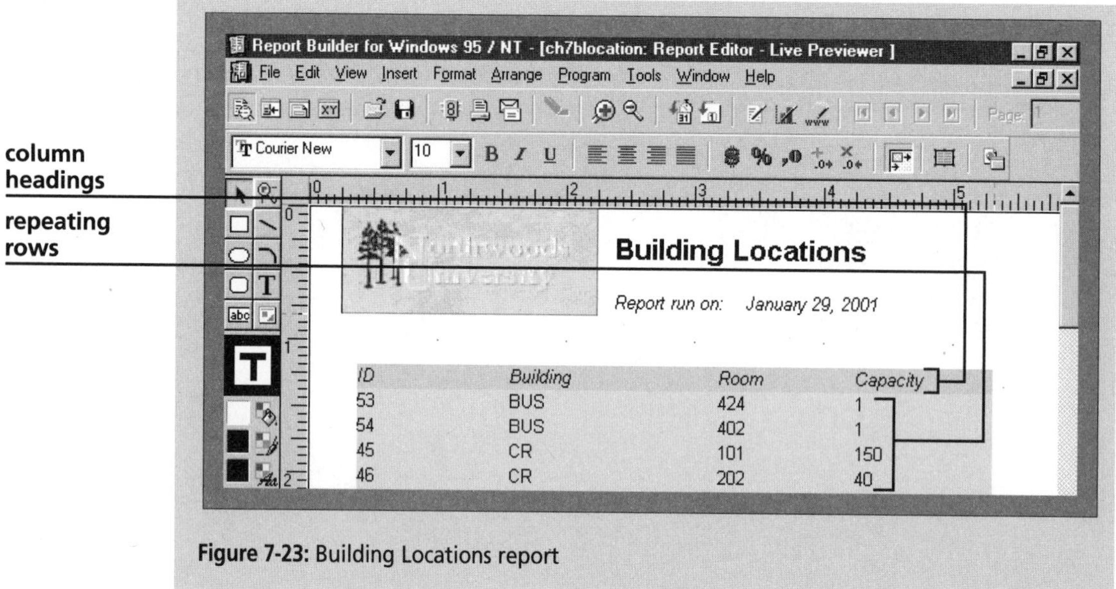

Figure 7-23: Building Locations report

When you examine the report, note that the report title and date heading are in the top margin. The body contains a row with the column headings, and multiple rows showing data from each record in the LOCATION table. The information in the report body is based on the report's Data Model, which is described in the next section.

The Report Data Model

When you create a report, you enter a SQL query to return all of the data records that will be displayed. Report Builder organizes the returned data fields into one or more **record groups**, which are sets of records with the same field headings, but with different data values, that represent the data retrieved by the query. The report's **data model** consists of the SQL query that retrieves the report data, and its associated record groups.

To open the Data Model Editor:

1 Click **View** on the top menu bar, then click **Data Model.** The Data Model Editor opens.

The Data Model Editor has a tool palette for creating queries and report data fields, and a toolbar for navigating to other windows within Report Builder, and for opening and saving files and running and printing reports. The navigation buttons include the Live Previewer button, for moving to the Live Previewer;

the Data Model button, which is currently pressed, for moving to the Data Model; the Layout Model button, for moving to the Layout Model; and the Parameter Form button, for moving to the report parameter form. (Parameter forms will not be covered in this book.)

The navigation buttons are displayed on the toolbar in the Data Model, Live Previewer, and Layout Model windows.

help

You might need to click View on the menu bar, then click Status Bar or Tool Palette if some of the Data Model window objects are not currently displayed.

Presently there are two objects in the report's Data Model: Q_1, which represents the SQL query that was typed in the Report Wizard when the report was created, and G_LOCID, which represents the record group resulting from the query. First you will open the query and modify it by adding an ORDER BY clause to order the records by LOCID.

To modify the query:

1. Double-click **Q_1** in the Data Model Editor. The SQL Query Statement dialog window is displayed, showing the SQL query for the report.
2. Modify the query as shown in Figure 7-24, then click **OK**. The Data Model Editor window is displayed again.

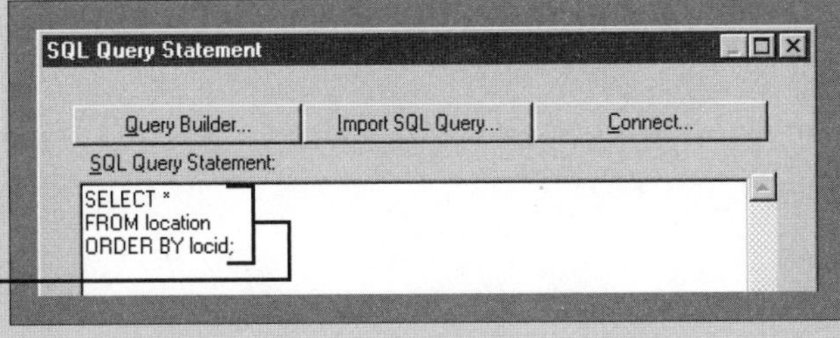

Figure 7-24: Modifying the report query in the Data Model Editor

You could also modify the SQL query by using the Report Wizard.

3 Click the **Live Previewer** button to view the result of your change in the Live Previewer. The data records are now sorted by location ID.

You could also view the report in the Live Previewer by clicking the Run button .

4 Click the **Data Model** button to return to the Data Model Editor.

The Q_1 query returns the LOCID, BLDG_CODE, ROOM, and CAPACITY fields, and the G_LOCID record group represents these fields in the report, as shown in Figure 7-25. The general format for the name of a record group is: `G_<name of first field returned by the record group query>`. In a record group, the individual fields are called **columns**. Note that the icon in front of each column name indicates the column's data type. Now you will open the Property Palette for the LOCID column in the G_LOCID record group, and view its properties.

To open the Property Palette for the LOCID column in the record group:

1 Select the LOCID column in the G_LOCID record group. The column is displayed with a black background to show that it is selected, as shown in Figure 7-25.

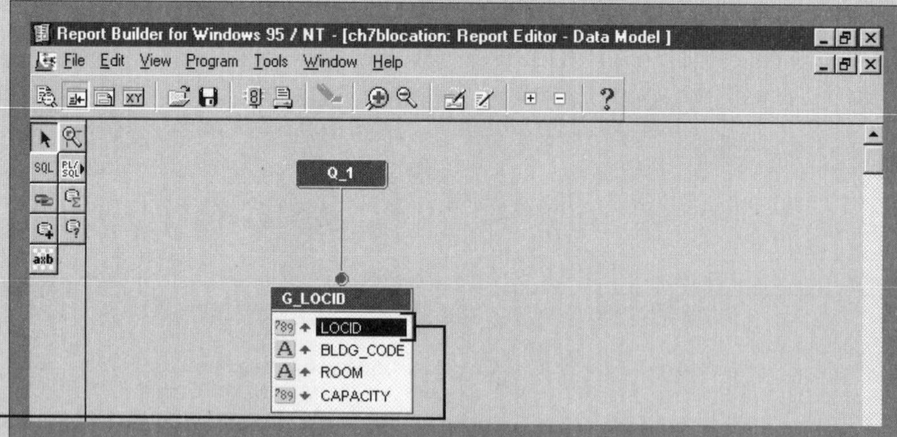

Figure 7-25: Selecting the LOCID column

2 Right-click, then click **Property Palette.** The LOCID column's Property Palette is displayed.

The **Name** property (**LOCID**) is the same as the corresponding database field. The **Column Type** property describes the type of data that the column displays. The report column types that you will use in this book are described in Figure 7-26.

Column Type	Contents
Scalar	Discrete data value retrieved from a database table
Summary	Data value that is the sum of other report data columns
Formula	Data value that is calculated from another report data column

Figure 7-26: Report column types

The current Column Type is Database—Scalar, which indicates that this column displays a discrete value from a database table. The **Datatype** and **Width** properties describe the data type and maximum width of the data column, respectively. The **Value if Null** property allows the developer to substitute a different value for the data field if the retrieved data value is NULL. The **Break Order** property is used to specify the order in which different group field levels appear in the report. (Since this report only has one group field level, changing this property has no impact on how the data are ordered in the report.)

Now you will close the LOCID column's Property Palette, and view the report in the Live Previewer again.

To close the Property Palette and view the report in the Live Previewer:

1. Close the Property Palette.
2. Click the **Live Previewer** button to display the report in the Live Previewer.

Layout Frames

Next, you will examine how the report layout relates to the data model. When you examine the report in the Live Previewer (see Figure 7-23), note that it has a row that contains the column headings, and then multiple rows that show the data from each record in the LOCATION table. These data rows are called **repeating** rows, since each one shows the same data fields (LOCID, BLDG_CODE, ROOM, CAPACITY), but with different data values. Now you will view the report layout in the Layout Model, which shows the report layout items as objects, rather than as actual data values.

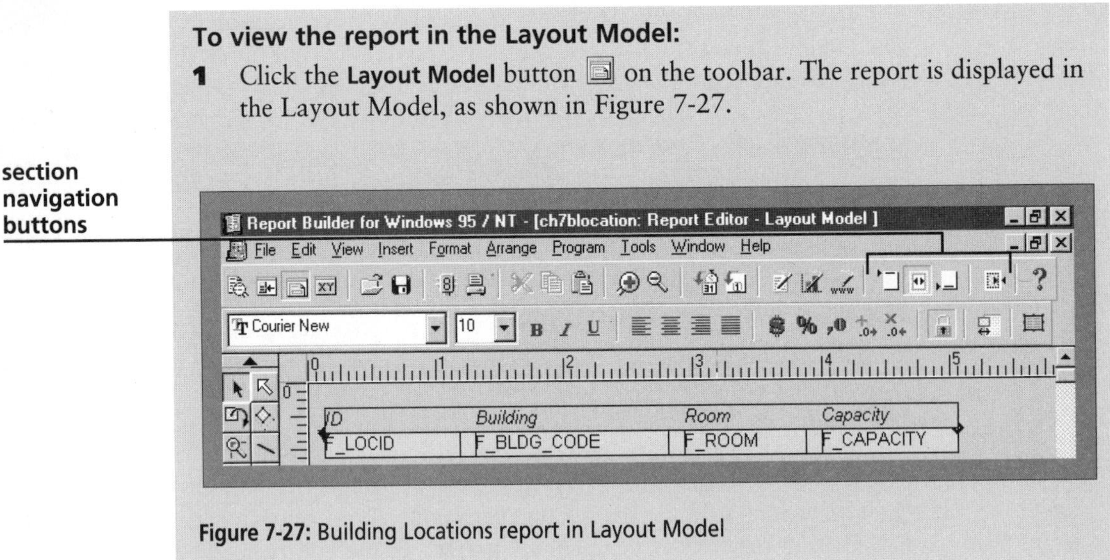

To view the report in the Layout Model:

1 Click the **Layout Model** button on the toolbar. The report is displayed in the Layout Model, as shown in Figure 7-27.

Figure 7-27: Building Locations report in Layout Model

The Layout Model is similar to the Layout Model Template Editor and to the Form Builder Layout Editor: it has a tool palette, toolbar, painting region, and status line. In addition, the top toolbar has four navigation buttons that allow you to work in different sections of the report.

A report has four sections: header, body, margins, and trailer. The **header** is an optional page (or pages) that appears at the beginning of the report and can contain text, graphics, data, and computations to introduce the report. The report **body** usually has multiple pages, and contains the report's data and computations. The report body has **margins** on each page, where you can put text such as titles and page numbers. In the two reports you have created so far, the company logo, report title, date, and date label were in the report margin. The report **trailer**, like the header, is an optional page (or pages) that appears at the end of the report. The trailer can include summary data, or for a report with hundreds of pages, it could contain text that indicates the end of the report.

The first button in the section navigation button group is the Header button, which enables you to make the report header the active section of the report, making it available for editing. The second button is the Body (or Main Section) button, which enables you to activate the report body. The third button is the Trailer button, which allows you to activate the report trailer. The fourth button is the Margin (or Edit Margins) button, which allows you to toggle the margins portion of the active report section on or off. When the Margin button is clicked, you can view and edit the margins of the report section that is currently active.

Currently the Body button is clicked, so this is the active report section and is available for editing. The Margin button is not clicked, so the margins in the report body are not visible or available for editing. Now you will click so you can edit the margins of the report body. You will delete the page number from the bottom margin, and make the page number appear in the top margin.

To edit the report body margins:

1. Be sure that the **Body** button is clicked, then click the **Margin** button to view the margins of the report body. The top margin of the report body is displayed.
2. Scroll to the bottom of the screen, select the report page number (**Page &<Page Number> of &<Total Pages>**), then press the **Delete** key to delete the page number.
3. Scroll back to the top of the report, click the **Page Number** button, select **Top Right** from the page number position list, make sure the **Page Number Only** option button is selected, then click **OK**. The new page number label and field are displayed on the right side of the top margin.
4. Reposition the page number so that it is aligned horizontally with the date fields, as shown in Figure 7-28.

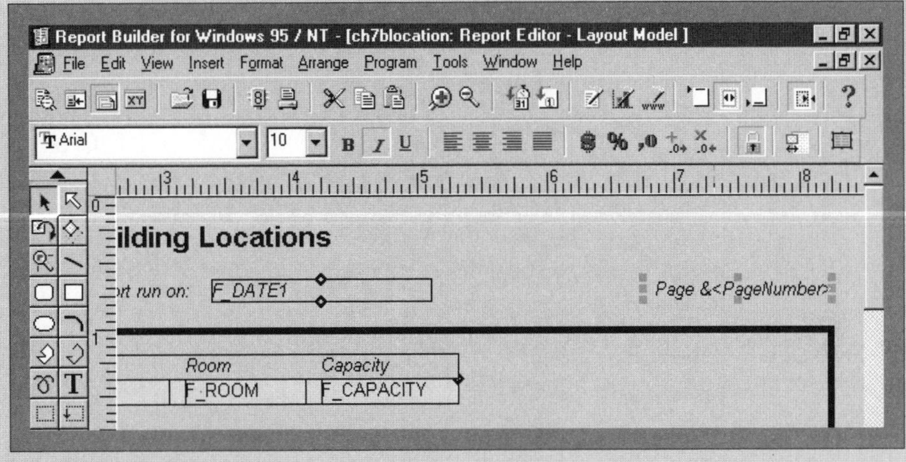

Figure 7-28: Repositioning the report page number

5. Change the page number font to **10-point Arial italic.**
6. Save the report.

Now you will examine the components of the report body. In the Layout Model, you can see that there are boxes around both the repeating fields and the column headings. Actually, there are several boxes, or frames, on the report. **Frames** are containers for grouping related report items so that you can set specific properties for a group of items, rather than having to set the property individually for each item. For example, all of the report column headings might be placed in a frame so that you can apply the same background color to all of the headings.

In the Layout Model, individual frames are not always visible, because they are on top of each other. Figure 7-29 shows a schematic representation of the Building Locations report's frames and objects.

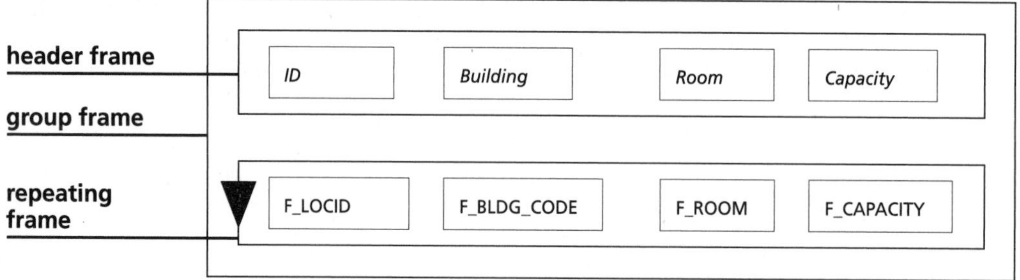

Figure 7-29: Report objects and frames

Reports can have three kinds of frames: header frames, repeating frames, and group frames. A **header frame** contains all of the column headings for a record group. In Figure 7-29, the column headings (ID, Building, Room, Capacity) are enclosed in a header frame. A **repeating frame** contains all of the data fields in a record group. A repeating frame is designated with a down-pointing arrow on its left side. In the Layout Model, individual data fields are given the default name `F_<database field name>`. Figure 7-29 shows that the record group's data fields (F_LOCID, F_BLDG_CODE, F_ROOM, and F_CAPACITY) are enclosed in a repeating frame. Repeating frames have variable sizing, and can shrink or stretch depending on how many data records are displayed in the report. For example, if the Building Locations report query retrieves 13 records, then the repeating frame stretches so that all 13 records are displayed. A **group frame** contains a record group's header frame and repeating frame.

If an object within a frame is moved outside of its frame, an error message will be generated by Report Builder when you view the report in the Live Previewer. For example, if you move the F_LOCID field outside of its repeating frame, an error message will appear. Furthermore, objects in a frame must be totally enclosed by their surrounding frames, or an error message will appear when you view the report in the Live Previewer. For example, in Figure 7-29, the header frame and repeating frame must be completely enclosed by the group frame. If one of these frames is outside of the group frame or if its borders overlap with the borders of the group frame, an error message will appear.

Frame names are derived from the name of their associated record groups. Figure 7-30 shows the general format used by the Report Wizard for naming report frames, as well as the names of the frames in the Building Locations report. Recall that in the Building Locations report, the record group name is G_LOCID.

Frame Type	General Name Format	Frame Name in Building Locations Report
Group frame	M_<record group name>_GRPFR	M_G_LOCID_GRPFR
Header frame	M_<record group name>_HDR	M_G_LOCID_HDR
Repeating frame	R_<record group name>	R_G_LOCID

Figure 7-30: Report frame types and default names

Now you will toggle the margins off so you can edit the report body. You will select different frames in the Building Locations report and examine their names. The easiest way to select a specific report frame in the Layout Model is to select an item in the frame, and then click the Select Parent Frame button on the Layout Model toolbar. A **parent frame** is the frame that directly encloses an object. In Figure 7-29, the header frame directly encloses the individual column headings, so the header frame is the parent frame of each column heading. Similarly, the group frame directly encloses the header frame, so the group frame is the parent frame of the header frame.

To select the frames and examine the frame names in the report body:

1. Click the **Margins** button to toggle the report body margins off and make the report body available for editing. The margins disappear, and the body becomes the active report section.
2. Click the **ID** column heading so that handles appear around its edges, and then click the **Select Parent Frame** button on the toolbar to select the ID column heading's parent frame, which is the header frame. Selection handles appear around all of the column headings, as shown in Figure 7-31.

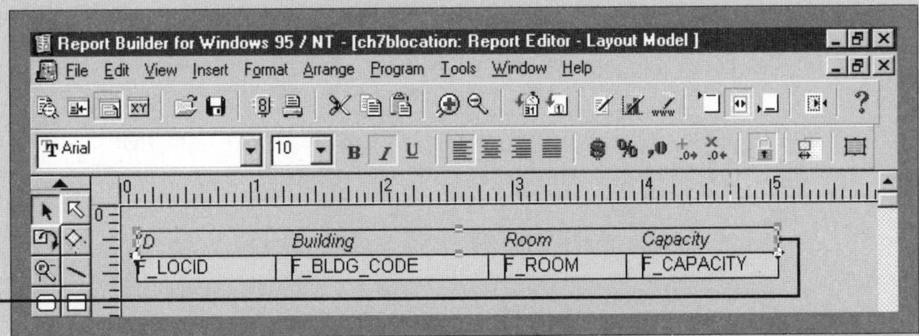

Figure 7-31: Selecting the header frame

3 Click **Tools** on the menu bar, then click **Property Palette.** The Property Palette for the header frame opens, and the frame name property is M_G_LOCID_HDR. Close the Property Palette.

Recall that the group frame is the parent frame of the header frame. Since the header frame is currently selected, clicking the **Select Parent Frame** button will select the frame that directly encloses it, which is the group frame. Next, you will select the group frame and open its Property Palette to examine its properties.

To select the group frame and open its Property Palette:

1 Click to select the header frame's parent frame, which is the group frame. Selection handles appear around all of the report objects, as shown in Figure 7-32.

group frame

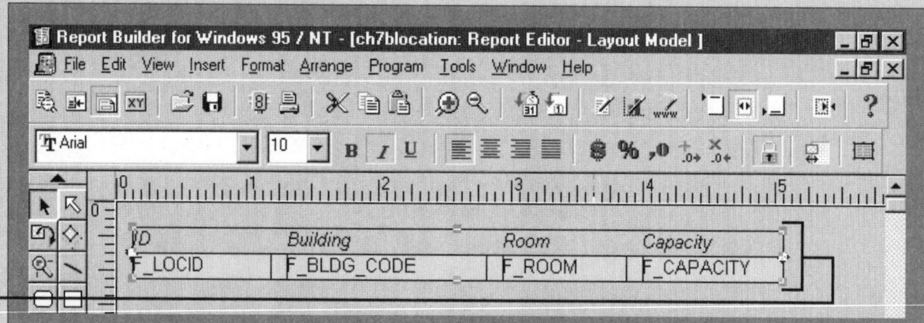

Figure 7-32: Selecting the group frame

2 Click **Tools** on the menu bar, then click **Property Palette.** The Property Palette for the group frame opens. The frame's name is M_G_LOCID_GRPFR. Close the Property Palette.

Finally, you will select the report's repeating frame and open its Property Palette. To do this, you will select one of the report data fields, and then click the Select Parent Frame button to select its parent frame, which is the repeating frame.

To select the report's repeating frame:

1 Click **F_LOCID** to select it. Selection handles appear around the data field.

2 Click to select the repeating frame. Selection handles appear around all of the data fields, as shown in Figure 7-33.

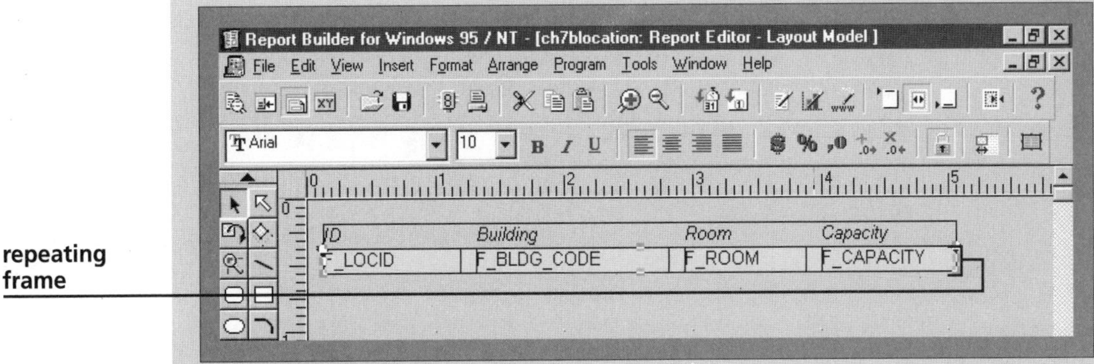

Figure 7-33: Selecting the repeating frame

3 Click **Tools** on the menu bar, then click **Property Palette.** Notice that the repeating frame's name (R_G_LOCID) is displayed at the top-left corner of the Property Palette sheet. The Source property is G_LOCID, which is the record group that is the source of the repeating frame's data.

You can change properties on a frame's Property Palette to specify the frame's background color, whether the frame should begin on a new page, or special printing instructions, such as leaving a set amount of blank space between rows. Now you will modify some of the properties of the R_G_LOCID repeating frame. First, you will modify the frame so that only five records appear per page. Then you will increase the vertical spacing between each record so that 0.25 inches of blank space appears between each row of data.

To modify the repeating frame properties:

1 On the R_G_LOCID repeating frame Property Palette, change the Maximum Records per Page property to **5**.

2 Change the Vert. Space Between Frames to **.25**. Click on another property to save the change, then close the Property Palette.

3 Click the **Live Previewer** button to view the report in the Live Previewer. The first five records appear on the first report page, with 0.25 inches of blank space between each row, as shown in Figure 7-34.

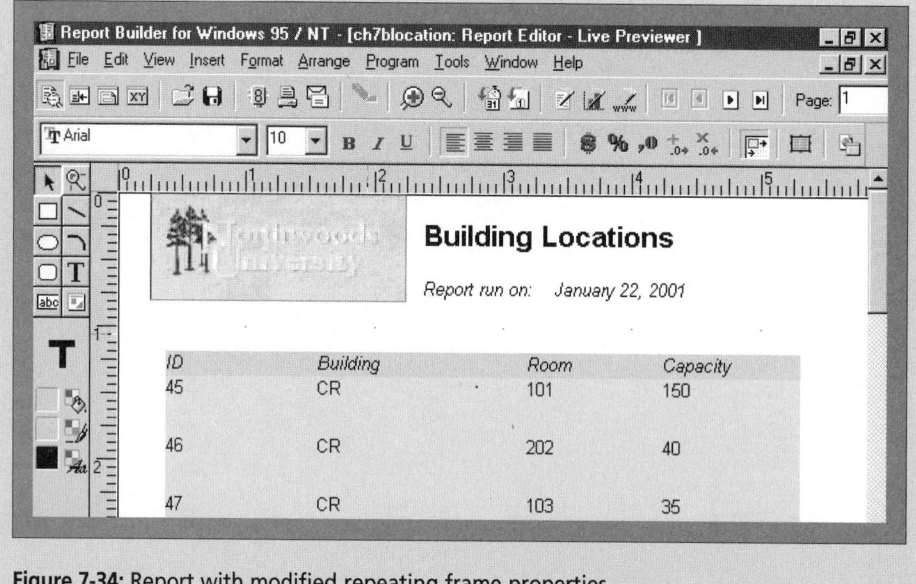

Figure 7-34: Report with modified repeating frame properties

4 Click the **Next Page** button on the toolbar. The next five records are displayed.

5 Click . The last three records are displayed.

6 Click the **First Page** button to navigate back to the first page of the report.

7 Close the report, and click **Yes** to save your changes.

Components of a Master-Detail Report

The Building Locations report had only one level of data and did not have a master-detail relationship. Reports with master-detail relationships have multiple record groups, and multiple group, header, and repeating frames. Next, you will open a report with a master-detail relationship, view the report's data model and frames, and modify the report's formatting to make it more readable.

To open the master-detail report:

1 Open **classlist.rdf** from the Chapter7 folder on your Student Disk. The file opens in the Object Navigator. Save the report as **ch7bclasslist.rdf**.

2 Double-click the **Live Previewer** icon to display the report in the Live Previewer (see Figure 7-35).

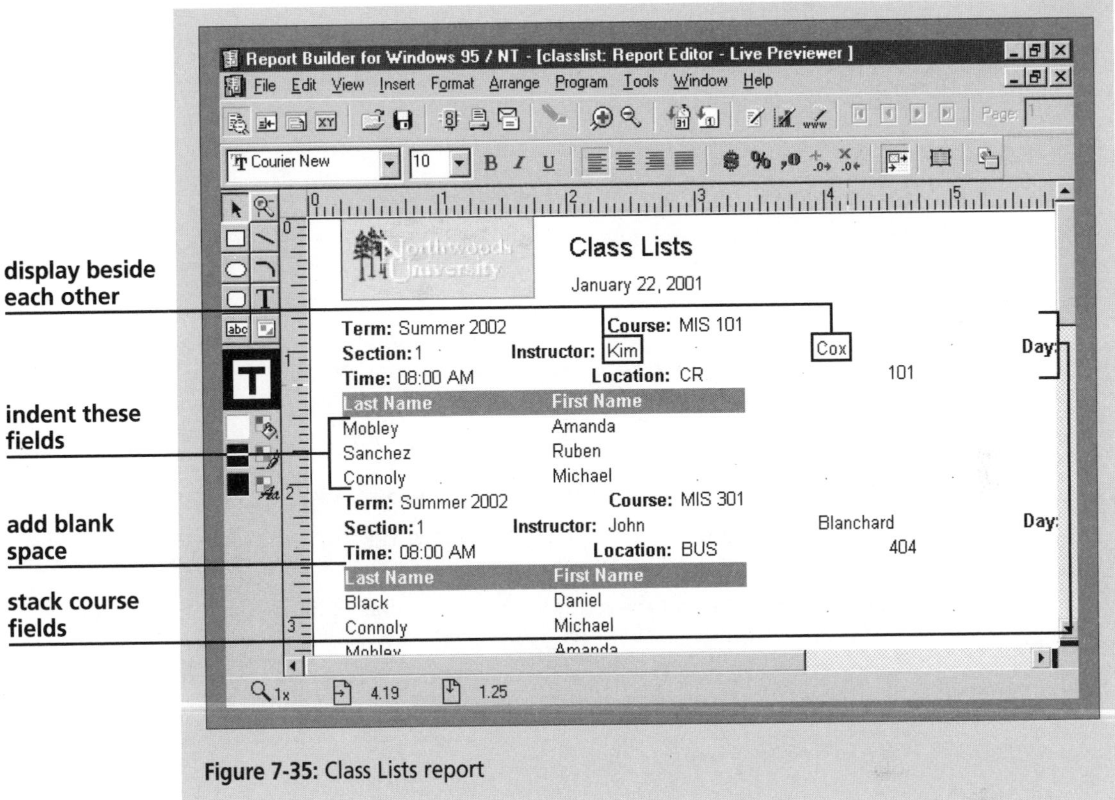

Figure 7-35: Class Lists report

The Class Lists report shown in Figure 7-35 provides class lists for all of the classes offered during the Summer 2002 term at Northwoods University. The report's output includes the term description, course call ID, section number, instructor's first and last names, day, time, and building code and room for each course offered during the term. It also shows the last and first names of each student enrolled in the course.

The figure highlights some of the problems with the report's current format. Since the class list needs to be distributed to individual instructors, the information for each course section should be printed on a separate page. The report would be easier to understand if the course data fields were stacked on top of each other vertically, rather than spread horizontally across the page. Fields that should appear adjacent to each other, such as the instructor's first and last name, should be anchored together instead of being spaced to accommodate the largest possible field width. Also, the report would look better if there were some blank space between the course section and student information, and if the student fields were indented on the page. To make these changes, you will need to edit the data fields and frames generated by the Report Wizard.

Understanding the Report Structure

Before you can edit the report, you need to understand its underlying structure. First you will examine the report's data model.

To examine the data model:

1 Click the **Data Model** button on the toolbar to open the data model.

In the data model, some of the record groups are not currently visible. First, you need to move the record groups so that they are all visible on the screen. To move a record group, you select it and then drag it to the desired position.

To move the record groups so that they are all visible:

1 Select **Q_1**, and move it to the top-left corner of the Data Model painting region, as shown in Figure 7-36.

2 Select **G_tdesc**, and move it so that it is to the right and slightly below Q_1, as shown in Figure 7-36.

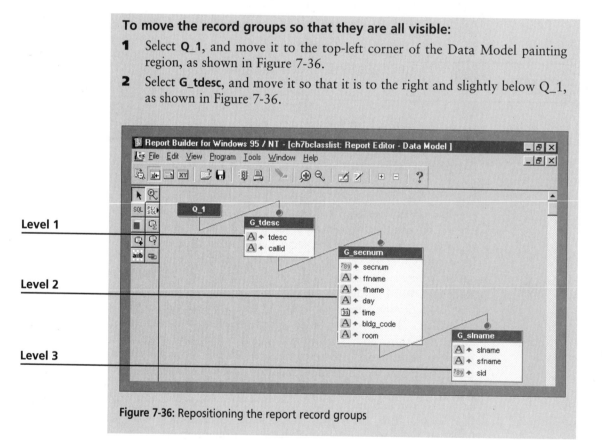

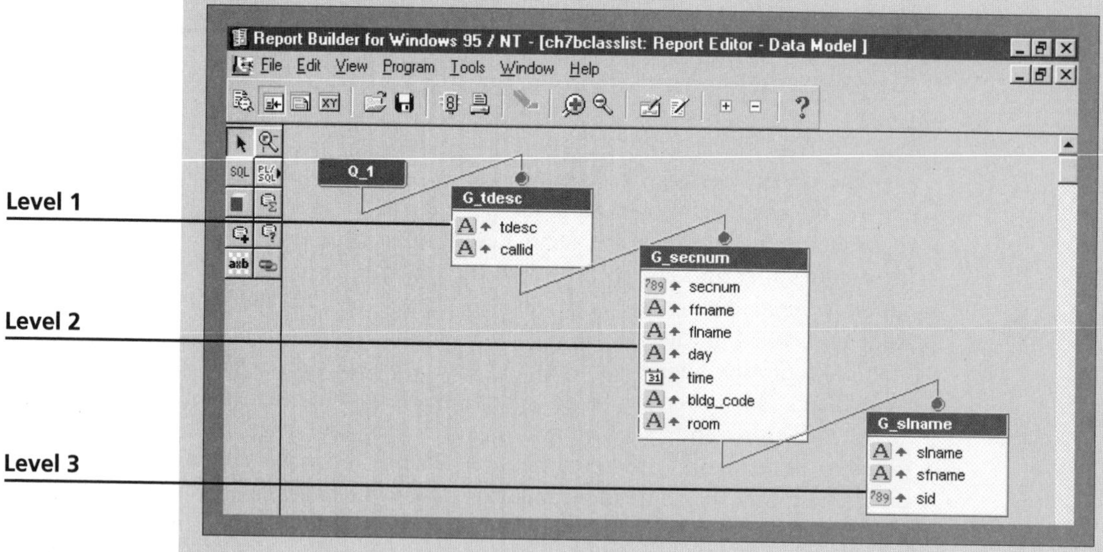

Figure 7-36: Repositioning the report record groups

3 Move the other record groups so that they are positioned as shown in Figure 7-36.

help

If a scroll bar appears on the right edge of one of the record groups, select the record group by clicking it, and then drag the bottom-center handle down so that all fields are displayed and the scroll bar disappears. If a scroll bar appears on the bottom or right side of one of the record groups, it means that the record group is not large enough to display all of its data. Select the record group by clicking it, and then drag the right-center handle horizontally to the right or vertically to the bottom so that all fields are displayed and the scroll bar disappears.

Note that there are three record groups: G_tdesc, G_secnum, and G_slname. Report data fields are grouped according to master-detail relationships. Since term description and course call ID will appear on each report page, these fields are in the Level 1(most general) G_tdesc group. One course call ID might have several section numbers, so all of the data fields that are unique for a given course section (SECNUM, FFNAME, FLNAME, DAY, TIME, BLDG_CODE, ROOM) appear in the Level 2 G_secnum record group. Each section number might have several students, so the student fields appear in the Level 3 (most detailed) G_slname record group.

Making the report formatting modifications highlighted in Figure 7-35 will require you to enlarge some of the report frames so that you can reposition the data fields and modify properties of individual frames. Now you will view the report frames in the Layout Model so you can select individual frames and modify their properties.

To view the report in the Layout Model:

1 Click the Layout Model button on the toolbar. The report layout is displayed.

help

If the report body margins are displayed, click the Margins button to toggle the margins off.

It is difficult to understand the report's frame relationships in the Layout Model, because many of the frames are stacked on top of each other. You will begin by examining a series of schematic diagrams that explain the frame structure of the report. Then, you will examine the individual report frames in the Layout Model.

First you will examine the layout of the G_slname record group's student fields. Recall that a header frame encloses a record group's column headings, a repeating frame encloses the record group's data fields, and a group frame encloses the record group's header frame and repeating frame. Figure 7-37 shows a schematic diagram of the frames that were created for the G_slname record group. (These frames exist in the Layout Model in Report Builder, but are hard to see because the header frame and repeating frame are stacked directly on top of each other, and the group frame is directly on top of these two frames.)

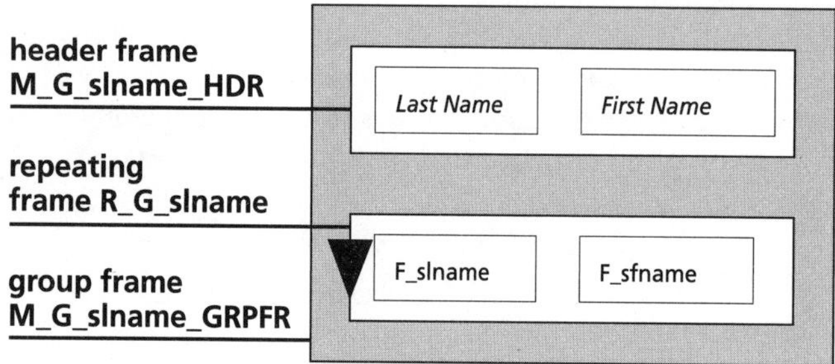

Figure 7-37: G_slname frames (group frame is shaded)

Figure 7-37 shows that the data fields that display the student last name and first name (F_slname and F_sfname) are surrounded by a repeating frame named R_G_slname. The column headings associated with these data fields (Last Name and First Name) are surrounded by a header frame named M_G_slname_HDR. Both the repeating frame and header frame are surrounded by a group frame named M_G_slname_GRPFR, which is shaded in the figure.

In a report with multiple record group levels, frames for higher-level (more detailed) record groups are nested *inside* frames for lower-level (less detailed) record groups. Recall that in this report, the Level 1 record group (G_tdesc) shows information for a term and course call ID, the Level 2 record group (G_secnum) shows information for a particular course section associated with that term and course call ID, and the Level 3 record group (G_slname) lists the students that are in a specific course section. The Level 1 record group's group frame contains the Level 2 record group's group frame, and the Level 2 record group's group frame contains the Level 3 record group's group frame. Figure 7-38 shows a schematic representation of these frames. To contrast the relationships between repeating frames and group frames, the repeating frames are filled in white, and the group frames are shaded.

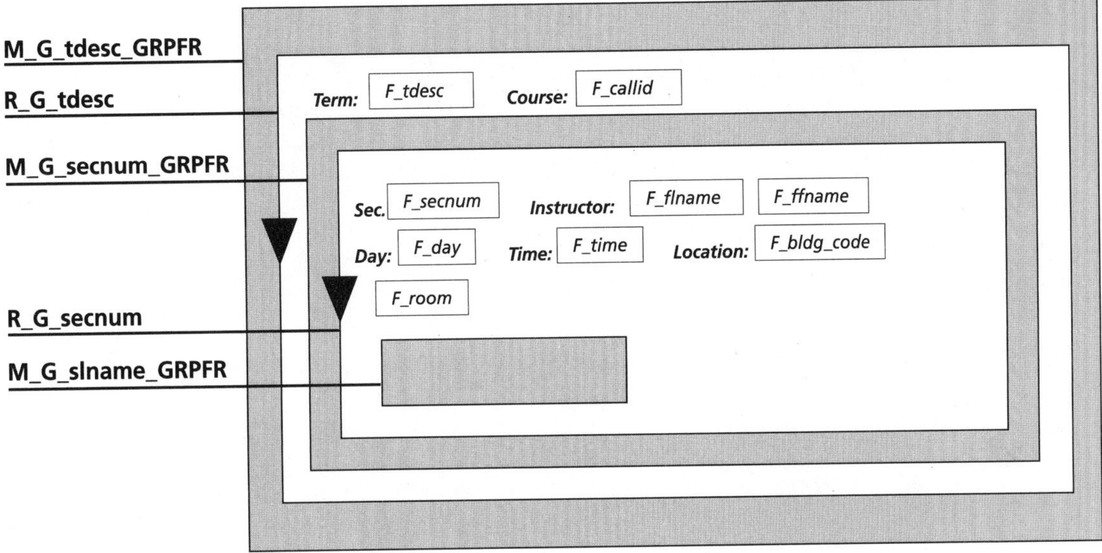

Figure 7-38: G_secnum and G_tdesc record group frames (repeating frames are white and group frames are shaded)

The G_secnum data fields and column headings and the G_slname group frame are enclosed by the G_secnum repeating frame (R_G_secnum) and the G_secnum group frame (M_G_secnum_GRPFR). There is no header frame for the G_secnum record group, since the column headings are displayed horizontally across the report. A header frame is created only when data fields are displayed directly under column headings in a columnar fashion. The term and course call ID fields and the G_secnum repeating frame are enclosed by the G_tdesc repeating frame (R_G_tdesc) and the G_tdesc group frame (M_G_tdesc_GRPFR).

Recall that these frame relationships are hard to see in the Layout Model, because many of the frames are displayed directly on top of each other. Now you will examine the frames in the Layout Model so that you become familiar with the frame relationships and gain experience selecting specific frames. First, you will examine the frames associated with the G_tdesc record group. Recall that the best way to select a frame is to select an item inside the frame, and then click the Select Parent Frame button ▣. Since sometimes a frame and its parent frame are directly on top of each other, there might be no visual change when you click ▣. To keep track of which frame is currently selected, you should always open each frame's Property Palette and check its name immediately after you select the frame.

To examine the G_tdesc record group frames:

1. Select the **F_tdesc** data field, then click the **Select Parent Frame** button to select its parent frame, which is the G_tdesc repeating frame (R_G_tdesc, as shown in Figure 7-38). Since this repeating frame surrounds all of the report items, selection handles appear around all of the report items. Now you will open the currently selected frame's Property Palette to confirm that its name is R_G_tdesc.
2. Click **Tools** on the menu bar, then click **Property Palette**. Confirm that the selected frame's name is R_G_tdesc, then close the Property Palette.

help

If R_G_tdesc is not the name of the current frame, close the Property Palette and repeat Steps 1 and 2.

3. Click to select the repeating frame's parent frame, which is the G_tdesc group frame (M_G_tdesc_GRPFR, as shown in Figure 7-38). There is no visible change in the selection handles that are displayed in the Layout Model, because the group frame is displayed directly on top of the repeating frame. Now you will open the current frame's Property Palette to confirm that its name is M_G_tdesc_GRPFR.
4. Click **Tools** on the menu bar, then click **Property Palette**. Confirm that the frame's name is M_G_tdesc_GRPFR, then close the Property Palette.

Now you will examine the frames associated with the G_secnum record group. Again, you will select a frame, and then open its Property Palette to determine the name of the selected frame.

To examine the G_secnum frames:

1. Click **F_secnum** to select it, then click the **Select Parent Frame** button to select its parent frame, which is the G_secnum repeating frame (see Figure 7-38). Selection handles appear around the G_secnum repeating frame.
2. Open the selected frame's Property Palette to confirm that its name is R_G_secnum, then close the Property Palette.
3. Click again to select the repeating frame's parent frame, which is the G_secnum group frame (M_G_secnum_GRPFR, as shown in Figure 7-38). The selection handles do not change in the Layout Model, because the group frame is displayed directly on top of the repeating frame.
4. Open the selected frame's Property Palette to confirm that its name is M_G_secnum_GRPFR, then close the Property Palette.

Finally, you will examine the frames associated with the G_slname record group. Since this record group's data are displayed in columns, it has a header frame that encloses the column headings. It also has a repeating frame that encloses the data fields, and a group frame that encloses the header frame and repeating frame.

To examine the G_slname record group frames:

1. Click the **Last Name** column heading, then click to select the column heading's parent frame, which is the header frame (M_G_slname_HDR, see Figure 7-37). Selection handles appear around both column headings.
2. Open the frame's Property Palette to confirm that its name is M_G_slname_HDR, then close the Property Palette.
3. Click the **F_slname** data field to select it, then click to select its parent frame, which is the G_slname repeating frame (R_G_slname, as shown in Figure 7-37). Selection handles appear around the repeating frame.
4. Open the selected frame's Property Palette to confirm that the selected frame's name is R_G_slname, then close the Property Palette.
5. Click again to select the repeating frame's parent frame, which is the G_slname group frame (M_G_slname_GRPFR, see Figure 7-37). Open the selected frame's Property Palette to confirm that its name is M_G_slname_GRPFR, then close the Property Palette.

Printing Report Records on Separate Pages

The first change you will make to the report's format is to make the data for each course section print on a separate page. To create a page break between sets of repeating records, you must open the Property Palette for the repeating frame that contains the records that you want on each page, and change its Maximum Records per Page Property to 1. For this report, you want the report to print separate pages for each course section number, containing the term description, course call ID, course section data, and student data. To do this, you will select the outermost repeating frame (R_G_tdesc), open its Property Palette, and change its Maximum Records per Page property to 1. Then you will view the report in the Live Previewer to see the result.

Sometimes you have to experiment to determine which repeating frame's Maximum Records per Page property must be changed to make report page breaks appear as desired.

To make the data for each course section number print on a separate page:

1. Click the **F_tdesc** data field to select it, then click the **Select Parent Frame** button to select the R_G_tdesc repeating frame.

2 Open the frame's Property Palette and confirm that you have selected R_G_tdesc. Change the Maximum Records per Page property to **1**, click on another property in the Property Palette to save the change, then close the Property Palette.

3 View the report in the Live Previewer. The report output is displayed as shown in Figure 7-39, with the course section data and class list for MIS 101 Section 1 displayed on the first page.

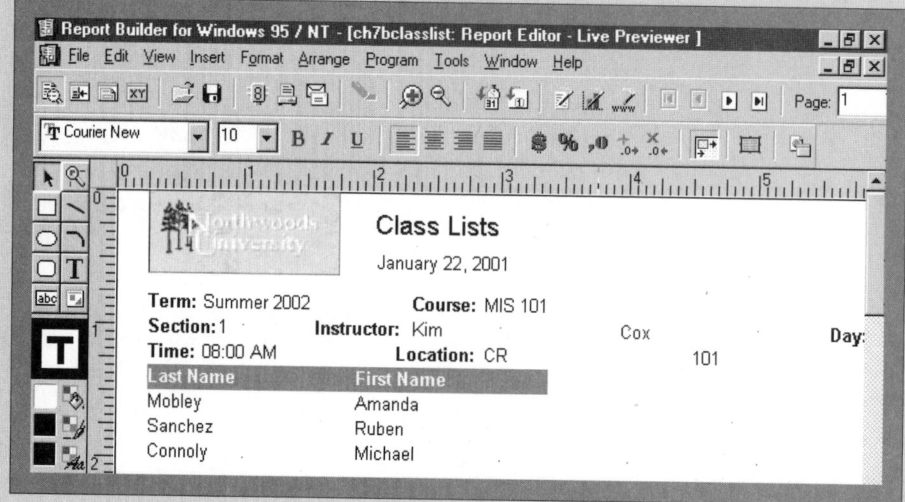

Figure 7-39: Class Lists report with data for each course section displayed on a separate page

4 Click the **Next Page** button to view the next class list. The class list for MIS 301 Section 1 is displayed.

5 Click . The class list for MIS 441 Section 1 is displayed.

6 Click the **First Page** button . The first page of the report is displayed again.

7 Save the report.

Repositioning the Report Objects

The next formatting task is to move the fields and headings in the R_G_secnum repeating frame so that they are stacked on top of each other. Sometimes when you move objects in the Layout Model or Live Previewer, you might find that you cannot move or drag a data field or label outside of its enclosing frame. This is because **confine mode** is on. When confine mode is off, you can freely drag objects anywhere on the Layout Model painting region. You can turn confine mode on by clicking the Confine Mode button on the Layout Model toolbar so that it appears selected. You can turn confine mode off by clicking again.

Another way to toggle confine mode on is to right-click anywhere in the Layout Model, then click Confine Mode on the menu so that it is checked. To turn confine mode off, right-click anywhere in the Layout Model, then click Confine Mode on the menu so that it is no longer checked.

It is safest to leave confine mode on. Recall that if a frame does not completely enclose all of the objects inside it, you will get an error message when you view the report in the Live Previewer.

Now you will examine the effects of turning confine mode on and off. You will see what happens when you try to move an object outside of its enclosing frame with confine mode on, and view the error message that is generated when you try to view the report in the Live Previewer.

To examine the effects of turning confine mode on and off:

1 Open the Layout Model, right-click in the blank area of the Layout Model painting region below the report objects, and make sure that **Confine Mode** has a check mark in front of it. Click anywhere on the painting region to close the menu, but do not click **Confine Mode** on the menu, or you will turn confine mode off.

help

If Confine Mode does not have a check mark, right-click anywhere in the Layout Model, click Confine Mode, then repeat Step 1.

2 Save the report.

3 Click the F_slname data field to select it, then try to drag it down so that it is below the F_sfname data field. Note that you can drag F_slname to the left and right, but you cannot drag it outside of its parent frame (R_G_slname).

4 Click **Edit** on the menu bar, then click **Undo** to return F_slname to its original position.

5 Save the file.

6 Click the **Confine Mode** button on the Layout Editor toolbar to turn confine mode off. Right-click in the painting region below the report objects, and confirm that Confine Mode is no longer checked.

7 Select F_slname, and drag it so that it is directly below F_sfname. This time you should succeed, and your screen should look like Figure 7-40.

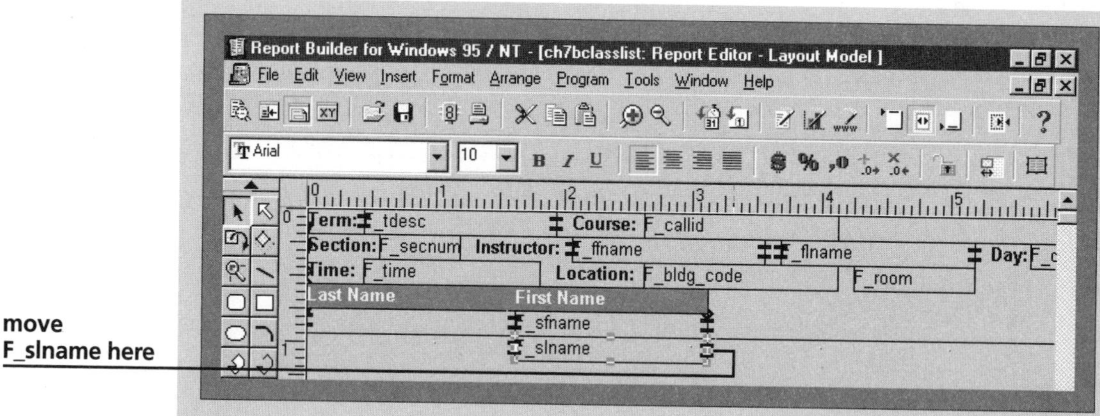

move
F_slname here

Figure 7-40: Repositioning F_slname with confine mode off

8 View the report in the Live Previewer. A warning dialog box opens with the message "REP-1213: 'Field F_slname' references column 'slname' at a frequency below its group." This error occurs because the F_slname is no longer in the repeating frame associated with its record group, but is in a frame at a lower group level.

9 Click **OK**. The error message is displayed again in the Live Previewer window. Click the **Layout Model** button to return to the Layout Model.

10 Click **File** on the menu bar, then click **Revert** to revert back to the report as it was the last time you saved it. Click **Yes** to undo your changes and revert back to the last saved copy of the report. The Object Navigator window is displayed.

11 Double-click the **Layout Model** icon to return to the Layout Model.

It is best to leave confine mode on at all times to avoid this error. If you do not leave confine mode on, you must always make sure that enclosing frames are large enough to let you move report objects where you want them, while keeping them enclosed in their parent frames. An alternative is to use **flex mode**, which automatically makes an enclosing frame larger when you move an enclosed object beyond its enclosing frame's boundary. To turn flex mode on, you click the **Flex Mode** button on the Layout Model toolbar. (You can find out if flex mode is on or off by right-clicking on the Layout Model painting region, and seeing if **Flex Mode** is checked on the menu.) To turn flex mode off, you click so the button is not pressed. Flex mode overrides confine mode; when flex mode is on, frames are automatically resized regardless of whether confine mode is on or off.

A problem with using flex mode is that when you move a report field, flex mode automatically resizes all of the surrounding frames. Sometimes this causes the outermost frames to extend beyond the boundaries of the report body, which generates an error when you view the report in the Live Previewer. Flex mode works well when you need to make a frame longer, because there is usually sufficient extra space on the length of the report page to resize all of the surrounding frames. However, when you need to make a frame wider, it is best to leave flex mode off and resize the frame manually, so you can ensure that report frames do not extend beyond the report body boundaries. Now you will examine the effects of using flex mode to resize report frames. First, you will move a report column lower on the report to make the report frames longer.

To examine the effects of using flex mode to make the report frames longer:

1. If necessary, click the **Flex Mode** button on the Layout Model toolbar so that it appears pressed.
2. Right-click in the blank area of the Layout Model painting region below the report objects, and confirm that **Flex Mode** is on and has a check mark before its name on the menu.

help

If Flex Mode already has a check mark, do not click Flex Mode again or you will turn it off.

3. Save the report.
4. Click **F_slname,** press the **Shift** key, then click **F_sfname** while keeping the Shift key pressed to select both columns as a group. Drag these columns down until their bottom edges are 1.5 inches from the top of the report, as shown in Figure 7-41. You will notice that as you drag the columns down, their parent frame (R_G_slname) automatically resizes to accommodate the new column locations.

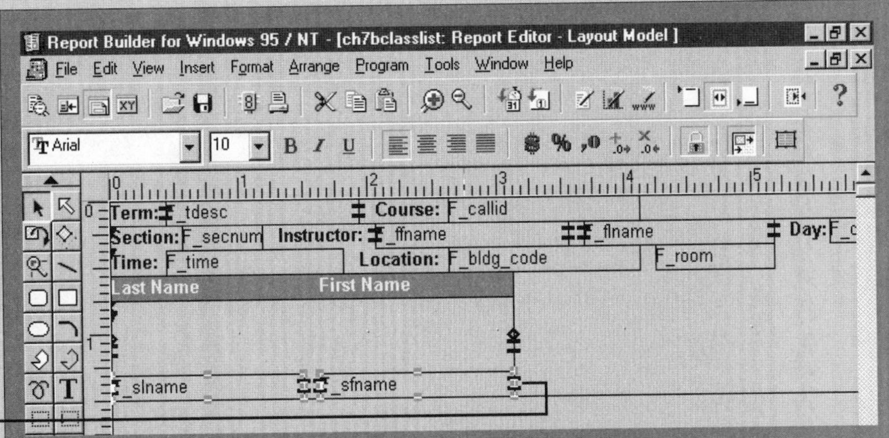

Figure 7-41: Using flex mode to make a report frame longer

5 View the report in the Live Previewer, and note the new positions of the F_slname and F_sfname data fields. There is now about 0.5 inches of white space between each student record as a result of vertically enlarging the repeating frame and moving the student columns lower in the frame.

6 Click **File** on the menu bar, then click **Revert** to revert back to the report as it was the last time you saved it. Click **Yes** to undo your changes. The Object Navigator window is displayed.

Now you will examine the effects of using flex mode to make report frames wider. You will move a report column to the right edge of the report, which causes the enclosing frames to be extended horizontally. Recall that this sometimes causes the report frames to extend beyond the report body boundary, as you will see next.

To examine the effects of using flex mode to make report frames wider:

1 Double-click the **Layout Model** icon to return to the Layout Model.

2 If necessary, click the **Flex Mode** button to turn flex mode on.

3 Click **F_sfname** to select it, then drag it to the right edge of the report so that the right edge of F_sfname is 5 inches from the left edge of the report, as shown in Figure 7-42. Note that the frames around F_sfname are resized as you drag the data column to the right edge of the report.

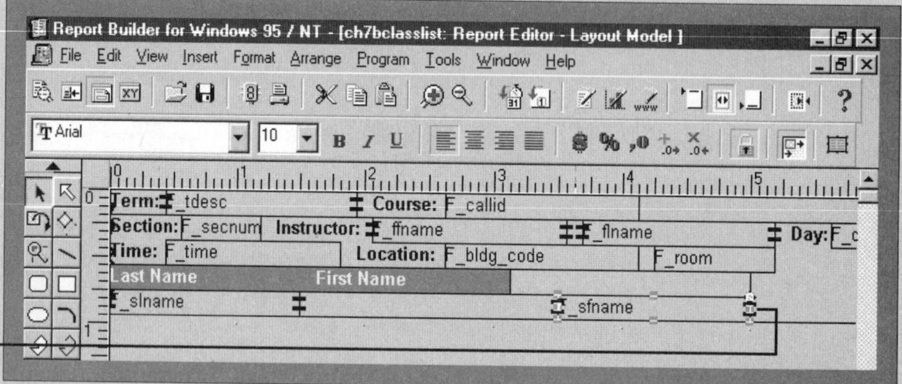

Figure 7-42: Moving F_sfname to the right edge of the report

4 View the report in the Live Previewer. The error message "REP-1212: Object 'Body' is not fully enclosed by its enclosing object 'callid'." is displayed. This error message indicates that some of the report objects extend beyond the edges of the report body.

5 Click **OK**, then open the Layout Model and scroll to the right edge of the report. You will see that some of the outer report frames now extend beyond the report body boundaries, as shown in Figure 7-43.

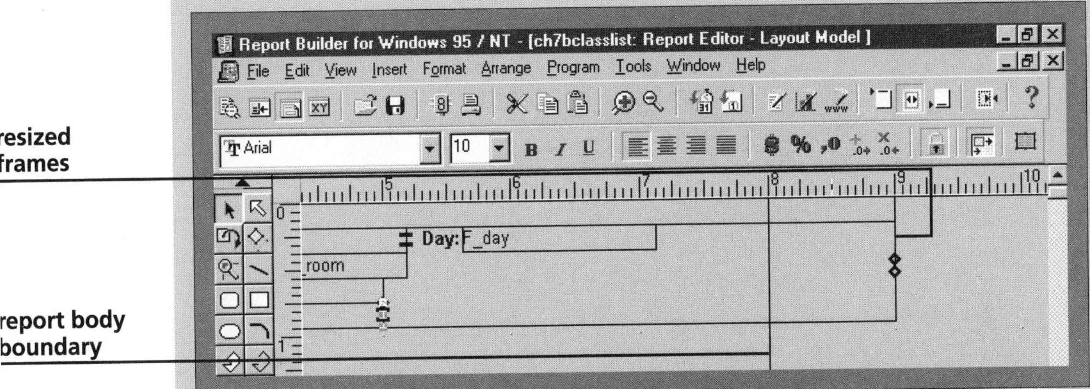

Figure 7-43: Viewing resized frames that extend beyond the report body boundary

6 Click **File** on the menu bar, click **Revert**, then click **Yes** to undo your changes.

Now that you understand the intricacies and peculiarities of confine mode and flex mode, you are ready to move the report objects. First, you will need to make the R_G_secnum repeating frame and its surrounding frames longer, so that there is room to stack the course section fields vertically. You will do this with flex mode on, since flex mode works well when you are making frames longer.

To resize the R_G_secnum repeating frame and its surrounding frames:

1 Open the Layout Model, and click the **Flex Mode** button to turn flex mode on.

2 Click **F_secnum** to select it, then drag it to the lower edge of the report so that the bottom edge of the outermost frame is about 2.5 inches from the top of the report, as shown in Figure 7-44.

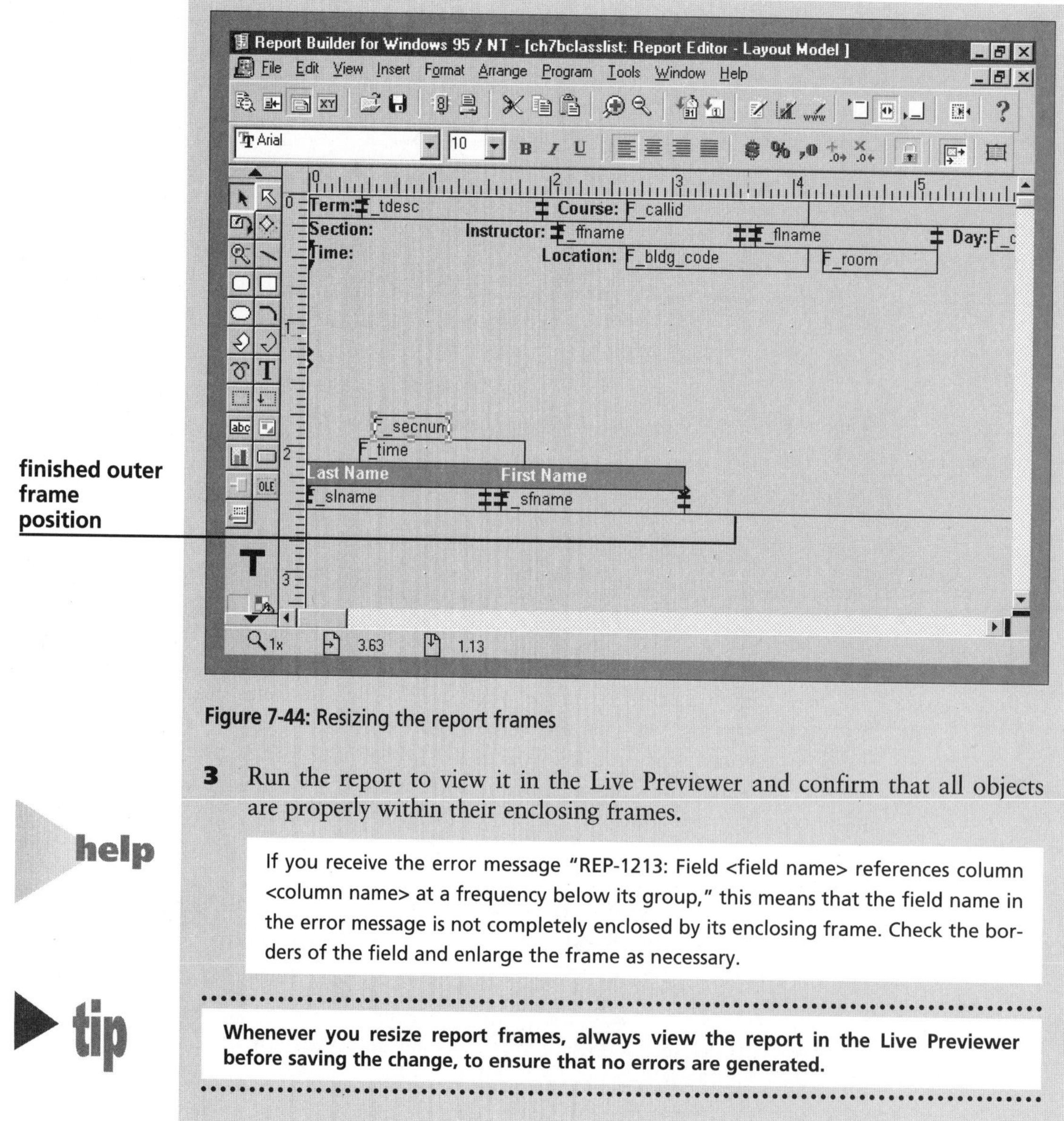

Figure 7-44: Resizing the report frames

3 Run the report to view it in the Live Previewer and confirm that all objects are properly within their enclosing frames.

help

If you receive the error message "REP-1213: Field <field name> references column <column name> at a frequency below its group," this means that the field name in the error message is not completely enclosed by its enclosing frame. Check the borders of the field and enlarge the frame as necessary.

tip

Whenever you resize report frames, always view the report in the Live Previewer before saving the change, to ensure that no errors are generated.

Now you will reposition the fields and labels in the R_G_secnum repeating frame so that the student records are indented, and the course section records are stacked vertically on top of each other. First you will select the M_G_slname_GRPFR group frame, and move it to the right so that the student fields are indented. Then you will adjust the course section headings and data fields.

To reposition the report fields and labels:

1. Open the Layout Model, select **F_slname**, and click the **Select Parent Frame** button. Click **Tools** on the menu bar, then click **Property Palette**, confirm that the **R_G_slname** frame is selected, then close the Property Palette.
2. Click again, and confirm that the **M_G_slname_GRPFR** group frame is selected. Then, use the arrow keys on the keyboard to move the group frame so that its top edge is about 2 inches from the top of the report body, and its left edge is indented about 0.25 inches from the left edge of the report body, as shown in Figure 7-45.
3. Open the Layout Model, turn flex mode off if necessary, and then move the course section fields and labels to the positions shown in Figure 7-45.

field elasticity indicator

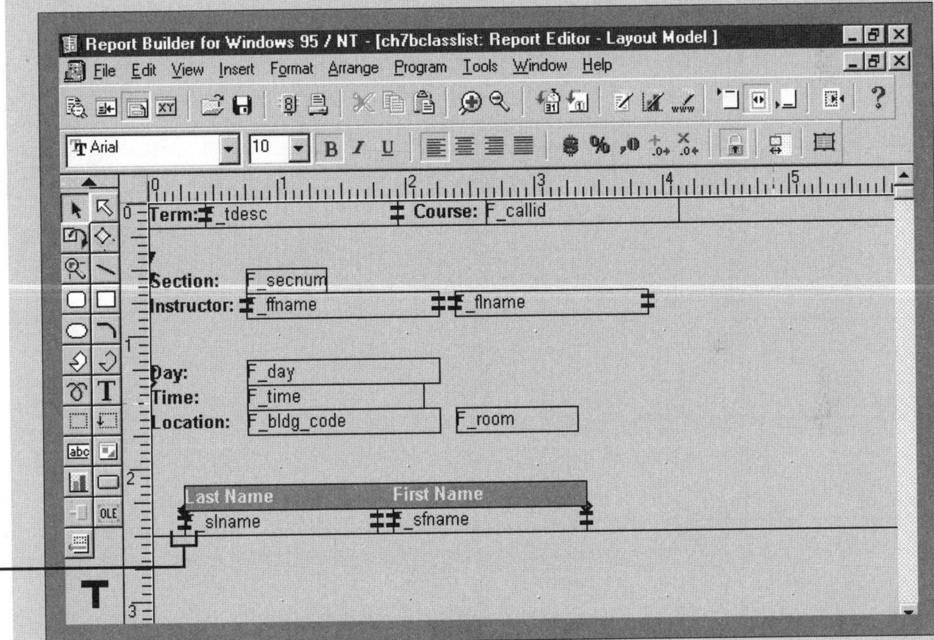

Figure 7-45: Repositioning the report fields and labels

6. Run the report to make sure it runs correctly, with no field referencing errors.
7. Save the report.

Adjusting the Spacing Between Report Columns

The final formatting task is to adjust the spacing between report columns so that data values that should appear right next to each other (such as the instructor's first and last name) are not separated by blank space. You have probably noticed that some of the report columns have horizontal tick marks on their right and left edges, as indicated in Figure 7-45. These marks reflect the column's **elasticity**, which determines whether a column's size is fixed on the printed report, or whether it can expand or contract automatically, depending on the height and width of the retrieved data value. Figure 7-46 illustrates the different elasticity options and associated markings on report column horizontal and vertical borders.

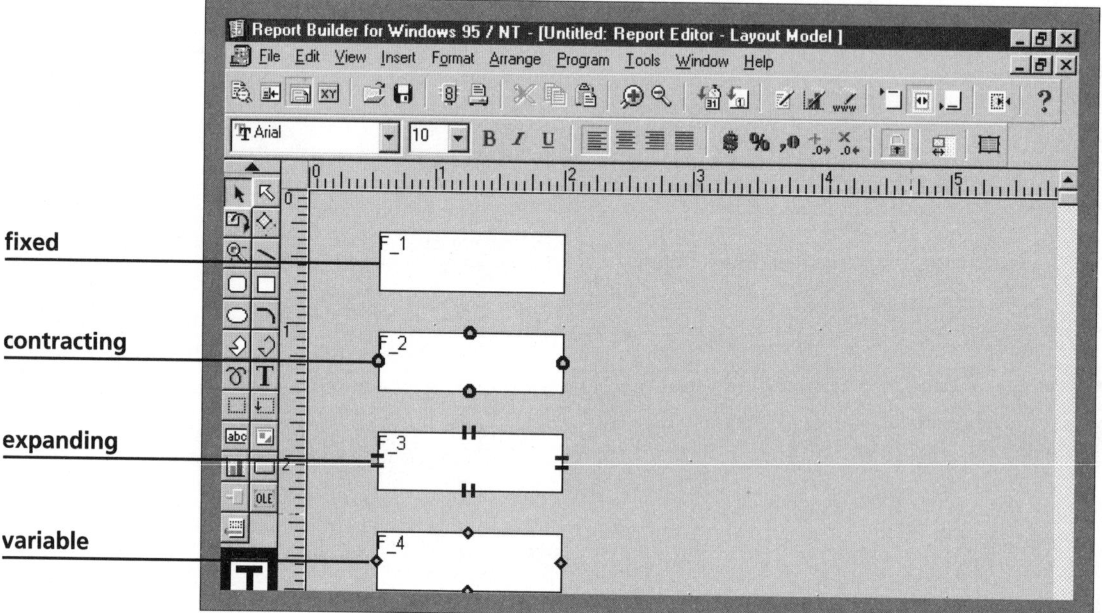

Figure 7-46: Report column elasticity indicators

The markings on the left and right edges specify the vertical elasticity property, or how tall the column can be, and the markings on the top and bottom specify the horizontal elasticity, or how wide the column can be. **Fixed** elasticity means that the column will have the size shown in the layout, and extra text will be truncated.

Contracting elasticity means that the column will contract if the data value is smaller than the layout size, but wider data will be truncated. **Expanding** elasticity means that the column will expand automatically to accommodate wider data values, but narrower data values will still occupy the entire space shown in the layout, and extra blank space will be displayed. **Variable** elasticity means that the report column will contract or expand as needed to fit the data value. A column can have **mixed** elasticity, meaning that the vertical and horizontal elasticity can be different. By default, character data columns created using the Report Wizard have horizontally expanding and vertically fixed elasticity, while columns with NUMBER and DATE data types are horizontally and vertically fixed.

To make adjacent report columns, such as the instructor first and last name, appear directly next to each other on the report regardless of the width of the individual data fields, you must adjust the columns' horizontal elasticity to variable. Now you will change the horizontal elasticity for the F_ffname, F_flname, F_bldg_code, and F_room columns to variable.

To adjust the column elasticity:

1 If necessary, open the Layout Model window.

2 Click **F_ffname**, press the **Shift** key, then click **F_flname**, **F_bldg_code**, and **F_room** to select the columns as an object group.

3 Click **Tools** on the menu bar, then click **Property Palette** to open the group Property Palette. Change the Horizontal Elasticity property to **Variable**, and then click anywhere on the Property Palette to save the change.

4 Close the Property Palette. Note that a diamond symbol now appears on the top and bottom edges of all the selected data fields to indicate that the horizontal elasticity is now variable.

help

If diamond symbols do not appear on the top and bottom edges of all four columns, open the Property Palette of each column that does not have the diamond shape at the top and bottom and change its Horizontal Elasticity property to Variable.

5 Run the report. The formatted report is displayed in the Live Previewer, as shown in Figure 7-47.

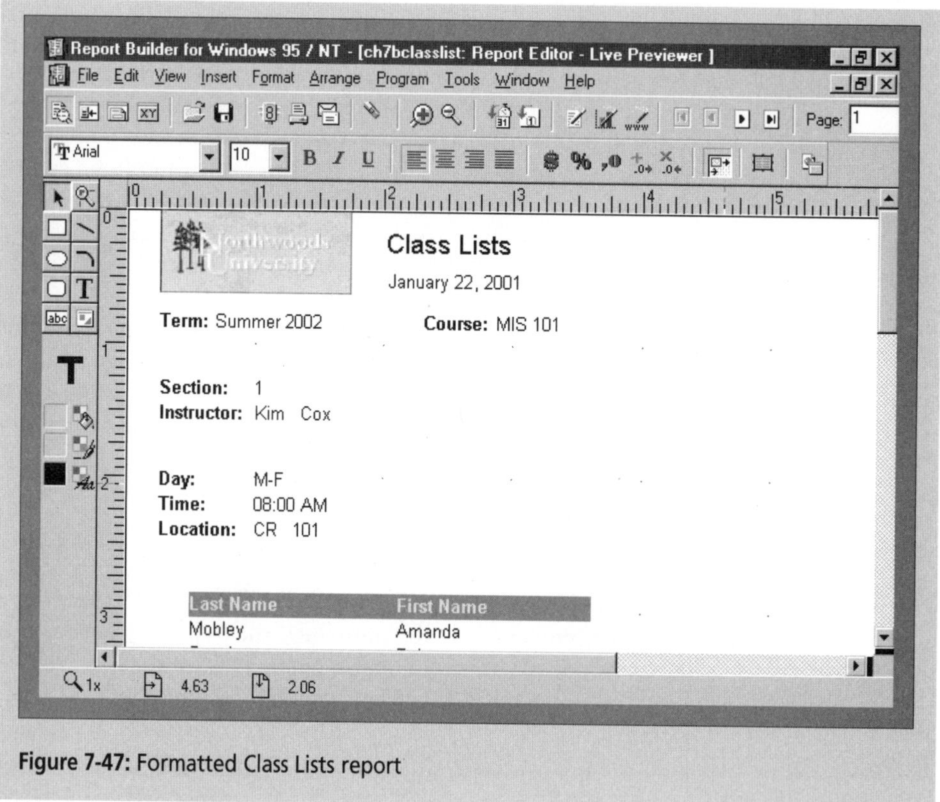

Figure 7-47: Formatted Class Lists report

Creating a User Parameter

Currently the Class Lists report shows class lists for only the Summer 2002 term. To make the report more flexible, users should be able to select from a list of term descriptions in the TERM table, and the report should display the associated class lists for any term. To implement this, you will create a **user parameter** that displays a list of possible report input values, which will be the term descriptions from the TERM table. Then you will modify the SQL query in the data model to use the selected term as a search condition.

It is always advisable to use number fields rather than character fields for search conditions in user parameters, because sometimes Oracle applications add blank spaces to pad out character fields, making it difficult to get an exact match. Therefore, you will use TERMID instead of TDESC for the search condition in the report's SQL query. However, the user will select from a list of term descriptions, and the associated term ID will be used as the search condition in the report. First, you will create the user parameter and specify the SQL query to display all of the term descriptions from the TERM table. After you create the user parameter, you will change its name. This name is used on the display presented to the user, so it should describe the contents of the parameter list. You will also change the user parameter's data type, which must match the data type of the TERMID field used in the report query search condition.

To create the user parameter:

1 Click **Window** on the menu bar, then click **Object Navigator.**

2 Click [+] beside **Data Model**, then click **User Parameters.**

3 Click the **Create** button on the Object Navigator toolbar to create a new user parameter. A new parameter is displayed.

4 Click the new parameter, then click it again if necessary so that the background of the parameter name turns blue. Change the name of the new user parameter to **TERM_DESCRIPTION.** This is the name that will appear beside the list of term descriptions that the user will select from, so it should describe the list contents. When the selection list is presented to the user, this parameter will appear as **Term Description.** It will be displayed using mixed case letters, and the underscore will be treated as a space.

5 Open the **TERM_DESCRIPTION** parameter's Property Palette by double-clicking the **User Parameter** icon beside TERM_DESCRIPTION. You will change the user parameter's properties to match those of the TERMID field in the TERM table, since this is the search field that will be passed to the report's SQL query.

6 Make sure the Data Type is Number, and change the Width property to **5** to match the properties of the TERMID field in the TERM table.

7 Click the space next to the List of Values property, then click the button that appears. The Parameter List of Values dialog box is displayed.

The Parameter List of Values dialog box lets you specify the list of values that the user can choose from for the report input. You can enter a static list of predetermined input values, or you can enter a SQL query that returns a list of values. You will use a SQL query that retrieves term IDs and term descriptions. The first field returned by the query must be the user parameter field (TERMID) that will be used as a search condition in the report's data model query. You will select the option of not displaying the TERMID in the user parameter list of values, so the selection list presented to the user will display only the second field returned by the query, which is the term description (TDESC). Now you will specify the user parameter list of values.

To specify the user parameter list of values:

1 Click the **SELECT Statement** option button.

2 Type the query shown in Figure 7-48 in the SQL Query Statement text box.

type this query

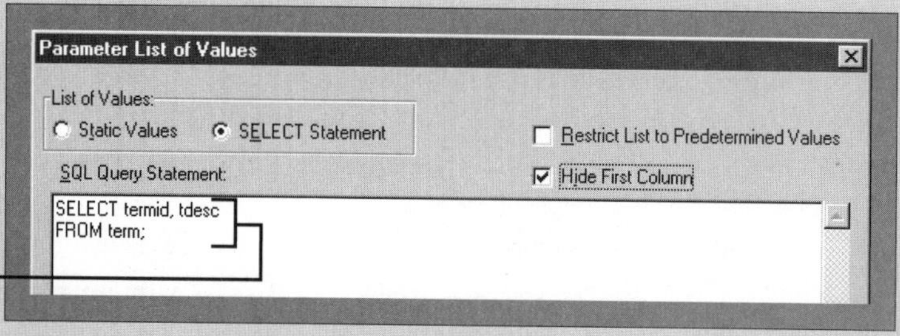

Parameter List of Values

List of Values: Static Values SELECT Statement

Restrict List to Predetermined Values

SQL Query Statement:

Hide First Column

```
SELECT termid, tdesc
FROM term;
```

Figure 7-48: User parameter SQL query

3 Clear the **Restrict List to Predetermined Values** check box, because you want the list to change dynamically as new terms are added to the TERM table.

4 Check the **Hide First Column** check box to hide the term ID in the list of values. Your completed Parameter List of Values dialog box should look like Figure 7-48.

5 Click **OK**, close the Property Palette, and then save the report.

Now you need to modify the report query to retrieve only the records for the term selected by the user and assigned to the user parameter. You will modify the SQL command by modifying the search condition to specify that TERMID equals the value of the user parameter, rather than the value "Summer 2002." To reference an input parameter in a query, you preface the parameter name with a colon, using the following general format `:<parameter name>`. The TERM_DESCRIPTION user parameter will be referenced as `:TERM_DESCRIPTION`.

To modify the report query:

1 In the Object Navigator window, double-click the **Data Model** icon to open the Data Model, and then double-click **Q_1** to view the data model SELECT statement.

2 Modify the SELECT statement as shown in Figure 7-49 so that it uses the TERMID value specified by the TERM_DESCRIPTION user parameter as the search condition value.

modify this line

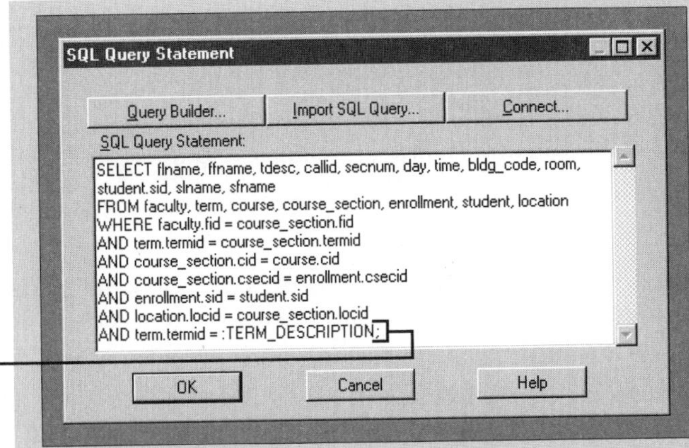

Figure 7-49: Modifying the data model query

3 Click **OK**.

4 Save the report and then view it in the Live Previewer. The Parameter Form window displays the user parameter input screen with a list arrow for selecting term descriptions.

5 Select the **Fall 2001** term and then click the **Run Report** button to confirm that the correct class lists are generated.

help

No reports will be generated for the Fall 2000, Spring 2001, and Summer 2001 terms, because there are no course section records for these terms in the database.

6 Close the report.

7 Exit Report Builder.

SUMMARY

- When you create a report, Report Builder organizes the data fields returned by the SQL query into one or more record groups, which are sets of records with the same field headings, but different data values.
- A new record group is created wherever there is a master-detail relationship in the records returned by a query.
- When you create a report using the Report Wizard, the default name for a record group is G_<field name of the first field in the group>.
- Frames are containers for grouping report record group data items and column headings.
- Reports can have header frames, repeating frames, and group frames. Header frames

group all of the column headings for a record group, repeating frames group the data items in a record group that display the actual data values, and group frames group associated header and repeating frames together.

- A header frame is only created when report data are displayed in columns with column headings on the first row and multiple records displayed below them.
- To avoid Report Builder errors, make sure that frames completely enclose the objects inside them.
- An object's parent frame is the frame that directly encloses the object. A good way to select a frame is to select one of its enclosed objects, then click the Select Parent Frame button.
- Frame names are derived from the record groups they enclose. The general format for the name of a group frame is M_<record group name>_GRPFR. The general format for the name of a header frame is M_G_<record group name>_HDR. The general format for the name of a repeating frame name is R_G_<record group name>.
- In a report with multiple group levels, frames for higher-level (more detailed) record groups are nested inside frames for lower-level (less detailed) record groups.
- To create a page break between sets of repeating records, you must open the Property Palette for the repeating frame that contains the records that you want on each page, and change its Maximum Records per Page Property to 1.
- When a report is in confine mode, you cannot move an object out of its enclosing frame.
- Flex mode automatically makes an enclosing frame larger when you move an enclosed object beyond the enclosing frame's boundary.
- Flex mode works well for making report frames longer. It does not work well for making report frames wider, because it enlarges all of the enclosing frames, and might make some of the frames extend beyond the report body boundaries, causing an error.
- Elasticity determines whether a field's size is fixed on the printed report, or whether it can expand or contract automatically, depending on the size of the retrieved data value.
- To make data values, such as the instructor first name and last name, appear directly next to each other on the report regardless of the width of the individual data fields, you must adjust the column horizontal elasticity to variable.
- User input parameters allow users to specify inputs to generate specific reports. A user input parameter list of values can be a static list of predetermined values or a list based on a database query.

REVIEW QUESTIONS

1. List the names of the record groups that might be created when you use the following queries to create master-detail reports using the Report Wizard in Report Builder. (Note: some queries might have more than one record group.)

a. SELECT category, invid, itemsize, color, qoh
   FROM item, inventory
   WHERE item.itemid = inventory.itemid;
b. SELECT fid, flname, ffname, bldg_code, room
   FROM faculty, location
   WHERE faculty.locid = location.locid;
c. SELECT custid, last, first, cadd, city, state, zip, orderdate, cust_order.orderid, inventory.invid, itemdesc, quantity
   FROM customer, cust_order, orderline, item, inventory
   WHERE customer.custid = cust_order.custid
   AND cust_order.orderid = orderline.orderid
   AND orderline.invid = inventory.invid
   AND inventory.itemid = item.itemid;

2. List the names of each repeating frame created by the queries in Question 1.
3. List the names of each group frame created by the queries in Question 1.
4. True or false: every repeating frame has its own group frame.
5. What is the difference between confine mode and flex mode?
6. When flex mode is on, does confine mode have to be on also?
7. What is the difference between expanding elasticity and variable elasticity?
8. When should report columns have fixed elasticity?
9. How do you create a relationship between a report's user parameter and the report's SQL query?
10. How do you determine what data type and width to assign to a user parameter?
11. How is the Name property of a user parameter displayed in a report? How does Report Builder format the Name parameter when it is displayed?

PROBLEM-SOLVING CASES

Run the northwoo.sql script to refresh the Northwoods University database, and run the clearwat.sql script to refresh the Clearwater Traders database. The script files are stored in the Chapter7 folder on your Student Disk.

1. In this exercise you will create a report that lists each item category in the Clearwater Traders database and the corresponding inventory ID, item description, size, color, current price, and quantity on hand.
   a. Create a report using the Report Wizard that lists item categories and descriptions, and then lists individual inventory IDs, sizes, colors, prices, and quantities on hand, as shown in Figure 7-50. Apply the clearwater.tdf custom template or another template to the report, and save the report as Ch7bEx1.rdf.

Inventory Items by Category

22 January 2001
Page 1

Category: Children's Clothing **Description: Children's Beachcomber Sandals**

Inventory ID	*Size*	*Color*	*Price*	*QOH*
11820	10	Blue	15.99	121
11821	11	Blue	15.99	111
11822	12	Blue	15.99	113
11825	11	Red	15.99	137
11826	12	Red	15.99	134
11824	10	Red	15.99	148
11823	1	Blue	15.99	121
11827	1	Red	15.99	123

Category: Men's Clothing **Description: Men's Expedition Parka**

Inventory ID	*Size*	*Color*	*Price*	*QOH*
11845	S	Navy	199.95	114
11848	XL	Navy	209.95	12
11847	L	Navy	209.95	0
11846	M	Navy	199.95	17

Figure 7-50

b. Create a schematic diagram of this report's objects and frames that is similar to the ones in Figures 7-37 and 7-38 for the Class Lists report. Label the data fields, repeating frames, header frames, and group frames using the default names provided by the Report Wizard.

c. Move the frames enclosing the inventory records down so there is about 0.25 inches of blank space between the Category and Description fields and the inventory column headings.

d. Format the report so there is about 0.25 inches of blank space between each group of Category/Description records. (*Hint*: Select the outermost repeating frame, and change its Vertical Spacing between Frames property to .25.) Your completed report should look like Figure 7-50.

2. In this exercise you will create a report that lists each inventory item in the Clearwater Traders database that is currently back ordered, and shows the corresponding shipping and back-order information.

   a. Use the Report Wizard to create a report that lists the inventory ID, description, size, and color for every item in the BACKORDER table that has not yet been received. The report should list the backorder ID, date expected, quantity expected, and date received and quantity received. The values for date received and quantity received will be NULL, since the report only lists unreceived back orders. Apply the clearwater.tdf template or another template to the report, and save the report as Ch7bEx2.rdf. Your report output should look like the report shown in Figure 7-51.

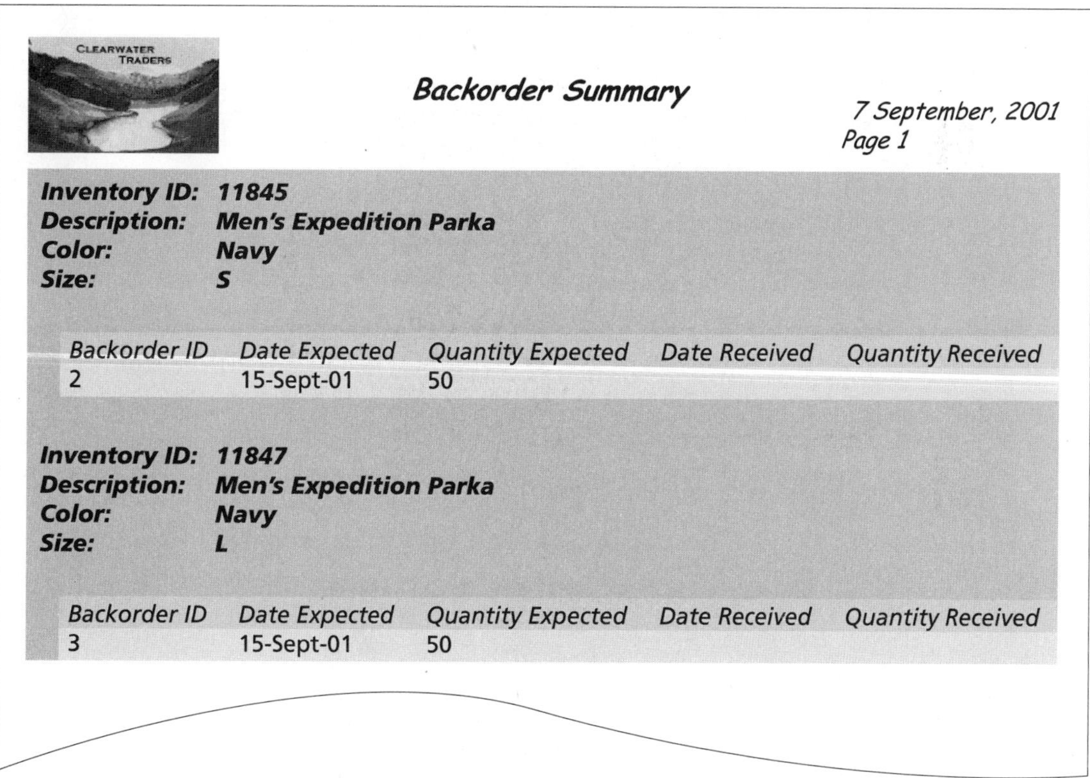

Figure 7-51

   b. Create a schematic diagram of this report's objects and frames that is similar to the ones in Figures 7-37 and 7-38 for the Class Lists report. Label the data fields, repeating frames, header frames, and group frames, using the default names provided by the Report Wizard.

   c. Format the report so there is about 0.25 inches of blank space between each inventory record. (*Hint*: Select the outermost repeating frame, and change its Vertical Spacing between Frames property to .25.)

   d. Indent the back order data about 0.25 inches from the left edge of the report.

   e. Enlarge the report frames and stack the inventory columns so the report output looks like Figure 7-51.

3. In this exercise you will create a report that lists each building code and room number in the Northwoods University database, and shows the room usage according to course sections during each term.

a. Use the Report Wizard to create a report like the one shown in Figure 7-52 that displays building codes and room numbers followed by term descriptions, and then displays the course call IDs, section numbers, days, and times that the room is used during the term. Apply the clearwater.tdf template or another template to the report, and save the report as Ch7bEx3.rdf.

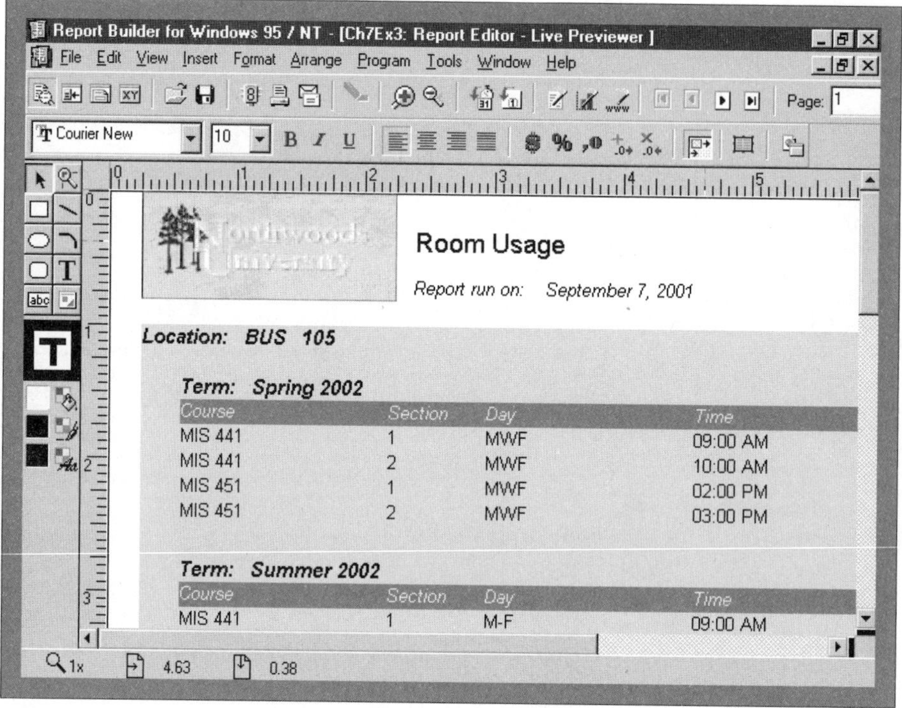

Figure 7-52

b. Format the report so that the data for each room appear on a separate page.

c. Change the elasticity of the building code and room fields so that they appear directly next to each other.

d. Modify the other components of the report so that the output looks like Figure 7-52.

4. Modify the report you created in Exercise 3 by creating a user parameter so that the user can select a specific building code and room combination, and view only the output for selection.

a. Open Ch7Ex3.rdf, and save it as Ch7bEx4.rdf.

b. Create a user parameter named LOCATION that displays building codes and room numbers for all locations in the LOCATION table, and returns LOC_ID as the parameter value.

c. Modify the report query to use the LOCATION parameter value as the search condition for LOCID.

5. Create a report that provides a schedule of the courses that are offered at Northwoods University. The report should list the term, then show each course call ID, name, and credits, and then show the section ID, section number, instructor, day, time, building code, and room for each associated course section. Create the report using the Report Wizard, and then format the report so that the output looks like Figure 7-53, with the following specifications:
    a. The output for each term should start on a new page.
    b. There should be about 0.25 inches of blank space between the data for each course.
    c. The location columns (building code and room) should appear directly beside each other.
    d. Save the report in a file named Ch7bEx5.rdf in the Chapter7 folder on your Student Disk.

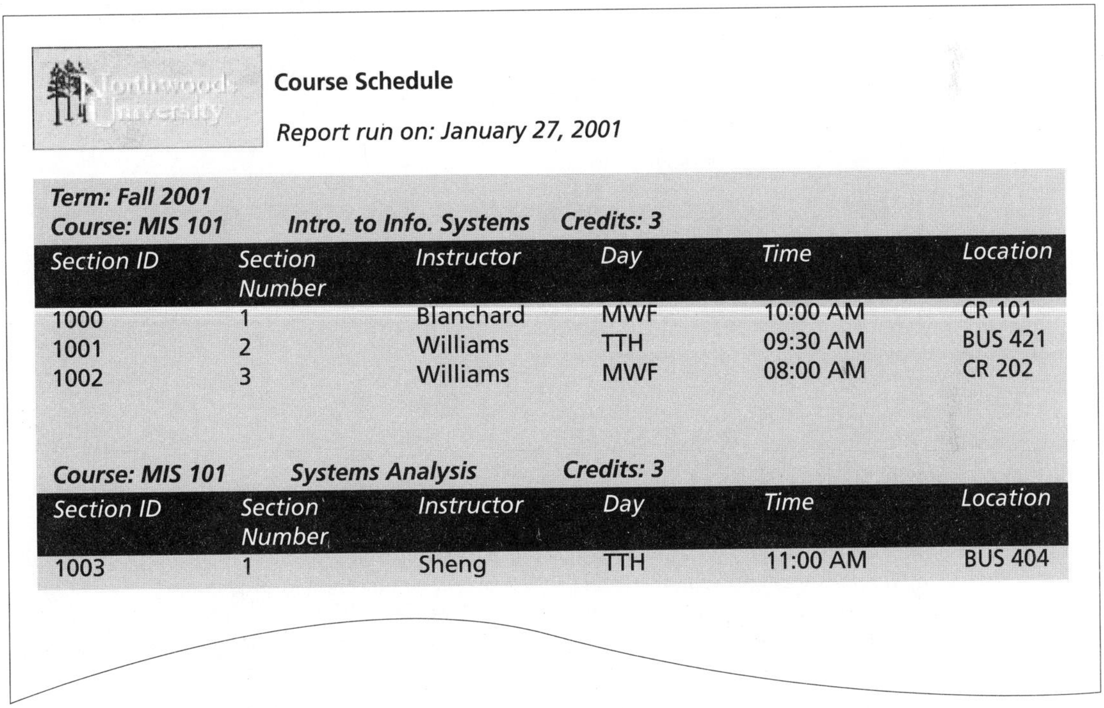

Course Schedule

Report run on: January 27, 2001

Term: Fall 2001

Course: MIS 101 ***Intro. to Info. Systems*** ***Credits: 3***

Section ID	*Section Number*	*Instructor*	*Day*	*Time*	*Location*
1000	1	Blanchard	MWF	10:00 AM	CR 101
1001	2	Williams	TTH	09:30 AM	BUS 421
1002	3	Williams	MWF	08:00 AM	CR 202

Course: MIS 101 ***Systems Analysis*** ***Credits: 3***

Section ID	*Section Number*	*Instructor*	*Day*	*Time*	*Location*
1003	1	Sheng	TTH	11:00 AM	BUS 404

Figure 7-53

6. Create a grade report for students at Northwoods University that lists the student's complete name and address, and then shows the term description, course call ID, course name, and grade. Only include courses where a grade has been assigned. Create the report using the Report Wizard, and format the report so that the output looks like Figure 7-54, with the following specifications:
    a. The output for each student should start on a new page.
    b. There should be about 0.25 inches of blank space between the records for each term.
    c. The student first name, middle initial, and last name fields should appear directly adjacent to each other. Similarly, the student's city, state, and ZIP code should appear directly adjacent to each other.

d. Save the report in a file named Ch7bEx6.rdf in the Chapter7 folder on your Student Disk.

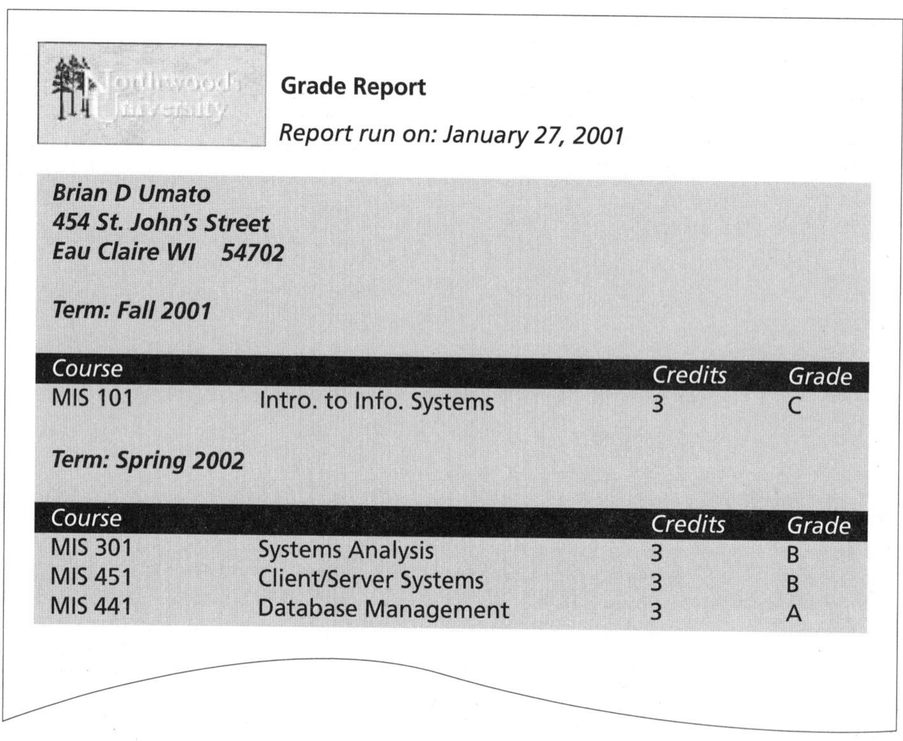
Grade Report

Report run on: January 27, 2001

Brian D Umato
454 St. John's Street
Eau Claire WI 54702

Term: Fall 2001

Course		*Credits*	*Grade*
MIS 101	Intro. to Info. Systems	3	C

Term: Spring 2002

Course		*Credits*	*Grade*
MIS 301	Systems Analysis	3	B
MIS 451	Client/Server Systems	3	B
MIS 441	Database Management	3	A

Figure 7-54

7. Modify Exercise 6 by creating a user parameter to allow the report to be run for only one student at a time. Use SID as the search condition, and display the student's first and last name in the parameter list of values. Order the student names alphabetically by last name. Save the report in a file named Ch7bEx7.rdf in the Chapter7 folder on your Student Disk.

LESSON C

objectives

- Create summary and formula columns in a report
- Create an executable report file that users can run independently of Report Builder
- Run a report from a form and pass parameters from the form to customize report output

The Northwoods University Student Transcript Report

In this lesson, you will work with a report that displays transcript information for students at Northwoods University, as shown in Figure 7-55.

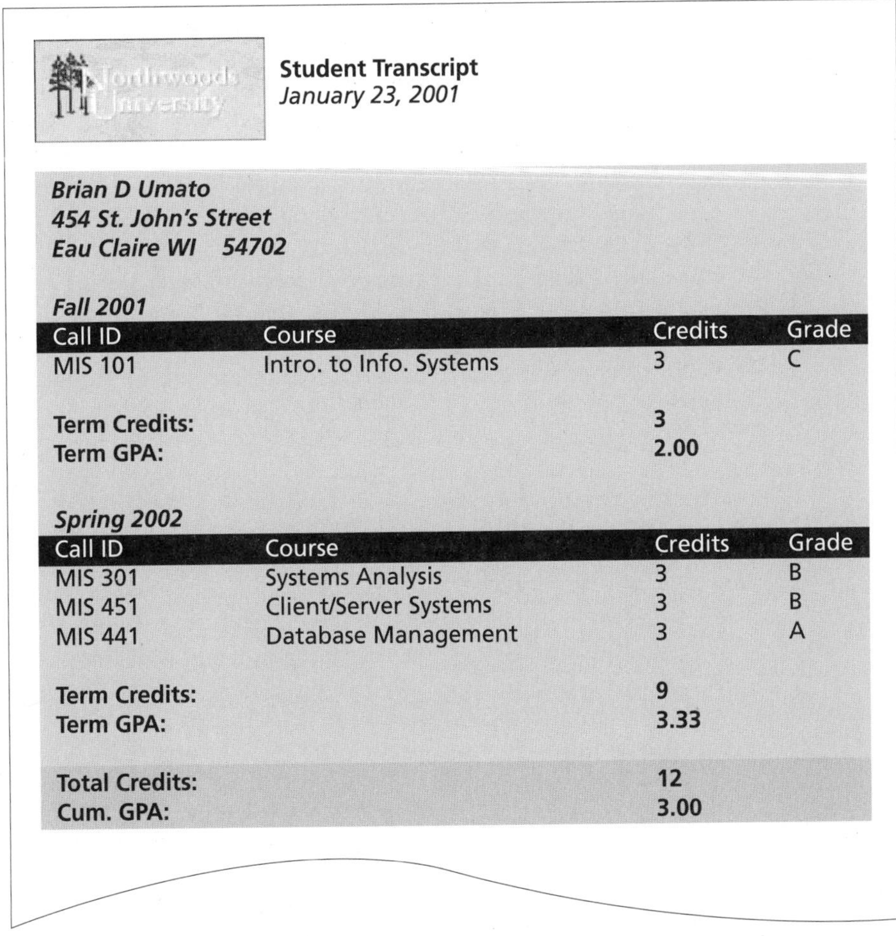

Student Transcript
January 23, 2001

Brian D Umato
454 St. John's Street
Eau Claire WI 54702

Fall 2001

Call ID	Course	Credits	Grade
MIS 101	Intro. to Info. Systems	3	C
Term Credits:		**3**	
Term GPA:		**2.00**	

Spring 2002

Call ID	Course	Credits	Grade
MIS 301	Systems Analysis	3	B
MIS 451	Client/Server Systems	3	B
MIS 441	Database Management	3	A
Term Credits:		**9**	
Term GPA:		**3.33**	
Total Credits:		**12**	
Cum. GPA:		**3.00**	

Figure 7-55: Student transcript report

The student's name and address are displayed at the top of the report, followed by the term description, and a listing of the courses, course credits, and grade received for each course. Course credits are summed for each term, and the student's total credits are also displayed. The student's grade point average (GPA) is calculated for each term, and the student's cumulative GPA is also displayed. First you will refresh your Northwoods University database tables by running the northwoo.sql script in SQL*Plus. Then you will start Report Builder and open the student transcript report.

To refresh your database tables and open the report:

1. Start SQL*Plus and run the **northwoo.sql** script file from the Chapter7 folder on your Student Disk.
2. Start Report Builder, and open **transcript.rdf** from the Chapter7 folder on your Student Disk. Save the file as **ch7ctranscript.rdf**. Run the report, connecting to the database as usual.

Currently the transcript report displays the student name and address, the term description, and the course and course grade data. Currently the student's first and last names are not adjacent to each other, and the student address overlaps with the course section header labels. The final report formatting has not been done yet, because whenever you make a modification using the Report Wizard, custom formatting—such as changing the size and spacing of fields in frames or adjusting the elasticity property of data fields—is lost. Whenever you create a report, it is best to make sure that all of the report data are displayed correctly before performing the final formatting. In this lesson, you will create the fields to sum the total credits and calculate the GPA. First you will view the report's data model.

To view the data model:

1. In the Live Previewer window, click the **Data Model** button on the toolbar. The Data Model Editor opens.
2. Rearrange the data model objects so that they are all visible on the screen, as shown in Figure 7-56.
3. Save the report.

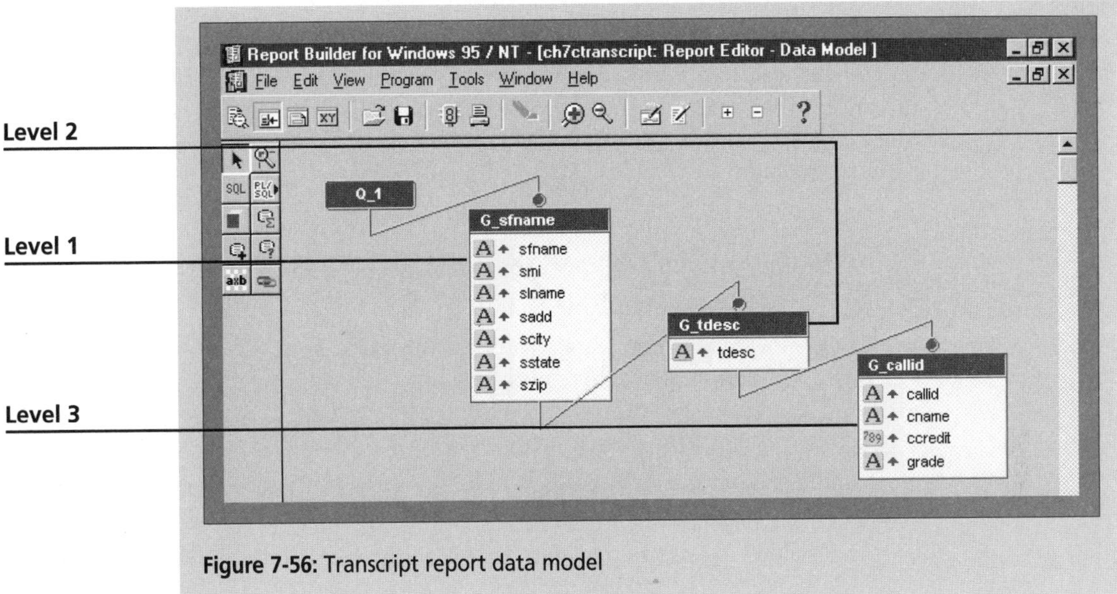

Figure 7-56: Transcript report data model

The Data Model shows that the report has three record groups. The Level 1 group, G_sfname, contains the student information. The Level 2 group, G_tdesc, shows the term description, and the Level 3 group, G_callid, displays the course and grade information.

Creating a Report Summary Column

The first modification you will make is to create a summary column to sum the term and overall credits. Recall that a **summary column** is a report column that summarizes data from other report columns by computing totals, such as the sum and average of a series of repeating data fields. You can easily create a summary column for any numerical report data field, using the Totals page in the Report Wizard. Now you will open the Report Wizard in re-entrant mode and examine the Totals page.

To examine the Totals page:

1. Click **Tools** on the menu bar, and then click **Report Wizard** to open the Report Wizard in re-entrant mode. The Report Wizard is displayed with tabs for each page across the top of the window.
2. Click the **Totals** tab. The Totals page is displayed.

The Totals page displays all of the report fields in the Available Fields column. To create a report summary column that summarizes the data that are displayed in a report field, you select the field from the Available Fields list, and then click the button that describes the function you would like to apply to the selected field: Sum, Average, Count, Minimum, Maximum, or % Total. Now you will create the summary column that sums the student credits.

To create the summary column to sum student credits:

1. Scroll down in the Available Fields list, and then click **ccredit**.
2. Click the **Sum >** button. Sum(ccredit) is displayed in the Totals list, indicating that a summary column to sum the ccredit field has been created.
3. Click **Finish** to close the Report Wizard and save the change. Make sure you are on Page 1 of the report, then scroll down if necessary to view the new summary columns. You will see totals for each term and for each student.
4. Select the new totals' labels and columns, and change the font to **10-point Arial bold.** Your formatted totals should look like Figure 7-57.

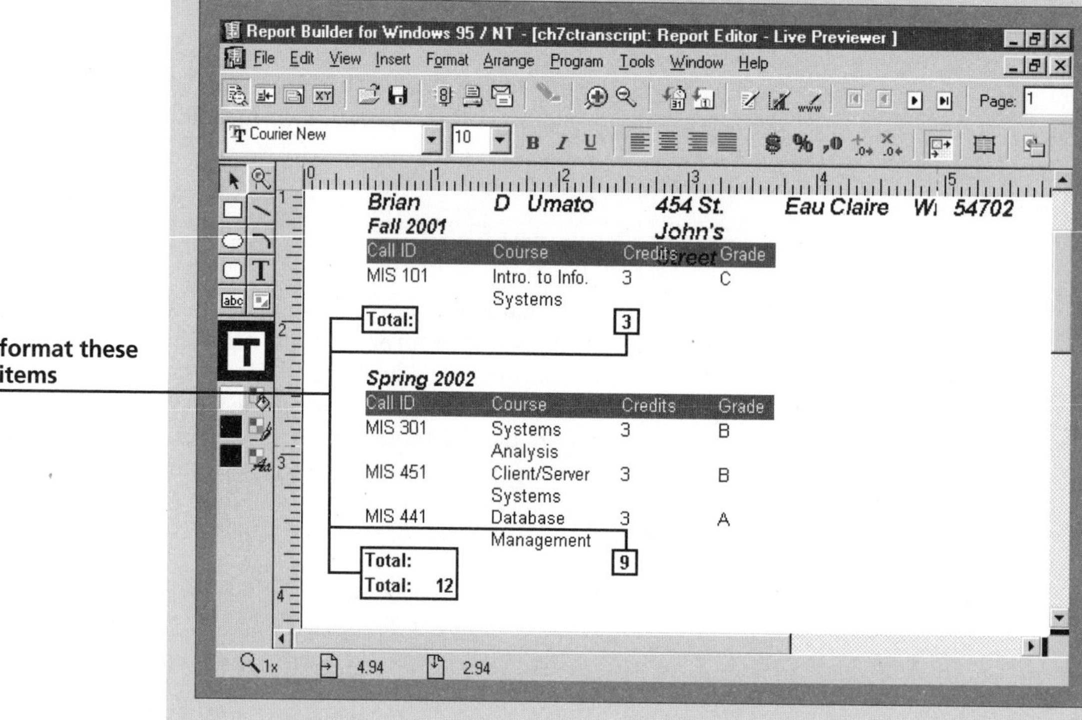

Figure 7-57: Formatting the summary columns and labels

5. Click the **Last Page** button and note the credits summary column that totals all of the credits for the entire report, as shown in Figure 7-58.

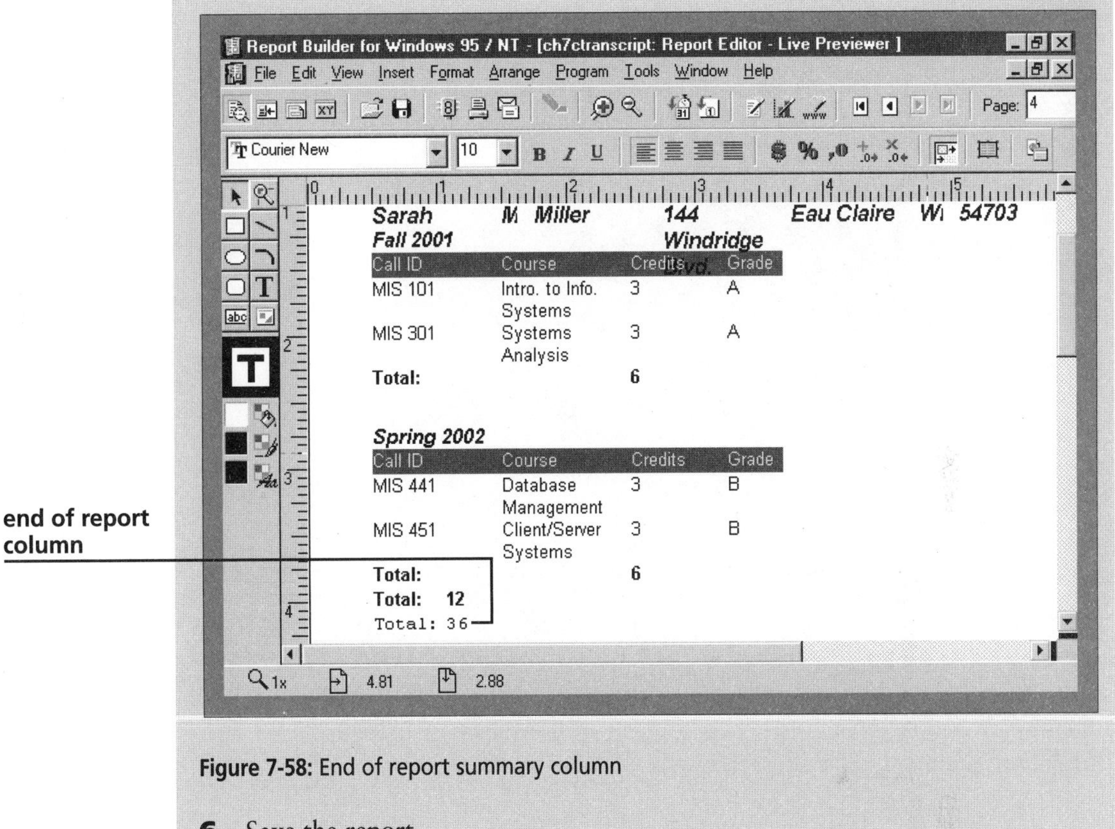

Figure 7-58: End of report summary column

6 Save the report.

When you use the Report Wizard to create a summary column, it creates a column that totals the selected field for each record group level *above* the field's record group in the report, and a column that totals the selected field for the entire report. In this report, the field being summed (CCREDIT) is in the G_callid (Level 3) record group. A new column that totals the CCREDIT values for each term is created in the G_tdesc (Level 2) record group, and a new column that totals the CCREDIT values for each student is created in the G_sfname (Level 1) record group. A new column is also created that is independent of any record group and that sums the field for the entire report.

The general format for the names of the new summary columns for each record group is `F_Sum<name of field being summed>Per<first record in associated record group>`. In this report, the summary column that sums the credits for each term is named `F_SumccreditsPertdesc`. It is displayed once per term, thus explaining the logic behind the name "Pertdesc." The totals column that sums the credits for each student is named `F_SumccreditsPersfname`, and is displayed once per student on the report. The name of the summary column for the entire report is `F_SumccreditsPerReport`, and it is only displayed once in the report. Now you will examine the report's data model and view the new summary columns.

To view the summary columns in the data model:

1 Open the Data Model and resize and reposition the record groups, if necessary, so that the names of the new summary columns are visible, as shown in Figure 7-59.

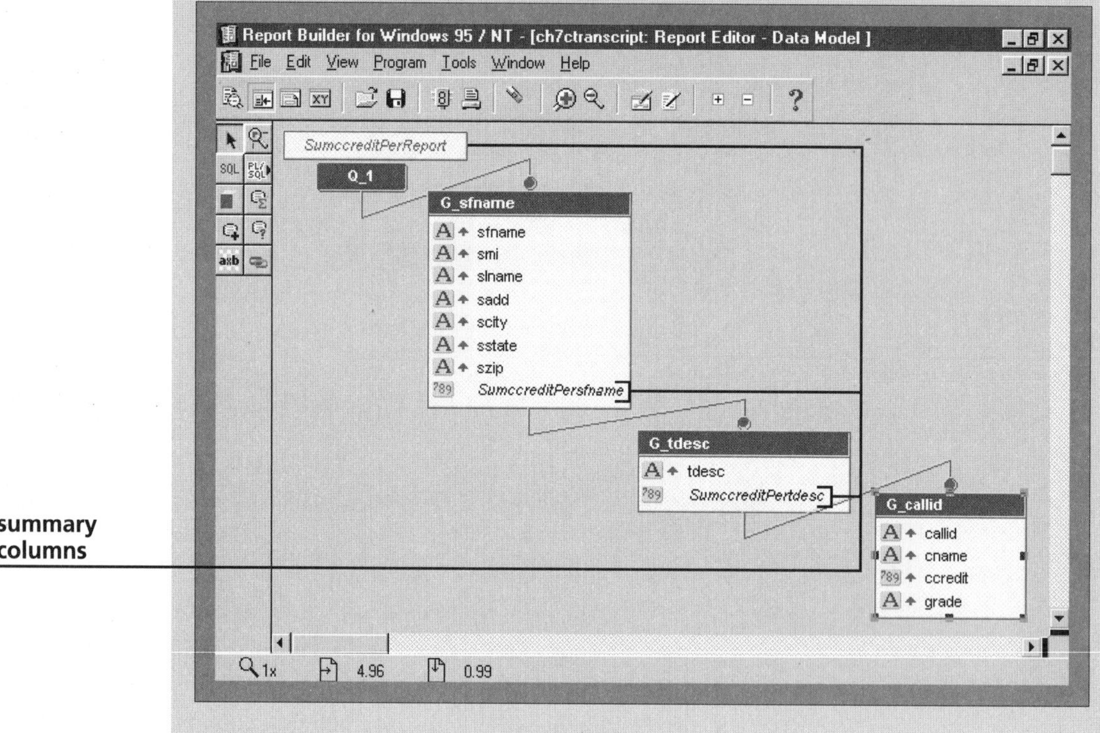

Figure 7-59: Data Model with summary columns

The Data Model shows that the per report summary column is displayed above the report query Q_1. The per sfname summary column is within the G_sfname record group, and the per tdesc summary column is within the G_tdesc record group. The most detailed record group, G_callid, does not have a summary column because this record group contains the CCREDIT field that is being summed.

Recall that record groups have repeating frames, which enclose the record group's data columns, and header frames, which enclose the record group's column headings. A record group's repeating frame and header frame are enclosed by a group frame. Summary columns are placed in a new kind of frame, called a footer frame. A record group's **footer frame** contains the summary columns that summarize the data fields in the record group and their associated labels. In this report, the call ID record group's footer frame encloses the G_tdesc summary column (SumccreditsPertdesc), which sums the CCREDIT column in the call ID record group. The G_tdesc record group's footer frame encloses the G_sfname summary column (SumccreditPersfname), and the G_sfname record group's footer frame encloses the SumccreditPerReport summary column. The general format for the name of a footer frame is `M_<record group name>_FTR`.

Now you will examine the per report and per student summary columns and footer frames in the Object Navigator to understand their relationships.

To examine the summary columns and footer frames in the Object Navigator:

1. Switch to the Object Navigator.
2. Select **CH7CTRANSCRIPT**, then click the **Collapse All** button to collapse all of the report objects.
3. Click + next to CH7CTRANSCRIPT, click + next to Layout Model, click **Main Section**, then click the **Expand All** button to expand all of the report body components. The report body components and their hierarchical relationships are displayed, as shown in Figure 7-60.

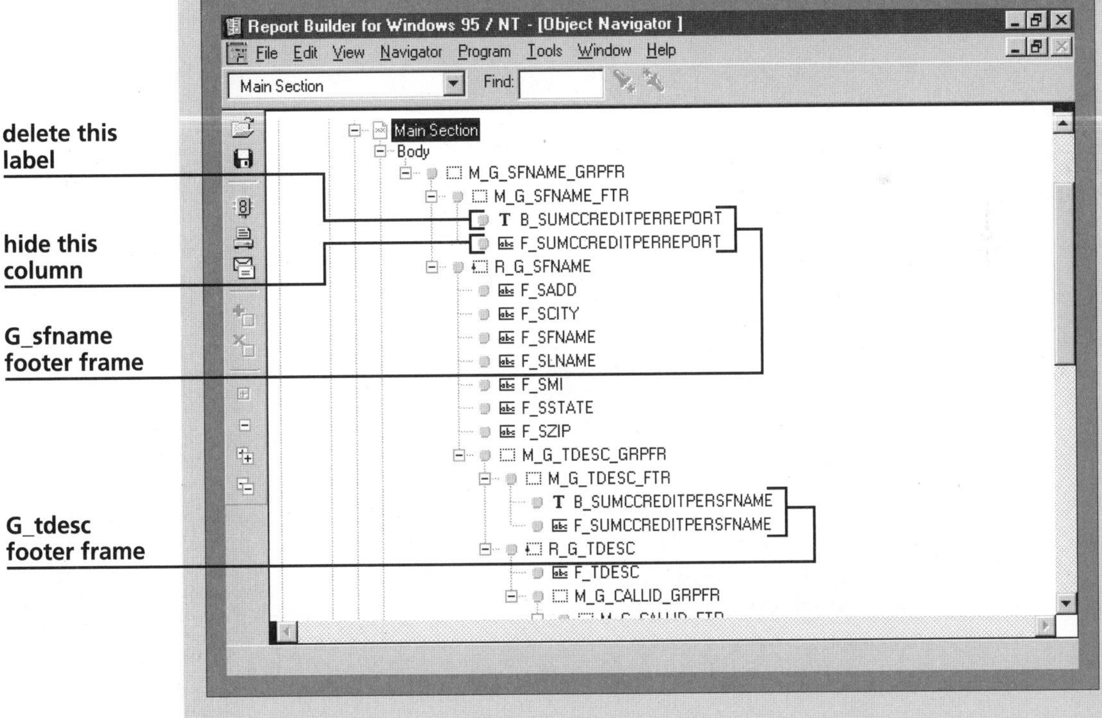

Figure 7-60: Transcript report footer frames

In the Report Builder Object Navigator, different object types can be identified by the icon that is displayed beside the object's name. [icon] denotes a group frame, [icon] denotes a repeating frame, [icon] denotes a boilerplate label (such as a column heading), and [icon] denotes a report column that displays data. Figure 7-60 shows that the G_sfname footer frame contains the per report summary column (F_SUMCCREDITPERREPORT) and its associated label (B_SUMCREDITPERREPORT), which is displayed as "Total:" in the Live Previewer. This frame is within the outermost group frame on the report, M_G_SFNAME_GRPFR. The report-level summary column sums the credits for all students shown on the report, and is displayed only once per report.

The G_tdesc footer frame contains the per student summary column (F_SUMCCREDITPERSFNAME) and its associated label. This summary column summarizes the credits for all terms for a particular student, and it is displayed once for every student. Now you will scroll down in the Object Navigator window to examine the summary column that summarizes the credits per term.

To examine the credits summary per term field:

1. Scroll down in the Object Navigator window until the G_callid footer frame (M_G_CALLID_FTR) is displayed.

The G_callid footer frame displays the "per term" summary column (F_SUMCCREDITPERTDESC) and its associated label. These items appear once for every term. Since the "per report" summary column, which sums the credits for all students, is not needed in this report, you will hide it by opening its Property Palette in the Object Navigator and setting its Visible property to No. Then you will delete its label.

To hide the per report summary column and delete its label:

1. Scroll up in the Object Navigator window, select **F_SUMCREDITPERREPORT** (see Figure 7-60), right-click, then click **Property Palette**.
2. Set the Visible property to **No**, click another property to save the change, and then close the Property Palette.
3. Click the **B_SUMCCREDITPERREPORT** boilerplate label item (see Figure 7-70), then click the **Delete** button [icon] to delete the label. Click **Yes** to confirm that you deleted the label.
4. Click the **Run** button [icon] to view the report in the Live Previewer. Navigate to the last report page and confirm that the per report total report totals column and label are no longer displayed.
5. Save the report.

Creating Formula Columns

Formula columns display the value returned by a function that performs mathematical computations on report data columns. You will use formula columns to calculate the student's grade point average for each term. Grade point average is calculated as follows:

$$\frac{\sum(\text{Course Credits} * \text{Course Grade Points})}{\sum(\text{Course Credits})}$$

Course grade points are awarded as follows: A = 4 grade points, B = 3, C = 2, D = 1, F = 0. First, you will create a formula column to calculate the total course credit points for each term, which is equal to course credits * (multiplied by) course grade points. For example, if a student received an A in a 3-credit course, the credit points would be 3 (the course credits) * 4 (the grade points for an A), or 12. To create a formula column in the report's data model, you first create the new formula column in the record group that contains the data values used in the calculations, and then you write the formula function using PL/SQL. Since the course credit points are calculated using the CCREDIT and GRADE data fields, this formula will be placed in the G_callid record group.

To create the formula column to calculate course credit points:

1. Open the Data Model.
2. Click the **Formula Column** tool on the Data Model tool palette. The pointer changes to + when you move it into the window.
3. Click the **G_callid** record group. A new formula column named *CF_1* appears in the record group, as shown in Figure 7-61. Resize the record group if necessary so that all of the data columns are displayed.

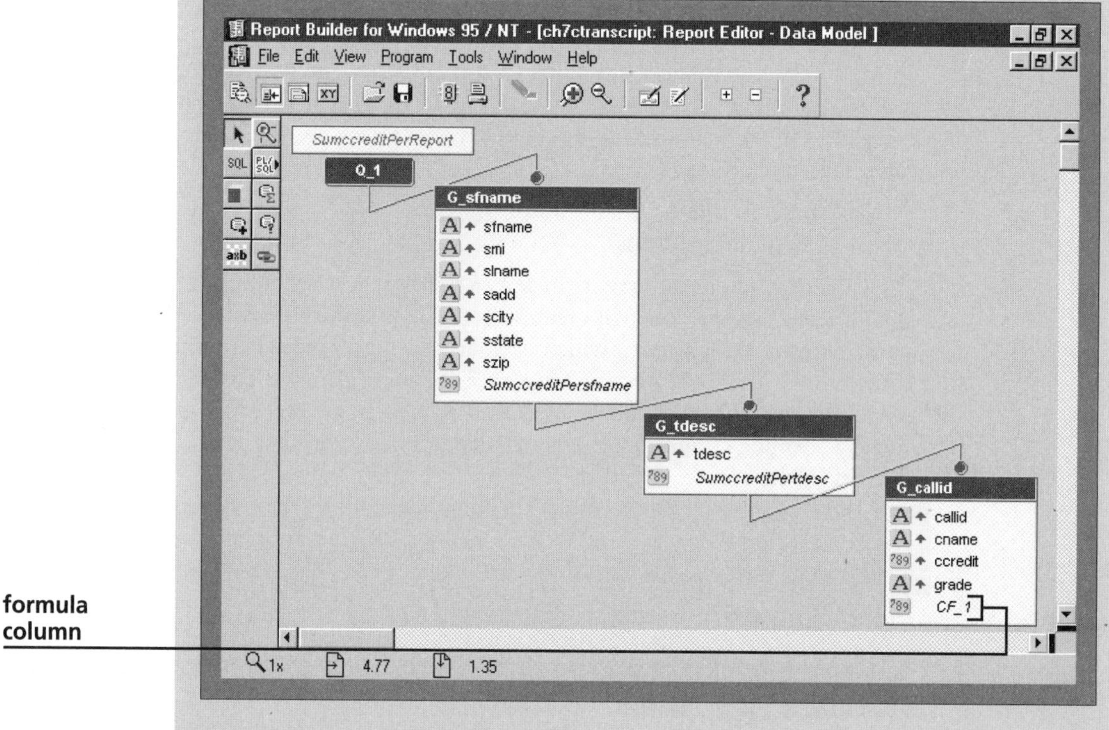

Figure 7-61: Creating a new formula column

4 Click **CF_1**, right-click, then click **Property Palette.** Change the Name property to **CF_credit_points**.

5 Click the **PL/SQL Formula** property so that a button appears, then click the button. The PL/SQL Editor opens.

Every formula column has an associated function written in PL/SQL. The function can reference values in the report's data fields, summary columns, and other formula columns. To reference a field in the PL/SQL formula column function, you simply preface the field's name (as it appears in the data model) with a colon.

Recall that a function always returns a value. To create a formula column function, you must declare a local variable in the function, perform the required calculations so that the variable is assigned to the value to be displayed in the formula column, and then use the RETURN command to instruct the function to return this value. Now you will write the PL/SQL function to calculate the course credit points.

To write the PL/SQL function to calculate the course credit points:

1 Type the code shown in Figure 7-62 in the PL/SQL Editor to return the credit points value for a specific course/grade combination.

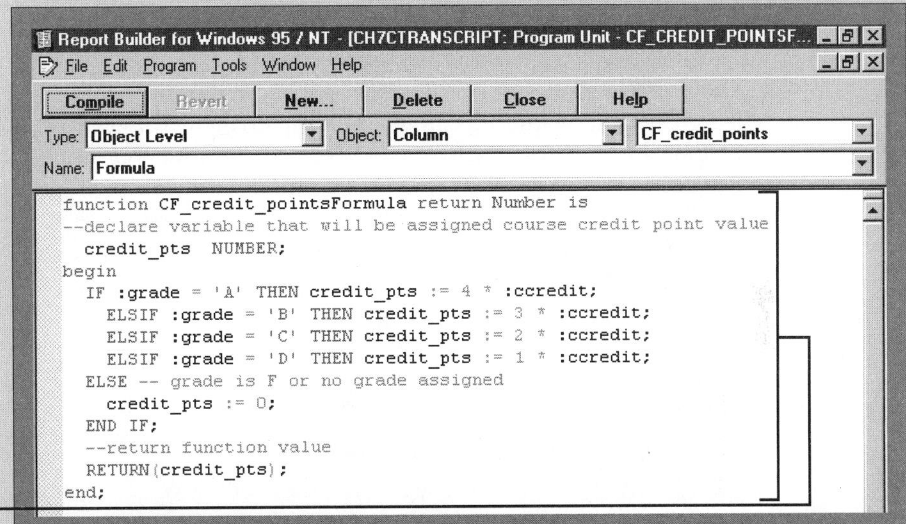


```
function CF_credit_pointsFormula return Number is
--declare variable that will be assigned course credit point value
  credit_pts  NUMBER;
begin
  IF :grade = 'A' THEN credit_pts := 4 * :ccredit;
    ELSIF :grade = 'B' THEN credit_pts := 3 * :ccredit;
    ELSIF :grade = 'C' THEN credit_pts := 2 * :ccredit;
    ELSIF :grade = 'D' THEN credit_pts := 1 * :ccredit;
  ELSE -- grade is F or no grade assigned
    credit_pts := 0;
  END IF;
  --return function value
  RETURN(credit_pts);
end;
```

Figure 7-62: Function to calculate course credit points

2 Compile the code, correct any syntax errors, close the PL/SQL Editor, and then close the Property Palette.

3 Save the report.

Currently, the formula column exists in the data model but is not displayed anywhere on the report. To display a formula column on a report, you must draw a new report data field in the Layout Model, using the **Field** tool, and then set the new field's Source property to the formula column.

Next, you will create a special kind of a field called a holding field to hold the value of CF_credit_points for each course record. A **holding field** is not displayed on the final report, but serves to hold the value of a calculated or summed column that will be used in another computation, so that you can check its value and confirm that it is calculated correctly. Fields that display formula column values must be in the same repeating frame as the data value in the formula that is in the highest-level record group. The data values used in the formula are GRADE and CCREDIT, which are in the G_callid (Level 3) record group. Therefore, the formula column must be in the R_G_callid repeating frame also. Now you will open the report Layout Model, make the R_G_callid repeating frame wider, draw a holding field on the painting region, and then assign its data source as the formula column you just created. First, you will make the R_G_callid repeating frame wider.

Recall that when you made a report frame longer in a previous lesson, you used flex mode to automatically enlarge the surrounding frames. However, recall that flex mode does not work very well when you need to make a frame wider, because it might make the surrounding frames extend beyond the boundaries of the report body, causing an error message to appear when you view the report in the Live Previewer. To make the R_G_callid repeating frame and the frames that enclose it wider, you will widen each frame manually, so you can ensure that the enclosing frames stay within the report body boundary.

The R_G_callid repeating frame is surrounded by the term description group frame (M_G_tdesc_GRPFR), the term description repeating frame (R_G_tdesc), and the call ID group frame (M_G_callid_GRPFR). You will select these frames individually in the layout model and make them wider one at a time. First you will make the outermost frame (the term description group frame) wider. Then you will view the report in the Live Previewer to make sure that no error messages are generated as a result of resizing the frame.

To make the term description frame wider:

1. Open the Layout Model.
2. Right-click below the report objects, and make sure that confine mode is on and flex mode is off.
3. Click **F_grade**, then click the **Select Parent Frame** button to select the R_G_callid repeating frame. Open the Property Palette to confirm that R_G_callid is the name of the selected frame, then close the Property Palette.
4. Click again to select the R_G_callid frame's parent frame, which is the call ID group frame (M_G_callid_GRPFR). Open the Property Palette to confirm that this is the name of the selected frame, then close the Property Palette.
5. Click again to select the M_G_callid_GRPFR frame's parent frame, which is the term description repeating frame (R_G_tdesc). Open the Property Palette to confirm that this is the name of the selected frame, then close the Property Palette.
6. Click again to select the R_G_tdesc frame's parent frame, which is the term description's group frame (M_G_tdesc_GRPFR). Open the Property Palette to confirm that this is the name of the selected frame, then close the Property Palette.
7. Make the M_G_tdesc_GRPFR about 1 inch wider by selecting its lower-right selection handle, and then dragging it toward the right edge of the screen. Your layout should look like Figure 7-63.

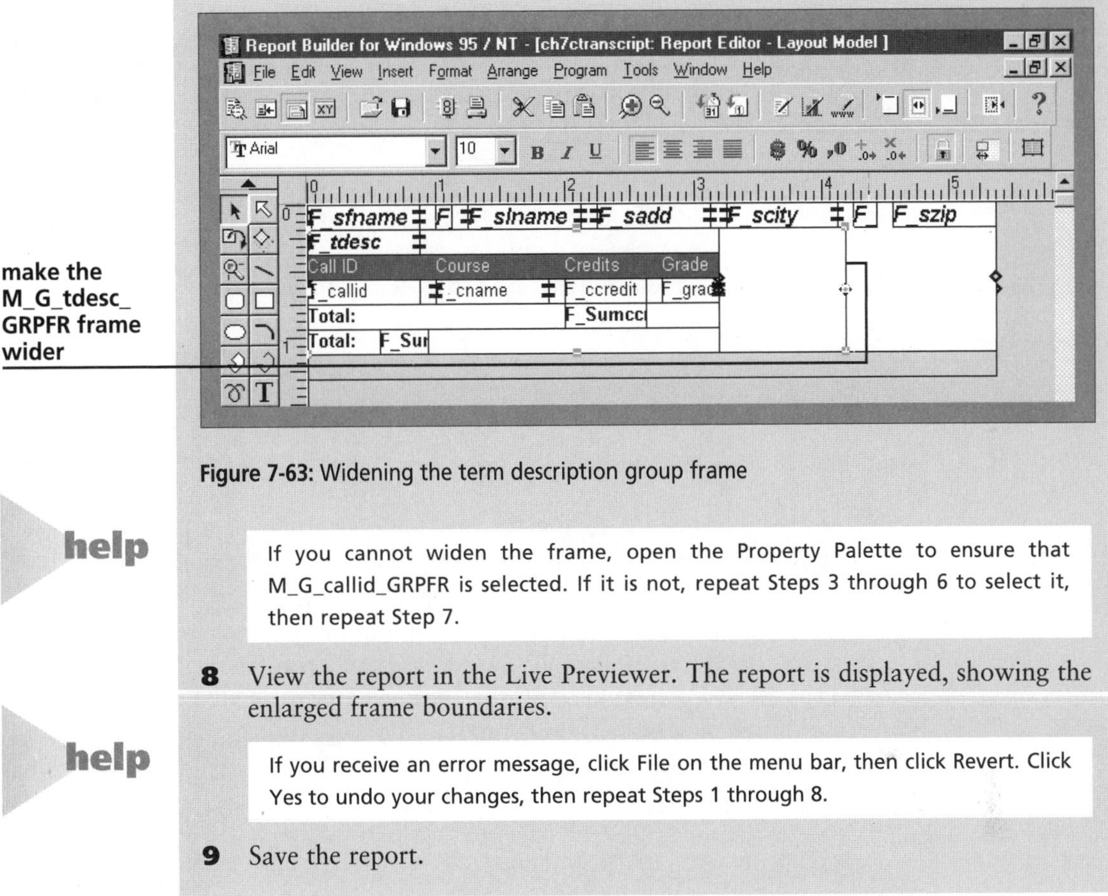

Figure 7-63: Widening the term description group frame

help

If you cannot widen the frame, open the Property Palette to ensure that M_G_callid_GRPFR is selected. If it is not, repeat Steps 3 through 6 to select it, then repeat Step 7.

8 View the report in the Live Previewer. The report is displayed, showing the enlarged frame boundaries.

help

If you receive an error message, click File on the menu bar, then click Revert. Click Yes to undo your changes, then repeat Steps 1 through 8.

9 Save the report.

Now you need to make the term description repeating frame, call ID group frame, and call ID repeating frames wider. You will do this next.

To widen the frames:

1 Open the Layout Model, and then repeat Steps 3 through 5 in the previous set of steps to select the term description repeating frame (R_G_tdesc).

2 Make the R_G_tdesc repeating frame about 1 inch wider by selecting its lower-right selection handle, and then dragging it to the right edge of the screen. Make sure that the term description repeating frame (R_G_tdesc) is still enclosed by and does not overlap the term description group frame (M_G_tdesc_GRPFR). Your layout should look like Figure 7-64.

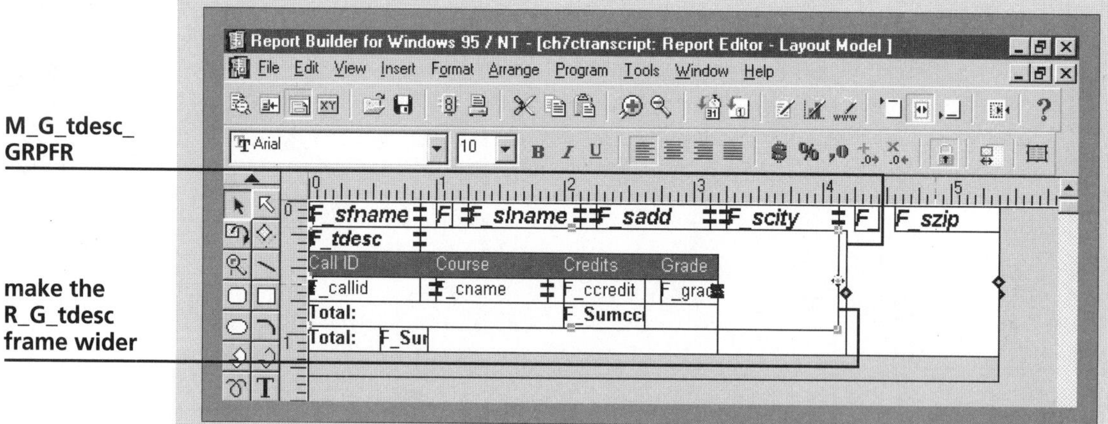

Figure 7-64: Widening the term description repeating frame

3 View the report in the Live Previewer and confirm that no error messages are generated as a result of resizing the frame, then save the report.

4 Open the Layout Model, select the call ID group frame (M_G_callid_GRPFR), using the process described in Steps 3 and 4 in the previous set of steps, and then make the frame about 1 inch wider. View the report in the Live Previewer to confirm that no error messages are generated, then save the report.

5 Open the Layout Model, select the call ID repeating frame (R_G_callid), and make it about 1 inch wider. Your Layout Model should look like Figure 7-65.

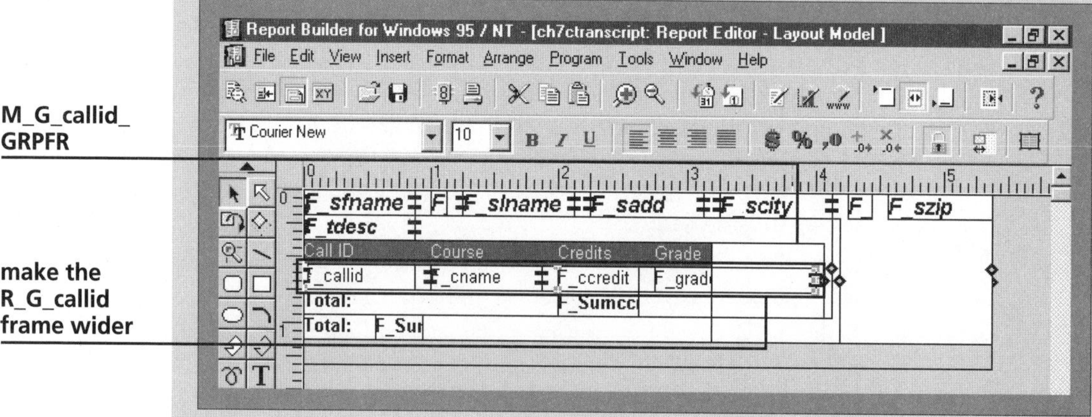

Figure 7-65: Widening the call ID group frame and repeating frame

6 View the report in the Live Previewer to confirm that no error messages are generated, then save the report.

Now you will draw the holding field in the R_G_callid repeating frame that will display the value of the credit points for each course. You will draw a data field on the painting region in the repeating frame, and then assign its data source as the credit_points formula column.

1 Open the Layout Model, click the **Field** tool [abc] on the tool palette, and draw a new field inside the R_G_callid frame, as shown in Figure 7-66.

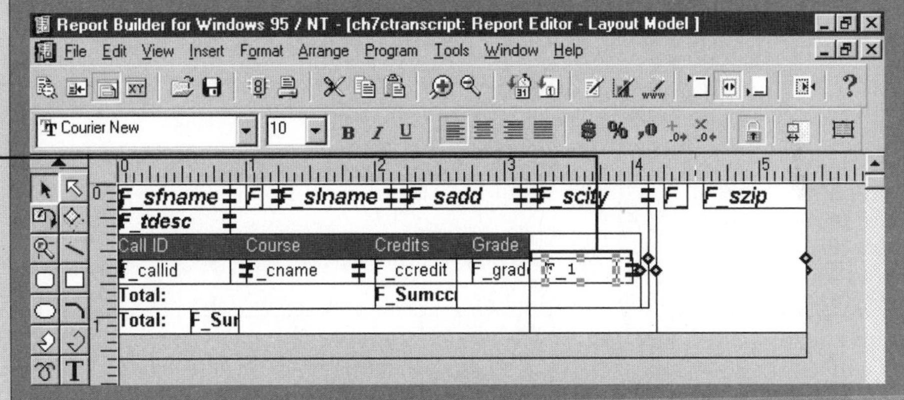

Figure 7-66: Drawing the holding field

2 Right-click on the new field, and open its Property Palette. Change the Name property to **F_credit_points.**

3 Click the **Source** property, then scroll down the list and select **CF_credit_points** as the field's data source. Note that there is a Visible property on the Property Palette sheet. Don't change it yet—you need to view the credit points value first to make sure the value is calculated correctly.

4 Click on another property to save the change, then close the Property Palette.

5 Save the report, and then view the report in the Live Previewer. Brian Umato's transcript on page 1 of the report looks like Figure 7-67, and displays the credit points formula column values correctly.

calculated credit point values (your font style might be different)

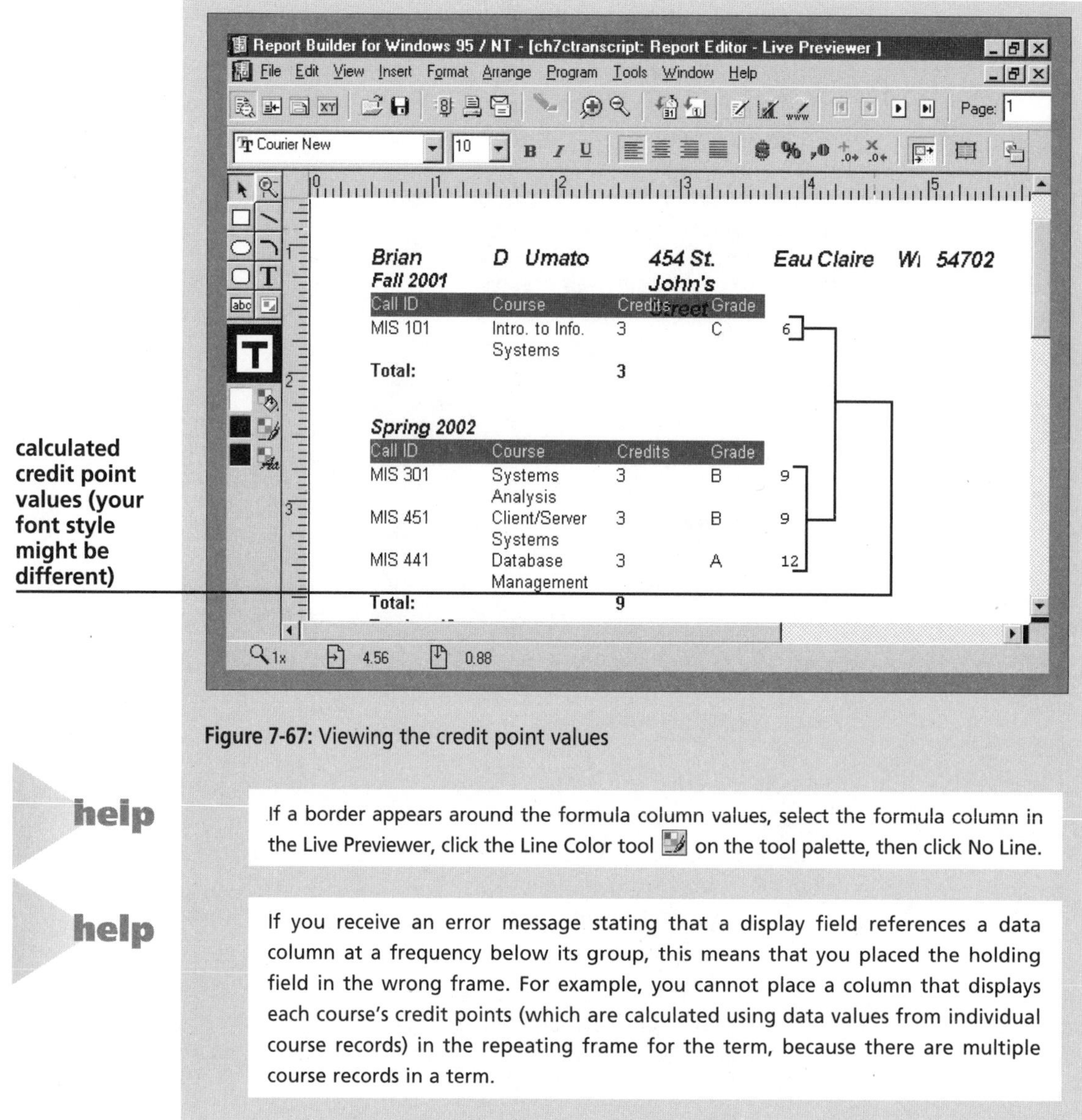

Figure 7-67: Viewing the credit point values

help If a border appears around the formula column values, select the formula column in the Live Previewer, click the Line Color tool on the tool palette, then click No Line.

help If you receive an error message stating that a display field references a data column at a frequency below its group, this means that you placed the holding field in the wrong frame. For example, you cannot place a column that displays each course's credit points (which are calculated using data values from individual course records) in the repeating frame for the term, because there are multiple course records in a term.

The next step is to sum the credit points for each term. You can create a summary column to sum values in a formula column just as you can sum values in a data column. Recall that when you use the Report Wizard to create a summary column, totals are automatically generated for all repeating groups in the report, and new footer frames are created. This time you will create the summary column manually so that you can control its placement and frequency. To create a summary column manually, you first create the new summary column in the data model, and then you draw a field on the report layout and set the field's source property as the new summary column. Since the sum of the credit points per term will be displayed once per term, the summary column will be placed in the G_tdesc record group. Summary columns have a Reset At property, which specifies the record group level at which the summary column is reset to zero and begins summing again. You want to calculate the sum of credit points for each term, so you will set the summary column's Reset At property to G_tdesc. Now you will create the summary column in the data model.

To create the summary column manually:

1 Open the Data Model, then click the **Summary Column** tool on the tool palette. The pointer changes to + as you move it across the window.

2 Click the pointer at the bottom of the G_tdesc record group. A new summary column named *CS_1* is displayed in the G_tdesc record group. If necessary, select the new summary column and drag it down so that it is the last field in the record group.

3 Open the new summary column's Property Palette, and change the name of the new summary column to **SumCF_credit_pointsPertdesc.**

4 Click the **Source** property, and select **CF_credit_points.**

5 Since you want the summary column to calculate the total credit points for every term, set the Reset At property to **G_tdesc.**

6 Click another property to save the change, then close the Property Palette.

7 Save the report.

Next, you will draw a holding field on the report layout to display the value of the credit points per term. You will draw the holding field so that it is displayed in the M_G_callid_FTR frame directly beside the summary column that sums the credits per term.

To draw the holding field:

1 Open the Layout Model, select the **F_SumccreditPertdesc** summary column, then click the **Select Parent Frame** button to select the call ID footer frame (M_G_callid_FTR). Open its Property Palette to confirm you have selected the correct frame, then close the Property Palette.

2 Make the frame wider by dragging its right edge about 0.75 inches to the right, so it is the same width as the R_G_callid frame (see Figure 7-68.)

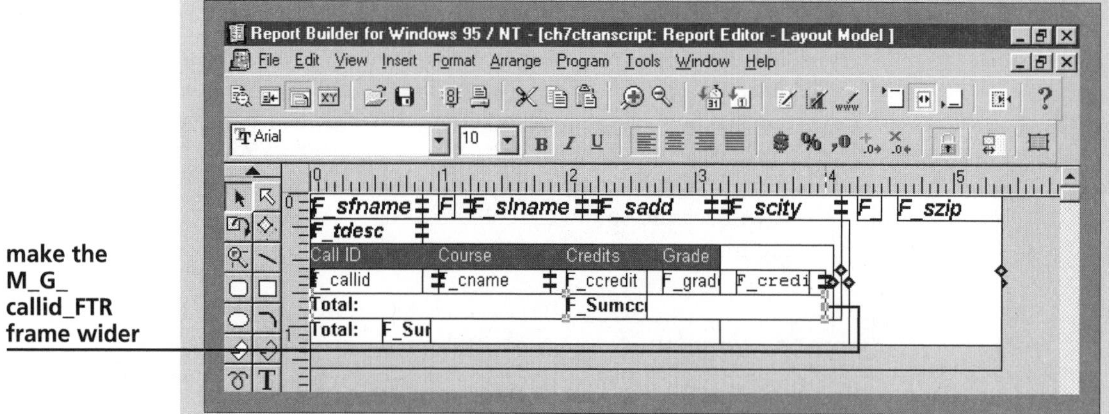

Figure 7-68: Widening the call ID footer frame

3 View the report in the Live Previewer to confirm that the frame is still enclosed by its parent frame, which is the call ID group frame, then save the report.

4 Open the Layout Model, click the **Field** tool, and draw a new field beside F_SumccreditPertdesc and directly below the F_credit_points holding field you drew before. The position of the new field is shown in Figure 7-69.

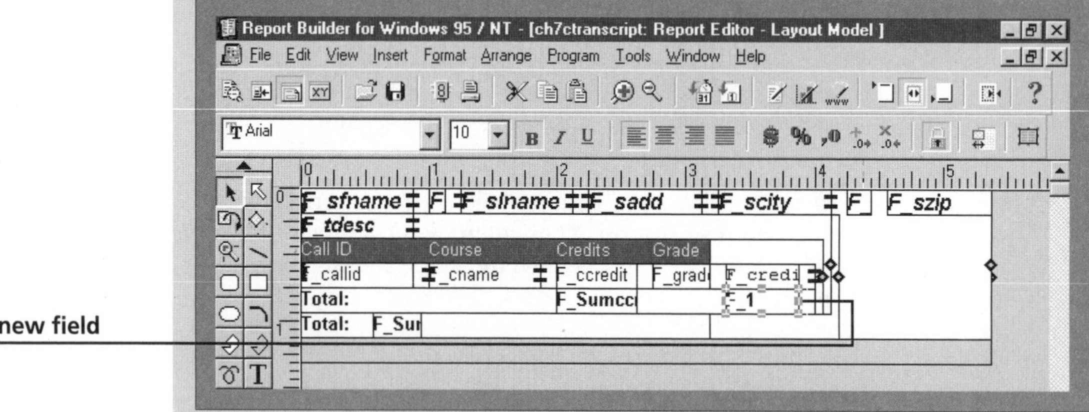

Figure 7-69: Drawing a field to display summed credit points per term

5 Open the new field's Property Palette, and change its name to **F_SumCF_credit_pointsPertdesc.**

6 Change the field's Source property to **SumCF_credit_pointsPertdesc**, click another property to save the change, then close the Property Palette.

7 View the report in the Live Previewer to confirm that the new summary column sums the credit point values correctly. The summary column should be displayed once per term, and the total credit points for Brian Umato should be 6 for the Fall 2001 term, and 30 for the Spring 2002 term, as shown in Figure 7-70.

help

If your summed credit point fields are outlined, select the fields, click the Line Color tool, then click No line.

summed credit points per term (your font style might be different)

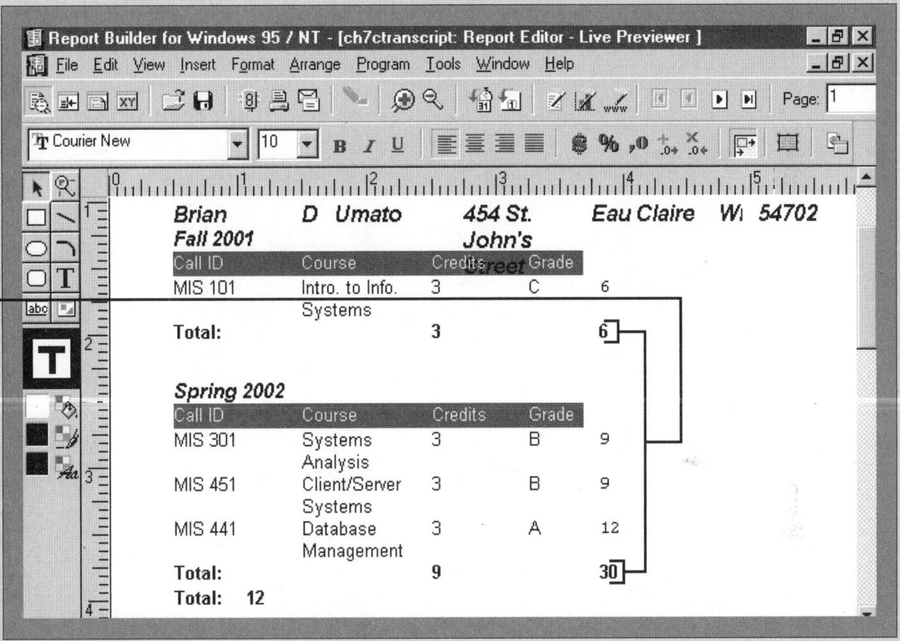

Figure 7-70: Viewing the summed credit points per term

8 Save the report.

Finally, you need to create a formula column to calculate the student grade point average for each term. In the report, this requires dividing the course credit points summary column (SumCF_credit_pointsPertdesc) by the term credits summary column (SumccreditPerTdesc). This value will be calculated once per term, so the formula column will be placed in the G_tdesc record group. Remember that to reference data model columns in a formula function, you preface the column name with a colon. Now you will create a formula column to calculate the term GPA.

To create the term GPA formula column:

1. Open the Data Model, click the **Formula column** tool, and create a new formula column in the G_tdesc record group. If necessary, make the G_tdesc record group larger so that all of the columns are displayed.
2. Open the new formula column's Property Palette, and name the column **CF_term_gpa**.
3. Click the **PL/SQL Formula** property, then click the button to open the PL/SQL Editor.
4. Type the code shown in Figure 7-71 to calculate the student GPA. Compile the code, correct any syntax errors, close the PL/SQL Editor, then close the Property Palette.

type this code

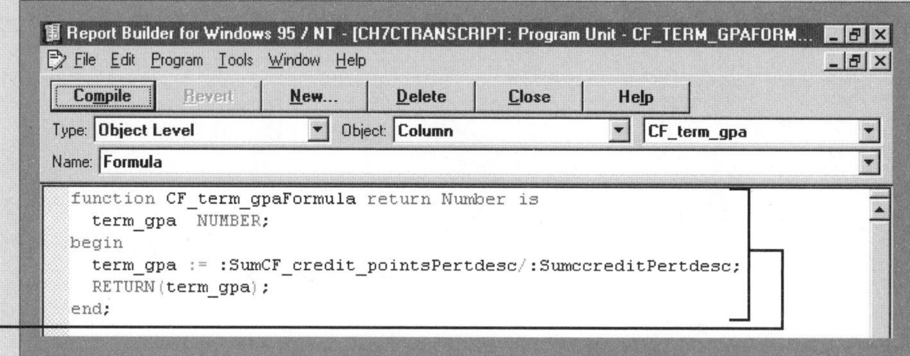
Report Builder for Windows 95 / NT - [CH7CTRANSCRIPT: Program Unit - CF_TERM_GPAFORM...
File Edit Program Tools Window Help
Compile
Revert
New...
Delete
Close
Help
Type: Object Level
Object: Column
CF_term_gpa
Name: Formula
function CF_term_gpaFormula return Number is
term_gpa NUMBER;
begin
term_gpa := :SumCF_credit_pointsPertdesc/:SumccreditPertdesc;
RETURN(term_gpa);
end;

Figure 7-71: Function to calculate term GPA

Now, on the report layout, you need to draw the report field that will display the term GPA. You will draw this field so that it is displayed directly below the summary column showing total term credits, which is in the call ID footer frame. You will need to make the call ID footer frame (M_G_callid_FTR) longer first. You will do this with flex mode on, which automatically resizes all of the footer frame's surrounding frames.

To lengthen the term description footer frame and draw the report field to display the term GPA:

1. Open the Layout Model, and, if necessary, click the **Flex Mode** button on the toolbar to turn flex mode on.
2. Click **F_SumccreditPertdesc**, then click the **Select Parent Frame** button to select the M_G_callid_FTR frame. Open the Property Palette to confirm that you have selected the correct frame, then close the Property Palette.
3. Drag the frame's center selection handle down so that it is sized as shown in Figure 7-72.

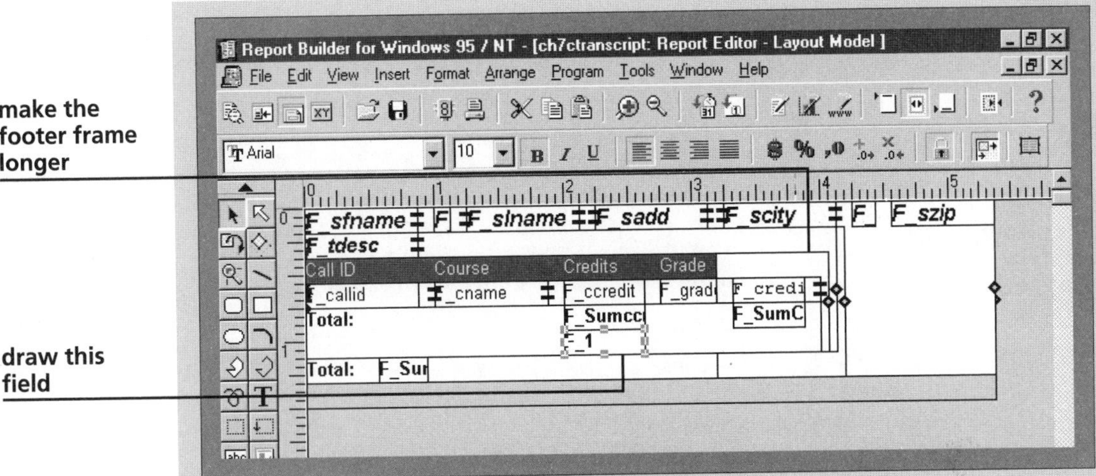

Figure 7-72: Lengthening the footer frame and drawing the term GPA field

4 Click the **Field** tool [abc] on the tool palette and draw a new field, as shown in Figure 7-72.

5 Open the new field's Property Palette, and change the Name property to **F_term_gpa**. Set the Source property to **CF_term_gpa**, and select **N.00** as the format mask, to make the field display to a precision of two decimal places, even if the value is a whole number. If N.00 is not among your list of format mask choices, type N.00 as the format mask.

6 Click on another property to save the change, then close the Property Palette.

7 Click the **Text** tool [T] on the tool palette and create a new label with the text **Term GPA:**. Place the label on the left edge of the report, aligned horizontally with the term GPA data field.

8 View the report in the Live Previewer to confirm that the GPA is calculated correctly. Your display of Brian Umato's transcript should look like Figure 7-73.

term GPA values (your font style might be different)

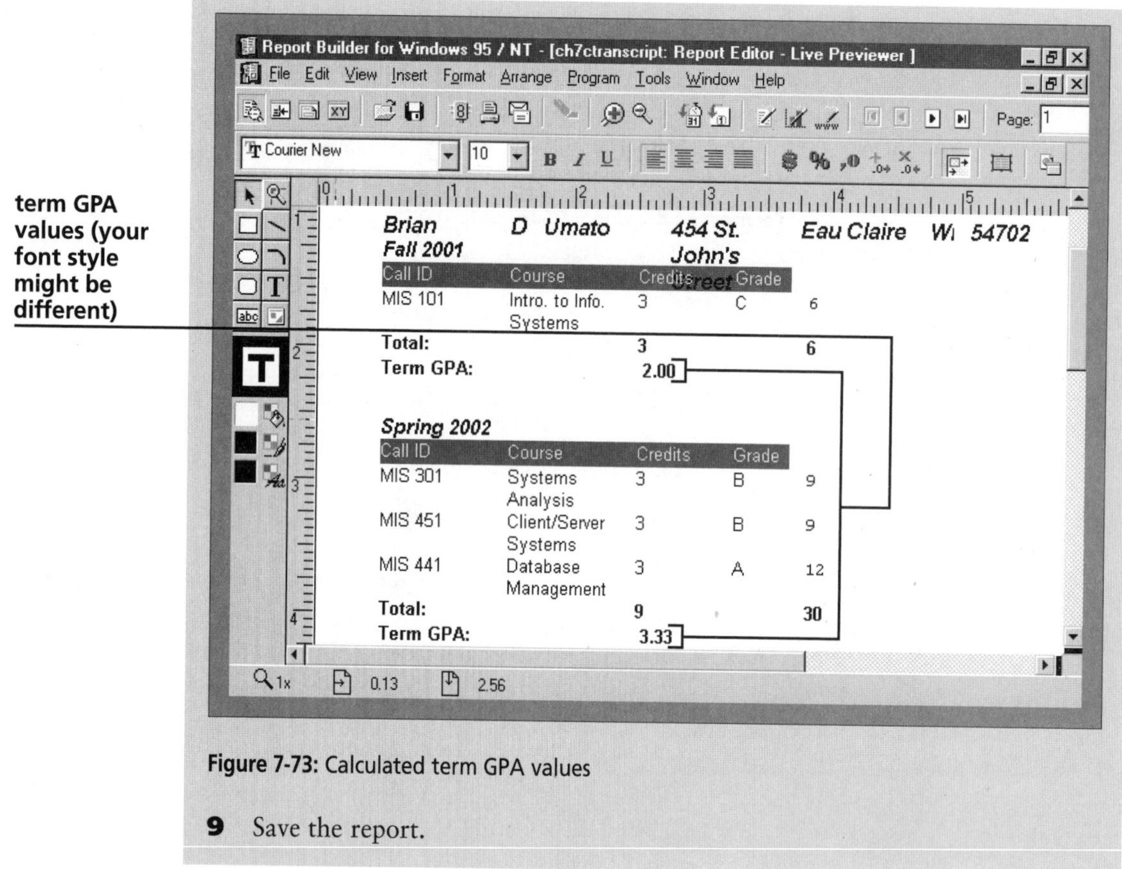

Figure 7-73: Calculated term GPA values

9 Save the report.

Since you have confirmed that the term GPA is being calculated correctly, you can now hide the holding fields that you don't want to appear on the finished report.

To hide the holding fields in the report display:

1 If necessary, open the Live Previewer, and then click one of the credit point fields to select the credit point fields in each repeating group. Open their Property Palette, set the Visible property to **No**, click another property to save the change, and then close the Property Palette. The credit points are no longer visible.

2 Repeat Step 1 for the sum of the credit points per term field.

3 Save the report. Final report formatting and calculating the student's overall GPA are left as an end-of-chapter exercise.

Creating and Running an Executable Report File

Recall from Chapter 5 that when users run a form created in Form Builder, they do not run it in the Form Builder environment, but run it directly from the operating system by double-clicking on the form's .fmx file in Windows Explorer. For the form to run successfully, the .fmx file type must be a registered file type on the user's workstation and must be associated with the Forms Runtime application, which is an executable file named IFrun60.exe that is located in the ORAWIN95\BIN\ or ORANT\BIN\ folder on the user's workstation. Similarly, when users run a report created in Report Builder, they run it directly by double-clicking the report's executable file in Windows Explorer. The report executable file has a .rep extension. Files with a .rep extension must be registered on the user's workstation, and must be associated with the Reports Runtime application, which is an executable file named Rwrun60.exe and is also located in the ORAWIN95\BIN\ or ORANT\BIN\ folder on the user's workstation. This file registration is created when Report Builder is installed on a user's workstation. Now you will learn how to generate an executable report file that can be run independently of Report Builder. Then you will open Windows Explorer and run the transcript report.

To generate the report executable file and run the transcript report from Windows Explorer:

1. Click **File** on the menu bar, point to **Administration**, and then click **Compile Report**. The Compile Report dialog box opens. The default filename is the same as the report filename, but with a .rep extension.
2. Confirm that you are saving the .rep file to the Chapter7 folder on your Student Disk, accept the default filename (ch7ctranscript.rep), then click **Save**.
3. Start Windows Explorer, navigate to the Chapter7 folder on your Student Disk, and double-click ch7ctranscript.rep.
4. When the Connect dialog window is displayed, log on to the database as usual. The report is displayed in the Reports Runtime Previewer window.

help

If the Open With dialog box opens instead of the Connect dialog box, it means that your workstation is not configured to run executable Oracle Reports files (which have .rep extensions), using the Oracle Reports Runtime application. For Windows 95 users, click Other, click Browse, click the Rwrun60.exe file in the ORAWIN95\BIN\ directory, click Open, and then click OK. For Windows NT users, click Other, click the R30run60.exe file in the \ORANT\BIN\ folder, click Open, then click OK. Now the Connect dialog box will open automatically when you double-click .rep filenames.

help

If the Connect dialog window does not appear, then you probably associated the .rep file with the wrong executable file. To delete an association, click View on the Windows Explorer menu bar, and then click Folder Options. Click the File Types tab, and then scroll down in the file list until you find an entry called REP Files. Click REP Files, click Remove, and then click Yes to confirm that you want to remove the association. Click the Close button to close the Folder Options dialog box, and then repeat the steps to create the file association.

The Reports Runtime Previewer window is similar to the Report Builder Live Previewer window. The report display is the same, but the window's toolbar buttons are limited to operations that allow the user to redirect the report output (print, e-mail, or configure the printer), restart or close the window, adjust the current zoom setting of the window, and step through the report pages. Now you will step through the report pages and print the report.

To step through the report and print it:

1. Click the **Next Page** button. The transcript information for student Daniel Black is displayed.
2. Click multiple times to step through all of the report pages, then click the **First Page** button to return to the first page of the report.
3. Click the **Print** button on the toolbar. The Print dialog box opens.
4. Click the **Pages** option button in the Print Range option button group, specify to print Pages from 1 to 1, then click **OK**. The first page of the transcript report should print on your printer.
5. Click the **Close Previewer** button to close the Reports Runtime Previewer.

Running a Report from a Form

Reports are often run from Oracle Form Builder applications so that the user can specify a parameter in a form, and then display and/or print the related report information. Figure 7-74 shows a form that is used to specify a specific term and student. When the user clicks the Run Transcript button, the form runs the transcript report and passes it the student ID and term ID values that the user selected on the form. When these values are received by the report, the transcript output is displayed for the selected term and student.

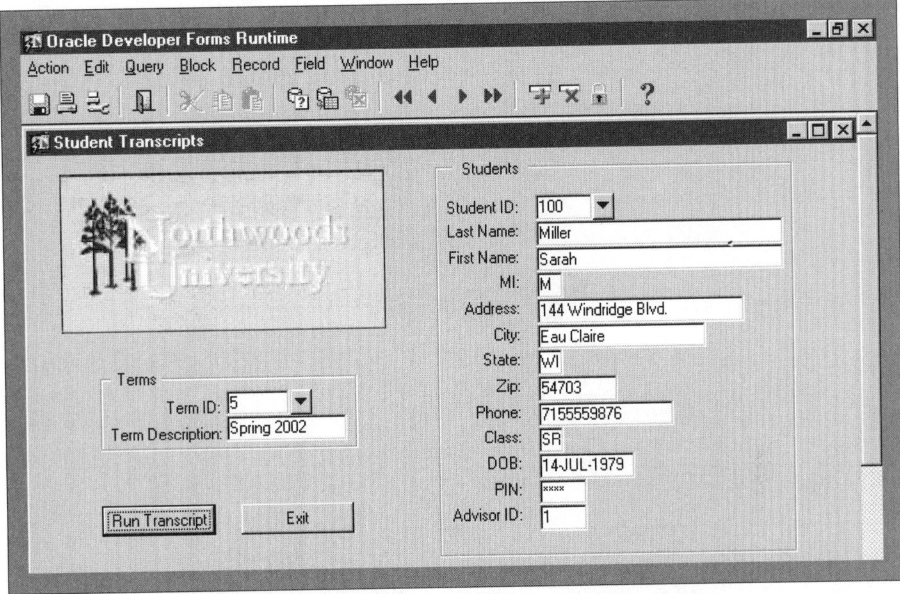

Figure 7-74: Form to run transcript report

To implement the form/report combination, you must add commands to the form to create a parameter list that contains the selected values for student ID and term ID. Then you must call the report and pass it the parameter list. You must also modify the report query so that it will accept the parameter list correctly. First you will modify the report query.

Modifying the Report Query

To pass one or more variable values from a form to a report, you must create bind parameters in the report query. **Bind parameters** are parameters whose values are passed from a form to a report. To create a bind parameter, you replace one or more expressions in the report's SQL query with a variable value preceded by a colon, and Report Builder automatically creates the parameter. Now you will modify the report query by adding search clauses for SID and TERMID, and setting these values equal to bind parameters whose values will be assigned when the form calls the report and passes the selected values for term and student. The bind parameters should be given descriptive names, because they will be displayed on the Parameter Form window when the report runs.

To create the bind parameters in the report query:

1. Save the report as **ch7cformtranscript.rdf.**
2. Open the Data Model, and then double-click **Q_1** to open the SQL query statement. Modify the query as shown in Figure 7-75, then click **OK.**

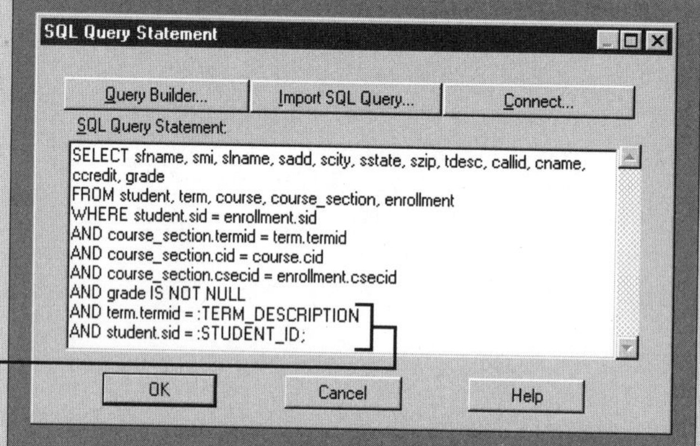

Figure 7-75 Adding bind parameters to the report query

3. The informational message "Note: The query 'Q_1' has created the bind parameter(s) 'TERM_DESCRIPTION', 'STUDENT_ID'." confirms that Report Builder created the bind parameters. Click **OK.**
4. Open the Object Navigator, and click [+] beside **User Parameters** under Data Model. The bind parameters STUDENT_ID and TERM_DESCRIPTION are displayed.
5. Save the report.

Parameter Lists

Next, you will open the form file in Form Builder, create the parameter list that will pass the term ID and student ID to the report, and add the RUN_PRODUCT command to the form. A **parameter list** is a list of variables that is used to pass data values from a Form Builder application to a report. Parameter lists are created in triggers or program units using PL/SQL. To create a parameter list, you first must declare a variable with the data type PARAMLIST to hold the list's internal identifier in the DECLARE section of the PL/SQL code, using the following general code format:

```
DECLARE
    <list ID variable name> PARAMLIST;
```

The next step is to create the list. This is done in the body of the PL/SQL procedure, using the following general code format:

```
BEGIN
   <list ID variable name> := CREATE_PARAMETER_LIST(<list
   name variable name>);
```

The `<list name variable name>` value can be any text string. After creating the parameter list, you use the ADD_PARAMETER procedure to add the parameter data values to the list. A single parameter list can be used to pass multiple parameters. Each parameter is added to the list separately, using its own ADD_PARAMETER command. Parameters passed using a parameter list can only be text (character) data or record group data. If you want to pass a date or number field, you must first convert it to a character using the TO_CHAR data conversion function.

The ADD_PARAMETER command has the following general format:

```
ADD_PARAMETER(<LIST>, <KEY>, <PARAMTYPE>, <VALUE>);
```

The command requires the following items:

- **LIST** is the name of the list ID variable for the parameter list that you defined when you created the parameter list.
- **KEY** is the name of the bind parameter as it was defined in the report that is receiving the parameter. KEY is passed as a character string, so it must be enclosed in single quotation marks. For the transcript report, the parameter keys are TERM_DESCRIPTION and STUDENT_ID, because these are the names of the bind parameters that were used in the report query (see Figure 7-75).
- **PARAMTYPE** can have a value of either TEXT_PARAMETER (for parameters that are character data) or DATA_PARAMETER (for parameters that are record groups).
- **VALUE** is the actual parameter character string or record group name. In the form that will call the transcript report, the values for term ID and student ID will be used. Since both of these values have a NUMBER data type, they must first be converted to character strings before they are added to the parameter list.

Now you will open the form shown in Figure 7-74 and create a trigger for the Run Transcript button. In the DECLARE section, the trigger declares the parameter list variable and the character variables that will be used when TERMID and SID are converted to character strings. In the BEGIN section, the trigger initializes the parameter list, then adds the TERMID and SID parameter values to the list. You will not compile the trigger yet, because it is not complete.

To open the form and initialize the parameter list:

1 Start Form Builder, and open the **studentreport.fmb** file from the Chapter7 folder on your Student Disk. Save the form as **ch7cstudentreport.fmb.**

2 In the Object Navigator, change the form name to **CH7CSTUDENTREPORT.**

3 Open the Layout Editor, click the **Run Transcript** button, right-click, point to **Smart Triggers**, then click **WHEN-BUTTON-PRESSED** to create a new trigger for the Run Transcript button. Type the code shown in Figure 7-76 to declare and initialize the parameter list. DO NOT compile the trigger yet.

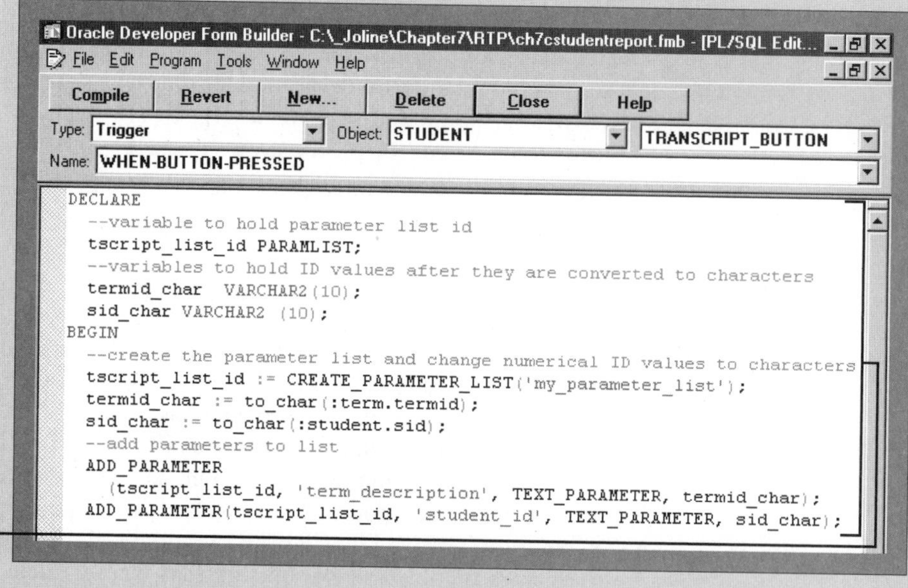
Oracle Developer Form Builder - C:_Joline\Chapter7\RTP\ch7cstudentreport.fmb - [PL/SQL Edit...
File Edit Program Tools Window Help
Compile
Revert
New...
Delete
Close
Help
Type: Trigger
Object: STUDENT
TRANSCRIPT_BUTTON
Name: WHEN-BUTTON-PRESSED
DECLARE
--variable to hold parameter list id
tscript_list_id PARAMLIST;
--variables to hold ID values after they are converted to characters
termid_char VARCHAR2(10);
sid_char VARCHAR2 (10);
BEGIN
--create the parameter list and change numerical ID values to characters
tscript_list_id := CREATE_PARAMETER_LIST('my_parameter_list');
termid_char := to_char(:term.termid);
sid_char := to_char(:student.sid);
--add parameters to list
ADD_PARAMETER
(tscript_list_id, 'term_description', TEXT_PARAMETER, termid_char);
ADD_PARAMETER(tscript_list_id, 'student_id', TEXT_PARAMETER, sid_char);
type this code

Figure 7-76: Creating the Run Transcript button trigger

After you finish using a parameter list, you must destroy it, or it will remain in the workstation's memory and take up space. The general format of the DESTROY command is `DESTROY_PARAMETER_LIST(param_list_id);`. You will add the command to destroy the parameter list after you add the command to run the report.

Using the RUN_PRODUCT Procedure to Run the Report

The RUN_PRODUCT procedure is a built-in Oracle procedure that is used to start the Reports Runtime application, specify the name of the report file that will be displayed, and pass the parameter list that contains data values that customize the report output. The general format of the RUN_PRODUCT procedure is:

```
RUN_PRODUCT(<PRODUCT>, <DOCUMENT>, <COMMUNICATION MODE>,
<EXECUTION MODE>, <LOCATION>, <PARAMETER LIST ID>, <DISPLAY>);
```

This procedure requires the following items:

- **PRODUCT** specifies which Oracle Developer application you want to run. RUN_PRODUCT can be used to run either Oracle Report Builder or Graphics Builder applications, so its value will be either REPORTS or GRAPHICS.
- **DOCUMENT** specifies the complete path and filename of the report .rdf file, which is specified as a character string enclosed in single quotation marks. The filename and folder path cannot contain any blank spaces.
- **COMMUNICATION MODE** can be either SYNCHRONOUS or ASYNCHRONOUS. **SYNCHRONOUS** specifies that control returns to the form only after the called product (Reports or Graphics) has closed. **ASYNCHRONOUS** specifies that control returns to the form while the product is still running, and you can multitask between the form and report applications.
- **EXECUTION MODE** can be either BATCH or RUNTIME. RUNTIME specifies that the called product's runtime environment (such as Reports Runtime) runs, while BATCH specifies that the called product just generates an executable file that is then displayed in the form. Use **RUNTIME** when running a Report Builder application so that control is transferred to the Reports Runtime environment. Use **BATCH** when running a Graphics application when a graphic is displayed directly in a form.
- **LOCATION** can be either FILESYSTEM or DB (database), and specifies where the document to be run is stored. In the exercises in this book, all documents will be stored in the file system on the user's workstation.
- **PARAMETER LIST ID** is the ID of the parameter list that contains the data values that are passed to the called product. If there is no parameter list, this value must be passed as NULL.
- **DISPLAY** is only used when you are running a Graphics chart. (Graphics charts will be discussed in Chapter 8.) This property specifies the name of the form image that will display the chart on the form. The DISPLAY value always must be passed as NULL when you are calling a report.

Now you will finish the Run Transcript button trigger. You will add the RUN_PRODUCT command to run the report, and then you will destroy the parameter list.

To finish the Run Transcript button trigger:

1 Add the commands at the end of the Run Transcript button trigger code shown in Figure 7-77.

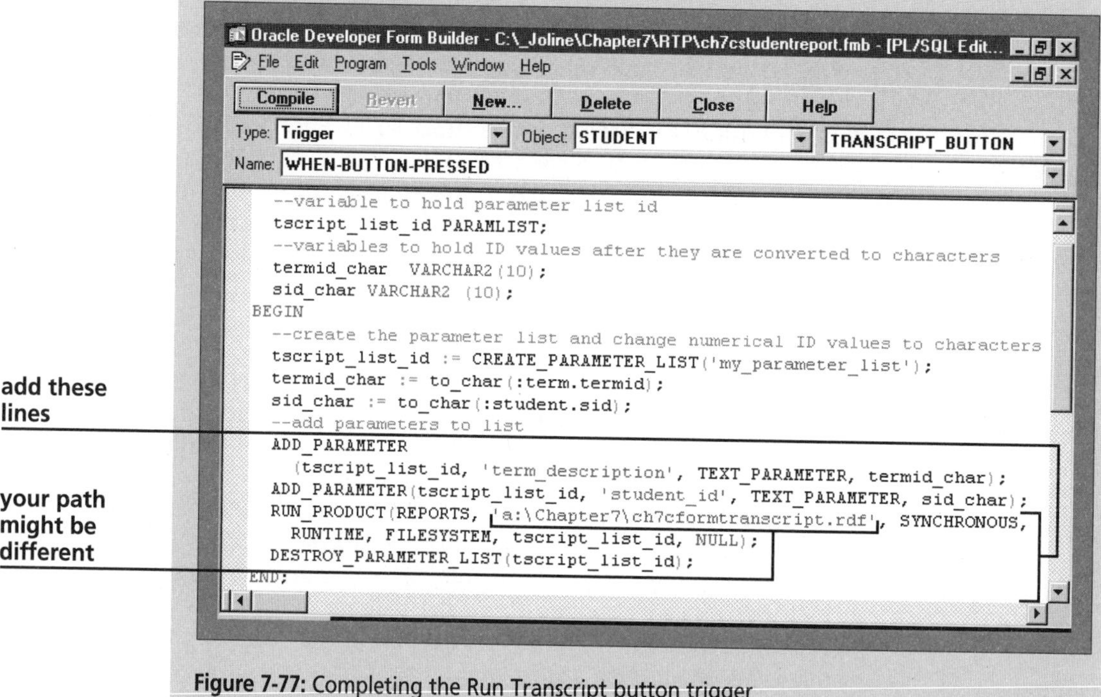
Oracle Developer Form Builder - C:_Joline\Chapter7\RTP\ch7cstudentreport.fmb - [PL/SQL Edit...
File Edit Program Tools Window Help
Compile Revert New... Delete Close Help
Type: Trigger Object: STUDENT TRANSCRIPT_BUTTON
Name: WHEN-BUTTON-PRESSED
--variable to hold parameter list id
tscript_list_id PARAMLIST;
--variables to hold ID values after they are converted to characters
termid_char VARCHAR2(10);
sid_char VARCHAR2 (10);
BEGIN
--create the parameter list and change numerical ID values to characters
tscript_list_id := CREATE_PARAMETER_LIST('my_parameter_list');
termid_char := to_char(:term.termid);
sid_char := to_char(:student.sid);
--add parameters to list
ADD_PARAMETER
(tscript_list_id, 'term_description', TEXT_PARAMETER, termid_char);
ADD_PARAMETER(tscript_list_id, 'student_id', TEXT_PARAMETER, sid_char);
RUN_PRODUCT(REPORTS, 'a:\Chapter7\ch7cformtranscript.rdf', SYNCHRONOUS,
RUNTIME, FILESYSTEM, tscript_list_id, NULL);
DESTROY_PARAMETER_LIST(tscript_list_id);
END;
add these lines
your path might be different

Figure 7-77: Completing the Run Transcript button trigger

2 Compile the trigger, correct any syntax errors, then close the PL/SQL Editor.

3 Save the form.

Now you will test the form/report combination. You will run the form, and select the Spring 2002 term and student Sarah Miller. When you click the Run Transcript button, the transcript report for the selected term and student should appear.

To test the form/report combination:

1 Click the **Run Form Client/Server** button to run the form, and connect to the database if necessary.

2 Click the LOV command button beside Term ID, and select **Spring 2002**.

3 Click the LOV command button beside Student ID, and select **Sarah Miller**.

4 Click the **Run Transcript** button. The Runtime Parameter Form appears with the selected values from the report displayed. Click the **Run** button to display the report. The report is displayed for the values for the selected student and term.

help If the report does not appear, check to make sure that the report is named ch7cformtranscript.rdf, and that the drive and path specification are correct. If you have saved the file under a different name, then change the PL/SQL string in the `RUN_PRODUCT` command accordingly.

5 Click the **Close Previewer** button. The Student Transcripts form is displayed. Click **Exit.**

6 Close all open Oracle applications.

SUMMARY

- Every time you make a modification using the Report Wizard, all custom formatting is lost. Therefore, it is best to make sure that the desired data and calculations are correctly displayed on the report, and then perform the final formatting as the last task.
- A summary column performs actions such as computing the sum and/or average of a series of repeating data fields. Summary columns can also find the minimum or maximum value in a data series, or find the number of records retrieved by a query.
- When you use the Report Wizard to create a summary column, it creates a summary column for the selected field for each group level above the field's record group in the report, and a summary column for the field for the entire report.
- Every summary column is placed in a footer frame that is in the same group frame as the one that encloses the repeating records being summarized.
- Formula columns display a value that is returned by a function that performs mathematical computations on report data columns. Formula columns must be displayed in the same frame as the most detailed data column in their formula.
- To create a formula column, you create the column in the data model, then write its computational logic using a PL/SQL function.
- To reference report columns in a PL/SQL formula column function, preface the column name as it appears in the data model with a colon.
- A holding column is a data model column that is not displayed in the final report, but serves to hold the value of a calculated or summed column that is not displayed, but will be used in another computation, so you can confirm that its value is correct.

- To create a summary column manually, you create the summary column in the data model, draw a field on the report layout to display the summary column value, and then set the new field's source property as the summary column.
- Users can run reports as standalone executable programs, or from Form Builder applications.
- Executable report files have a .rep extension, and can be run by double-clicking on the file name in Windows Explorer, or by creating a Windows shortcut icon.
- To pass variable values from a form to a report, you must create bind parameters in the report query. To create a bind parameter, you replace one or more expressions in the report's SQL query with a variable value preceded with a colon.
- Bind parameters should be given descriptive names, because they will appear on the Parameter Form window as labels when the report runs.
- A parameter list is a list of variables that is used to pass data values from Form Builder applications to reports.
- Parameters passed using parameter lists can only be character data or record group data. If you want to pass a date or number field, you must first convert it to a character using the TO_CHAR data conversion function.
- A single parameter list can pass multiple parameters. Each parameter value must be added to the parameter list using a separate ADD_PARAMETER command.

REVIEW QUESTIONS

1. List the name of the summary column field that would be calculated if you used the Report Wizard to calculate a summary column that sums the specified field in each of the following queries:

Query	Summary Column
a. SELECT locid, bldg_code, room, capacity FROM location;	CAPACITY
b. SELECT itemid, itemdesc, invid, qoh FROM item, inventory WHERE item.itemid = inventory.itemid;	QOH
c. SELECT tdesc, callid, secnum, currenrl FROM term, course, course_section WHERE term.termid = course_section.termid AND course.cid = course_section.cid;	CURRENRL

2. List the names of the footer frames that would be generated for the queries in Question 1.
3. True or false: you can create a formula column using the Report Wizard.
4. Which of the following describes where holding fields that display formula columns must be drawn?
   a. in the frame that contains the data values used in the formula column function
   b. in the frame that contains the summary columns used in the formula column function
   c. in the frame that is the outermost group frame in the report
   d. there is no good way to tell.
5. What is the difference between. .rdf and .rep report files?
6. True or false: when you call a form from a report, you must specify the complete path to the report's .rep file, and the path cannot contain any spaces.
7. Which of the following data field values from the CUST_ORDER table could be passed directly in a parameter list without using any data type conversion functions?
   a. ORDERID
   b. ORDERDATE
   c. METHPMT
   d. ORDERSOURCE
8. Modify the following report queries so that bind parameters would be created for the specified columns:

Query	Bind Parameter Column(s)
a. SELECT flname, ffname, bldg_code, room, ffphone FROM faculty, location WHERE faculty.locid = location.locid;	FID
b. SELECT itemdesc, category, itemsize, color, curr_price FROM item, inventory WHERE item.itemid = inventory.itemid;	ITEMID, CATEGORY
c. SELECT tdesc, callid, secnum, day, time FROM term, course, course_section WHERE term.termid = course_section.termid AND course.cid = course_section.cid;	TERMID, CALLID

9. Why should bind parameters always be given descriptive names?

PROBLEM-SOLVING CASES

Run the northwoo.sql script to refresh the Northwoods University database, and run the clearwat.sql script to refresh the Clearwater Traders database. The script files are stored in the Chapter7 folder on your Student Disk.

1. In this exercise, you will modify the transcript report you created in the tutorial lessons so that its formatting and function match the report shown in Figure 7-55.
   a. Open the ch7ctranscript.rdf file from the Chapter7 folder on your Student Disk, and save it as Ch7cEx1.rdf.
   b. Calculate and display the student's cumulative GPA for all terms. (*Hint*: Sum all course credit points and divide the result by the sum of all course credits.)
   c. Revise the formatting so that the summary and formula column labels match the ones shown in Figure 7-55; the student first name, middle initial, and last name are adjacent to each other; and the student's city, state, and ZIP code are adjacent to each other.
2. In this exercise, you will modify the ch7cstudentreport.fmb form file by adding a button with the label "Run All Terms." When the user clicks this button, the form passes the selected student ID to the transcript report, and the report displays the student's records for all terms. To do this, you will create another copy of the transcript report that only has a single bind parameter.
   a. Open ch7ctranscript.rdf from the Chapter7 folder on your Student Disk, save it as ch7cEx2.rdf, and create a bind parameter for SID.
   b. Open the ch7cstudentreport.fmb file from the Chapter7 folder on your Student Disk, save it as Ch7cEx2.fmb, and then add a "Run All Terms" button. Make a trigger for the button that creates a parameter list to pass the SID value to the Ch7cEx2.rdf report, and then calls the report.
3. Create a report that displays building codes at Northwoods University, and then shows all related location IDs, rooms, and capacities, in order of room number. Create a summary column that sums the capacity of all of the rooms in a specific building, and also shows the capacity of all rooms in the LOCATION table. Format the report as shown in Figure 7-78, and save the file as Ch7cEx3.rdf.

Building Capacity Report

Report run on: January 27, 2001

Building: BUS

Room	*Capacity*
105	42
211	55
402	1
404	35
421	35
424	1
433	1
Building Capacity:	170

Building: CR

Room	*Capacity*
101	150
103	35
105	35
202	40
Building Capacity:	260

Figure 7-78

4. In this exercise, you will create a form with a LOV that displays the building codes at Northwoods University. The user can select a specific building and click a button labeled "Show Capacity," and the report created in Exercise 3 is displayed for only the selected building.

a. Open the Ch7cEx3.rdf report file that you created in Exercise 3, and save the file as Ch7cEx4.rdf.

b. Modify the report query so that it has a bind parameter for the BLDG_CODE variable, and so that it no longer displays the total capacity for all buildings. Create the parameter so that the label of the parameter on the parameter form is BUILDING_CODE.

c. Create a form named Ch7cEx4.fmb that has fields that display the LOCID and BLDG_CODE fields from the LOCATION table.

d. Create a LOV in the form that displays each BLDG_CODE in the LOCATION table only once. (*Hint*: Use the DISTINCT qualifier.)
e. Create a button on the form labeled "Show Capacity" that creates a parameter list with the selected building code, and then calls the Ch7cEx4.rdf report from the Chapter7 folder on your Student Disk. The report should display capacity information for only the selected building.
f. Create an Exit button to exit the form.

5. Create a report that shows item descriptions from the Clearwater Traders database, and associated inventory IDs, sizes, colors, current prices, and quantities on hand, as shown in Figure 7-79. Save the report as Ch7cEx5.rdf.

CLEARWATER TRADERS

Inventory Valuation Report

27 January 2001
Page 1

Description: 3-Season Tent

Inventory ID	*Size*	*Color*	*Price*	*QOH*	*Value*
11668		Sienna	269.99	15	$ 4,049.85
11669		Forest	269.99	9	$ 2,429.91
				Total Item Value:	$ 6,479.76

Description: Children's Beachcomber Sandals

Inventory ID	*Size*	*Color*	*Price*	*QOH*	*Value*
11820	10	Blue	15.99	121	$ 1,934.79
11822	12	Blue	15.99	113	$ 1,806.87
11821	11	Blue	15.99	111	$ 1,774.89
11823	1	Blue	15.99	121	$ 1,934.79
11825	11	Red	15.99	137	$ 2,190.63
11827	1	Red	15.99	123	$ 1,966.77
11826	12	Red	15.99	134	$ 2,142.66
11824	10	Red	15.99	148	$ 2,366.52
				Total Item Value:	$16,117.92

Figure 7-79

a. Create a formula column that computes the value of each inventory item, which is equal to current price times quantity on hand.
b. Create a summary column that displays the total value of each inventory item.

c. Create a summary column that displays the total value for all inventory items on the report.

d. Format the report as shown in Figure 7-79.

6. Create a report that displays customer order information for Clearwater Traders, as shown in Figure 7-80. The top of the report should display the customer name and address, as well as the order date, order ID, and payment method. The detail section should show the item ID, item description, size, color, order price, and quantity, and calculate the extended total. The summary section at the bottom of the report should show a subtotal that sums all of the extended totals, calculates sales tax and shipping and handling, and sums the subtotal, tax, and shipping and handling to create a final order total.

Customer Invoice

Alissa R White
987 Durham Rd.
Sister Bay WI 54234

Order Date *April 1, 1999*
Order ID *1660*
Payment Method *CHECK*

Item ID	*Description*	*Size*	*Color*	*Price*	*Quantity*	*Extended Total*
559	Men's Expedition Parka	S	Navy	199.95	2	$ 399.90
786	3-Season Tent		Forest	269.99	1	$ 269.99

Subtotal: $ 669.89
Tax: $ 40.19
Shipping & Handling: $ 7.50

ORDER TOTAL: $ 717.58

Figure 7-80

a. Create the report using the Report Wizard to display the required data fields. Each customer order should appear on a separate report page. Save the report as Ch7cEx6.rdf.

b. Create a formula column to calculate the extended total, which is the order price times quantity.
c. Create a summary column to calculate the subtotal, which is the sum of all of the extended totals.
d. Create a formula column that calculates the tax as 6 percent of the subtotal for Wisconsin residents, and as 0 for residents of all other states.
e. Create a formula column to calculate shipping and handling as follows: If the order subtotal is less than $25, then the shipping cost is $3.50. If the subtotal is $25 or more, but less than $75, then the shipping cost is $5.00. If the subtotal is $75 or more, then the shipping cost is $7.50.
f. Create a column to calculate the order total, which is the sum of the subtotal, tax, and shipping and handling.
g. Format the report so the output looks like Figure 7-80.

CHAPTER 8

Using Graphics Builder

objectives

- Use Graphics Builder to create pie and bar charts based on database data
- Link two charts using a drill-down relationship
- Create a Form Builder application that receives user inputs and passes them to a chart
- Create a Form Builder application that displays a chart directly on a form
- Create a Report Builder application that displays a chart that corresponds to report data

Introduction Graphics Builder is the Oracle Developer application used for creating graphical displays of database data such as pie charts, bar charts, and line charts. Oracle Graphics Builder applications can run as stand-alone applications, or they can be embedded within forms or reports. In this chapter, you will learn how to create a variety of charts and how to display charts using Form Builder and Report Builder applications. You will also learn how to create an interactive drill-down relationship between two related charts that allows you to select a value in one chart and to view related detail data in a second chart.

Displaying Data in Charts

Numerical data displayed in a graphical format using bar, pie, and line charts are faster and easier to interpret and understand than data shown in a textual or numerical format. Graphics Builder lets you retrieve database data and display their values graphically using a variety of chart types. Before you create a chart, it is important to consider the type of data you want to chart and the purpose of the chart. Figure 8-1 lists the chart types you can create using Graphics Builder, describes what each chart is designed to illustrate, and provides an example of an application.

Chart Type	Description	Example
Column	Shows discrete values using vertical columns	Display current quantities on hand for different inventory items
Bar	Shows discrete values using horizontal bars	Same as column
Gantt	Shows task or project scheduling or progress information	Display work schedule information by showing hours of the day on the X-axis and names of employees on the Y-axis, with horizontal bars showing each employee's scheduled start and stop times
High-Low	Shows multiple Y-axis values for a single X-axis point	Display a student's highest, lowest, and average GPA
Line	Shows data values as points connected by lines to show trends over time	Display total order revenues for each item category for each month of the past year
Mixed	Combines a column and line chart	Display total order revenues for each month of the past year as columns, with a line connecting the top of the columns to show sales trends
Pie	Shows how individual data values contribute to a total amount	Illustrate what percentage of sales each item category (Women's Clothing, Men's Clothing, etc.) contributes to total sales revenue

Figure 8-1: Oracle Graphics Builder chart types and uses

Chart Type	Description	Example
Scatter	Presents two sets of unrelated data for identifying potential trends	Plotting student age versus GPA
Table	Shows text data in a table	Show a listing of all Northwoods University course call IDs, descriptions, days, times, and locations
Double Y-charts	Shows a line chart with two different sets of Y-axis data for the same X-axis values	Illustrate how revenue dollars and units sold have varied over the same period

Figure 8-1: (continued) Oracle Graphics Builder chart types and uses

Figure 8-2 shows an example of a column chart created using the data from the INVENTORY table in the Clearwater Traders database to show the current quantity on hand of the Women's Fleece Pullover in different sizes and colors.

One of the most common mistakes people make when creating charts is to plot discrete data points using a line chart. Figure 8-3 shows the same data as Figure 8-2, but using a line chart format. Since the quantity on hand for each color/size combination is discrete (e.g., the quantity for size S, color Teal has no relationship or bearing on the quantity on hand for Size M, color Teal), the chart makes no sense.

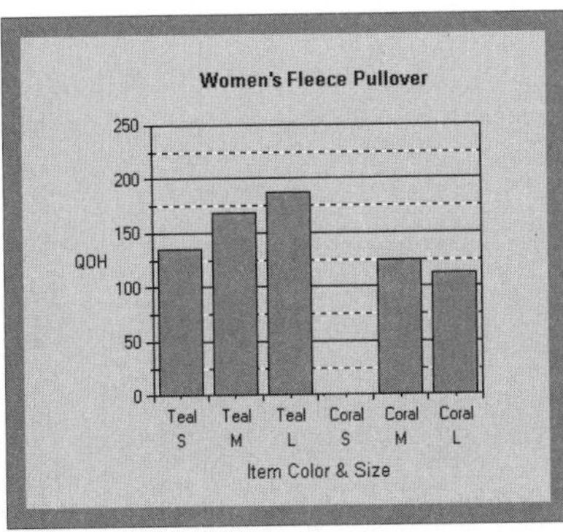

Figure 8-2: Column chart

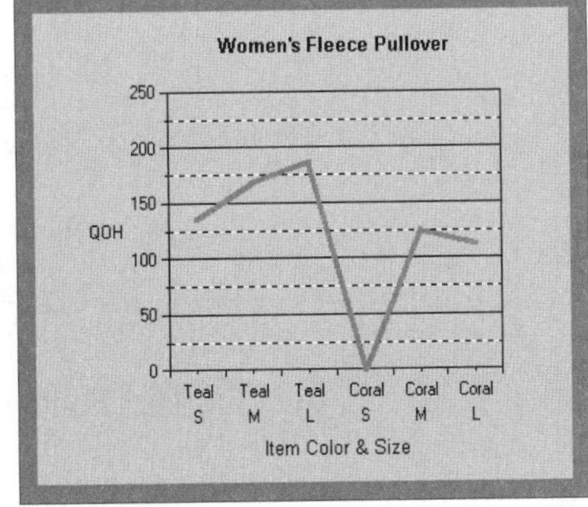

Figure 8-3: Inappropriate use of a line chart

Creating a Chart

The management at Clearwater Traders wants to create a chart to display the portion of its total order revenue that comes from specific catalogs and from its Web site. A pie chart is appropriate because it is used to compare individual components (such as proportions of revenue from each order source) to an overall total (such as total order revenue). Figure 8-4 shows a design sketch for the Order Revenue Source pie chart. If Clearwater Traders had $1,000 in total revenue and one-third came from Catalog 152, one-third came from Catalog 153, and one-third came from the Web site, each of these three sources would contribute one-third to the total revenue, so each would represent one-third of the pie. The chart labels show the source name as well as the value that each contributes to the total revenue. Now you will start Graphics Builder and then create the Order Revenue Sources pie chart.

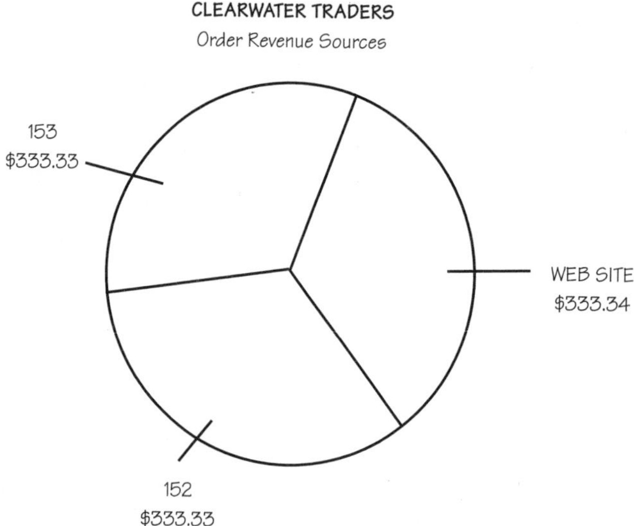

Figure 8-4: Design sketch for pie chart

Starting Graphics Builder

Just as with any other Developer application, you must start Graphics Builder and then log on to the database. After starting Graphics Builder, you will run the SQL script to refresh your Clearwater Traders database tables.

To start Graphics Builder and run the SQL script:

1. Click the **Start** button on the Windows taskbar, point to **Programs**, point to **Oracle Developer 6.0**, and then click **Graphics Builder**.
2. Click **File** on the menu bar, click **Connect**, and then connect to the database in the usual way. Graphics Builder opens in the Object Navigator window, as shown in Figure 8-5. Maximize the Object Navigator window if necessary.

tip

Another way to start Graphics Builder is to open Windows Explorer, change to the ORAWIN\BIN (or ORANT\BIN) folder, and then double-click the file Rwbld60.exe.

tip

If you are using Graphics Builder with a Personal Oracle8 database on your local machine, you will need to create a database alias and enter a connect string to connect to the database. Follow the instructions in the Instructor's Manual, or ask your instructor or technical support person for help.

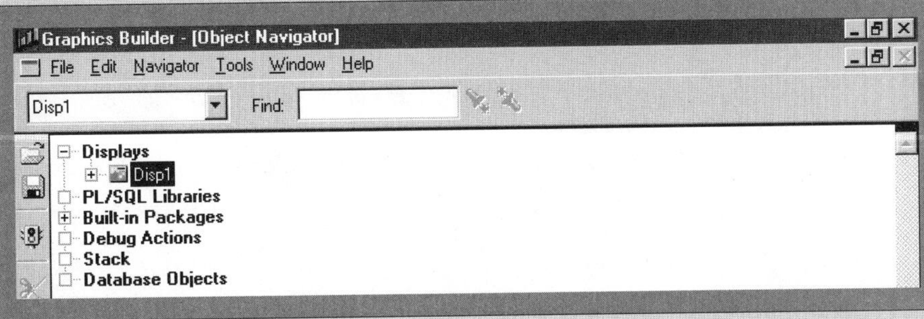

Figure 8-5: Oracle Graphics Builder Object Navigator window

help

The Graphics Builder [Graphics: Disp1: Layout Editor] window might open when you start Graphics Builder. If this happens, click Window on the menu bar, click Object Navigator, and then maximize the Object Navigator window.

3. Start SQL*Plus, log on to the database, and run the clearwat.sql script from the Chapter8 folder on your Student Disk to refresh your database tables.

   Note: If you are storing your Student Disk files on a floppy disk, you will need to use two floppy disks for this chapter. When your first disk becomes full, save your solution files to the second disk.
4. Exit SQL*Plus.

A Graphics Builder chart object is called a **display**. When you start Graphics Builder, a new chart object named Disp1 is created automatically. The Object Navigator display shows the possible components of a chart, as well as the external Oracle resources that you can access within Graphics Builder. First you will save the chart.

To save the chart:

1. In the Object Navigator window, click **Disp1**.
2. Click the **Save** button on the Object Navigator toolbar. The Save dialog box opens.
3. If necessary, click the **File System** option button, and then click **OK**. The Save As dialog box opens.
4. Navigate to the Chapter8 folder on your Student Disk, type **ch8ordersource** in the File Name text box, and then click **Save**. The display name in the Object Navigator window changes to the file name.

To create the chart, you will use the Chart Genie in the Layout Editor to automate the chart creation process. The Chart Genie is similar to the wizards you used in Form Builder and Report Builder.

To start the Chart Genie:

1. In the Object Navigator, click **Window** on the menu bar, click **Graphics: ch8ordersource.ogd:Layout Editor** to open the Layout Editor, and then maximize the Layout Editor window if necessary. The Graphics Layout Editor window is similar to the Form Builder Layout Editor.
2. Click **Chart** on the menu bar, and then click **Create Chart** to open the Chart Genie - New Query dialog box.

Creating a new chart is a two-step process: first you create the query to define the data that you want to display in the chart, and then you define the chart properties. The following sections will describe this process.

Creating the Chart Query

The data that the chart query retrieves must include the numerical data that are displayed in the pie slices of a pie chart, or on the X- and Y-axes of a column, bar, or line chart. The query must also retrieve the textual or numerical data that are displayed as labels on the pie slices, or on the X- and Y-axes of the other chart types. When writing the query to retrieve the chart data, you must first visualize what the chart will look like. Then you can determine which data fields the query needs to retrieve. In order for your chart to display the different proportions of total order revenue from different order sources, you will retrieve the total revenue amount for each order source. (Graphics Builder will automatically calculate the proportion that each source contributes to the total.) Since the chart will be a pie chart, the query will also retrieve the data fields that will be displayed as the pie slice labels, which are the names of the different order sources. Therefore, you need to create a query in which one column shows the names of the different order sources, and another column shows the total revenue (ORDER_PRICE multiplied by (*) QUANTITY from the ORDERLINE table) from each source. Since the total revenue column is a calculated value, the SQL command must create an alias for this column. Recall that an alias is an alternate name for a column retrieved in a SELECT command. When you use a calculated value in a query for a Graphics chart, you must create an alias for the column with calculated values, or some Graphics Builder functions will not work correctly.

To define your chart query, you can enter a SQL statement, import a SQL query from a text file, or import the data from a variety of other source file types, such as Microsoft Excel or Lotus 1-2-3. Now you will define the query by entering the SQL statement directly.

To define the chart query:

1 Click in the Name text box, and change the query name to **ordersource_query**.

2 Click the **Type** list arrow, and make sure that **SQL Statement** is selected. Since you will directly enter a SQL query to retrieve the data for the pie chart, the File text box is disabled.

Click in the SQL Statement text box, and then type the query shown in Figure 8-6.

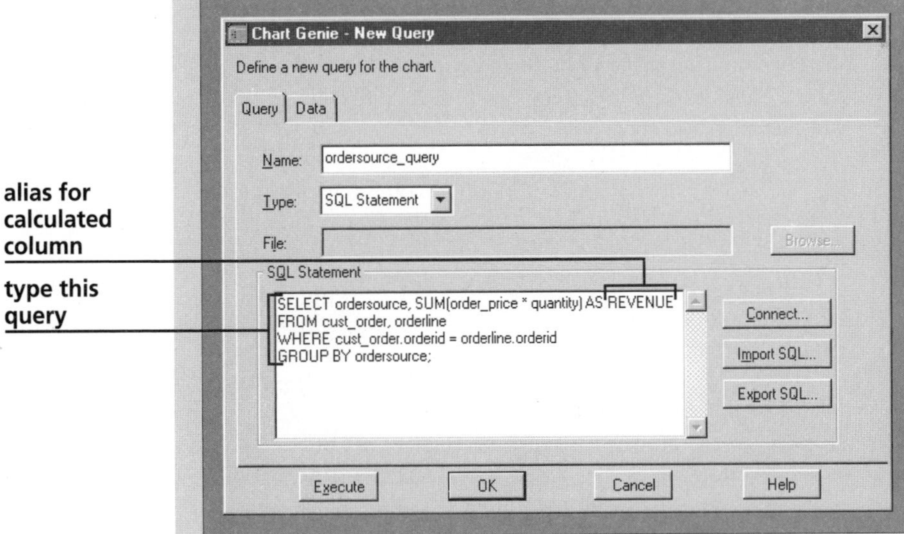

Figure 8-6: SQL command for ordersource_query

4 Click **Execute** to execute the SQL statement. The Data tab of the Chart Genie - New Query dialog box displays the query's output, as shown in Figure 8-7.

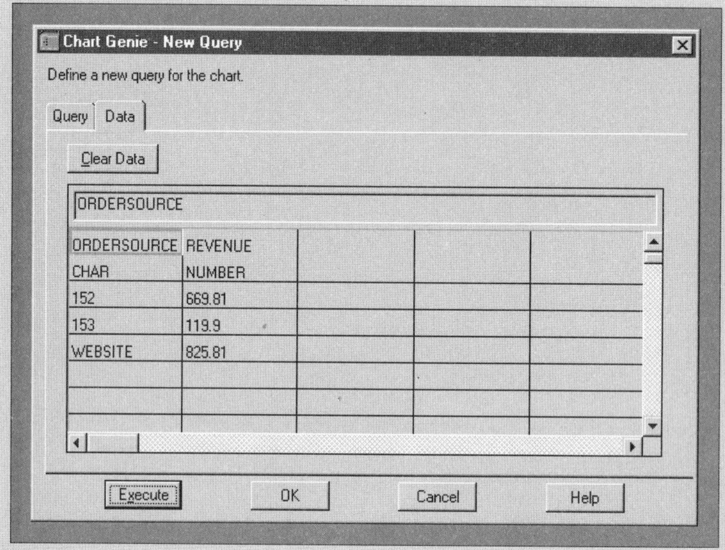

Figure 8-7: Data tab for ordersource_query

help

If you receive a DBMS (ORA-) error message when you try to execute a query, or if your query does not return the data shown, then you probably made a SQL error, such as forgetting a join clause. Use Query Builder to generate the query, or create the query in a text editor and then test it using SQL*Plus. When the command is correct, copy the tested SQL code into the SQL Statement text box on the Query tab and execute the query again.

5 Click **OK** to accept the data and open the Chart Properties dialog box, as shown in Figure 8-8.

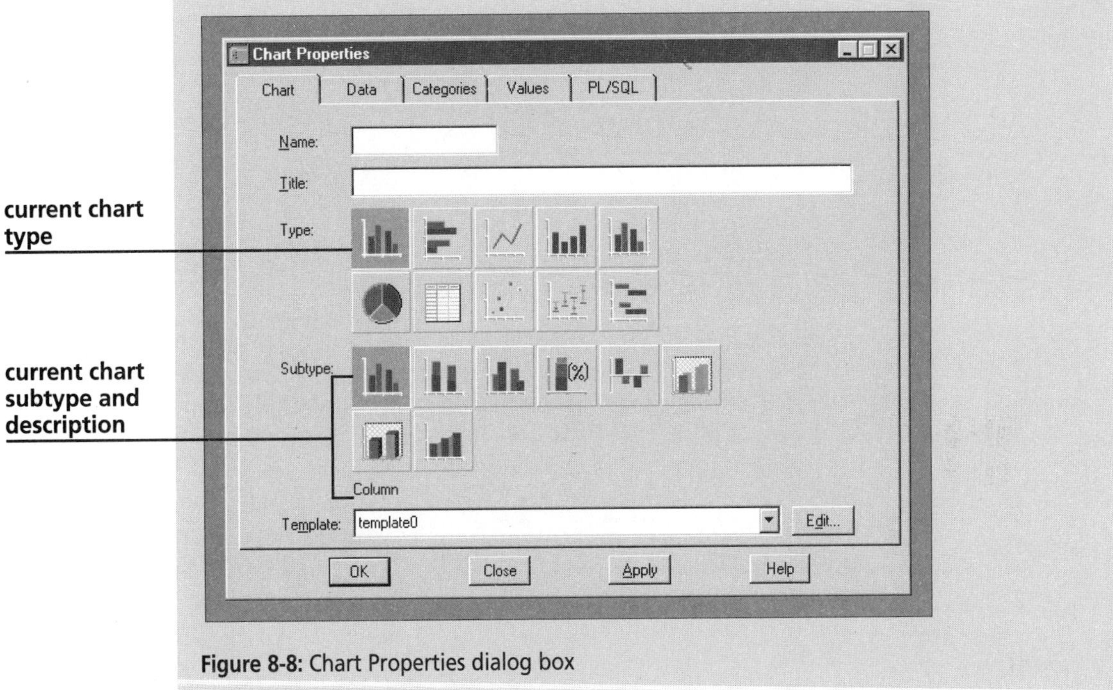

Figure 8-8: Chart Properties dialog box

Defining the Chart Properties

The Chart Properties dialog box lets you configure the chart's appearance. You use the Chart tab of the Chart Properties dialog box to specify the chart name, title, type, subtype, and template, which allows you to apply a predefined style to the chart. The **chart name** should be different from the query name, because you might have several charts that use the same query to define the data that they display. The **chart title** is displayed at the top of the chart. The **chart type** defines the type of chart you want to create—pie, column, line, etc. The **subtype** defines the different display and shading options for the selected chart type. The default chart type is column, which is currently selected. The default chart subtype is also column. The name of the current subtype appears above the Template list box. The Template list box lets you select a predefined template for the chart, which is useful if you want to create many charts using the same chart properties.

To name the chart and change its properties:

1. Click in the Name text box, and then type **ordersource_chart**.
2. Click in the Title text box, and then type **Order Revenue Sources**.

3 Click the **Pie** chart type (the first chart type in the second row). Note that the Subtype options change to show variations of the pie chart.

4 Click the **Pie with depth** subtype (the third subtype from the left). Do NOT click OK.

help

If you accidentally click OK, right-click on the gray chart background within the selection handles that define the chart's perimeter, then click Properties to redisplay the Chart Properties dialog box.

5 Click the **Data** tab. The Edit button enables you to edit the SQL query. The Filter Function list enables you to select a predefined PL/SQL function that filters the query data so that only certain rows are returned. The Mapping option buttons are used to specify how the data are mapped when a Gantt or High-Low chart type is used. The Data Range option buttons allow you to specify that only a set number of records are returned. This is useful for speeding up the performance of queries that might return hundreds or thousands of records.

6 Click the **Categories** tab. This tab allows you to specify the chart labels from a list of the data fields returned by the SQL query. The Chart Genie selected the ORDERSOURCE data field for the Chart Categories data source, since this was the only character data field returned by the query. This is the data field that you want to use for the pie slice labels (see Figure 8-4), so you will not change the Chart Categories selection.

7 Click the **Values** tab. This tab enables you to specify the query column that will be used to determine the pie slice proportions for the chart. Since REVENUE is the only numerical column in the query SELECT statement, the Chart Genie selected it as the default choice for the chart values.

8 Click the **PL/SQL** tab. This is used to create a PL/SQL trigger that will execute when the user performs a specific action, such as clicking the mouse button. This is an advanced chart application, and will not be covered in this book.

9 Click **OK** to close the Chart Properties dialog box and display the chart in the Layout Editor, as shown in Figure 8-9.

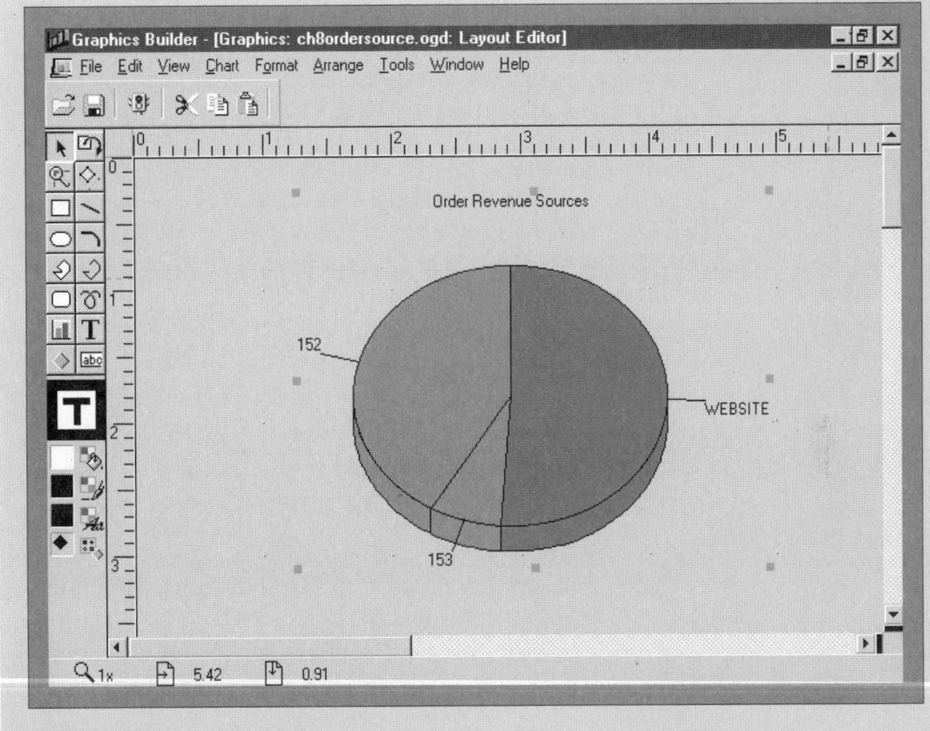

Figure 8-9: Chart in Layout Editor

10 Click the **Save** button on the toolbar to save the chart.

Editing and Formatting the Chart Display

The next step is to edit and reformat the chart display. Currently, each slice of the pie chart shows the approximate proportion of total revenue for each of the different sources, but it would be more informative if it listed the actual dollar figures along with the source names. You can re-enter the Chart Properties dialog box and edit the chart style, SQL query, categories, and values. Now you will open the Chart Properties window and edit the display to show the actual revenue values along with the order source values for the pie slice labels.

To edit the displayed data:

1. If necessary, click anywhere in the pie chart to select it. Selection handles appear around a square perimeter of the chart, as shown in Figure 8-9, to indicate that the chart is selected.

If selection handles appear around the individual pie slices as well as the chart, you selected the chart's individual components rather than the entire chart. Click anywhere outside the chart, and then click the chart to select it.

2 Within the selected area of the pie, right-click anywhere on the gray background, and then click **Properties** on the menu to reopen the Chart Properties dialog box.

tip

You also can double-click anywhere on the background area around the chart to open the Chart Properties dialog box.

3 Click the **Categories** tab, click **REVENUE** in the Query Columns list box, click **Insert** to copy REVENUE into the Chart Categories list, and then click **OK** to close the Properties dialog box and save your changes. The chart now displays revenue amounts in the pie slice labels, as shown in Figure 8-10.

revenue values

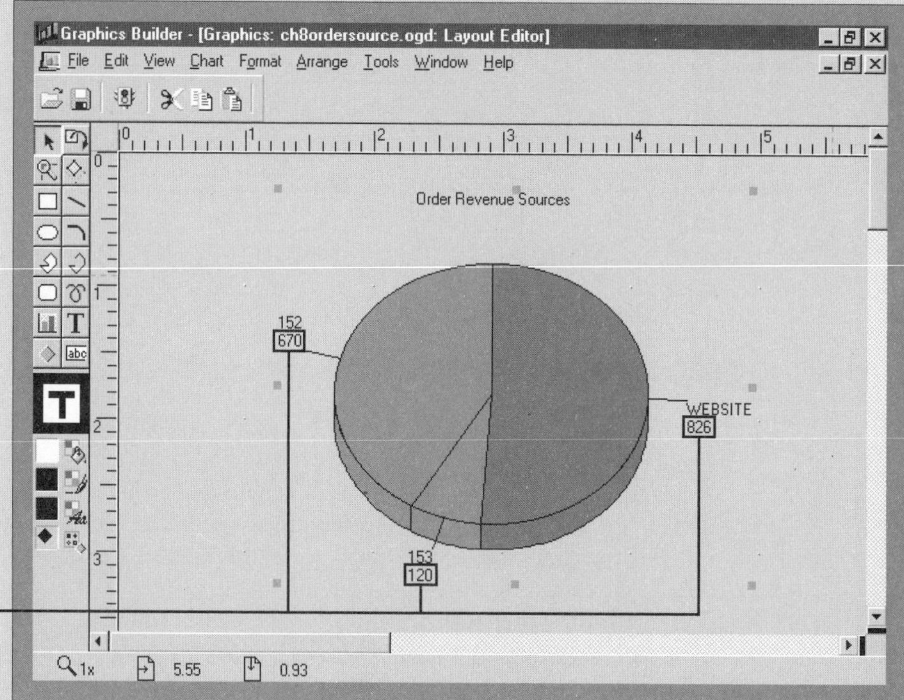

Figure 8-10: Updated chart display with revenue values

Now the chart includes the revenue amounts, but they are displayed as numbers instead of as currency, which makes it hard to differentiate them from the catalog numbers. You can apply a currency format mask to the revenue amounts so that they will appear with dollar signs and two decimal places. Then you will reposition the chart and change the chart labels to make them more prominent and enhance the chart's appearance.

In Graphics Builder, you can apply format masks only to NUMBER and DATE data types.

To format the revenue amounts as currency:

1 If necessary, click the chart to select it, and then click any one of the revenue values to select them as a group. Selection handles should appear around each revenue value, as shown in Figure 8-11.

selection handles

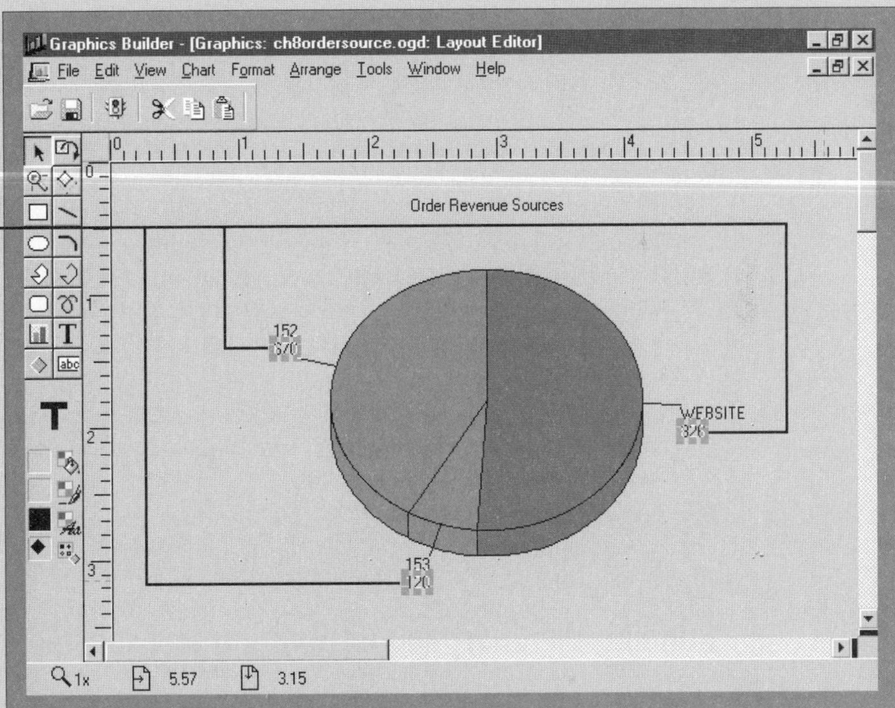

Figure 8-11: Chart revenue values selected

2 With the revenue values selected, click **Format** on the menu bar, click **Number** to open the Number Format dialog box, click **$999,990.99**, and then click **OK** to apply the selected format mask to the selected objects. The currency values now should be formatted correctly in the Layout Editor window.

help

If the format mask you want to use does not appear in the list of available format masks, click Format on the menu bar, and then click Number. Type the format mask you want to use in the Format text box, click Add, then click OK.

Now the revenue values are displayed as currency with two decimal places. Next, you will reposition and resize the chart, and then format the chart labels.

To reposition and resize the chart, and format the labels:

1. Select the chart so that selection handles appear around the entire chart area. Drag the chart to the top-left corner of the Layout Editor, then click the lower-right selection handle, and use the rulers to resize the chart so it is about **3.9** inches wide and **3.0** inches tall. The icons in the bottom-left corner of the Layout Editor window show the exact dimensions of the chart as you resize it.

   help

   If the rulers are not displayed at the top and left edges of the Layout Editor window, click View on the menu bar, and then click Rulers to display them.

2. Select the entire chart, and then click the middle of the chart and drag it so the upper-left corner of the chart is **0.2** inches from the top and **0.2** inches from the left edge of the Layout Editor painting region.
3. Click the chart title so selection handles appear around the title, click **Format** on the menu bar, point to **Font**, click **Font**, change the font to **10-point Arial bold**, and then click **OK**. (If your system does not have Arial, substitute another font.)
4. Click any one of the order source labels to select them as a group, press the **Shift** key, click any one of the revenue figures to select them as a group, and then change the font for both labels to **8-point Arial bold**.
5. Save the chart.

Running the Chart and Generating an Executable Chart File

Now that the chart looks the way you want it to in the Layout Editor, you need to run it in the Graphics Runtime application to see how it looks when run. Like the other Oracle Developer applications, Graphics Builder allows you to save Graphics charts both as design files (with an .ogd extension) and as executable files (with an .ogr extension). As with the other Developer applications, to run .ogr files directly from Windows Explorer, .ogr files must be a registered file type on your workstation. When you run a chart in the Graphics Builder environment, it is displayed in the Graphics Debugger window, which is used for previewing graphics files during development. When users run the chart directly from Windows Explorer, the window title is "Graphics Runtime." Now you will run the chart in the Graphics Debugger window, and then create an executable .ogr file and run it from Windows Explorer.

To run the chart and create an executable file:

1 Click the **Run** button on the Layout Editor toolbar to run your chart in the Graphics Debugger window. Close the Graphics Debugger window by clicking **File** on the menu bar, and then clicking **Close**.

tip

You can also close the Graphics Debugger window by clicking ☒ on the inner window in the Graphics Debugger window. If you click ☒ on the outer window, you will close Graphics Builder.

2 Click **File** on the menu bar, point to **Administration**, point to **Generate**, and then click **File System**. The Save As dialog box opens.

3 Make sure that the Chapter8 folder on your Student Disk is selected in the Save in text box and that the filename is **ch8ordersource.ogr**, and then click **Save**. The executable file is saved on your Student Disk.

Now you will run the executable file by starting Windows Explorer and double-clicking the filename in Windows Explorer.

To run the ch8ordersource.ogr file from Windows Explorer:

1 Start Windows Explorer, and navigate to the Chapter8 folder on your Student Disk.

2 Double-click **ch8ordersource.ogr**. If the Connect dialog box is displayed, then .ogr files are registered correctly on your workstation, and you do not need to follow the steps in the Help paragraphs that follow. If the chart is displayed in the Graphics Runtime window, you can proceed to Step 3.

help

If the Open With dialog box opens instead of the Connect dialog box, your workstation is not configured to run executable Graphics Builder files (which have .ogr extensions) using the Oracle Graphics Runtime application. For Windows 95 users, click Other, click Browse, click the gorun60.exe file in the ORAWIN95\BIN\ directory, click Open, and then click OK. For Windows NT users, click Other, click the gorun60.exe file in the \ORANT\BIN\ folder, click Open, then click OK. Now the Connect dialog box will open automatically when you double-click files with .ogr extensions.

help

If the Connect dialog box is not displayed after you make the file association, then the .ogr file is associated with the wrong executable file. To edit a file registration, click View on the Windows Explorer menu bar, and then click Folder Options. Click the File Types tab, and then scroll down in the file list until you find an entry called Graphics Builder Executable. Click Graphics Builder Executable, click Edit, click Edit again, click Browse, and then navigate to the \ORAWIN95\BIN\gorun60.exe file. Click OK, click Close to save the new registration, then click Close to close the Folder Options dialog box.

3 When the Connect dialog box opens, connect to the database as usual, view the chart in the Graphics Runtime window, then click ☒ on the window title bar to close the Graphics Runtime window.

4 If necessary, click **Graphics Builder** on the Windows taskbar to return to Graphics Builder.

Creating a Column Chart

Next you will create a column chart to display a detailed analysis of the revenue generated by different items sold through the Web site order source. Specifically, you want to see how much revenue was generated by each item in the ITEM table (Women's Hiking Shorts, Women's Fleece Pullover, etc.). You will not display how much revenue was generated by each individual color and size combination for the items. Figure 8-12 shows the design sketch for the revenue detail column chart. Each item is shown as a separate column on the X-axis, and total revenue determines each column's height on the Y-axis. You will display this chart beside the Order Revenue Sources pie chart that you just created. For this chart, you will create the query and layout components manually instead of using the Chart Genie.

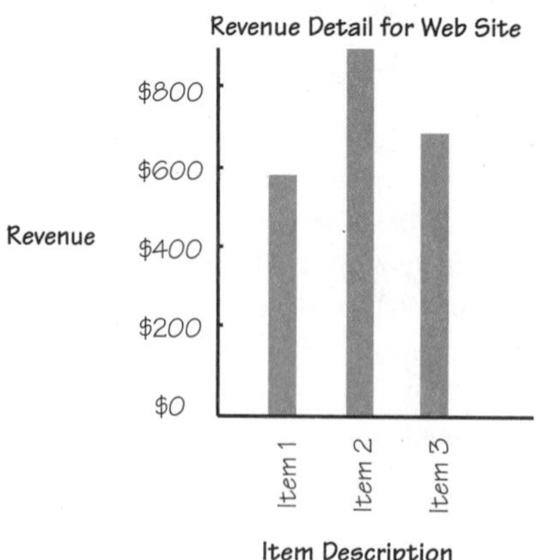

Figure 8-12: Design sketch for the revenue detail column chart

Creating a Chart Manually

The only difference between creating a chart manually and creating a chart using the Chart Genie is that the Chart Genie automatically displays the Create Query window and the Chart Properties window. When you create the chart manually, you must explicitly open these windows. You will create the new chart so it is displayed beside the pie chart you created earlier. First you will save the ch8ordersource.ogd file with a new filename, so your original file remains intact.

To save the chart with a new filename:

1 If necessary, start Graphics Builder and open the **ch8ordersource.ogd** file that you created in the previous exercise from the Chapter8 folder on your Student Disk.

2 Click **File** on the menu bar, point to **Save As**, and then click **File System**. If necessary, navigate to the Chapter8 folder on your Student Disk. Type **ch8revdetail.ogd** in the File Name text box, and then click **Save**.

3 If necessary, click **Window** on the menu bar, and then click **Object Navigator** to open the Object Navigator window.

Now you will create the chart query. For this chart, you want to display the item descriptions on the X-axis and show the total revenue generated by each item as a column value on the Y-axis. Therefore, the query will need to return item descriptions and associated total revenue for each item. You will calculate total revenue for each item as quantity times (*) price for all orders placed using the Web site as the source.

To create the chart query manually:

1 Click the **Queries** subcategory under the ch8arevdetail.ogd display (see Figure 8-13), and then click the **Create** button on the Object Navigator toolbar. The Query Properties dialog box opens.

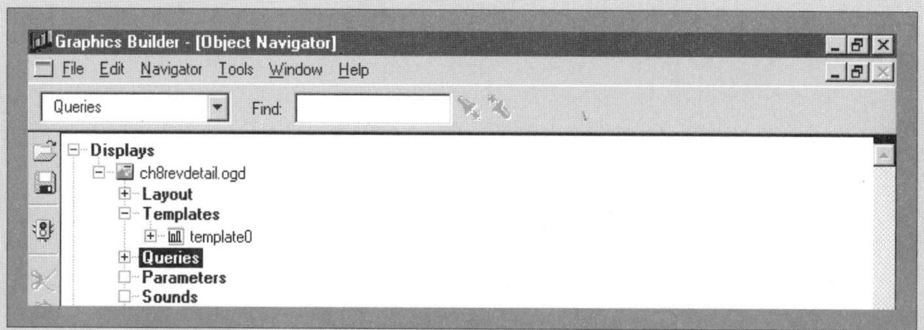

Figure 8-13: Creating a new chart query manually

2 Name the new query **revdetail_query.**

3 If necessary, click the **Type** list arrow, click **SQL Statement**, and then click in the SQL Statement text box and type the query shown in Figure 8-14.

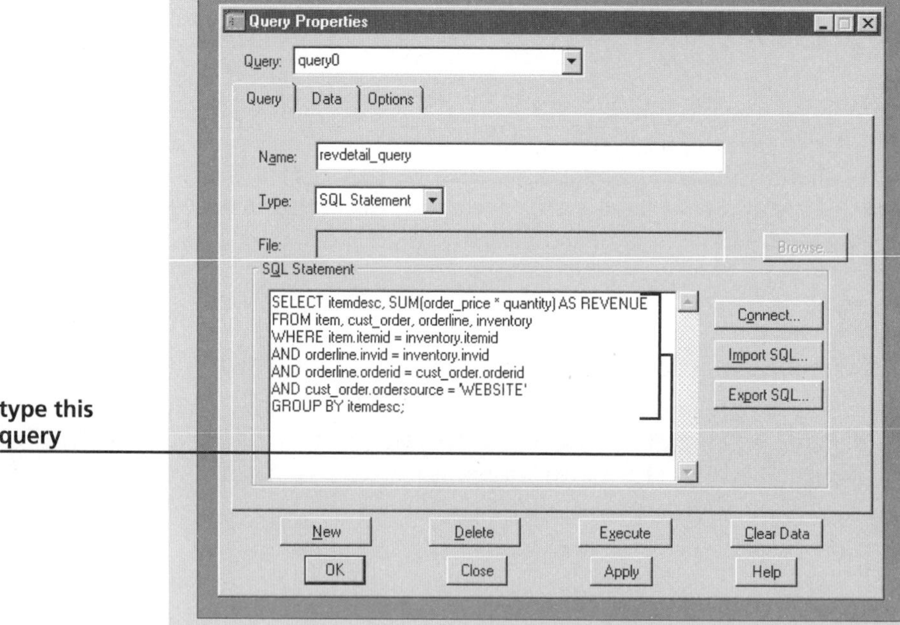

Figure 8-14: SQL command for revdetail_query

4 Click **Execute** to execute the SQL statement and display the Data tab, as shown in Figure 8-15. If your data are different, double-check to make sure you typed the SQL command correctly.

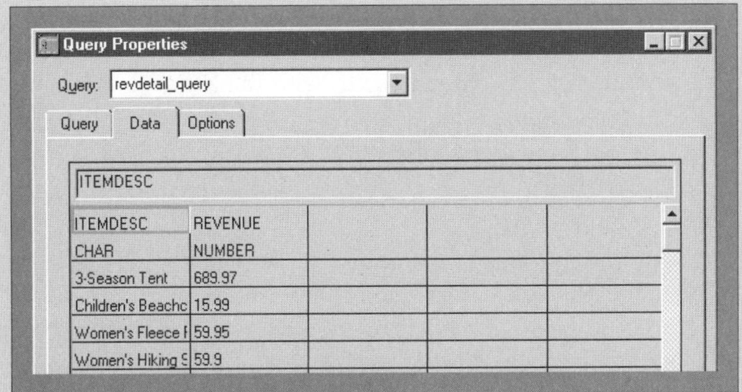

Figure 8-15: Data for revdetail_query

5 Click **OK** to accept the data and to close the Query Properties dialog box.

Creating a Layout Manually

Next you will open the Layout Editor where the pie chart is displayed. You will create the layout for the column chart in the area to the right of the pie chart.

To create the layout for the column chart:

1 Click **Window** on the menu bar, and then click **Graphics: ch8revdetail.ogd: Layout Editor** to open the Layout Editor.

2 Click **View** on the menu bar, and then click **Zoom Out**. Zoom out again if necessary until you can see at least five inches of blank space to the right of the pie chart.

3 Click the **Chart** tool on the Tool palette, and then click anywhere in the layout area and to the right of the pie chart. The Chart Genie dialog box opens and prompts you for a query source.

4 Make sure that the **Existing Query** option button is selected and that **revdetail_query** is displayed in the query list, and then click **OK**. The Chart Properties dialog box opens.

5 Type **revdetail_chart** as the chart name and **Revenue Detail for Web Site** as the chart title.

6 Be sure that the **Column** chart type (the first chart in the first type row) and the **Column with depth** chart subtype (the first chart in the second subtype row) are selected.

7 Click the **Categories** tab. Recall that this property specifies the labels that correspond to the displayed data. ITEMDESC is selected as the chart category for each bar, which agrees with the X-axis labels on your chart design sketch in Figure 8-12.

8 Click the **Values** tab. REVENUE is listed as the source for the Y-axis columns, which is also correct. Click **OK** to close the dialog box and create the column chart.

9 Move the column chart if necessary so that it is on the screen next to the pie chart. Click **View** on the menu bar, and then click **Zoom In** so you can see the chart in the normal view mode. Move the column chart so that its top edge is **0.2 inches** from the top edge of the layout screen and so that it does not overlap the pie chart, as shown in Figure 8-16.

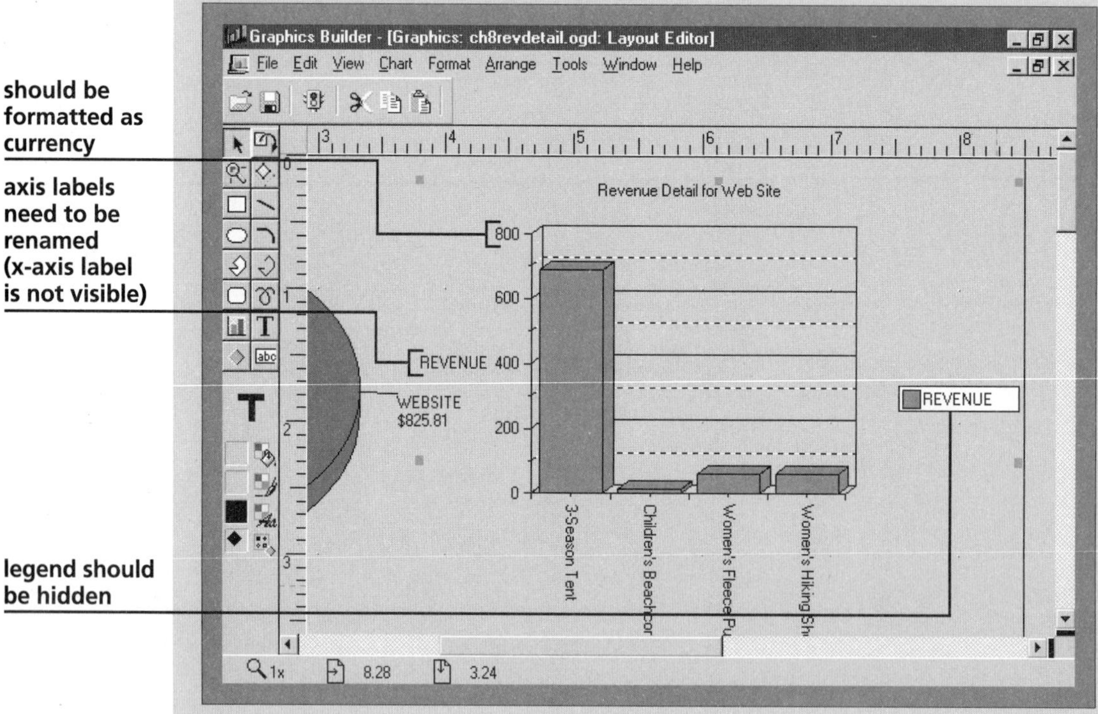

Figure 8-16: Positioning the column chart in the Layout Editor

10 Save the chart.

Formatting the Layout

Figure 8-16 highlights some of the problems with the present layout format. Currently the axis labels are the same as the query column titles—ITEMDESC and REVENUE. You will create custom axis labels so the labels match the design sketch in Figure 8-16. You also will format the **tick labels**, which are the labels on the individual tick mark items on the axes. You will add a currency format mask for the Y-axis revenue values and change the font size and style for the chart title and axis labels. And, since there is only one data series, the chart legend will be hidden. First you will open the Axis Properties window to modify the axis labels and tick labels.

To format the chart axes:

1. Select the column chart, right-click, and then click **Axes**. The Axis Properties dialog box opens, as shown in Figure 8-17. You use this dialog box to select either the X- or Y-axis and then specify the axis display properties, such as the label, direction, tick mark style, tick label rotation, and number of tick marks per interval. You will change the Y-axis first.

The Y-axis is labeled Y1 because you could have two Y-axes in a double-Y chart.

click to select axis

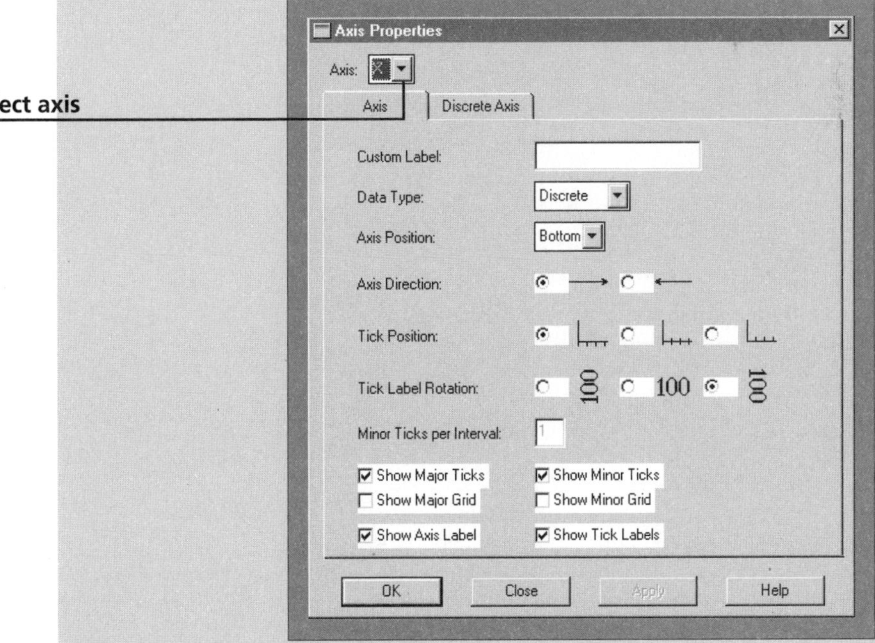

Figure 8-17: Axis Properties dialog box

2 Click the **Axis** list arrow, and then click **Y1**.
3 Click in the Custom Label text box, and type **Revenue**.
4 Click **Apply** to save your changes but leave the Axis Properties dialog box open.

▶ tip

When you click Apply, the changes are saved, but the dialog box remains open. When you click OK, the changes are saved, and the dialog box closes.

5 Click the **Axis** list arrow, and then click **X**.
6 Click in the Custom Label text box, and type **Item Description**.
7 Click the first **Tick Label Rotation** option button so the tick labels are displayed vertically and parallel to the right edge of the page or screen. Click **OK** to save your changes and close the Axis Properties dialog box. The X-axis labels are displayed vertically in the chart layout.

Next you will modify the format mask of the Y-axis tick labels, and change the font sizes and styles for the chart labels. You will do this directly within the Layout Editor.

To modify the format mask and chart label fonts:

1 Click any one of the Y-axis revenue value labels to select the labels as a group, click **Format** on the menu bar, click **Number**, click **$999,990** in the Number Format dialog box, and then click **OK**. The labels now are displayed as currency values.
2 Select the column chart title, click **Format** on the menu bar, point to **Font**, click **Font**, and then select **10-point Arial bold**.
3 Change the X- and Y-axis labels (Revenue and Item Description) to an **8-point Arial bold** font, and change the X- and Y-axis tick labels (currency values and item descriptions) to an **8-point Arial regular** font. Your chart display should look like Figure 8-18.

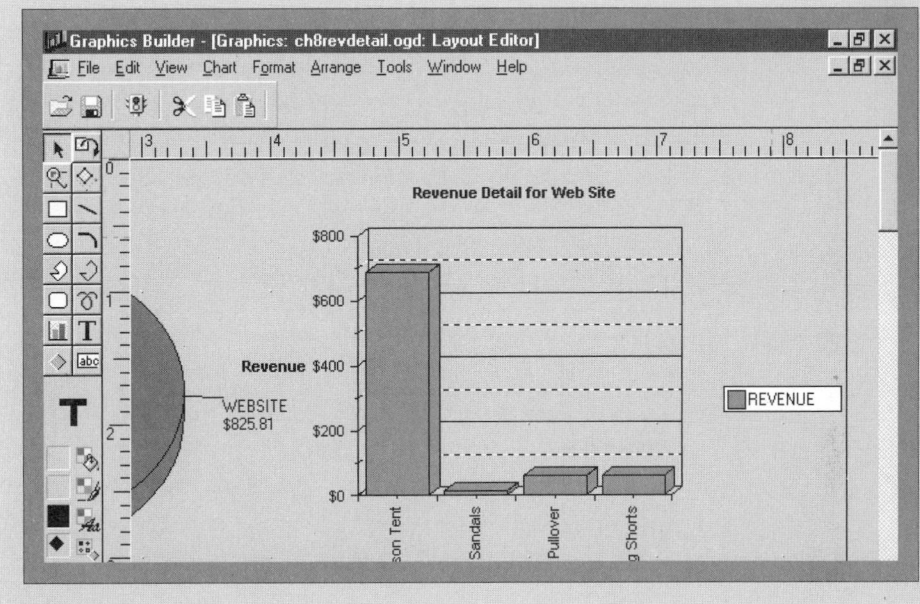

Figure 8-18: Formatted column chart

4 Save the chart.

By default, the chart displays a legend (see Figure 8-16) that identifies the current data in the columns. There is only one data series (REVENUE) in the chart, so you will hide the legend, using the Frame properties dialog box. This dialog box allows you to customize the appearance of aspects of the chart frame, such as the shadow depth and legend display.

To hide the chart legend:

1 Select the column chart, right-click, and then click **Frame** to open the Frame Properties dialog box.

2 Clear the **Show Legend** check box to hide the legend, and then click **OK**. The legend no longer is displayed in the Layout Editor window.

Currently, the columns are the same color on their front, side, and top surfaces. To give two-dimensional objects a three-dimensional appearance, you create the illusion of light shining on one surface with the other surfaces in shadow. You can enhance the appearance of the three-dimensional columns if you choose a light shade for the front surface and a slightly darker shade of the same color for the side and top surfaces.

To enhance the appearance of the column bars:

1. Click the front bar surface on one of the column bars in the column chart, click the **Fill Color** button on the Tool palette, and then change the fill color to **light blue.**
2. Click the side surface of any column bar to select it, click , and then change the fill color to a slightly **darker blue** shade as used on the front surface of the bars.

To achieve the best three-dimensional effect, choose colors that are directly beside each other on the Fill Color palette.

You cannot select the front, side, and top surfaces of a column bar as a group.

3. Click the top surface of any column bar to select it, click, and then change the fill color to the same **darker blue** shade you used in Step 2.
4. Reposition the pie and column charts so that you can see both the pie and column charts in the Layout Editor window at normal size, as shown in Figure 8-19.

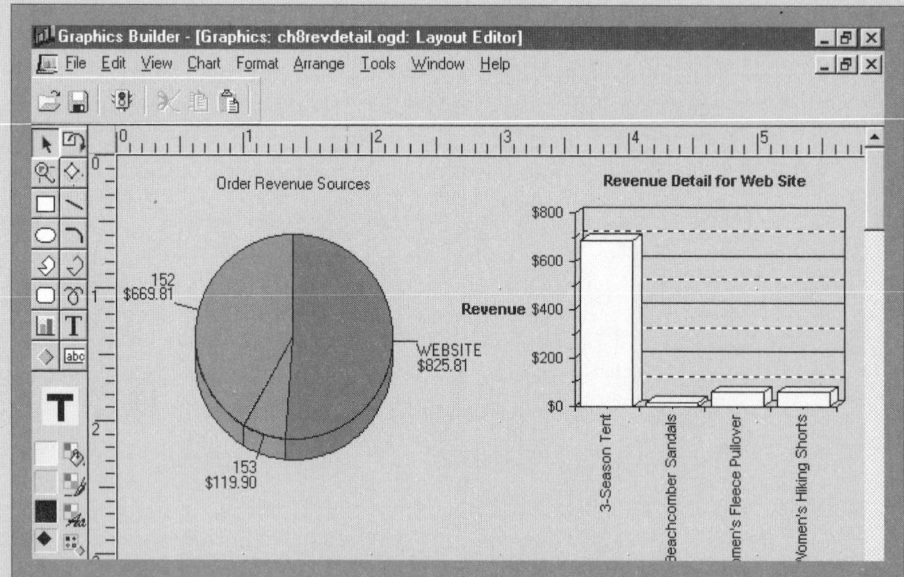

Figure 8-19: Resizing and repositioning the pie and column charts

5. Save the chart and then click the **Run** button to view your charts in the Graphics Debugger window. Adjust the chart sizes and positions if necessary. Close the Graphics Debugger window.

Creating a Chart Drill-Down Relationship

Next you will see how to create a drill-down relationship between two charts, using the two charts that you just created. A **drill-down relationship** allows the user to select a data series on one chart, and then see detail information about the selected data series on a second chart. For example, when the user selects the catalog 152 pie slice in the pie chart, the detail information about all catalog 152 orders will automatically be displayed in the associated column chart. You can create a drill-down relationship with any two charts that have a master-detail relationship where the first (master) chart specifies values that the user can select as search conditions for the second (detail) chart's query.

The first step in creating a drill-down relationship between the pie chart and bar chart is to modify the chart layout as shown in Figure 8-20. You will modify the column chart title to be more general, because data for different order sources will be displayed. You also will need to add instructions to direct the user to click a pie slice to view the associated revenue detail information.

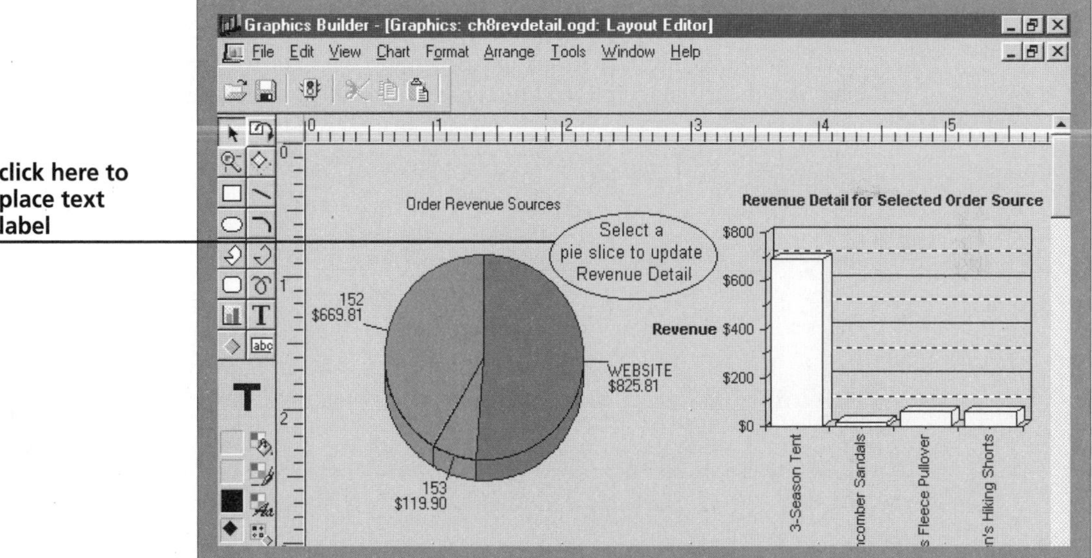

Figure 8-20: Modifying the chart layout for the drill-down relationship

To modify the chart layout for the drill-down relationship:

1 In the Layout Editor, select the column chart, right-click, and then click **Properties.** If necessary, click the **Chart** tab, then change the chart title to **Revenue Detail for Selected Order Source.** Click **OK** to save the change and close the Properties sheet.

2 Click the **Text** tool [T] on the Tool palette, click in the location shown in Figure 8-20, and then create a text label with the following text: **Select a pie slice to update Revenue Detail**. Divide the text into three lines as shown in Figure 8-20 by pressing the **Enter** key at the end of each line.

3 Drag the pointer over the new text to select it, then format the text as **10-point Arial regular**.

4 Click the text so that selection handles appear around it, then click **Format** on the menu bar, point to **Alignment**, then click **Center** to center the text.

5 Click the **Ellipse** tool on the Tool palette, and then draw an ellipse to enclose the user instructions as shown in Figure 8-20. If necessary, set the fill color to **No Fill**, and select **black** for the line color. Your modified chart layout should look like Figure 8-20.

Next, you need to modify the chart properties to create the drill-down relationship. You must create a **chart parameter** first, which is a variable whose value is assigned when the user clicks one of the pie chart slices to select a specific order source. This parameter value is passed to the column chart query, and the values displayed on the column chart change to display the data for the selected parameter. You also must change the column chart query so that the retrieved data are for the parameter value rather than its current search condition, which is WEBSITE. First you will create the parameter for the slices in the pie chart.

To create the drill-down parameter:

1 In the Layout Editor, click the pie chart, and then click any one of the pie slices so that all the slices are selected as individual objects.

2 Right-click anywhere in the selected objects, and then click **Properties** on the menu. The Object Properties dialog box opens for the REVENUE_slices object.

3 Click the **Drill-down** tab, and then click **New** to the right of the Set Parameter text box. The Parameters dialog box opens.

help

If you created a parameter previously in this chart, the New button to the right of the Set Parameter text box will be named Edit. Clicking the Edit button will allow you to create a new parameter.

4 Type **selected_source** for the parameter name. Since ORDERSOURCE is a character data field, make sure that **Char** is the parameter data type.

5 Type **WEBSITE** in the Initial Value text box. (This value will change when the chart runs, but an initial value must be entered.) This text is used as a search condition in the chart query, so it is case sensitive.

6 Click **OK** to close the Parameters dialog box.

Next, you need to specify the query and query field where the parameter will be used. The parameter will be the value for the ORDERSOURCE field in the column chart query, which is the revdetail_query.

To specify the new parameter query and query field:

1 In the Object Properties dialog box, click the **To Value of** list arrow, and then click **ORDERSOURCE**. This specifies that the parameter will take the ORDERSOURCE value of the pie slice that is selected by the user.

2 To specify that the parameter will be used in the column chart revdetail_query, click the **Execute Query** list arrow, and then click **revdetail_query**. Do not click the OK button yet.

Finally, you need to edit the revdetail_query so that the parameter is used for the search condition, instead of the WEBSITE value that you entered when you originally created the query. You will edit the SQL query and specify that the order source is equal to the parameter value (:selected_source). As in other applications, the parameter name is always preceded by a colon.

To edit the revdetail_query:

1 In the Object Properties dialog box, click the **Edit** button to the right of the Execute Query text box. The Query Properties dialog box opens. If necessary, click the **Query** tab so you can edit the SQL query.

2 Modify the query as shown in Figure 8-21.

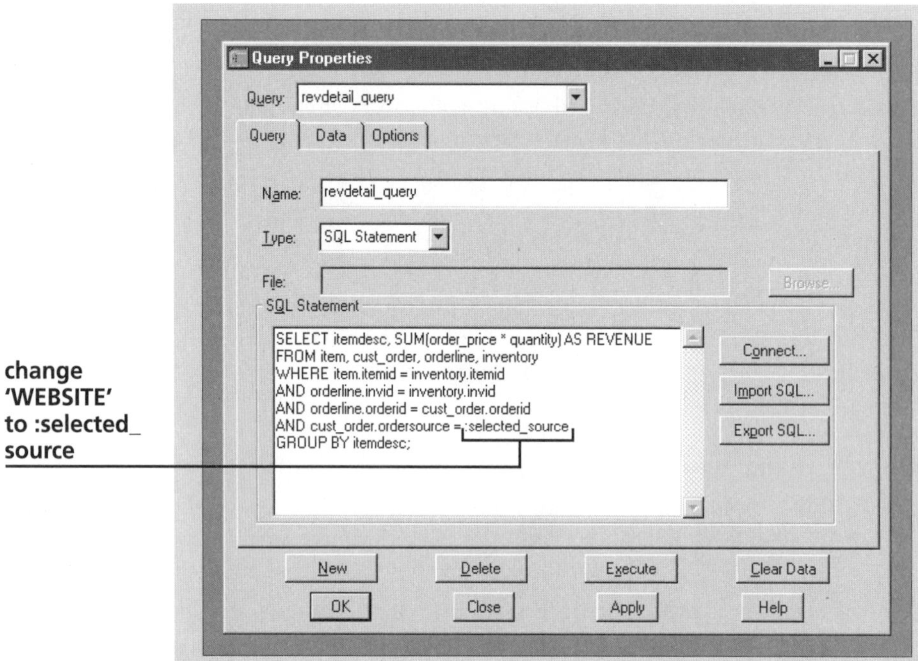

Query Properties

Query: revdetail_query

Query | Data | Options

Name: revdetail_query

Type: SQL Statement

File:

SQL Statement

```
SELECT itemdesc, SUM(order_price * quantity) AS REVENUE
FROM item, cust_order, orderline, inventory
WHERE item.itemid = inventory.itemid
AND orderline.invid = inventory.invid
AND orderline.orderid = cust_order.orderid
AND cust_order.ordersource = :selected_source
GROUP BY itemdesc;
```

Browse... | Connect... | Import SQL... | Export SQL...

New | Delete | Execute | Clear Data

OK | Close | Apply | Help

change 'WEBSITE' to :selected_ source

Figure 8-21: Modifying the revdetail_query

3 Click **OK** to save the changes and close the Query Properties dialog box.

4 Click **OK** to close the Object Properties dialog box.

5 Save the chart.

Now you can run the chart to test if the drill-down relationship works correctly.

To test the drill-down relationship:

1 Run the chart. When the chart first is displayed in the Graphics Debugger window, the WEBSITE data are displayed in the column chart, because WEBSITE was specified as the parameter's initial value.

2 Click the pie slice for catalog **152**. The column chart now displays the data for order source 152.

3 Click the pie slice for catalog **153**. Again, the values displayed in the column chart are dynamically updated.

help

If your drill-down relationship does not work correctly, close the Debugger window, open the Object Navigator, find the parameter you just created, and then delete it. Repeat the steps to create the parameter. If the drill-down relationship still does not work, ask your instructor or technical support person for help.

4 Close the Graphics Debugger window.

5 Open the Object Navigator window, click **ch8revdetail.ogd** under Displays, click **File** on the menu bar, then click **Close** to close the chart file.

Creating a Form That Calls a Graphics Chart

Oracle Graphics Builder charts often are called from within a Form Builder application, or displayed directly on a form. In this exercise you will create a form that allows a user to select an item ID from a list of items in the Clearwater Traders Database. After the user selects the item ID, the form displays a chart that shows the quantity on hand for each inventory item associated with the selected item ID. It does this by passing the item ID to the chart using a parameter list. Figure 8-22 shows the form that displays information about different items in the Clearwater Traders database.

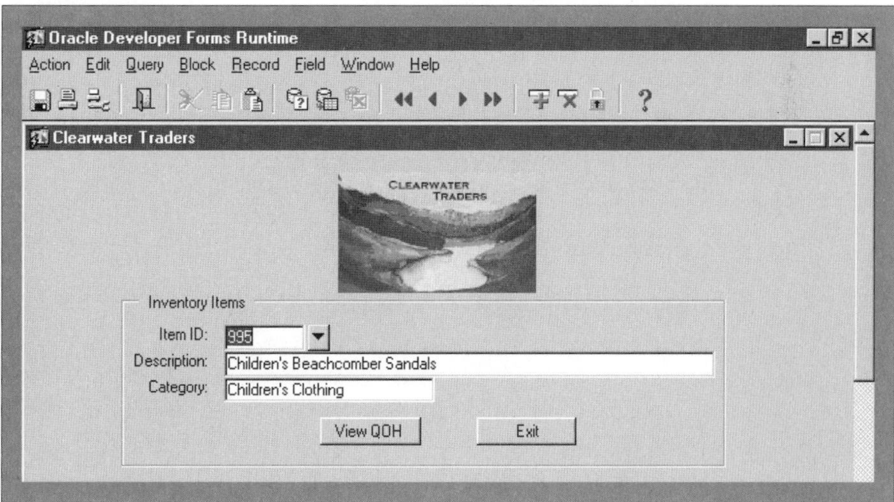

Figure 8-22: Form to display chart

When the user clicks the LOV command button beside the Item ID field, a LOV display presents a list of the database inventory items. When the user selects a specific item—in this case, Children's Beachcomber Sandals—and then clicks the View QOH button, a chart is displayed that shows the inventory quantity on hand for each inventory item with size and color corresponding to the selected item, as shown in Figure 8-23.

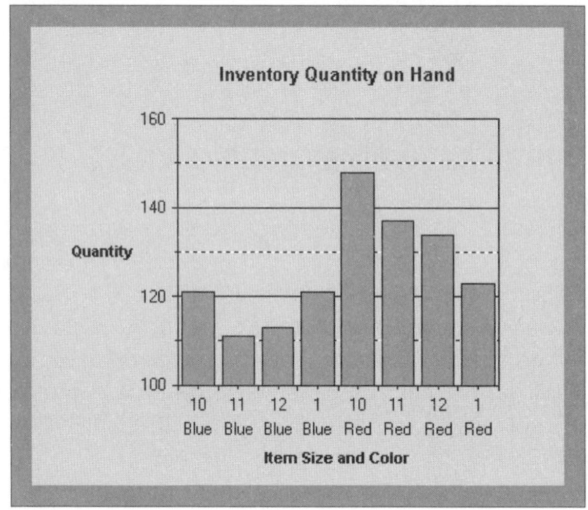

Figure 8-23: Chart displaying quantity on hand for selected item ID

Recall from Chapter 7 that when you passed a parameter from a form to a report, the parameter was passed in a parameter list. In this application, the form button trigger must create a parameter list that contains the selected item ID, and then execute the RUN_PRODUCT procedure. This time, RUN_PRODUCT starts a Graphics Builder chart instead of a report. First, you will modify the chart query so it uses a parameter instead of a specific search condition value for Item ID. Then you will modify the form so it creates a parameter list containing the selected item ID, and then calls the chart.

Modifying the Chart Query

The query in the current chart display is coded so that the search condition is for Item ID 995 (Children's Beachcomber Sandals). To pass the item ID variable values from the form to the chart, you must create a parameter in the chart, just as you created a parameter in the report in Chapter 7 when you called a report from a form. To create a parameter in a chart, you create the parameter in the Object Navigator window, and then replace the search condition in the chart's SQL query with the parameter's name, preceded by a colon. First you will start Graphics Builder, open the chart display, and create a parameter.

To open the chart display and create a parameter:

1. If necessary, start Graphics Builder.
2. Open the **inventory.ogd** file from the Chapter8 folder on your Student Disk, and save it as **ch8inventory.ogd** in the Chapter8 folder on your Student Disk.
3. If necessary, open the Object Navigator window, click the **Parameters** object under ch8inventory.ogd, then click the **Create** button. The Parameters dialog box is displayed.

The Parameters dialog box allows you to specify the parameter name, data type, and initial value. Since the parameter will be assigned to the value of ITEMID in the ITEM table, its data type will be NUMBER. Its initial value can be any valid ITEMID value. After you create the parameter, you will modify the chart query so that the chart displays the data for the item ID value that is assigned to the parameter. Now you will create the parameter and modify the chart query.

To create the parameter and modify the query:

1. Complete the Parameters dialog box as shown in Figure 8-24, then click **OK**. The new parameter is displayed in the Object Navigator window.

your default parameter name might be different

type or select these values

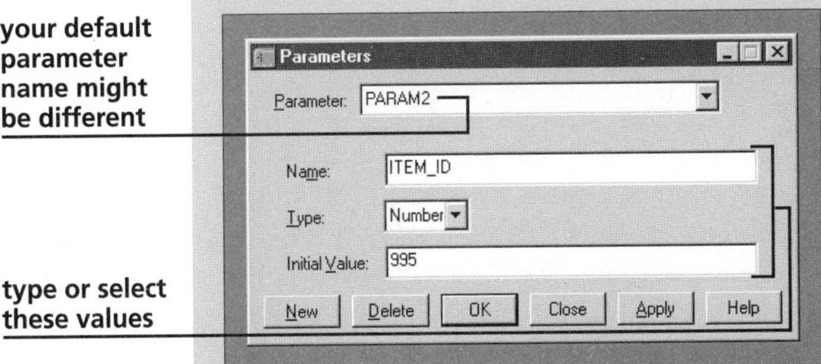

Figure 8-24: Chart parameter specification

2. If necessary, click beside Queries to display the chart queries, then double-click the **Queries** icon beside qoh_query to open the Query Properties dialog box.
3. Modify the query search condition as shown in Figure 8-25, then click **OK**.

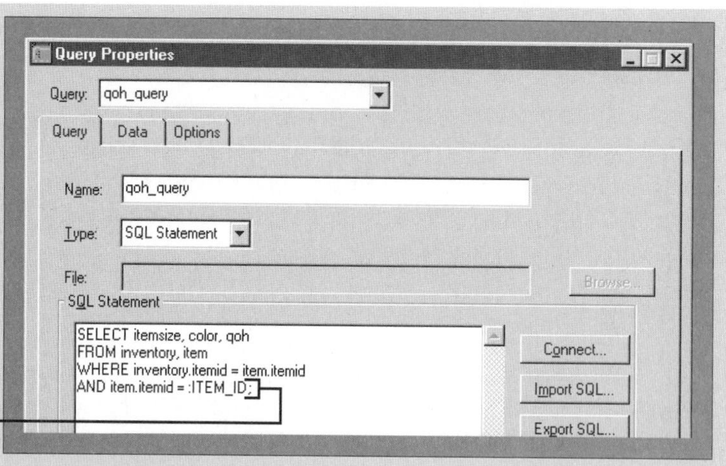

Figure 8-25: Modifying the chart query

4 Save the chart.

5 Close ch8inventory.ogd

Modifying the Form

Next, you will open the form file and create the trigger for the View QOH button. The View QOH button's trigger will create the parameter list that will pass the selected item ID to the chart, and execute the RUN_PRODUCT command to run the chart. When the form executes the RUN_PRODUCT command, it starts Graphics Runtime and displays the chart in the Graphics Runtime window.

You will create the parameter list and use the RUN_PRODUCT procedure the same way you created the parameter list that was passed to the report in Lesson C in Chapter 7, except that this time the product specification will be Graphics instead of Reports.

To open the form and create the View QOH button trigger:

1 Start Form Builder, open the **item.fmb** file from the Chapter8 folder on your Student Disk, and save the file as **ch8item.fmb.**

2 Open the Layout Editor, select the **View QOH** button, right-click, point to **Smart Triggers,** then click **WHEN-BUTTON-PRESSED** to create a trigger for the View QOH button. Type the code shown in Figure 8-26 to create the parameter list and run the chart.

type this code

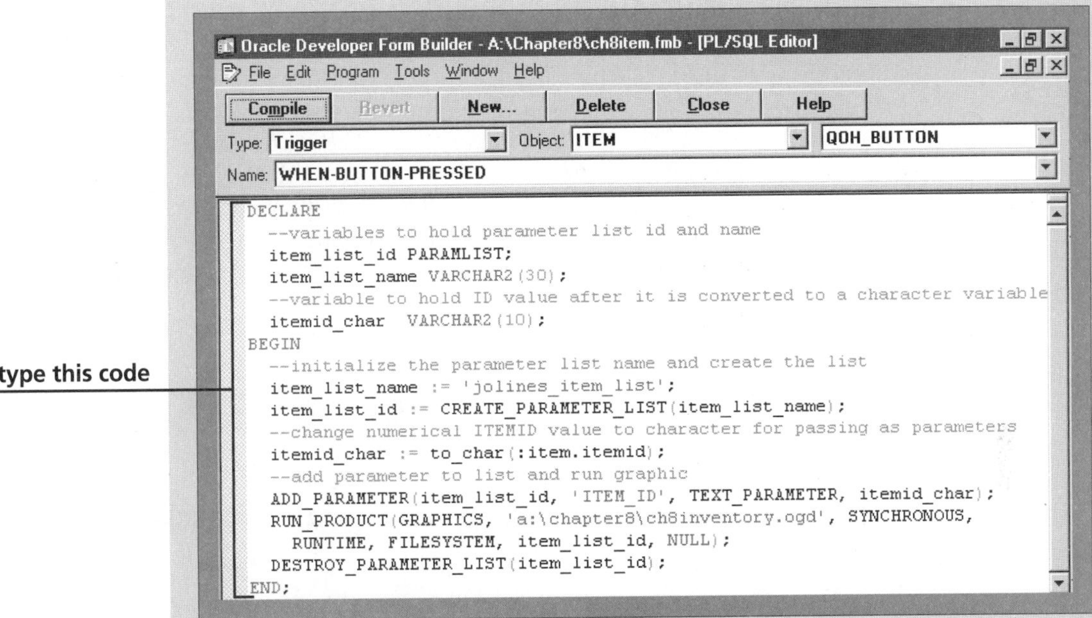

```
DECLARE
  --variables to hold parameter list id and name
  item_list_id PARAMLIST;
  item_list_name VARCHAR2(30);
  --variable to hold ID value after it is converted to a character variable
  itemid_char  VARCHAR2(10);
BEGIN
  --initialize the parameter list name and create the list
  item_list_name := 'jolines_item_list';
  item_list_id := CREATE_PARAMETER_LIST(item_list_name);
  --change numerical ITEMID value to character for passing as parameters
  itemid_char := to_char(:item.itemid);
  --add parameter to list and run graphic
  ADD_PARAMETER(item_list_id, 'ITEM_ID', TEXT_PARAMETER, itemid_char);
  RUN_PRODUCT(GRAPHICS, 'a:\chapter8\ch8inventory.ogd', SYNCHRONOUS,
    RUNTIME, FILESYSTEM, item_list_id, NULL);
  DESTROY_PARAMETER_LIST(item_list_id);
END;
```

Figure 8-26: View QOH button trigger

3 Compile the code, correct any syntax errors, then close the PL/SQL Editor.

4 Save the form. Now you will run the form and confirm that it successfully displays the chart and passes the selected Item ID.

5 Run the form, click the **LOV** command button next to Item ID, select item **559** (Men's Expedition Parka), then click **OK**.

6 Click the **View QOH** button. The inventory quantity on hand for each size of the selected item is displayed in the Graphics Runtime window. This may take a few moments to load and display.

help If the chart does not appear, check to make sure the chart .ogd file is named ch8inventory.ogd, and that it is in the Chapter8 folder on your Student Disk.

7 Click the **Close** button [X] on the window title bar to close the Graphics Runtime window, then click the Exit button to close Forms Runtime.

8 Click **File** on the Form Builder menu bar, then click **Close** to close the form file in Form Builder.

Creating a Form with an Embedded Graphics Chart

Another way to display a chart using a form is to display the chart directly on the form. Now you will create a form that allows the user to select an item from the Clearwater Traders' ITEM table using a LOV. The corresponding inventory colors, sizes, and quantities on hand are displayed in a column chart directly on the form, as shown in Figure 8-27. When a chart is displayed directly on a form, a Graphics Builder application named Oracle Graphics Batch starts when the form is run. This application dynamically updates the chart when its query data change. After you exit the form, Oracle Graphics Batch continues running and must be closed manually.

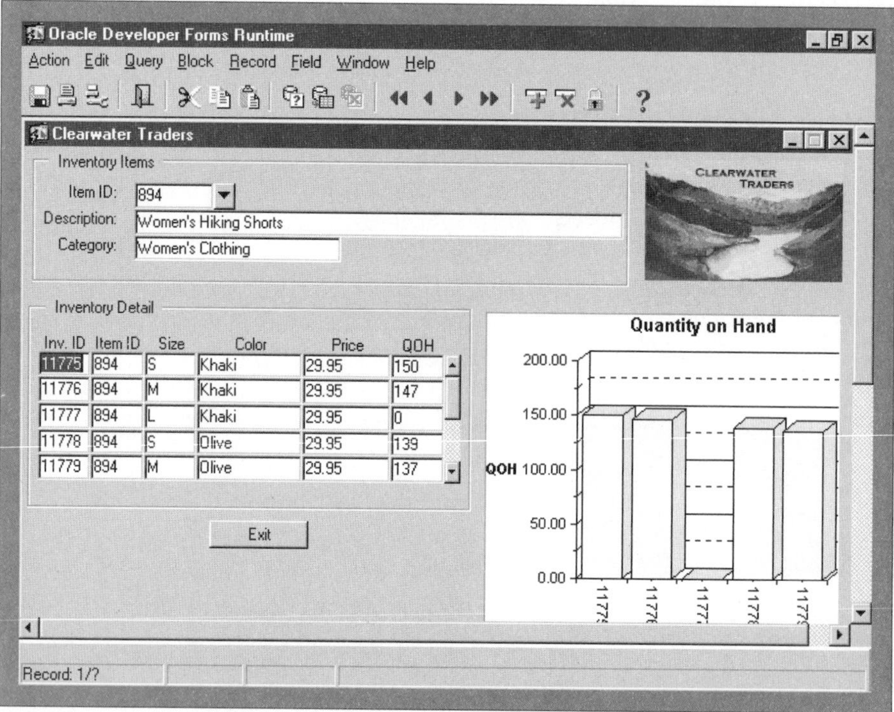

Figure 8-27: Chart displayed directly on form

You will use the Chart Wizard in Form Builder to create the chart item on the form. The Chart Wizard has the following pages to lead you through the chart creation process:

- **Type** page, which allows you to specify the chart title, type (bar, column, line, etc.), and subtype options

- **Block** page, where you specify the name of the form data block that is the source of the chart data
- **Category** page, where you select the form data field that provides the data for the X-axis
- **Value** page, where you select the form data field that provides the data for the Y-axis
- **File Name** page, where you specify the file that stores the chart .ogd file. You can modify the chart properties by opening this file in Graphics Builder.

Like the other Oracle wizards, the Chart Wizard is re-entrant, meaning that you can use it to modify an existing chart created with the Wizard. Now you will open the form and create the chart.

To open the form and create the chart:

1. Switch to Form Builder if necessary, open the **item1.fmb** file from the Chapter8 folder on your Student Disk, and save it as **ch8item1.fmb**.
2. Open the Layout Editor, and click the **Chart Wizard** button on the toolbar. The Chart Wizard Welcome page is displayed. Click **Next.** The Type page is displayed.
3. Type **Quantity on Hand** in the Title box, be sure that **Column** is the selected chart type, and select **Depth** as the subtype. Click **Next**. The Block page is displayed. The form has two blocks, ITEM and INVENTORY. The chart data come from the INVENTORY block, so the chart will be part of this block.
4. Select **INVENTORY** as the block name, then click **Next**. The Category page is displayed.
5. Inventory ID numbers will appear on the X-axis, so select **INVID** from the Available Fields list, click the **Select** button so that INVID is displayed in the Category Axis list, then click **Next**. The Value page is displayed.
6. Quantity on hand values will appear on the Y-axis, so select **QOH** from the Available Fields list, click so that QOH is displayed in the Value Axis list, then click **Next**. The File Name page is displayed.
7. Click **Save As**, navigate to the Chapter8 folder on your Student Disk, type **ch8form** in the File name box, and click **Save**. (You do not need to type the file extension, because it will be added automatically.) The filename is displayed in the File Name page, as shown in Figure 8-28. Click **Finish**. The new chart is displayed on the form.

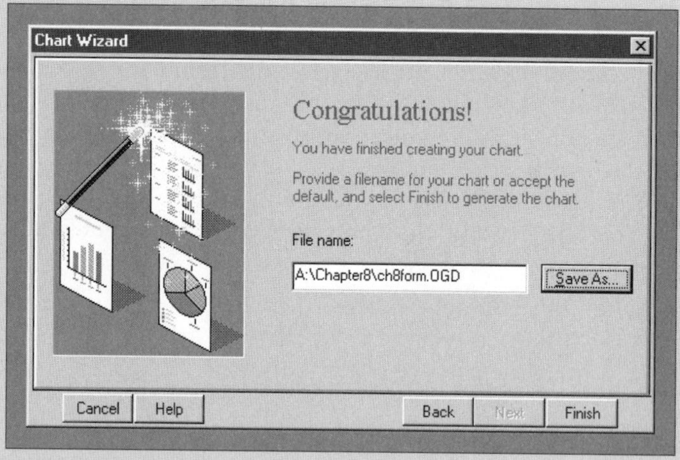

Figure 8-28: Specifying the chart filename

You must specify the filename as shown in Figure 8-28, or the chart will be placed in the default chart directory, and will be difficult to find later.

8 Click the chart so that selection handles appear around its edges. Resize and reposition the chart so your form looks like Figure 8-29.

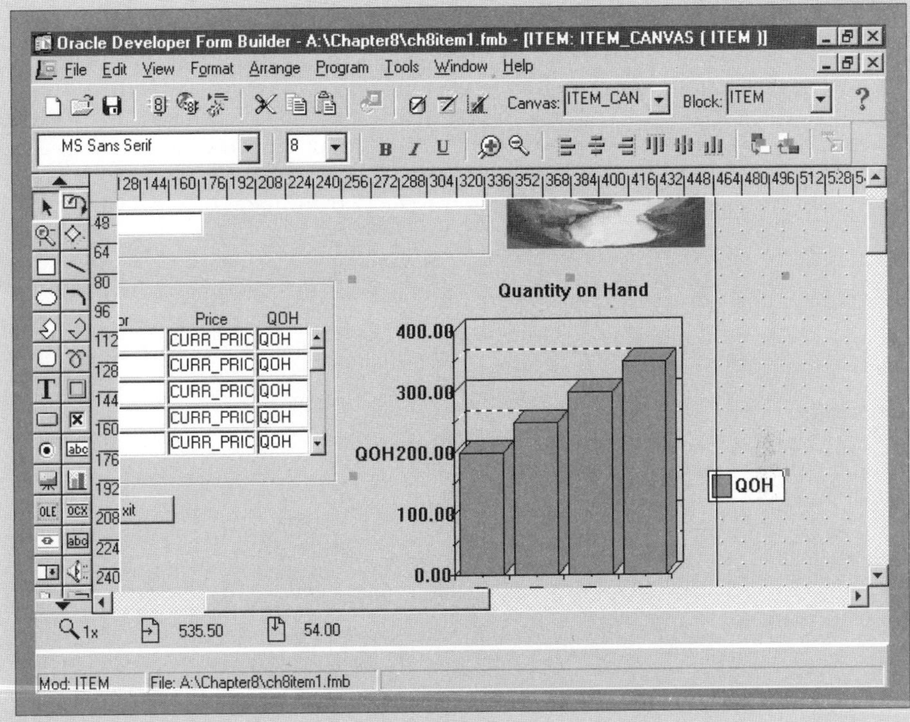

Figure 8-29: Resizing and repositioning the chart

help

If a white box labeled CHART_ITEM appears rather than the template that looks like a column chart, an error occurred with the Chart Wizard. Exit Form Builder, open the form again, and repeat Steps 1–8. If the white box appears again, ask your instructor or technical support person for help.

9 Save the form.

Next, you need to modify some of the chart's properties. You cannot do this directly in Form Builder, so you will open the chart file in Graphics Builder. You will hide the chart legend, modify the X- and Y-axis label fonts, and create a custom label for the X-axis.

To modify the chart properties:

1 Click **Graphics Builder** on the Windows taskbar.

2 Open the **ch8form.ogd** chart file from the Chapter8 folder on your Student Disk. Since the chart is based on a form data block rather than a query, the X-axis labels appear as generic "INVID" values. The actual values will be inserted when the chart is run.

3 Select the chart, right-click, then click **Frame** to open the Frame Properties dialog box. Clear the **Show Legend** check box, then click **OK**.

help

If the legend still appears in the Layout Editor after you make the change, click Chart on the menu bar, then click Update Chart to refresh the layout.

4 Select the chart, right-click, then click **Axes** to open the Axis Properties dialog box. Select the X-axis, type **Inventory ID** as the custom label for the X-axis, then click **OK**.

5 Select the X-axis tick labels, and change the font to **8-point Arial regular.**

6 Select the Y-axis tick labels, and change the font to **8-point Arial regular.**

7 Change the font of the X- and Y-axis labels to **8-point Arial bold.**

8 Change the font of the chart title to **10-point Arial bold.**

9 Change the fill color of the front of the columns to **pale yellow.**

10 Change the fill color of the sides and tops of the columns to a **slightly darker yellow** shade.

11 Change the format mask of the Y-axis tick labels (QOH) to **999,990.**

12 Save the chart. Your formatted chart should look like Figure 8-30.

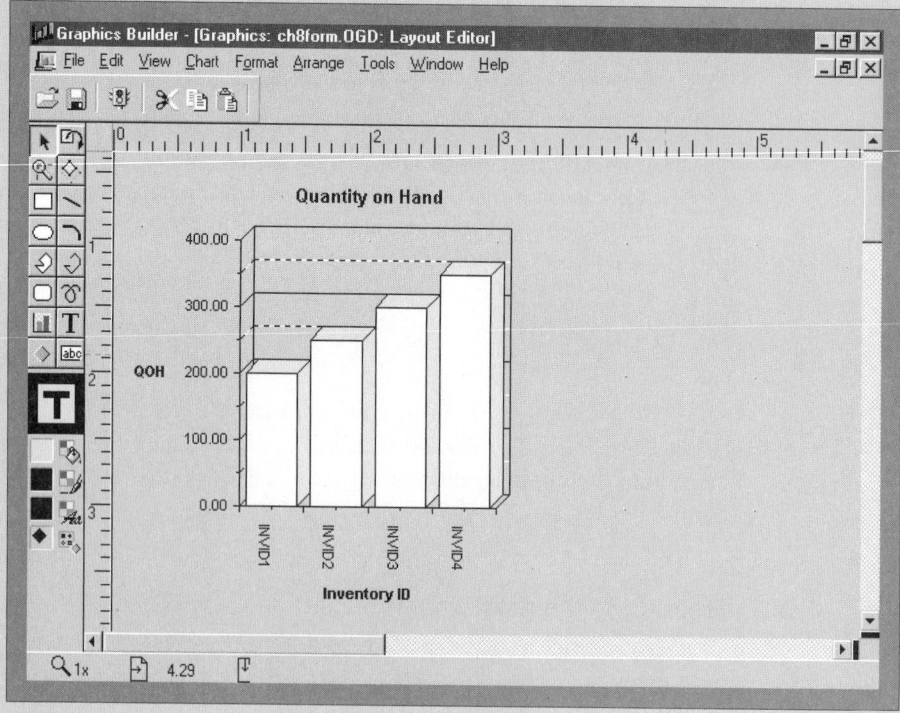

Figure 8-30: Modifying the chart properties

13 Close ch8form.ogd.

Now you will switch back to Form Builder, and run the form to view the chart and to determine if any other modifications are necessary.

To switch to Form Builder and run the form:

1 Click **Form Builder** on the Windows taskbar.

help When you view the chart in the Form Builder Layout Editor, the chart image will still look the same. This is because when you create a chart using the Chart Wizard, a chart template is created to show the chart as a symbol, so you can adjust its size and position. This chart template is updated when you close the form in Form Builder and then reload it.

2 Click the **Run Form Client/Server** button to run the form.

help If an error message is displayed saying that some of the form items are not on the canvas, open the Layout Editor and resize and reposition the chart so it is entirely on the canvas. Then run the form again.

3 Click the **LOV** command button beside Item ID, and select Item **894** (Women's Hiking Shorts). The associated inventory records are displayed in the Inventory Detail frame and on the chart, as shown in Figure 8-31.

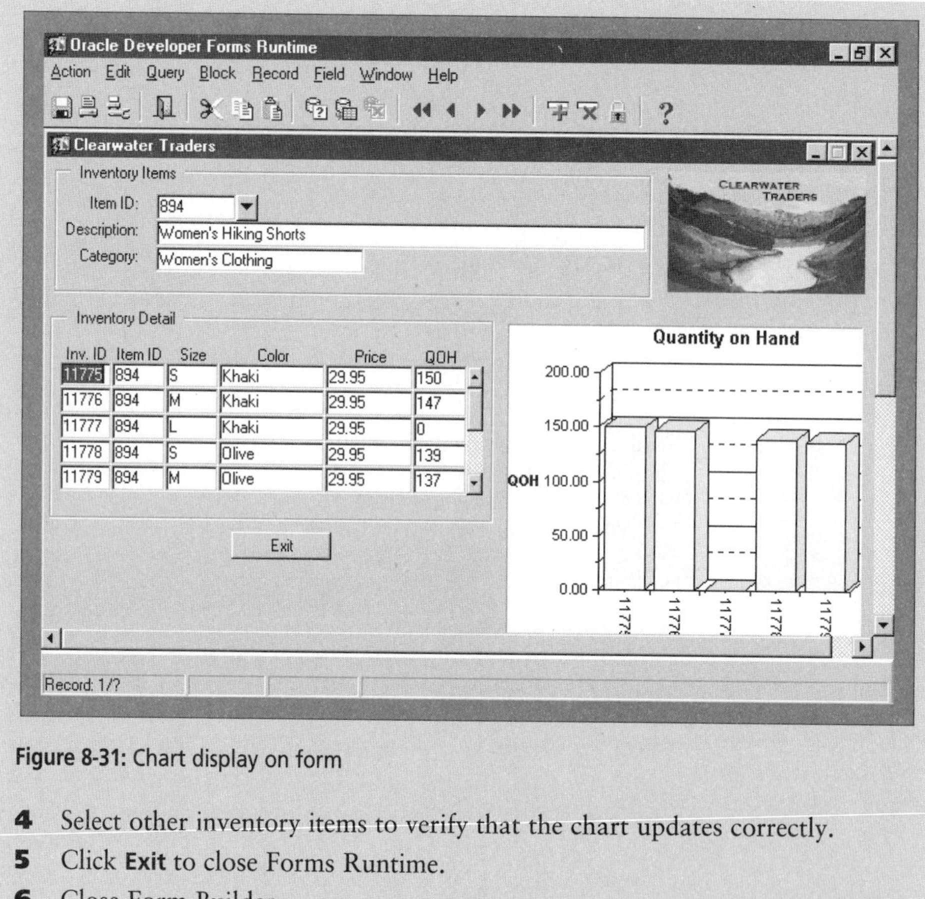

Figure 8-31: Chart display on form

4 Select other inventory items to verify that the chart updates correctly.
5 Click **Exit** to close Forms Runtime.
6 Close Form Builder.
7 Click the **Oracle Graphics Batch** application button on the Windows taskbar, and then close the application.

Creating a Report with an Embedded Graphics Chart

Next, you will create a report that contains a graphics chart that displays some of the report data visually. Clearwater Traders needs an inventory report, similar to the form you just created, that shows each item ID and description and lists all of the associated inventory numbers, sizes, colors, current prices, and quantities on hand. The report also will include a chart that shows the associated quantity on hand for each inventory item. The design sketch for this report is shown in Figure 8-32.

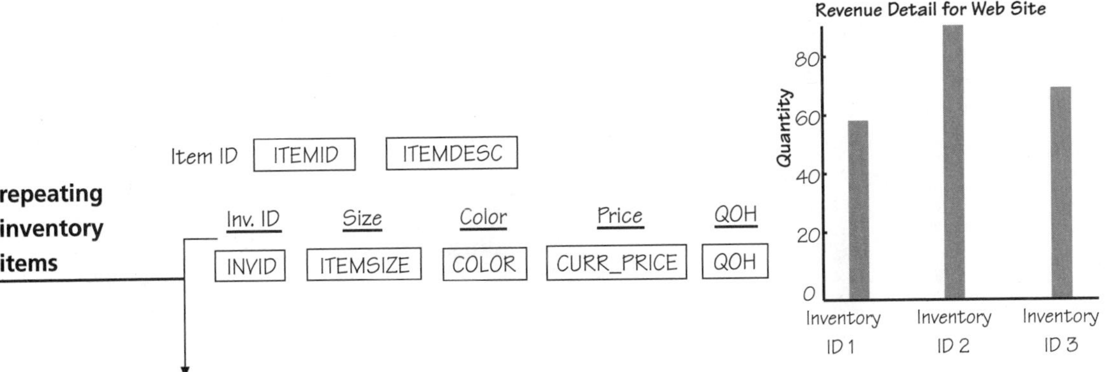

Figure 8-32: Design sketch for report with chart display

The report data appear in a master-detail layout on the left side of the page, with the item information as the master record and the inventory information as repeating detail records. The chart shows the inventory ID and quantity on hand information. To create this report, you will create a chart in the report using the Report Builder Chart Wizard.

The Report Builder Chart Wizard has the same pages as the Form Builder Chart Wizard, except that it also includes a Breaks page that specifies the frequency with which the chart will appear in the report. A chart can appear once at the beginning of the report, once at the end, or once within any of the report's repeating frames. For this report, you want the chart to appear once for each item ID that is displayed. Now you will start Report Builder, open the report, and create the chart using the Chart Wizard.

To start Report Builder and create the chart:

1. Start Report Builder, open the **inventoryqoh.rdf** file from the Chapter8 folder on your Student Disk, and save the report as **ch8inventoryqoh.rdf**.
2. Click **File** on the menu bar, then click **Connect**, and connect to the database as usual.
3. Double-click the **Layout Model** icon to view the report in the Layout Model, then click the **Chart Wizard** button on the button bar. The Chart Wizard Welcome page is displayed.
4. Click **Next**. The Type page is displayed.
5. Type **Quantity on Hand** in the Title box, and be sure that **Column** is the selected chart type and **Plain** is the subtype. Click **Next**. The Category page is displayed.

6 Inventory ID numbers will appear on the X-axis, so select **invid** from the Available Fields list, click the **Select** button [>] so that invid is displayed in the Category Axis list, then click **Next**. The Value page is displayed.

7 Quantity on hand values will appear on the Y-axis, so select **qoh** from the Available Fields list, click [>] so that qoh is displayed in the Value Axis list, then click **Next**. The Break page is displayed.

8 Click **once per Item ID: , itemdesc (R_G_itemid)** in the Chart list box to specify that the chart will be displayed once for every item ID. Click **Next**. The File Name page is displayed.

9 Click **Save As**, navigate to the Chapter8 folder on your Student Disk, type **ch8report** in the File name box, and click **Save**. The filename is displayed on the File Name page. Click **Finish**. The new chart is displayed on the report below the inventory records.

10 Save the report.

Next, you will move the chart so that it appears to the right of the inventory records. Since the chart is inside the R_G_item repeating frame, you will need to turn on flex mode so that the repeating frame is automatically resized when you drag the chart to the right edge of the report. You will do this next.

To move the chart:

1 Click the **Flex Mode** button on the button bar to activate flex mode.

2 Click the chart to select it, then drag the chart to the right edge of the report and then up so that it is displayed directly to the right of the inventory records, as shown in Figure 8-33.

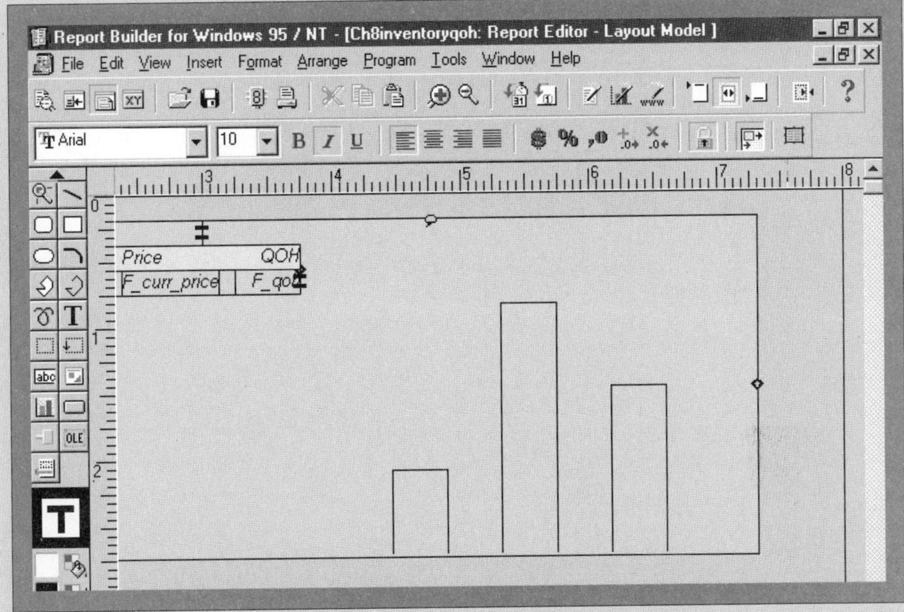

Figure 8-33: Repositioning the report chart

3 Save the file.

4 Click the **Run** button to view the chart in the Live Previewer. Your report should look like Figure 8-34.

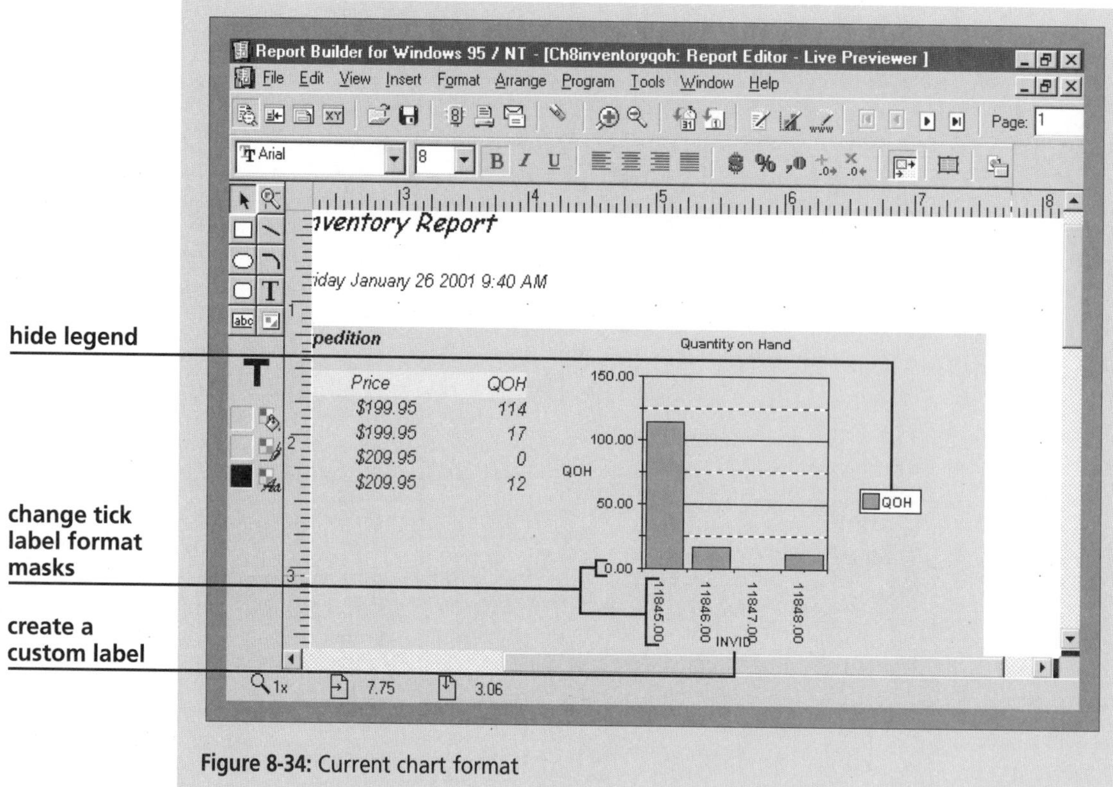

Figure 8-34: Current chart format

Figure 8-34 shows that you need to modify the chart to hide the legend, create a custom label for the X-axis, and change the tick label (inventory ID) format masks so that the numbers do not have a decimal point. You will switch to Graphics Builder, open the report chart file, and modify the chart properties.

To modify the chart properties:

1. Click **Graphics Builder** on the Windows taskbar, and open the **ch8report.ogd** chart file from the Chapter8 folder on your Student Disk. Like the chart on the form, the chart is displayed as a template rather than with actual data values, because the actual data values come from the report.
2. Select the chart, right-click, then click **Frame** to open the Frame Properties dialog box. Clear the **Show Legend** check box, then click **OK.**
3. Click **Chart** on the menu bar, then click **Update Chart** to refresh the chart display, if necessary.
4. Select the chart, right-click, then click **Axes** to open the Axis Properties dialog box. Type **Inventory ID** as the custom label for the X-axis, then click **OK.**

5 Select the X-axis labels, click **Format** on the menu bar, then click **Number**. The Number Format dialog box is displayed. Type **999999** in the Number text box, click **Add** to add the new format mask, then click **OK** to apply the new format mask to the chart.

6 Select the Y-axis labels, click **Format** on the menu bar, click **Number**, select **999,990** from the Number Format list, then click **OK** to apply the format mask to the chart.

7 Save the chart.

8 Click **Report Builder** on the taskbar.

9 Click the **Run** button to refresh the Live Previewer display. Your finished chart display should look like Figure 8-35.

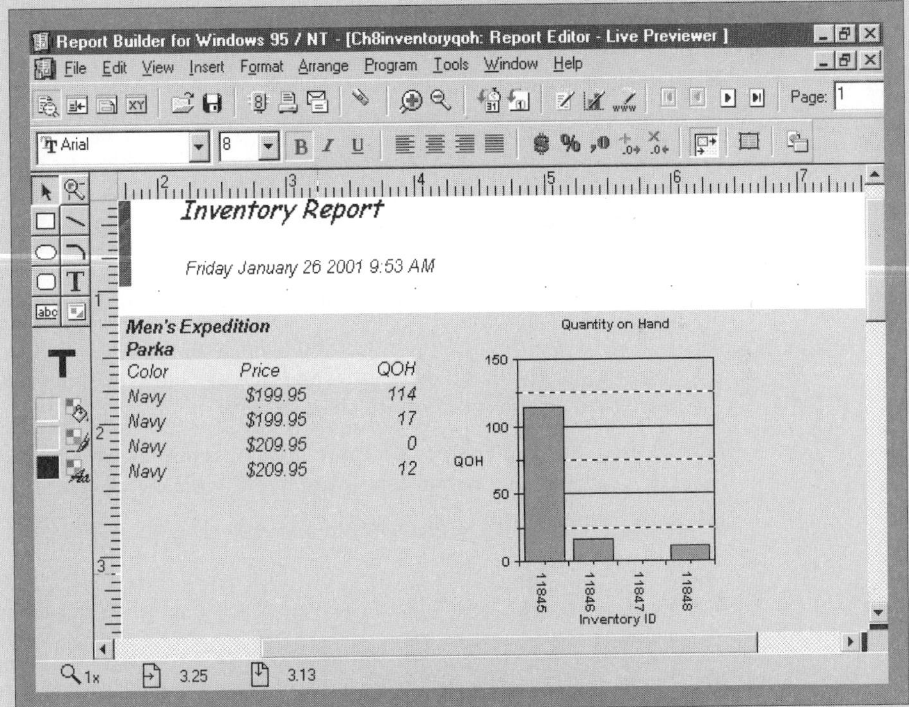

Figure 8-35: Finished chart display

10 Save the report.

11 Exit Report Builder and all open Oracle applications.

SUMMARY

- Data are easier to understand when displayed in graphical forms using bar, pie, and line charts rather than in textual or tabular formats, but it is important to consider the type of data you want to chart and the purpose of what you want to illustrate in your chart.
- One of the most common mistakes people make when creating charts is to plot discrete data points using a line chart.
- To create a new chart, create the query to define the chart, and then define the chart display properties.
- For pie charts, the different pie slice labels will be the values of one query data column, and the corresponding percentage of the whole pie will be the values in the second query column.
- For column charts, the X-axis item labels will be the values of one query data column, and their Y-axis values will be the values in another query column.
- When creating calculated column values using SQL arithmetic or aggregate functions in Graphics Builder, you should assign the calculated column an alias using the SQL AS clause.
- Graphics Builder allows you to save charts both as design files (with an .ogd extension) and as finished executable files (with an .ogr extension).
- A drill-down relationship can be created for any two charts that have a master-detail relationship where one chart specifies different values that the user can choose as a search condition for the second chart's query.
- To pass a data value from a form to a graphics display, you must modify the chart so that its query has a parameter instead of a specific search condition.
- You can use the RUN_PRODUCT procedure to pass a parameter list from a form to a Graphics Builder chart and then display the chart.
- You can use the Chart Wizard to create charts that are displayed directly on forms and reports.
- To change the appearance of a chart created using the Chart Wizard, open the chart file in the Graphics Builder and modify its properties.

REVIEW QUESTIONS

1. Specify the chart type that best illustrates the following information:
   a. how the total enrollment for each term has varied over all past terms in the Northwoods University database
   b. the total dollar amount of orders that each customer has placed using the Clearwater Traders database
   c. the proportion of faculty members who hold each of the different faculty ranks in the Northwoods University database
2. True or false: a line chart is used to show data trends with continuous values.
3. True or false: in Graphics Builder, the chart object name is always the same as the associated chart file in the file system.
4. Write the SQL query to create a chart to illustrate the following information:
   a. the proportion that the current enrollment for each course call ID for the Spring 2002 term contributes to the overall term enrollment
   b. how current enrollments have varied in MIS 301 over all terms
   c. current enrollment and maximum enrollment for all course section numbers of MIS 101 during the Fall 2001 term
5. Graphics Builder only allows you to apply format masks to ______________ and ______________ data types.
6. To create a drill-down relationship between two charts, the data in the charts must have a ______________ relationship.

PROBLEM-SOLVING CASES

Run the northwoo.sql script to refresh the Northwoods University database, and run the clearwat.sql script to refresh the Clearwater Traders database. The script files are stored in the Chapter8 folder on your Student Disk.

1. Create a pie chart named Ch8Ex1.ogd that shows each building code in the LOCATION table of the Northwoods University database as a pie slice, with the total capacity of the building's rooms as the proportion of the pie. List the total capacity of each building as a label on the pie chart. Change the chart title to "Building Capacities."
2. Open the Ch8Ex1.ogd file you created in Exercise 1, and save it as Ch8Ex2.ogd. Then create a bar chart on the layout that shows building room numbers on the X-axis and room capacities on the Y-axis. Create a drill-down relationship so that when you select a given building's pie slice, you see the selected building's room numbers and capacities. Provide a descriptive title for the chart, add custom axis labels, and delete the legend. (*Hint*: To make the values for the LIB building appear correctly, click the Continuous Axis tab on the Axis Property sheet, and specify 20 as the Step Size. Do not show minor tick marks on the Y-axis.)
3. Create a column chart named Ch8Ex3.ogd that shows the faculty ranks (ASSO, FULL, INST, ASST) from the Northwoods University FACULTY table on the X-axis, and shows how many faculty members there are in each rank as column values on the Y-axis. Give the chart a descriptive title and custom axis labels, and delete the legend. (*Hint*: Use the SQL COUNT function, and modify the Y-axis on the Continuous Axis tab on the Axis Properties sheet so that 0 is the axis's Minimum Value, and 1 is the Step size.)
4. Open the Ch8Ex3.ogd file you created in Exercise 3, and save it as Ch8Ex4.ogd. Add a table chart to the layout that shows faculty member last and first names and middle initials. Create a drill-down relationship so that when you select a given faculty rank on the column chart, the data for faculty members with the selected rank are displayed. Create descriptive labels and titles and format the charts attractively.
5. Create a column chart named Ch8Ex5.ogd with two data series that shows how the maximum enrollment compares with the current enrollment for all courses offered during the Summer 2002 term at Northwoods University. Show course call ID values on the X-axis, and enrollment values on the Y-axis. Create a descriptive chart title and custom axis labels, and format the chart attractively.

6. Create a line chart named Ch8Ex6.ogd to show how total enrollments in all courses have varied over all terms. (*Hint*: Sum CURRENRL values for each course.) Show the term description on the X-axis, and the enrollment figures on the Y-axis. Give the chart a descriptive title, hide the legend, create custom axis labels, and format the chart attractively.
7. Open the Ch8Ex6.ogd file you created in Exercise 6, and save it as Ch8Ex7.ogd. Add a column chart to the layout that shows the total enrollment for all sections of each individual course call ID for the selected term. Create a drill-down relationship so that when you select a specific term description on the line chart, the data for the individual courses for the selected term are displayed. Create descriptive labels and titles and format the charts attractively.
8. In this exercise, you will create a form that allows the user to select a term description. The form then calls a graphics display that shows how the maximum enrollment compares with the current enrollment for all courses offered during the selected term. (This exercise modifies the chart that was created in Exercise 5.)
   a. Open the Ch8Ex5.ogd chart you created in Exercise 5, and save it as Ch8Ex8.ogd.
   b. Create a parameter named TERM_ID for the chart search condition, and modify the query so the term description search condition uses TERMID.
   c. Create a form named Ch8Ex8.fmb that has a control block that displays term IDs and term descriptions for Northwoods University. Create a LOV to allow users to select a specific term by displaying the term ID and term description. Use a form-style layout, and only display one term record at a time.
   d. Create a command button on the form with the label Show Enrollments. When the user clicks the Show Enrollments button, the form should display the Ch8Ex8.ogd chart in the Graphics Runtime window. The chart should show maximum enrollments and current enrollments for each course call ID for the selected item.
9. In this exercise, you will create a form that displays course information and directly displays a chart showing the current enrollment figures for all course call IDs during a selected term. Create a form named Ch8Ex9.fmb.
   a. Create a control block that displays term IDs and term descriptions for Northwoods University, and an associated LOV to allow users to select a specific term. Use a form-style layout and display only one term record at a time.
   b. Create a view named CSEC_VIEW that includes course call IDs, term IDs, term descriptions, section numbers, and maximum and current enrollments.
   c. Create a data block for the CSEC_VIEW view. Use a tabular layout, display five records at a time, and include a scroll bar. Modify the data block's WHERE property so the layout displays records for the selected term in the form's TERM block. (*Hint*: There are no course sections for term IDs 1, 2, 3, or 4.)
   d. Use the Chart Wizard to create a column chart that shows the current enrollment figure for all course call IDs for the selected term. Display the call ID on the X-axis, and the current and maximum enrollment on the Y-axis. Save the chart as Ch8Ex9.ogd in the Chapter8 folder on your Student Disk. Provide a descriptive chart title for the chart, add custom axis labels, delete the legend, and format the chart attractively.

10. Create a report named Ch8Ex10.rdf that lists all building codes, room numbers, and associated capacities. Use a Group Above layout style, and display the rooms and associated capacities for each building code in a separate record group. Use the Chart Wizard to create a chart on the report that show each building's room numbers on the X-axis, and the associated capacities on the Y-axis. Name the chart Ch8Ex10.ogd, and store the chart in the Chapter8 folder on your Student Disk. Provide a descriptive chart title for the chart, add custom axis labels, delete the legend, and format the chart attractively. Position the chart so it is displayed on the right edge of the report, beside the associated records.

CHAPTER 9

Creating an Integrated Database Application

Introduction▶ Now that you have created a variety of individual forms, reports, and charts, you can integrate them into a database application that allows users to access all of these items from a single entry point. The user interface will be a new Form Builder form module that will serve as the launching point. After you have created the form module that integrates all of the individual applications, you will learn how to use Project Builder to manage all of the individual application files.

LESSON A

objectives

- Design a database application interface
- Create a global path variable to specify the location of form, report, and graphics files
- Use timers in a Form Builder application to create a splash screen

Designing an Integrated Database Application

When you are developing a new database system, you must first identify the business processes that the database system will support, and the data items required to support these processes. Then you create the database tables, and develop the forms, reports, and charts to manage the data and support the business processes. Finally, you must design an integrated database application that lets users access the different forms, reports, and chart files. An integrated database application can have a front-end entry screen, or **splash screen**, to introduce the application. Then, the application displays the main system screen. The main system screen often displays a **switchboard**, which consists of command buttons or iconic buttons that enable users to quickly and easily access commonly used forms, reports, or charts. All forms, reports, and charts should also be accessible through pull-down menus for users who prefer to use the keyboard rather than the mouse pointer. Pull-down menus provide access to features that are used less often than the ones on the switchboard, and to features that have multiple levels of choices. For example, you could have a pull-down menu titled Reports, and then the next-level menu would list the different reports that the system could generate.

The integrated database application that will be developed in this chapter is for the Clearwater Traders database. Figure 9-1 shows the design for the main system screen for the Clearwater Traders order-processing system. The top part of the design shows the pull-down menu structure, and the bottom part of the design shows the system switchboard.

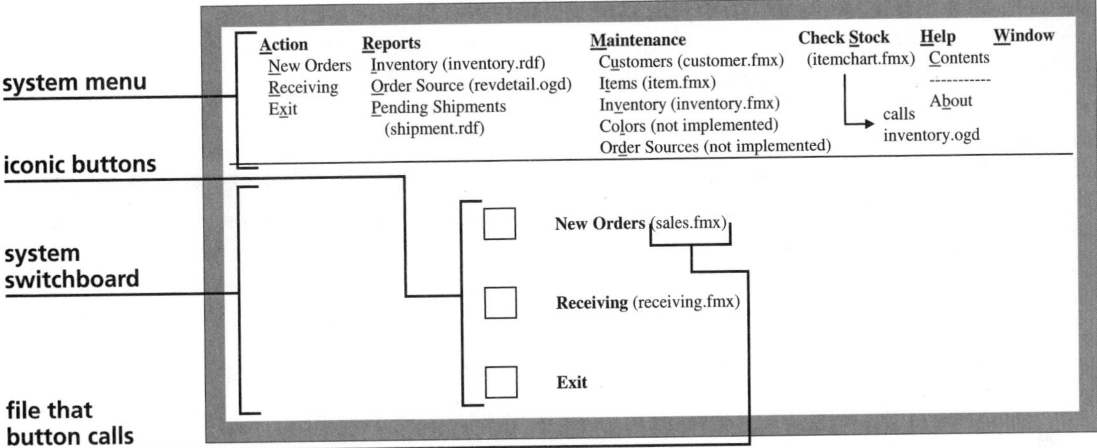

Figure 9-1: Main system screen design

The form applications to support the two main business processes—processing new orders and receiving incoming shipments—have switchboard access using iconic buttons with descriptive labels beside them. The associated .fmx files that each button will call are shown beside the labels.

The first pull-down menu selection in Figure 9-1 is Action. This menu contains selections for processing new orders and incoming shipments, and an Exit command. The Reports pull-down menu provides access to the Inventory, Order Source, and Pending Shipments reports. The Maintenance pull-down menu provides access to the basic forms that let users insert, update, and delete records from each database table. Note that forms for two of the selections (Colors and Order Sources) have not been implemented yet. You will learn how to develop placeholder elements called **stubs**, which are programs that handle undeveloped system features, in Lesson B of this chapter. The Check Stock pull-down menu provides access to a combined form/chart application that allows users to select a given item number, and then see a chart of the stock levels of the associated inventory items. The Check Stock selection is included as a top-level menu selection because a sales representative might want to access the form while processing a new order. The final two selections on the Clearwater Traders database application menu are Help and Window. The Help selection has two second-level selections: Contents, which provides access to the Help search engine, and About, which gives details about the application. The Window menu selection allows users to move between windows in a multiple-window application.

Most Windows applications have Help as the last menu choice, but Form Builder automatically places Window as the last pull-down menu selection.

Figure 9-2 summarizes all of the forms, reports, and chart files that will be used in the project. All of the form design (.fmb) files are included in the CWPROJECT subfolder in the Chapter9 folder on your Student Disk. Note that some of the forms call other application files.

Design File Location/Name	Description
\cwproject\sales.fmb	Custom form that records information about new customer sales orders in the CUST_ORDER, ORDERLINE, and INVENTORY tables
\cwproject\cust_order.fmb	Custom form that allows the user to add a new customer or update information about an existing customer during the sales order process
\cwproject\receiving.fmb	Custom form to record information about new inventory item shipments in the INVENTORY, SHIPPING, and BACKORDER tables
\cwproject\customer.fmb	Data block form that displays and edits CUSTOMER data records
\cwproject\item.fmb	Data block form that displays and edits ITEM data records
\cwproject\inventory.fmb	Data block form that displays and edits INVENTORY data records and displays associated ITEM foreign key records
\cwproject\itemchart.fmb	Custom form that allows the user to select a specific inventory item and then displays an associated column chart showing inventory quantities on hand
\cwproject\inventory.ogd	Column chart displayed by itemchart.fmb showing quantities on hand of different inventory items for a selected product
\cwproject\inventory.rdf	Report that displays ITEM and INVENTORY information
\cwproject\revdetail.ogd	Drill-down graphics application that shows total revenue from different order sources in a pie chart and related inventory detail information in a column chart
\cwproject\shipment.rdf	Report that displays incoming shipment information

Figure 9-2: Summary of items to be used in the integrated database application

Note: The application created in this chapter requires integrating many different files. It is a good idea to copy the Chapter9 folder and its contents from your floppy drive to a hard drive and work from there to improve application performance. The tutorial exercises will illustrate that the Chapter9 folder has been copied to the root directory of the E drive.

When you create a project, you should place all of the application files in the same folder so they are easier to access and maintain. If you use the same application file in two different projects (such as using the same form in multiple projects), it is best to create multiple copies of the file, because the developers in one project might make changes to the application file that would prevent it from working correctly in the other project. For complex projects with many different applications, it is a good idea to use separate form files rather than combining all of the project forms into a single .fmb file with many different canvases. With separate form files, the project is more **modular**, meaning that it is broken up into modules that are independent of one another. Each module can be developed, tested, and debugged individually, and then integrated into the overall application. Smaller modules are easier to work with, and make it easier for project teams with multiple team members to split up the development effort, since different team members can simultaneously work on different form files.

Creating a Global Path Variable

When you create an integrated database application, it is useful to create a global variable that is assigned to a text string representing the path of the home drive and directory where all of the individual application files are stored. For example, if the project files are stored in the \Chapter9\cwproject subfolder on the E drive of your computer, the global path variable, named `:GLOBAL.project_path`, would be assigned to the text string 'e:\Chapter9\cwproject\'. Whenever you call a form, report, or chart, you can concatenate this path variable with the specific file name being called, to specify the complete path to the called application. For example, if you want to call the customer.fmx file that is in the \Chapter9\cwproject subfolder, you would use the command `CALL_FORM (:GLOBAL.project_path||'customer.fmx');`. This way, if the home directory where the project files are stored changes, you simply change the path specification in the global path variable assignment statement, instead of having to change the path in each individual CALL_FORM and RUN_PRODUCT procedure call.

Now you will start Form Builder and create the new form that will integrate the database applications. You will create a PRE-FORM trigger to automatically maximize the form at runtime, and assign the global project path to a text string specifying the drive letter and folder path containing the project application files. If your Chapter9 folder is in the root directory of a floppy disk, the global path will be a:\Chapter9\cwproject\. If the folder is on another drive, you must specify that drive letter. If the project files are stored in a subfolder within another folder, you must set this variable equal to the complete path specification, including all subfolders.

To create the form and create the PRE-FORM trigger:

1. Start Form Builder, select the **Build a new form manually** option button, and then click **OK**.
2. Save the new form as **main.fmb** in the \Chapter9\cwproject subfolder.
3. If necessary, change to Ownership View, create a form-level PRE-FORM trigger, and type the code shown in Figure 9-3 to maximize the window and create the global project path. Set the global project path variable equal to the string representing the drive and complete path to your \Chapter9\cwproject subfolder.

type this code

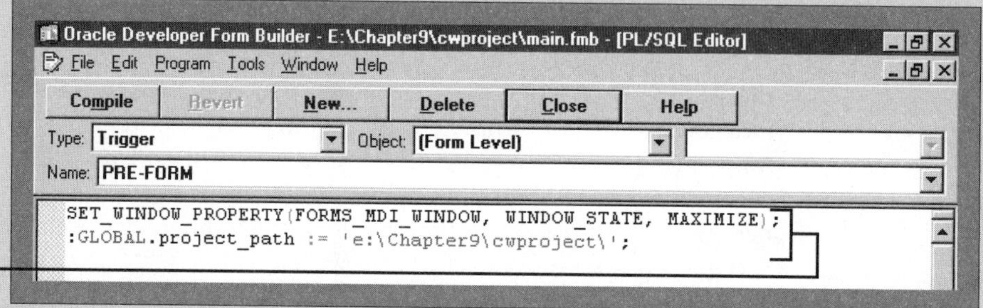

Figure 9-3: PRE-FORM trigger to create global project path variable

4. Compile the code, correct any syntax errors, and then close the PL/SQL Editor.
5. Save the form.

Creating a Splash Screen

Recall that a splash screen is the first image that appears when you run an application. It introduces the application and usually identifies the system author(s) and copyright information. To implement the splash screen, you will have two separate windows within the main application form. When the user starts the program, the first window will display the splash screen window shown in Figure 9-4 for a few seconds. Then the application will switch to the second window and display the main application screen. To implement this, you will modify the PRE-FORM trigger to load the splash screen image and then set a timer. The splash screen window will appear first, and when the time set by the timer runs out, a form-level trigger named WHEN-TIMER-EXPIRES will execute. This trigger sets the application's focus to an item on the main application window, hiding the splash screen window and causing the main application window to appear.

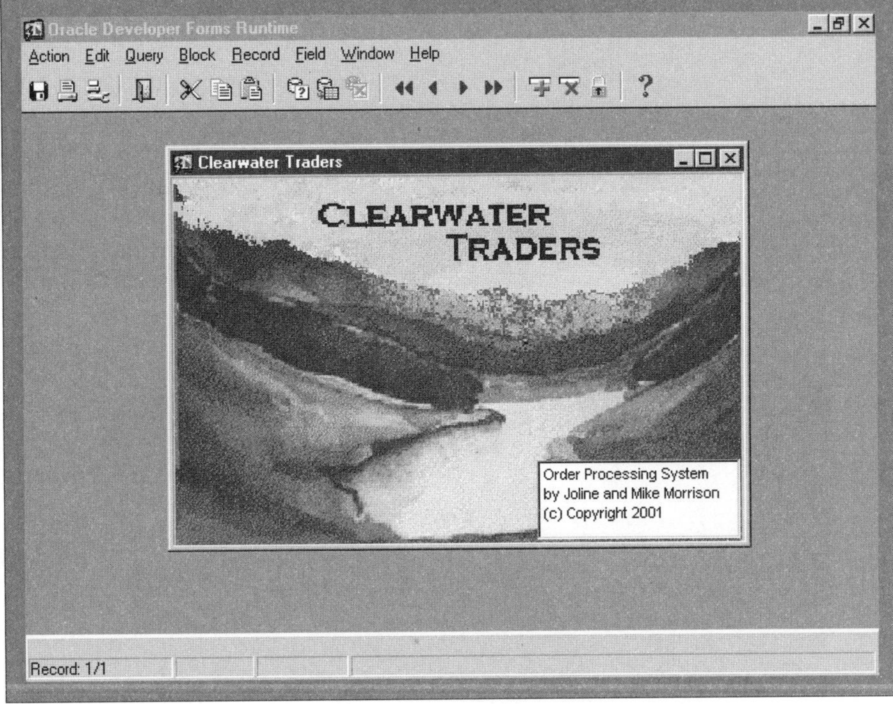

Figure 9-4: Splash screen

The first step in developing the splash screen is to create the windows, canvases, and blocks for the splash screen and main application screen. You will do this next.

To create the windows, canvases, and blocks for the splash screen and main application screen:

1. Open the Object Navigator window in Ownership View if necessary, and click ⊞ beside Windows. Rename WINDOW1 **SPLASH_WINDOW.**
2. Select **Windows,** then click the **Create** button to create a second window. Name the second window **MAIN_WINDOW.**
3. Switch to Visual View, click ⊞ beside SPLASH_WINDOW, and click **Canvases,** then click to create a new canvas. Name the new canvas **SPLASH_CANVAS.**
4. Click ⊞ beside MAIN_WINDOW, and create a new canvas in MAIN_WINDOW named **MAIN_CANVAS.**

5 Switch to Ownership View, click **Data Blocks**, then click [icon] to create a new data block. Click the **Build a new data block manually** option button, click **OK**, then rename the new block **SPLASH_BLOCK.** Create a second new data block, and name it **MAIN_BLOCK**. Your MAIN form module should contain the data blocks, canvases, and windows shown in Figure 9-5.

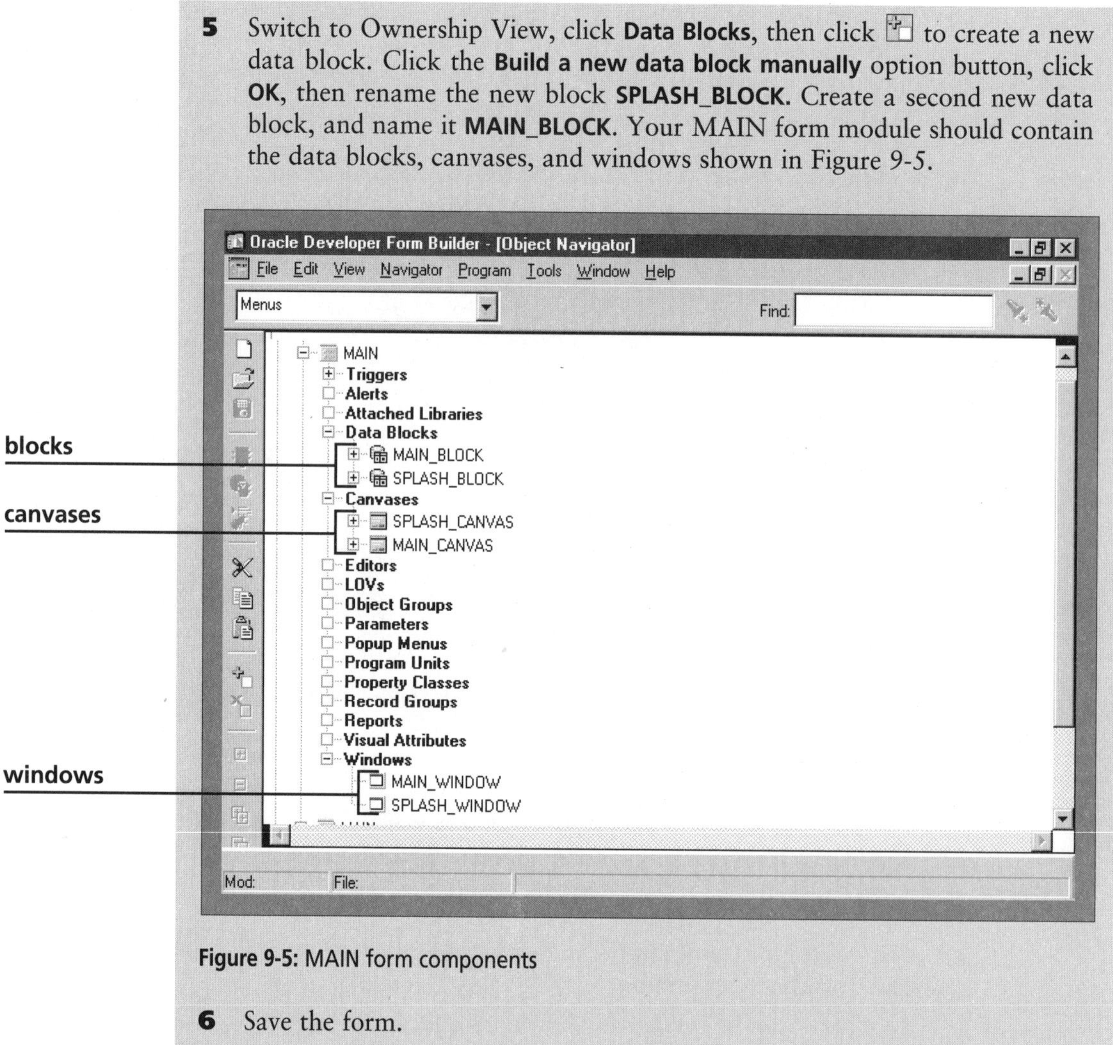

Figure 9-5: MAIN form components

6 Save the form.

Configuring the Splash Screen Window

Windows applications are displayed in a window that has a title bar and buttons that allow the user to minimize, maximize, and close the window. This window can run in its normal state, in which windows for multiple applications are visible on the screen display, or in a maximized state, in which the window for a single application fills the screen display. When a Form Builder form is running, it is displayed in the Forms Runtime window. So far, the windows that you have made in Form Builder applications have been **document windows**, which are displayed inside the Forms Runtime window. If the user resizes the Forms Runtime window so that it is smaller than the document window inside it, the document window is clipped and scroll bars appear. Windows within a Form Builder application can also be **dialog windows**, which are displayed in their own window frame, rather than in the Forms Runtime window. You will configure the splash screen window as a dialog window so that it displays only the window title bar, and does not display the Forms Runtime pull-down menu selections or toolbar. You also will modify the window's Hide on Exit property so that the window disappears from the screen as soon as the application focus switches to the main application window. Splash screen windows typically do not fill the entire screen display, so you will also change the window size to 320 by 200 pixels (the Clearwater Traders splash screen logo looks best in a window that is this size).

To configure the splash screen window:

1. Open the SPLASH_WINDOW Property Palette, and type **Clearwater Traders** as the Title property.
2. Scroll to the Window Style property, click on the space next to the property title, and select **Dialog** from the list.
3. Change the Hide on Exit property to **Yes.**
4. Change the Move Allowed and Resize Allowed properties to **No**. Change the Width property to **320**, and the Height property to **200.**
5. Change the X- and Y-Position property so the window is centered on the screen when it is displayed. To calculate the X- and Y-positions to center the window, use the following formulas: X-position = [(screen resolution width) - 320]/4; Y-position = [(screen resolution height) - 200]/4. For example, if your screen resolution is 640 by 480, your X-position would be (640 – 320)/4, or 80, and Y-position would be (480 - 200)/4, or 70. Close the SPLASH_WINDOW Property Palette.
6. Open the SPLASH_CANVAS Property Palette, and change the Width to **320**, and the Height to **200.** Close the Property Palette.
7. Save the form.

Creating an Image Item in the Splash Window

The splash screen image is a graphic art file that will appear on the SPLASH_CANVAS. There are two approaches to displaying graphic art on a form. The first approach is to create the image as a boilerplate image, as you did when you imported the Clearwater Traders and Northwoods University logos into forms in previous chapters. This approach stores the image inside the .fmb file when you design the form. However, this approach will not work for a splash screen because the image is the only item on the canvas, and a canvas with only boilerplate objects and no items under the Item category in the Object Navigator will not be displayed by the Forms Runtime application.

You will use the second approach, which is to create an image item on the SPLASH_CANVAS. An **image item** is an empty item container in the .fmb design file, and the actual image is loaded into the image item from the file system at run-time. This method must be used when you display a canvas with no items other than a graphic image. Now you will create the image item, adjust its size so that it is the same size as the canvas, and adjust its X- and Y-positions so that it is in the top-left corner of the canvas.

To create an image item for the SPLASH_CANVAS and adjust its properties:

1 Double-click the **Canvas** icon beside the SPLASH_CANVAS to view the canvas in the Layout Editor. If necessary, select SPLASH_BLOCK from the block list.

2 Zoom out so you can see the entire canvas.

3 Click the **Image Item** tool on the Tool palette, and then draw a box that is about the same size as the canvas and just inside the canvas borders, as shown in Figure 9-6. The fill color of your image item might be different.

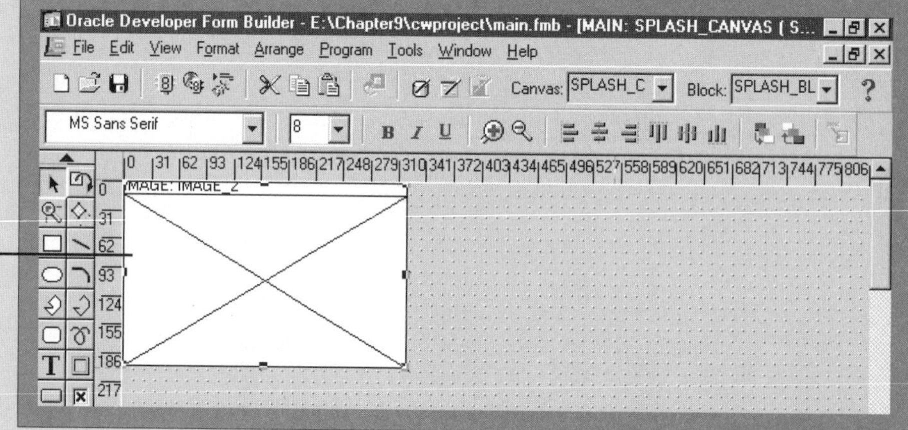

Figure 9-6: Splash screen image item

4 Right-click the new image item, then click **Property Palette** to open its Property Palette. Change the Name property to **SPLASH_IMAGE**. Next, you will adjust the position and size properties for SPLASH_IMAGE so that it is the same size as the canvas and positioned in the upper-left corner of the canvas.

5 Click in the space next to the Sizing Style property, and select **Adjust**. This automatically adjusts the graphic's size so it is the same size as the image item on the canvas.

6 Change the X- and Y-position properties for SPLASH_IMAGE to **0**, if necessary.

7 Change the Width property to **320**, and the Height property to **200**, making it the same size as the canvas.

8 Close the Property Palette and then save the form.

Now you will create a **display item** on the image. A display item is a form item that displays text that the user cannot change. You will draw the display item on the canvas, and change its properties so that it displays the name of the system, the system developer, and the copyright date.

To create the splash screen display item:

1 In the Layout Editor, click **View** on the menu bar, then click **Normal Size** to view SPLASH_CANVAS at its normal size. Scroll up and to the left so all of the SPLASH_IMAGE is displayed.

2 Click the **Display Item** tool, and draw a display item in the lower-right corner of the canvas, as shown in Figure 9-7.

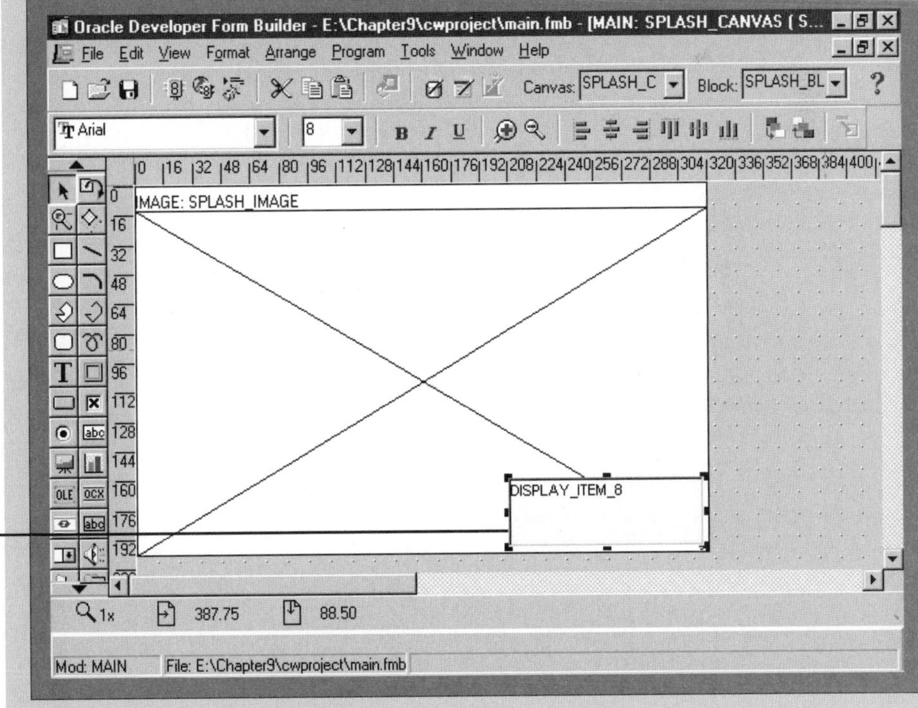

Figure 9-7: Display item on splash image

3 Select the new display item, click the **Fill Color** tool on the tool palette, and change the display item's fill color to **white**.

4 Click the **Text Color** tool on the tool palette, and change the text color to **black**, if necessary.

5 Change the font to **8-point Arial regular**.

6 Open the display item's Property Palette, and change the Name property to **SPLASH_ITEM** and the Maximum Length to **500**. The Maximum Length property specifies the maximum length of the text string that can be displayed in the display item.

7 Close the Property Palette and then save the form.

Next, you will modify the PRE-FORM trigger to load the graphic image file into the image item before the form appears, assign a text string to be displayed in the SPLASH_ITEM display item, and create a timer that determines how long the splash screen appears on the screen. To load an image file into an image item, you use the READ_IMAGE_FILE procedure, which has the following general format: `READ_IMAGE_FILE(<FILE_NAME>, <FILE_TYPE>, <ITEM_NAME>);`. This command requires the following parameters:

- **FILE_NAME** is the complete path and file name of the graphic art image file, and is passed as a character string in single quotation marks. If the image file is named cwsplash.tif and stored in the \Chapter9\project subfolder, the parameter will be `:GLOBAL.project_path || 'cwsplash.tif'`.
- **FILE_TYPE** is the type of image file being used, and is passed as a character string in single quotation marks. Legal values are the following file types: .BMP, .PCX, .PICT, .GIF, .CALS, .PCD, and .TIF. For the TIF file used in the previous example, the parameter would be passed as 'TIFF'.

The different image file types depend on the graphics art application used to create the file, and how the image is compressed to make it take less file space. Bitmap (.BMP) files are usually uncompressed, while .PCX, .GIF, and .TIF files use different compression methods. .PICT and .PCD files are made by specific graphics applications, and .CALS files are used to compress black and white images for fax transmissions. Most popular graphics applications support one of these types of files.

- **ITEM_NAME** is the name of the image item where the file will be displayed, and is passed as a character string in single quotation marks using the format `'block name.item name'`. For the image item you just created, the parameter would be `'splash_block.splash_image'`.

To create a timer, you use the CREATE_TIMER function, which has the following general format: `TIMER_ID := CREATE_TIMER(<TIMER_NAME>, <MILLISECONDS>,<ITERATE>);`. This command requires the following parameters:

- **TIMER_ID** is a previously declared variable of data type TIMER.
- **TIMER_NAME** can be any character string up to 30 characters long. TIMER_NAME is entered as a character string in single quotation marks. An example of a timer name is 'splash_timer'.
- **MILLISECONDS** is a numeric value that specifies the time duration, in milliseconds, until the timer expires. (When the timer expires, it calls a form-level trigger named WHEN-TIMER-EXPIRED, which you will create later in this lesson.)
- **ITERATE** specifies whether the timer should be reset immediately after it expires. Valid values are **REPEAT** (meaning it should be reset and start counting down again) and **NO_REPEAT** (meaning it should stay expired). The REPEAT option could be used to create animated graphics that are displayed repeatedly. The splash screen will be displayed only once, so you will use the NO_REPEAT option.

Now you will modify the PRE-FORM trigger to load the image into the SPLASH_WINDOW, create the splash display text, and create the timer. You will modify the splash display text to display your name as the system developer, and to show the current year for the copyright year.

To modify the PRE-FORM trigger:

1 Switch to the Object Navigator, click [+] beside Triggers, then double-click the **Trigger** icon beside PRE-FORM to open the PRE-FORM trigger in the PL/SQL Editor. Modify the PRE-FORM trigger using the code shown in Figure 9-8.

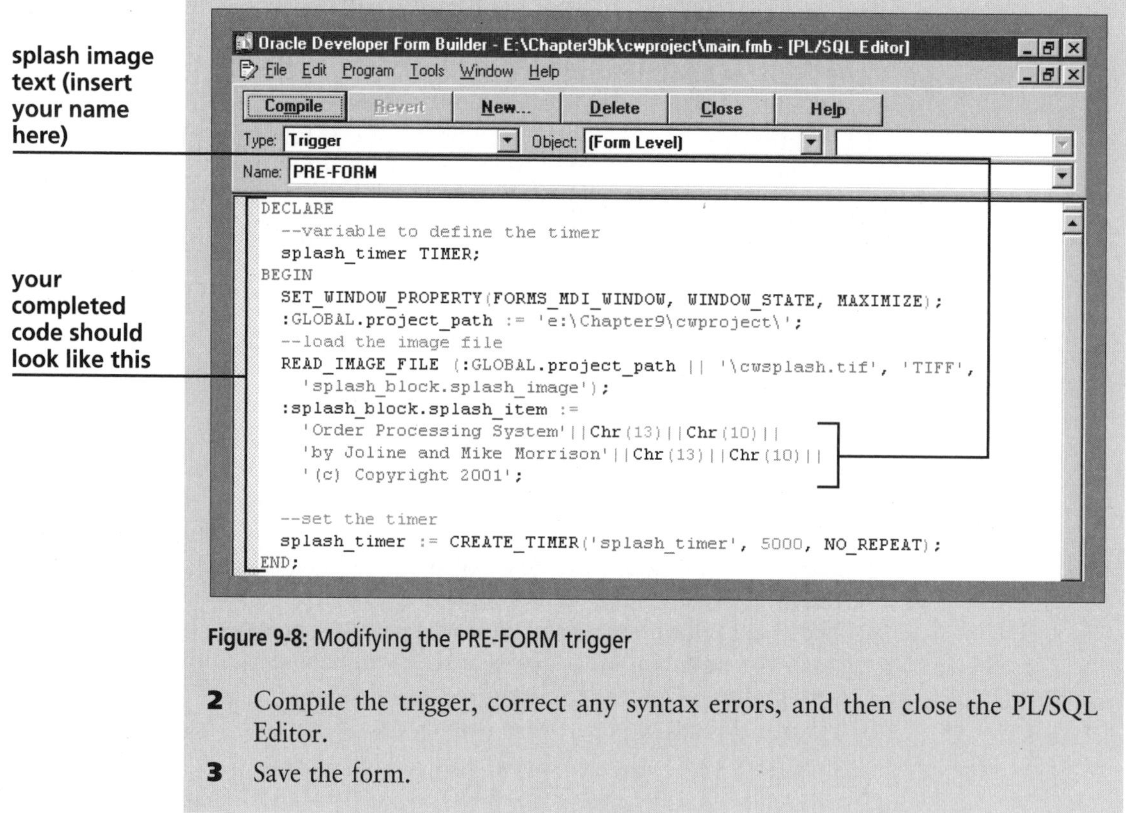

splash image text (insert your name here)
your completed code should look like this
Oracle Developer Form Builder - E:\Chapter9bk\cwproject\main.fmb - [PL/SQL Editor]
File Edit Program Tools Window Help
Compile
Revert
New...
Delete
Close
Help
Type: Trigger
Object: (Form Level)
Name: PRE-FORM
DECLARE
--variable to define the timer
splash_timer TIMER;
BEGIN
SET_WINDOW_PROPERTY(FORMS_MDI_WINDOW, WINDOW_STATE, MAXIMIZE);
:GLOBAL.project_path := 'e:\Chapter9\cwproject\';
--load the image file
READ_IMAGE_FILE (:GLOBAL.project_path || '\cwsplash.tif', 'TIFF',
'splash_block.splash_image');
:splash_block.splash_item :=
'Order Processing System'||Chr(13)||Chr(10)||
'by Joline and Mike Morrison'||Chr(13)||Chr(10)||
'(c) Copyright 2001';
--set the timer
splash_timer := CREATE_TIMER('splash_timer', 5000, NO_REPEAT);
END;

Figure 9-8: Modifying the PRE-FORM trigger

2 Compile the trigger, correct any syntax errors, and then close the PL/SQL Editor.

3 Save the form.

Creating the WHEN-TIMER-EXPIRED Trigger

When a form timer expires, the form-level WHEN-TIMER-EXPIRED trigger is called. If no WHEN-TIMER-EXPIRED trigger exists, nothing happens. If such a trigger does exist, then the code in the trigger executes.

The next step is to create a WHEN-TIMER-EXPIRED trigger that displays the main application window. To show a window in a multiple-window application, you first show the window using the SHOW_WINDOW command, and then move the form insertion point to an item on that window using the GO_ITEM command. Therefore, you first need to create an item in the main application window so you can move the form's insertion point to that item.

To create an item in the MAIN_WINDOW:

1 Double-click the **Canvas** icon beside MAIN_CANVAS to display the MAIN_CANVAS in the Layout Editor.

2 If necessary, select **MAIN_BLOCK** from the Block list box in the Layout Editor. You will create a button that will become the iconic button that will call the sales.fmx file. You will adjust the button size and position later.

3 Click the **Button** tool on the Tool palette, and draw a button anywhere on the canvas. Open the button's Property Palette, and change the button's name to **NEW_ORDERS_BUTTON.** This will be the iconic button that is beside the New Orders label in Figure 9-1.

4 Close the Property Palette.

5 Save the form.

Next you will create the WHEN-TIMER-EXPIRED trigger. This trigger will show the MAIN_WINDOW and switch the application's focus to the NEW_ORDERS_BUTTON.

To create the WHEN-TIMER-EXPIRED trigger:

1 Open the Object Navigator window in Ownership View.

2 Under the MAIN module, select **Triggers**, and then click the **Create** button to make a new form-level trigger.

3 Create a WHEN-TIMER-EXPIRED trigger using the code shown in Figure 9-9.

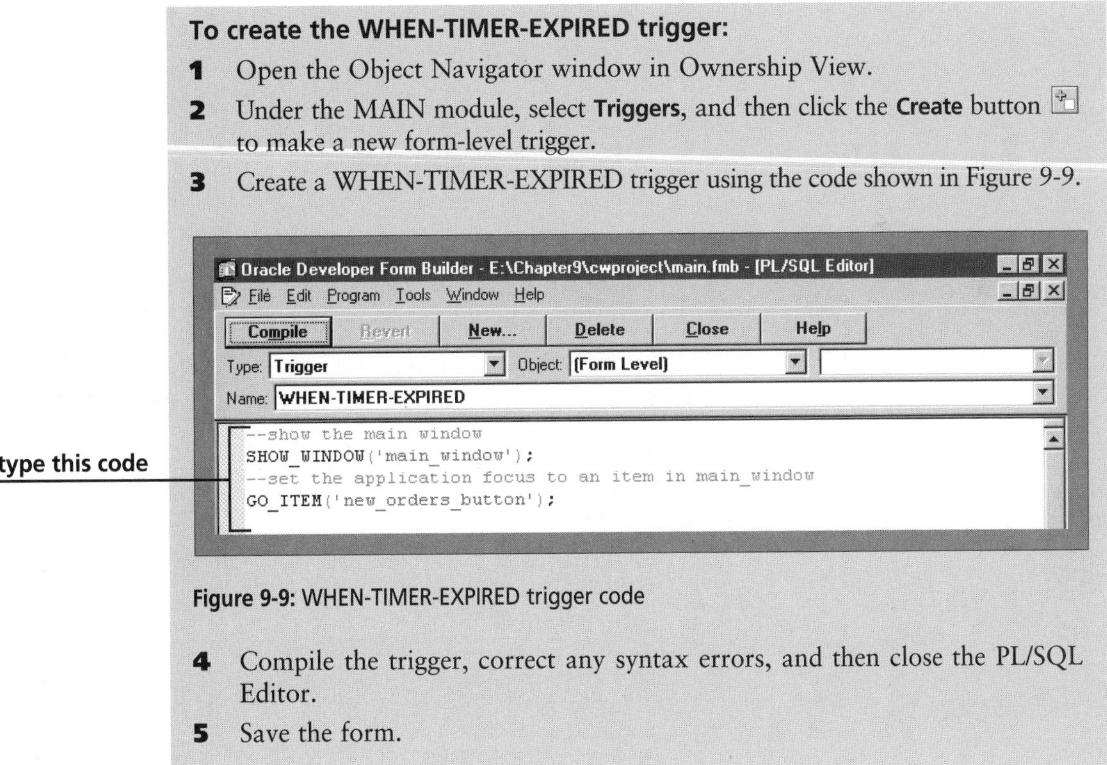

Figure 9-9: WHEN-TIMER-EXPIRED trigger code

4 Compile the trigger, correct any syntax errors, and then close the PL/SQL Editor.

5 Save the form.

When a Form Builder application runs, the Forms Runtime application displays the window that contains the item that is first in the form's navigation sequence. Recall that the form navigation sequence is determined by the order in which the blocks and items are listed in the Object Navigator. Therefore, SPLASH_BLOCK must be the first block listed, and SPLASH_ITEM must be the first item in the block. Next you will double-check the navigation sequence to make sure that the window with the splash screen will be the first window displayed.

To check the navigation sequence:

1 If necessary, open the Object Navigator window in Ownership View.

2 Confirm that the navigation sequence looks like the one shown in Figure 9-10. If it does not look like Figure 9-10, drag the objects into the correct positions, and verify that SPLASH_IMAGE is in SPLASH_BLOCK and NEW_ORDERS_BUTTON is in MAIN_BLOCK.

3 Save the form.

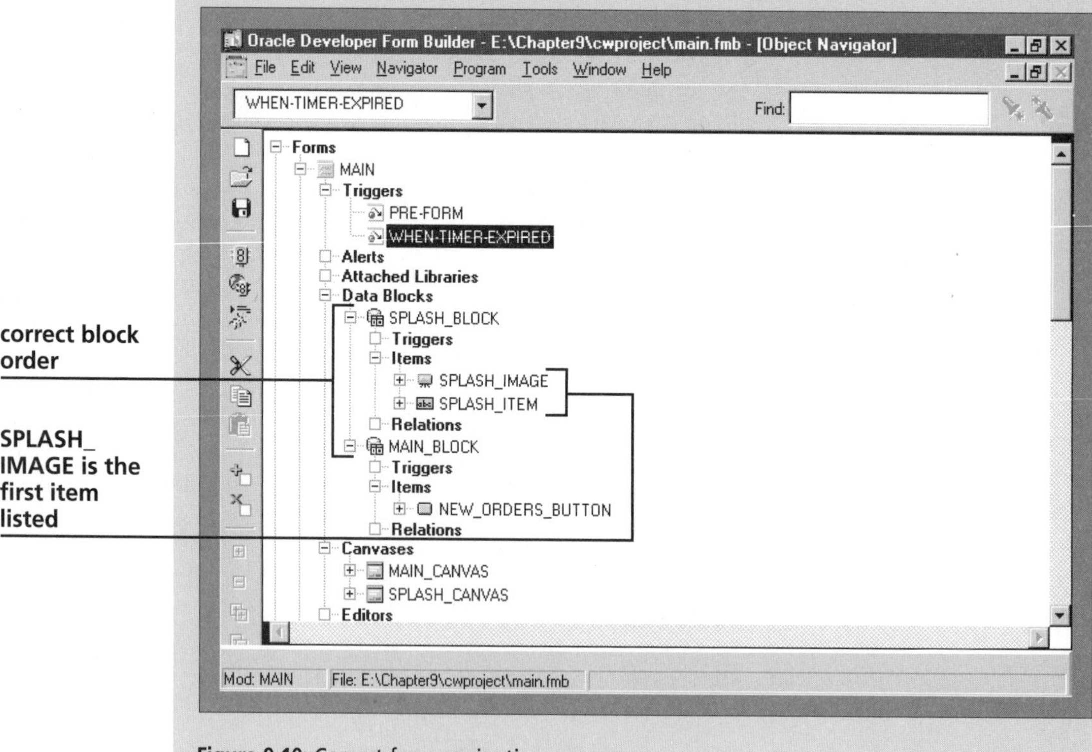

Figure 9-10: Correct form navigation sequence

Now you can test the splash screen.

To test the splash screen:

1 Click the **Run Form Client/Server** button to run the application. Connect to the database if necessary. The splash screen should be displayed as shown in Figure 9-4, for five seconds and then the main window should be displayed. The splash window should be centered in the screen display.

help

If the SPLASH_WINDOW is displayed but no image appears, double-check to make sure that the cwsplash.tif image file is in the \Chapter9\cwproject subfolder, and that you correctly assigned the global path text string to the location of your \Chapter9\cwproject\ subfolder.

2 Close Forms Runtime.

3 If necessary, adjust the size of the SPLASH_ITEM on the canvas so that all of the text is displayed and your splash screen looks like Figure 9-4.

Creating the Iconic Buttons

Next, you will format the MAIN_WINDOW and create the iconic buttons and labels shown in Figure 9-11. Iconic buttons are like LOV command buttons—the picture on the button comes from an icon (.ico) file specified in the button's Property Palette. From a design standpoint, it is important to use an icon image that relates to the button's function, and to provide a Tooltip, which is a text tip that appears when the user places the mouse pointer on the button and that describes the button's function in a small text box under the button. Now you will format the MAIN_WINDOW and create the iconic buttons. Recall that to specify the location of an icon file on an iconic button, you must type the complete path (including drive letter) to the folder that contains the icon folder, and that you must omit the .ico extension from the icon filename.

The icon files used in the steps are stored in the Chapter9 folder. You can display the icon files directly from the Chapter9 folder, or you can copy the files to the folder specified for Oracle application icons in the system registry on your client workstation. Instructions for creating the folder to store the Oracle application icons and for modifying the system registry are provided in the Instructor's Manual.

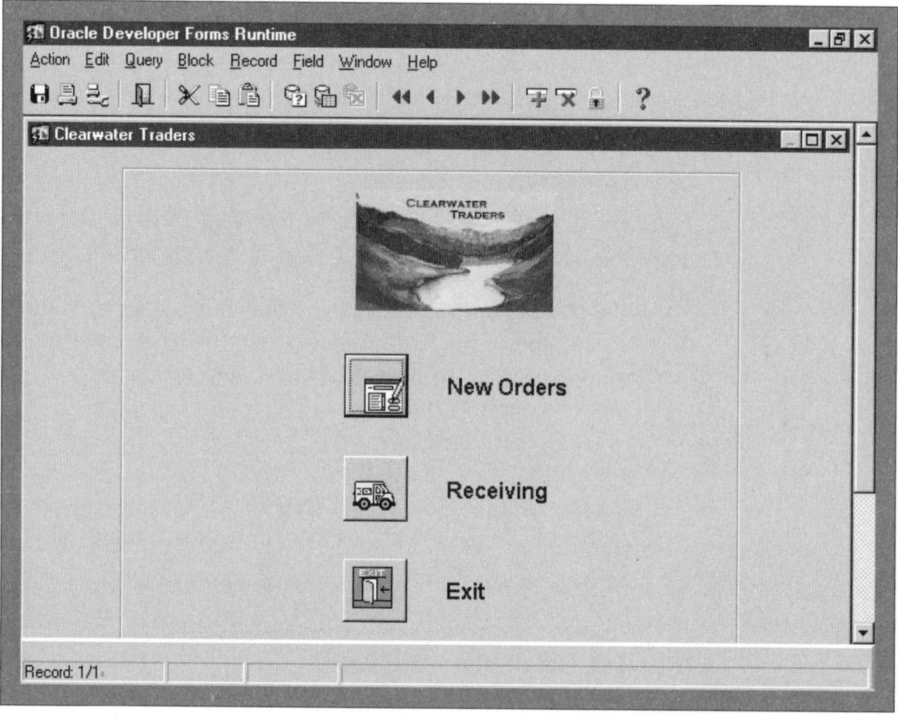

Figure 9-11: Main window layout

To format the MAIN_WINDOW and create the iconic buttons:

1 Open the Property Palette for the MAIN_WINDOW, and change the title to **Clearwater Traders.** Change the Move Allowed, Resize Allowed, and Maximize Allowed properties to **No.**

2 Change the Width and Height properties of the MAIN_WINDOW so that the window is displayed full-screen width on your computer, then close the MAIN_WINDOW Property Palette.

3 Open the Property Palette for the MAIN_CANVAS, and change the canvas's Width and Height properties so that they are the same width and height as the MAIN_WINDOW. Close the Property Palette.

4 Double-click the **Canvas** icon to open the MAIN_CANVAS in the Layout Editor. Select the **New Orders** button that you created earlier, and confirm that MAIN_BLOCK is displayed in the block list at the top of the Layout Editor.

help

If the New Orders button is not in the MAIN_BLOCK, open the Object Navigator, and move the button into the MAIN_BLOCK.

5 Right-click the **New Orders** button and click **Property Palette**, and then change the following properties. (The "sales" icon should appear on the NEW_ORDERS_BUTTON after you close the button's Property Palette.)

Property	Value
Label	**<deleted>**
Iconic	**Yes**
Icon Filename	**<drive letter where your Chapter9 folder is located> \Chapter9\sales**
Width	**36**
Height	**35**
Tooltip	**Create sales order**

6 Close the button's Property Palette.

help If the icon does not appear, check that you correctly specified the path to the folder where the icon file is, that the icon file name is spelled correctly, and that the file exists in the specified folder.

help This button size assumes you are using a 640 by 480 screen resolution. If your screen resolution is different, you might need to change the size of the buttons to make the icons appear correctly.

7 Copy the button and paste it two times to make the next two iconic buttons.

8 Modify the properties of the two new buttons as follows:

Name	Icon Filename	Tooltip
RECEIVING_BUTTON	**<drive letter where your Chapter9 folder is located> \Chapter9\truck02h**	**Receive shipment**
EXIT_BUTTON	**<drive letter where your Chapter9 folder is located> \Chapter9\exit**	**Exit**

9 Position the buttons and use the Text tool [T] to create the associated labels, as shown in Figure 9-11. Format the labels using a **12-point Arial bold** font.

10 Click **File** on the menu bar, point to **Import**, and then click **Image**. Click **Browse**, navigate to the **Chapter9** folder, select **clearlogo.tif**, click **Open**, and then click **OK**. Resize and reposition the logo as necessary so that your form looks like Figure 9-11.

11 Click the **Frame** tool on the tool palette, and draw a frame around the buttons and logo, as shown in Figure 9-11. Change the frame properties as follows:

Property	**Value**
Name	**MAIN_FRAME**
Update Layout	**Manually**
Bevel	**Inset**
Frame Title	**<deleted>**

12 Save the form.

13 Run the form and then adjust the item positions and sizes if necessary.

Creating the Button Triggers

Now you need to create the WHEN-BUTTON-PRESSED triggers for the iconic buttons. Recall that to call one form from another, you use the CALL_FORM procedure, which has the following general format: `CALL_FORM(<FORM MODULE NAME>, <DISPLAY>, <SWITCH_MENU>);`. This command requires the following parameters:

- **FORM MODULE NAME** is the full path and filename, including the drive letter, to the called form's .fmx file. Recall that all of the forms and reports that are called from the main application form are located in the CWPROJECT subfolder of the Chapter9 folder. You will use the `:GLOBAL.project_path` variable to specify the drive letter and folder path and then concatenate this to a string specifying the path, application's filename. For example, the form module name parameter to call the sales.fmx file will be `:GLOBAL.project_path || 'sales.fmx'`.
- **DISPLAY** specifies whether the calling form is hidden or not hidden by the called form. Valid values are HIDE and NO_HIDE. Usually the calling form is hidden and the DISPLAY parameter value is HIDE. The only time the DISPLAY parameter value would be NO_HIDE is when you want the two forms to appear side by side on the screen.
- **SWITCH_MENU** specifies whether the called form's standard pull-down menus are inherited from the calling form or if custom menus specified by the programmer will replace the called form's pull-down menus. Valid values are **NO_REPLACE** (pull-down menus are inherited from the calling form) and **DO_REPLACE** (pull-down menus are specified by the programmer). You will learn how to replace pull-down menus in the next lesson. For now, the calling form will use the standard Forms Runtime menu selections, and you will use the NO_REPLACE value.

Currently, only the .fmb (design) files are in the cwproject subfolder. Before you can test the trigger and see if the button successfully opens the Sales form, you must generate the .fmx files for the sales.fmb and receiving.fmb form files in the cwproject subfolder.

To generate the executable (.fmx) form files:

1. Open the **sales.fmb** file from the \Chapter9\cwproject subfolder.
2. Make sure that the **SALES** form is selected in the Object Navigator, click **File** on the menu bar, point to **Administration**, and then click **Compile File** to create the sales.fmx file in the cwproject subfolder. The message "Module built successfully." is displayed in the message area when the .fmx file is successfully generated.
3. Close the SALES form. Click **No** if you are asked if you want to save your changes.
4. Open the **receiving.fmb** form file, and repeat Step 2 with the RECEIVING form to create the receiving.fmx file in the cwproject subfolder.

help

If you receive an error message when you compile the RECEIVING form, it might be because your database account does not have a view named SHIP_VIEW, which you should have created in Chapter 6. If you did not create SHIP_VIEW in Chapter 6, you can create the view by opening the shipview.txt file from the Chapter9 folder on your Student Disk, copying the code into SQL*Plus, and executing the command.

5. Close the RECEIVING form.

Now you will create the button trigger to call the Sales form in the main form application.

To create the button triggers:

1. In the Object Navigator window, double-click the Canvas icon beside MAIN_CANVAS in the MAIN form. Select the **New Orders** button, right-click, point to **Smart Triggers**, and then click **WHEN-BUTTON-PRESSED** to create a new button trigger. Type the following code in the PL/SQL Editor: `CALL_FORM (:GLOBAL. project_path || 'sales.fmx', HIDE, DO_REPLACE);`.
2. Compile the trigger, correct any syntax errors, and then close the PL/SQL Editor.
3. Save the form.

Now you will run the main application form module, and test to see if the New Orders button trigger works correctly. When you click the New Orders button, the Customer Orders (sales.fmx) form should be displayed.

To test the New Orders button:

1. Click the **Run Form Client/Server** button to run the form.
2. Click the **New Orders** iconic button. The Customer Orders (sales.fmx) form should be displayed.

help

If the sales form does not appear when you click the New Orders button, check to be sure that you entered the path for the global path variable in the PRE-FORM trigger correctly. Also make sure you entered the path to the .fmx file correctly in the New Orders button trigger, and confirm that the sales.fmx file is in the \Chapter9\cwproject\ subfolder.

3. Click **Exit** to exit the Customer Orders form, and then click ☒ to exit Forms Runtime.

Now you will create the button triggers for the Receiving and Exit iconic buttons. The Receiving button will call the Receiving (receiving.fmx) form from the cwproject subfolder, and the Exit button will exit the main application form module.

To create the triggers for the Receiving and Exit buttons:

1. Create a WHEN-BUTTON-PRESSED trigger for the Receiving iconic button, and type the following code in the PL/SQL Editor: **`CALL_FORM (:GLOBAL.project_path || 'receiving.fmx', HIDE, DO_REPLACE);`**.
2. Create a WHEN-BUTTON-PRESSED trigger for the Exit button with the following code: **`EXIT_FORM;`**.
3. Save the form and then run the form to confirm that the new button triggers work correctly.
4. Exit Forms Runtime.
5. Exit Form Builder.

S U M M A R Y

- All form, report, and chart applications within an integrated Oracle database application should be available through pull-down menus, and the most commonly used features should have easy switchboard access through iconic buttons.
- When you create an integrated Oracle database application, you should place all of the form, report, and chart files in the same folder.

- If you use the same form, report, or chart file in two different projects, you should create multiple copies of the file, because the developers in one project might make changes to the application file that would prevent it from working correctly in another project.
- Stubs are programs that handle application access to other programs that have not yet been implemented.
- When you create an integrated database application, set a global variable equal to a text string representing the home drive and folder path where all of the individual form, report, and chart files are stored. If the drive or folder path changes, change the path specification in the global path variable instead of in each individual CALL_FORM and RUN_PRODUCT procedure call.
- A splash screen is the first image that appears when you run an application. It introduces the application and usually identifies the system author(s) and copyright information.
- To implement the splash screen, you have two separate windows in the form. The splash screen is displayed in a dialog window, which does not have pull-down menus. The main application window is displayed in a document window, which is displayed in its own window frame and has pull-down menus.
- If an imported graphic art image is the only image on a canvas, you must create an image item to store the graphic, or the canvas will not be displayed at runtime.
- Boilerplate graphic art image items are loaded at design time, while image item graphic images are loaded at runtime using the READ_IMAGE_FILE procedure.
- A timer is a system-based object that waits a set period of time before it expires. When a timer expires, the WHEN-TIMER-EXPIRED trigger is called. If no WHEN-TIMER-EXPIRED trigger exists, nothing happens. If the trigger does exist, then the code in the trigger is executed.
- To show a different window in a form, use the SHOW_WINDOW command, and then change the form's insertion point to an item in the window, using the GO_ITEM command.
- When you use iconic buttons in a user interface, it is important to use an icon image that relates to the button's function and to provide a Tooltip that appears at the bottom of the form when the user positions the mouse on the button.
- When a Form Builder application runs, the Forms Runtime application displays the window that contains the item that is first in the form's navigation sequence.

REVIEW QUESTIONS

1. In an integrated database application, what applications should be accessed using the switchboard, and what applications should be accessed using pull-down menus?
2. What is the advantage to creating a global path variable in an integrated database application, rather than specifying the path to each individual application?

3. What is a splash screen, and when is it displayed?
4. What is the difference between the dialog and document window styles?
5. What are the two methods for displaying an imported graphic art image file in a form?
6. You are creating an integrated Oracle database application, and have stored all of the form, report, and graphics files in a folder named \Oracle\myproject that is on the D drive of your computer.
   a. Write the assignment statement for the :GLOBAL.path variable.
   b. Write the CALL_FORM command that you will use to call a form file named inventory.fmx that is stored in the d:\Oracle\myproject subfolder. Specify the parameters so that the calling form will be hidden, and the new form's menus will be replaced.
   c. You decide to move all of the project files to the f:\Databaseprojects\project1 subfolder on your network server. What do you need to change in the code of your integrated database application?
7. How do you display a different window in a multiple-window form?
8. Which window is displayed first when you start a form module?

PROBLEM-SOLVING CASES

1. Figure 9-12 shows the main screen for a project application for the Northwoods University database. In this exercise, you will create the main application form that will call the form applications shown. The form files (faculty.fmb, student.fmb, course_sec.fmb) are stored in the \Chapter9\nwproject1 subfolder on your Student Disk.

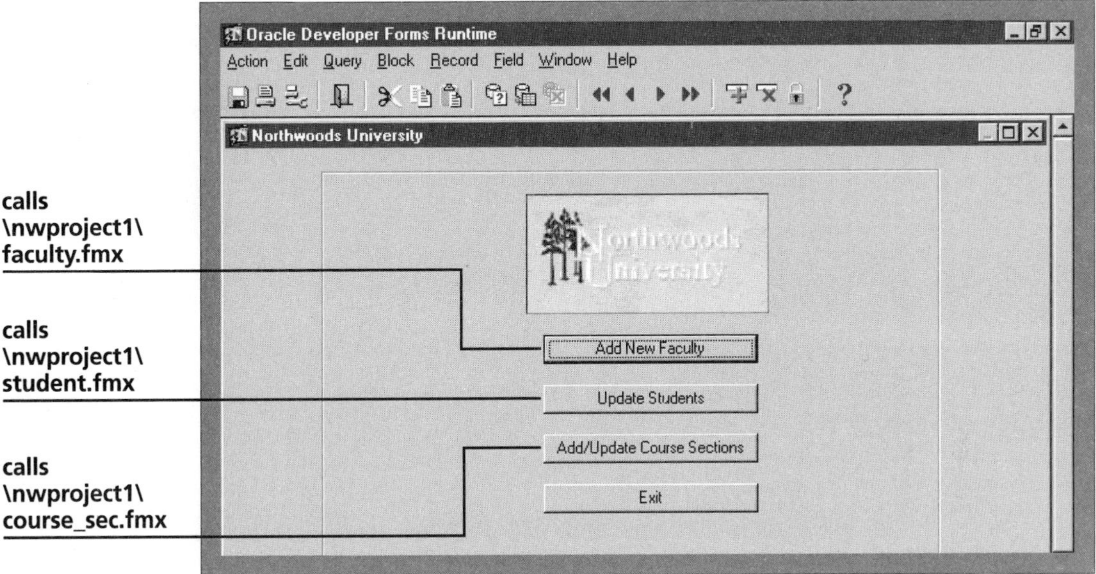

Figure 9-12

a. Create a form named Ch9aEx1.fmb that will be the main application form that calls the individual application files. Create a PRE-FORM trigger to maximize the window automatically, and to create a global path variable to specify the location of the project files in the \Chapter9\nwproject1 subfolder on your Student Disk.
b. Create a splash screen in the form that displays the nwsplash.tif image stored in the \Chapter 9\nwproject1 subfolder. Add a display item on the image item that provides copyright name and date information.
c. Create four command buttons, as shown in Figure 9-12, to provide switchboard access to the application form files.
d. Create a trigger for the first three buttons so that they call the associated .fmx file using the global project path variable. Generate the required .fmx files, and test the application to make sure that the buttons call the forms correctly. (*Hint*: In order to successfully compile the course_sec.fmb file, you might need to create a new sequence named csecid_sequence that starts with 2000 and has no maximum value, if you have not done this in a previous exercise.)

2. In this exercise, you will create an integrated database application for the Northwoods University database that allows students to log on to the system using their student ID and PIN. When a student successfully logs on, the application allows her to view her personal information and course grades, and register for courses.

a. Create a form named Ch9aEx2.fmb that will be the form that calls the individual application files. This form will have two canvases. The first canvas will look like Figure 9-13, and will allow the user to enter her student ID and PIN. If the student enters a valid SID and SPIN value on this canvas, then the main canvas (see Figure 9-14) will be displayed. (*Hint*: Use an explicit cursor and the %FOUND operator to determine if the student ID and PIN exist in the STUDENT table. If they do, save the SID as a global variable named: GLOBAL.sid, then display the main canvas.)

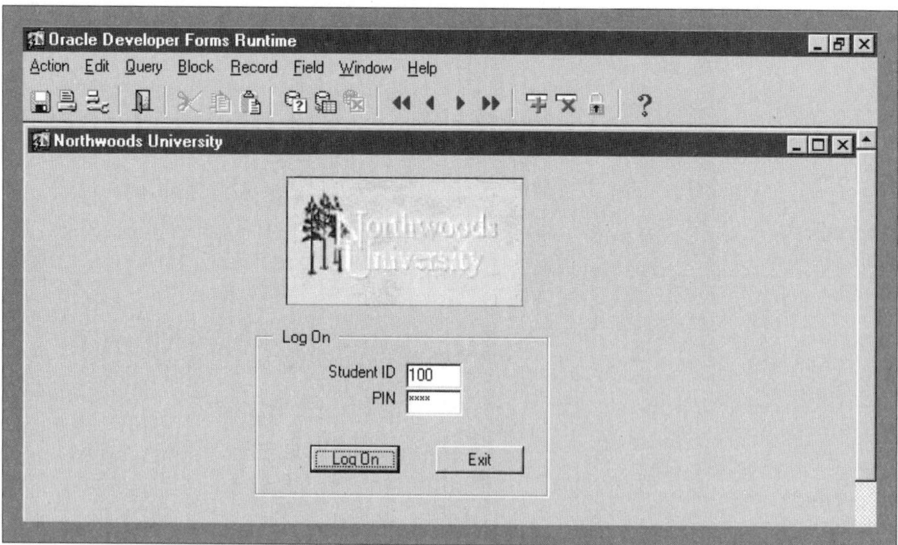

Figure 9-13

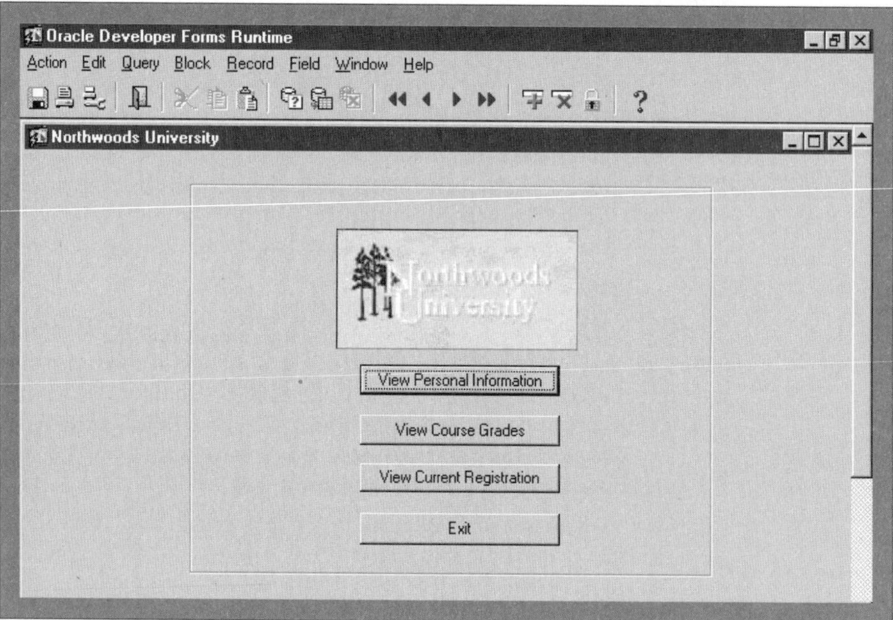

Figure 9-14

b. Create a PRE-FORM trigger that maximizes the window and initializes the global project path variable to specify the location of the project in the \Chapter9\nwproject2 subfolder on your Student Disk.

c. Create a splash screen in the form that displays the nwsplash.tif image stored in the \Chapter9\nwproject2 subfolder. Add a display item on the image item that provides copyright name and date information.
d. Create a form named STUDENT that is displayed when the user clicks the View Personal Information button. This form should display the fields shown in Figure 9-15. Create a data block form based on the fields in the STUDENT table, and modify the WHERE property of the data block so that the block sid = :GLOBAL.sid. The student will need to click the Execute Query button to display her personal information. Save the form as student.fmb in the \Chapter9\nwproject2 subfolder.

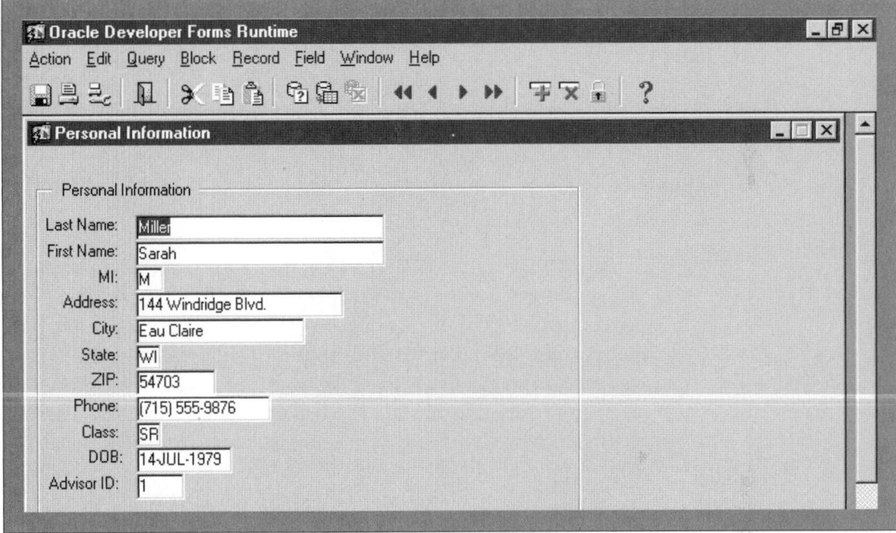

Figure 9-15

e. Create a form named GRADES that displays the SID, TERMID, TDESC, CALLID, and GRADE fields for the selected student when the user clicks the View Course Grades button, as shown in Figure 9-16. Create a view named GRADE_VIEW that contains the fields shown in Figure 9-16. Then, create a data block with the GRADE_VIEW view as its source. Modify the WHERE property of the data block so that only the grades for the current student are displayed. Format the form so it looks like Figure 9-16. The student will need to click the Execute Query button to display the grades. Save the form in the \Chapter9\nwproject2 subfolder as grades.fmb.

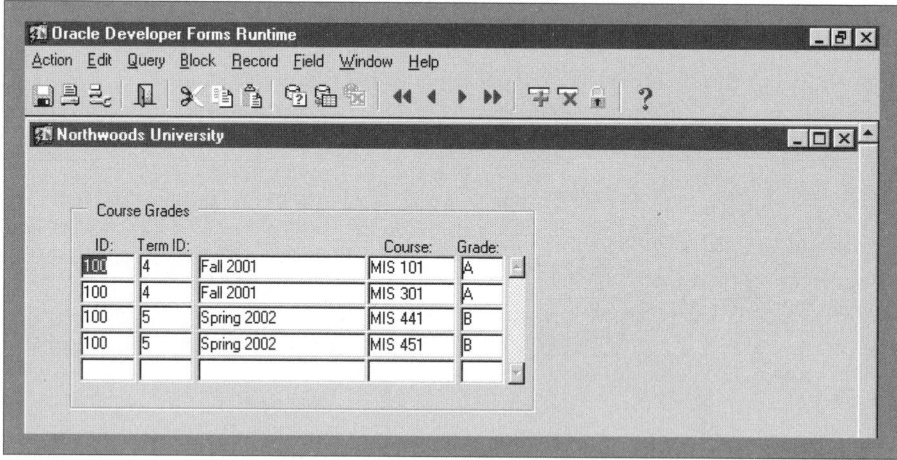

Figure 9-16

f. Create a form named REGISTER that is displayed when the user clicks the Register for Courses button. Save the form as register.fmb in the Chapter9\ nwproject2 subfolder. This form (see Figure 9-17) will allow the student to view her current schedule. Courses that the student is currently enrolled in but has not yet received a grade for are displayed in the Current Schedule frame fields when the form first is displayed. To display this data create a database view named ENROLLMENT_VIEW containing the SID and the Current Schedule frame fields. The view should only include records for which a grade value has not yet been assigned. Then, create a data block based on ENROLLMENT_VIEW. Do not display the SID in the layout. Modify the data block's WHERE property so that only the schedule information for the current student is displayed. The student will need to click the Execute Query button when the form first is displayed, to display her current schedule. (*Hint*: Be sure to test the form using a student ID that has current enrollment records for which no grade has been assigned.)

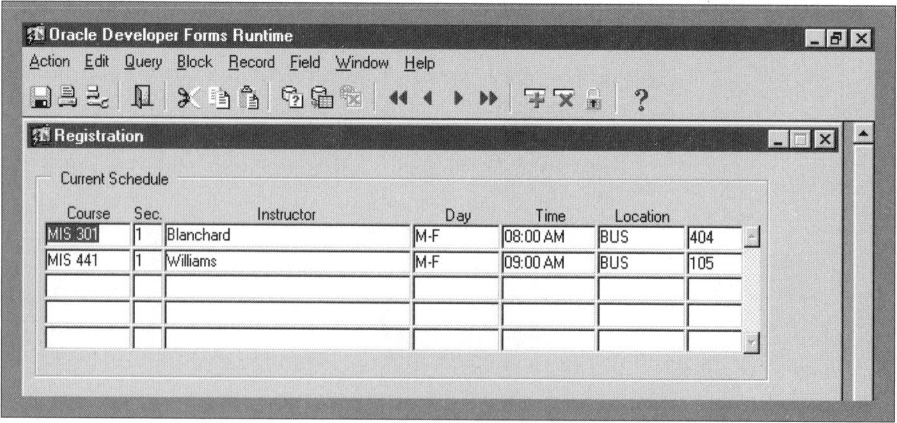

Figure 9-17

LESSON B

objectives

- Use Project Builder to manage files in an integrated database application
- Create custom pull-down menus
- Display different custom pull-down menus on specific forms

Using Project Builder to Organize and Access Project Files

Project Builder is an Oracle application that allows database developers to organize and manage the individual forms, reports, and graphics that make up an integrated database application. It provides a central point where multiple software developers can view and edit all the application files in the Form Builder, Report Builder, or Graphics Builder development environment, and then recompile changed files so that the files work together correctly. In this lesson you will learn how to create a new project, and work with project files using Project Builder.

Creating a New Project

Now you will use Project Builder to create a project for the Clearwater Traders Order Processing System. You will use the Project Wizard, which guides the project creation process using the following pages:

- **Project filename page**, which specifies the name of the project registry file. When you create a new project, Project Builder creates a **project registry file**, which is a file with a .upd extension that contains a series of pointers, or references, that specify the locations of individual project applications.
- **Project definition page**, which specifies the project title and the default drive and folder path where the project files are stored.
- **Default database connection page**, which specifies the project's database connection information (username, password, and connect string), so you can access the individual project components without having to create a new connection each time.
- **User information page**, which specifies the project author and contains comments about the project, such as the date last modified.
- **Finish page**, which determines whether to just create the project registry file, or to open a dialog box that allows you to immediately add files to the project.

Now you will start Project Builder and create the project.

To create the project:

1 Click **Start** on the Windows taskbar, point to **Programs**, point to **Oracle Developer 6.0**, and then click **Project Builder**. The Welcome to Oracle Developer dialog box is displayed.

You can also start Project Builder by starting Windows Explorer, changing to the ORAWIN\BIN directory, and then double-clicking the PJ60.exe file to start Project Builder. (Windows NT users can change to the ORANT\BIN directory, and then double-click the PJ60.exe file.)

2 Make sure the **Use the Project Wizard** option button is selected, and then click **OK**. The Welcome to the Project Wizard page is displayed. Click **Next**. If you have previously created a project, then the Project definition page is displayed. Make sure the **Create a standalone project** option button is selected, then click **Next**. The Project filename page is displayed.

3 Click **Browse**, then navigate to the \Chapter9\cwproject subfolder. Type **clearwater** in the File name text box, and click **Save** to save the Project Registry file. The complete path to the Project Registry file is displayed in the Project Registry Filename text box. If your project is stored on the E drive in the \Chapter9\cwproject subfolder, the path and filename would be e:\Chapter9\cwproject\clearwater.

4 Click **Next**. The Project definition page is displayed.

5 Delete the current text in the Title text box, then type **Order Processing System** for the project title. If necessary, in the Project Directory text box, type the drive letter and path name where your Chapter9 folder is located plus **\Chapter9\cwproject**. Your completed Project definition page should look like Figure 9-18.

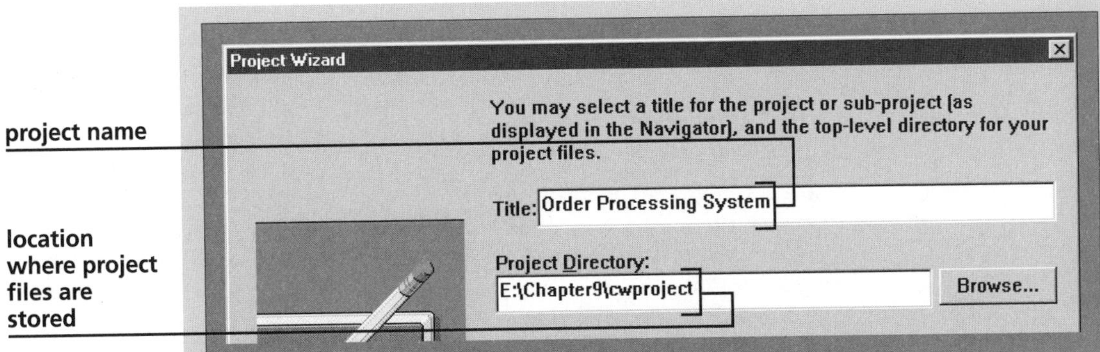

Figure 9-18: Project definition page

7 Click **Next.** The Default database connection page is displayed. This page allows you to connect to the database. You can select a predefined (existing) connection, use the current connection, or create a new connection. Connect to the database by typing in your usual connection information. (This will not create a predefined connection that you can use when you create other projects, so you will create a predefined connection later in the lesson.)

8 Click **Next.** The User Information page is displayed. Type your name in the Author text box. Type **Project created on <current date>** in the Comments box, and then click **Next.** The Finish page is displayed.

9 Make sure the **Select files to add to the project** option button is selected, and then click **Finish.** The Add Files to Project dialog window is displayed.

Now you will specify the files that are included in the project. You will select all of the form .fmb files in the \Chapter9\cwproject subfolder.

To add the form files to the project:

1 Navigate to the \Chapter9\cwproject subfolder. Click the **cust_order.fmb** file, press the **Ctrl** key, and then click the rest of the .fmb files one at a time, while keeping the Ctrl key pressed, until all eight of the project's .fmb files are selected, as shown in Figure 9-19. (Be sure to use the main.fmb file that you successfully completed in Lesson A of this chapter.) Note that the project registry file (clearwater.upd) appears in the window, confirming that it was created.

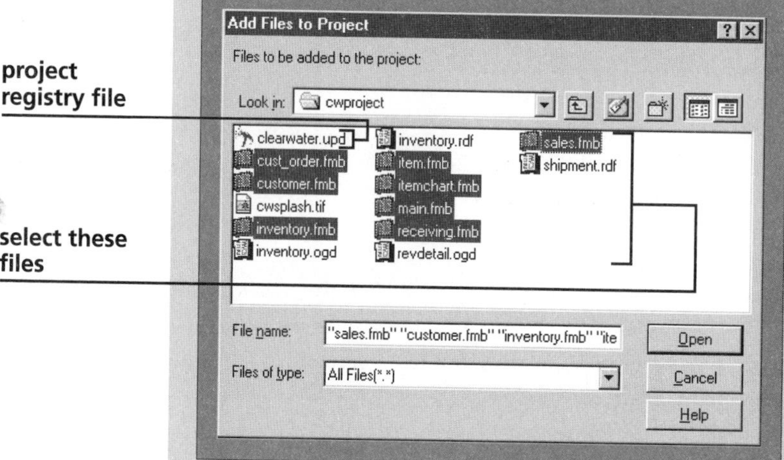

Figure 9-19: Selecting the project .fmb files

help

If the file extensions are not displayed in the Add Files to Project window, start Windows Explorer, click View on the menu bar, then click Options. Click the View tab, scroll in the list box until you find the Hide file extensions for known file types check box, and then clear the check box to display file extensions.

2 Click **Open.** The Project Builder – Project View window is displayed, as shown in Figure 9-20. If necessary, maximize the window.

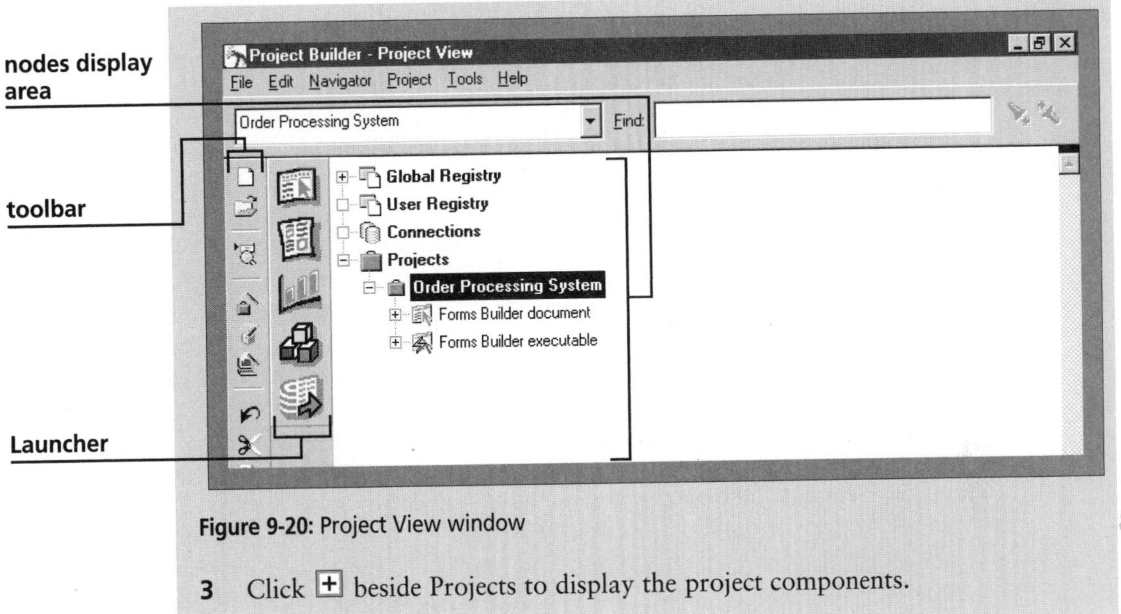

Figure 9-20: Project View window

3 Click [+] beside Projects to display the project components.

The Project View window, also called the Project Navigator, has three main components: a toolbar that provides access to common Project Builder commands; the Launcher, which is a secondary toolbar on the left edge of the window that provides access to the Developer applications (Form Builder, Report Builder, Graphics Builder, Procedure Builder, and Query Builder); and the nodes display area, which shows Project Builder components and provides a hierarchical view of the components in individual projects.

Each project has the following components: Global Registry, User Registry, Connections, and Projects. The **Global Registry** specifies the different types of files that can be included in a project. These files can include Oracle applications (such as Form Builder and Report Builder) and other applications, such as Excel spreadsheets and HTML files. The **User Registry** stores configuration information (such as user ID, password, and user name) for each individual user. Each user registry inherits information about the types of applications that can be included in a project from the Global Registry, and users can also define additional file types in their individual user registries. A user registry also stores environment and preference settings. **Connections** specify user names, passwords, and database connect strings, and can be assigned across projects and to different projects so that users don't have to log on each time they launch a project application. The **Projects** node displays each project and its components.

Viewing Project Dependencies

Now you will examine the components of the Order Processing System project. Project components have **dependencies**, which are conditional relationships based on the order in which project components are defined. Executable files, called **targets**, depend on their associated **source code** files (uncompiled design files created in Form Builder, Report Builder, or Graphics Builder), called **inputs.** Project Builder automatically detects these dependencies, and deduces that any target files in your project are dependent on the associated input files. If an input file changes, it must be recompiled to create a new target file. When you add files to a project, only add the input files. Project Builder will automatically display the associated target files, because they are implied by the existence of the input files. Project Builder always assumes that all project target files are stored in the project default directory.

You can view the project files in the Project Navigator in two different views. The **Project view** shows project items organized by item types and project-to-subproject relationships. The individual projects are organized alphabetically by project filename, then alphabetically by category, and then by filename. The Project view is useful for tracking the different types of files in a project. The **Dependency view** shows files in the Project Navigator in the order of dependency; project nodes are at the highest point in the hierarchy, followed by target nodes, followed by input components. For example, a target .fmx form file depends on its input .fmb file. In either view, input filenames are always shown in boldface type.

Now you will view the current project files and their dependencies. By default, the Project View window opens in Project view. First, you will examine the Order Processing System project's project files and dependencies in the Project Navigator window.

To examine the project files and dependencies:

1 Click [+] beside **Forms Builder document**. A listing of the Form Builder input (.fmb) files is displayed, as shown in Figure 9-21.

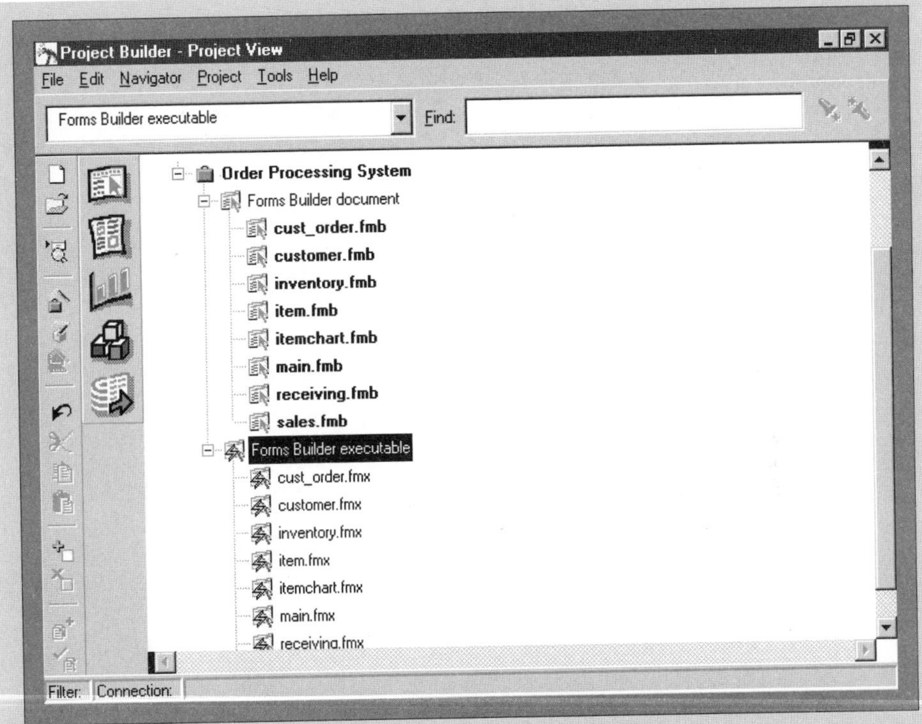

Figure 9-21: Project files in Project view

2 Click ⊞ beside **Forms Builder executable.** A listing of the Form Builder target (.fmx) files is displayed, as shown in Figure 9-21.

3 Click **Navigator** on the menu bar, and then click **Dependency View.** Only the target (.fmx) files are displayed under the Order Processing System project, because they are at the top dependency level.

4 Click ⊞ beside **sales.fmx.** Its associated input file (sales.fmb) is displayed below it (see Figure 9-22), showing that the sales.fmx file is dependent on the sales.fmb file.

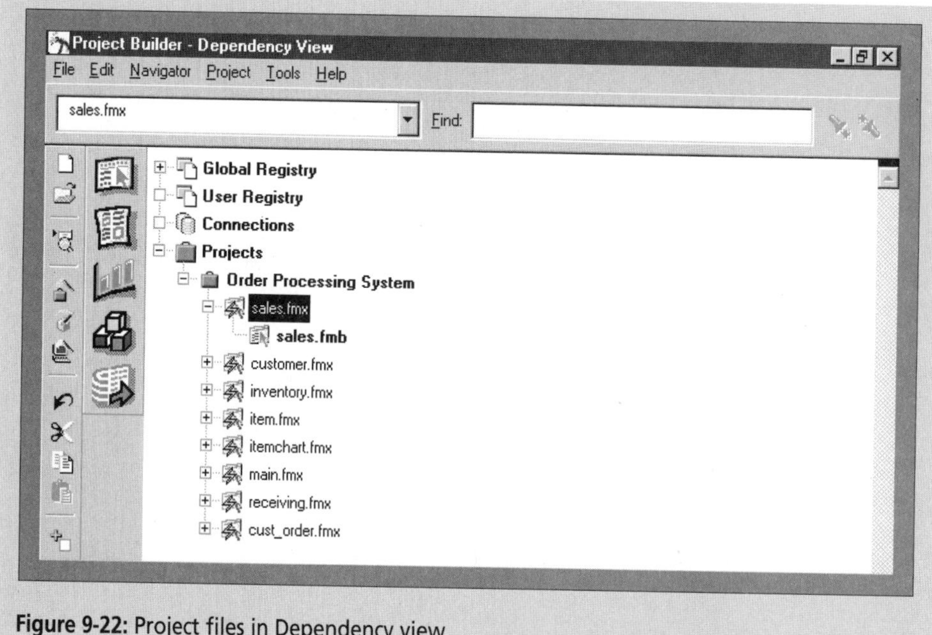

Figure 9-22: Project files in Dependency view

Adding Files to a Project

Now you will add the rest of the files to the project. Figure 9-1 shows that the project has two report files (inventory.rdf and shipment.rdf), and two graphics files (revdetail.ogd and inventory.ogd). After you add the files, it is not necessary to save the project—Project Builder project files are automatically saved every time you make a change to the project.

To add the report and graphics files to the project:

1. Click **Navigator** on the menu bar, and then click **Project View** to return to the Project view.
2. Click **Order Processing System** in the Project Navigator window to select it. If you do not select the project before adding files, the new files will be added to whatever item is currently selected.
3. Click the **Add Files To Project** button on the toolbar. The Add Files To Project dialog box opens.

Another way to add files to a project is by clicking Project on the menu bar, then clicking Add Files To Project.

4 If necessary, navigate to the \Chapter9\cwproject subfolder, select **inventory.ogd,** press the **Ctrl** key, and then select **inventory.rdf, revdetail.ogd,** and **shipment.rdf,** while keeping the Ctrl key pressed.

5 Click **Open**. The Reports Builder and Graphics Builder files are displayed in the Project Navigator window.

6 Click [+] beside **Graphics Builder document, Graphics Builder executable,** and **Reports Builder document.** The report and graphics files are displayed. There are no Reports Builder executable files because the report .rdf files are both the input and target files, and do not need to be compiled.

Creating a Predefined Connection

In Project Builder, you can define a connection that specifies a database username, password, and connect string. You can then assign this connection to different projects that use the same database connection. When you start different form, report, and chart applications directly from the Project Navigator window, the database connection is automatically made in Form Builder, Report Builder, or Graphics Builder. Now you will create a predefined project connection.

To create a predefined project connection:

1 Click the **Connections** node to select it, right-click, then click **New Connection**. The New Connection dialog box is displayed.

2 Type **Clearwater Connection** in the title text box, and then type your usual connection information using your username, password, and database connect string. Click **OK** to save the connection. The new connection is displayed in the Project Navigator window.

tip

When you create a connection in Project Builder, your password is saved in the project registry file. This file could be exported to another file format and someone could view your password, so it is a good idea to always keep project files in a secure location.

After you have created a connection, you can assign it to a project. To do this, you drag the Connection icon and drop it on the project in the Project Navigator window. Now you will assign the Clearwater Connection to the Order Processing System project.

To assign the connection to the project:

1 Click **Clearwater Connection**, then drag and drop it onto Order Processing System.

Creating New Projects, Reopening Projects, and Deleting Projects

Recall that you do not need to explicitly save project files, because all changes are automatically saved as soon as they are performed. When you exit Project Builder, your projects will automatically appear in the Project Navigator window the next time you start Project Builder, as long as the project registry files have not been deleted or moved. Now you will create a second project using the Project Wizard, exit Project Builder, and then start Project Builder again to observe that the project files are still displayed in the project window.

Like the wizards in the other Oracle utilities, the Project Wizard is re-entrant, meaning that it can be used to edit the properties of an existing project, if an existing project is selected when the Project Wizard is started. When you want to create a new project, you must be sure that an existing project is not selected, so that the Project Wizard is not opened in re-entrant mode.

To create a new project using the Project Wizard:

1 Click **Projects** in the Project Navigator window, so no objects in the existing project are selected and the Project Wizard is not opened in re-entrant mode.

2 Click **Tools** on the menu bar, then click **Project Wizard**. The Welcome to the Project Wizard page is displayed. Click **Next**.

3 Make sure the **Create a standalone project** option button is selected, then click **Next**.

4 Click **Browse**, navigate (if necessary) to the \Chapter9\cwproject subfolder, type **sample** as the File name, click **Save**, then click **Next**.

5 Accept the default title and project directory, then click **Next**.

6 Open the Predefined connection list box, and select **Clearwater Connection**, then click **Next**.

7 You will not specify any information on the User information page, so click **Next**.

8 Click the **Create an empty project** option button, then click **Finish**. The sample project is displayed under the Order Processing System project in the Project Navigator window.

Now you will exit Project Builder, and start it again. When you start Project Builder, the projects that were displayed when you last exited the program are automatically redisplayed in the Project Navigator.

To exit Project Builder and start it again:

1. Click **File** on the menu bar, then click **Exit** to exit Project Builder.
2. Start Project Builder again, and click **Cancel** when the Welcome to Oracle Developer dialog box is displayed. The Order Processing System and Sample Project projects are displayed in the Project Navigator window.

When you delete a project in the Project Navigator window, a dialog box opens to give you the option of deleting the project registry (.upd) file in the file system also. Even if you do not delete the project registry file from the file system, the project will not appear in the Project Navigator window the next time you start Project Builder. Now you will delete the sample project you created.

To delete the project:

1. If necessary, click **sample** to select it, then click the **Delete** button on the toolbar. The Project Builder – Remove Items dialog window is displayed. This window gives you the option of deleting the project registry (.upd) file from the file system as well as removing the project from the Project Navigator window.
2. Click **Yes** to delete the project, then click **Yes** again to delete its registry file. The sample project no longer appears in the project navigator.

To delete a project, you could also select the project, click Navigator on the menu bar, then click Delete, or press the Delete key on the keyboard.

Opening, Editing, and Compiling Input Files

When you create a project using Project Builder, you can open individual project files, edit them, and then recompile them from the Project Navigator window. To make the Order Processing System application work correctly, you need to modify the forms that call other forms, so that their CALL_FORM commands use the global path variable and call all forms from the \Chapter9\cwproject subfolder using the `:GLOBAL.project_path` variable. Now you will open the sales.fmb file in Form Builder from the Project Navigator window, and edit the Create/Edit Customer button trigger that calls the cust_order.fmx file. (Modifying the rest of the project forms that call other form or chart files will be left as an end-of-chapter exercise.) Since the database connection information is attached to the project, the database connection will already be made, and you will not need to explicitly connect to the database as you usually do when you start Form Builder.

To open sales.fmb in Form Builder and modify the CALL_FORM command:

1. If necessary, click ⊞ beside Order Processing System, then click ⊞ beside Forms Builder document files to display the project's .fmb files.
2. Double-click the **Form Builder Document** icon beside sales.fmb. The Sales form opens in the Form Builder Object Navigator window.

Another way to open sales.fmb is to click the Form Builder button on the Launcher toolbar to start Form Builder, and then navigate to the \Chapter9\cwproject subfolder and open sales.fmb. If you open the form this way, you will have to click File on the menu bar, then click Connect and explicitly connect to the database.

3. In Form Builder, click **Tools** on the menu bar, then click **Layout Editor**. Select the **CUST_ORDER_CANVAS**, then click **OK**.
4. Right-click the **Create/Edit Customer** button, click **PL/SQL Editor**, and edit the CALL_FORM command as follows:

```
CALL_FORM(:GLOBAL.project_path || 'customer.fmx');
```

5. Compile the trigger, correct any syntax errors, and close the PL/SQL Editor.
6. Save the form, then close Form Builder.

Currently, only the sales.fmx and receiving.fmx project executable files exist in the \Chapter9\cwproject subfolder. Now you must create all of the target form (.fmx) and graphics (.ogr) files for the project by compiling (which is called building in Project Builder) all of the input (.fmb and .ogd) files. There are three options for building input files in Project Builder: **Build Selection**, which only builds a selected target file; **Build Incremental**, which builds all target files whose associated input files have been changed but have not yet been rebuilt; and **Build All**, which builds all of the project's input files. To build a selection, you must have the selection's target file selected in the Project Navigator window. For example, to rebuild sales.fmb, you must select sales.fmx in the Project Navigator window. To incrementally build the project files or to build all of the project files, you must have the project selected in the Project Navigator window. Now you will build all of the project files, then run the main project application file (main.fmx), click the New Orders button to call sales.fmx, then click the Create/Edit Customer button to confirm that the cust_order.fmx file is called correctly from sales.fmx.

To build all of the project files, and test sales.fmx:

1. In the Project Navigator window, click **Order Processing System** to select the project so you can build all of the project files.
2. Click **Project** on the menu bar, then click **Build All**. You will see various forms and program elements flash on your computer screen as Project Builder builds the files. When the Building confirmation dialog box is displayed, confirming that all 10 project items were successfully built, click **OK**.
3. Double-click the **Form Builder executable** icon beside main.fmx to run the main application form. Note that when the form opens, you are automatically logged on to the database.
4. Click the **New Orders** button to run sales.fmx. The New Orders form is displayed.

help If the New Orders form does not appear, be sure that you are using the main.fmb file that you successfully created in Lesson A of this chapter.

5. Click **Create/Edit Customers**. The Customers form is displayed. Click **Exit** to exit the Customers form, click **Exit** again to exit the Sales form, then click **Yes** to confirm closing the form. Click the **Exit** iconic button to close main.fmx.

help If the Customers form does not appear, make sure that you modified the CALL_FORM procedure correctly in the Create/Edit Customers button trigger, and that you recompiled the sales.fmb file.

Adding Pull-Down Menus to Forms

Currently, the default Forms Runtime pull-down menu choices are Action, Edit, Block, Field, Record, Query, Window, and Help. To replace these default menu choices with custom menu choices, like the ones shown in Figure 9-1, you will need to create a menu module. A **menu module** is a separate Oracle application that you create in Form Builder. It can contain individual menus and other objects such as program units and parameters. A menu module is saved in the file system as a design file with an .mmb extension, and as an executable file with an .mmx extension. The executable (.mmx) menu file can be attached to any form module. After you create a menu module, you create one or more **menu items** that contain customized sets of pull-down menu choices. Now you will start Form Builder from the Project Navigator window using the Launcher toolbar. Then you will create a new menu module and a menu item for your integrated database application.

To create the menu module and menu item:

1. Click the **Form Builder** button on the Launcher toolbar. The Welcome to Form Builder window is displayed. Click **Cancel**. The Object Navigator window is displayed with a new form (MODULE1) loaded. If necessary, maximize the Object Navigator window.
2. If necessary, click **MODULE1** to select it, then click the **Delete** button to delete it.
3. In the Object Navigator window, click the top-level **Menus** object, and then click the **Create** button on the Object Navigator toolbar to create a new menu module. Rename the new menu module **CW_MENU**.
4. Click the **Menus** item under CW_MENU, and then click to create a new menu item in the menu module. Rename the new menu item **MAIN_MENU**. Your Object Navigator window should look like Figure 9-23.

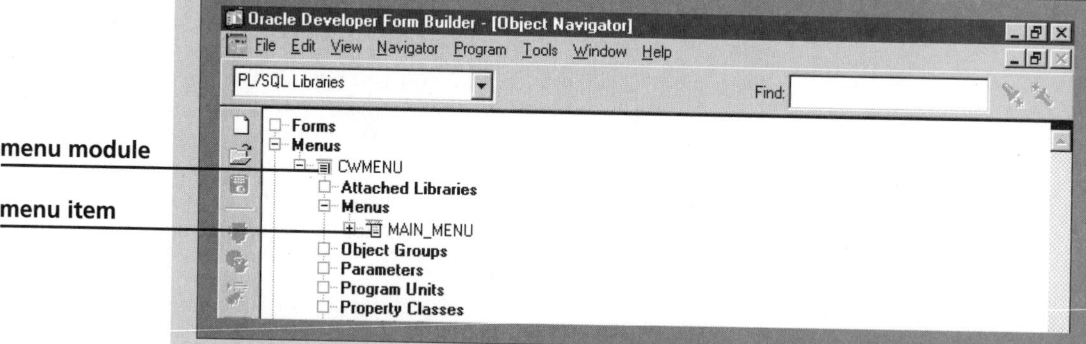

Figure 9-23: Creating a new menu module and menu item

5. Double-click the **Menu Item** icon beside MAIN_MENU to open the Menu Editor. Maximize the Menu Editor window if necessary.

You also can select MAIN_MENU, click Tools on the menu bar, then click Menu Editor to open the Menu Editor.

6. Click **File** on the menu bar, and then click **Save**, and save the menu module as **CW_MENU.mmb** in the \Chapter9\cwproject subfolder.

Using the Menu Editor

The Menu Editor window (see Figure 9-24) enables you to build a pull-down menu structure and define the underlying action triggers to carry out the user selections. Like other Oracle applications, it has a toolbar and status line at the bottom of the screen. By default, a new menu item contains a default menu selection named <New_Item>. Before you create any menus, it is important to configure the Menu Editor environment so that the menu design file is saved automatically when you generate an executable file. You will do this next.

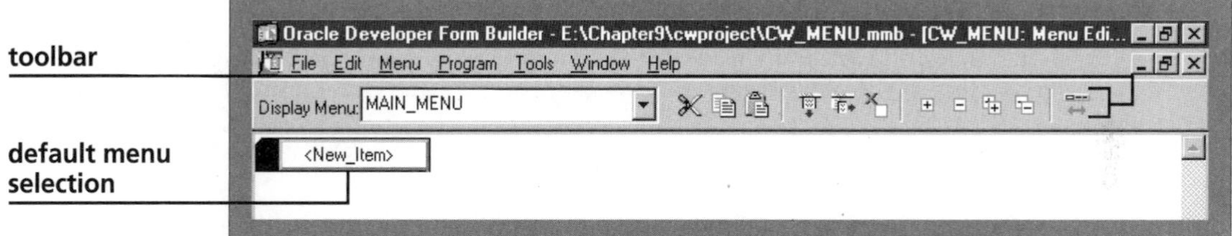

Figure 9-24: Menu Editor window

To configure the Menu Editor:

1 Click **Tools** on the menu bar in the Menu Editor window, and then click **Preferences** to open the Preferences dialog window.

2 Make sure that the Save Before Building and Build Before Running check boxes are checked. Leave the other check boxes cleared, and click **OK** to save your changes.

Now you can create the menu items. The top-level menu choices on the menu bar are called **parent menu items** and the lowest-level selections that cause an action to occur are called **child menu items.** When the user selects a child menu item, a program called an **action trigger** runs that issues commands to perform actions such as calling another form, calling a report, or clearing the form. For example, when a user clicks Action and then clicks New Orders, an action trigger executes that calls the New Orders form. You can use the Menu Editor to create the menu items visually, by creating the parent menu items across the screen and the child menu items below their associated parent selections. Now you will create the parent and child menu items shown in Figure 9-1.

To create the menu items:

1 Click the **<New Item>** box, then click it again if necessary, so the insertion point appears in the menu item label. Delete the current label (<New Item>, including the angle brackets), and type **Action** to rename the first parent menu item.

2 Click the **Create Down** button on the Menu Editor toolbar to create a child menu item under the Action menu item.

3 Type **New Orders** to change the <New Item> menu item label to New Orders.

4 Click again to create another child menu item, and change the new menu item label to **Receiving.**

5 Click again to create another child menu item, and change the new menu item label to **Exit.**

6 Press the ↑ key three times to select the Action menu item.

tip

You can also select the Action menu item by clicking it with the mouse pointer.

7 Click the **Create Right** button on the Menu Editor toolbar to create a new parent menu item to the right of the Action menu item.

8 Type **Reports** to rename the new menu item.

9 Repeat the previous steps to create the menu items shown in Figure 9-25. Use the **Create Down** button and the **Create Right** button on the Menu Editor toolbar until you have created the entire menu structure.

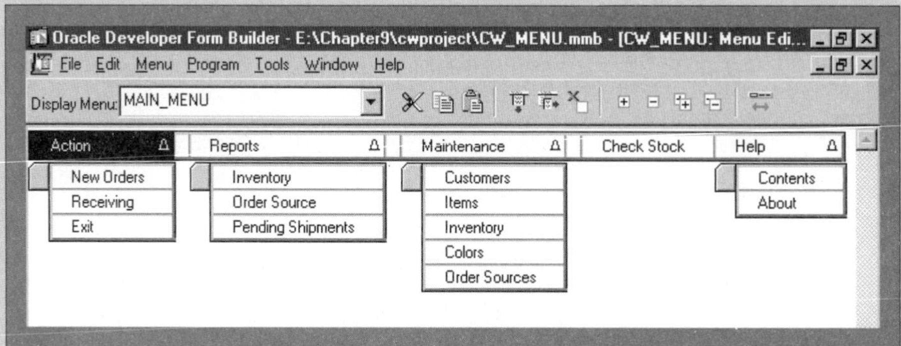

Figure 9-25: Project menu structure

10 Click **File** on the menu bar, then click **Save** to save the menu module.

help

If you accidentally create a menu item that you don't need, click on the menu item you want to delete, click the Delete button on the Menu Editor toolbar, and then click Yes to confirm the deletion.

help

Do not create the Window menu selection shown in Figure 9-1, because it will be added automatically in the Forms Runtime window.

Your menu structure is almost finished—you are missing the Window menu selection, which the system automatically adds at runtime, so you don't need to create it on your menu structure. Now you will create the separator bar in the Help submenu that separates Contents and About in Figure 9-1.

To create the separator bar:

1. Select the **Contents** child-level menu item, and then click the **Create Down** button on the Menu Editor toolbar to insert a new child menu item under Contents.
2. Right-click the new menu item, placing the mouse pointer on the right edge of the menu item, to the right of the <New Item> text, and then click **Property Palette** to open the Property Palette for the new menu item.
3. Change the name of the menu item to **SEPARATOR_BAR.**
4. Click the **Menu Item Type** property selection, select **Separator** from the list, and then close the Property Palette.
5. Click **File** on the Menu Editor menu bar, then click **Save** to save the menu module.

Closing and Reopening Menu Module Files

Menu modules are stored separately from form applications. Therefore, to open a menu module, you need to retrieve it from the file system independently of the form it will be used in. To see how this is done, let's suppose you decide to take a break and close your file and work on it later. You will close the menu module and then reopen it.

To close the menu module file then reopen it:

1. Click **File** on the menu bar, then click **Close** to close the CW_MENU menu module. The Object Navigator window is displayed, and the CW_MENU menu module object no longer is displayed under the Menus object.
2. Click the **Open** button on the Object Navigator toolbar, click the **Files of type** list arrow, click **Menus (*.mmb)**, select **CW_MENU.mmb** from the file list, then click **Open** to open the file.
3. To return to the Menu Editor, double-click the **Menu** icon beside CW_MENU. Notice that each parent menu item with underlying selections has a down-pointing tab beside the label name. To open a parent item, click its tab.

4 Click the **Action** tab ▼ to display the child menu items. Note that the Action tab now points upward ▲, indicating that its child items now are displayed. Also, note that the child items do not have tabs beside them.

5 Click the **Action** tab ▲ to close the Action menu.

Creating the Access Key Selections

In Figure 9-1, each menu selection has an underlined letter. This letter is called the menu item's **access key**. Users can press Alt plus the access key to open the top-level menu selections on the menu bar without using the mouse. Once a menu selection is opened, the user can select a submenu selection by simply pressing the submenu's access key, without pressing the Alt key. In a Form Builder menu module, the first letter of each menu item label is the default access key. If two menu selections have the same first letter, then the access key will only work for the first selection. Therefore, it is sometimes necessary to change the default access key, because it already has been used or because another key seems more intuitive. For example, in the Maintenance menu selection, two items begin with the same letter (Customers and Colors). Also, the access key for Exit might be *x* rather than *E* because *x* sounds more like Exit. To override the default access key choice, type an ampersand (&) before the desired access key menu letter. The ampersand will not appear on the menu when it is running.

If you want an ampersand to appear in a menu choice, type the ampersand twice. For example, to make an ampersand the access key for a menu selection of Research & Development, you would enter it as Research && Development.

To change the menu access keys:

1 Open the **Action** menu item so Exit is displayed.

2 Select the **Exit** child item by clicking it, and then click it again to go into text editing mode.

3 Type **&** before the x so the label now reads E&xit.

4 Modify all other menu items whose access key is not the first letter of the label so your menu labels look like Figure 9-26.

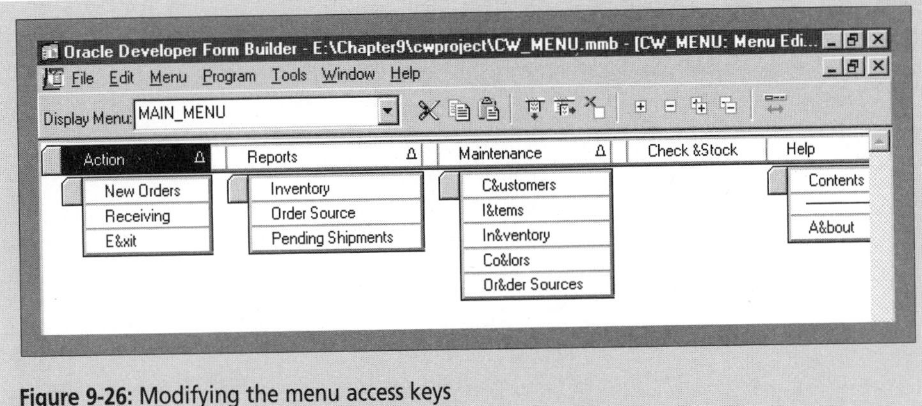

Figure 9-26: Modifying the menu access keys

5 Save the menu module.

Modifying the Iconic Button Triggers and Creating the Menu Item Action Triggers

Every child menu item must have an associated action trigger that executes when the user selects the item. You will create the action trigger for the New Orders menu selection next. It will have exactly the same code as the NEW_ORDERS_BUTTON trigger, so you will open the main.fmb form file, and copy that code and then paste it into the menu item's action trigger.

To create the New Orders menu item action trigger:

1 In Form Builder, click **Window** on the menu bar in the Menu Editor window, then click **Object Navigator** to open the Object Navigator window.

2 Click the top-level **Forms** object in the Object Navigator window, then click the **Open** button, and open **main.fmb** from the \Chapter9\cwproject subfolder.

3 Open the MAIN_CANVAS in the Layout Editor, right-click the **New Orders** button, and then click **PL/SQL Editor** to view the trigger code. Maximize the PL/SQL Editor window if necessary. Select all of the code, click **Edit** on the menu bar, click **Copy**, and then close the PL/SQL Editor.

4 To reopen the Menu Editor window, click **Window** on the menu bar, and then click **CW_MENU: Menu Editor**.

5 Right-click the **New Orders** menu selection, and then click **PL/SQL Editor** on the menu. An action trigger for the menu choice is created automatically, and you can enter the code that will execute when the menu item is selected. Maximize the PL/SQL Editor window if necessary.

6 Click **Edit** on the menu bar, and then click **Paste** to paste the copied trigger code into the Source code pane.

7 Compile the trigger, correct any syntax errors, and then close the PL/SQL Editor.

Next, you will create the triggers for the Receiving and Exit menu items. For the Receiving item, you will copy the code from the form's Receiving button trigger, and then paste it into the action trigger for the Receiving menu item.

To create the action triggers for the Receiving and Exit menu items:

1. Open the MAIN_CANVAS in the Layout Editor, right-click the **Receiving** button, and then click **PL/SQL Editor** to view the trigger code. Copy the code and then close the PL/SQL Editor.
2. Click **Window** on the menu bar, and then click **CW_MENU: Menu Editor** to open the Menu Editor window.
3. Right-click the **Receiving** menu selection, and then click **PL/SQL Editor** on the menu. Paste the copied trigger code into the Source code pane, compile the trigger, and close the PL/SQL Editor.
4. Right-click the **Exit** menu selection, and then click **PL/SQL Editor** to open the Editor again.
5. Type **EXIT_FORM;** in the PL/SQL Editor.
6. Compile the trigger, correct any syntax errors, and then close the PL/SQL Editor.
7. Save the menu module.

The next child menu item that needs an action trigger is Inventory (under the Reports menu), for calling the inventory chart report. Recall that to call a report from a form, you use the RUN_PRODUCT procedure. Since you do not pass a parameter to this report, you will not need to create a parameter list. However, the RUN_PRODUCT procedure generates a compile error unless you create a parameter list variable and pass this variable as a parameter in the RUN_PRODUCT command. Therefore, you will declare a parameter list variable named "dummy _list," but you will not initialize it or add any parameters to it. Now you will create the Inventory menu item's action trigger to run the report.

To create the Inventory menu item's action trigger:

1. In the Menu Editor window, select the **Inventory** child item selection under Reports, right-click, and then open the PL/SQL Editor. Type the code shown in Figure 9-27.

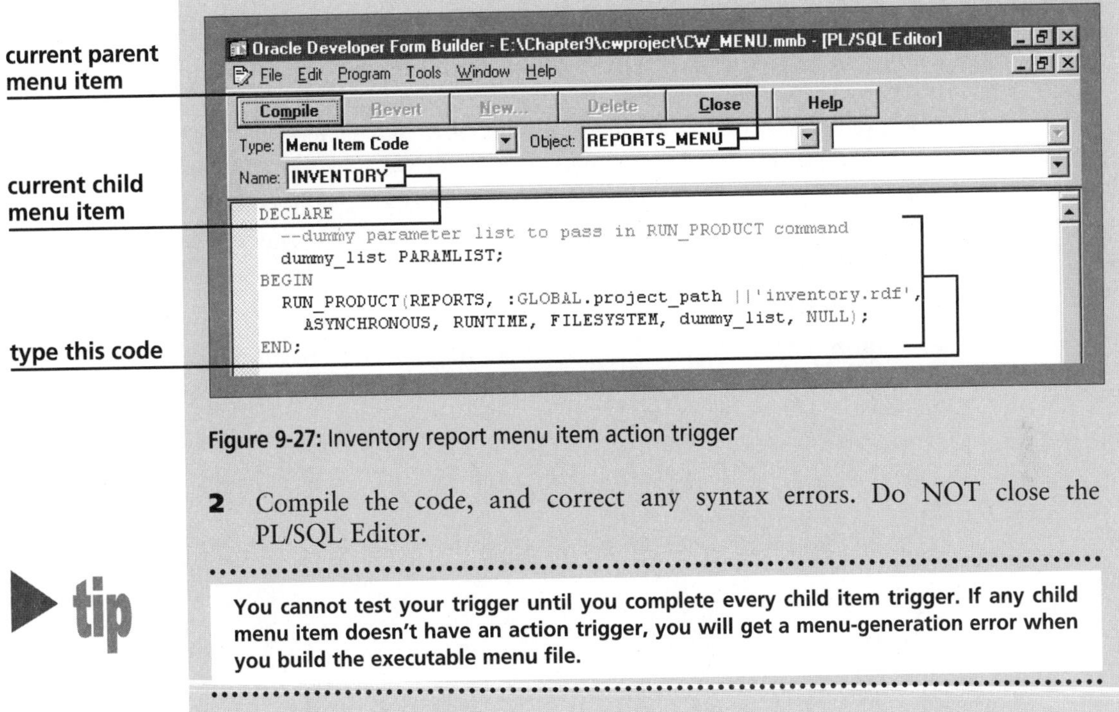

Figure 9-27: Inventory report menu item action trigger

2 Compile the code, and correct any syntax errors. Do NOT close the PL/SQL Editor.

tip

You cannot test your trigger until you complete every child item trigger. If any child menu item doesn't have an action trigger, you will get a menu-generation error when you build the executable menu file.

The next menu selection under Reports is Order Source, which will display the revdetail.ogd chart application. This action trigger will use the RUN_PRODUCT procedure just like the previous action trigger, except that you will substitute GRAPHICS for REPORTS as the product name, and change the filename. You can easily move to another menu item in the PL/SQL Editor window. Figure 9-27 shows that the Object list displays the current parent menu item, and the Name list displays the current child menu item. Since the Order Source menu item's parent item is also REPORTS_MENU, you can select the ORDER_SOURCE child menu item from the Name list and open its PL/SQL Editor. Now you will copy the current code in the INVENTORY menu item, move to the ORDER_SOURCE menu item, and paste the code and modify it to create the ORDER_SOURCE action trigger. Then you will repeat the process to create the PENDING_SHIPMENTS action trigger.

To create the Order Source and Pending Shipments action triggers:

1. Select all of the code in the INVENTORY action trigger, click **Edit** on the Object Navigator menu bar, then click **Copy**.
2. Open the Name list, then click ORDER_SOURCE. The ORDER_SOURCE PL/SQL Editor window opens.
3. Click in the PL/SQL Editor window to make it the current window, click **Edit** on the menu bar, then click **Paste** to paste the copied code to the ORDER_SOURCE PL/SQL Editor window.
4. Modify the code's RUN_PRODUCT command as follows:

```
RUN_PRODUCT(GRAPHICS, :GLOBAL.project_path ||
'revdetail.ogd', ASYNCHRONOUS, RUNTIME,
FILESYSTEM, dummy_list, NULL);
```

5. Compile the trigger and correct any syntax errors. Do NOT close the PL/SQL Editor.
6. Repeat Steps 2 and 3 to open the PL/SQL Editor window for the PENDING_SHIPMENTS action trigger, and paste the trigger code.
7. Modify the PENDING_SHIPMENTS action trigger's RUN_PRODUCT command as follows:

```
RUN_PRODUCT (REPORTS, :GLOBAL.project_path ||
'shipment.rdf', ASYNCHRONOUS, RUNTIME,
FILESYSTEM, dummy_list, NULL);
```

8. Compile the trigger and correct any syntax errors. Close the PL/SQL Editor.
9. Save the menu module.

The first three Maintenance menu selections (Customers, Items, and Inventory) and the Inventory Display selection under Check Stock all require action triggers to call .fmx files. These triggers will use the CALL_FORM command just as the iconic button triggers you created did. You will create these action triggers next.

To create the action triggers to call the .fmx files:

1. In the Menu Editor window, right-click the **Customers** child menu item under Maintenance, and then click **PL/SQL Editor.**
2. Type the following trigger code: `CALL_FORM(:GLOBAL.project_path || 'customer.fmx', HIDE, NO_REPLACE);`. Compile the code and correct any syntax errors.

3 Create and compile similar triggers for the Items, Inventory, and Check Stock child menu items, using the following trigger code:

Child Menu Item	Trigger Code
ITEMS	**CALL_FORM (:GLOBAL.project_path || 'item.fmx', HIDE, NO_REPLACE);**
INVENTORY	**CALL_FORM (:GLOBAL.project_path || 'inventory.fmx', HIDE, NO_REPLACE);**
CHECK_STOCK	**CALL_FORM (:GLOBAL.project_path || 'itemchart.fmx', HIDE, NO_REPLACE);**

4 Close the PL/SQL Editor.

5 Save the menu module.

Next, you need to create the stubs for the menu selections (Color and Order Sources under Maintenance, and About under Help) that are not implemented yet. To do this, you will create an alert to inform the user that the feature has not been implemented, and then create the action triggers to display the alert.

To create the alert and action triggers for the unimplemented menu items:

1 Open the Object Navigator window in Ownership View, select **Alerts** under the MAIN form module, then click the **Create** button to create a new alert. Rename the alert **NOT_DONE_ALERT**.

2 Open the NOT_DONE_ALERT Property Palette, and change the following properties:

Title	**Not Implemented**
Message	**Feature not yet implemented.**
Alert Style	**Stop**
Button1	**OK**
Button2	**<deleted>**

3 Close the Property Palette, and save the form.

4 Open the Menu Editor, and create action triggers for the Colors and Order Sources menu selections (under Maintenance) and for the About menu selection (under Help) to display the NOT_DONE_ALERT, using the following trigger code:

```
DECLARE
    alert_button NUMBER;
BEGIN
    alert_button := SHOW_ALERT('NOT_DONE_ALERT');
END;
```

5 Compile the triggers and correct any syntax errors, then close the PL/SQL Editor.

6 Save the menu module.

Finally, you will create an action trigger for the Contents menu selection (under Help) to activate the Forms Designer Help screen.

To create the action trigger for Contents:

1 In the Menu Editor, right-click **Contents** (under Help), then click **PL/SQL Editor** to create an action trigger with the following code: **DO_KEY('help');**. Compile the code, correct any syntax errors, then close the PL/SQL Editor.

2 Save the menu module.

Compiling and Debugging the Executable Menu File

Before you attach a menu to a form, you must compile the menu design (.mmb) file into an executable (.mmx) menu file. Every time you modify your menu module, you must recompile the .mmb file into a new .mmx file. Otherwise, your form will use your old .mmx file and will not show your recent changes. It is best to compile your menu module in Form Builder the first time you compile it, so you can correct errors immediately. When you make incremental changes later on, you can recompile the menu module directly in Project Builder. Now you will compile the menu module and generate the executable menu file.

To generate the executable menu file:

1 In the Menu Editor window, click **File** on the menu bar, point to **Administration**, and then click **Compile File.** If you have not logged on to the database yet, you will be prompted to do so. The message "Module generated successfully" on the status line indicates that the executable file was created successfully.

You can compile a menu module in the Object Navigator by clicking File, pointing to Administration, and then clicking Compile File as long as one of the menu items is selected. If a form item is selected, this command will compile the current form rather than the menu module.

help

If you receive a menu compile error message like the one shown in Figure 9-28, it indicates that you forgot to create an action trigger for one of your child menu items. The "No PL/SQL source code in menu item CONTENTS" error indicates that the Contents menu item is missing its action trigger. Recall that every child menu item must have an action trigger. If you encounter a similar error, add the necessary trigger code under the menu selection indicated in the error message and then recompile the menu file.

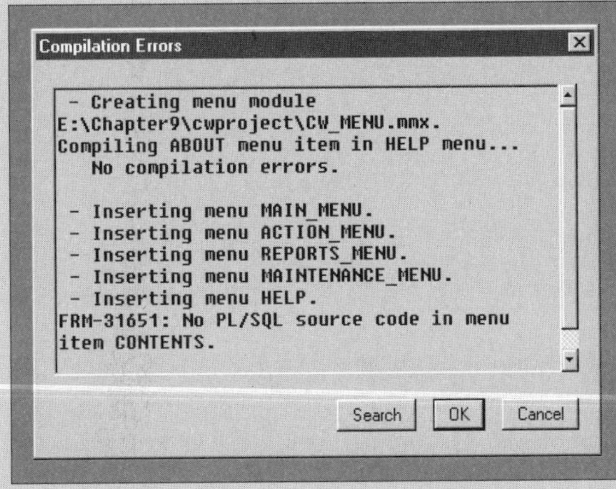

Figure 9-28: Menu compile error

2 Close Form Builder. Click **Yes** if you are asked if you want to save your changes.

Adding the Menu Module to the Project

Now you need to add the menu module files to the project, attach the menu module to the form in the MAIN module's Property Palette, and test the menus. First you will add the menu module design (.mmb) and executable (.mmx) files to the project.

To add the menu module files to the project:

1 If necessary, click **Project Builder** on the taskbar to open Project Builder.

2 If necessary, click **Order Processing System** to select the project, then click the **Add Files to Project** button. The Add Files to Project dialog box opens.

3 Click **CW_MENU.mmb**, press the **Ctrl** key, then click **cw_menu.mmx** while keeping the Ctrl key pressed to select the menu module design and executable files. Click **Open** to add the files to the project. The menu file types are displayed in the Project Navigator.

4 Click [+] beside Forms Builder menu document and Form Builder menu executable to view the menu module files.

Next, you will modify the main.fmb input file so that it displays the custom menu. To do this, you will open main.fmb and change the form's Menu Module property on the form's Property Palette to specify the full folder path to the cw_menu.mmx file, including the drive letter.

To display the custom menu on the main project form:

1 In the Project Navigator window, click [+] beside the Forms Builder document to display the project's .fmb files if necessary. Double-click the **Form Builder Document file** icon beside main.fmb to open main.fmb in Form Builder.

2 In the Form Builder Object Navigator window, open the MAIN module Property Palette by double-clicking the **Forms** icon next to MAIN. Confirm that the Menu Source property value is **File.**

3 Scroll down to the Menu Module property. The current value should be DEFAULT&SMARTBAR. The smartbar is the Forms Runtime toolbar that is used for working with data block forms. Since the main.fmb form is not a data block form, you will not display the smartbar. Change the value to the full path and filename of your newly generated menu module. If the Chapter9 folder were stored on the E drive, the full path would be **e:\Chapter9\cwproject\cw_menu.mmx.**

You cannot specify the menu module location using the :GLOBAL.project_path variable, so if you change the location of the project files, you will need to change the menu module path in the form Property Palette also.

4 Close the form's Property Palette, save the form, then close Form Builder and return to the Project Navigator window in Project Builder.

Now you must recompile the main.fmb file. To do this, you will use the Build Selection option in Project Builder. Whenever you make a change to a single project file, always build only the changed file, rather than rebuilding all of the project files, because it is time-consuming to rebuild files that do not need to be rebuilt. To build a single file, select the *target* (.fmx) file in the Project Navigator window, and then click the Build Selection button on the Project Navigator toolbar. After you rebuild the file, you will run the form to confirm that the custom menus are displayed.

To rebuild the main.fmb file and test the form:

1 In the Project Navigator window, click [+] beside Forms Builder executable if necessary, then click **main.fmx** to select it.

2 Click the **Build Selection** button on the toolbar to build the selected file. Click **OK** when the "1 item built." message is displayed.

3 Double-click **main.fmx** under Forms Builder executable to run the form. Your custom pull-down menus are displayed on the main application's menu bar, as shown in Figure 9-29. Note that the Forms Runtime toolbar is not displayed.

custom pull-down menus

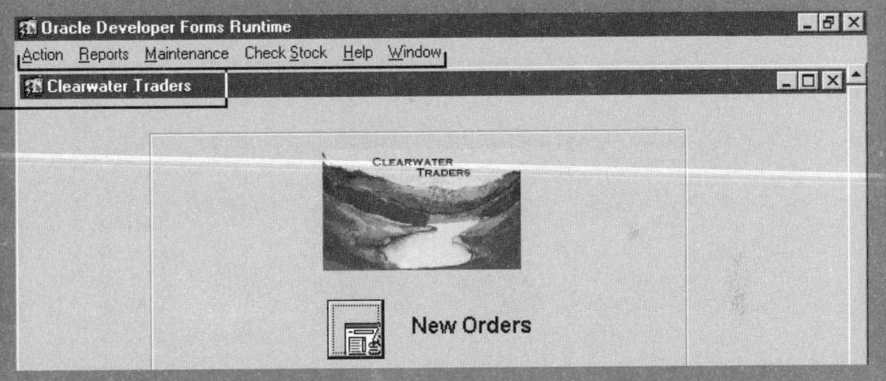

Figure 9-29: Form with custom menu

help

If you receive the error message "FRM-10221: Cannot read file <path to .mmx file>," it is because the name or path of the menu executable file is not specified correctly in the form's Menu Module property, or because the correct file is not in the specified folder.

4 Test all of your pull-down menu choices to confirm that they call the correct form, report, or graphic, and that the NOT_DONE_ALERT is displayed for unimplemented selections.

help

The data block forms (Customers, Items, Inventory) will not work correctly because they require the default Forms Runtime menu module. You will learn how to restore the default menus to these forms later in the lesson.

help

The chart will not appear in the Check Stock menu selection when you select an item, because the path information in its triggers needs to be modified. You will correct this in an end-of-chapter exercise.

help

If you edit a menu action trigger, save the menu file, then run the form and find that nothing has changed, it means that you forgot to recompile the modified .mmb file into a new .mmx file. After making any changes in the menu action triggers, always remember to recompile your menu file before you run the form to test the change.

5 Test all of the pull-down menu choices using the access keys to confirm that the access keys work correctly. Recall that to open a top-level menu selection, you must press the Alt key plus the access key, and that to execute a child-level menu selection, you simply press the access key.

6 Click the **Exit** iconic button to exit the application.

Replacing Called Form Pull-Down Menus

Every time you call a form from main.fmb, the CALL_FORM command uses the NO_REPLACE option. As a result, each time a new form is called, the called form inherits the CW_MENU.mmb menu module from main.fmb. In the Order Processing System application, this is a problem for the Customer, Item, and Inventory forms displayed under the Maintenance menu, because these are data block forms, and they should display the default Forms Runtime window pull-down menus.

Currently, the Menu Module property for the Customer, Item, and Inventory forms is set to DEFAULT&SMARTBAR, which means they will display the default Forms Runtime pull-down menu and toolbar when they are called using the DO_REPLACE option. Now you will edit the action triggers for the pull-down menu items that call these data block forms, so they use the DO_REPLACE option.

To edit the action triggers for the menu items that call the Customer, Item, and Inventory forms:

1 In the Project Navigator window, double-click the **Menu** icon beside CW_MENU.mmb to open the menu module for editing in Form Builder.

2 In the Form Builder Object Navigator, double-click the **Menu Module** icon beside CW_MENU. The Menu Editor opens.

3 Right-click the **C&ustomers** child menu item under Maintenance, then click **PL/SQL Editor** to open the menu item's action trigger.

4 Modify the action trigger as follows:

```
CALL_FORM(:GLOBAL.project_path || 'customer.fmx',
HIDE, DO_REPLACE);
```

5 Compile the trigger, and correct any syntax errors. Do NOT close the PL/SQL Editor.

6 Open the Name list box to view the other child menu items under the Maintenance parent menu item, click **Items** to display its action trigger in the PL/SQL Editor window, then modify its action trigger so it also uses the DO_REPLACE option in the CALL_FORM command. Compile the trigger and correct any syntax errors.

7 Repeat Step 6 for the Inventory action trigger, then close the PL/SQL Editor.

8 Save the menu module.

9 Exit Form Builder.

Now you will rebuild the menu module in Project Builder, run the main.fmx form from Project Builder, and test the menu module to make sure that the default data block menus are displayed for the Customer, Item, and Inventory forms.

To rebuild the menu module, run the application, and test the modified menu module:

1 In the Project Navigator window, click **CW_MENU.mmx** under Forms Builder menu executable, then click the **Build Selection** button to rebuild the modified menu module. Click **OK** when the "1 item built." message is displayed.

help If you receive a message stating "No items were built," open the CW_MENU.mmb file in Form Builder and compile the menu and generate the new CW_MENU.mmx file in Form Builder.

2 Double-click the **Forms Runtime Executable** icon beside main.fmx to run the application. The main application window is displayed.

3 In the Clearwater Traders database application, click **Maintenance** on the menu bar, then click **Customers** to display the customer form. The default Forms Runtime pull-down menu and toolbar are displayed.

help If the Customers form still inherits the menu selections from main.fmb, make sure you modified the CALL_FORM procedure to use the DO_REPLACE option, and that you rebuilt the edited menu module's .mmb file.

4 Click **Exit** to exit the Customer form.

5 Repeat Steps 3 and 4 to confirm that the Items and Inventory forms have the default Forms Runtime menus. Do not exit the application yet.

6 Click the **Exit** iconic button to close the form.

7 Close Project Builder and all other Oracle applications.

SUMMARY

- Project Builder provides a central point from which project team members can view and edit all project files in their Form Builder, Report Builder, or Graphics Builder development environments, and then recompile changed files so that they work together correctly.
- When you create a new project, Project Builder creates a project registry file that contains a series of pointers specifying the locations of individual project applications.
- A project database connection specifies a user name, password, and database connect string that can be assigned to projects, so developers don't have to log on each time they open a different Developer project application.
- In Project Builder, executable files are called targets, and design files are called inputs. Target files depend on their associated input files, and if an input file changes, it must be rebuilt (or recompiled) to create a new target file.
- Project Builder project registry files are automatically saved every time you make a change to the project. When you exit Project Builder, your projects will automatically appear in the Project Navigator window the next time you start it, as long as the project registry files have not been deleted or moved.
- When you rebuild input files in Project Builder, you can choose Build Selection, which only builds a selected target file; Build Incremental, which compiles all target files whose associated input files have been changed but not yet rebuilt; or Build All, which builds all of the project's input files.
- To replace the Forms Runtime default menu with a custom menu, you must create a new menu module. Menu modules are saved in the file system as uncompiled (design) files with an .mmb extension and as executable files with an .mmx extension. The executable (.mmx) menu file can be attached to any form module.
- Main menu items with underlying selections are called parent menu items. Menu choices without submenu choices under them are called child menu items.
- Form Builder automatically places Window as the last pull-down menu selection. In a menu module, the Window menu selection is added automatically at runtime, so you don't need to create it.

- An action trigger is the procedure that is executed when a user selects a pull-down menu item. Every child menu item must have an associated action trigger, which is executed when the user selects the item.
- The default access key for menu selections is the first letter of the selection label. To override the default, type an ampersand (&) before the desired new access key.
- Every time you modify your menu module, you must recompile the .mmx file.
- To display a different menu module in a called form, specify the new menu module filename in the form's Menu Module property, and then call the form using the DO_REPLACE option in the CALL_FORM command.

REVIEW QUESTIONS

1. True or false: Project Builder creates the application form that provides the switchboard interface to the database application.
2. The project registry file is saved in:
   a. the database
   b. the ORAWIN\BIN or ORANT\BIN folder
   c. the default project folder
   d. the Global Registry
3. What is the difference between the project registry and the user registry?
4. What is the purpose of creating a database connection and applying it to a project in Project Builder?
5. ______________ files are dependent on ______________ files.
   a. .fmx, .fmb
   b. .ogr, .ogd
   c. .mmx, .mmb
   d. all of the above
6. Describe the difference between an input and a target in Project Builder.
7. Describe the circumstances in which you would use the Build All, Build Incremental, and Build Selection options.
8. When do you use the Launcher toolbar in Project Builder?
9. What is the difference between an .mmb and an .mmx file?
10. True or false: to attach a custom menu module to a form, you set the form's Menu Module property equal to the full path to the menu module's .mmb file.
11. What is a parent menu item? What is a child menu item?
12. What is the default access key for pull-down menu selections? How do you override the default access key?

13. Why must you recompile your menu module every time you modify it?
14. When do you use the DO_REPLACE parameter in the CALL_FORM command?

PROBLEM-SOLVING CASES

1. Make the following modifications to the Clearwater Traders Order Processing System so that it works correctly. Save your modified files in the \Chapter9\cwproject subfolder.
   a. Update the LOV command button trigger in the itemchart.fmb file so that it loads the inventory.ogd graphics display from the \Chapter9\cwproject subfolder using the global path variable.
   b. Create a new menu module named EXIT_MENU.mmb that only displays Exit on the menu bar, and contains the trigger code EXIT_FORM. Then, modify the sales.fmb, receiving.fmb and intemchart.fmb project forms so that the forms display the EXIT_MENU.mmx menu module rather than the CW_MENU.mmx menu module. (Hint: You will need to modify the cw_menu.mmb menu module so that the New Orders, Receiving, and Check Stock forms are called using the DO_REPLACE option in the CALL_FORM command. Then, you will need to modify the Menu Module property in these forms so that it displays the EXIT_MENU.mmx menu module.)
   c. Rebuild all needed project files and test your application to make sure the new menu module is displayed.
2. Create a new menu module named EXIT_MENU.mmb that only displays Exit on the menu bar, and contains the trigger code EXIT_FORM. (You might have already created this menu module in Exercise 1.) Then, create a new form to serve as the About window for the Clearwater Traders Order Processing System. Look at other Windows-based applications to help design the form. Modify the CW_MENU module so the new form is called from the pull-down menu, and display the EXIT_MENU.mmx menu module in the form. Add the form to your clearwater.upd project in Project Builder, and save the form as about.fmb in the \Chapter9\cwproject subfolder.
3. Create a new project named Northwoods University Database System, based on the project application you created in Chapter 9, Lesson A, Exercise 1. Name the project registry file northwoods1.upd, and specify \Chapter9\nwproject1 as the project's default directory. Specify yourself as the project author, and add a comment stating the date the project was created.
   a. If necessary, copy the ch9aex1.fmb form file that you created in Chapter 9, Lesson A, Exercise 1, to the \Chapter9\nwproject1 subfolder, and add it to the project. Also, add the faculty.fmb, student.fmb, and course_sec.fmb form files (located in the \Chapter9\nwproject1 subfolder) to the project.
   b. Add the location.rdf and classlist.rdf report files (located in the \Chapter9\nwproject1 subfolder) to the project.
   c. Create a database connection named "Northwoods Connection," and assign it to the project.

d. Create a menu module named Ch9bEx3main.mmb with the menu choices shown in Figure 9-30. Add action triggers so that the menu selections call the indicated form or report, and create "Not implemented yet" stubs for the About menu selection. Use the DO_REPLACE option in the CALL_FORM commands. Add the menu module to the project.

Action	**Reports**	**Help**	**Window**
Add New Faculty (faculty.fmb)	Locations (location.rdf)	Contents	
Update Students (student.fmb)	Class Lists (classlist.rdf)	-----------	
Add/Edit Course Sections (course_sec.fmb)		About	
Exit			

Figure 9-30

e. Modify the Ch9aEx1.fmb form file so it displays the Ch9bEx3main.mmx menu module.

f. Create a new menu module named EXIT_MENU.mmb that only displays Exit on the menu bar, and contains the trigger code EXIT_FORM. (You might have already created this menu module in Exercise 1.) Copy the EXIT_MENU.mmb file to the \Chapter9\nwproject1 subfolder, then add it to the project. Modify the faculty.fmb, student.fmb, and course_sec.fmb form files so that they display the EXIT_MENU.mmb menu module.

4. Create a new project named Northwoods University Student Registration System, based on the project application you created in Chapter 9, Lesson A, Exercise 2. Name the project registry file northwoods2.upd, and specify \Chapter9\nwproject2 as the project's default directory. List yourself as the project author, and add a comment stating the date the project was created.

a. Add the following forms to the project that you created in Chapter 9, Lesson A, Exercise 2: Ch9aEx2.fmb, student.fmb, grades.fmb, and register.fmb form files.

b. Add the transcript.rdf report file (located in the \Chapter9\nwproject2 subfolder on your Student Disk) to the project.

c. Create a menu module named Ch9bEx4main.mmb with the menu choices shown in Figure 9-31. Add action triggers so that the menu selections call the indicated form or report, and create a "Not implemented yet" stub for the About menu selection. Add the menu module to the project.

Action	**Run Transcript**	**Help**	**Window**
View Personal Information (student.fmb)	(transcript.rdf)	Contents	
View Course Grades (grades.fmb)		-----------	
View Current Schedule (register.fmb)		About	
Exit			

Figure 9-31

d. Call the forms in the project so that they display the default Forms Runtime menu and toolbar.
e. Modify the command that calls the transcript.rdf report file so that the report is displayed only for the student who logs on to the application. (*Hint*: You will need to pass the value of the student's SID as a parameter to the report.)
f. Modify the transcript.rdf report so that it accepts the parameter created in the previous step.
g. Modify the Ch9aEx2.fmb form file so it displays the Ch9bEx4main.mmx menu module.

CHAPTER 10

Creating Web Applications Using the Oracle Application Server

Introduction The **World Wide Web (WWW)** has become synonymous with the Internet. Commercial Web sites are riding high on Wall Street, and Web-generated sales could reach hundreds of billions of dollars in the next few years. Selling products from Web sites requires more than advertising, however. Customers need to be able to submit order information and receive responses to inquiries. The Oracle Web Application Server, also called the Oracle Application Server, provides a variety of methods for creating Web pages that interact with an Oracle database. This chapter shows how to use PL/SQL and Oracle WRB (Web Request Broker) commands to generate dynamic Web pages. We also introduce techniques for creating LiveHTML, JavaScript applications.

LESSON A

objectives

- Learn about Web addressing
- Understand the differences between static and dynamic Web pages
- Learn how dynamic Web pages are created
- Turn static Web pages into dynamic Web pages with LiveHTML and PL/SQL

Web Basics

A **Web page** is a file with an .htm or .html extension that contains Hypertext Markup Language (HTML) tags and text. **HTML** is a document-layout language with hypertext-specification capabilities. HTML is not a programming language, although it can contain embedded programming commands. HTML's primary task is to define the structure and appearance of Web pages and to enable Web pages to have embedded hypertext links to other documents. HTML consists of tags embedded in text. A **tag** is a marker that indicates how a particular line or section of a Web page is displayed. HTML tags are normally used in pairs, enclosed in angle brackets (< >), and usually take the form of `<tag name>text formatted by tag</tag name>`. For example, the title of a Web page named "Mike's Web Page" uses the `<title>` tag and would be formatted as:

```
<title>Mike's Web Page</title>
```

> Note: In this chapter, text that is inserted in commands by the user will be placed in square brackets ([ ]) rather than in angle brackets (< >) to avoid confusion with HTML tag formats.

Users access and display Web pages using special applications called **Web browsers** that display HTML files as Web pages. Two popular browsers are Netscape Navigator and Microsoft Internet Explorer. Tags are not displayed in a Web page—they just define how the text will appear when displayed in a Web browser. Here is the skeleton of a basic HTML document:

```
<html>
<head>
<title>[Web page title text]</title>
</head>
<body>
<h2>[Web page header text]</h2>
[Rest of Web page document text]
</body>
</html>
```

Notice the beginning `<html>` and ending `</html>` paired tags. These tags tell a Web browser that the enclosed text is to be treated as an HTML document. The `<head>...</head>` tags enclose the document's header section. The `<title>...</title>` tags delimit text that will be displayed in the title bar of the browser window. The `<body>...</body>` tags enclose the text (and other tags) that compose the actual Web page. Web pages can also have embedded **hyperlinks,** which are references to other Web pages. When you click a hyperlink, the Web page (and corresponding .htm or .html file) specified by the hyperlink tag is displayed.

Web Communications and Protocols

A **Web server** is a computer that is connected to the Internet and runs a special software process called a **listener** that "listens" for requests from Web browsers. When a browser sends a request for a Web page to a Web server, the listener forwards the request to the Web server, which reads it and then sends the requested HTML file across the network back to the browser. This file contains a formatted HTML document that is displayed by the user's browser as a Web page.

Communication protocols are agreements between a sender and a receiver regarding how data will be sent, and how data are interpreted by the receiver. The Internet is built upon two network protocols: the **Transmission Control Protocol (TCP)** and the **Internet Protocol (IP).** Both protocols are required for any Internet activity, so they are usually referred to collectively, as **TCP/IP**. Any computer connected to a network using the TCP/IP protocols loads software into its memory for processing TCP/IP network traffic. The computer might be used to send e-mail, browse the Web, or act as a Web server. In all situations, TCP/IP processing software will be loaded into the computer's main memory when the computer is booted.

Internet messages, such as Web pages and e-mail messages, are broken into **packets**, or small chunks of data that can be routed independently of one another through the Internet. TCP is responsible for reassembling network packets into complete messages. IP specifies how messages are addressed.

IP addresses are generally expressed as four numbers, (each ranging in value from 0 to 255) and separated by periods (or decimal points). An example of an IP address is 137.28.224.5. Numbers of this type are difficult to remember, so an IP address can also be represented by a **domain name,** which is a name that has meaning and therefore is easier to remember. Examples of domain names are www.oracle.com and www.software-expert.com. **Domain name servers** are computers that maintain tables with domain names matched to their corresponding numerical IP addresses.

Information on the World Wide Web is usually transferred using a communications protocol called the **Hypertext Transfer Protocol** (**HTTP**). Older Internet protocols such as the File Transfer Protocol (FTP) also are supported by most Web browsers. FTP is still the primary protocol used for transferring files.

Web Addresses

A **Web address,** also called a **Uniform Resource Locator (URL)**, is a string of characters, numbers, and symbols that specifies the communications protocol (such as HTTP or FTP), the domain name or IP address of a Web server, and, optionally, the folder path (such as /www/pub), and the name of an HTML Web page file. If the protocol is not specified, Web browsers by default assume an HTTP protocol. The standard notation assumes that the starting folder is the Web server's root document folder. For example, if c:\orant\ows\new is the server's root document folder, an HTML document named example1.htm located in the c:\orant\ows\new\examples\ folder on a Web server with the IP address 137.28.224.5 is addressed as http://137.28.224.5/examples/example1.htm. If you do not specify an HTML filename in a URL, (such as example1.htm), then your Web browser displays the Web server's default home page. Although a Web administrator can select any file name as a default home page, default Web home page files are usually named index.htm, default.htm, or home.htm.

A URL can also specify a **file URL**, which is an HTML file that is accessible by the Web browser, by omitting the protocol and server address and just specifying a system drive letter (such as C: or D:) and a folder path and filename. A file URL is only used when you are developing new Web pages and want to see how they look when they are displayed in a browser. The file folder path and filename must be on the local workstation. For example, if you want to display a file named index.html that is stored on the C drive in the \Webdocs folder of your computer, the file URL is `file///c:\Webdocs\index.html`.

Any time you specify a domain name (such as www.uwec.edu) as part of a Web address, your Web browser sends a message to a domain name server requesting the IP numeric address corresponding to the domain name. After receiving the IP address from the domain name server, your browser will try to contact the server listening on that address. If you know the IP address of a Web page, you can enter the IP address rather than the domain name. Using a domain name slows response time by the amount of time it takes your browser to learn the IP address from the domain name server. Despite the potential for slower response times, domain names generally should be used if they exist, because an administrator might decide to change a server's IP address while retaining the same domain name.

Running Multiple Web Server Listeners on the Same Computer

Most Web servers run an FTP server process as well as a Web server listener process, to allow Web developers to send new HTML files to the Web server when old page files become outdated. Some Web servers run a second Web server listener to respond to and process administrator requests, which might be used to perform tasks such starting or shutting down the Web server, adding new administrators, and installing new Web server programs. This is managed through the use of **ports.**

In general, a port is a place where information goes into and/or comes out of a computer. For example, a modem is usually connected to the serial port on a personal computer. When you are using the TCP/IP protocol on a network server, a port is a number that specifies how incoming network messages are handled. When the TCP/IP software is loaded onto a server, a table is loaded that lists port numbers and associated server programs. Incoming messages addressed to the server specify a server port number. This number is associated with the server program that will be used to process the incoming message. By default, the Web server process is assigned to port 80, so incoming requests to a Web server are usually addressed to port 80. However, an administrator can change the Web server port to any other port. If he does, incoming messages will have to include the new port number in the URL by appending it to the IP address, using a colon and the port number. For example, the address of a Web server with an IP address of 137.28.224.5 whose Web server process is listening on port 81 would be: `http://137.28.224.5:81/`.

Some port numbers are reserved for other common TCP/IP server processes. An FTP server usually listens on port 21. An Internet e-mail server usually runs two listening processes—one on port 25 for receiving and sending messages and one on port 110 that enables users to read their mail. The default port for the Oracle Application Server administration listener is port 8888. Assuming this port assignment isn't changed, an Oracle Application Server administrator can access a site remotely and modify the Web server's configuration, as well as perform a limited number of administrative tasks for Oracle databases, by accessing a URL such as `http://[Web server's address]:8888/`.

Static and Dynamic Web Pages

In a **static Web page**, tags and text are fixed at the time the page is created. Each time a static page is accessed, it will display the same information. Static Web pages are useful for displaying information that doesn't change very often. In a **dynamic Web page**, also called an **interactive Web page**, the page content varies according to user inputs and/or database queries. User inputs to dynamic pages might be "I want to purchase 100 shares of Technology Inc. at $99 per share" or "What is my current checking account balance?" The ability to respond to these requests has created a new system of commerce. So, how can a Web server interactively read and respond to user requests?

Currently, the technologies supporting interactive Web pages can be grouped into four basic categories. The first uses HTML **forms**, which are enhanced HTML documents designed to collect user inputs (like paper or Oracle forms) and send them to a Web server. HTML forms allow users to input data using text boxes, option buttons, and lists. When the form is submitted to the Web server, a program running on the Web server processes the form inputs and dynamically composes a Web page reply. This program, called the servicing program, uses the **Common Gateway Interface (CGI)** protocol. The problem with using CGI-based servicing programs is that each form submitted to a Web server starts its own copy of the servicing program. A busy Web server is likely to run out of memory when it services many forms simultaneously. As interactive Web sites have gained popularity, Web server vendors have developed proprietary technologies to process form inputs without starting a new copy of the servicing program for every form. Examples of these new technologies include Oracle's Web Request Broker (WRB), Netscape's Netscape Service Application Programming Interface (NSAPI), Microsoft's Internet Server Application Programming Interface (ISAPI), and Allaire's Cold Fusion Server.

The second approach uses a technology called **Server Side Includes (SSIs)**. An SSI is a command that is sent to a Web server from within an otherwise static Web page. An SSI can display a variety of outputs, ranging from simple Web page hit counters and "current time" displays, to page contents tailored to the type of browser the user has, to database and file information. Examples of SSIs will be provided later in the lesson. The Web server must be signaled as part of the page's URL that the Web page includes SSIs, and the Web server must be configured to handle the SSIs correctly. The output from the SSI is then inserted into the static Web page content. The Oracle Application Server implementation of SSIs is called LiveHTML.

The third approach downloads compiled executable programs stored on a Web server to a local Web browser and runs them on the local computer. The user's browser and operating system must have the ability to run the executable program file. This program interacts with the user and sends and retrieves data from other servers as needed. Examples of technologies that use this approach are Java and ActiveX. The advantages of this approach are that more complex user interfaces can be created and user inputs are validated and responded to from the local Java or ActiveX program, rather than from a remote Web server.

The final approach allows uncompiled code in languages such as JavaScript or VBScript to be typed into the HTML document along with the static HTML text. Special tags indicate to the user's browser that this text is code, and if the user's browser has the capability of recognizing and interpreting the code, the code is run by the browser. As with Java or ActiveX, more complex user interfaces are possible with this approach.

It is not clear if any one of these approaches will dominate the others. HTML forms are the most widely used, but their network traffic and server processing requirements are demanding and have Web developers actively investigating other approaches. Many older Web browsers don't support the third and fourth approaches, so for now, HTML forms are the most popular choice for supporting the widest number of users.

The Oracle Application Server

The Oracle Application Server maintains a listener process on a Web server that listens for user requests for Web pages. Requests that reference PL/SQL stored program units are forwarded to the **Web Request Broker** (**WRB**). (Recall from Chapter 4 that a stored program unit is a PL/SQL program unit that is stored in the database, and runs on the database server.) The WRB retrieves the stored program unit from an Oracle database. The database can be on the same server as the Oracle Application Server application, or on a database server located somewhere else on the network. The WRB runs the stored procedure and sends the resulting HTML-encoded Web page response back to the user's browser.

For example, suppose a customer enters his customer ID and payment information onto a form and submits the form to the following Oracle Web server URL: `http://137.28.224.5/owa_default_service/owa/startorder`. The Web listener receives this information and determines, by reading and interpreting its address, that it should be forwarded to the WRB. (The `/owa_default_service/owa/` part of the address is what tells the Web listener that the message is intended for the WRB.) A stored PL/SQL program unit named `startorder` is then called by the WRB to process the information and prepare a response.

Oracle's WRB enables one copy of a Web program in main memory to service many simultaneous requests. The WRB uses **cartridges**, which are programs that run on the server and interpret HTML form inputs from Web browsers, then generate dynamic Web page responses. Some examples of Oracle cartridges are:

- **LiveHTML** Allows the output of any executable program supported by the Web server's operating system to be included in Web pages
- **PL/SQL Cartridge** Processes PL/SQL stored procedures. The PL/SQL Cartridge handles database access better than Oracle's Java Cartridge, but lacks some of Java's functionality.
- **Java Cartridge** Allows Java execution on the server and enables Java server applications to use PL/SQL
- **Perl Cartridge** Runs Perl scripts. Perl is a command language commonly used in UNIX environments.
- **ODBC Cartridge** Enables you to connect to relational databases that support the ODBC (Open Database Connectivity) interface

Creating Dynamic Web Pages

Note: To perform the LiveHTML, PL/SQL, and Developer exercises in this chapter, you will need an account on an Oracle database that your Web server is configured to access. In addition, when doing the LiveHTML and Developer exercises, you must be able to copy Web page files to a folder that is accessible to your Web server (not your local client workstation). Finally, your Web administrator must have configured your Web server to recognize LiveHTML, PL/SQL, and Developer pages. Instructions for creating these configurations are provided in the Installation Instructions in the Instructor's Manual.

Note: The exercises in this lesson require Microsoft Internet Explorer (Version 3.01 or higher) or Netscape Navigator (Version 3 or higher). In this chapter, Web browser commands and illustrations are provided using Windows Internet Explorer. If you are using Netscape Navigator or another browser, your commands might be slightly different.

Creating a LiveHTML Web Page

A Web server must **parse**, or translate and interpret, each command on a Web page that uses LiveHTML. Since parsing can significantly slow down the Web server and sometimes incorrectly translate and interpret non-HTML documents, methods have been devised to signal to a Web server when a URL refers to an HTML document containing SSIs. For example, the following HTML tag is placed in a static Web page to create a link to a second Web page that might be parsed for LiveHTML (SSI) commands:

```
<a href="http://137.28.224.5/livehtml/hello.html">Hello World</a>
```

The text "Hello World" is the highlighted hyperlink text that the user clicks to move to the referenced Web page. The Oracle Application Server knows to process this page for SSIs because `/livehtml/` is embedded in the URL as part of the path to the hello.html document file. An **actual path** is the drive letter and folders that specify the location of a file on a server or client workstation. A **virtual path** is a name that is assigned to the value of an actual path. Along with a file location, a virtual path can also specify how a file or program referenced by the virtual path is to be processed. If a Web administrator has specified "livehtml" as a virtual path, this can signal the Oracle Application Server to parse SSIs in the hello.html document and tell the Oracle Application Server to substitute an actual path (probably not `/livehtml/`) for the `/livehtml/` virtual path. The actual path where the LiveHTML Web page is located can be on any drive or directory accessible to the Web server. Oracle Application Server allows a Web administrator to create as many virtual paths as needed. If a virtual path indicates how a file or program is to be processed, it will be associated with a specific Oracle Web cartridge, such as the PL/SQL or LiveHTML Cartridge.

Note: To do the following exercise, a virtual Web server path named ClassSSI (which is case sensitive) must have been created and associated with the LiveHTML Cartridge and with a folder on your Web server. If a virtual path with a different name has been created for your LiveHTML exercise, your instructor must tell you its name. In addition, you must be able to send or copy an HTML file to your personal subfolder with the path referenced by ClassSSI.

First you will create a static Web page file named `call_ssi.html` that has a hyperlink reference to a dynamic Web page file named hello.html. (You will create hello.html later.) To create a reference from one Web page to another, you use the following general tag: `<a href="http://[IP address or domain name of your Web server>]/[<path and file name (on the Web server) of html file that you wish to link to]"></a>`. In this exercise, your tag will look like:

```
<a href="http://[IP address or domain name of your Web server]/
ClassSSI/[your Oracle username]/hello.html></a>
```

ClassSSI is a virtual path that tells the Web server to process the hello.html Web page for SSIs, and it specifies the root of an actual path on the Web server where hello.html will be stored. You will store the call_ssi.html file (which calls hello.html) on your Student Disk.

To create the call_ssi.html Web page file:

1 Start Notepad (or any other text or HTML editor), and type the commands shown in Figure 10-1. Save the file as **call_ssi.html** in the Chapter10 folder on your Student Disk.

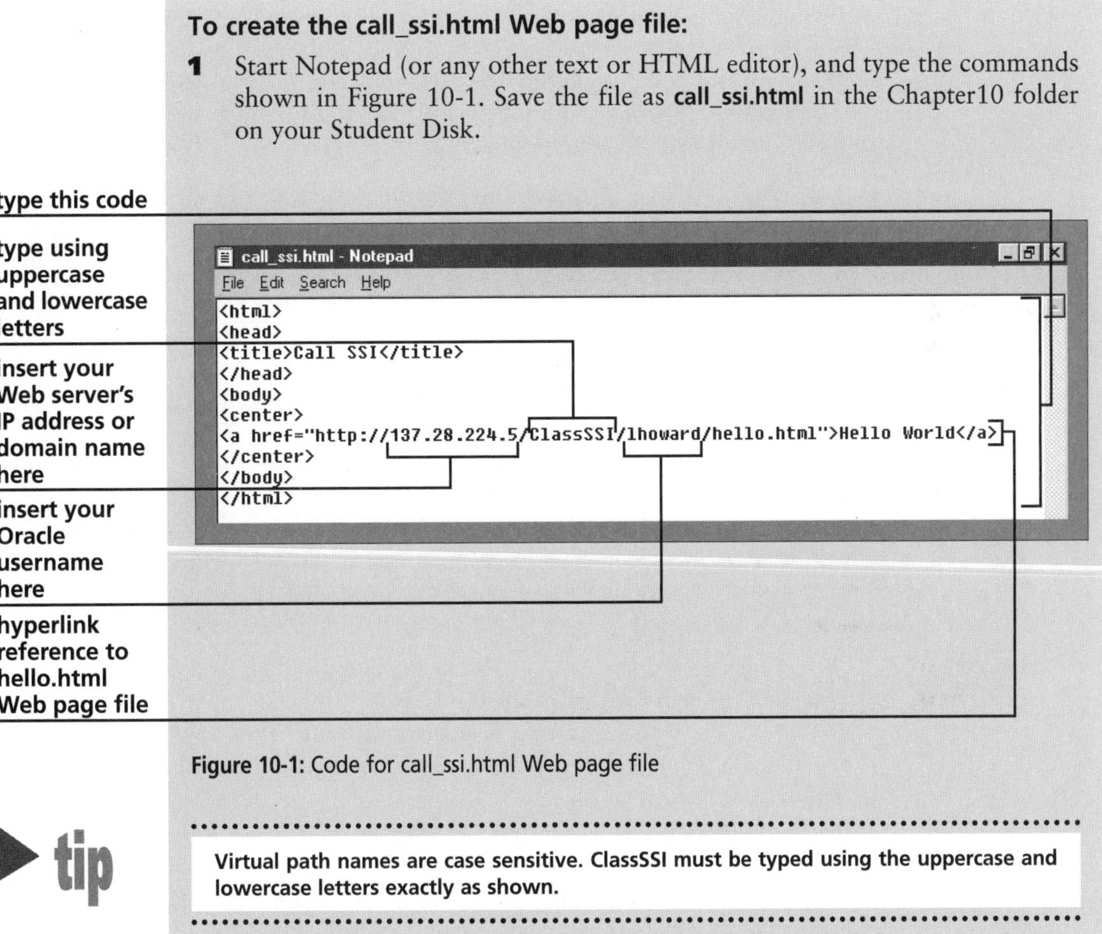
type this code
type using uppercase and lowercase letters
insert your Web server's IP address or domain name here
insert your Oracle username here
hyperlink reference to hello.html Web page file
call_ssi.html - Notepad
File Edit Search Help
<html>
<head>
<title>Call SSI</title>
</head>
<body>
<center>
<a href="http://137.28.224.5/ClassSSI/lhoward/hello.html">Hello World</a>
</center>
</body>
</html>

Figure 10-1: Code for call_ssi.html Web page file

tip

Virtual path names are case sensitive. ClassSSI must be typed using the uppercase and lowercase letters exactly as shown.

There are two ways to view a Web page file. To view a Web page file that is saved on a Web server, you type the Web server URL, and the path to the file, which might include a virtual path, in the Address box in the Web browser. To view a Web page file that is saved on your local workstation, you load the file in the Web browser by clicking File on the Web browser menu bar, clicking Open, and then specifying the file's name and location. Now you will view call_ssi.html with your Web browser using a file URL.

To view call_ssi.html in your Web browser:

1 Start your Web browser, click **File** on the menu bar, click **Open,** click **Browse,** navigate to the **call_ssi.html** file in the Chapter10 folder on your Student Disk, click **Open**, and then click **OK**. The file is displayed in your browser as a simple Web page with a single hyperlink to Hello World, as shown in Figure 10-2. Note that the file URL is displayed in the Web browser Address box. *Do not* click the hyperlink, because you have not yet created the hello.html file that it references.

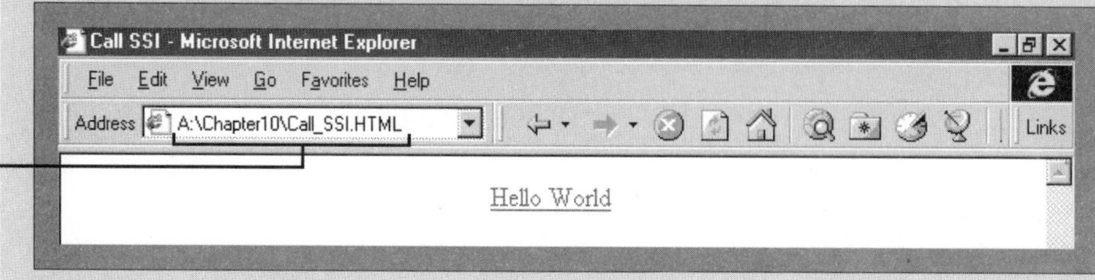

Figure 10-2: call_ssi.html Web page display

Note: This text displays Web pages using Microsoft Internet Explorer Version 4.0. Your browser screens and toolbars might appear different, depending on your browser, browser version, and individual configuration.

help

If your Web page does not look like the one in Figure 10-2, or if you receive an error message, make sure your code exactly matches the code shown in Figure 10-1. When you find the error, correct your Notepad file, save the corrected file, and then refresh the display in your Web browser by clicking View on the menu bar, then clicking Refresh.

Now you will create the hello.html Web page file that is referenced by the Hello World hyperlink. This Web page file contains LiveHTML commands, which are SSIs. LiveHTML commands must be enclosed in HTML comment tags, which are normally used by Web developers to document Web pages. They are usually not processed by the Web browser. This convention avoids errors if a browser reads an SSI file that hasn't been processed by a Web server. Comment tags have the following general format: `<!--[Comment text]-->`.

When a Web page is parsed for LiveHTML commands, the file is searched for HTML comments. Within a comment, the character following the exclamation point and two hyphens `<!--` in the comment tag is checked to see if it is the pound (#) character. If it is, then the word following the pound (#) character is checked to see if it is a valid LiveHTML command. If it is, the command is executed and passed

any parameters that follow the command. Oracle limits LiveHTML commands to one command per line in an HTML file. LiveHTML commands have the following general format:

```
<!--#[command [param1="value1" param2="value2" …]] -->
```

Figure 10-3 describes common LiveHTML (SSI) commands.

Command	Description
Config	Sets parameters for how the included files or scripts are to be parsed. If used, it is usually the first LiveHTML command in a file.
Include	Specifies that a file is to be included in the generated HTML page at this point
Echo	Displays the value of an environmental variable
Fsize	Displays the size of a file
LAST_MODIFIED	Displays the last modification date of the file
Exec	Executes a program or script. This command is sometimes disabled for security reasons.
Request	Oracle-specific command that sends a request to another cartridge using inter-cartridge exchange (ICX)

Figure 10-3: Common LiveHTML (SSI) commands

Each of these commands uses one or more parameters that affect the command's action. You can find more information and examples of these and other SSI commands in books specifically addressing Web programming, and from Web sites using this technology.

Now you will create the hello.html Web page file. It will contain the `:echo` LiveHTML command, and display environmental variables. (An **environmental variable** is a variable that is stored on a workstation and is available to any program running on the workstation.) The LiveHTML commands and corresponding environmental variables that will be displayed on the hello.html Web page file are: DATE_LOCAL (current date and time from the Web server), PATH_TRANSLATED (actual path to the Web page file), and DOCUMENT_URI (virtual path to the current Web page file.) After you create the hello.html file, you will send or copy the hello.html file to the folder on your Web server that has been created for SSI files by your instructor. Then you will view hello.html by clicking the Hello World hyperlink on the call_ssi.html Web page in your Web browser.

To create the hello.html Web page:

1 Switch to your text editor, and create a new file. Type the code shown in Figure 10-4 and save the file as **hello.html** in the Chapter10 folder on your Student Disk. The <p>...</p> tags indicate that the enclosed text is within a single paragraph, and should have blank lines above and below it. The <br> tag indicates that the preceding text should be followed by a line break.

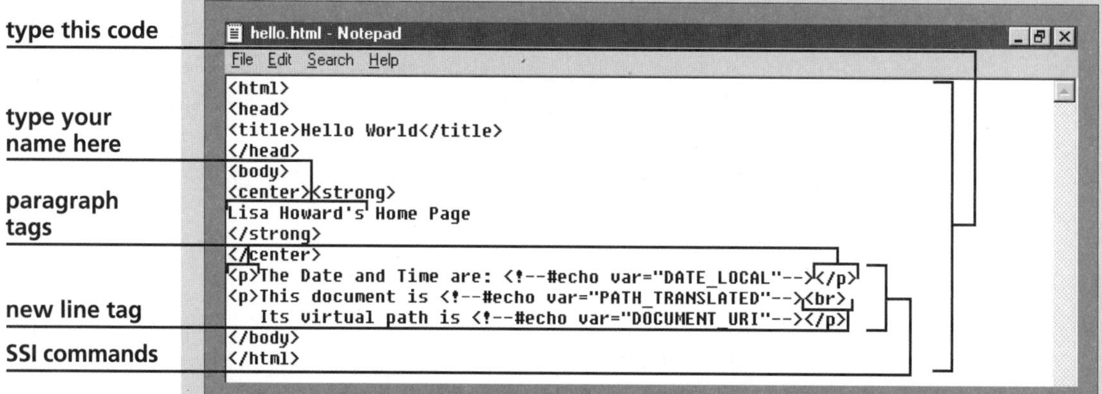
type this code
type your name here
paragraph tags
new line tag
SSI commands
hello.html - Notepad
File Edit Search Help
<html>
<head>
<title>Hello World</title>
</head>
<body>
<center><strong>
Lisa Howard's Home Page
</strong>
</center>
<p>The Date and Time are: <!--#echo var="DATE_LOCAL"--></p>
<p>This document is <!--#echo var="PATH_TRANSLATED"-->
Its virtual path is <!--#echo var="DOCUMENT_URI"--></p>
</body>
</html>

Figure 10-4: Code for hello.html Web page file

2 Using FTP or a mapped network drive letter, transfer or copy hello.html to the Web server folder assigned for your LiveHTML files.

tip

Your instructor will tell you the name of this folder and the method you should use to copy the file to the folder.

3 Switch back to your Web browser, and click the **Hello World** hyperlink on the call_ssi.html Web page. The hello.html Web page is displayed, as shown in Figure 10-5.

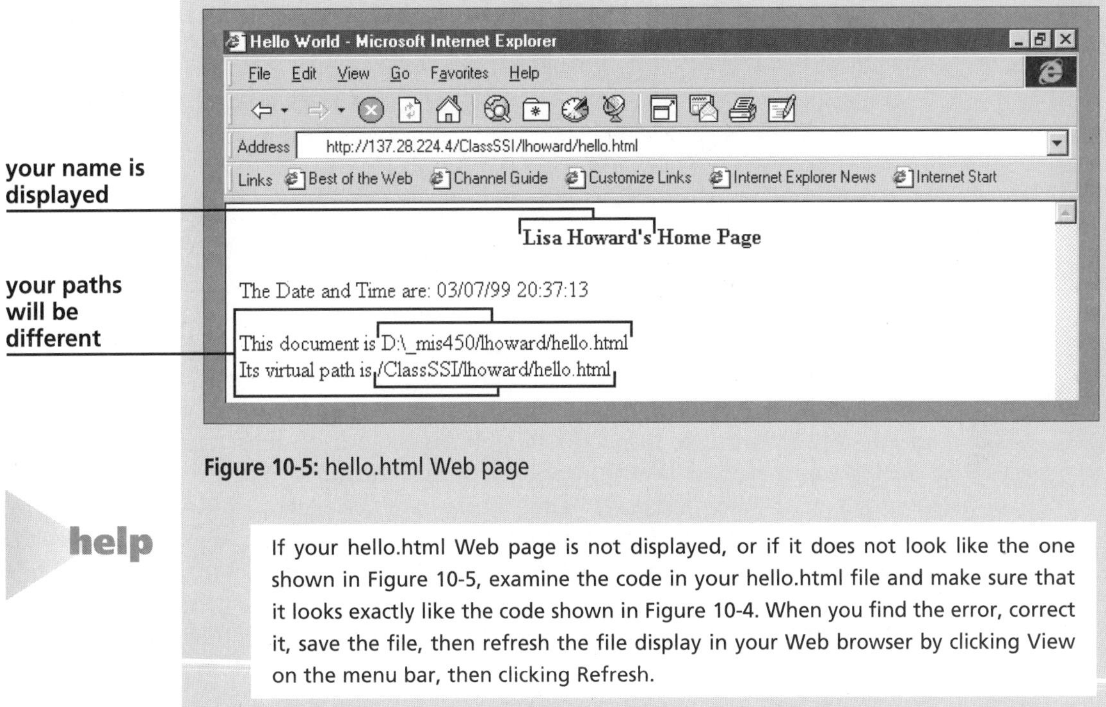

Figure 10-5: hello.html Web page

help

If your hello.html Web page is not displayed, or if it does not look like the one shown in Figure 10-5, examine the code in your hello.html file and make sure that it looks exactly like the code shown in Figure 10-4. When you find the error, correct it, save the file, then refresh the file display in your Web browser by clicking View on the menu bar, then clicking Refresh.

This LiveHTML Web page uses the `echo` command to display environmental variables. A more powerful use of LiveHTML is to display database query information. Now you will create a LiveHTML page that calls a PL/SQL stored program unit and retrieves database information from the Northwoods University database. The PL/SQL program unit will be executed using the Oracle PL/SQL Cartridge called from your LiveHTML page. The PL/SQL program unit will display a listing of information about all Northwoods University database course sections that students can enroll in, which are course sections that are in a term in which the current enrollment status is OPEN, and the course section's current enrollment is less than its maximum allowable enrollment. First, you will refresh your Northwoods University database tables by running the northwoo.sql script in SQL*Plus. Then you will modify the TERM table in the Northwoods University database so the STATUS field for the Spring 2002 term is OPEN rather than CLOSED, to allow your query to display more open courses.

To refresh the Northwoods University database tables and update the STATUS field:

1. Start SQL*Plus, and run the **northwoo.sql** script from the Chapter10 folder on your Student Disk.

2 Type the following commands to change the status of the Spring 2002 term to OPEN:

```
UPDATE term
SET status = 'OPEN'
WHERE termid = 5;
COMMIT;
```

3 Exit SQL*Plus.

Recall that you use Oracle Procedure Builder to create PL/SQL stored program units. Next you will start Procedure Builder and confirm that it is configured to develop PL/SQL Web applications. Whenever you create PL/SQL stored program units that will be displayed as Web pages, your database must have a special user account named WWW_USER that the Oracle Application Server uses to interact with the database. You must have privileges to execute some of WWW_USER's stored program units, and you must give the WWW_USER the necessary privileges to run your stored program units. In addition, WWW_USER must have a stored program unit named HTP. This is a package specification containing an important procedure used to send HTML tags and text back to the user's Web browser.

To verify that your database is configured for developing PL/SQL Web applications:

1 Start Procedure Builder, and maximize the Object Navigator window.

2 Click **File** on the menu bar, and then click **Connect** and connect to the database in the usual way.

3 Click [+] beside **Database Objects**, scroll down if necessary and click [+] beside **WWW_USER**, and then click [+] beside **Stored Program Units.** The HTP package specification should be displayed.

help If the WWW_USER or HTP package specifications are not displayed, ask your instructor or technical support person for help. The Oracle Application Server files that are used in this chapter might not be installed on your server.

Next you will create a Web page file named open_courses1.html that displays outputs from a PL/SQL stored program unit named OPEN_COURSES1. This stored program unit displays the results of a query showing all courses in the Northwoods University database in which the ENROLLMENT field for the course's term ID has the value OPEN, and the course's current enrollment is less than its maximum allowable enrollment. To run the OPEN_COURSES1 stored program unit, you will use the `request` SSI command. To notify the Web server that OPEN_COURSES1 is a PL/SQL program, you will use **class_plsql** as the virtual path to the stored program unit. This tells the Web server to process the program using the Oracle PL/SQL Cartridge. Now you will create the open_courses1.html Web page file, store it on your Student Disk, and then transfer it to the folder on your Web server that your instructor has told you to use for LiveHTML Web page files.

To create the open_courses1.html Web page file:

1 Switch to Notepad, create a new file, type the commands shown in Figure 10-6, and save the file as **open_courses1.html** in the Chapter10 folder on your Student Disk. Note that the filename ends with the number one (1), not the lower case letter L (l).

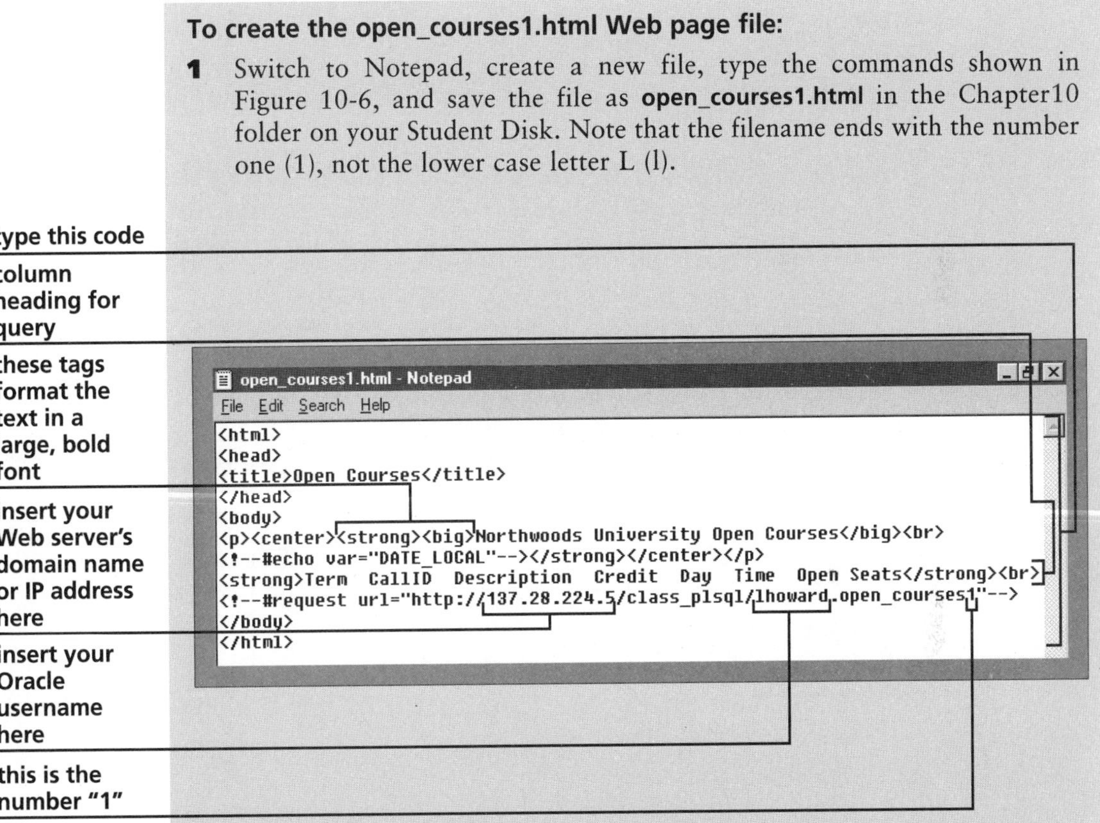
type this code
column heading for query
these tags format the text in a large, bold font
insert your Web server's domain name or IP address here
insert your Oracle username here
this is the number "1"
open_courses1.html - Notepad
File Edit Search Help
<html>
<head>
<title>Open Courses</title>
</head>
<body>
<p><center><strong><big>Northwoods University Open Courses</big>
<!--#echo var="DATE_LOCAL"--></strong></center></p>
<strong>Term CallID Description Credit Day Time Open Seats</strong>
<!--#request url="http://137.28.224.5/class_plsql/lhoward.open_courses1"-->
</body>
</html>

Figure 10-6: Code for open_courses1.html Web page file

2 Transfer or copy open_courses1.html to the folder on your Web server where you store your LiveHTML Web page files.

Next you will use Procedure Builder to create the OPEN_COURSES1 stored program unit. This stored program unit declares a cursor that returns the term description, course call ID, course name, course credits, day, time, and number of open seats for all courses for which the term enrollment status is OPEN. You will process this cursor using a cursor FOR loop, and use a procedure named **p** that is in the HTP package specification within the WWW_USER database account. This procedure prints a character string, and is used to create outputs that contain HTML tags within PL/SQL programs. To call the procedure, you must preface the procedure name (p) with its user name (WWW_USER), a period, and the name of its package specification (HTP) plus a period. The program unit creates an HTML file whose body contains a text string for each row returned by the cursor. Each row consists of the cursor output fields concatenated together onto a single line, with two blank spaces between each field.

To create the open_courses1 stored program unit:

1 Switch to Procedure Builder, click **Stored Program Units** under your username in the Object Navigator window, click the **Create** button, type **open_courses1** in the Name text box, be sure the **Procedure** option button is selected, and then click **OK**.

2 Type the code shown in Figure 10-7 to create the DECLARE section of the program unit. Do not compile the program unit yet.

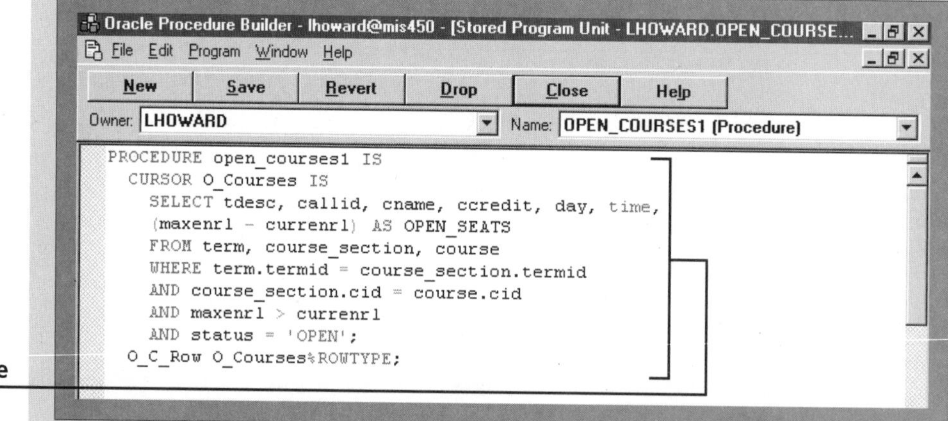

Figure 10-7: DECLARE section of OPEN_COURSES1 program unit

3 Type the code shown in Figure 10-8 directly under the DECLARE section you just typed to create the body of the program unit. Click **Save** to compile the code, correct any syntax errors, and then close the PL/SQL Editor.

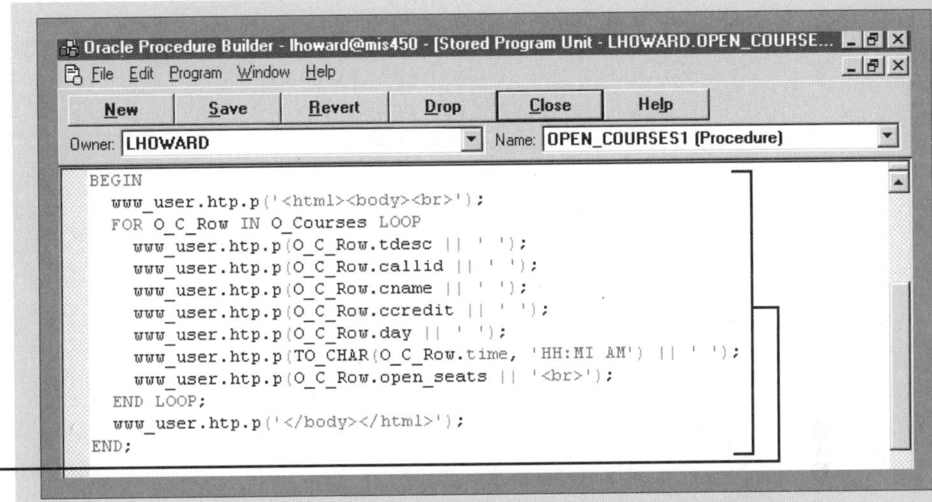

type this code under DECLARE section

Figure 10-8: Body of OPEN_COURSES1 program unit

Whenever you use a looping operation in a PL/SQL program (or any other program that is called within a dynamic Web page), it is very important to test the program to make sure that the loop is terminating properly *before* you attempt to run the program on the Web server. If you create a program with an infinite loop, it will consume most of the processing capabilities of the Web server's central processing unit, and will cause the Web server to be very slow in servicing requests from other users. It might even consume all of the Web server's virtual memory, and cause the Web server to crash. Now you will test your stored program unit to verify that the program is terminating properly. You will run the OPEN_COURSES1 stored program unit from the PL/SQL command prompt, and confirm that it is terminating properly.

To test the OPEN_COURSES1 stored program unit:

1. Open the PL/SQL Interpreter window.
2. Type **open_courses1;** at the PL/SQL> prompt, then press the **Enter** key. The program unit should terminate successfully and the insertion point should appear beside a new PL/SQL> prompt, as shown in Figure 10-9. This indicates that the program unit terminated properly.

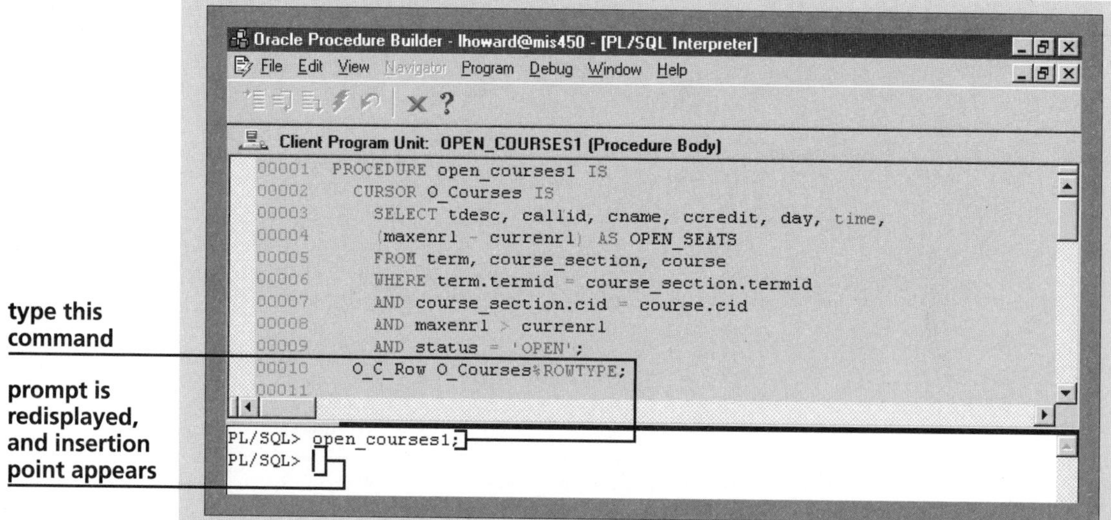

Figure 10-9: Testing the OPEN_COURSES1 program unit

help

Figure 10-10 shows what happens when a loop in a program unit does not terminate, but loops indefinitely. The program code has been modified so that the cursor is processed using a LOOP ... EXIT WHEN loop, which requires an explicit EXIT WHEN command that signals to exit the loop when the last cursor row is processed. In Figure 10-10, this command has been commented out. The program unit compiles successfully, but when it runs, the program loops indefinitely, and the insertion point "hangs" under the PL/SQL> prompt. If this happens when you test your program unit, exit the program by clicking on any other system application on the taskbar, then returning to Procedure Builder. You might need to press Ctrl+Alt+Delete on the keyboard, select Procedure Builder on the task list, and then click End Task to exit Procedure Builder. Start Procedure Builder again if necessary, debug the program, and test it again to confirm that the loop terminates properly.

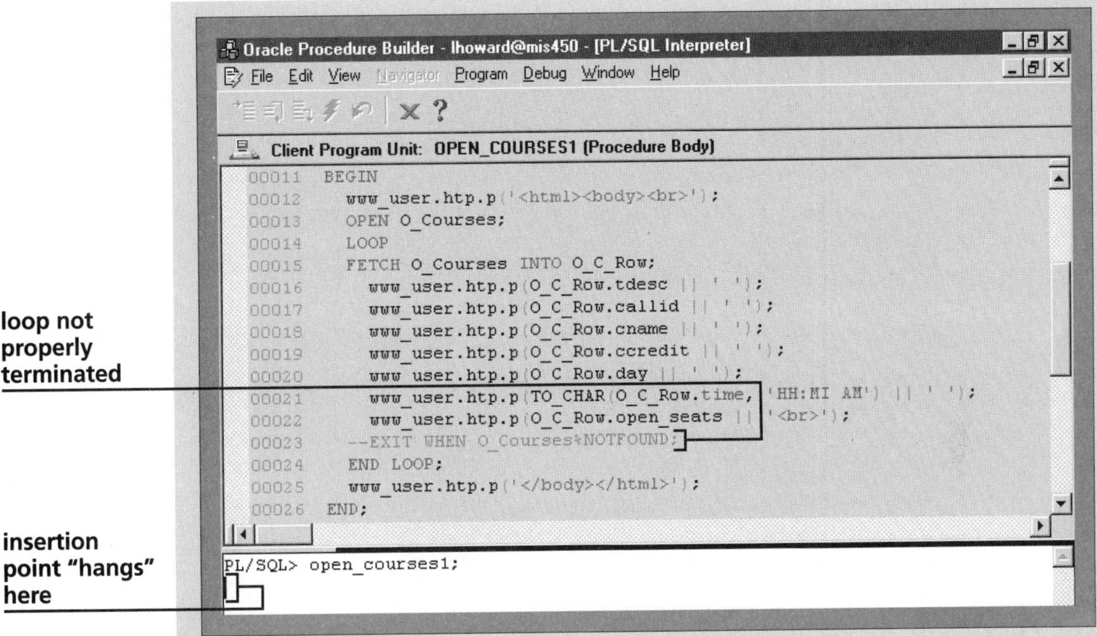

Figure 10-10: Example of program unit with infinite loop

Recall that you had to grant specific privileges to allow other users to insert, update, delete, or select records in your database tables. Like database tables, stored program units are database objects that belong to a specific user. In order to allow the Oracle Application Server to execute your open_courses1 program unit, you must grant the WWW_USER account this privilege. The general format for granting the EXECUTE privilege to another user on a stored program unit is: `GRANT execute ON [stored program unit name] TO [user name];`. Now you will grant the EXECUTE privilege to the WWW_USER account at the PL/SQL> prompt in Procedure Builder.

To grant the EXECUTE privilege to WWW_USER on the OPEN_COURSES1 program unit:

1. If necessary, open the Interpreter PL/SQL> prompt pane.
2. Type the following command at the PL/SQL> prompt:

```
GRANT execute
ON open_courses1
TO www_user;
```

Now you are ready to view the open_courses1.html Web page file using your Web browser. You will switch to your Web browser, and type the URL for the open_courses1.html file by specifying the Web server domain name or IP address, and the virtual path to the file.

To view the open_courses1.html Web page file:

1 Switch to your Web browser, and type the following URL in the Address text box:

```
http://[Your Web Server Address]/ClassSSI/[Your Oracle
user name]/open_courses1.html
```

Note that you must substitute your Web server's domain name or IP address and your user name to specify your subfolder on the Web server. Don't forget that the ClassSSI part of the above URL is case sensitive. (The rest of the URL is not case sensitive.) Your Web page should appear as shown in Figure 10-11.

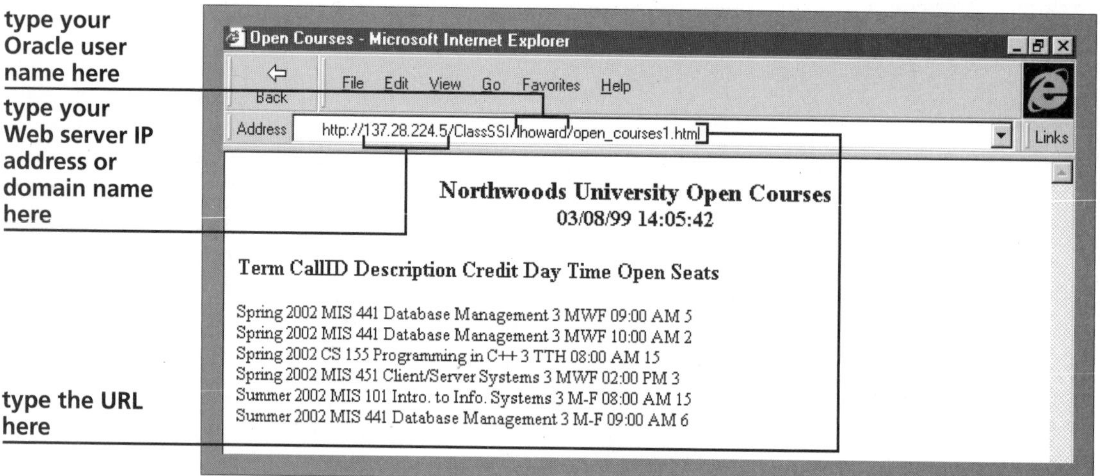

Figure 10-11: Web page that returns database information

Aligning Text Using HTML Tables

To improve the appearance of data on a Web page, you can use an **HTML table** to align text and form inputs on the Web page. An HTML table is enclosed between `<table>...</table>` tags, and individual table rows are indicated with `<tr>...</tr>` tags. You can add as many rows as you want to a table. Within a single table row, you use the `<td>...</td>` tags to define table columns and delimit the contents of individual table cells.

You can modify these table tags to specify the table's appearance. For example, modifying a table tag to `<table width="95%">...</table>` tells the Web browser to size the table to span 95 percent of the available screen width. Modifying the first row of a table's column/cell tag to `<td width="20%">...</td>` instructs the browser to size the column to span 20 percent of the total table width. The sum of all of the column widths should equal 100 percent, even when the `<table>` tag width is less than 100 percent of the screen width. Since the column widths are specified in relation to the table, you include the width attribute in the `<td>...</td>` tags only in the first table row, and this controls column widths for the entire table. Similarly, the first row determines the number of columns in a table, so you should include the same number of `<td>...</td>` tags in every row of the table.

By default, the text in each table field is left-justified. To center the text in a field, you would modify the field tag as follows: `<td align="center">...</td>`. To make the text in a field right-justified, you would modify the field tag as follows: `<td align="right">...</td>`. To modify the alignment of a column, you must modify the alignment for each field in the column, not just the column heading.

Figure 10-12 shows the HTML code for creating a table that displays the first four rows in the LOCATION table in the Northwoods University database, and Figure 10-13 shows the table generated by this code displayed in the Internet Explorer Web browser.

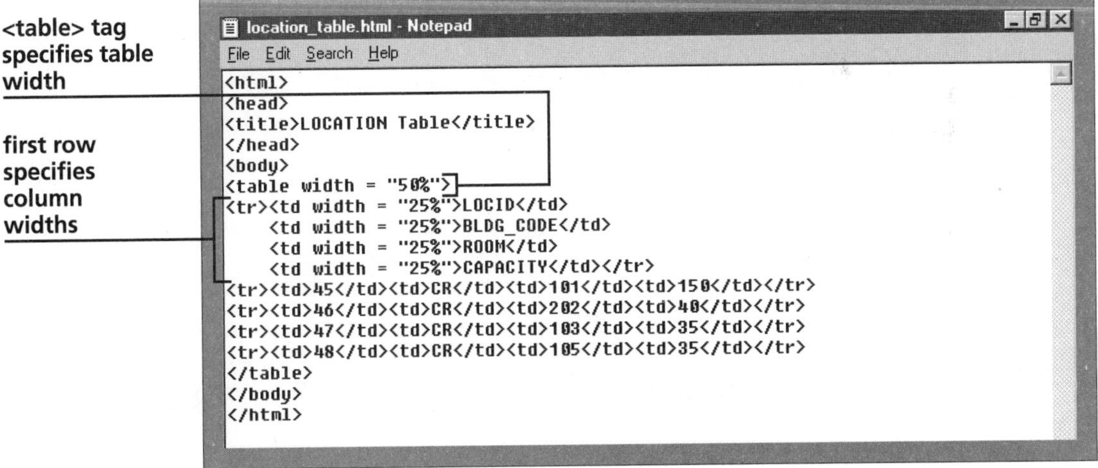


```
<html>
<head>
<title>LOCATION Table</title>
</head>
<body>
<table width = "50%">
<tr><td width = "25%">LOCID</td>
    <td width = "25%">BLDG_CODE</td>
    <td width = "25%">ROOM</td>
    <td width = "25%">CAPACITY</td></tr>
<tr><td>45</td><td>CR</td><td>101</td><td>150</td></tr>
<tr><td>46</td><td>CR</td><td>202</td><td>40</td></tr>
<tr><td>47</td><td>CR</td><td>103</td><td>35</td></tr>
<tr><td>48</td><td>CR</td><td>105</td><td>35</td></tr>
</table>
</body>
</html>
```

Figure 10-12: Code for HTML table

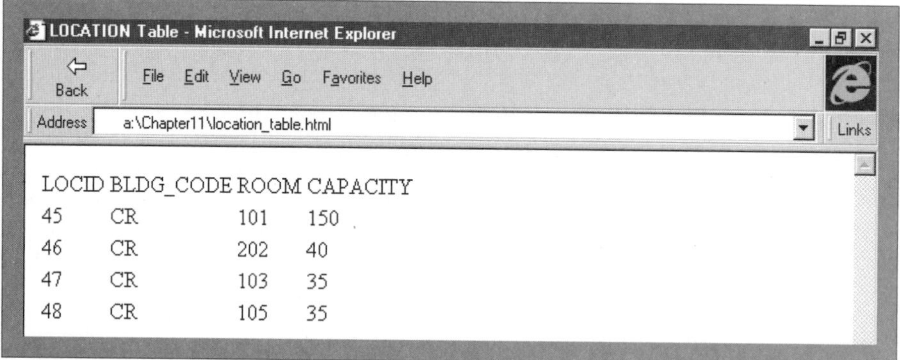

Figure 10-13: HTML table display

The `<table width = "50%">` tag specifies that the table will be displayed using 50 percent of the screen width. The first table row specifies that each column heading will take 25 percent of the table width. Note that the column width specifications are only included for the heading row, and are omitted for the data rows. All columns accept the default left justification.

When you create an HTML table, it is very important to make sure that every row starts with a `<tr>` tag and ends with a `</tr>` tag, and that every field starts with a `<td>` tag and ends with a `</td>` tag. Otherwise, the data in the browser will not be displayed correctly.

To improve the appearance of the data retrieved in the Web page in Figure 10-11, you will reformat the output so the data are displayed in an HTML table. First you will modify the open_courses1.html file so it displays the column headings using a table.

To format the open_courses1 Web page file output in an HTML table:

1. Switch to Notepad, and if necessary, open the open_courses1.html file from the Chapter10 folder on your Student Disk. Save the file as **open_courses2.html**.
2. Modify the file as shown in Figure 10-14.

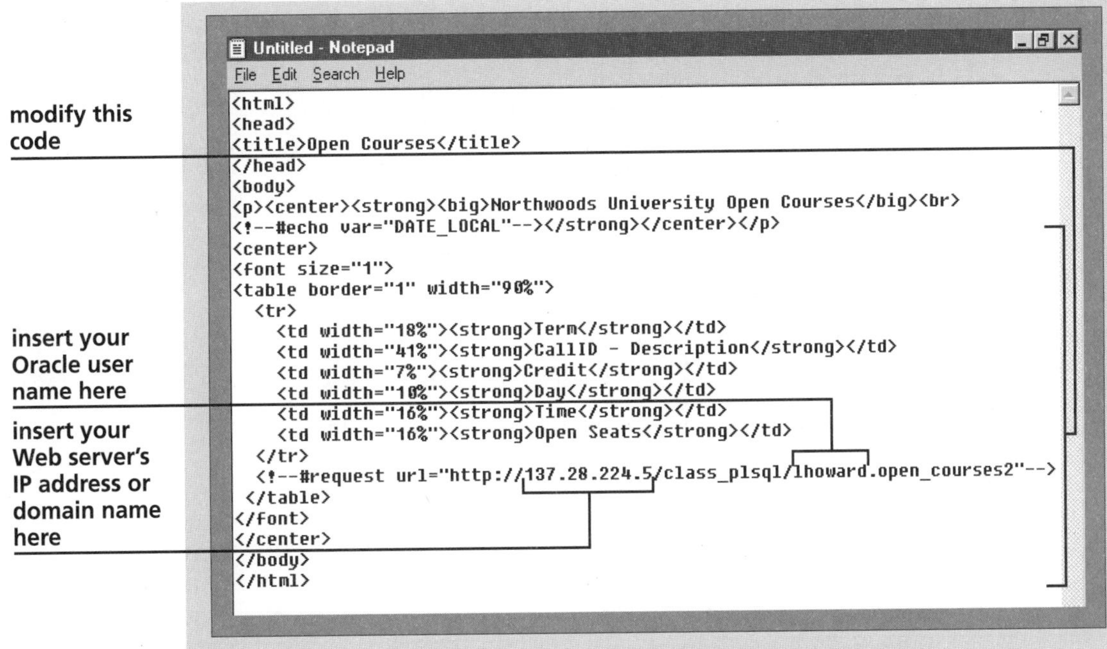

Figure 10-14: Code to format open_courses2.html Web page file as a table

3 Transfer the open_courses2.html file to the Web server folder where you keep your LiveHTML files.

Next, you will create a new stored program unit named OPEN_COURSES2. You will copy the code from the OPEN_COURSES1 stored program unit, and paste the copied code into the new OPEN_COURSES2 program unit. Then, you will modify the code so the retrieved database fields are formatted using an HTML table.

To create the OPEN_COURSES2 stored program unit:

1 Switch to Procedure Builder, open the Object Navigator if necessary, and double-click the icon beside the **OPEN_COURSES1** stored program unit under your user name to make the program unit available for editing in the PL/SQL Editor.

2 Copy the text of the program unit between the BEGIN and END commands.

3 Click the **New** button on the PL/SQL Editor button bar. Type **OPEN_COURSES2** for the new program unit name, make sure the Procedure option button is selected, then click **OK**.

4 Click between the BEGIN and END commands in the new program unit, and paste the text you copied from the OPEN_COURSES1 program unit.

5 Modify the code in the body of the program unit as shown in Figure 10-15.

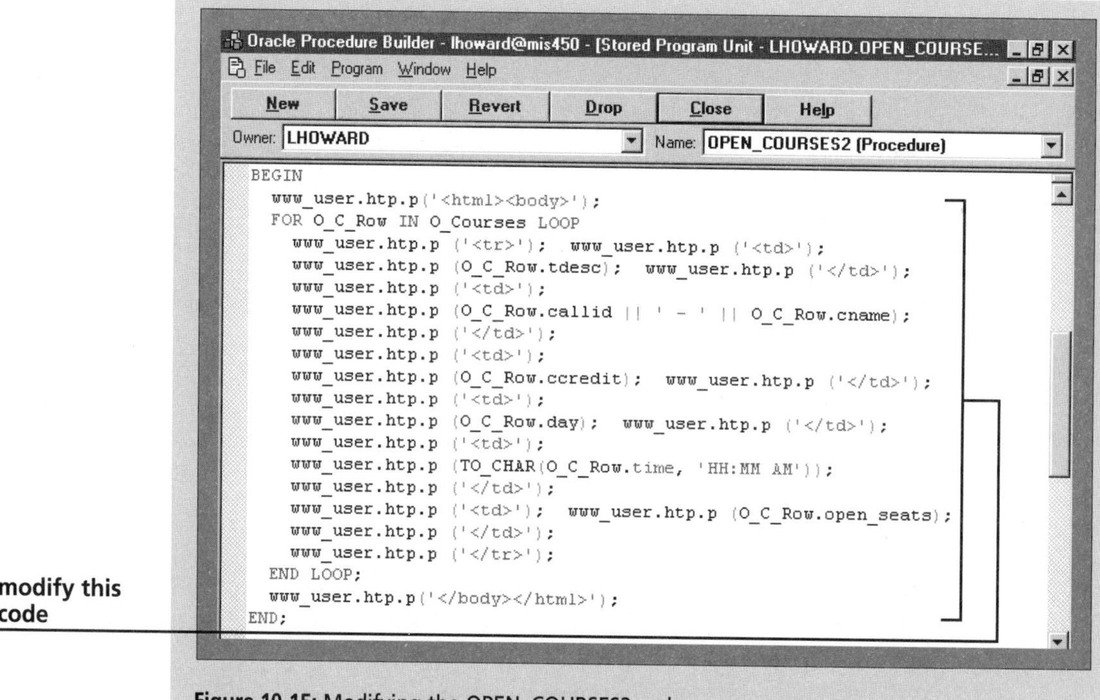

Figure 10-15: Modifying the OPEN_COURSES2 code

6 Compile the code and correct any syntax errors, then close the PL/SQL Editor.

Next, you will test the new program unit in the Procedure Builder to make sure that the loop terminates properly. Then you will store the program unit as a stored program unit in the database.

To test and store the new program unit:

1 Type **open_courses2;** at the PL/SQL> prompt and confirm that the program unit terminates properly, and that the PL/SQL> prompt is redisplayed. If the program unit does not terminate properly, debug and retest it in the Procedure Builder until it terminates properly.

The final step is to grant execute privileges to the WWW_USER account for the OPEN_COURSES2 stored program unit. You will do this next.

To grant privileges for the new stored program unit:

1. Click **Program** on the menu bar, then click **PL/SQL Interpreter** to open the PL/SQL Interpreter window and PL/SQL> prompt pane.
2. Type the following command at the PL/SQL> prompt, then press the **Enter** key to execute the command:

```
GRANT EXECUTE
ON open_courses2
TO www_user;
```

Now you will switch to your Web browser, open the open_courses2.html Web page file, and test to make sure the data retrieved by the stored program unit are formatted correctly.

To test the new program unit formatting:

1. Switch to your Web browser, and type the following URL in the Address text box:

```
http://[your Web server address]/ClassSSI/[your Oracle
user name]/open_courses2.html
```

2. Press the **Enter** key to display the Web page. Your Web page should be displayed with the data formatted in a table, as shown in Figure 10-16.

Open Courses - Microsoft Internet Explorer

Address http://137.28.224.5/ClassSSI/lhoward/open_courses2.html

Northwoods University Open Courses
03/08/99 20:07:32

Term	CallID - Description	Credit	Day	Time	Open Seats
Spring 2002	MIS 441 - Database Management	3	MWF	09:07 AM	5
Spring 2002	MIS 441 - Database Management	3	MWF	10:07 AM	2
Spring 2002	CS 155 - Programming in C++	3	TTH	08:07 AM	15
Spring 2002	MIS 451 - Client/Server Systems	3	MWF	02:07 PM	3
Summer 2002	MIS 101 - Intro. to Info. Systems	3	M-F	08:07 AM	15
Summer 2002	MIS 441 - Database Management	3	M-F	09:07 AM	6

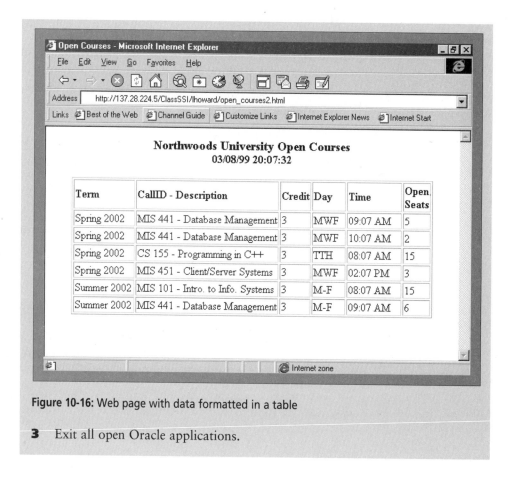

Figure 10-16: Web page with data formatted in a table

3 Exit all open Oracle applications.

SUMMARY

- A Web page is a file with an .htm or .html extension that contains Hypertext Markup Language (HTML) tags and text.
- HTML is a document-layout language with additional hypertext-specification capabilities.
- A tag is a marker that indicates how a particular line or section of a Web page is displayed. HTML tags are used in pairs, enclosed in angle brackets (< >), and usually take the form of <tag name>text formatted by tag</tag name>.
- Web pages can have hyperlinks, which are references to other Web pages. When you click a hyperlink, the Web page (and corresponding .htm or .html file) specified by the hyperlink tag is displayed.
- A Web server is a computer that is connected to the Internet and runs special software called a listener that listens for requests from users who wish to connect to it. When a user connects to a Web server, the server sends an HTML file across the network to the user.

- Communication protocols are "agreements" between a sender and a receiver regarding how data will be sent, and how data are interpreted by the receiver. The Internet is built on two network protocols: the Transmission Control Protocol and the Internet Protocol (TCP/IP).
- IP addresses are generally expressed as four numbers (ranging in value from 0 to 255) separated by periods (or decimal points). Since numbers of this type are difficult to remember, IP addresses can also be represented by domain names, which are names that have meaning.
- Information on the World Wide Web is most commonly transferred using a communications protocol named the Hypertext Transfer Protocol (HTTP). Older Internet protocols such as the File Transfer Protocol (FTP) also are supported by most Web browsers.
- The location of a Web page on a Web server is called a Uniform Resource Locator (URL), and it specifies the IP address or domain name of the Web server, as well as the path specification to the HTML file.
- In static Web pages, tags and text are fixed at the time the page is created and do not change over time. In dynamic Web pages, the HTML file is dynamically created in response to user inputs.
- Current technologies used to create dynamic Web pages include the Common Gateway Interface (CGI), Server Side Includes (SSIs), sending an executable program to the user when a Web page is requested, and writing uncompiled code directly into an HTML document and sending it to the user's computer to execute.
- An SSI is a command that is sent to a Web server from within a static Web page. The Web server must be configured to correctly process the SSI. The output from the SSI is inserted onto the Web page that is displayed to the user.
- The Oracle Application Server maintains a listener process on a Web server that listens for user requests for Web pages. Requests that reference PL/SQL stored program units are forwarded to the Web Request Broker (WRB).
- The WRB uses cartridges, which are programs that run on the server and interpret HTML form inputs from Web browsers and generate dynamic Web page responses.
- LiveHTML is Oracle's SSI implementation. Live HTML commands are always enclosed in HTML comment tags.
- To create PL/SQL programs that create output for Web pages, your database must have a special user account named WWW_USER that the Oracle Application Server uses to interact with the database. The WWW_USER account must have a stored program unit named HTP, which is a package specification that contains procedures that are used to write PL/SQL programs that can interact with the Oracle Application server and be displayed as Web pages.
- To allow the Oracle Application Server to execute a stored program unit, you must grant the WWW_USER account this privilege for the stored program unit using the GRANT command.
- To improve the appearance of data on a Web page, you can use an HTML table to align text and form inputs.

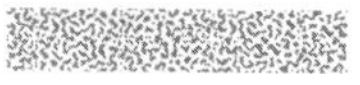

REVIEW QUESTIONS

1. How is an IP address determined from a domain name?
2. What server process is normally assigned to TCP/IP port 80?
3. How does the Oracle Application Server know that a Web address refers to a LiveHTML Web page?
4. What is an HTML tag?
5. If a protocol isn't specified in a URL, what protocol is automatically assumed and used by a Web browser?
6. What Oracle Web listener is commonly found running at port 8888?
7. Describe four approaches used to add dynamic content to Web pages.
8. What is a cartridge?
9. Explain how an SSI adds dynamic content to an otherwise static Web page.
10. How does the Oracle Application Server know that a Web address refers to a PL/SQL program?
11. What information can be relayed to a Web server by a virtual path?
12. Why should you place LiveHTML (SSI) commands within HTML comments?
13. How does a Web server know that an SSI command is enclosed within an HTML comment?
14. What privileges must you grant to WWW_USER before your stored program units may be accessed from the Web?
15. What privileges must be granted to you from WWW_USER before you can create Web-enabled PL/SQL stored program units?

PROBLEM-SOLVING CASES

1. Modify your LiveHTML hello.html Web page so that it displays the text "Date Last Modified," and shows the date the file was last modified. Save the modified HTML file as Ch10aEx1.html.
2. In this exercise, you will create a static Web page that uses a LiveHTML command to call a PL/SQL stored procedure that uses a SQL query to retrieve faculty member last names, first names, and phone numbers.
   a. Use Procedure Builder to create a stored program unit named FACULTY1. Use a loop to read through the retrieved records and output them to the Web page, using the WWW_USER.HTP.P procedure. Be sure to test your program unit to make sure your loop terminates properly before you run the program unit from a Web page.

b. Create an HTML file named Ch10aEx2.html that uses a LiveHTML command to call the FACULTY1 stored program unit. Provide an informative Web page title and column labels. Do not format the program unit output using an HTML table.

3. Modify the FACULTY1 program unit and Ch10aEx2.html Web page file that you created in Exercise 2 so that the stored program unit and HTML file use HTML table tags to present the output in a table format. Save the revised stored program unit as FACULTY2, and save the revised HTML file as Ch10aEx3.html.

4. In this exercise, you will create a static Web page that uses a LiveHTML command to call a PL/SQL stored procedure that uses a SQL query to retrieve the item description, size, color, current price, and quantity on hand for all items in the Clearwater Traders INVENTORY table that are currently in stock (QOH is not equal to 0).

   a. Use Procedure Builder to create a stored program unit named INSTOCKINVENTORY that retrieves the required records. Use a loop to read through the retrieved records and output them to the Web page, using the WWW_USER.HTP.P procedure. Format the program unit output using an HTML table. Be sure to test your program unit to confirm that your loop terminates properly before you run the program unit from a Web page.

   b. Create an HTML file named Ch10aEx4.html that uses a LiveHTML command to call the INSTOCKINVENTORY stored program unit. Create an informative Web page title and column headings, and display the program unit output using an HTML table.

5. In this exercise, you will create a static Web page that uses a LiveHTML command to call a PL/SQL stored procedure that uses a SQL query to retrieve the item description, size, color, order price, quantity, extended total (quantity times price), and total order amount for all items in order ID 1061 in the Clearwater Traders database.

   a. Use Procedure Builder to create a stored program unit named ORDERDETAIL that retrieves the required records. (You will insert the value 1061 directly into the query as a search condition.) Use a loop to read through the retrieved records and output them to the Web page, using the WWW_USER.HTP.P procedure. The output should display columns for item description, size, color, order price, quantity, and current order total. The current order total column should show a running order total by summing the extended totals, including the current line. Be sure to test your program unit to confirm that your loop terminates properly before you run the program unit from a Web page. Format the output using an HTML table. (*Hint*: You can calculate the extended total and order total in the SQL query and reference the calculated fields using an alias.)

   b. Create an HTML file named Ch10aEx5.html that uses a LiveHTML command to call the ORDERDETAIL stored program unit. Include a descriptive Web page title and column headings, and format the program unit output using an HTML table.

6. In this exercise, you will create a static Web page that uses a LiveHTML command to call a PL/SQL stored procedure that uses a SQL query to retrieve the last name, first name, middle initial, ZIP code, order ID, and total order amount for every order placed using the Web site as its order source. The Web page display will be a form with columns for the customer name, customer ZIP code, order ID, and order amount.
   a. Use Procedure Builder to create a stored program unit named WEBORDERS that retrieves the required records. (You will insert the value "WEBSITE" directly into the query as a search condition.) Use a loop to read through the retrieved records and output them to the Web page, using the WWW_USER.HTP.P procedure. Display the customer names in a single field, formatted as follows: "Edwards, Mitch M." Calculate the total order amount as the sum of the quantity times (*) order price for each order line in the order. Be sure to test your program unit to confirm that your loop terminates properly before you run the program unit from a Web page. (*Hint*: Group the records by order ID and the other CUSTOMER data fields.)
   b. Create an HTML file named Ch10aEx6.html that uses a LiveHTML command to call the WEBORDERS stored program unit. Format the program unit output using an HTML table.

LESSON B

objectives

- Use a PL/SQL program to process user inputs from a Web form
- Learn how to avoid displaying form parameters in a URL
- Create a dynamic Web page using JavaScript

Using PL/SQL to Process Form Inputs

So far, the dynamic Web pages you have created have not accepted user inputs. Now you will create a Web page that uses an HTML form to collect user inputs. These inputs will be processed by a PL/SQL stored program unit that will return specific data values based on the inputs. The Web page will allow a student at Northwoods University to enter her last name and PIN. When she submits these values, a PL/SQL stored procedure will generate the Web page response, which will be a list of all classes she has taken and her grades in these courses.

HTML Form Commands

An HTML form allows users to input data. To create a form within an HTML file, you use the following general code:

```
<form action = "http://[address of Web server]/[virtual path name
specifying the type of cartridge that will process the form inputs]/
[name of the program receiving the form inputs]>
<input type="[type]" name="[variable name] [additional parameters]>
<input type="[type]" name="[variable name] [additional parameters]>
</form>
```

A form can contain standard controls used in graphical user interfaces such as text boxes, command buttons, option buttons, and lists. You can create a text box for collecting input using the following code:

```
<input type="text" name="[parameter variable name that input is
assigned to]" size = "[maximum width of parameter text]">
```

Form inputs are submitted when the user clicks a standard form command button, called a **submit button**. A submit button is created using the following general code: `<input type="submit" value = "[button caption]">`. Another standard button found on many HTML forms is a **reset button**, which clears the text in all of the form text boxes. The standard code for creating a reset button is: <input type="reset" value = "[button caption]">. To place a blank space between two form buttons, you use the HTML command . You can use multiple   commands, one after another, to create a larger blank space.

Now you will create a static Web page that contains a form, and then you will display the Web page in your Web browser. The form will have input boxes for the student name and PIN, and Submit and Reset buttons with a blank space between the buttons. The form controls will be displayed in a table so that they are aligned in columns. You will use class_plsql as the virtual path to tell the Web server to use the PL/SQL Web cartridge to process the form, and you will pass the form inputs to a stored procedure named TRANSCRIPT that you will create later.

To create the form Web page and display it in your Web browser:

1 Open Notepad and type the code shown in Figure 10-17. Save the file as **northwoods.html** in the Chapter10 folder on your Student Disk.

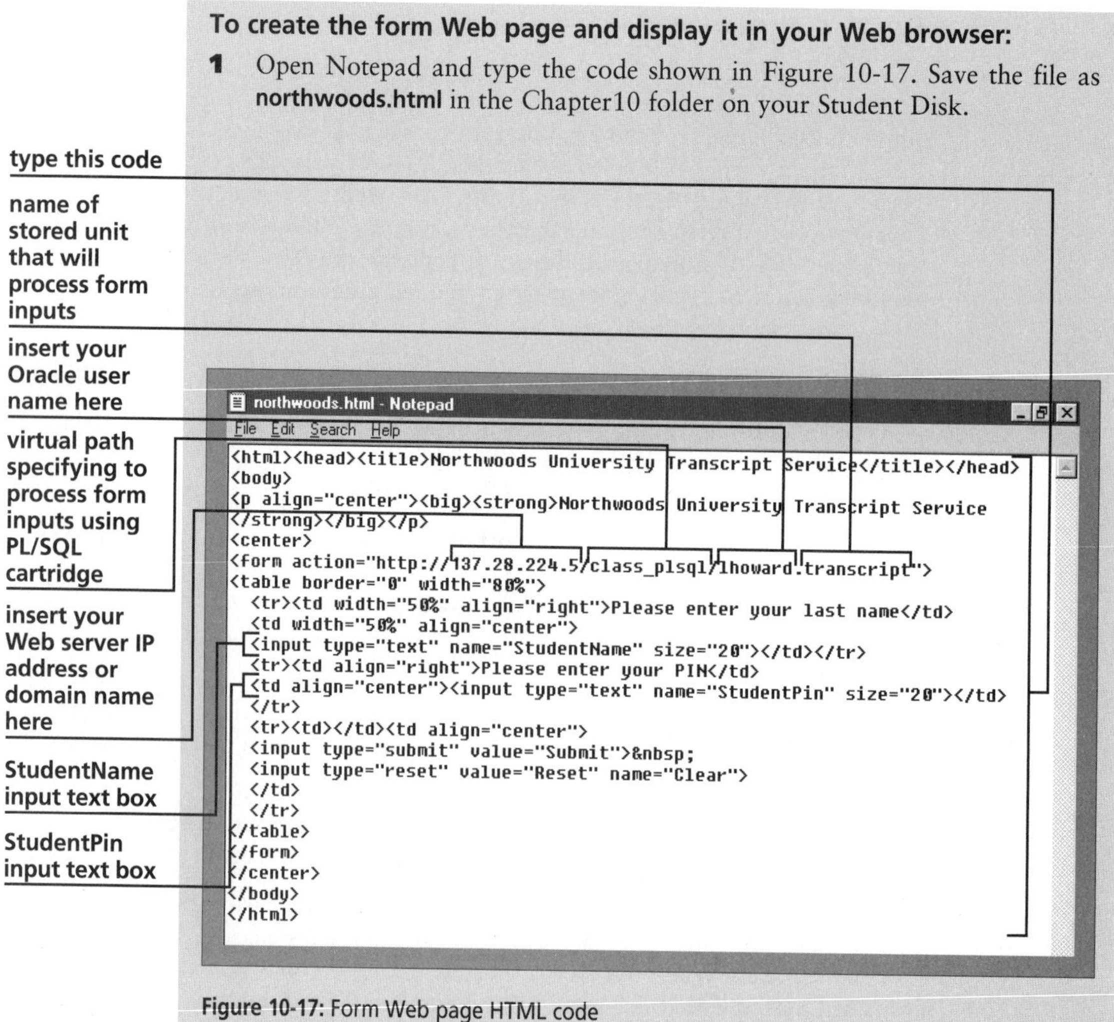


```
<html><head><title>Northwoods University Transcript Service</title></head>
<body>
<p align="center"><big><strong>Northwoods University Transcript Service
</strong></big></p>
<center>
<form action="http://137.28.224.5/class_plsql/lhoward.transcript">
<table border="0" width="80%">
  <tr><td width="50%" align="right">Please enter your last name</td>
  <td width="50%" align="center">
  <input type="text" name="StudentName" size="20"></td></tr>
  <tr><td align="right">Please enter your PIN</td>
  <td align="center"><input type="text" name="StudentPin" size="20"></td>
  </tr>
  <tr><td></td><td align="center">
  <input type="submit" value="Submit"> 
  <input type="reset" value="Reset" name="Clear">
  </td>
  </tr>
</table>
</form>
</center>
</body>
</html>
```

Figure 10-17: Form Web page HTML code

2 Open your Web browser, click **File** on the menu bar, click **Open**, click **Browse**, and then open the **northwoods.html** file from the Chapter10 folder on your Student Disk. Your form should appear as shown in Figure 10-18.

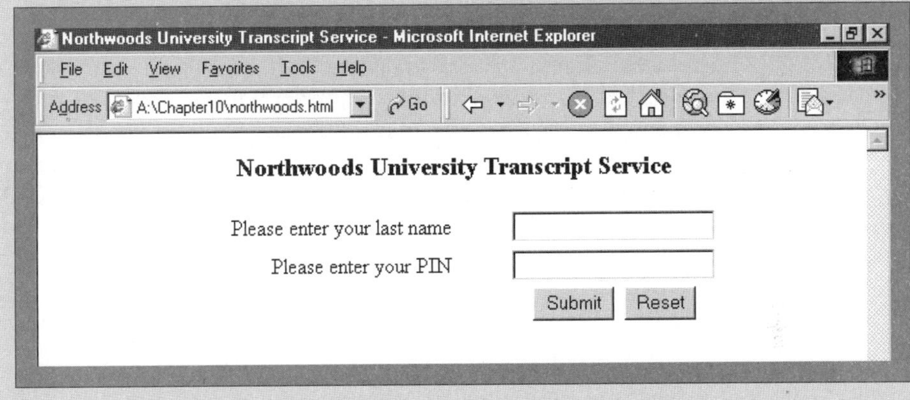

Figure 10-18: Student Transcript input form

Now you will use Procedure Builder to create a stored program unit named TRANSCRIPT. This program unit accepts the student last name and PIN as input parameters. It verifies that the input student last name and PIN match the values for a corresponding record in the STUDENT table, and then it uses an explicit cursor to retrieve the student's term descriptions, course call IDs, course names, credits, and grades. The EXCEPTION section notifies the user if an error was made when entering the last name or PIN. Since the code for this program unit is rather long, you will copy the code from a text file named transcript.txt that is in the Chapter10 folder on your Student Disk, and paste it into the PL/SQL Editor window.

To create the transcript stored program unit:

1 Start Procedure Builder and connect to the database if necessary.

2 In the Object Navigator window, click [+] beside Database Objects, click [+] beside your username, click **Stored Program Units**, click the **Create** button, and create a new program unit named **TRANSCRIPT**.

3 Delete the current program unit header text in the PL/SQL Editor.

4 Switch to Notepad, and open **transcript.txt** from the Chapter10 folder on your Student Disk. Click **Edit** on the menu bar, then click **Select All** to select all of the text. Click **Edit** on the menu bar, and then click **Copy**.

5 Switch back to Procedure Builder, and paste the copied text into the PL/SQL Editor. Save the program unit, then close the PL/SQL Editor.

Next, you will test your new program unit to make sure that it does not have any infinite loops.

To test the program unit and save it as a stored program unit:

1 Open the PL/SQL Interpreter, and type the following at the PL/SQL> prompt:

```
transcript('Miller', '8891');
```

If the program unit successfully terminates, the PL/SQL> prompt should reappear. If the PL/SQL> prompt does not reappear, indicating an infinite loop, close the program unit and debug and test it until it works correctly.

Next you will need to grant execute privileges on the transcript program unit to the WWW_USER database account. You will do this from the PL/SQL> prompt in Procedure Builder.

To grant execute privileges for the transcript program unit:

1 At the PL/SQL> prompt, type the following command:

```
GRANT execute ON transcript TO www_user;
```

Next, you will test the northwoods.html file and TRANSCRIPT stored program unit together. First you will open the northwoods.html file in your Web browser, and enter a student name and PIN.

To test the files:

1 Switch to your Web browser, click **File** on the menu bar, click **Open**, then open the **northwoods.html** file from the Chapter10 folder on your Student Disk.

2 Type **Miller** in the student name input text box and **8891** in the PIN input text box, and then click **Submit**. The Web page shown in Figure 10-19 should be displayed. Note that the form parameter values are displayed in the URL after the Web page address.

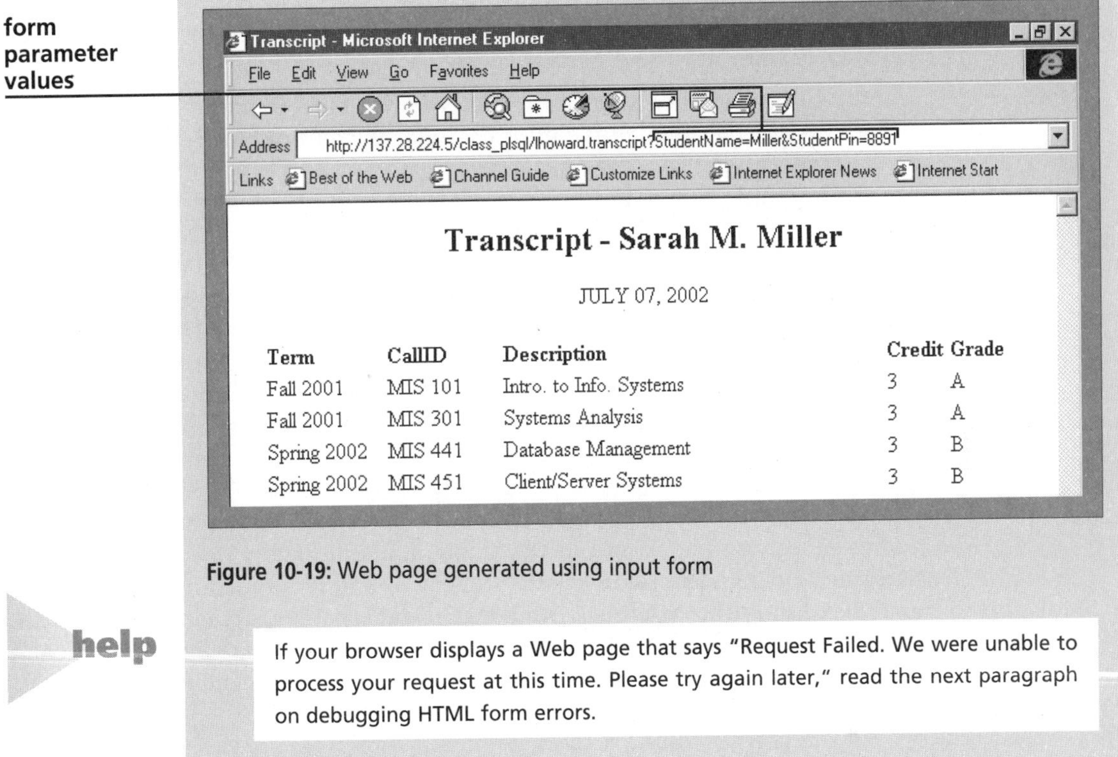

Figure 10-19: Web page generated using input form

help If your browser displays a Web page that says "Request Failed. We were unable to process your request at this time. Please try again later," read the next paragraph on debugging HTML form errors.

Debugging HTML Forms

Recall that when you create an HTML form, you use a command that includes a "form action" attribute. This form action attribute determines the Web program that will process the form's inputs. In the northwoods.html Web page, the form action attribute refers to the TRANSCRIPT stored program unit. When you create an input text box on a form, the values inserted into the input boxes are passed to the program listed in the form's action attribute. The processing program expects to receive these parameters, and if it doesn't receive them, it quits, and the Web server displays the message "Request Failed. We were unable to process your request at this time. Please try again later."

The northwoods.html Web page has the following form input controls:

```
<input type=text name=StudentName size="20">
<input type=text name=StudentPin size="20">
```

Recall that the heading of the transcript program unit is as follows:

```
PROCEDURE transcript (StudentName VARCHAR2, StudentPin VARCHAR2) IS
```

The transcript program unit expects StudentName and StudentPin to be passed to it in this order and with these names. Any of the following errors in the HTML file will cause the "Request Failed" error:

- Misspelling *name* as *nmae*: <input type=text nmae=StudentName size="20">
- Misspelling *StudentName* as *StudntName*: <input type=text name= StudntName size="20">
- Omitting an input box
- Placing the input boxes in the wrong order on the form

Avoiding Displaying Form Parameters in a URL

In Figure 10-19, the student's name and PIN are displayed in the transcript Web page URL. Does this pose a security risk? You might think that because the output is displayed for the student only after she enters these numbers, it shouldn't matter if this information is displayed in the URL. However, most Web browsers maintain a history list of Web sites that have been recently visited that includes addresses like the one shown in the Address line in Figure 10-19. A subsequent user can view a history list, enter the student ID and PIN found in the history list, and then display the supposedly confidential student transcript information of any student who has used the browser recently. Form parameter information could potentially include other sensitive data, such as credit card numbers and Social Security numbers.

This is probably not a serious problem for individuals who access Web sites from their home computers (unless other people in their homes pose a security risk). Business environments pose a higher risk—employee workstations are often in unlocked workspaces, and security is not ensured. In educational computer laboratories, the security risk is very high: anyone might sit down at any time in front of any of the lab workstations and view sensitive information that has been passed on Web forms. The solution is to prevent form inputs from appearing in a URL.

Web forms have two primary methods for sending their inputs to a Web server: the default method, also called the GET method, which was used in the previous form application, and the POST method. With the **GET** method, parameters are passed as command line parameters in the URL to the Web server, and are displayed on the URL address line. After receiving form inputs, a GET request is passed as an environmental variable to the program processing the request. There are two problems with the GET method: first, the parameter values are displayed in the URL address, which poses a security risk. Second, environmental variables typically are limited to 255 characters in length, which puts a severe limitation on the length of form parameters. To avoid these problems, most Web forms use the POST method. After receiving form inputs using the **POST** method, the Web server passes the inputs to the processing program as though they were directly typed into the program from the keyboard. As a result, form inputs are not displayed in the URL, and there are no restrictions on the length of these inputs.

Now you will modify your northwoods.html file so the form inputs are processed using the POST method. To do this, you will insert the following command as the first command in the <form...> tag:

```
method="POST"
```

To change the Web program so that it uses the POST method to process form inputs:

1 Switch to Notepad, modify the <form...> tag to specify the POST method, as shown in Figure 10-20, then save the file.

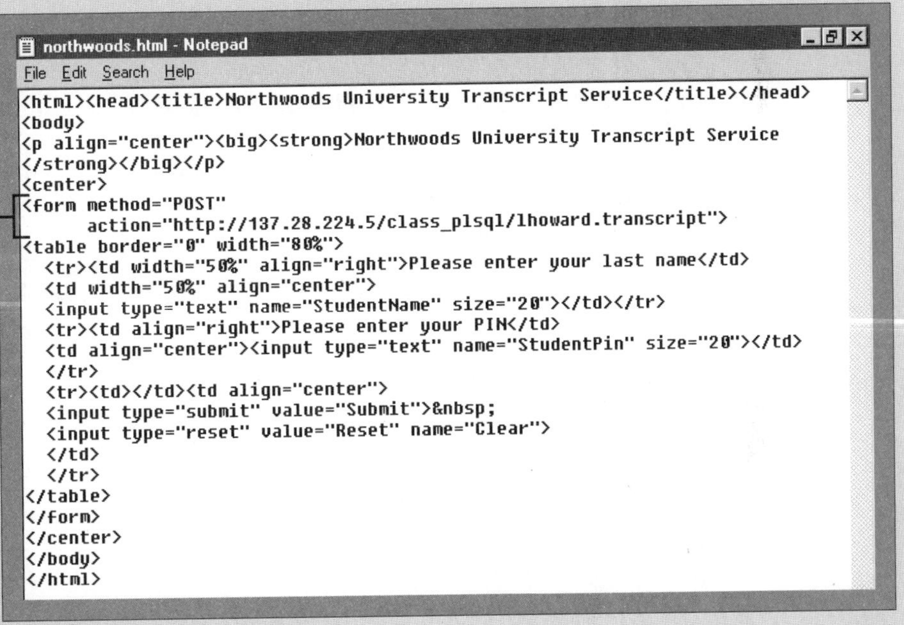

modify these lines

```
northwoods.html - Notepad
File Edit Search Help
<html><head><title>Northwoods University Transcript Service</title></head>
<body>
<p align="center"><big><strong>Northwoods University Transcript Service
</strong></big></p>
<center>
<form method="POST"
      action="http://137.28.224.5/class_plsql/lhoward.transcript">
<table border="0" width="80%">
  <tr><td width="50%" align="right">Please enter your last name</td>
  <td width="50%" align="center">
  <input type="text" name="StudentName" size="20"></td></tr>
  <tr><td align="right">Please enter your PIN</td>
  <td align="center"><input type="text" name="StudentPin" size="20"></td>
  </tr>
  <tr><td></td><td align="center">
  <input type="submit" value="Submit"> 
  <input type="reset" value="Reset" name="Clear">
  </td>
  </tr>
</table>
</form>
</center>
</body>
</html>
```

Figure 10-20: Specifying the POST method in northwoods.html

2 Switch to your Web browser, and open **northwoods.html** from the Chapter10 folder on your Student Disk, so that it is displayed in the browser.

3 When the input form is displayed, type **Miller** for the student name, and **8891** for the student PIN, then click **Submit**. The transcript is displayed, but the student name and PIN no longer appear on the URL.

The Oracle Application Server supports a variety of encryption standards to ensure security. Although data encryption is beyond the scope of this book, you can investigate these capabilities using other references if you need to transmit passwords and other sensitive information across a network.

Using JavaScript for Client-Side Form Processing

Recall that one of the approaches for creating interactive Web pages is writing uncompiled JavaScript or VBScript code directly into an HTML document that is sent across the network to the user's computer. This allows a Web page to perform client-side form processing. In client-side form processing, the Web page responds to user inputs from HTML form elements—such as text input boxes, command buttons, and option buttons—by processing the inputs on the user's workstation, rather than sending inputs back to the Web server for processing. For security reasons, client-side JavaScript and VBScript applications cannot launch applications on a client computer, read or write files on a client computer, or write files on a Web server.

Now you will create a client-side JavaScript program. Figure 10-21 shows an HTML file containing JavaScript code to display the current date and time, the date the document was last modified, and the name of the user's Web browser application. The JavaScript code within the HTML file is **interpreted**, or translated to machine-readable language, and executed one line at a time by the JavaScript interpreter that is built into most newer Web browsers.

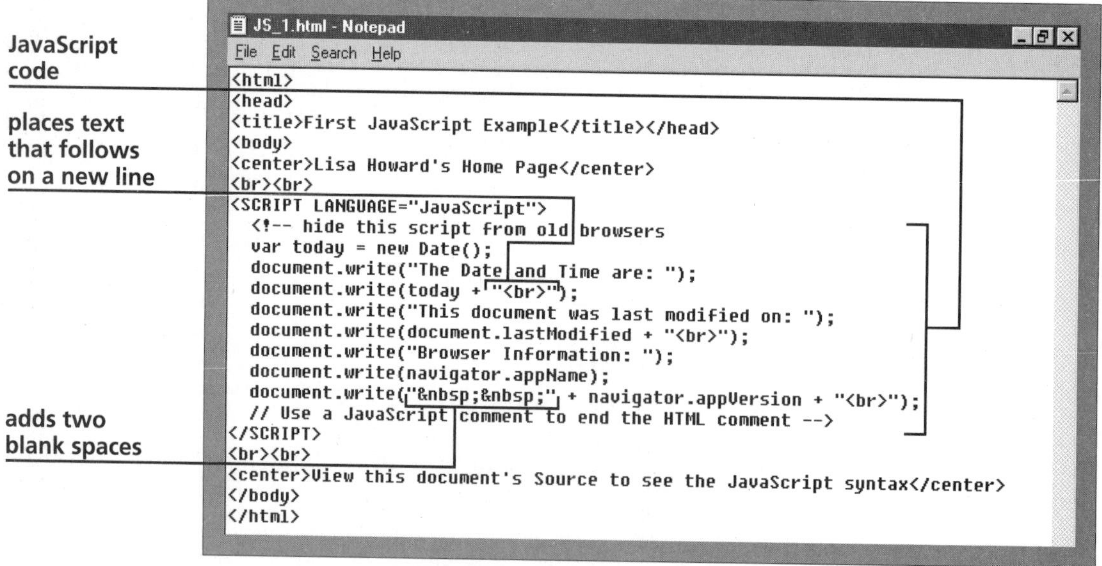

JavaScript code
places text that follows on a new line
adds two blank spaces
JS_1.html - Notepad
File Edit Search Help
<html>
<head>
<title>First JavaScript Example</title></head>
<body>
<center>Lisa Howard's Home Page</center>

<SCRIPT LANGUAGE="JavaScript">
<!-- hide this script from old browsers
var today = new Date();
document.write("The Date and Time are: ");
document.write(today + "<br>");
document.write("This document was last modified on: ");
document.write(document.lastModified + "<br>");
document.write("Browser Information: ");
document.write(navigator.appName);
document.write(" " + navigator.appVersion + "<br>");
// Use a JavaScript comment to end the HTML comment -->
</SCRIPT>

<center>View this document's Source to see the JavaScript syntax</center>
</body>
</html>

Figure 10-21: HTML file containing JavaScript

In an HTML file, JavaScript code must be enclosed within the following tags: `<SCRIPT LANGUAGE="JavaScript">…</SCRIPT>`. The LANGUAGE parameter tells the Web browser which type of scripting language is being used. Since some older Web browsers are not able to interpret JavaScript, JavaScript commands are always enclosed in HTML comment tags. The comment tags keep these older browsers from displaying JavaScript code as text in the body of the Web page. The

JavaScript interpreter ignores the first line of all HTML comments, but processes the rest of the lines. The line signaling the end of the JavaScript code must be prefaced with the JavaScript comment indicator (//), and must include the HTML end-of-comment indicator (-->).

The `document.write` command instructs the JavaScript interpreter to display the text that follows in the HTML document. The text may be inserted directly and enclosed in double quotes, or it may be text returned from another JavaScript command or variable. For example, in Figure 10-21, the **today** variable returns the current date, the **document.lastModified** variable returns the date the HTML file was last modified, and the **navigator.appName** and **navigator.appVersion** variables return the name and version of the user's Web browser. Now you will create and view the HTML file containing JavaScript.

To create the JavaScript html file:

1. Switch to Notepad, and create a new file. Type the code shown in Figure 10-21, and save the file as **js_1.html** in the Chapter10 folder on your Student Disk.
2. Switch to your Web browser, and open js_1.html. Your JavaScript Web page should be displayed as shown in Figure 10-22.

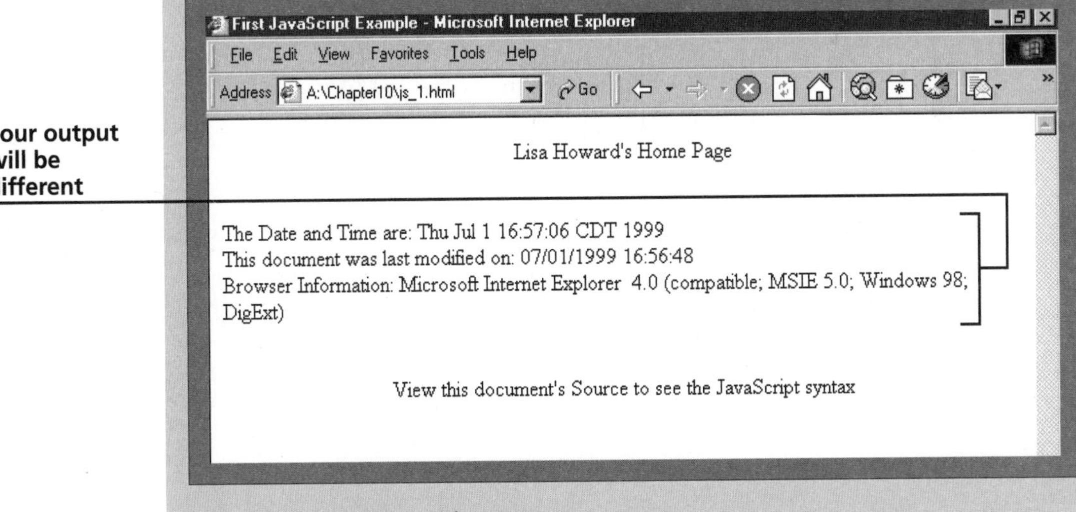

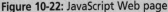

Figure 10-22: JavaScript Web page

JavaScript is a rich language with many commands other than the ones used in this example. There are several books devoted entirely to JavaScript commands, as well as many Web sites that provide tutorials and examples of JavaScript.

Using JavaScript to Validate Form Inputs

Next, you are going to create a Web page that uses JavaScipt commands to validate user form inputs. To do this, you will first have to learn some basic JavaScript language syntax.

Functions and Alerts Recall that a function is a program unit that returns a specific value to the calling program. The general format for a JavaScript function is:

```
function [function name]([parameter1, parameter2, …]) {
       [function program steps]
       return [value returned by function]
       }
```

The function header contains the **function** keyword and a list of the parameters passed to the function in parentheses. Each parameter in the parameter list is delimited by commas. After the header, the left curly brace (**{**) signifies the beginning of the function body, and the right curly brace (**}**) signifies the end of the function body. The right curly brace must be on a line by itself immediately following the function body. Variables are declared within the function body. The `return` command specifies the value that is returned by the function. This can be a data value, or it can also be one of the logical values true or false, which are always typed in lowercase letters. When the `return` command is executed, the function is immediately exited and program control returns to the calling program.

Figure 10-23 shows an example of an HTML program with a JavaScript function.

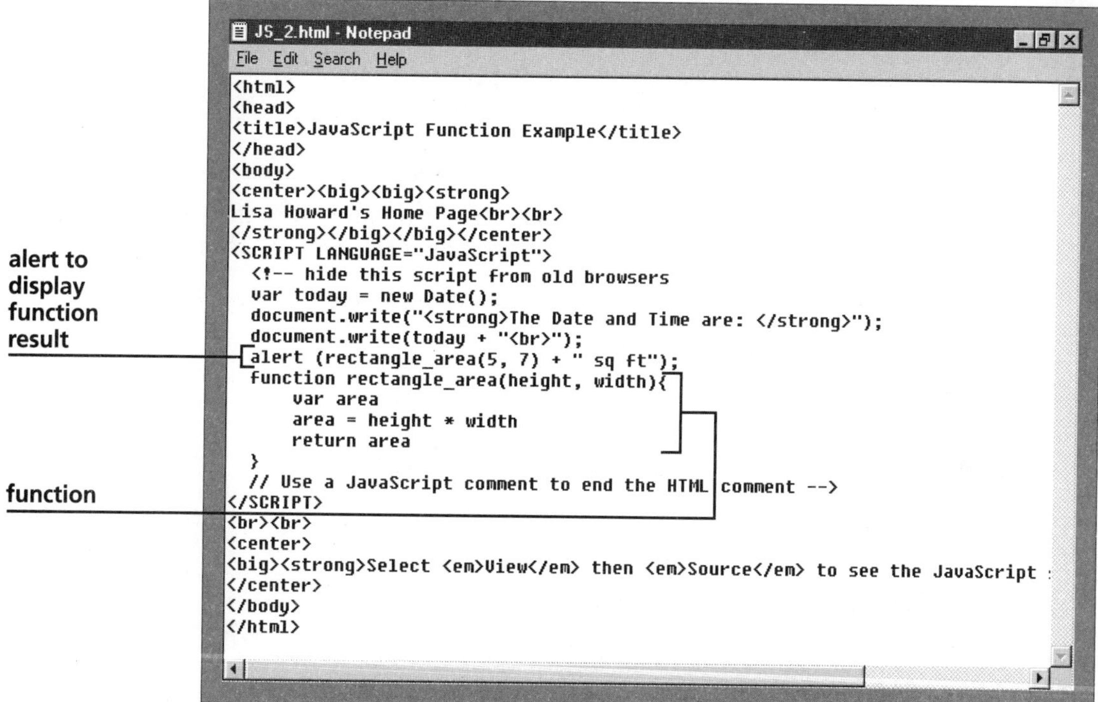


```
<html>
<head>
<title>JavaScript Function Example</title>
</head>
<body>
<center><big><big><strong>
Lisa Howard's Home Page<br><br>
</strong></big></big></center>
<SCRIPT LANGUAGE="JavaScript">
  <!-- hide this script from old browsers
  var today = new Date();
  document.write("<strong>The Date and Time are: </strong>");
  document.write(today + "<br>");
  alert (rectangle_area(5, 7) + " sq ft");
  function rectangle_area(height, width){
      var area
      area = height * width
      return area
  }
  // Use a JavaScript comment to end the HTML comment -->
</SCRIPT>
<br><br>
<center>
<big><strong>Select <em>View</em> then <em>Source</em> to see the JavaScript :
</center>
</body>
</html>
```

Figure 10-23: HTML program with JavaScript function

The function name is rectangle_area, and it receives two parameters: height and width. The **var** keyword following the function header is used to declare a variable, named area, that is assigned to the rectangle area, which is the product of the height and width. Although you can omit declaring the variable, and the function will still work, it is a good practice to always explicitly declare variables within functions, since this makes the function easier to understand and maintain. The function in Figure 10-23 returns the calculated value for the rectangle area.

The **alert** command calls the JavaScript message function that displays a dialog box containing the text specified in parentheses. Text to be displayed by an alert is enclosed in quotation marks, and the contents of variables are displayed by specifying the variable name without quotation marks. In the example, the alert will display the area calculated by the rectangle_area function, followed by the text "sq ft." The plus (+) operator can also be used to concatenate strings into a single larger string of characters for display in an alert. Notice that the alert command calling the rectangle_area function is in the JavaScript code before the function is declared. The browser's JavaScript interpreter will process the alert command, see that rectangle_area refers to a function, and then call the function and return the result.

IF and IF/ELSE Structures The general format for the JavaScript IF control structure is:

```
If (condition) {
[Program statements]
}
```

If the condition is true, the program statements between the curly braces are executed. If the condition is false, these program statements are skipped.

The general format for the JavaScript IF/ELSE control structure is:

```
If (condition) {
[Program statements]
} else {
[Program statements]
}
```

If the condition is true, the program statements between the first set of curly braces are executed. If the condition is false, the program statements between the second set of curly braces are executed.

Assignment and Comparison Operators The JavaScript assignment operator that is used to assign a value to a variable is a single equal sign (=). This is equivalent to the := assignment operator in PL/SQL. The JavaScript comparison operator that is used to determine if a condition is true or false is two equal signs with no space between them (==). This is equivalent to the = comparison operator in PL/SQL. The following code illustrates how these operators are used:

```
var number1 = 2
var number2 = 3
If (number1 == number2) {
 alert("number1 is equal to number2")
} else {
 alert("number1 is not equal to number2")
}
```

Notice that you can declare a variable with the "var" keyword and assign a value to the variable on the same line.

The AND, OR, and NOT Logical Operators In some cases, you will want to test multiple conditions with a single, more complex expression. Figure 10-24 shows the JavaScript logical operators that are used to combine conditions using AND, OR, and NOT.

Operation	JavaScript Operator
AND	&&
OR	||
NOT	!

Figure 10-24: JavaScript logical operators

As with any programming language, if two conditions are joined using the AND (&&) operator, both conditions must be true for the overall expression to be true. If two conditions are joined using the OR (||) operator, only one condition must be true for the overall expression to be true. The NOT (!) operator negates (returns the opposite value of) the condition. Here are some examples using these operators. Note that the two slashes (//) start a JavaScript comment.

```
5 > 2 && 34 > 5              // result is true
5 > 2 && 34 < 5              // result is false
5 > 2 || 34 > 5              // result is true
5 > 2 || 34 < 5              // result is true
5 < 2 || 34 < 5              // result is false
!true                        // result is false
!(6 > 5)                     // result is false
!(6 < 5)                     // result is true
```

Now you are ready to add JavaScript code to validate form inputs in the northwoods.html Web page file that you created earlier in the lesson. This application received a student's name and PIN using an HTML form, and then displayed the student's transcript information using a PL/SQL program unit. Recall that to create a form within an HTML file, you used the following general code:

```
<form action = "http://[address of Web server]/[virtual path
name specifying the type of cartridge that will process the
form inputs]/[name of the program receiving the form inputs]>
```

To validate form inputs using a JavaScript function, you need to add two additional parameters to modify this `<form...>` tag. Currently, the `<form...>` tag contains the method and action parameters. You will add the form **name** and **onsubmit** parameters so that the `<form...>` tag will have the following general format:

```
<form name="[form name]"
   method="POST"
   action=[same as before]
   onsubmit="return [JavaScript function name]">
```

The form name parameter specifies how the form is referenced in the JavaScript function. In this example, the form name is login, so the form input text fields will be referenced as `document.login.StudentName` and `document.login.StudentPIN`. The onsubmit parameter specifies the name of the JavaScript function that will be used to validate the form inputs. In your application, this function will be named `validate( )`, and you will write it later. Now you will modify the `<form...>` tag.

To modify the `<form...>` tag:

1 Switch to Notepad, and open the **northwoods.html** file from the Chapter10 folder on your Student Disk. Save the file as **northwoods2.html.**

2 Add the commands shown in Figure 10-25 to modify the <form...> tag.

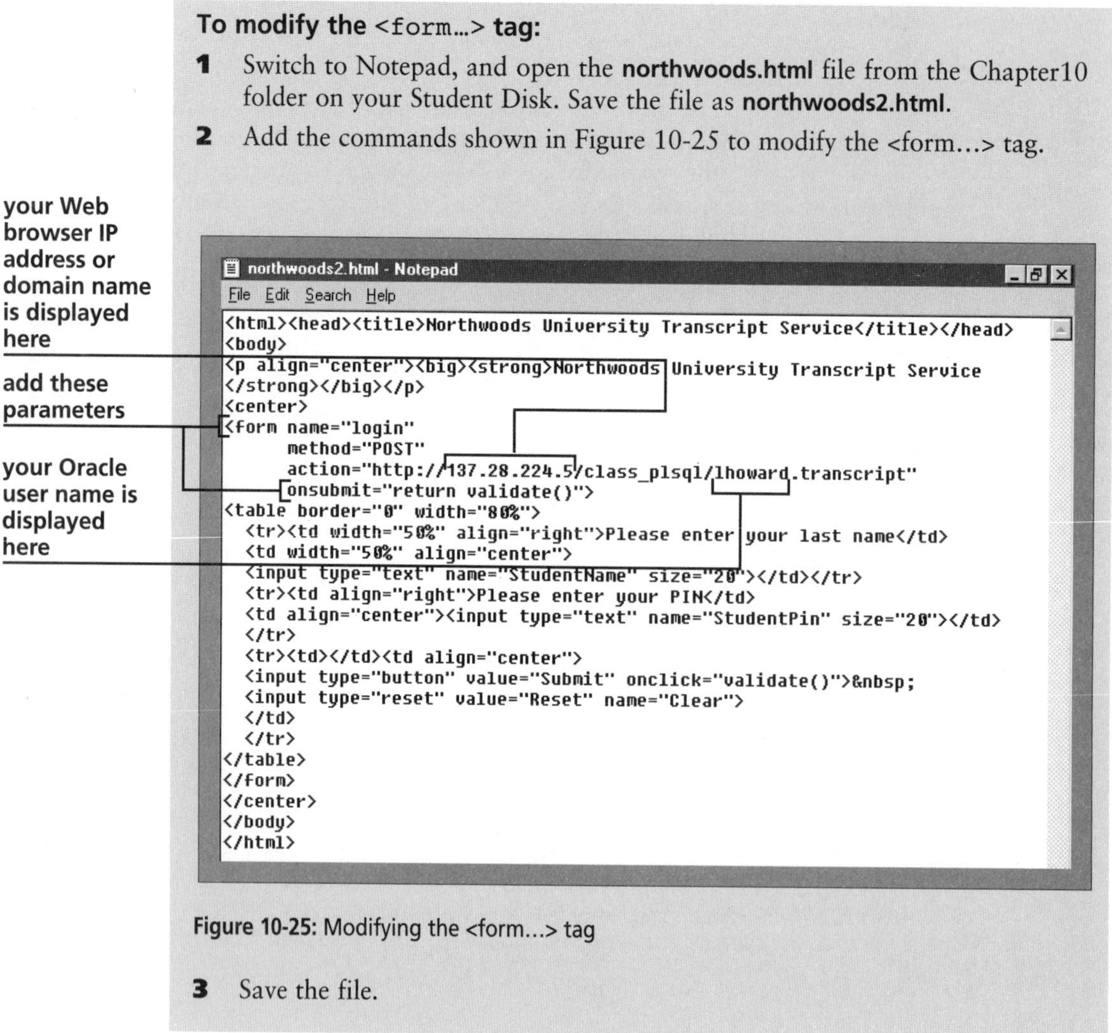


```
<html><head><title>Northwoods University Transcript Service</title></head>
<body>
<p align="center"><big><strong>Northwoods University Transcript Service
</strong></big></p>
<center>
<form name="login"
      method="POST"
      action="http://137.28.224.5/class_plsql/lhoward.transcript"
      onsubmit="return validate()">
<table border="0" width="80%">
  <tr><td width="50%" align="right">Please enter your last name</td>
  <td width="50%" align="center">
  <input type="text" name="StudentName" size="20"></td></tr>
  <tr><td align="right">Please enter your PIN</td>
  <td align="center"><input type="text" name="StudentPin" size="20"></td>
  </tr>
  <tr><td></td><td align="center">
  <input type="button" value="Submit" onclick="validate()"> 
  <input type="reset" value="Reset" name="Clear">
  </td>
  </tr>
</table>
</form>
</center>
</body>
</html>
```

Figure 10-25: Modifying the <form...> tag

3 Save the file.

Next you will write the JavaScript function that validates that the user has entered a value for the student name and PIN. This function will test to see if the value entered for either the student name or PIN is a blank string (" "). If either the student name or PIN is blank, the function will return the logical value false. If neither value is a blank string, the function will return the logical value true.

To create the JavaScript data validation function:

1 Add the commands shown in Figure 10-26 to northwoods2.html to create the function that validates that the user has entered a value for the student name and PIN.

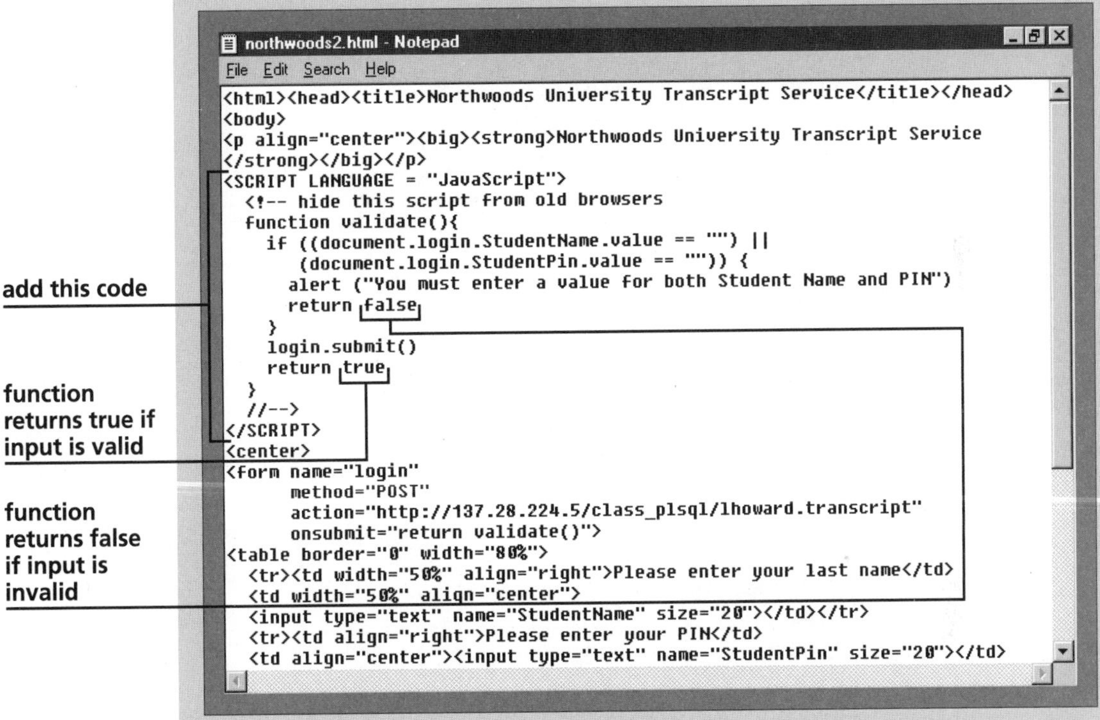

northwoods2.html - Notepad

File Edit Search Help

```
<html><head><title>Northwoods University Transcript Service</title></head>
<body>
<p align="center"><big><strong>Northwoods University Transcript Service
</strong></big></p>
<SCRIPT LANGUAGE = "JavaScript">
  <!-- hide this script from old browsers
  function validate(){
    if ((document.login.StudentName.value == "") ||
       (document.login.StudentPin.value == "")) {
      alert ("You must enter a value for both Student Name and PIN")
      return false
    }
    login.submit()
    return true
  }
  //-->
</SCRIPT>
<center>
<form name="login"
      method="POST"
      action="http://137.28.224.5/class_plsql/lhoward.transcript"
      onsubmit="return validate()">
<table border="0" width="80%">
  <tr><td width="50%" align="right">Please enter your last name</td>
  <td width="50%" align="center">
  <input type="text" name="StudentName" size="20"></td></tr>
  <tr><td align="right">Please enter your PIN</td>
  <td align="center"><input type="text" name="StudentPin" size="20"></td>
```

add this code

function returns true if input is valid

function returns false if input is invalid

Figure 10-26: JavaScript function to validate form inputs

2 Save the file.

Finally, the form's Submit button must be changed to an ordinary button with an **onclick** parameter that specifies the name of the JavaScript function that will be called for data validation when the button is clicked. This is done using the following general code to specify the Submit button:

```
<input type="button" value="Submit" onclick="[JavaScript function name]">
```

To change the Submit button:

1 Modify the Submit button specification in northwoods2.html as shown in Figure 10-27.

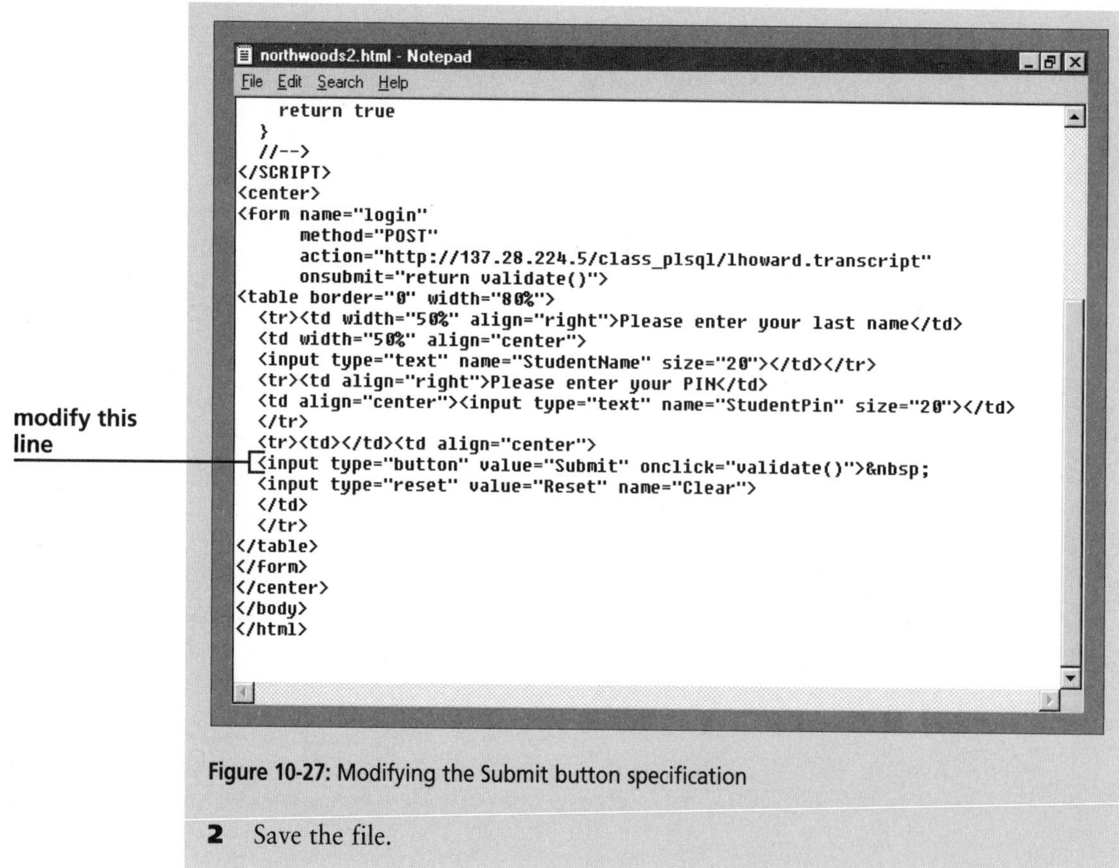


```
    return true
  }
  //-->
</SCRIPT>
<center>
<form name="login"
      method="POST"
      action="http://137.28.224.5/class_plsql/lhoward.transcript"
      onsubmit="return validate()">
<table border="0" width="80%">
  <tr><td width="50%" align="right">Please enter your last name</td>
  <td width="50%" align="center">
  <input type="text" name="StudentName" size="20"></td></tr>
  <tr><td align="right">Please enter your PIN</td>
  <td align="center"><input type="text" name="StudentPin" size="20"></td>
  </tr>
  <tr><td></td><td align="center">
  <input type="button" value="Submit" onclick="validate()"> 
  <input type="reset" value="Reset" name="Clear">
  </td>
  </tr>
</table>
</form>
</center>
</body>
</html>
```

Figure 10-27: Modifying the Submit button specification

2 Save the file.

The next step is to test your JavaScript program by opening the northwoods2.html file in your Web browser. It will look identical to your earlier Web page input form, but if either text box is empty when the Submit button is clicked, an error message will be displayed to tell the user to enter a value for both student name and PIN.

To test the JavaScript data validation program:

1 Switch to your Web browser, and open **northwoods2.html** from the Chapter10 folder on your Student Disk. The input form is displayed.

2 Do not enter any inputs in the form fields—just click the **Submit** button. The error message shown in Figure 10-28 is displayed.

help

If you receive an error message stating that the script did not run successfully, go to the line specified in the error message and make sure your code exactly matches the code shown in Figures 10-25, 10-26, and 10-27. When you find the error, correct your Notepad file, save the corrected file, and refresh the display in your Web browser by clicking View on the menu bar, then clicking Refresh.

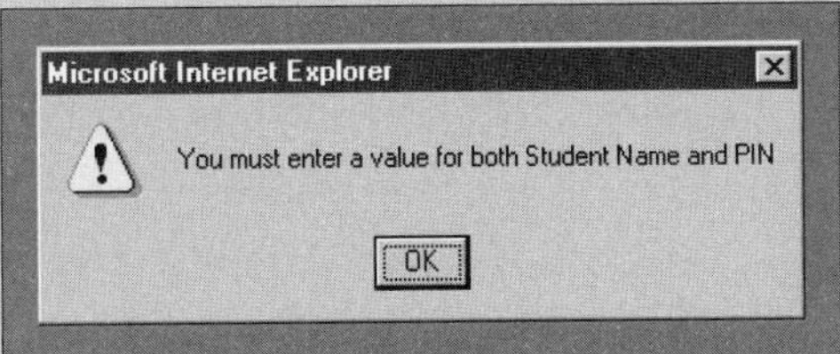

Figure 10-28: Error message generated by JavaScript function

3 Click **OK** to close the dialog box.

SUMMARY

- HTML forms allow users to submit inputs to Web pages using an interface consisting of Windows controls such as text boxes, command buttons, and option buttons.
- Form inputs are submitted when the user clicks a standard form command button called a Submit button. Another standard button found on many HTML forms is a Reset button, which clears the text in all of the form text boxes.
- HTML forms can be used to gather inputs for PL/SQL stored program units.
- Form parameters can be passed using the GET method, which stores the parameters as environmental variables, and displays them as part of the URL. Form parameters can also be passed using the POST method, which submits them directly to the form and avoids displaying them as part of the URL.
- The name, capitalization, and order of the variables defined by form input tags must match the name, capitalization, and order of the parameters declared for the executable program processing the form.
- Scripting languages such as JavaScript are mainly used to validate user form inputs before submitting them to a Web server.
- Scripting languages perform client-side form processing, whereby the Web page responds to user inputs from HTML form elements—such as text input boxes, command buttons, and option buttons—by processing the inputs on the user's workstation rather than sending inputs back to the Web server for processing.
- For security reasons, scripting languages cannot read or write to files on a client computer, and they cannot launch other applications on a client computer.
- JavaScript code is enclosed in HTML files in comment tags so that it will not be displayed as text by older Web browsers that do not process JavaScript code.

REVIEW QUESTIONS

1. What is the difference between an HTML form and a Developer form?
2. Give two reasons for creating a Web database application rather than a standard Developer database application.
3. Assume you have a form with three input items named LNAME, FNAME, and ADDRESS and that these inputs are displayed in that order. Will a stored procedure with the parameters LASTNAME, FIRSTNAME, and ADDRESS (in that order) successfully receive this form's inputs?
4. What does an HTML form's Reset button do?
5. When should you use the GET method for passing parameters to a Web page, and when should you use the POST method?
6. What are scripting languages, such as JavaScript, usually used for?
7. What is the difference between Java and JavaScript?

PROBLEM-SOLVING CASES

1. Create a Web page with an HTML form to allow a user to enter an item ID, size, and color for Clearwater Traders merchandise. Save the Web page file as Ch10bEx1.html. When the user clicks the Submit button, this Web page should call a PL/SQL stored program unit named INVENTORYITEMS that you will create using Procedure Builder. The INVENTORYITEMS stored program unit will display a Web page showing the item's category, description, current price, and quantity on hand. (*Hint*: Review the sections on implicit and explicit cursors in Chapter 4 for help on how to query the database for this information in the PL/SQL program unit.) Format the returned Web page using appropriate HTML tags and text to make it understandable.
2. Use JavaScript to create a function that checks the Web page you created in Exercise 1 to confirm that the user entered values for item ID, size, and color. Save your modified HTML file as Ch10bEx2.html.
3. Create a Web page with an HTML form to allow a user to enter a Term ID. Include text on the Web page that indicates that Term ID 5 corresponds to the Spring 2002 term, and Term ID 6 corresponds to the Summer 2002 term. (Note that there are no course section records for the other terms.) Save the Web page file as Ch10bEx3.html. When the user clicks the Submit button, this Web page should call a PL/SQL stored program unit named COURSE_SECTIONS that you will create using Procedure Builder. This stored program unit will display a Web page showing the course call ID, section number, faculty last name, day, time, building code, and room for every course offered during the selected term. (*Hint*: Review the sections on implicit and explicit cursors in Chapter 4 for help on how to query the database for this information in the PL/SQL program unit.) Format the returned Web page using appropriate HTML tags and text to make it understandable.
4. Use JavaScript to create a function that checks the Web page you created in Exercise 3 to confirm that the user entered a value for Term ID. Save your modified HTML file as Ch10bEx4.html.

Index

Special Characters

A

B

D

E

F

I

J

K

L

M

N

O

Q

R

S

T

U

V